Lecture Notes in Computer Science 5981

Commenced Publication in 1973
Founding and Former Series Editors:
Gerhard Goos, Juris Hartmanis, and Jan van Leeuv

Hiroyuki Kitagawa   Yoshiharu Ishikawa
Qing Li   Chiemi Watanabe (Eds.)

Database Systems
for Advanced Applications

15th International Conference, DASFAA 2010
Tsukuba, Japan, April 1-4, 2010
Proceedings, Part I

 Springer

Volume Editors

Hiroyuki Kitagawa
University of Tsukuba, Graduate School of Systems and Information Engineering
Tennohdai, Tsukuba, Ibaraki 305–8573, Japan
E-mail: kitagawa@cs.tsukuba.ac.jp

Yoshiharu Ishikawa
Nagoya University, Information Technology Center
Furo-cho, Chikusa-ku, Nagoya 464-8601, Japan
E-mail: ishikawa@itc.nagoya-u.ac.jp

Qing Li
City University of Hong Kong, Department of Computer Science
83 Tat Chee Avenue, Kowloon, Hong Kong, China
E-mail: itqli@cityu.edu.hk

Chiemi Watanabe
Ochanomizu University, Department of Information Science
2-1-1, Otsuka, Bunkyo-ku, Tokyo 112-8610, Japan
E-mail: chiemi@is.ocha.ac.jp

Library of Congress Control Number: 2010922033

CR Subject Classification (1998): H.2, H.3, H.4, H.5, C.2, J.1

LNCS Sublibrary: SL 3 – Information Systems and Application, incl. Internet/Web
and HCI

ISSN   0302-9743
ISBN-10   3-642-12025-3 Springer Berlin Heidelberg New York
ISBN-13   978-3-642-12025-1 Springer Berlin Heidelberg New York

springer.com

© Springer-Verlag Berlin Heidelberg 2010
Printed in Germany

Typesetting: Camera-ready by author, data conversion by Scientific Publishing Services, Chennai, India
Printed on acid-free paper   06/3180

Message from the DASFAA 2010 Chairs

It is our great pleasure to welcome you to the proceedings of the 15th International Conference on Database Systems for Advanced Applications (DASFAA 2010). DASFAA is an international forum for academic exchange and technical discussions among researchers, developers, and users of databases from academia, business, and industry. DASFAA is a leading conference in the areas of databases, large-scale data management, data mining, and the Web. We are delighted to have held the 15th conference in Tsukuba during the cherry blossom season – the very best season of the year.

The call for papers attracted 237 research submissions from 25 countries / regions (based on the affiliation of the first author). Among them, 55 regular papers and 16 short papers were selected for presentation after a thorough review process by the Program Committee. The Industrial Committee, chaired by Hideko S. Kunii and Umesh Dayal, selected 6 industrial papers for presentation from 15 submissions and organized an industrial invited talk. The conference program also included 22 demo presentations selected from 33 submissions by the Demo Committee, chaired by Takahiro Hara and Kian-Lee Tan.

We are proud to have had two distinguished keynote speakers: Gerhard Weikum (Max-Planck Institute for Informatics) and Raghu Ramakrishnan (Yahoo! Research). Their lectures were the highlight of this conference. Tutorial Co-chairs, Kazutoshi Sumiya and Wookey Lee organized three tutorials by leading experts: Mining Moving Objects, Trajectory and Traffic Data (by Jiawei Han, Zhenhui Li, and Lu An Tang), Querying Large Graph Databases (by Yiping Ke, James Cheng, and Jeffrey Xu Yu), and Introduction to Social Computing (by Irwin King). A stimulating panel was organized by Panel Co-chairs Yasushi Kiyoki and Virach Sornlertlamvanich. This rich and attractive conference program boasts conference proceedings that span two volumes in Springer's *Lecture Notes in Computer Science* series.

Beyond the main conference, Masatoshi Yoshikawa and Xiaofeng Meng, who chaired the Workshop Committee, put together workshops that were of interest to all. The workshop papers are included in a separate volume of proceedings also published by Springer in its *Lecture Notes in Computer Science* series.

DASFAA 2010 was jointly organized by the University of Tsukuba and the Database Society of Japan (DBSJ). It received in-cooperation sponsorship from the KIISE Database Society of Korea, the China Computer Federation Database Technical Committee, ARC Research Network in Enterprise Information Infrastructure, Asian Institute of Technology (AIT), "New IT Infrastructure for the Information-explosion Era," MEXT Grant-in-Aid for Scientific Research on Priority Areas, Japan, Information Processing Society of Japan (IPSJ), the Institute of Electronics, Information, and Communication Engineers (IEICE), Japan PostgreSQL Users Group, MySQL Nippon Association, and the Japanese

Firebird Users Group. We are grateful to the sponsors who contributed generously to making DASFAA 2010 successful. They are Beacon Information Technology Inc., Mitsubishi Electric Corporation, National Institute for Materials Science (NIMS), KDDI R&D Laboratories Inc., National Institute of Advanced Industrial Science and Technology (AIST), Ricoh Co., Ltd., NTT DATA Corporation, Hitachi, Ltd., Ricoh IT Solutions Co., Ltd., SRA OSS, Inc., Japan, and Nippon Telegraph and Telephone Corporation. We also appreciate financial support from the Telecommunications Advancement Foundation and Kayamori Foundation of Informational Science Advancement.

The conference would not have been possible without the support of many colleagues. We would like to express our special thanks to Honorary Conference Chair Yoshifumi Masunaga for his valuable advice on all aspects of organizing the conference. We thank Organizing Committee Chair Masaru Kitsuregawa and Vice Chair Miyuki Nakano, DBSJ Liaison Haruo Yokota, Publicity Co-chairs Jun Miyazaki and Hyoil Han, Local Arrangements Committee Chair Toshiyuki Amagasa, Finance Chair Atsuyuki Morishima, Publication Chair Chiemi Watanabe, and Web Chair Hideyuki Kawashima. We are grateful for the strong support from the DASFAA 2010 Geographical Area Chairs: Bonghee Hong (Korea), Li-Zhu Zhou (China), Jeffrey Xu Yu (Hong Kong), Ming-Syan Chen (Taiwan), Stéphane Bressan (Singapore), Vilas Wuwongse (Thailand), Krithi Ramamritham (India), James Bailey (Australia), Chen Li (America), and Peer Kröger (Europe). Our thanks go to all the committee members and other individuals involved in putting it all together.

Finally, we thank the DASFAA Steering Committee, especially the immediate past Chair, Kyu-Young Whang, and current Chair, Katsumi Tanaka, for their leaderships and encouragement.

April 2010   Hiroyuki Kitagawa
                                                Yoshiharu Ishikawa
                                                Qing Li

Organization

Honorary Conference Chair

Yoshifumi Masunaga   Aoyama Gakuin University, Japan

General Conference Chair

Hiroyuki Kitagawa   University of Tsukuba, Japan

Organizing Committee Chair

Masaru Kitsuregawa   The University of Tokyo, Japan

Organizing Committee Vice Chair

Miyuki Nakano   The University of Tokyo, Japan

DBSJ Liaison

Haruo Yokota   Tokyo Institute of Technology, Japan

Program Committee Co-chairs

Yoshiharu Ishikawa   Nagoya University, Japan
Qing Li   City University of Hong Kong, China

Industrial Committee Co-chairs

Hideko S. Kunii   Ricoh IT Solutions Co., Ltd., Japan
Umesh Dayal   HP Labs, USA

Tutorial Co-chairs

Kazutoshi Sumiya   University of Hyogo, Japan
Wookey Lee   Inha University, Korea

Panel Co-chairs

Yasushi Kiyoki   Keio University, Japan
Virach Sornlertlamvanich   NECTEC, Thailand

Demo Committee Co-chairs

Takahiro Hara   Osaka University, Japan
Kian-Lee Tan   National University of Singapore, Singapore

Workshop Committee Co-chairs

Masatoshi Yoshikawa   Kyoto University, Japan
Xiaofeng Meng   Renmin University, China

Publicity Co-chairs

Jun Miyazaki   Nara Institute of Science and Technology, Japan
Hyoil Han   LeMoyne-Owen College, USA

Local Arrangements Committee Chair

Toshiyuki Amagasa   University of Tsukuba, Japan

Finance Chair

Atsuyuki Morishima   University of Tsukuba, Japan

Publication Chair

Chiemi Watanabe   Ochanomizu University, Japan

Web Chair

Hideyuki Kawashima   University of Tsukuba, Japan

Geographical Area Chairs

Korea: Bonghee Hong   Pusan National University, Korea
China: Li-Zhu Zhou   Tsinghua University, China
Hong Kong: Jeffrey Xu Yu   Chinese University of Hong Kong, China
Taiwan: Ming-Syan Chen   National Taiwan University, Taiwan
Singapore: Stéphane Bressan   National University of Singapore, Singapore
Thailand: Vilas Wuwongse   Asian Institute of Technology, Thailand
India: Krithi Ramamritham   Indian Institute of Technology at Bombay,
                                      India
Australia: James Bailey   The University of Melbourne, Australia

America: Chen Li	University of California, Irvine and BiMaple, USA
Europe: Peer Kröger	Ludwig-Maximilians-Universität München, Germany

Best Paper Committee Co-chairs

Katsumi Tanaka	Kyoto University, Japan
Kyu-Young Whang	Korea Advanced Institute of Science and Technology (KAIST), Korea
Jianzhong Li	Harbin Institute of Technology, China

DASFAA Awards Committee

Tok Wang Ling (Chair)	National University of Singapore, Singapore
Kyu-Young Whang	Korea Advanced Institute of Science and Technology (KAIST), Korea
Katsumi Tanaka	Kyoto University, Japan
Kirthi Ramamirtham	Indian Institute of Technology at Bombay, India
Jianzhong Li	Harbin Institute of Technology, China
Dik Lun Lee	Hong Kong University of Science & Technology, China
Arbee L.P. Chen	National Chengchi University, Taiwan

Steering Committee

Katsumi Tanaka (Chair)	Kyoto University, Japan
Ramamohanarao Kotagiri (Vice Chair)	University of Melbourne, Australia
Kyu-Young Whang (Advisor)	Korea Advanced Institute of Science and Technology (KAIST), Korea
Yoshihiko Imai (Treasurer)	Matsushita Electric Industrial Co., Ltd., Japan
Kian-Lee Tan (Secretary)	National University of Singapore, Singapore
Yoon Joon Lee	Korea Advanced Institute of Science and Technology (KAIST), Korea
Qing Li	City University of Hong Kong, China
Krithi Ramamritham	Indian Institute of Technology at Bombay, India
Ming-Syan Chen	National Taiwan University, Taiwan
Eui Kyeong Hong	Univerity of Seoul, Korea
Hiroyuki Kitagawa	University of Tsukuba, Japan
Li-Zhu Zhou	Tsinghua University, China
Jianzhong Li	Harbin Institute of Technology, China
BongHee Hong	Pusan National University, Korea

Organizing Committee

Shuji Harashima	Toshiba Corporation, Japan
Atsushi Iizawa	Ricoh IT Solutions Co., Ltd., Japan
Minoru Inui	Beacon IT Inc., Japan
Tatsuo Ishii	SRA OSS, Japan
Hiroshi Ishikawa	Shizuoka University, Japan
Kyoji Kawagoe	Ritsumeikan University, Japan
Yutaka Kidawara	National Institute of Information and Communications Technology (NICT), Japan
Hajime Kitakami	Hiroshima City University, Japan
Isao Kojima	National Institute of Advanced Industrial Science and Technology (AIST), Japan
Kazunori Matsumoto	KDDI Lab., Japan
Masataka Matsuura	Fujitsu Ltd., Japan
Hirofumi Matsuzawa	IBM Japan, Japan
Shojiro Nishio	Osaka University, Japan
Makoto Okamato	Academic Resource Guide, Japan
Tetsuji Satoh	University of Tsukuba, Japan
Jun Sekine	NTT DATA Corporation, Japan
Shigenobu Takayama	Mitsubishi Electric Cooporation, Japan
Takaaki Tasaka	SGI Japan, Ltd., Japan
Yoshito Tobe	Tokyo Denki University, Japan
Masashi Tsuchida	Hitachi, Ltd., Japan
Masashi Yamamuro	NTT Corporation, Japan
Kazumasa Yokota	Okayama Prefectural University, Japan

Program Committee

Toshiyuki Amagasa	University of Tsukuba, Japan
Masayoshi Aritsugi	Kumamoto University, Japan
James Bailey	University of Melbourne, Australia
Ladjel Bellatreche	Poitiers University, France
Boualem Benatallah	University of New South Wales, Australia
Sourav Bhowmick	Nanyang Technological University, Singapore
Athman Bouguettaya	CSIRO, Australia
Chee Yong Chan	National University of Singapore, Singapore
Lei Chen	Hong Kong University of Science and Technology, China
Ming-Syan Chen	National Taiwan University, Taiwan
Reynold Cheng	University of Hong Kong, China
Gao Cong	Aalborg University, Denmark
Bin Cui	Peking University, China
Alfredo Cuzzocrea	ICAR-CNR / University of Calabria, Italy
Zhiming Ding	Chinese Academy of Sciences, China
Gill Dobbie	University of Auckland, New Zealand

Guozhu Dong	Wright State University, USA
Jianhua Feng	Tsinghua University, China
Ling Feng	Tsinghua University, China
Sumit Ganguly	Indian Institute of Technology Kanpur, India
Yunjun Gao	Zhejiang University, China
Lukasz Golab	AT&T Labs, USA
Vivekanand Gopalkrishnan	Nanyang Technological University, Singapore
Stéphane Grumbach	INRIA, France
Wook-Shin Han	Kyungpook National University, Korea
Takahiro Hara	Osaka University, Japan
Kenji Hatano	Doshisha University, Japan
Wynne Hsu	National University of Singapore, Singapore
Haibo Hu	Hong Kong Baptist University, China
Seung-won Hwang	POSTECH, Korea
Mizuho Iwaihara	Waseda University, Japan
Ramesh C. Joshi	Indian Institute of Technology Roorkee, India
Jaewoo Kang	Korea University, Korea
Norio Katayama	National Institute of Informatics, Japan
Yutaka Kidawara	National Institute of Information and Communications Technology, Japan
Myoung Ho Kim	Korea Advanced Institute of Science and Technology (KAIST), Korea
Markus Kirchberg	Institute for Infocomm Research, A*STAR, Singapore
Hajime Kitakami	Hiroshima City University, Japan
Jae-Gil Lee	IBM Almaden Research Center, USA
Mong Li Lee	National University of Singapore, Singapore
Sang-goo Lee	Seoul National University, Korea
Sang-Won Lee	Sungkyunkwan University, Korea
Wang-Chien Lee	Pennsylvania State University, USA
Cuiping Li	Renmin University, China
Jianzhong Li	Harbin Institute of Technology, China
Ee-Peng Lim	Singapore Management University, Singapore
Xuemin Lin	University of New South Wales, Australia
Chengfei Liu	Swinburne University of Technology, Australia
Jiaheng Lu	Renmin University, China
Sanjay Madria	University of Missouri-Rolla, USA
Nikos Mamoulis	University of Hong Kong, China
Weiyi Meng	Binghamton University, USA
Xiaofeng Meng	Renmin University, China
Jun Miyazaki	Nara Institute of Science and Technology, Japan
Yang-Sae Moon	Kangwon National University, Korea
Yasuhiko Morimoto	Hiroshima University, Japan
Miyuki Nakano	University of Tokyo, Japan
Wolfgang Nejdl	L3S / University of Hannover, Germany

Masato Oguchi	Ochanomizu University, Japan
Tadashi Ohmori	University of Electro-Communications, Japan
Makoto Onizuka	NTT Cyber Space Laboratories, NTT Corporation, Japan
Satoshi Oyama	Hokkaido University, Japan
HweeHwa Pang	Singapore Management University, Singapore
Jian Pei	Simon Fraser University, Canada
Wen-Chih Peng	National Chiao Tung University, Taiwan
Evaggelia Pitoura	University of Ioannina, Greece
Sunil Prabhakar	Purdue University, USA
Tieyun Qian	Wuhan University, China
Krithi Ramamritham	Indian Institute of Technology Bombay, India
Uwe Röhm	University of Sydney, Australia
Shourya Roy	Xerox India Innovation Hub, India
Yasushi Sakurai	NTT Communication Science Laboratories, NTT Corporation, Japan
Simonas Saltenis	Aalborg University, Denmark
Monica Scannapieco	University of Rome, Italy
Markus Schneider	University of Florida, USA
Heng Tao Shen	University of Queensland, Australia
Hyoseop Shin	Konkuk University, Korea
Atsuhiro Takasu	National Institute of Informatics, Japan
Kian-Lee Tan	National University of Singapore, Singapore
David Taniar	Monash University, Australia
Egemen Tanin	University of Melbourne, Australia
Jie Tang	Tsinghua University, China
Yufei Tao	Chinese University of Hong Kong, China
Vincent S. Tseng	National Cheng Kung University, Taiwan
Anthony K.H. Tung	National University of Singapore, Singapore
Vasilis Vassalos	Athens University of Economics and Business, Greece
Guoren Wang	Northeastern University, China
Jianyong Wang	Tsinghua University, China
Jiying Wang	City University of Hong Kong, China
Wei Wang	University of New South Wales, Australia
Raymond Chi-Wing Wong	Hong Kong University of Science and Technology, China
Vilas Wuwongse	Asian Institute of Technology, Thailand
Jianliang Xu	Hong Kong Baptist University, China
Haruo Yokota	Tokyo Institute of Technology, Japan
Ge Yu	Northeastern University, China
Jeffrey Xu Yu	Chinese University of Hong Kong, China
Rui Zhang	University of Melbourne, Australia
Aidong Zhang	University at Buffalo, SUNY, USA
Yanchun Zhang	Victoria University, Australia
Aoying Zhou	East China Normal University, China

Industrial Committee

Rafi Ahmed	Oracle, USA
Edward Chang	Google, China and University of California Santa Barbara, USA
Dimitrios Georgakopoulos	CSIRO, Australia
Naoko Kosugi	NTT Corporation, Japan
Kunio Matsui	Nifty Corporation, Japan
Mukesh Mohania	IBM Research, India
Yasushi Ogawa	Ricoh Co. Ltd., Japan
Makoto Okamoto	Academic Resource Guide, Japan
Takahiko Shintani	Hitachi, Ltd., Japan

Demo Committee

Lin Dan	Missouri University of Science and Technology, USA
Feifei Li	Florida State University, USA
Sanjay Kumar Madria	Missouri University of Science and Technology, USA
Pedro Jose Marron	University of Bonn, Germany
Sebastian Michel	Max-Planck-Institut für Informatik, Germany
Makoto Onizuka	NTT CyberSpace Laboratories, NTT Corporation, Japan
Chedy Raissi	National University of Singapore, Singapore
Lakshmish Ramaswamy	The University of Georgia, Athens, USA
Lidan Shou	Zhejiang University, China
Lei Shu	Osaka University, Japan
Tomoki Yoshihisa	Osaka University, Japan
Koji Zettsu	National Institute of Information and Communications Technology, Japan
Xuan Zhou	CSIRO, Australia

Workshop Committee

Qiang Ma	Kyoto University, Japan
Lifeng Sun	Tsinghua University, China
Takayuki Yumoto	University of Hyogo, Japan

Local Arrangements Committee

Kazutaka Furuse	University of Tsukuba, Japan
Takako Hashimoto	Chiba University of Commerce, Japan
Yoshihide Hosokawa	Gunma University, Japan
Sayaka Imai	Sagami Women's University, Japan

Kaoru Katayama	Tokyo Metropolitan University, Japan
Shingo Otsuka	National Institute for Materials Science, Japan
Akira Sato	University of Tsukuba, Japan
Tsuyoshi Takayama	Iwate Prefectural University, Japan
Hiroyuki Toda	NTT Corporation, Japan
Chen Han Xiong	University of Tsukuba, Japan

External Reviewers

Sukhyun Ahn	Qingsong Jin	Ardian Kristanto
Muhammed Eunus Ali	Kaoru Katayama	Poernomo
Mohammad Allaho	Yoshihiko Kato	Yinian Qi
Parvin Asadzadeh	Hideyuki Kawashima	Meena Rajani
Seyed Mehdi	Hea-Suk Kim	Harshana Randeni
Reza Beheshti	Kyoung-Sook Kim	Gook-Pil Roh
Stéphane Bressan	Georgia Koloniari	Jong-Won Roh
Xin Cao	Susumu Kuroki	Seung Ryu
Ding Chen	Injoon Lee	Sherif Sakr
Keke Chen	Jinseung Lee	Jie Shao
Shiping Chen	Jongwuk Lee	Zhitao Shen
Yi-Ling Chen	Ki Yong Lee	Wanita Sherchan
Hong Cheng	Ki-Hoon Lee	Reza Sherkat
Van Munin Chhieng	Sanghoon Lee	Liangcai Shu
Taewon Cho	Sunwon Lee	Chihwan Song
Jaehoon Choi	Yutaka I. Leon-Suematsu	Kazunari Sugiyama
Tangjian Deng	Kenneth Leung	Keiichi Tamura
Pham Min Duc	Guoliang Li	Takayuki Tamura
Takeharu Eda	Jianxin Li	Masashi Toyoda
Yuan Fang	Jing Li	Mayumi Ueda
Chuancong Gao	Lin Li	Muhammad Umer
Shen Ge	Xian Li	Daling Wang
Nikos Giatrakos	Yu Li	Yousuke Watanabe
Kazuo Goda	Bingrong Lin	Ling-Yin Wei
Jian Gong	Xin Lin	Jemma Wu
Yu Gu	Yimin Lin	Hairuo Xie
Adnene Guabtni	Xingjie Liu	Kefeng Xuan
Rajeev Gupta	Yifei Liu	Masashi Yamamuro
Tanzima Hashem	Haibing Lu	Zenglu Yang
Jenhao Hsiao	Min Luo	Jie (Jessie) Yin
Meiqun Hu	Jiangang Ma	Peifeng Yin
Guangyan Huang	Chris Mayfield	Tomoki Yoshihisa
Oshin Hung	Debapriyay	Naoki Yoshinaga
Rohit Jain	Mukhopadhyay	Weiren Yu
Bin Jiang	Tiezheng Nie	Kun Yue
Lili Jiang	Sarana Nutanong	Dana Zhang

Rong Zhang	Xiaohui Zhao	Xuan Zhou
Shiming Zhang	Bin Zhou	Gaoping Zhu
Wenjie Zhang	Rui Zhou	Ke Zhu
Geng Zhao	Xiangmin Zhou	Qijun Zhu

Organizers

University of Tsukuba

The Database Society of Japan (DBSJ)

In Cooperation with

KIISE Database Society of Korea
The China Computer Federation Database Technical Committee
ARC Research Network in Enterprise Information Infrastructure
Asian Institute of Technology (AIT)
"New IT Infrastructure for the Information-explosion Era", MEXT (Ministry of Education, Culture, Sports, Science and Technology) Grant-in-Aid for Scientific Research on Priority Areas, Japan
Information Processing Society of Japan (IPSJ)
The Institute of Electronics, Information, and Communication Engineers (IEICE)
Japan PostgreSQL Users Group
MySQL Nippon Association
The Japanese Firebird Users Group

Sponsoring Institutions

Platinum Sponsors

BeaconIT, Japan

MITSUBISHI ELECTRIC
CORPORATION, Japan

Gold Sponsors

National Institute for
Materials Science (NIMS),
Japan

KDDI R&D Laboratories
Inc., Japan

National Institute of
Advanced Industrial
Science and Technology
(AIST), Japan

Silver Sponsors

HITACHI
Inspire the Next

Ricoh Co., Ltd., Japan

NTT DATA
CORPORATION, Japan

Hitachi, Ltd., Japan

Bronze Sponsors

Ricoh IT Solutions Co.,
Ltd., Japan

SRA OSS, Inc., Japan

Table of Contents – Part I

XML Search and Matching

Graphs

Spatial Databases

XML Technologies

Time Series and Streams

Advanced Data Mining

Query Processing

Web

Sensor Networks and Communications

Information Management

Communities and Web Graphs

Table of Contents – Part II

Data Streams

Similarity Search and Event Processing

Storage and Advanced Topics

Industrial

Demo

Tutorials and Panels

Knowledge on the Web: Robust and Scalable Harvesting of Entity-Relationship Facts

Gerhard Weikum

Max-Planck Institute for Informatics
Saarbruecken, Germany
`weikum@mpi-inf.mpg.de`

Abstract. The proliferation of knowledge-sharing communities like Wikipedia and the advances in automatic information extraction from semistructured and textual Web data have enabled the construction of very large knowledge bases. These knowledge collections contain facts about many millions of entities and relationships between them, and can be conveniently represented in the RDF data model. Prominent examples are DBpedia, YAGO, Freebase, Trueknowledge, and others.

These structured knowledge collections can be viewed as "Semantic Wikipedia Databases", and they can answer many advanced questions by SPARQL-like query languages and appropriate ranking models. In addition, the knowledge bases can boost the semantic capabilities and precision of entity-oriented Web search, and they are enablers for value-added knowledge services and applications in enterprises and online communities.

The talk discusses recent advances in the large-scale harvesting of entity-relationship facts from Web sources, and it points out the next frontiers in building comprehensive knowledge bases and enabling semantic search services. In particular, it discusses the benefits and problems in extending the prior work along the following dimensions: temporal knowledge to capture the time-context and evolution of facts, multilingual knowledge to interconnect the plurality of languages and cultures, and multimodal knowledge to include also photo and video footage of entities. All these dimensions pose grand challenges for robustness and scalability of knowledge harvesting.

H. Kitagawa et al. (Eds.): DASFAA 2010, Part I, LNCS 5981, p. 1, 2010.

Cloud Data Management @ Yahoo!

Raghu Ramakrishnan

Yahoo! Research, USA
ramakris@yahoo-inc.com

Abstract. In this talk, I will present an overview of cloud computing at Yahoo!, in particular, the data management aspects. I will discuss two major systems in use at Yahoo!–the Hadoop map-reduce system and the PNUTS/Sherpa storage system, in the broader context of offline and online data management in a cloud setting.

Hadoop is a well known open source implementation of a distributed file system with a map-reduce interface. Yahoo! has been a major contributor to this open source effort, and Hadoop is widely used internally. Given that the map-reduce paradigm is widely known, I will cover it briefly and focus on describing how Hadoop is used at Yahoo!. I will also discuss our approach to open source software, with Hadoop as an example.

Yahoo! has also developed a data serving storage system called Sherpa (sometimes referred to as PNUTS) to support data-backed web applications. These applications have stringent availability, performance and partition tolerance requirements that are difficult, sometimes even impossible, to meet using conventional database management systems. On the other hand, they typically are able to trade off consistency to achieve their goals. This has led to the development of specialized key-value stores, which are now used widely in virtually every large-scale web service.

Since most web services also require capabilities such as indexing, we are witnessing an evolution of data serving stores as systems builders seek to balance these trade-offs. In addition to presenting PNUTS/Sherpa, I will survey some of the solutions that have been developed, including Amazon's S3 and SimpleDB, Microsoft's Azure, Google's Megastore, the open source systems Cassandra and HBase, and Yahoo!'s PNUTS, and discuss the challenges in building such systems as "cloud services", providing elastic data serving capacity to developers, along with appropriately balanced consistency, availability, performance and partition tolerance.

H. Kitagawa et al. (Eds.): DASFAA 2010, Part I, LNCS 5981, p. 2, 2010.

Distributed Cache Indexing for Efficient Subspace Skyline Computation in P2P Networks

Lijiang Chen[1], Bin Cui[1], Linhao Xu[2], and Heng Tao Shen[3]

[1] Department of Computer Science & Key Laboratory of High Confidence Software
Technologies (Ministry of Education), Peking University
{clj,bin.cui}@pku.edu.cn
[2] IBM China Research Lab, China
xulinhao@cn.ibm.com
[3] School of ITEE, University of Queensland, Australia
shenht@itee.uq.edu.au

Abstract. Skyline queries play an important role in applications such as multi-criteria decision making and user preference systems. Recently, more attention has been paid to the problem of efficient skyline computation in the P2P systems. Due to the high distribution of the P2P networks, the skyline computation incurs too many intermediate results transferred between peers, which consumes mass of the network bandwidth. Additionally, a large number of peers are involved in the skyline computation, which introduces both heavy communication cost and computational overhead. In this paper, we propose a novel *Distributed Caching Mechanism* (DCM) to efficiently improve the performance of the skyline calculation in the structured P2P networks, using a *Distributed Caching Index* (DCI) scheme and an advanced cache utilization strategy. The DCI scheme is employed to efficiently locate the cache that can properly answer a future skyline query. Exploring the property of *entended skyline*, we can optimize the utilization of the cached results for answering future skyline queries. Extensive evaluations on both synthetic and real dataset show that our approach can significantly reduce both bandwidth consumption and communication cost, and greatly shorten the response time.

1 Introduction

Peer-to-peer (P2P) computing and its applications have attracted much attention recently. Due to its advanced features like scalability, flexibility, computing and storage capability, it has been widely employed in various applications such as resource sharing, distributed data management [9]. In the field of data management, P2P systems have been successfully exploited to support different types of queries, such as conventional SQL queries [9] and range queries [10]. Presently, skyline computation against a large-scale multi-dimensional data in P2P networks has gained increasing interests [3,5,15], since the skyline operator [2] has widely been applied in data mining, multi-criteria decision making and user preference systems.

A skyline query over a set of d-dimensional data, selects the points which are not dominated by any other point in the database. A point p_1 dominates another point p_2, if

H. Kitagawa et al. (Eds.): DASFAA 2010, Part I, LNCS 5981, pp. 3–18, 2010.

p_1 is no worse than p_2 in any dimension and is better than p_2 in at least one dimension. Generally, the skyline query is evaluated based on every dimension of the database. However, in the real applications, users may have specific interests in different subsets of dimensions. Thus skyline queries are often performed in an arbitrary subspace according to users' preferences. We refer to this type of query as *Subspace Skyline Query*, and focus on the problem about the subspace skyline query cache mechanism in this paper.

There is a long stream of research on solving the skyline query problem, and many algorithms have been developed for this purpose on either centralized, distributed or fully-decentralized P2P environment [2,3,5,15]. However, no one has utilized any caching mechanism to improve the performance of the skyline computation. The caching mechanisms are widely used to facilitate the query processing in the P2P networks [10,14]. The query results are cached at peers and used to answer future skyline queries, which can significantly reduce the response time and communication cost. However, the skyline query is very different from the conventional SQL query and similarity query, which can be split into sub-queries and whose results can be combined from sub-queries or overlapped queries. Taking range query as example, we can find the caches covering or overlapping the query range to obtain the answer. On the contrary, it is difficult to share the result sets with the skyline queries on different subspaces, because of the characteristic of the skyline query. In other words, the cache for a certain query result can only be used to answer the query which is exactly same as the cached one (on same subspace), which abates the efficiency of the existing caching methods tremendously.

On the other side, we can expect that the use of the cache mechanism for our P2P skyline computation can gain significant benefits for the following two reasons. First, in the skyline computation, almost all of peers in the P2P network will be involved. Even some progressive algorithms still have to access a huge number of peers. Therefore, the cache mechanism has potential to enhance the skyline computation in the P2P networks. Second, given a d-dimensional dataset, there are $2^d - 1$ different skyline queries over any of the non-empty subspaces of the whole space. Each skyline query demands an individual result with a set of skyline points, i.e., there are $2^d - 1$ different skyline results in total respecting to d-dimensional space. Assume that all these results have been cached, then any query can be answered with a marginal cost to locate the cache and fetch the answer. The scalability and huge storage capability features of P2P systems make it possible for the assumption.

In this paper, we propose a novel approach to efficiently improve the performance of the skyline computation in a structured P2P network using *Distributed Caching Mechanism* (DCM). Specifically, we design a *Distributed Caching Index* (DCI) method to manage the distributed caches. Thus the query peer can locate the promising cache efficiently and an advanced cache utilization strategy can exploit the caches containing answers when no exactly matched cache is found. Our approach has two advantages: (1) our DCM approach is overlay independent, as the index value for a certain query result is unique for the skyline query, and this value can be used to locate the peer which indexes the value. The index value is single dimensional and hence can be easily supported by any type of structured P2P overlay; (2) our caching mechanism is orthogonal to the query algorithm. The purpose of the query algorithm here is to conduct the skyline query when the cache which contains the answer is not available. Videlicet, any skyline computation algorithm can be adapted to our DCM in P2P networks.

In this work, we focus on providing a general solution for caching skyline query in a structured P2P overlay and contribute to its advancements with the following:

- We propose a novel *Distributed Caching Mechanism*, which explores the *Distributed Caching Index* technique to maintain the distributed result caches. The DCI is well adapted to the structured P2P network and can locate the cached result progressively.
- A concept of *extended skyline* [13] which is the superset of skyline, is used for cache in our caching mechanism. Exploring the property of the *extended skyline*, we propose an advanced cache utilization strategy to optimize the DCM by maximizing the utilization of the existing caches. We are able to use the cached query results to the best of its abilities when no exactly matched cache is found. Thus the expensive skyline query processing can be avoided as much as possible.
- We conduct extensive experiments on both real and synthetic datasets to evaluate the performance of the proposed caching mechanism. Particularly, we exploit the SSP [15] and TDS [16] skyline query algorithm as the baseline approaches. We adopt the BATON [8] structure, which is used in SSP [15], as the P2P overlay. The same experimental platform can fairly evaluate the effect of our caching mechanism. Our results show that the DCM can greatly reduce the network traffic consumption and shorten the response time.

The rest of this paper is organized as follow. The preliminaries of the work are reviewed in Section 2. Section 3 presents our *Distributed Caching Mechanism*, including the challenges, index strategy and skyline calculation algorithms. The extensive experimental study is reported in Section 4, and finally we conclude the paper in Section 5.

2   Preliminaries

In this section, we first discuss the related work, followed by the notations and definitions in this paper.

2.1   Related Work

Skyline query processing was first introduced into database systems by Borzonyi et al. [2], with the Block Nested Loop (BNL) and Divide-and-Conquer algorithms. Most early researches on the skyline query are focused on the traditional centralized database. Chomicki et al. [4] proposed a Sort-Filter-Skyline algorithm as a variant of BNL. Godfrey et al. [7] provided a comprehensive analysis of those aforementioned algorithms without indexing support, and proposed a new hybrid method Linear Elimination Sort for Skyline. Yuan et al. [16] investigated the subspace skyline computation. It utilizes two sharing strategy based algorithms, Bottom-Up and Top-Down algorithms, to compute all possible non-empty subset skyline points of a given set of dimensions. The skyline computation in both distributed and fully-decentralized P2P systems has attracted more attention recently. Wang et al. [15] proposed the Skyline Space Partitioning (SSP) approach to compute skylines on the BATON [8]. Vlachou et al. [13] proposed a threshold based algorithm called SKYPEER, to compute subspace skyline queries on a super-peer architecture P2P network. It use extended skyline to share subspace skyline results from all peers. But every peer in the network should pre-computes all extended subspace skyline results of its data. In this paper, we are the first to combine the cache mechanism and extended skyline. All peers in our network need not to do pre-computation as SKYPEER did. And our aim is to minimize the skyline computation processing at peers, which is a heavy and costly task. What's more, our approach can be applied to any structured P2P networks.

Another area relevant to our work is caching query results in the P2P systems. The caching mechanism is essential to improve the query performance in the P2P systems, as it can significantly shorten the response time and reduce the communication cost. For decentralized unstructured P2P systems, Wang et al. [14] proposed a distributed caching mechanism that distributes the cache results among neighboring peers. For structured P2P systems, Saroiu et al. [12] investigated a content caching mechanism for KaZaA [1] P2P network, which shows a great effect on the search performance. Sahin et al. [10] employed caching to improve range queries on the multidimensional CAN [11] P2P system, based on Distributed Range Hashing (DRH). To the best of our knowledge, there is no previous research work on the skyline query caching in the P2P networks, and we cannot simply adopt any existing cache mechanism in this scenario due to the characteristic of the skyline query.

2.2 Notations and Definitions

Without loss of generality, we make two assumptions: 1) all values are numeric; and 2) the larger values on each dimension are preferred by users in their skyline queries. In practice, we can easily apply our proposal to the cases where the smaller values are preferred by adding a negative sign to each values on the relevant dimensions. All notations used throughout this paper are listed in Table 1.

Table 1. Notations Used in Discussions

Symbol	Description
d	Data space dimensionality
Ω	Unit hypercube $[0, 1]^d$
P	A peer in the P2P system
$\mathcal{D}$	Dataset stored in the P2P system
O	A data object in $\mathcal{D}$

Definition 1. (***Dominating Relationship***) *Let the data space be a d-dimensional hypercube* $\Omega = [0, 1]^d$, *and the dataset* $\mathcal{D}$ *be a set of points in* Ω. *For any two points* $u, v \in \mathcal{D}$, *u dominates v if* $u_i \geq v_i$ *where* $1 \leq i \leq d$ *and there exists at least one dimension j such that* $u_j > v_j$.

Definition 2. (***Skyline Point***) *A data point u is a skyline point if u is not dominated by any other point v in* $\mathcal{D}$.

Consider a running example as follow: a travel agency online has a list of hotels, each has three attributes, listed in *Table 2*, i.e. *grade (star)*, *occupancy rate* and *discount*. H_2 is a skyline point as it has the largest value on the dimension *occupancy rate* and is not dominated by any other object.

Table 2. Hotel Attributes and Values

	Grade (Star)	Occupancy Rate	Discount
H_1	5	0.6	0.1
H_2	4	0.8	0.2
H_3	2	0.7	0.4
H_4	5	0.4	0.5
H_5	3	0.5	0.3

Definition 3. (*Subspace Skyline Point*) *Given a subspace S and a data object $\mathcal{O}$ in dataset $\mathcal{D}$, the projection of $\mathcal{O}$ in the subspace S, denoted by $\mathcal{O}_S$, is an $|S|$-tuple $(\mathcal{O}.d_{s1}, \ldots, \mathcal{O}.d_{s|S|})$, where $d_{s1}, \ldots, d_{s|S|} \in S$. $\mathcal{O}_S$ is the subspace skyline point if it is not dominated by the projection of any other object in $\mathcal{D}$ on the subspace S. Thus we call $\mathcal{O}$ a subspace skyline point on the subspace S.*

Definition 4. (*Query Containing Relationship*) *Suppose two subspace skyline queries $\mathcal{Q}_p$ and $\mathcal{Q}_q$ belong to subspaces S_p and S_q respectively. Let S_i ($1 \leq i \leq |S_p|$) be the s_i-th dimension of d, and S_j ($1 \leq j \leq |S_q|$) be the s_j-th dimension of d, then we define the query containing relationship ($\subset$) as: $\mathcal{Q}_p \subset \mathcal{Q}_q \Leftrightarrow \{(\forall s_i(\in \mathcal{Q}_p)=1, s_i(\in \mathcal{Q}_q)=1) \wedge (\exists s_j(\in \mathcal{Q}_q)=1, s_j(\in \mathcal{Q}_p)=0)\}$. Correspondingly, we can also say that $\mathcal{Q}_p$ is a contained skyline query of $\mathcal{Q}_q$.*

For example, based on *Definition 4*, if $\mathcal{Q}_p=(0,1,1)$ and $\mathcal{Q}_q=(1,1,1)$ then we have $\mathcal{Q}_p \subset \mathcal{Q}_q$, as $d_2, d_3(\in \mathcal{Q}_p)=1$ and $d_2, d_3(\in \mathcal{Q}_q)=1$, while $d_1(\in \mathcal{Q}_q)=1$ and $d_1(\in \mathcal{Q}_p)=0$.

3   Distributed Caching Mechanism

This section will introduce our *Distributed Caching Mechanism* algorithm, which utilizes a well-adapted *Distributed Caching Index* technique to index the distributed skyline caches and route the skyline query to the promising peers.

Before presenting our approach, we specify the concept of *cache*. The term *cache* here refers to a set of skyline points, which are the result of a certain skyline query. In this paper, we divide cache into two categories: *local cache* and *global cache*. The local cache is the skyline result of the local dataset at a peer, while the global cache is the final skyline result of all peers in the network. All local caches are thought to be stored at their owner peers. In practice, we can compute the local skyline in each peer on the fly as it is time efficient and does not incur any communication cost. However, the global cache management is a non-trivial task as we cannot store all the global caches at each peer due to the storage limitation at each peer and the maintenance cost over the network. Therefore, the purpose of our work is to design a method to manage and utilize the global cache in an effective way.

3.1   Indexing Cached Results on P2P Overlay

After the skyline results have been computed on the P2P network, the results are then cached at the query originator peer. To answer future skyline queries based on the cached results, an efficient index structure is designed on the structured P2P overlay for indexing the cached skylines, which is named as *Distributed Caching Index* (DCI). Each skyline result set is indexed by a unique value which is maintained by a certain peer whose index range contains the index value. We should notice that only the query originator caches the skyline results and the location of the cached skylines is published via the index value on the P2P network to facilitate the skyline computation. Therefore, the DCI distributes the indices proportionally over the P2P network, which can guarantee a balanced workload at each peer. In what follows, we present the proposed DCI method based on the BATON overlay and then discuss how to deploy the proposed DCI on DHT-based overlay at the end of this section.

Formally, for a d-dimensional data space, given a skyline query $\mathcal{Q}_i=(d_1,...,d_d)$, we define its value $V_{\mathcal{Q}_i}=d_1 \times 2^0 + ... + d_d \times 2^{d-1}$ and the maximum value of all skyline queries $V_{\mathcal{Q}_{max}}=2^0+2^1+...+2^{d-1}=2^d-1$, which is also equal to the number of different subspace skyline queries. Then, the index value V_{I_i} of $\mathcal{Q}_i$ is calculated as follows:

$$V_{I_i} = \mathcal{R}_{min} + \frac{V_{\mathcal{Q}_i}}{\lceil \frac{V_{\mathcal{Q}_{max}}}{\mathcal{R}_{max}-\mathcal{R}_{min}} \rceil} \tag{1}$$

where $\mathcal{R}_{max}$ and $\mathcal{R}_{min}$ indicate the maximum and minimum values of the index range of the BATON network respectively. Note that the cache index V_{I_i} satisfies the condition $V_{I_i} \in [\mathcal{R}_{min}, \mathcal{R}_{max})$, which guarantees the cache index can be found at a certain peer in the BATON network.

Publishing Cache Index: After a skyline query $\mathcal{Q}_i$ has been processed, the query originator P first calculates the cache index value V_{I_i} and then checks if its own index range contains V_{I_i}. If true, P publishes the cache index of $\mathcal{Q}_i$ in itself. Otherwise, P routes the cache index publishing message to the proper peer according to the BATON protocol [8].

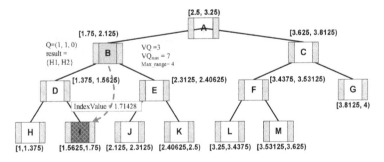

Fig. 1. An Example of Cache Index Publishing

Example 1. Consider the example shown in the Figure 1 and suppose the index range of the BATON is [1,4). If the peer B has processed a skyline query $\mathcal{Q}=(1,1,0)$ and obtains the result $\{H_1, H_2\}$, then B caches the result and publishes the result to the peer I. This is because the index value of $\mathcal{Q}=(1,1,0)$ is $3/\lceil 7/4 \rceil = 1.71428 \in [1.5625, 1.75)$, which belongs to the index range of the peer I.

Processing Query with Cache Index: If the cache index of a skyline query Q_i has been published on the P2P network and Q_i is issued by another peer afterward, then the peer can locate a proper skyline cache to answer Q_i. At the beginning, the query issuer P checks if its index range contains the cache index V_{I_i}. If true, then the cache index on P can be used directly to locate the cache. The query issuer then visits the peer who caches the skyline result using $addr$ parameter. After that, the query issuer publishes a new cache index pointing to itself. If V_{I_i} does not belong to the index range of P, Q_i will be routed to a proper peer according to the BATON protocol. According to the BATON protocol, the routing cost of the skyline query to a certain peer is $O(\log N)$, where N is the total peer population. Thus the publishing or searching cost of a cache index with the DCI is also $O(\log N)$.

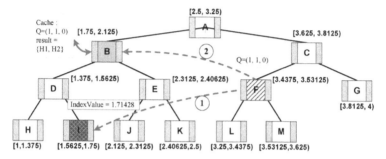

Fig. 2. A Query Example of using Cache Index

Example 2. The example in the Figure 2 shows how to process a skyline query by using the DCI technique. Suppose the peer B issued $Q=(1, 1, 0)$ before and cached the result. When a peer F issues the same query Q, it first tries to find if the results of Q have been cached. To this end, F visits the peer I since the index value 1.71428 falls into its index range. As such, F obtains the cached results from B.

The above algorithm can compute skyline efficiently with the cached results. Once a specific cache is established, it can be explored to answer the following identical skyline queries. However, due to the dynamicity of the P2P system, we should address the problem of the cache update. Our approach is similar to [10]. First, the cache may be changed when data update occurs. When some data object at a peer updates (insert or delete), we should check whether the update changes the local skylines. If the update does not change the skyline in any subspace, no further operation is needed. Otherwise, if the affected local skylines have been already cached, we send the skyline update to the corresponding global cache. We further check the global cache and see if we need to take further operations. Note that the local cache update may not incur the global cache update. If the global cache needs to be updated, we either update the global cache (for data insertion), or just remove the cache and corresponding index (for data deletion). In the deletion case, if the global cache is kept, we need re-examine it by comparing with the skylines from other peers, which is expensive and unacceptable.

Another possible factor affecting cache is node join or leave. If a new peer participates the P2P system carrying some new data objects, it informs all the index-maintaining peers and all those peers record the information. For the following queries, the indexing peer returns the cached result and notifies the originator peer to connect the new peer for the additional skyline candidates, and then it deletes the marked updates. The originator obtains the local skylines from the new peer and then merges them with the cached global skylines. If the global skylines are changed, the originator publishes the new cache index and disables the old ones. On the other side, the node leave may also trigger the cache update problem. When a peer leaves the network, one of its neighbors takes over its index range. If it has some data objects, it will hand over its data objects to its neighbor.

3.2   Optimizing Cache Utilization

While the proposed cache mechanism can utilize the cached answers for future skyline queries, the effectiveness is limited, as the cached results for a certain query can be only used to answer the identical future queries.

The skyline queries are often performed in an arbitrary subspace according to the user's preference. Therefore, we need to deal with 2^d-1 subspace skyline queries. An alternative caching mechanism is to pre-materialize a set of skyline points on various subspaces and then store them in the cache for the future skyline queries. Obviously, such approaches are infeasible as there are a large number of skyline answer sets and the maintenance cost could be very expensive. Recall the caching mechanism for the range query processing, the cached results can be reused for any overlapped range query, and thus the caches gain effective utilization. Therefore, the favorable solution for the skyline caching mechanism is to improve the utilization of the cached skyline answers for future queries.

Relationship of Subspace Skyline Results: In [13], Vlachou et al. analyzed the relationship between subspace skyline query results and proposed a new concept of extended skyline which based on the concept of extended domination. They are defined as follow:

Definition 5. (*Extended Domination*) *For any dimension set U, where $U \subseteq \mathcal{D}$, p extended dominates q if on each dimension $i \in U$, $p_i > q_i$.*

Definition 6. (*Extended Skyline*) *The extended skyline is a set of all points that are not extended dominated by any other data point in dataset $\mathcal{D}$.*

The extended domination is a strictly dominating relationship substantively. The extended skyline is a containing superset of normal skyline set. From the above definitions, an important property of subspace skyline relationship can be found [13]:

Theorem 1. *For two extended skyline queries, if $\mathcal{Q}^*_p \subset \mathcal{Q}^*_q$ then $\mathcal{R}^*_p \subseteq \mathcal{R}^*_q$.*

With the property of the extended skyline, we are able to cache the extended skyline results for the future skyline queries. As long as the results of a containing-query can be found, we just get the results from the cache and do not need to issue a new skyline query to the P2P system, which can significantly increase the cache hit rate and reduce the processing cost. Additionally, it is very easy to obtain the exact skyline points from the extended skyline points, as the result set is small and we only need to remove the dominated data points.

The Enhanced Caching Algorithm: In *Section 3.1*, only the cache of the identical skyline query can be used for a coming skyline query. If a skyline query was not is-sued before, there does not exist any matched cache. A naive method is to broadcast the skyline query to the whole P2P network and ask each peer to compute its local sky-line. Finally, the originator peer merges all returned local skylines to identify the global skylines. The progressive algorithms, such as the SSP [15], still involve a great num-ber of peers to participate the process of the skyline computation, with both expensive communication cost and data transferring cost. Therefore, the key point to improve the skyline computation performance is to increase the utilization ratio of the cached results for answering future queries. The proposed cache utilization strategy is designed to ex-ploit the result containment between skyline queries, and hence it can greatly alleviate the expensive skyline computation over the P2P network.

According to *Theorem 1*, we have a relationship between two skyline queries: $Q^*_p \subset Q^*_q \Rightarrow R^*_p \subseteq R^*_q$, if we conduct the extended skyline query. Obviously, when a skyline query Q_p is issued for the first time, no exactly matched cache can be found for this query. However, if we can find one cache for its containing-query Q_q which is available in the P2P network, then we can use R^*_q to answer Q_p since R^*_q contains all the skyline points of R_p. The communication cost of this method is $O(\log N)$ where N is the number of peers in the P2P network, while the data transferring cost is equal to the size of R^*_q. In this way, we can avoid computing a skyline query over the whole P2P network if there exists its containing query.

In order to find the cache of a containing-query of the skyline query Q_p, a naive solution is to orderly select every containing-query Q_q that $Q_p \subset Q_q$, until one desired cache is found. This method guarantees that if the cache of a containing-skyline-query exists, it then can be located and taken into use. However, in the worse case that no cache for any containing-query is available, the maximal communication cost for locating the cache is $(2^{d-|S_q|}) \times (\log N)$, where N is the number of peers in the P2P network. When the dimensionality is large and $|S_p|$ is small, it incurs a heavy cost overhead. The other solution is based on the random selection strategy, i.e., randomly adding one dimension as an active dimension to generate a minimal containing-query and search its cache. Suppose Q_p has $|S_p|$ active dimensions, we randomly select one containing-query Q_q at the first time, which satisfies Q_q has only one more active dimension than Q_p. If the desired cache was not found, we then recursively add another active dimension to Q_q and repeat the same process, until one cache of a containing-skyline-query is found or reaching the containing-query whose dimensions are all active. This method may skip some containing-queries existing in the system, but the maximal communication cost for locating the desired cache is only $(d - |S_p|) \times (\log N)$, where N is the number of peers in the P2P network.

Compared with the above approaches, the first approach can find cache as long as any containing-query cache exists in the network, thus it has a higher cache hit rate. However it may incur a heavy communication cost to locate the desired cache. On the contrary, the second approach has a lower cache hit rate but incurs less suffering in case that the desired cache is not available. So the overall optimal performance is a tradeoff between the cache hit rate and the communication cost when locating the desired cache. In the implementation, we find if there exist some caches for the skyline queries with nearly full dimensionality, the cache hit rate for the second approach is high, and hence yields better performance. In this work, we adopt the latter mechanism in our proposed DCM scheme.

Algorithm 1. dcmSkyline($\mathcal{Q}$)

---

1   $\mathcal{R}_\mathcal{Q}$=∅ /* $\mathcal{R}_\mathcal{Q}$ is the result of $\mathcal{Q}$ */;
2   $\mathcal{R}_{\mathcal{Q}_i}$=∅ /* $\mathcal{R}_{\mathcal{Q}_i}$ is the result of $\mathcal{Q}_i$ that is $\mathcal{Q}$'s containing skyline query */;
3   compute index value $V_{I_\mathcal{Q}}$ of $\mathcal{Q}$ according to the Formula 1;
4   find the peer P whose $V_{I_\mathcal{Q}} \in [P.minSubRange, P.maxSubRange)$;
5   **if** (*find $\mathcal{Q}$'s cache index in P*) **then**
6     $cacheIndex$ =< $\mathcal{Q}, addr$ >;
7     $\mathcal{R}_\mathcal{Q}$=fetchCache(*cacheIndex*);
8   **if** ($\mathcal{R}_\mathcal{Q} \neq$ ∅) **then**
9   indexPublish($\mathcal{Q}, addr$);
10   **return** $\mathcal{R}_\mathcal{Q}$;

11   **else**
12   randomly select $\mathcal{Q}_i$ that $\mathcal{Q} \subseteq \mathcal{Q}_i \wedge \mathcal{Q} \oplus \mathcal{Q}_i$==1;
13   **if** ($\mathcal{Q}_i$ *exists*) **then**
14       $\mathcal{R}_{\mathcal{Q}_i}$=dcmSkyline($\mathcal{Q}_i$);
15   **if** ($\mathcal{R}_{\mathcal{Q}_i} \neq$ ∅) **then**
16         $\mathcal{R}_\mathcal{Q}$=skyline($\mathcal{R}_{\mathcal{Q}_i}$);
17   indexPublish($\mathcal{Q}, addr$);
18   **return** $\mathcal{R}_\mathcal{Q}$;

19   **else**
20   process the skyline query $\mathcal{Q}$;

Algorithm 1 shows the process of the skyline query computation based on the enhanced cache utilization strategy. Given a skyline query $\mathcal{Q}$, the query originator peer computes the index value $V_{I_\mathcal{Q}}$, according to the *Formula 1*. Then the peer P whose index range covers $V_{I_\mathcal{Q}}$ is found (lines 3-4). If $V_{I_\mathcal{Q}}$ is in P, $\mathcal{Q}$ is then routed to the peer that keeps the desired cache and the cached results are returned to the query originator peer. After that, a new cache index is published (lines 5-10). However, if the cache for answering $\mathcal{Q}$ is not available in the P2P network, the query originator peer will randomly select one of its containing-query $\mathcal{Q}_i$ which satisfies the condition that $\mathcal{Q}_i$ is the smallest containing-query of $\mathcal{Q}$ (line 12). If $\mathcal{Q}_i$ is a valid query which has one more active dimension than $\mathcal{Q}$, the *Algorithm 1* will be called recursively to find the cache for the containing-query $\mathcal{Q}_i$, until one of these caches is found or reaching the largest containing-query (i.e., the skyline query with full dimensionality) (lines 13-14). If the containing cache is found, the query originator peer computes the skyline query $\mathcal{Q}$ based on that returned cache to get the final skyline results and publishes a new cache index to the P2P network (lines 15-18). At last, if no cache for any containing-query is available, the query originator peer conducts the skyline computation algorithm to get the answers (e.g., using the SSP algorithm to find the skyline points) (lines 19-20).

Note that, in all operations related to the cache, the extended skyline points are used, i.e., transferring the cached extended skyline points and caching the extended skyline points. Only when returning the skyline results to the end-user, the peer computes the exact skyline results from the extended skyline points locally. The correctness of this algorithm is guaranteed by *Theorem 1*.

4 An Experimental Study

In this section, we report the experimental results obtained from the extensive simulation implemented in Java. We study the performance of our skyline caching mechanism DCM on the BATON overlay with respect to three aspects: dimensionality, network size and data size. The proposed DCM is evaluated against the adapted TDS approach based on the advanced centralized skyline algorithm TDS [16] and the progressive P2P skyline algorithm SSP [15]. For two baseline algorithms, we examine the performance with and without cache. In the scenario that the cache mechanism is not available, the query originator computes the skyline using the skyline query algorithm. For example, in the TDS approach, the query originator peer floods the query to the whole P2P network; upon receiving a query request from the query originator, each peer computes its local skyline, returns the result back to the query originator peer, and finally these local skylines are merged to generate the global skyline. On the contrary, if the previous query results are cached, the query originator always tries to find the answer from the cache directly. We consider the following five performance metrics:

- **Cache hit probability:** the probability that the cache of the exactly match skyline query or the containing skyline queries is found.
- **Involved node number:** the node population visited by the skyline query, including originator nodes, nodes routing query, nodes delivering message and nodes conducting skyline query.
- **Message communication cost:** the total number of messages transferred in the network for skyline computation.
- **Response time:** the overall elapsed time to answer the skyline query.

The experimental results show that the above metrics include all processing cost in the skyline computation, i.e., queries delivering and computing, intermediate results transferring, cache and index maintaining.

The performances are measured through simulation experiments on a Linux Server with four Intel Xeon 2.80GHz processors and 2.0GB RAM. All experiments are repeated 10 times, and each of them issues 1000 skyline queries starting from a random node. We distribute the data points randomly into all peers in the BATON network. We use three kinds of different datasets: two synthetic datasets of independent and anti-correlated distribution, which have up to 8 million data points, and a real dataset of the NBA players' season statistics from 1949 to 2003 [6], which approximates a correlated data distribution. Our approach yields similar performance superiority for three datasets, as the performance is mainly affected by the cache hit rate. Although, more skyline points need to be transferred in the network for anti-correlated data, the performance improvement of anti-correlated dataset is almost similar to that of independent data. Therefore, we only present the results on the synthetic independent datasets here due to the space constraint. The parameters used in the experiments are listed in Table 3. Unless stated otherwise, the default parameter values, given in bold, are used.

Table 3. Parameters Used in Experiments

Parameter	Setting
Dimensionality	4, 5, ..., 10
Peer population	$2^7, ..., 2^{12}$
Data volume at each peer	50, 100, 200, 400, **800**, 8000

4.1 Cache-Hit Probability

We first study the cache-hit probability of our proposed DCM. The cache hit for the DCM we calculate is not only the probability of the exactly-matched cache hit, but also with the probability of the containing-query cache. The cache-hit probability is related to the queries which have been processed before, and has no relationship with both data size and network size. Therefore, only dimensionality is concerned as it determines the total number of distinct queries. The results in the Figure 3(a) show that the probability drops down when the dimensionality increases. It reaches 99% when the dimensionality is 4, because the total number of the skyline query results is $2^4 - 1$, which is very small compared with 1000 randomly generated subspace skyline queries and undoubtedly has many repeated queries. As expected, due to the number of different queries increases exponentially (as dimensionality increases), the hit rate drops accordingly. For example, the cache hit rate for the exact-match method is less than 40% when the dimensionality equals to 10. However, our DCM approach is very robust with respect to the dimensionality change. Even though the dimensionality reaches 10, the cache-hit probability remains about 90%. Enhanced with our optimized cache utilization strategy, we can get the query result from the caches of the containing-queries, not only from the cache of the identical query.

The experimental results shown in the Figure 3(b) demonstrate the different cache-hit performance between the DCM and the exact-match method. We randomly issue 1000 subspace queries in a 10-dimensional dataset and compute the cache hit rates for the

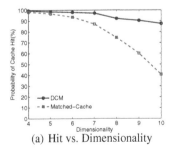
(a) Hit vs. Dimensionality

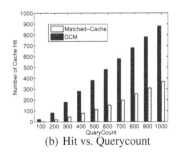
(b) Hit vs. Querycount

Fig. 3. Performance on Probability of Cache Hit

two approaches after every 100 queries have been finished. We can see that the DCM outperforms the exact-match method in all cases. The final probability of the DCM is nearly 90% with the help of the containing-query caches, while the probability of the exact-match method is no more than 40%.

Obviously, the proposed DCM can improve the cache-hit rate and hence it can greatly reduce the processing cost of the baseline skyline algorithms. Since the DCM yields a significant improvement against the exact-match method, we will omit the comparison with the exact-match method in the following experiments, as we concern how the caching mechanism can improve the performance of the baseline algorithms in case of without cache.

4.2 Number of Involved Nodes

The metric *involved node number* counts all nodes who participate in skyline query computation, even including the nodes which only relay the query messages. The results on Figure 4 show that our DCM involves much fewer nodes than the baseline SSP and TDS algorithms. Even when the network size increases, the DCM approaches still yield better performance, and the gap between the DCM approaches and the baseline algorithms remains wide. This is owing to the efficiency of our proposed DCI technique, and we only need $O(logN)$ cost to locate and fetch the cached results in most cases. Therefore, our DCM can reduce the involved peer population and process the skyline query efficiently.

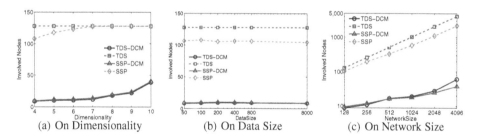

(a) On Dimensionality   (b) On Data Size   (c) On Network Size

Fig. 4. Performance on Number of Involved Nodes

The idea of the TDS [16] cannot work effectively in the P2P environment, because it can get only the partial answers from the caches of the containing queries and have to flood messages to all peers in order to locate the missing answers. Actually, the cache hit rate of the TDS is the same as other centralized approaches as only the cache of the exactly matched query is reusable. On the contrary, with the extended skyline, our DCM approach can obtain the answers from the caches of any containing skyline query directly, and hence the cache-hit rate (or cache utilization rate) is improved.

4.3 Message Communication Cost

The next performance metric we consider is the message communication cost. From Figure 5, we can find that the message cost of the enhanced DCM approaches is much lower than the baseline SSP and TDS algorithms, and is steady with the increase of dimensionality, network size and data size. In the DCM, if the cache was found, the main cost is to locate and return the cache to the query originator. Note that, in case that the cache could be found, our DCI-based approach has only $O(logN)$ cost to locate the peer with the cache.

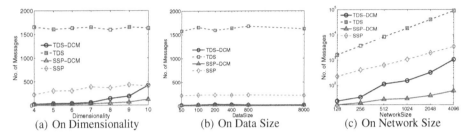

(a) On Dimensionality   (b) On Data Size   (c) On Network Size

Fig. 5. Performance on Message Communication Cost

4.4   Response Time

The response time indicates the average elapsed time for the query originator to obtain the global skylines. The results are shown in Figure 6. Like the previous cases, the DCM outperforms the baseline SSP and TDS approaches. However, the difference is that the SSP performs worse than the TDS. The reason is because the TDS may find the partial skylines from the caches of the containing queries, and it broadcasts the query to all peers to get all "missing" skylines simultaneously; while the SSP must find a certain region according to the partition histories and compute the skyline in that region, and then visits other regions progressively if necessary.

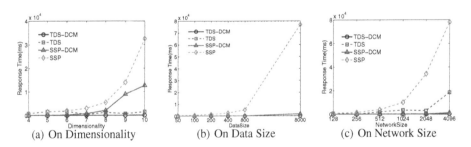

(a) On Dimensionality   (b) On Data Size   (c) On Network Size

Fig. 6. Performance on Response Time

4.5   Effect of Updates

Finally we show the performance of the algorithms on peer updates. We randomly select 1024 peers and 1 million 6-dimensional data points and vary the peer update rate from 5% to 20% within the period of 1000 query executions. The type of the node update behaviors (i.e., node join, node leave, data insert, data deletion and data update) is randomly chosen. We use the response time to evaluate the efficiency of different approaches. In Figure 7, the performance of all DCM-based approaches is better than that of the baseline methods, although the DCM mechanism is deteriorated nearly 4 times comparing to the case without updates, as some caches may be disabled due to the updates. The updates have no impact on the baseline SSP, because it calculates each skyline query in the originated peer and propagates to the promising peers progressively regardless of update. The performance of the TDS is marginally deteriorated by the reason that the intermediate result can not be used when peers update. But the effect is very limited because the crucial consumption of the response time is the network communication, and TDS has to access all the peers to retrieve the missing skylines.

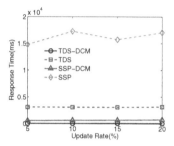

Fig. 7. Performance on Updates

5   Conclusion

In this paper, a novel query-driven caching mechanism DCM is proposed for computing skyline on a structured P2P overlay. The proposed method can be easily applied to other structured P2P overlays and is compatible with any distributed skyline computation algorithm. The DCM uses a distributed indexing strategy to locate the cache and an enhanced cache utilization strategy is introduced to improve the cache performance. We conducted the extensive experiments on various datasets and the experimental results demonstrate the superiority of our approach.

Acknowledgement

This research was supported by the National Natural Science foundation of China under Grant No.60603045 and 60873063.

References

1. http://www.kazaa.com/
2. Borzsonyi, S., Kossmann, D., Stocker, K.: The skyline operator. In: Proc. ICDE, pp. 421–430 (2001)
3. Chen, L., Cui, B., Lu, H., Xu, L., Xu, Q.: isky: Efficient and progressive skyline computing in a structured p2p network. In: Proc. ICDCS (2008)
4. Chomicki, J., Godfrey, P., Gryz, J., Liang, D.: Skyline with presorting. In: Proc. ICDE, pp. 717–719 (2003)

5. Cui, B., Lu, H., Xu, Q., Chen, L., Dai, Y., Zhou, Y.: Parallel distributed processing of constrained skyline queries by filtering. In: Proc. ICDE (2008)
6. Data, B. (2008), http://databasebasketball.com/
7. Godfrey, P., Shipley, R., Gryz, J.: Maximal vector computation in large data sets. In: Proc. VLDB, pp. 229–240 (2005)
8. Jagadish, H.V., Ooi, B.C., Vu, Q.H.: Baton: A balanced tree structure for peer-to-peer networks. In: Proc. VLDB, pp. 661–672 (2005)
9. Ooi, B., Tan, K., Zhou, A., Goh, C., Li, Y., Liau, C., Ling, B., Ng, W., Shu, Y., Wang, X., Zhang, M.: PeerDB: Peering into personal databases. In: Proc. SIGMOD (2003)
10. Sahin, O.D., Gupta, A., Agrawal, D., Abbadi, A.E.: A peer-to-peer framework for caching range queries. In: Proc. ICDE, p. 165 (2004)
11. Ratnasamy, M.R.S., Francis, P., Shenker, S.: A scalable content-addressable network. In: Proc. SIGCOMM, pp. 161–172 (2001)
12. Saroiu, R.S.S., Gummadi, K.P., Levy, H.M.: An analysis of internet content delivery systems. In: Proc. OSDI (2002)
13. Vlachou, A., Doulkeridis, C., Kotidis, Y., Vazirgiannis, M.: Skypeer: Efficient subspace skyline computation over distributed data. In: Proc. ICDE, pp. 416–425 (2007)
14. Wang, C., Xiao, L., Liu, Y., Zheng, P.: Distributed caching and adaptive search in multilayer P2P networks. In: Proc. ICDCS, pp. 219–226 (2004)
15. Wang, S., Ooi, B.C., Tung, A.K.H., Xu, L.: Efficient skyline query processing on peer-to-peer networks. In: Proc. ICDE, pp. 1126–1135 (2007)
16. Yuan, Y., Lin, X., Liu, Q., Wang, W., Yu, J.X., Zhang, Q.: Efficient computation of the skyline cube. In: Proc. VLDB, pp. 241–252 (2005)

iDISQUE: Tuning High-Dimensional Similarity Queries in DHT Networks

Xiaolong Zhang[1], Lidan Shou[1], Kian-Lee Tan[2], and Gang Chen[1]

[1] College of Computer Science, Zhejiang University, China
{xiaolongzhang,should,cg}@cs.zju.edu.cn
[2] School of Computing, National University of Singapore, Singapore
tankl@comp.nus.edu.sg

Abstract. In this paper, we propose a fully decentralized framework called iDISQUE to support tunable approximate similarity query of high dimensional data in DHT networks. The iDISQUE framework utilizes a distributed indexing scheme to organize data summary structures called iDisques, which describe the cluster information of the data on each peer. The publishing process of iDisques employs a locality-preserving mapping scheme. Approximate similarity queries can be resolved using the distributed index. The accuracy of query results can be tuned both with the publishing and query costs. We employ a multi-probe technique to reduce the index size without compromising the effectiveness of queries. We also propose an effective load-balancing technique based on multi-probing. Experiments on real and synthetic datasets confirm the effectiveness and efficiency of iDISQUE.

1 Introduction

In many applications, objects (documents, images, etc.) are characterized by a collection of relevant features which are represented as points in a high-dimensional space. Given a query point, a similarity search finds all data points which are nearest (most similar) to the query point. While there have been numerous techniques proposed for similarity search in high-dimensional space, these are mostly studied in the context of centralized architectures [8,13]. Unfortunately, due to the well-known dimensional curse problem, search in a high-dimensional space is considered as a "hard" problem. It has been suggested that since the selection of features and the choice of a distance metric in typical applications is rather heuristic, determining approximate nearest neighbors should suffice for most practical purposes.

The relentless growth of storage density and improvements in broad-band network connectivity has fueled the increasing popularity of massive and distributed data collections. The Peer-to-Peer (P2P) systems, as a popular medium for distributed information sharing and searching, have been gaining increasing interest in recent years. There is strong demand for similarity search in peer-to-peer networks as well.

Unfortunately, most of the existing P2P systems are not designed to support efficient similarity search. In unstructured P2P networks, there is no guarantee on the completeness and quality of the answers. On the other hand, while structured DHT-based P2P systems [19,16,17] offer an efficient lookup service, similarity search still cannot be

H. Kitagawa et al. (Eds.): DASFAA 2010, Part I, LNCS 5981, pp. 19–33, 2010.

resolved readily. The reason is that the DHT networks usually employ consistent hashing mechanisms which destroy data locality. To facilitate similarity search in DHT networks, we need locality-preserving lookup services, which can map similar data objects in the original data space to the same node in the overlay network.

We advocate a fully decentralized index to organize the high-dimensional data in a DHT network. Our framework is motivated by the following ideas. First, we need a locality-preserving mapping scheme to map the index entries into DHT network. The *locality sensitive hash* (LSH) has been proved to be effective for mapping data with spatial proximity to the same bucket [12]. It has been utilized in [8] to support high dimensional similarity search in centralized setting. By carefully designing the mapping scheme in DHT, we can realize high-dimensional locality among the peers. Second, mapping all data objects of each peer to the DHT would result in a huge distributed index. To reduce the size of the distributed index and the cost of index maintenance, we should publish data summaries to the DHT network instead of the individual data objects. For this purpose, each peer needs a summarization method to derive representative summaries from the data objects. We note that clustering is a typical summarization approach for organizing high-dimensional data. For example, in centralized environment, the *iDistance* [13] approach indexes high-dimensional data in clusters, and provides effective similarity search based on a mapping scheme for these clusters. Although it is difficult to cluster the entire data set in a P2P environment, we can exploit local clustering in each peer to construct a distributed index.

In this paper, we propose a practical framework called iDISQUE to support tunable approximate similarity queries of high-dimensional data in DHT networks. The contributions of our work are summarized as follow:

- We propose a fully decentralized framework called iDISQUE to handle approximate similarity queries, which is based on a novel locality-preserving mapping scheme. We also present a tunable query algorithm for approximate similarity search. The accuracy of query results can be tuned by both the indexing and query costs.
- We present a distributed query multi-probe technique to allow for reducing the size of the indices without compromising the effectiveness of queries. In addition, we introduce a load-balancing technique based on multi-probing.
- We conduct extensive experiments to evaluate the performance of our proposed iDISQUE framework.

The rest of the paper is organized as following: In Section 2, we present the related work. In Section 3, we describe the preliminaries. Next we give an overview of the iDISQUE framework in Section 4. Section 5 presents the locality-preserving indexing in iDISQUE. Section 6 describes the query processing method in iDISQUE. In Section 7, we present the techniques to handle load imbalance. We present the experiment study in Section 8, and finally conclude the paper in Section 9.

2   Related Work

We are aware of a few previous works which propose techniques to support similarity search in P2P environments. These systems can be divided into three main categories.

The first category includes attribute-based systems, such as MAAN [4] and Mercury [3]. In such systems, the data indexing process and similarity search process are based on single dimensional attribute space, and their performances are poor even in low dimension.

The second category including systems such as Murk [7] and VBI-tree [14] is based on multiple dimension data and space partitioning schemes, and maps the specific space regions to certain peers. Although these systems generally perform well at low dimensionality, their performance deteriorates rapidly as the dimension increases due to the "dimensionality of curse".

The third category is metric-based P2P similarity search systems. The examples include [6] [18] and [11]. The SIMPEER framework [6] utilizes the *iDistance* [13] scheme to index high-dimensional data in a hierarchical unstructured P2P overlay, however this framework is not fully decentralized, making it vulnerable to super-peer failure. In [18] the author defines a mapping scheme based on several common reference points to map documents to one dimensional chord. However, their scheme is limited to applications in the document retrieval.

Perhaps the most similar work to ours is [11]. In [11] the author has proposed a algorithm to approximate K-Nearest Neighbor queries in structured P2P utilizing the Locality Sensitive Hashing [8] scheme. However, we argue that their scheme is not a fully decentralized scheme, since their scheme relies on a set of *gateway peers*, where it would incur the "single-failure" problem when the workload of gateway peers is large. More over, the effectiveness of their mapping scheme and load balance scheme are largely dependent on the *percomputed* global statistics, which are difficult to collected in a fully distributed way. While our approach is fully decentralized, and no percomputed global statistics are required. It makes our approach more practical in reality.

3   Preliminaries

In this section, we briefly introduce the basic mechanisms used in our iDISQUE framework, namely the *locality sensitive hashing* and the *iDistance* indexing scheme.

3.1   Locality Sensitive Hashing

The basic idea of locality sensitive hashing (LSH) is to use a certain set of hash functions which map "similar" objects into the same hash bucket with high probability [12]. LSH is by far the basis of the best-known indexing method for approximate nearest-neighbor (ANN) queries. A LSH function family has the property that objects close to each other have higher probabilities of colliding than those that are far apart. For a domain S of the point set with distance measure D, a LSH family is defined as: a family $\mathcal{H} = \{h : S \rightarrow U\}$ is called (r_1, r_2, p_1, p_2)-sensitive for D if for any $v, q \in S$

- if $D(v, q) \leq r_1$ then $Pr_{\mathcal{H}}[h(v) = h(q)] \geq p_1$;
- if $D(v, q) \geq r_2$ then $Pr_{\mathcal{H}}[h(v) = h(q)] \leq p_2$,

where $Pr_{\mathcal{H}}[h(v) = h(q)]$ indicates the *collision probability*, namely the probability of mapping point v and q into the same bucket.

To utilize LSH for approximate similarity (K-nearest-neighbor (KNN)) search, we should pick $r_1 < r_2$ and $p_1 > p_2$ [8]. With these choices, nearby objects (those within distance r_1) have a greater chance (p_1 vs. p_2) of being hashed to the same value than objects that are far apart (those at a distance greater than r_2 away).

3.2 The iDistance Indexing Scheme

The *iDistance* [13] scheme is an indexing technique for supporting similarity queries on high-dimensional data. It partitions the data space into several clusters, and selects a reference point for each cluster. Each data object is assigned a one-dimensional iDistance value according to its distance to the reference point of its cluster. Therefore, all objects in the high-dimensional space can be mapped to the single-dimensional keys of a B+-tree. Details of the similarity query processing in iDistance can be found in [13].

4   Overview

In this section, we first give an overview of the iDISQUE framework. Then, we present a simple indexing solution.

4.1 The iDISQUE Framework

The iDISQUE framework comprises a number of peers that are organized into a DHT network. In a DHT network, the basic **lookup** service is provided. Given a key, the lookup service can map it to an ID denoted by ***lookup(key)***, which can be used to find a peer responsible for the key. Without loss of generality, we use Chord [19] as the DHT overlay in our framework.

A peer sharing high-dimensional data is called a *data owner*, while a peer containing index entries of the data shared by other peers is called an *indexing peer*. For sharing and searching the data among the peers, the iDISQUE framework mainly provides the following two services:

– **The index construction service**
  When a peer shares its data, a service called index construction is invoked. The index construction service consists of the following four steps. First, the data owner employs a local clustering algorithm, to generate a set of data clusters. For simplicity in presentation and computation, we assume that all data clusters are spherical in the vector space. Second, the data owner employs the *iDistance* scheme [13] to index its local data in clusters, using the cluster centers created in the previous step as the reference points. Third, for each data cluster being generated in step one, the data owner creates a data structure called *iDisque*(iDIStance QUadruplE), denoted by $C*$ as follows:

$$C* =< C, r_{min}, r_{max}, IP >$$

  where C is the center of the cluster, r_{min} and r_{max} define the minimum and maximum distances of all cluster members to the center, IP is the IP address of the data owner. The above *iDisque* is a compact data structure which describes the data cluster information. Therefore, the data owner is able to capture the summary of its own data via a set of iDisques. Fourth, for each iDisque being generated, the data owner publishes its replicas to multiple peers using a mapping scheme which maps the center of the iDisque to a certain number (denoted by L_p) of indexing peers, then each iDisque is replicated to these indexing peers. In the rest of paper, L_p is called the *publishing mapping degree*. The indexing peers receiving the iDisque will insert it into their local index structures.

- **The querying processing service**
  A peer can submit a similarity query to retrieve the data objects most relevant to the given query. When a similarity (KNN) query is issued, it is first sent to a specified number (denoted by L_q) of indexing peers which may probably contain candidate iDisques, using the same mapping scheme as described above. In the rest of paper, L_q is referred to as the *query mapping degree*. Second, each indexing peer receiving the query will then look up its local part of the distributed index to find colliding iDisques. The query will then be sent to the data owners of these candidate clusters. Third, the data owners of the candidate clusters process the query utilizing their local iDistance indexes, and return their local K-nearest-neighbors as the candidate query results. Finally, the querying node sorts the data in the candidate query result sets by distance to the query point, and produces the final results.

4.2   A Naive Indexing Scheme

For a Chord overlay, a straightforward indexing scheme can be implemented as follows: For each iDisque, we replicate it for L_p copies and map them evenly to the Chord ring ranging from 0 to 2^m-1 at a constant interval of $P = 2^m/L_p$. Constant P is called the *publishing period*. Therefore, any two replicas rep_i and rep_j are mapped to two Chord keys so that

$$|ChordKey(rep_i) - ChordKey(rep_j)| = n \cdot P,$$

where n is an integer in $(1, \ldots, \lfloor L_p/2 \rfloor)$. Such replication scheme guarantees that an interval of length greater than P in the Chord key space must contain at least one replica of an iDisque. Therefore, any interval of length greater than P must contain all iDisques of the entire system. When a query is issued, we randomly choose an interval of length P in the Chord ring, and randomly select L_q keys in this interval ($L_q \leq P$). The query is then delivered to the indexing peers in charge of these L_q keys.

In the above naive indexing scheme, the iDisques are randomly distributed in the network. Therefore, its index storage space is uniformly allocated among the indexing peers, and the query load is balanced too. We denote by iDISQUE-Naive the above naive indexing scheme. We note that if L_q is large enough to cover all peers in the interval, the query is able to touch all iDisques in the system and is therefore accurate. However, a typical L_q value is much smaller than P in a large network. Therefore, the naive scheme cannot achieve high accuracy without a large L_q value. To improve the query accuracy at a low cost, we shall look at a locality-preserving indexing scheme in the iDISQUE framework.

5   Locality-Preserving Index Scheme

In this section, we introduce a locality-preserving indexing scheme for iDISQUE. In the remaining sections, we will only assume the locality-preserving indexing scheme for the iDISQUE framework unless explicitly stated. We begin by introducing a locality-preserving mapping scheme, and then present the index construction method.

5.1   Locality-Preserving Mapping

To facilitate locality-preserving mapping, a family of LSH functions is needed. Without loss of generality, we assume the distance measure to be the $\mathcal{L}_2$ norm. Therefore, we can use the family of LSH functions $\mathcal{H}$ for $\mathcal{L}_p$ norms, as proposed by Datar et al. in [5], where each hash function is defined as

$$h_{a,b}(v) = \lfloor \frac{(a \cdot v + b)}{W} \rfloor,$$

where a is a d-dimensional random vector with entries chosen independently from a p-stable distribution, and b is a real number chosen uniformly from the range $[0, W]$. In the above family, each hash function $h_{a,b} : R^d \to Z$ maps a d-dimensional vector v onto a set of integers. The p-stable distribution used in this work is the Gaussian distribution, which is 2-stable and therefore works for the Euclidean distance.

To resolve similarity queries, the locality-preserving mapping scheme in iDISQUE has to be able to map similar objects (both data and query points) in the high-dimensional space to the same Chord key. Our proposed mapping scheme consists of the following consecutive steps:

First, to amplify the gap between the "high" probability p_1 and "low" probability p_2 (refer to section 3.1), we define a function family $\mathcal{G} = \{g : R^d \to U^k\}$ such that $g(v) = (h_1(v), \ldots, h_k(v))$, where $h_j \in \mathcal{H}$. Each g function produces a k-dimensional vector. By concatenating the k LSH functions in g, the collision probability of far away objects becomes smaller (p_2^k), but it also reduces the collision probability of nearby objects(p_1^k).

Second, to increase the collision probability of nearby objects, we choose L independent g functions, $g_1, \ldots, g_L$, from $\mathcal{G}$ randomly. Each function g_i ($i = 1, \ldots, L$) is used to construct one hash table, resulting in L hash tables. We can hash each object into L buckets using functions $g_1, \ldots, g_L$. As a result, nearby objects are hashed to at least one same bucket at a considerably higher probability, given by $1 - (1 - p_1^k)^L$.

Third, to map each k-dimensional vector, $g_i(v)$, to the Chord key space as evenly as possible, we multiply $g_i(v)$ with a k-dimensional random vector $R_i = [a_{i,1}, \ldots, a_{i,k}]^T$:

$$\rho_i(v) = g_i(v) \cdot [a_{i,1}, \ldots, a_{i,k}]^T \bmod M$$

where M is the maximum Chord key $2^m - 1$. Each element in R_i is chosen randomly from the Chord key space $[0, 2^m - 1]$. We call each ρ_i a LSH-based mapping function, and denote by Ψ_L the function set $\{\rho_i, \ldots, \rho_L\}$. Therefore, given $\Psi_L = \{\rho_1, \ldots, \rho_L\}$, we can map a point $v \in R^d$ to L Chord keys $\rho_1(v), \ldots, \rho_L(v)$. A query at point q is said to collide with data point v if there exists a function ID $i \in \{1, \ldots, L\}$, so that $\rho_i(q) = \rho_i(v)$.

5.2   Index Construction

We now propose the detailed process of index construction in iDISQUE. Given the four steps of the index construction service as described in the previous section, we shall focus on the first and the fourth steps, namely *clustering local data* and *publishing iDisques*, since the other two steps are already clearly described.

Clustering local data: When clustering its local data, a data owner must take two important requirements into consideration. On one hand, as the iDisque structures are published and stored in the DHT network, we need the number of iDisques (clusters) published by a data owner to be as small as possible so that the cost of constructing and storing the distributed index can be limited. On the other hand, the dimension (geometric size) of each cluster must be small enough to guarantee the query accuracy.

In order to balance the number of clusters and the query accuracy, we can tune the entire system using a parameter δ, which specifies the maximum radius allowed for each data cluster. A large δ value reduces the number of iDisques published and maintained in the system, while impairing the accuracy. In contrast, a small δ does the reverse. Due to the restriction of cluster size, one data point might be "singular" as its distance from its nearest neighbor is more than δ. These points are called *singularities*. We shall create a *singular cluster* for each singularity.

It is worth mentioning that although there are several existing clustering algorithms to cluster local data to data cluster with a maximum radius δ, e.g., BIRCH [20], CURE [10], in the implementation we adopt BIRCH [20] to cluster local data due to its popularity and simplicity.

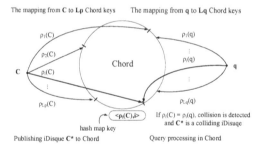

Fig. 1. Publishing and querying an iDisque

Publishing iDisques: In order to publish iDisques to peers, we need a LSH-based mapping function sequence $\Psi_{L_p} = \{\rho_1, \ldots, \rho_{L_p}\}$, whose length (cardinality) is equal to the publishing mapping degree L_p. All data owners use the same set of Ψ_{L_p}.

Given the above locality-preserving mapping scheme, we can publish iDisques in a straightforward approach. Figure 1 shows an example of the mapping and publishing process. First, for each iDisque denoted by $C*$ (meaning that it is centered at C), we map its cluster centroid C to L_p independent Chord keys $\rho_1(C), \ldots, \rho_{L_p}(C)$, using the aforementioned mapping scheme. Second, for each Chord key $\rho_i(C)$, we evoke the Chord **lookup** service to find the peer which owns the Chord key, and send the publishing message containing the respective function ID i, the Chord key $\rho_i(C)$, together with the iDisque, $C*$, to that peer. As multiple keys might be assigned to a same peer in the Chord protocol, the upper bound cost of publishing an iDisque is L_p messages.

When a publishing message is sent to an indexing peer, the recipient inserts the received data into its local data structure, i.e. a hash map indexed by composite key $(\rho_i(C),i)$. This data structure provides efficient local lookup given the Chord key and the mapping function ID.

6   Query Processing in iDISQUE

6.1   Basic Query Processing Scheme

Query processing is analogous to the publishing process as Figure 1. When a KNN query at point q is issued, we map it to L_q Chord keys, named *query Chord key set*, using a LSH-based mapping function sequence $\Psi_{L_q} = \{\rho_1, \ldots, \rho_{L_q}\}$ which is a prefix of sequence Ψ_{L_p} (the first L_q functions in the LSH-based mapping functions pre-defined in the system). Peers in charge of these Chord keys may probably contain *colliding iDisques* (those that are mapped to the same Chord key as the query point via the same mapping function ρ_i). We refer to such indexing peers as *candidate peers*. Then, the querying node distributes the query along with its Chord key and function ID to each respective candidate peer to look for colliding iDisques.

Based on the mapping scheme proposed in section 5.1, the probability of collision between an iDisque centroid C and the query q depends on two factors: (1) the distance between q and C, and (2) the number of function IDs on which the two points *may* collide. Assume the *publishing mapping degree* of an iDisque centered at C is L_p, the latter can be calculated as $min(L_p, L_q)$. Therefore, the probability of collision is estimated as $1 - (1 - p_1^k)^{min(L_p, L_q)}$. In the query process, we can tune the coverage of the query by varying L_q. If L_q is large enough, all iDisques close to the query point in the data space may have a chance to collide with the query on at least one function ID i, where $i \leq L_p$. In contrast, if L_q is small, the query cost is reduced, but its accuracy will inevitably be impaired.

The above procedure supports progressive query refinement during run-time query processing. The querying peer maintain a top K result *queue* sorted by their distances to query point, and distribute the query progressively to candidate peers in an order analogous to $\rho_1, \rho_2, \ldots, \rho_{L_q}$. Each time a new candidate iDisque is returned, the querying peer asks for the top-K data points from the data owners of the candidate iDisques, and merges the local results in the result queue. Meanwhile, the querying peer maintains a temporary list of data owners which it has already asked for data. Therefore, if replicas of the same iDisque are returned, they will simply be discarded.

The query processing in each candidate peer is straightforward. Upon receiving a query, the candidate peer looks up its local hash map structure (as described in section 5.2) for a colliding iDisque. If the query contains multiple function IDs and Chord keys (in case that multiple functions are mapped to the same peer), multiple lookups will be performed. Since some of the resultant colliding iDisques may not be really close to the query point (due to false positives introduced by LSH-mapping), all colliding iDisques will further undergo a distance check in data space. Then the candidate iDisques are sent back to the querying peer. The pseudo code of the basic query algorithm is presented algorithm 6.1.

Algorithm 6.1: QUERYPROCESSING(q, K, L_q)

---

input: q as the query object
$\quad\quad\quad$ K as the number of nearest neighbors
$\quad\quad\quad$ L_q as the query mapping degree
$queue := \emptyset$
Construct Ψ_{L_q} from Ψ_{L_p}
Create *query Chord key set* using Ψ_{L_q}
for each key_i in *query Chord key set*
$\quad$ **do** $\left\{ \begin{array}{l} \text{Search indexing peer } \textbf{lookup}(key_i) \\ \textbf{for} \text{ each } iDisque \text{ returned} \\ \quad \textbf{do} \left\{ \begin{array}{l} \text{Search data owner } iDisque.IP \\ \text{Merge the local top K results to } queue \end{array} \right. \end{array} \right.$
return $(queue)$

6.2 Multi-probing in iDISQUE

In the above basic query processing algorithm, one problem is that the query accuracy is bounded by $min(L_p, L_q)$. When the publishing mapping degree L_p is small, the query accuracy is restricted. Therefore, the publishing mapping degree has to be large to achieve high query accuracy, resulting in large index size and high cost of index publishing. A recent study on a new technique called multi-probing [15] has shed some light on this problem. The multi-probe query technique employs a probing sequence to look up multiple buckets which have a high probability of colliding the target data. As a result, the method requires significantly fewer hash functions to achieve the same search quality compared to the conventional LSH scheme. Inspired by this idea, we propose a distributed multi-probe technique in iDISQUE.

For each mapping function ρ_i, the multi-probe technique creates multiple query keys instead of one query key, which are also probable to collide with the keys of candidate iDisques. The multiple keys are generated in a query-directed method. For a query point q, we first obtain the *basic query keys* $\rho_1(q), \ldots, \rho_{L_q}(q)$. Second, for each basic query key $\rho_i(q)$, we employ the multi-probe method proposed in [15] to generate T number of *extended query keys*, $\rho_i^{(1)}(q), \rho_i^{(2)}(q), \ldots \rho_i^{(T)}(q)$, in descending order of their probability to contribute to the query. Parameter T determines the number of multi-probings for each basic query key. Therefore, the total number of query keys is $(T + 1) \cdot L_q$. Third, we look up the indexing peers of the basic and extended query keys in an order like following $\rho_1(q), \rho_2(q), \ldots, \rho_{L_q}(q), \rho_1^{(1)}(q), \rho_2^{(1)}(q), \ldots, \rho_{L_q}^{(1)}(q), \ldots, \rho_1^{(j)}(q),$ $\rho_2^{(j)}(q), \ldots, \rho_{L_q}^{(j)}(q), \ldots$ Users are allowed to determine how far the multi-probing should proceed according to the search accuracy and query cost.

Utilizing the above technique, we are able to achieve high accuracy even when the publishing mapping degree of iDisques is very small. In addition, we can always improve the search accuracy at the expense of more query keys (or higher query costs). Therefore, the search quality will not be restricted by the publishing mapping degree, which is specified by the data owners during index creation.

7   Load-Balancing

The iDISQUE framework can easily adopt load-balancing techniques. In this section, we propose a technique to handle load balancing. There could be two categories of data imbalance in iDISQUE. One is the imbalance of data storage among data owners, the other is the imbalance of index size among indexing peers. The former category can be addressed by a simple data migration or replication scheme such as LAR [9]. In iDISQUE, we shall focus on the problem of load imbalance among indexing peers.

In our method, the load of an indexing peer is defined as the number of iDisques it maintains. In the ideal case, all iDisques are distributed evenly across the entire network. However, in the locality-preserving mapping scheme, the assignment of iDisques might be skewed among different indexing peers. To address the problem, we restrict the maximum number of iDisques published to indexing peers by defining a variable *capacity threshold* τ for each indexing peer, according to its storage or computing capacity. If the number of iDisques maintained by an indexing peer reaches the threshold, the peer is overloaded and any request to insert a new iDisque to it will be rejected. The publishing peer (data owner) can then utilize the multi-probe technique to discover a new indexing peer which hopefully maintains fewer iDisques.

The discovering process is as following: For an iDisque denoted by $C*$ and a mapping function ρ_i, the *basic publishing key* $\rho_i(C)$ and a series of *extended publishing keys* are all created by the multi-probe technique in descending order of their similar probability to the *basic publishing key*. If the indexing peer mapped by the basic publishing key is overloaded, we probe one-by-one, in order of descending probability, the peers mapped by the extended keys, until an indexing peer accepts $C*$.

When processing similarity query, a load-balanced system must utilize the multi-probe technique. If a query misses on a basic publishing key, it will proceed to an extended key. However, the message cost of a query is determined by the query mapping degree L_q and parameter T. In the rest of this paper, we assume that multi-probing is enabled when the proposed load balancing technique works.

8   Experimental Results

In this section, we evaluate the performance of iDISQUE framework. We implement both the *iDISQUE-Naive* indexing scheme and the locality-preserving indexing scheme, which we refer to as *iDISQUE-LSH*.

For comparison, we implement a fully decentralized high-dimensional similarity search P2P system based on spatial partitioning. For a d-dimensional space, we split the data space evenly on each dimension into s equal-length segments. Therefore, we obtain a d-dimensional grid of s^d space partitions. For each partition we create an index entry containing the IPs of peers, which own data records in that partition. All partitions, in a Z-order, are assigned to the peers in the Chord overlay in a round-robin manner. This indexing scheme is called *SPP* (Spatial-partitioning). When a KNN query is issued to SPP, we generate a sphere in the d-dimensional space centered at the query point with a predefined radius. Indexing peers containing partitions overlapping the sphere are queried, and the results are returned to the querying peer. If the number of results of the query is smaller than K, the sphere is enlarged iteratively to overlap more partitions. The SPP scheme provides accurate query results.

We conduct the following experiments to evaluate iDISQUE: (1) The comparison among the three indexing schemes, namely SPP, iDISQUE-Naive, and iDISQUE-LSH; (2) Tuning various parameters in iDISQUE-LSH; (3) Load-balancing; and (4) Scalability.

8.1   Experiment Setup

The effectiveness of approximate similarity search is measured by recall and error rate, as defined in [8]. Given a query object q, let I(q) be the set of ideal answers (i.e., the k nearest neighbors of q), let A(q) be the set of actual answers, then recall is defined as: $\frac{|A(q) \cap I(q)|}{|I(q)|}$. The error rate is defined as: $\frac{1}{|Q|K} \sum_{q \in Q} \sum_{i=1}^{K} \frac{d_i^{\#}}{d_i^*}$, where $d_i^{\#}$ is the i-th nearest neighbor found by iDISQUE, and d_i^* is the true i-th nearest neighbor. Since error rate does not add new insight over relative recall and we do not report in our experiments due to the space limit.

One real dataset and one synthetical dataset are used to evaluate the iDISQUE framework. The real dataset is a subset (denoted by **Covertype**) containing 500k points selected randomly from the original Covertype dataset [2], which consists of 581k 55-dimensional instances of forest Covertype data. The synthetical dataset is generated from the Amsterdam Library of Object Images set [1] which contains 12000 64-dimensional vectors of color histogram. We create new data points by displacing by a small amount (0.005) the original data points on a few random directions, and obtain a synthetical dataset containing 100k points (denoted by *ALOI*). For both datasets, 1000 queries are drawn from the datasets randomly and for each query a peer initiator is randomly selected. As most experimental results show similar trends for these two datasets, unless otherwise stated, in the following experiment we will only present the results for the **Covertype** dataset due to the limit of space.

In our experiments, the default network size N is 1000. The dataset is horizontally partitioned evenly among the peers, and each node contains 100 data points by default. The splitting number of SPP on each dimension is $s = 4$. We have experimented with different parameter values for the locality-preserving mapping and picked the ones that give best performance. In our experiment, the parameters of locality-preserving mapping are $k = 15$, $W = 3.0$ for both datasets. We also conduct a lot of experiments with different K (K = 1,5,10) for the KNN queries, and the default K value is 5. In our experiments, for simplification, we assume each iDisque C* has the same publishing mapping degree, and the default publishing and querying mapping degree (L_p and L_q) is 10. Since the current iDISQUE framework is based on Chord, in the experiments the query processing cost is measured by the number of lookup messages for the *indexing peers*. All the queries are executed for 10 times, and the average results are presented.

8.2   Comparative Study - Experiment 1

In this experiment we first compare the total storage space of the proposed indexing schemes ($L_p = 1$ for the iDISQUE schemes). Note that the index size of iDISQUE-Naive is the same as that of iDISQUE-LSH. The results indicate that the storage space of iDISQUE is much larger than SPP, although the former is still acceptable compared to the size of the datasets. It must be noted that as the value of L_p increases, the storage cost will increase linearly. This justifies the need for the multi-probe technique.

Table 1. Results of the storage space

Dataset	Data size (KB)	SPP index size (KB)	iDISQUE index size (KB)
Covertype(500k)	214844	248	15877
ALOI(100k)	50000	132	4895

We now compare the distribution of the index storage space among peers for the three indexing schemes. For fairness in comparison, the results are presented in percentages. Figure 2(a) shows the cumulative distribution of index storage space among 1000 peers using the Covertype dataset. The figure indicates that iDISQUE-Naive provides the most uniform distribution in index storage space, while the result of iDISQUE-LSH is also satisfactory. However, the results of SPP indicate that nearly 95% of its index storage is assigned to 5% the nodes. This partly explains the small index size of SPP. As a result, the skewed index in SPP would inevitably cause serious problem of load imbalance. Therefore, a scheme based on spatial-partitioning is not truly viable for querying high-dimensional data.

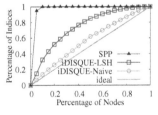

(a) Comparison of cumulative distributions of index size

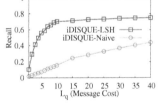

(b) Comparison of search effectiveness

Fig. 2. Results of Exp. 1

To compare the effectiveness and efficiency of the three schemes, we plot the accuracy results of iDISQUE-Naive and iDISQUE-LSH versus their message costs in Figure 2(b), given a publishing mapping degree of 40. The message cost of a query can be determined by the query mapping degree L_q. Specifically, L_q is equivalent to the number of messages caused by a query when multi-probe is disabled. Note that the recall of SPP is not plotted as they are always 1. The message cost of SPP is 35. Figure 2(b) indicates that the recall of iDISQUE-Naive increases almost linearly when the message cost grows. It reaches 44% at the cost of 40 messages. In contrast, the recall of iDISQUE-LSH increases rapidly at first, and reaches about 70% at 10 messages. This shows that iDISQUE-LSH's accuracy is acceptable. Since iDISQUE-LSH is much more effective than iDISQUE-Naive given the same message cost, we will focus on iDISQUE-LSH in the remaining experiments.

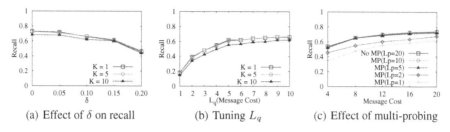

(a) Effect of δ on recall   (b) Tuning L_q (c) Effect of multi-probing

Fig. 3. Results of Exp. 2

8.3 Performance Tuning - Experiment 2

Figure 3(a) shows the effect of δ on recall. We can see that the smaller the δ, the higher the recall. This is because a smaller δ leads to finer iDisques, which can represent the actual data points with a better approximation. If $\delta = 0$, all iDisques are singular. In such case, the iDisques represent real data and the recall is 73% for $K = 1$. Similar trends are observed when K = 5 and K = 10. Such results confirm our analysis in Section 5.2. When δ is 0.1, it can strike a balance between the storage space of iDisques and the accuracy of queries. Therefore, the default value of δ is set to 0.1 (there are 9201 iDisques when $\delta = 0.1$).

Figure 3(b) shows the effect of L_q on recall for K = 1, 5, 10. It can be observed that when $L_q < 5$, the recall increases rapidly. When $L_q = 5$, the recall is nearly 60% for K = 1. These results confirm the effectiveness of our tunable approximate similarity query framework.

Figure 3(c) shows the accuracy of iDISQUE-LSH enabling the distributed multi-probe technique (MP means to enable multi-probing). The accuracy of a conventional (non-multiple-probing) scheme is also plotted. As shown in the figure, multi-probe with small publishing mapping degrees (L_p=1, 2, 5, 10) can achieve highly competitive accuracy at nearly the same cost as a conventional scheme, which has a larger publishing degree of 20. The recall of multi-probe with $L_p = 10$ is even marginally better than the conventional scheme when the message cost is greater than 12. As a smaller L_p value leads to smaller index size and also smaller index publishing costs, the multi-probe technique can reduce the index size without compromising the quality of queries.

8.4 Load-Balancing - Experiment 3

In this experiment, we study the effect of the proposed load-balancing technique, which is based on multi-probing. We vary the capacity threshold τ and illustrate the results of the publishing cost, load distribution, and query accuracy in Figure 4. The threshold τ is measured in units of the average load of all peers without load-balancing (i.e. a τ value of 2 means twice the average load). We also plot the same results for iDISQUE without load-balancing. The results indicate that as τ increases, the publishing cost reduces rapidly because less multi-probing is required. However, the load becomes more skewed as the restriction is being lifted. Meanwhile, as τ increases, more iDisques can be found by the extended query key, therefore producing higher accuracy. To obtain balanced load while producing quality results and without introducing high publishing cost, a τ value of 2.5 could be a good choice.

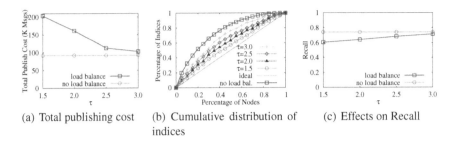

(a) Total publishing cost   (b) Cumulative distribution of   (c) Effects on Recall
                                     indices

Fig. 4. Results of Exp. 3

8.5   Scalability - Experiment 4

In this experiment, we evaluate the scalability of iDISQUE using five networks, where the number of peers ranges from 1000 to 5000. We also create five randomly selected subsets of the **Covertype** dataset with different sizes (100k, 200k, $\cdots$, 500k), and create an iDISQUE index for each of these subsets in a respective network. Figure 5 shows the accuracy at various query message costs (L_q) in different network scales. The results show that the network size has little impact on the accuracy of iDISQUE. Therefore, the proposed scheme is scalable in terms of effectiveness.

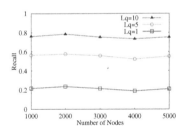

Fig. 5. Result of Scalability

9   Conclusion

In this work, we proposed a framework called iDISQUE to support tunable approximate similarity query of high dimensional data in DHT networks. The iDISQUE framework is based on a locality-preserving mapping scheme, and its query accuracy can be tuned by both the indexing and query costs. We also proposed a distributed multi-probe technique for iDISQUE to reduce its index size without compromising the effectiveness of queries. Load balancing among the indexing nodes was achieved by utilizing a novel technique based on multi-probing. The experimental results confirmed the effectiveness and efficiency of the proposed framework.

For future work, we would look at data update strategies in the iDISQUE framework. We would also consider dynamic load balancing techniques to handle skewness in queries.

Acknowledgment

This work was supported in part by the National Science Foundation of China (NSFC Grant No. 60803003, 60970124), and by Chang-Jiang Scholars and Innovative Research Grant (IRT0652) at Zhejiang University.

References

1. The amsterdam library of object images homepage (2008),
   `http://staff.science.uva.nl/~aloi/`
2. Uci kdd archive (2008), `http://www.kdd.ics.uci.edu`
3. Bharambe, A.R., Agrawal, M., Seshan, S.: Mercury: supporting scalable multi-attribute range queries. In: SIGCOMM (2004)
4. Cai, M., Frank, M.R., Chen, J., Szekely, P.A.: Maan: A multi-attribute addressable network for grid information services. In: GRID (2003)
5. Datar, M., Immorlica, N., Indyk, P., Mirrokni, V.S.: Locality-sensitive hashing scheme based on p-stable distributions. In: Symposium on Computational Geometry (2004)
6. Doulkeridis, C., Vlachou, A., Kotidis, Y., Vazirgiannis, M.: Peer-to-peer similarity search in metric spaces. In: VLDB (2007)
7. Ganesan, P., Yang, B., Garcia-Molina, H.: One torus to rule them all: multi-dimensional queries in p2p systems. In: WebDB, pp. 19–24 (2004)
8. Gionis, A., Indyk, P., Motwani, R.: Similarity search in high dimensions via hashing. In: VLDB (1999)
9. Gopalakrishnan, V., Silaghi, B., Bhattacharjee, B., Keleher, P.: Adaptive replication in peer-to-peer systems. In: ICDCS (2004)
10. Guha, S., Rastogi, R., Shim, K.: Cure: An efficient clustering algorithm for large databases. In: SIGMOD (1998)
11. Haghani, P., Michel, S., Aberer, K.: Distributed similarity search in high dimensions using locality sensitive hashing. In: EDBT (2009)
12. Indyk, P., Motwani, R.: Approximate nearest neighbors: Towards removing the curse of dimensionality. In: STOC (1998)
13. Jagadish, H.V., Ooi, B.C., Tan, K.-L., Yu, C., Zhang, R.: idistance: An adaptive b$^+$-tree based indexing method for nearest neighbor search. In: ACM TODS (2005)
14. Jagadish, H.V., Ooi, B.C., Vu, Q.H., Zhang, R., Zhou, A.: Vbi-tree: A peer-to-peer framework for supporting multi-dimensional indexing schemes. In: ICDE (2006)
15. Lv, Q., Josephson, W., Wang, Z., Charikar, M., Li, K.: Multi-probe lsh: Efficient indexing for high-dimensional similarity search. In: VLDB (2007)
16. Ratnasamy, S., Francis, P., Handley, M., Karp, R.M., Shenker, S.: A scalable content-addressable network. In: SIGCOMM (2001)
17. Rowstron, A.I.T., Druschel, P.: Pastry: Scalable, decentralized object location, and routing for large-scale peer-to-peer systems. In: Guerraoui, R. (ed.) Middleware 2001. LNCS, vol. 2218, p. 329. Springer, Heidelberg (2001)
18. Sahin, O.D., Emekçi, F., Agrawal, D., Abbadi, A.E.: Content-based similarity search over peer-to-peer systems. In: Ng, W.S., Ooi, B.-C., Ouksel, A.M., Sartori, C. (eds.) DBISP2P 2004. LNCS, vol. 3367, pp. 61–78. Springer, Heidelberg (2005)
19. Stoica, I., Morris, R., Karger, D.R., Kaashoek, M.F., Balakrishnan, H.: Chord: A scalable peer-to-peer lookup service for internet applications. In: SIGCOMM, pp. 149–160 (2001)
20. Zhang, T., Ramakrishnan, R., Livny, M.: Birch: A new data clustering algorithm and its applications. Data Min. Knowl. Discov. (1997)

Adaptive Ensemble Classification in P2P Networks

Hock Hee Ang, Vivekanand Gopalkrishnan,
Steven C.H. Hoi, and Wee Keong Ng

Nanyang Technological University, Singapore

Abstract. Classification in P2P networks has become an important research problem in data mining due to the popularity of P2P computing environments. This is still an open difficult research problem due to a variety of challenges, such as *non-i.i.d. data distribution, skewed or disjoint class distribution, scalability, peer dynamism* and *asynchronism*. In this paper, we present a novel P2P Adaptive Classification Ensemble (PACE) framework to perform classification in P2P networks. Unlike regular ensemble classification approaches, our new framework *adapts* to the test data distribution and *dynamically* adjusts the voting scheme by combining a subset of classifiers/peers according to the test data example. In our approach, we implement the proposed PACE solution together with the state-of-the-art linear SVM as the base classifier for scalable P2P classification. Extensive empirical studies show that the proposed PACE method is both efficient and effective in improving classification performance over regular methods under various adverse conditions.

1 Introduction

Distributed data mining is important and beneficial to a broad range of real-world applications [1] on distributed systems. Recent popularity of peer-to-peer (P2P) networks has also enabled them as excellent platforms for performing distributed data mining tasks, such as P2P data classification [2,3,4,5]. While its potential is immense, data mining in a P2P network is often considerably more challenging than mining in a centralized environment. In particular, P2P classification faces a number of known challenges [6] including *scalability, peer dynamism* and *asynchronism*, etc.

In the past few years, a number of distributed classification techniques have been proposed to perform classification in P2P networks [1,2,4,5]. Among these, ensemble approaches are the most popular due to their simple implementation and good generalization performance. The key idea is to build individual classifiers on the local training data of peers, and then combine *all* their predictions (e.g., weighted majority voting) to make the final prediction on unseen test data examples. Ensemble approaches are proven to perform well, provided the following assumptions are fulfilled: (1) outputs of individual classifiers are *independent*, and (2) generalization error rates of individual classifiers are smaller than 50%, i.e., individual classifiers do *not* perform worse than a random guessing approach.

H. Kitagawa et al. (Eds.): DASFAA 2010, Part I, LNCS 5981, pp. 34–48, 2010.

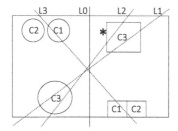

Fig. 1. Feature space for a 2-class problem

Table 1. Non i.i.d. class distributions

Peer	Class 1	Class 2	Class 3	Class 4
skewed				
P1	70	10	10	10
P2	10	70	10	10
P3	10	10	70	10
P4	10	10	10	70
disjointed				
P1	50	50	0	0
P2	50	50	0	0
P3	50	50	0	0
P4	0	0	50	50

Unlike in centralized environments, in a P2P learning environment, regular ensemble methods often cannot completely fulfill the above assumptions due to the dynamics of P2P networks. Below, we discuss some scenarios where regular ensemble approaches could fail to achieve satisfactory performance in P2P networks.

Scenario 1: disjoint data distribution: One typical challenge with P2P classification is the issue of *disjoint data distribution and bias* [2,7]. Figure 1 depicts a sample 2-D feature space for a two-class classification problem. The data space is represented by two symbols: circles and squares (each representing one class), and the solid line (L0) denotes the optimal decision plane/model. Labels within the symbols represent the peers/classifiers (C1, C2 and C3) that own the data, and their decision planes/models are represented by dotted lines (L1, L2 and L3) respectively. In a regular ensemble learning approach, if we assume that peers' training data are i.i.d., the ensemble solution should be close to L0. However, in this scenario, the ensemble model will be biased towards L1 and L2 causing its accuracy to suffer. Although the data distribution between the two classes is equal, the bias is still present due to the difference in number of votes. However, in reality, it is not possible to adjust the bias by simply using the data density.

Scenario 2: skewed class distribution: Skewed class distribution is very common in typical classification problems [8] and for a P2P system, the skewness can vary widely among peers. Table 1 presents non i.i.d. data and class distributions that may be present in a P2P learning environment. The numbers denote the percentage of the class data that is held by peers and peers may not hold the same amount of data locally. Ideally, peers' data should be i.i.d. and balanced, allowing the ensemble classifier to perform better than individual classifiers. In reality, however, skewed class distribution is usually unavoidable, and can considerably deteriorate performance of the ensemble classifier. For instance, according to a recent empirical study [2], the accuracy of an SVM ensemble classifier trained from an imbalanced two-class data set decreases when the skew between classes increases.

Scenario 3: disjoint class distribution: In extreme cases of *skewed class distribution*, some peers may have no data from certain classes. We refer to this special scenario as *disjoint class distribution* (Table 1). The regular ensemble approach that simply combines outputs from all classifiers could perform very poorly in this scenario. For example, since peers P1, P2, and P3 contain no training data from class 3, they will simply make a wrong prediction for a test sample from class 3. As a result, no matter how peer P4 performs, the ensemble classifier that combines all the four peers in a majority voting approach will always make a wrong prediction for any example from class 3.

From the above discussions, we observe that regular ensemble classification approaches have two shortcomings: (1) *all* the classifiers (peers) are engaged in the voting scheme to predict a test data example; (2) the ensemble is often *not aware* of test data distribution, so the same combination scheme is *universally* applied for any test data example. Given the settings of a P2P network, it will not be possible to obtain a representative testing dataset for estimating the generalization errors of the classifiers. Hence, one uses the training errors as the estimate. However, training errors are not always indicative of the generalization error especially as shown in the previous scenarios, where classifiers may in fact have generalization errors larger than 50% although not shown by their training errors. Hence, the first shortcoming is a clear violation of the principals of ensemble learning, which can deteriorate the accuracy of the entire ensemble. The second shortcoming often leads to a suboptimal combination scheme because it does not reward/penalise classifiers and hence the inability to deal with non-i.i.d. data distribution.

To overcome the above shortcomings, in this paper, we investigate a novel P2P ensemble classification framework that *adapts* to the test data distribution and engages only a *subset* of classifiers (peers) *dynamically* to predict an unseen test data example. This raises three challenges: (1) *how to effectively and efficiently choose a subset of classifiers (peers) according to a test data example dynamically?* (2) *how to develop an effective voting scheme to combine the outputs from the subset of classifiers (peers)?* (3) *how to minimize communication cost and interactions between peers towards an efficient and scalable solution?*

Inspired by the mixture of expert classification architecture [9] and the k Nearest Neighbor classifier [10] where both assume that the closer the training data are to the test data, the more appropriate the classifier will be, led us to take into consideration how well the training error of a classifier estimates its generalization error, which is the basis for constructing an adaptive ensemble.

This paper addresses these challenges, and makes the following contributions:

- We propose a novel <u>P2P</u> <u>A</u>daptive <u>C</u>lassification <u>E</u>nsemble (PACE) framework, which can be integrated with any existing classification algorithm. The PACE framework *adapts* to the test data distribution and adopts a *dynamic* voting scheme that engages only a *subset* of classifiers (peers).
- We implement an effective P2P classification algorithm based on the PACE framework with the state-of-the-art linear SVM algorithm as the base classifier. This enables highly efficient and scalable solutions in real large-scale applications.

– We conduct extensive empirical evaluations on both efficacy and efficiency of our approach. Results show that our new algorithm is comparable to competing approaches under normal conditions, and is considerably better than them under various adverse conditions.

The rest of this paper is organized as follows. Section 2 reviews existing work on P2P classification. Section 3 presents our proposed PACE approach. Section 4 shows our experimental results and Section 5 concludes this paper.

2   Related Work

Classification approaches for P2P systems can be generally classified into two categories: collaborative [1,2] and ensemble [4,5] approaches. While both categories of approaches use the divide-and-conquer paradigm to solve the classification problem, collaborative approaches only generate a single model for the classification task while ensemble approaches often combine multiple models/classifiers for predictions.

To take advantage of statistical property of SVMs, Ang et al. proposed a variant of the cascade SVM approach, which makes use of Reduced SVM (RSVM) to reduce the communication cost and improve classification accuracy. Although this approach claims to reduce the communication cost with RSVM, propagation of non-linear SVM models, made up of a number of support vectors, is very costly. In addition, the tasks of cascading are repeated in all peers, wasting computational resources.

On the contrary, Bhaduri et al. proposed to perform distribution decision tree induction [1]. This is a much more efficient approach which propagates only the statistics of the peers' local data, with the decision tree of each peer converging to the global solution over time.

Like traditional distributed systems, ensemble approaches are also very popular in P2P systems, which has several advantages for classification in P2P networks. First, voting ensemble is a loosely coupled algorithm, which means that it does not require high-level synchronization. Secondly, as it does not require all models to participate in the voting, it is able to give a partial solution anytime [4]. This also means that it is *fault tolerant*, as failures of a few peers only *slightly* affect the final prediction. The following are some examples of ensemble approaches in P2P classification.

Recently, Siersdorfer and Sizov [5] proposed to classify Web documents by propagating linear SVM models built from local data among neighboring peers. Predictions are then performed using only the collected models, incurring zero communication cost. However, collecting only from the neighboring peers restricts the representation of the ensemble of classifiers as the data on each peer is relatively small compared to the entire data in the network. This decreases the prediction accuracy of the ensemble. As their experiments were conducted using a small number of peers (16), this problem may be overlooked.

In another work, Luo et al. [4] proposed building local classifiers using Ivotes [11] and performed prediction using a communication-optimal distributed

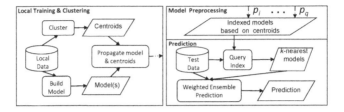

Fig. 2. Architectural framework of PACE. Dotted arrows represent network communications.

voting protocol that requires the propagation of unseen data to most, if not all peers. This incurs huge communication cost if predictions are frequent. In addition, their work assumes that the error rates of all classifiers are equal, which as previously discussed is not a valid assumption; besides it also does not address the limitations of majority voting that can happen in the P2P networks. Recently Ang *et al.* [2] showed that DIvotes in the P2P networks is sensitive to the effects of skewed class distribution.

3   Approach

PACE aims to maintain the advantages of regular voting ensemble solutions while effectively overcoming their drawbacks (c.f. Section 1). Unlike conventional ensemble approaches, PACE is novel in that it *adapts* to the test data distribution, and employs a *dynamic* voting scheme, which chooses only a subset of important classifiers/peers for assessing a test data example. To facilitate the selection of important classifiers with respect to test data distribution, we propose a cluster-driven approach, which provides an *efficient* way to examine how close a test data example is to a specific peer at the expense of slight increase in cost. Further, to combine outputs from the subset of selected classifiers, we evaluate several k nearest neighbor weighted voting approaches, which exploit various information towards an effective combination in the voting process.

The architectural framework of PACE is illustrated in Figure 2, and includes two major phases: (1) training and clustering, and (2) prediction. In the training and clustering phase, every peer builds classifier(s) on their local training data, after which, they cluster the data and then propagate the classifiers and cluster centroids to other peers. Each peer indexes its collected models (from other peers) using the corresponding centroids. In the prediction phase, peers use the index created earlier to select a subset of models from the set of all collected models that is most relevant to the given test instance. Voting is then performed using this subset of classifiers to produce the final prediction. Next, we provide the in depth details of the two phases.

3.1   Training and Clustering Phase

In the training phase, a peer builds a base classifier (or a set of classifiers using techniques such as bagging, boosting or Ivotes) from its local data. Based on the model propagation approach, the local classifiers are propagated to *all other peers*. The collected models are then used for performing prediction locally. Model propagation is a popular approach in P2P classification where prediction cost is a concern especially when training data are few and testing data are in abundance [1,2,5,7]. We reiterate here that PACE is a generic ensemble classification framework, which can be integrated with any existing classification algorithm as the base classifier. In our approach, we adopt the state-of-the-art linear SVM algorithm as the base classifier to exploit its high efficiency for training classification models.

As noted earlier, simply combining the ensemble of classifiers by majority voting is insufficient to guarantee satisfactory classification accuracy. However, without additional data it is not possible to perform model selection or use advanced model combination techniques. Unfortunately, considering the size of the P2P network and the communication cost, it is not possible to manipulate the data as required by existing ensemble model selection or advanced combination techniques. Assuming that the classifier's accuracy is correlated to the distance between the testing and training data, we have to somehow capture the distance between the testing and training data or the locality of the training data. Hence, we propose the use of clustering to capture the locality of the training data of each peer.

Clustering is performed on the training data of the peer to generate the set of centroids, which are representative of the training data of the classifier. The centroid serves as a summarization of a group of data examples. By using only a small number of centroids, the additional overheads on the communication cost will be reduced substantially. We empirically show that the inclusion of centriods ensures robust ensemble performance. Here, we employ the simple and efficient k-means clustering algorithms, in which we can specify the number of clusters, which is directly proportional to the communication cost.

Note that the training and clustering steps can be performed either concurrently or sequentially. However, in addition to the locality information, we also want to capture the classifier's classification accuracy on the particular centroid. This mainly aims to address the problem of *skewed class distribution*. Note that a classifier trained on an imbalanced dataset will often have a high error rate. Hence, even if the classifier is trained on data near to the test data, it still might be possible that the classifier's accuracy is not at an acceptable level. Hence, the accuracy of the centroid can be used as the balancing parameter. In order to obtain the error rate of the classifier on the cluster, we require both training and clustering to be completed. Once the cluster testing is completed, the classification model together with the centroids and their error rates are propagated to other peers.

By propagating the model(s) and centroids to all peers, we are duplicating the knowledge of the peers in the P2P network, which allows the knowledge of peers to remain in the network even when the owners have left or failed. This reduces the adverse effects of peers failing. In addition, predictions can be handled locally by peers without additional communication cost and waiting

---

Algorithm 1. Training and Clustering Phase.

---

    input : Local data $\mathbf{D}_i$, number of cluster g
    output: classification model $\mathbf{M}_i$, centroids $\mathbf{C}_i$, model error rate $\mathbf{Err}_i$
1 $\mathbf{M}_i \leftarrow trainClassifier(\mathbf{D}_i)$;
2 **Clusters** $\leftarrow clusterData(\mathbf{D}_i, g)$;
3 $\mathbf{C}_i \leftarrow computeCentroids(\mathbf{Clusters})$;
4 $\mathbf{Err}_i \leftarrow predictionTest(\mathbf{M}_i, \mathbf{D}_i)$;

---

time. However, there are two problems with such approaches, viz., the cost of data propagation and validity of the models.

With consideration to the cost of data propagation and efficiency of PACE, we choose to employ LIBLINEAR [12] for training linear SVM as the base classifier in this paper. LIBLINEAR is an implementation of linear SVM that performs a dual coordinate gradient descent for optimizing the linear SVM solution. LIBLINEAR reaches an ϵ-accurate solution in $O(log(1/\epsilon))$ iterations, and is one of the fastest linear SVM solutions. In addition, the linear SVM only produces a *single weight vector as its model* for a two-class problem (for multiclass problem, based on one against all strategy, the number of weight vectors is the number of classes minus one), which significantly reduces communication cost incurred for model propagation. Another problem is the validity of the models. Assuming that models get outdated (due to concept drift), full propagation and replication of the models will increase staleness of knowledge and degrade accuracy over time. One approach which can be used to handle this problem is to use aging schemes to invalidate or decrease the weights of the models as time passes. However, other than peers leaving, new peers may also join the network or old peers may receive new data. With new peers joining, their base classifiers can be propagated to other peers and used together with other collected models, which can be easily achieved using (weighted) majority voting. Whereas for old peers receiving new data, one simple approach is to create an additional base classifier (and centroids) with the new data and propagate it out. Alternatively, one can choose to update the old model (and centroids) with the new data and propagate them out to replace the old model (and centroids). The training and clustering phase is summarized in Algorithm 1.

3.2   Prediction Phase

Assuming not all classifiers fulfill the accuracy criteria required for ensemble classifiers, we need to filter out those irrelevant classifiers to prevent them from adversely affecting the ensemble's accuracy. Let us now examine Figure 1 that illustrates **scenario 1**. Given a test instance represented by *, we note that a regular ensemble consisting of all the classifiers $L1$, $L2$ and $L3$ incorrectly classifies the test instance. On the other hand, if we were to select a model (subset of all the models) which is trained on the data nearest to the test instance, e.g. $L3$ in this case, we are more likely to correctly predict the test instance. However, we note that the closest classifier may not be the optimal solution since it also depends on the error rate of the classifier.

Let $dist(D_i, T)$ denote the distance from test instance T to the training data D_i of peer p_i in the feature space, M_i denote the classification model of p_i, $Err_{emp}(M_i)$ denote the empirical error of M_i and $P_{err}(M_i, T)$ denote the confidence probability of M_i wrongly classifying T. We introduce the following lemma to show the relationship between two models on a test instance (proof omitted due to space constraints):

Lemma 1. *When $dist(D_i, T) > dist(D_j, T)$, then $P_{err}(M_i, T) > P_{err}(M_j, T)$ if $Err_{emp}(M_i) = Err_{emp}(M_j)$.*

Using Lemma 1 as the basis, we can provide a better estimation on the expected error $Err_e xp$ given a test instance T, which is ideal for weighing classifiers in the ensemble. In addition, given that the criteria for selecting classifiers for the ensemble is based on Lemma 1, we will be able to address scenarios 1, 2 and 3. However, Lemma 1 assumes equal empirical error among the classifiers, which as noted earlier is an incorrect assumption. Hence, using Lemma 1 as the basis, we propose to combine the empirical error $Err_{emp}(M_i)$ and the distance $dist(D_i, T)$ as the weight for the classifier in the ensemble.

$W(M_i, T) = 1 - P_{err}(M_i, T) = (1 - Err_{emp}(M_i)) * w(d)$ where $w(.)$ is an inverse distance function and $d = dist(D_i, T)$. However, it is not possible to perform the distance computation between all test data examples and all peers' training data. Hence, we approximate $dist(D_i, T)$ with the distance of test instance T to the nearest centroid c from the set of centroids C_i generated from clustering the local training data D_i ($dist(D_i, T) \approx dist_{min}(C_i, T) = \min_{c \in C_i} dist(c, T)$).

To ensure asymptotic increase in accuracy as the size of the ensemble increases, we have to ensure that the Err_{exp} of each individual classifiers is less than 0.5. Although we try to estimate Err_{exp} with W, it is obvious that W is less accurate, partly due to the fact that the distance measure uses centroids instead of the actual data points, causing loss in accuracy. Therefore, instead of choosing classifiers that meet the accuracy criteria, we perform a ranking and choose the top k classifiers. Since every classifier M_i minimizes $Err_{emp}(M_i)$, using the empirical error estimates for ranking may create unnecessary bias. Hence, all classifiers are ranked according to their distance to the test instance, regardless of their empirical error. Selecting only the top k models allows us to minimize of adverse effects of possible erroneous classifiers. However, as we are unable to determine the actual Err_{exp}, the choice of k only serves as an estimation. Hence, we empirically examine the choice of k and determine its effect on the classification accuracy.

To allow better flexibility, we relax the value of d in W allowing the value of either $dist_{min}(C_i, T)$ or $rank(M_i, T)$. Note that the main purpose of $w(.)$ is to increase the weight of the *closer* models (*research problem 2*). Hence, in this paper, we examine a variety of k nearest neighbor based weighting schemes as follows:

- $w_0(d) = 1$
- $w_1(d) = \frac{dist_{min}(C_{last}, T) - dist_{min}(C_{current}, T)}{dist_{min}(C_{last}, T) - dist_{min}(C_{first}, T)}$
- $w_2(d) = e^{-\lambda rank}$.

Algorithm 2. Prediction.

---

    input   : test instance T, set of all peers' classifiers $\mathbf{M}$, centroids of all peers'
             training data $\mathbf{C}$, prediction errors of all peers' models $\mathbf{Err}$, size of
             ensemble k;
    output: prediction y;
1 weighted votes counts $\mathbf{VC}$;
2 list of k classifier $\mathbf{TopK}$;
3 **while** $|VC| < k$ **do**
4   retrieve next nearest centroid C_{near} from index;
5   **if** *model M_i of $C_{near} \notin$ TopK* **then**
6          $\mathbf{TopK} \leftarrow M_i$;
7          $y_i \leftarrow predict(\mathbf{M}_i, T)$;
8          $increase\mathbf{VC}_{y_i} by((1 - \mathbf{Err}_i) * w(rank_i, dist_i))$
9 $y \leftarrow getClassWithMaxVote(\mathbf{VC})$;

Note that the ranking of models has to be performed for each test instance, because as their distribution varies so will their k nearest neighbors. A naïve approach is to compute the distance between all classifiers' centroids each and everytime a new instance arrives. However, this is too costly as each ranking incurs gN distance computations. Hence, to maintain high efficiency of PACE, we propose to use a distance aware indexing algorithm such as k-d tree [13] or locality-sensitive hashing (LSH) [14]. As the centroids are propagated with the models, every peer can create their own index locally. Given a centroid, we only need to index it once, and thereafter perform retrieval based on the index, unlike the naïve approach where we have to recompute the distance of the test instance to all collected centroids. Given appropriate parameters, a single lookup is sufficient to retrieve all the required *nearest* models. On the arrival of a model, we update the index using the centroids which allows us to retrieve the k-nearest classifiers.

Here, we summarize the preprocessing and prediction phase. First, when a new classifier and its centroids are received, we index the centroids and store the classifier. This indexing step is critical to maintaining high efficiency for the prediction phase. Next, when a test instance T arrives, we first retrieve k nearest models using the index. Note that there can be more than k retrievals since each model has more than one centroids. Next, we compute the centroid distance to the test instance or simply record the rank. Then the prediction of the classifier, multiplied by its training error and the kNN weight is stored. Given the votes of the k nearest classifiers, we compute the largest voted class that is output as the prediction of the ensemble. Pseudocode of the prediction phase is presented in Algorithm 2.

3.3 Complexity Analysis

Here, we provide a time complexity analysis of PACE. In the training phase, each peer builds a LIBLINEAR SVM classifier. The cost of building the linear SVM model is $O(\log(1/\epsilon)\ell_i d)$ for an ϵ-accurate solution where d is number of dimension of the dataset [12]. Other than model construction, peers also cluster the local data which costs $O(\ell_i k d)$ [15]. Once both model construction and clustering are completed, the training data is evaluated costing $O(\ell_i d)$. In addition, upon the arrival of a peer's model and centroids, the distance indexes for the models are updated. Since this is not a part of the prediction, we count it as a part of the training cost. As the datasets used in our experiments are all high dimensional, we use LSH [14] as the indexing algorithm. Given a $(1, c, p1, p2)$-sensitive hash function for $\mathbb{R}^d$, the cost of constructing the index is $O((d+\tau)(gN)^\rho \log_{1/p^2} gN)$, where τ is the time to compute the hash function and $\rho = \log(p_1/p_2)$. Hence, assuming the $k << \log(1/\epsilon)$, the worst case time complexity for training and clustering is $O(\log(1/\epsilon)\ell_i d)$.

In the prediction phase, when an instance arrives, we shall use the precomputed index to retrieve the top k nearest neighbors. As noted earlier, a single lookup would be sufficient to retrieve all k nearest neighbors (given an appropriate c value or perform lookup in an incremental manner). Hence, the cost of getting the k nearest neighbors is $O(gd(gN)^{1/c^2})$. Finally, prediction is performed for the top k models costing $O(kd)$. Hence, the worst case time complexity for prediction is $O(gd(gN)^{1/c^2})$.

Next, we provide a brief overview of the communication cost incurred by each peer. After the model construction, clustering and cluster validation, each peer propagates its model, centroids and centroids' accuracy to all other peers. Each peer has g centroids, which are all d-dimensional vectors. The model for a two class problem is in d-dimensional space. Including the centroids' accuracy, the communication cost is $O(Ngd)$ bytes. Since all models are available at every peer, the prediction phase does not require any communication.

4   Experiments and Analysis

In this section, we present experiments that demonstrate the cost-benefits of PACE on various distributions. In addition, we study the effect of parameters on classification accuracy, computation and communication cost of PACE.

4.1   Experiment Setup

To mimic P2P systems in our experiments, we employed some of the larger datasets available from the UCI repository [16] — Multiclass Covertype (581,012 instances, 54 features, 7 class labels and 500 peers), MNIST (70,000 instances, 780 features, 10 class labels and 100 peers) and (KDD) Census-Income (295,173 instances, 50 features, 2 class labels and 200 peers) datasets. In addition, we generated a two class (Binary) Covertype dataset by using class two against all other classes. All attributes were normalized to the range of [0,1]. The number of peers was chosen in accordance to the size of the dataset such that each peer has at least 500 instances. For the Binary and Multiclass Covertype, we conducted 10-fold cross validation. For the Census-Income and MNIST datasets, we tested using the provided testing data with 10 independent iterations.

Table 2. Classification accuracy in %

Dataset	Centralized Linear SVM	P2P Ivotes	Linear SVM Ensemble	PACE $k = 10$	PACE $k = 0.1N$
I.I.D. Data Distribution					
Census Income	94.60 ± 0.00	94.79 ± 0.02	94.32 ± 0.01	94.31 ± 0.02	94.32 ± 0.02
Binary Covertype	75.68 ± 0.15	79.83 ± 0.12	75.61 ± 0.15	75.51 ± 0.33	75.61 ± 0.23
Multiclass Covertype	71.28 ± 0.12	76.12 ± 0.16	70.83 ± 0.16	70.67 ± 0.25	70.80 ± 0.20
MNIST	79.82 ± 0.00	88.00 ± 0.25	86.65 ± 0.19	82.43 ± 0.92	82.43 ± 0.92
Disjoint Class Distribution					
Multiclass Covertype	N.A.	68.74 ± 0.12	65.99 ± 0.16	69.70 ± 0.32	69.81 ± 0.45
MNIST	N.A.	43.66 ± 3.36	53.94 ± 0.84	80.14 ± 1.48	80.14 ± 1.48

Table 3. Average computational cost per peer in msec (Training and Prediction)

Dataset	Centralized Linear SVM	P2P Ivotes	Linear SVM Ensemble	PACE $k = 10$	PACE $k = 0.1N$
I.I.D. Data Distribution					
Census Income	135 0	164 21200	20 11100	66 560	66 560
Binary Covertype	63 0	1102 49100	16 6112	40 224	40 820
Multiclass Covertype	200 0	2081 85889	27 14700	50 458	50 1790
MNIST	3670 0	2779 2300	66 3700	1422 320	1422 320
Disjoint Class Distribution					
Multiclass Covertype	N.A.	1237 55300	9 13300	31 490	31 2050
MNIST	N.A.	1129 1700	8 1900	1137 260	1137 260

We compare PACE to the following algorithms — Centralized Linear SVM, Ensemble of Linear SVM (LinSVME) with weighted majority voting and P2P Ivotes [4]. Centralized Linear SVM is used as the benchmark for accuracy achievable in a centralized environment. Since the main objective of this paper is to address the limitations of majority voting in the P2P networks, we compare with the two other (weighted) voting approaches. We used the LIBLINEAR [12] linear SVM package as the base classifier for PACE and LinSVME. In addition, we used Kmeans++ [15] as the clustering algorithm for PACE. P2P Ivotes uses the C4.5 Release 8 by Quinlan [17] as the base classifier. All algorithms are coded in C++. Default settings for the Linear SVM were used, and for P2P Ivotes, the bite size was set at 400 for the MNIST dataset and 800 for the rest of the other datasets and error threshold was set at 0.002.

4.2   Accuracy

Table 2 presents classification accuracies of all competing approaches under various distributions. Under the assumption that the data is independent and identically distributed among all peers (I.I.D. Data Distribution), we observe that PACE achieves higher accuracy than centralized Linear SVM on the MNIST dataset but slightly lower accuracy (less than 1%) on other datasets. However, note that in a real P2P environment, it is not possible to centralize all data to learn a classifier. Compared with P2P Ivotes, PACE yields lower accuracy on Binary, Multiclass Covertype and MNIST datasets, but is comparable on the Census dataset. Note that the base classifier for P2P Ivotes is essentially an ensemble of classifiers, because of which it performs better although at a much higher cost. Compared with LinSVME, PACE achieves comparable accuracy on all datasets except MNIST.

4.3   Disjoint Data Distribution

For this experiment, we distributed the multiclass datasets (Covertype and MNIST) among the peers such that each peer has local training data from only two different classes (an example of scenario 3). Classification accuracy of the P2P approaches for disjointed data distribution is also presented in Table 2. As the results show, accuracies for all approaches drop in comparison with the case of i.i.d. However, PACE achieves significantly higher accuracy than the competitors on all datasets, demonstrating that it is more resilient to the adverse effect of disjoint class distribution. Note that the significant drop in accuracy for P2P Ivotes and LinSVME on MNIST dataset could be due to the fact that the size distribution of the different classes in MNIST dataset are almost equal compared to the Multiclass Covertype dataset and hence the difference.

In addition, we present the average computational cost incurred for a single peer during the training and prediction (on entire test set) phase (c.f. Table 3). Note that for PACE, we exclude the cost for building the index and for index retrieval, since these are implementation dependent, and fall outside the compared phases. For instance, a hash-based index incurs negligible cost while testing, but more in the construction stage. Observe that P2P Ivotes incurs the highest cost on almost all datasets. This is because it dynamically builds additional classifiers depending on the difficulty of the local dataset. Hence, in addition to incurring higher training cost, as there are more classifiers involved, the prediction cost is also relatively higher. Compared with LinSVME, the training cost of PACE is higher due to cost of clustering the local training data. However, since training is not done as frequently and this is also within acceptable range (not more than a few seconds for each peer), it is not a big issue. However, PACE incurs significantly lesser prediction cost (even if index retrieval were to be added) than LinSVME. This is because LinSVME uses all classifiers for prediction, whereas PACE only uses a small number. From Table 3, we can see that PACE performs within acceptable time for varying number of peers (100–500) with varying dataset sizes and dimensions, thus demonstrating the efficiency and scalability of PACE. It is apparent that the communication cost of PACE increases linearly with the number of models (peers). Hence, given an appropriate choice of classification model (such as linear SVM that is represented with only a single vector), PACE is highly scalable in terms of communication cost (results omitted due to space constraints).

4.4   Skewed Class Distribution

Here, we examine the effect of skewed class distribution on classification accuracy of the P2P approaches. Using a two class classification problem as the base case, we experimented with the Binary Covertype dataset which has an even class distribution. We skewed the local training data of every peer such that

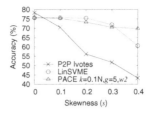

Fig. 3. Effect of skewed class distribution on accuracy (Binary Covertype Dataset)

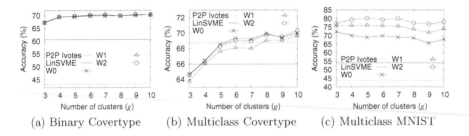

(a) Binary Covertype   (b) Multiclass Covertype   (c) Multiclass MNIST

Fig. 4. Effect of number of clusters on accuracy for skewed (a) and disjoint (b and c) class distributions; number of voters, $k = 10$

it is $s\%$ away from the natural distribution while maintaining equal size distribution (skews in both ways). From the results (Figure 3), we observe that for all approaches, as s increases, the accuracy decreases. However, we note that P2P Ivotes has the sharpest descent. While LinSVME initially performs better than PACE, its accuracy decreases sharply as s increases past 0.3. Although all approaches are affected by skew, we note that PACE has the smoothest and smallest descent in accuracy.

4.5   Parameter Sensitivity

Here, we examine the effects of the number of clusters g, number of voting peers k, and the inverse distance weighting scheme, and provide some insight towards their selection. Note that in Figures 4 and 5, lines denoted by "W" indicate the different weighting schemes for PACE.

Number of clusters g: First we study the effect of the number of clusters on classification accuracy for non i.i.d. data distribution (in Figure 4). Plots for i.i.d. data distribution are not presented because the variations are less than 0.5% and do not present any knowledge. We observe that in all cases, as the number of clusters increases, the classification accuracy also increases but the magnitude of increase also decreases. This happens because as the number of clusters increases, they become more compact and representative, thereby increasing the accuracy of the distance approximation and improving the prediction accuracy. However, computation and communication costs also increase proportionally. Hence, our approach is not very sensitive to the number of clusters and a smaller number

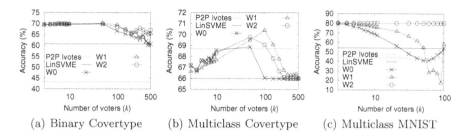

(a) Binary Covertype   (b) Multiclass Covertype   (c) Multiclass MNIST

Fig. 5. Effect of number of voting nearest neighbors on accuracy for skewed (a) and disjoint (b and c) class distributions; number clusters, $g = 5$

of clusters are preferred. Note that satisfactory accuracy can be achieved on all datasets with as few as 5 clusters.

Number of voting nearest neighbors k: Next, we study the effect of the number of voting nearest neighbors (size of ensemble) on classification accuracy (c.f. Figure 5). While the results for i.i.d. data distribution are not presented due to space constraints, we note that as the number of voting peers increases, the accuracy also increases. However, the increase quickly diminishes when the number of voting peers exceeds 20% of the total peers.

Whereas under skewed or disjoint class distributions (c.f. Figure 5), we observe that as k increases, the classification accuracy increases initially and then decreases. Note that this is also an expected result which justifies the rationale of our approach. The reason for the initial increase is because classifiers that were ranked higher have a higher probability of correctly classifying the test data, which also depends on training error. Hence we observe that the nearest neighbor may not have the highest accuracy. However, as k increases, we are gradually adding classifiers that are less probable of correctly classifying the test data and when the true error rate of these added classifiers falls below 0.5, the accuracy of the ensemble starts to deteriorate. In addition, we observe that for small values of k (lower 10%), PACE is able to outperform P2P Ivotes and LinSVME for both distributions.

5   Conclusions

In this paper, we studied the problem of distributed learning in P2P networks. We found potential pitfalls that arise due to data distributions in P2P networks, and presented several scenarios in which majority voting performs badly, demonstrating the significance of our work. To address these issues, we proposed a novel <u>P2P</u> <u>A</u>daptive <u>C</u>lassification <u>E</u>nsemble (PACE) framework which dynamically adapts to the distributions of the test data and selects only relevant classifiers to participate in the prediction vote. To validate the efficiency and effectiveness of our approach, we conducted extensive empirical evaluations. Results show that in the normally assumed environment, PACE performs similar to other existing P2P classification approaches. However, under varying conditions typical of real world environments, PACE outperforms existing approaches and minimizes the effects of adverse conditions.

In future, we intend to perform a more in-depth study on the relationship between test data and classifiers with respect to their proximity in the feature space. While the idea of our framework is to correlate data proximity and accuracy of the training error as an estimate for generalization error, its implementation may not be the most appropriate due to considerations such as time and communication cost. Hence, alternatives approaches will be explored in future.

References

1. Bhaduri, K., Wolff, R., Giannella, C., Kargupta, H.: Distributed decision-tree induction in peer-to-peer systems. Statistical Analysis and Data Mining 1(2), 85–103 (2008)
2. Ang, H.H., Gopalkrishnan, V., Hoi, S.C.H., Ng, W.K.: Cascade RSVM in peer-to-peer networks. In: Daelemans, W., Goethals, B., Morik, K. (eds.) ECML PKDD 2008, Part I. LNCS (LNAI), vol. 5211, pp. 55–70. Springer, Heidelberg (2008)
3. Gorodetskiy, V., Karsaev, O., Samoilov, V., Serebryakov, S.: Agent-based service-oriented intelligent P2P networks for distributed classification. In: Hybrid Information Technology, pp. 224–233 (2006)
4. Luo, P., Xiong, H., Lü, K., Shi, Z.: Distributed classification in peer-to-peer networks. In: KDD, pp. 968–976 (2007)
5. Siersdorfer, S., Sizov, S.: Automatic document organization in a P2P environment. In: ECIR, pp. 265–276 (2006)
6. Datta, S., Bhaduri, K., Giannella, C., Wolff, R., Kargupta, H.: Distributed data mining in peer-to-peer networks. IEEE Internet Computing, Special issue on Distributed Data Mining 10(4), 18–26 (2006)
7. Ang, H.H., Gopalkrishnan, V., Hoi, S.C.H., Ng, W.K., Datta, A.: Classification in P2P networks by bagging cascade RSVMs. In: VLDB Workshop on DBISP2P, pp. 13–25 (2008)
8. Chan, P.K., Stolfo, S.J.: Toward scalable learning with non-uniform class and cost distributions: A case study in credit card fraud detection. In: KDD, pp. 164–168 (1998)
9. Jordan, M.I., Xu, L.: Convergence results for the em approach to mixtures of experts architectures. Neural Networks 8(9), 1409–1431 (1995)
10. Cover, T., Hart, P.: Nearest neighbor pattern classification. IEEE Transactions on Information Theory 13(1), 21–27 (1967)
11. Breiman, L.: Pasting small votes for classification in large databases and on-line. Machine Learning 36(1-2), 85–103 (1999)
12. Hsieh, C.J., Chang, K.W., Lin, C.J., Keerthi, S.S., Sundararajan, S.: A dual coordinate descent method for large-scale linear SVM. In: ICML, pp. 408–415 (2008)
13. Berchtold, S., Ertl, B., Keim, D.A., Kriegel, H.P., Seidl, T.: Fast nearest neighbor search in high-dimensional space. In: ICDE, pp. 209–218 (1998)
14. Andoni, A., Indyk, P.: Near-optimal hashing algorithms for approximate nearest neighbor in high dimensions. In: FOCS, pp. 459–468 (2006)
15. Arthur, D., Vassilvitskii, S.: K-means++: the advantages of careful seeding. In: SODA, pp. 1027–1035 (2007)
16. Asuncion, A., Newman, D.: UCI machine learning repository (2007)
17. Quinlan, J.R.: C4.5: Programs for Machine Learning. Morgan Kaufmann, San Francisco (1993)

Mining Rare Association Rules in the Datasets with Widely Varying Items' Frequencies

R. Uday Kiran and P. Krishna Reddy

Center for Data Engineering,
International Institute of Information Technology-Hyderabad,
Hyderabad, Andhra Pradesh, India - 500032
`uday_rage@research.iiit.ac.in, pkreddy@iiit.ac.in`

Abstract. Rare association rule is an association rule consisting of rare items. It is difficult to mine rare association rules with a single minimum support (*minsup*) constraint because low *minsup* can result in generating too many rules in which some of them can be uninteresting. In the literature, *minimum constraint model* using "multiple *minsup* framework" was proposed to efficiently discover rare association rules. However, that model still extracts uninteresting rules if the items' frequencies in a dataset vary widely. In this paper, we exploit the notion of "item-to-pattern difference" and propose multiple *minsup* based FP-growth-like approach to efficiently discover rare association rules. Experimental results show that the proposed approach is efficient.

Keywords: rare association rules, frequent patterns, multiple minimum supports.

1 Introduction

Association rule mining is an important data mining technique which discovers interesting associations among the entities (or items) in a dataset. Since the introduction of association rules in [1], mining the association rules has been extensively studied in the literature [2,3]. The basic model of association rule is as follows:

Let $I = \{i_1, i_2, ..., i_n\}$ be a set of items. Let T be a set of transactions (dataset), where each transaction t is a set of items such that $t \subseteq I$. A pattern (or an itemset) X is a set of items $\{i_1, i_2, ..., i_k\}$, $1 \leq k \leq n$, such that $X \subseteq I$. Pattern containing k number of items is called k-pattern. An association rule is an implication of the form, $A \Rightarrow B$, where $A \subset I$, $B \subset I$ and $A \cap B = \emptyset$. The rule $A \Rightarrow B$ holds in T with *support s*, if s% of the transactions in T contain $A \cup B$. Similarly rule $A \Rightarrow B$ holds in T with *confidence c*, if c% of transactions in T that support A also support B. Given T, the problem of mining association rules is to discover all rules that satisfy user-specified minimum support (*minsup*) and minimum confidence (*minconf*) constraints. The patterns which satisfy *minsup* are called frequent patterns. The rules which satisfy both *minsup* and *minconf* constraints are called strong rules.

Example 1: Let our running example be the transaction dataset, T, shown in Figure 1(a). The set of items $I = \{bread, jam, ball, bat, bed, pillow\}$. Total number of transactions in T is 20. A set of items $\{bread, jam\}$ is a pattern. This pattern occurs

H. Kitagawa et al. (Eds.): DASFAA 2010, Part I, LNCS 5981, pp. 49–62, 2010.

in 5 transactions; therefore, its support, $S(bread \cup jam) = \frac{5 \times 100}{20} = 25\%$ (in support count, $S(bread \cup jam) = 5$).Let $jam \Rightarrow bread$ be an association rule which is derived from this pattern. The confidence of this rule, $C(jam \Rightarrow bread) = \frac{S(bread \cup jam)}{S(jam)} = \frac{25 \times 100}{40} = 62.5\%$. If $minsup = 25\%$ and $minconf = 50\%$ then the pattern $\{bread, jam\}$ is a frequent pattern and the rule $jam \Rightarrow bread$ is a strong rule. This rule says that 25% of customers buy $bread$ and jam together, and those who buy jam also buy $bread$ 62.5% of the time. Throughout this paper, we discuss this example in terms of support counts.

TID	Items	TID	Items
1	Bread, Jam	11	Bread, Ball
2	Bread, Ball, Pillow	12	Bread
3	Bread, Jam	13	Ball, Jam
4	Bread, Bed, Pillow	14	Ball, Bat
5	Bread, Jam, Bat	15	Ball, Jam
6	Bread, Jam	16	Ball, Bat
7	Bread, Bed, Pillow	17	Jam, Bat
8	Ball, Bat	18	Ball, Bat
9	Bread, Jam	19	Ball, Bat
10	Bread, Ball	20	Bread, Ball

(a)

Patterns	S	MIS	MMS
Bread	12	9	Y
Ball	11	8	Y
Jam	8	5	Y
Bat	7	4	Y
Pillow	3	2	Y
Bed	2	2	Y
Bread, Ball	4	-	N
Bread, Jam	5	-	Y
Bread, Bed	2	-	**Y**

Patterns	S	MMS
Bread, Pillow	2	**Y**
Ball, Jam	2	N
Ball, Bat	5	Y
Jam, Bat	2	N
Bed, Pillow	2	Y
Bread, Pillow, Bed	2	**Y**

(b)

Fig. 1. Rare item problem illustration. (a) Transaction dataset and (b) Frequent patterns generated at $minsup = 2$. The terms 'S', 'MIS' and 'MMS' are used as acronyms for support, $minimum$ $item$ $support$ and multiple $minsup$ framework. In Figure 1(b), second column represents the MIS values used for the items in multiple $minsup$ framework. In Figure 1(b), the frequent patterns generated in multiple $minsup$ framework are projected over the frequent patterns generated in single $minsup$ framework using terms 'Y' and 'N'. These terms represent 'frequent patterns generated' and 'frequent patterns not generated' in multiple $minsup$ framework respectively.

Given a transaction dataset consisting of a set of items, mining association rules generally involves two steps: (*i*) Discovering all frequent patterns and (*ii*) Generating all strong association rules from the set of frequent patterns.

Most of the real-world datasets are non-uniform in nature. That is, in a dataset, some items appear frequently, while others appear relatively infrequent or rare. A rare association rule is an association rule consisting of rare items. Rare association rules can provide useful knowledge [5]. However, it is difficult to mine rare association rules because single $minsup$ based association rule (or frequent pattern) mining approaches like Apriori [1] and Frequent Pattern-growth (FP-growth) [6] suffer from the dilemma called "rare item problem" [7]. The "rare item problem" can be described as follows. If $minsup$ is set too high, we miss the frequent patterns involving rare items because rare items fail to satisfy high $minsup$. To find frequent patterns consisting of both frequent and rare items, we have to set $minsup$ very low. However, this may cause combinatorial explosion and produce too many frequent patterns. In addition, some of the uninteresting patterns can be generated as frequent patterns.

Uninteresting patterns are of two types.

1. Patterns which have low support and contain only frequent items.
2. Patterns which have low support and contain *highly* frequent and rare items. Considering such patterns as interesting or uninteresting is a subjective matter which depends on user requirement, application type etc. In this paper, we consider such patterns as uninteresting.

Example 2: Continuing with Example 1, it can be observed that at high *minsup*, say *minsup* = 5 (in support count), the rare items *bed* and *pillow* fail to participate in generating frequent patterns because their support value is less than *minsup*. To find frequent patterns consisting of rare items, let us specify low *minsup*, say *minsup* = 2. The frequent patterns generated at *minsup* = 2 are shown in Figure 1(b). Among these generated frequent patterns, the patterns {bread, ball}, {ball, jam} and {jam, bat} (patterns shown in bold letters) are uninteresting because they contain frequent items occurring together in very less number of transactions. Also, the patterns {bread, pillow}, {bread, bed} and {bread, pillow, bed} (patterns shown in bold-italics letters) can also be considered as uninteresting because they contain frequent item *bread* occurring along with rare items *bed* and/or *pillow* in very less number of transactions.

To improve the performance of mining frequent patterns consisting of both frequent and rare items, efforts are being made to discover frequent patterns using "multiple *minsup* framework" [7,9,11,12]. Independent of the detailed implementation technique, the model used in these approaches is as follows.

1. Each item in the transaction dataset is specified with a support constraint called *minimum item support* (*MIS*).
2. A pattern is defined as frequent, if its support is greater than or equal to the minimal *MIS* value among all its items. In other words, *minsup* of a pattern is represented as the minimal *MIS* value among all its items.

Generally, items' *MIS* values are specified based on their respective support values. So, as compared with frequent items, rare items are specified with relatively lower *MIS* values. If a pattern contains only frequent items, it has to satisfy relatively high *minsup* value to be a frequent pattern. If a pattern contains rare items, it has to satisfy relatively low *minsup* value to be a frequent pattern. Thus, this model can efficiently prune those uninteresting patterns which have low support and contain only frequent items.

Even though this model improves the performance over single *minsup* framework, it still extracts uninteresting frequent patterns which have low support and contain *highly* frequent and rare items. Hence, this model is also insufficient to mine frequent patterns especially in the datasets where items' frequencies vary widely, because, in such datasets, users can consider those rules which have low support and contain both *highly* frequent and rare items as uninteresting rules. In this paper, we refer this model as *minimum constraint model*.

> **Example 3:** For the transaction dataset shown in Figure 1(a), let the *MIS* values for the items *bread, ball, jam, bat* and *pillow* be 9, 8, 5, 4, 2 and 2 respectively. In Figure 1(b), column titled "MMS" presents the frequent patterns generated under *minimum constraint model* with reference to the frequent patterns generated at *minsup* = 2. The following observations can be made from the discovered frequent patterns. First, this model has pruned all uninteresting patterns which have low support and contain only frequent items from becoming frequent patterns. Second, this model was unable to prune uninteresting patterns which have low support and contain *highly* frequent and rare items from becoming frequent patterns.

The generation of uninteresting frequent patterns in *minimum constraint model* is due to the reason that this model specifies *minsup* of a pattern by considering only the minimal frequent (or *MIS*) item within it. In this paper, we exploit the notion of "item-to-pattern difference" and, extend it to the *minimum constraint model* so that the proposed model can specify *minsup* of a pattern by considering both minimal and maximal frequent items within it. Thus, the proposed model prunes uninteresting patterns while mining frequent patterns in the datasets where items' frequencies vary widely. We call this model as *minimum-maximum constraint model*. For this model, we also discuss a pattern-growth approach which uses the prior knowledge regarding the items' *MIS* values and discovers frequent patterns with a single scan on the dataset. Experimental results show that the proposed model is efficient.

The rest of the paper is organized as follows. In Section 2, we summarize the existing approaches for mining rare association rules. In Section 3, we describe the proposed approach. In Section 4, experimental results conducted on synthetic and real world datasets are presented. In Section 5, we discuss conclusions and future work.

2   Related Work

To address "rare item problem", *minimum constraint model* (discussed in Section 1), which uses "multiple minimum support framework" was discussed and, extended to Apriori approach to discover complete set of frequent patterns [7]. This multiple *minsup* based Apriori-like approach suffers from the performance problems like generating huge number of candidate patterns and multiple scans on the transactional dataset. Therefore, an effort has been made to extend *minimum constraint model* to FP-growth approach as it does not suffer from the performance problems as those of Apriori [9].

The above two approaches assume that items' *MIS* values will be specified by the user prior to their execution. However, in the datasets where there exists numerous items, it is mostly difficult for the user to specify items' *MIS* values. Therefore, a method shown in Equation 1 has been proposed to specify *MIS* values for the items dynamically [11].

$$MIS(i_j) = S(i_j) - SD \quad when \ (S(i_j) - SD) > LS \qquad (1)$$
$$= LS \qquad\qquad otherwise$$

where, *LS* refers to user-specified "least support" and *SD* refers to support difference. *SD* can be either user-specified or derived using Equation 2.

$$SD = \lambda(1 - \beta) \tag{2}$$

where, λ represents the parameter like mean, median of the item supports and $\beta \in [0, 1]$.

In [12], an effort has been made to improve the performance of [9] by efficiently identifying only those items which can generate frequent patterns.

The approaches discussed in [7,9,11,12] are based on *minimum constraint model*. Therefore, these approaches can efficiently prune uninteresting patterns which have low support and contain only frequent items. However, they cannot prune uninteresting patterns which have low support and contain both *highly* frequent and rare items. The reason is *minimum constraint model* do not specify *minsup* of a pattern by considering both minimal and maximal frequent items within it, instead specifies *minsup* of a pattern by considering only the minimal frequent item within it.

A stochastic mixture model known as negative binomial (NB) distribution has been discussed to understand the knowledge of the process generating transaction dataset [8]. This model along with a user-specified precision threshold, finds local frequency thresholds for groups of patterns based on which algorithm finds all NB-frequent patterns in a dataset. It considers highly skewed data (skewed towards right) with the underlying assumption of Poisson processes and Gamma mixing distribution. Hence, the model can effectively be implemented in the datasets like general web logs etc., which are exponentially distributed. However, this approach is not effective on other datasets like general super markets datasets which are generally not exponentially distributed. The reason is frequent items will distort the mean and the variance and thus will lead to a model which grossly overestimates the probability of seeing items with high frequencies. If we remove items of high frequencies as suggested, we may miss some interesting rules pertaining to frequent items.

An approach has been suggested to mine the association rules by considering only infrequent items i.e., items having support less than the *minsup* [10]. However, this approach fails to discover associations between frequent and rare items.

An Apriori-like approach which tries to use a different *minsup* at each level of iteration has been discussed [13]. This model still suffers from "rare item problem" because it uses a single *minsup* constraint at each iteration. Also, this approach being an Apriori-like approach suffers from the performance problems like generating huge number of candidate patterns and multiple scans on the dataset.

The model proposed in this paper is different from the *minimum constraint model* which was used in [7,9,11,12]. The proposed approach specifies *minsup* of a pattern by considering both minimal and maximal frequent item within it, whereas in *minimum constraint model*, *minsup* of a pattern is specified by considering only the minimal frequent item within it. The approaches discussed in [8] and [10] deal with frequent patterns consisting of rare items. The proposed approach extracts frequent patterns consisting of both frequent and rare items. The approach discussed in [13] do not specify *minsup* for each pattern, instead specifies *minsup* at each iteration or level of frequent pattern mining. In the proposed approach, independent to the iteration, each pattern has to satisfy a *minsup* depending upon the items within it.

3   Proposed Approach

3.1   Basic Idea

In the datasets where items' frequencies vary widely, the *minimum constraint model* (discussed in Section 1) generates uninteresting frequent patterns which have low support and contain *highly* frequent and rare items. The main issue is to develop a model to filter such uninteresting frequent patterns. One of the characteristic feature of an uninteresting frequent pattern generated in *minimum constraint model* is that the support of a pattern is much less than the support of maximal frequent item within it.

The basic idea of the proposed approach is as follows. Uninteresting patterns are filtered by limiting the difference between the support of a pattern, and the support of the maximal frequent item in that pattern. To explain the basic idea, we define the following terms: item-to-pattern difference (*ipd*) and maximum item-to-pattern difference (*mipd*).

Definition 1. *Item-to-pattern difference* (ipd). *Given a pattern $X = \{i_1, i_2, \cdots, i_k\}$, where $S(i_1) \leq S(i_2) \leq \cdots \leq S(i_k)$ and $MIS(i_1) \leq MIS(i_2) \leq \cdots \leq MIS(i_k)$, the ipd of pattern X, denoted as $ipd(X) = S(i_k) - S(X)$.*

> **Example 5:** In the transactional dataset shown in Figure 1(a), $S(bread) = 12$ and $S(bread, jam) = 5$. The *ipd* value of the pattern $\{bread, jam\}$ i.e., $ipd(bread, jam) = 7 (= 12 - 5)$.

The metric *ipd* provides the information regarding the difference between the support of a pattern with respect to support of the maximal frequent item within it. If *ipd* value is less for a pattern, it means support of the pattern is close to the the support of the maximal frequent item within it. A high *ipd* value for a pattern means support of the respective pattern is very less (or away) from the support of the maximal frequent item within it.

Definition 2. *Maximum item-to-pattern difference* (mipd). *The mipd is a user-specified maximum ipd value. A pattern X is an interesting pattern if and only if $ipd(X) \leq mipd$. Otherwise, the pattern X is an uninteresting pattern.*

> **Example 6:** Continuing with the Example 5, if $mipd = 7$ then the pattern $\{bread, jam\}$ is an interesting pattern because $ipd(bread, jam) \leq mipd$. The pattern $\{bread, bed\}$ is an uninteresting pattern because $ipd(bread, bed) > mipd$.

The value for *mipd* can be specified to a certain percentage of a transactional dataset. If *mipd* value is high, it discovers both interesting and uninteresting patterns as in *minimum constraint model*. If *mipd* value is less, there is a danger of filtering out interesting patterns.

The rules to extract interesting patterns under the proposed framework are as follows. Let $X = \{i_1, i_2, \cdots, i_k\}$, where $S(i_1) \leq S(i_2) \leq \cdots \leq S(i_k)$, be a pattern. Let $MIS(i_j)$ be the *minimum item support* for an item $i_j \in X$. The pattern X is frequent if

(i) $S(X) \geq minimum(MIS(i_1), MIS(i_2), \cdots, MIS(i_k))$, and
(ii) $ipd(X) \leq mipd$.

Example 7: For the transaction dataset shown in Figure 1(a), let the *MIS* values for the items *bread, ball, jam, bat* and *pillow* be 9, 8, 5, 4, 2 and 2 respectively. Let user-specified *mipd* = 7. The interesting pattern {*bread, jam*} is a frequent pattern because $S(bread, jam) \geq minimum(9,5)$ and $ipd(bread, jam) \leq mipd$. The uninteresting pattern {*bread, bed*} is an infrequent pattern, because, even though $S(bread, bed) \geq minimum(9,2)$ its $ipd(bread, bed) > mipd$.

The items' *MIS* values and *mipd* value should be specified by the user depending upon the type of application. Also, the items' *MIS* values can be specified using Equation 1.

Calculating *minsup* for a pattern X: Based on the above two requirements, for the pattern X to be a frequent pattern, its *minsup*, denoted as $minsup(X)$ can be calculated using Equation 3.

$$minsup(X) = maximum(minimum(MIS(i_1), \cdots, MIS(i_k)), (S(i_k) - mipd)) \qquad (3)$$

The value, "$S(i_k) - mipd$" can be derived by substituting "$ipd(X) = S(i_k) - S(X)$" in the equation "$ipd(X) \leq mipd$". Thus, the proposed approach specifies *minsup* of a pattern by considering both minimal and maximal frequent item within it.

Problem Definition: In a transactional dataset T, given items' *MIS* values and *mipd* value, discover complete set of frequent patterns that satisfy (*i*) lowest *MIS* value among all its items and (*ii*) *ipd* value less than or equal to the user-specified *mipd*.

The frequent patterns discovered using this model follow "sorted closure property". The "sorted closure property" says, if a *sorted k-pattern* $\langle i_1, i_2, ..., i_k \rangle$, for k $\geq$ 2 and MIS(i_1) $\geq$ MIS(i_2) $\geq ... \geq$ MIS(i_k), is frequent, then all of its subsets involving the item having lowest MIS value (i_1) need to be frequent; however, other subsets need not necessarily be frequent patterns.

Example 8: Consider a transaction dataset having three items i_1, i_2 and i_3. Let the *MIS* values for these items be 5%, 10% and 20% respectively. If a sorted 3-pattern {i_1, i_2, i_3} has support 6% then it is a frequent pattern because $S(i_1, i_2, i_3) \geq min(5\%, 10\%, 20\%)$. In this frequent pattern, all supersets of i_1 i.e., {{i_1}, {i_1, i_2}, {i_1, i_3}} are frequent because of *apriori property* [1]. However, the supersets of items i_2 and i_3, {i_2, i_3}, can be still an infrequent pattern with $S(i_2, i_3) = 9\%$ which is less than the its required *minsup* ($min(10\%, 20\%)$).

Therefore, the user should not consider all subsets of a frequent pattern as frequent. Instead, user can consider all subsets of the minimal frequent (or *MIS*) item within a frequent pattern as frequent. This property is elaborately discussed in [7].

3.2 Algorithm

Given the transactional dataset T, items' *MIS* values and *mipd* value, the proposed approach utilizes the prior knowledge regarding the items' *MIS* values and discovers frequent patterns with a single scan on the transactional dataset. The approach involves the following three steps.

1. Construction of a tree, called *MIS*-tree.
2. Generating *compact MIS-tree* from *MIS*-tree.

3. Mining *compact MIS-tree* using *conditional pattern bases* to discover complete set of frequent patterns.

The first two steps i.e., construction of *MIS*-tree and *compact MIS-tree* are similar to those in [9,12]. However, mining frequent patterns from the *compact MIS-tree* is different from [9,12].

Structure of *MIS*-tree: The structure of *MIS*-tree includes a prefix-tree and an item list, called *MIS*-list, consisting of each distinct item with frequency (or support), *MIS* value and a pointer pointing to the first node in prefix-tree carrying the respective item.

Construction of *MIS*-tree: Initially, the items in the transactional dataset are sorted in descending order of their *MIS* values. Let this sorted order of items be *L*. Next, *MIS*-list is populated with all the items in *L* order. The support values of the items are set to 0. The *MIS* values of the items are set with their respective *MIS* values. A *root* node labeled "null" is created in the prefix-tree. Next, each transaction in the transactional dataset is scanned and *MIS*-tree is updated as follows. (*i*) Items in the respective transaction are ordered in *L* order. (*ii*) For these items, their frequencies (or supports) are updated by 1 in the *MIS*-list. (*iii*) In *L* order, a branch which consists of these items is created in the prefix-tree. The construction of a branch in the prefix-tree of *MIS*-tree is same as that in FP-tree. However, it has to be noted that FP-tree is constructed with support descending order of items and *MIS*-tree is constructed with *MIS* descending order of the items. To facilitate tree traversal, an item header table is built so that each item points to its occurrences in the tree via a chain of node-links.

We now explain the construction of *MIS*-tree using the transactional dataset shown in Figure 1(a). Let the *MIS* values for the items *bread, ball, jam, bat, pillow* and *bed* be 9, 8, 5, 4, 2 and 2 respectively. Let user-specified *mipd* = 7. The sorted-list of items in descending order of their *MIS* values is *bread, ball, jam, bat, pillow* and *bed*. Let this sorted order be *L*. In *L* order, all items are inserted into *MIS*-list by setting their support and *MIS* values to 0 and their respective *MIS* values. In the prefix-tree, a root node is created and labeled as "null". Figure 2(a) shows the *MIS*-tree constructed before scanning the transactional dataset. In the first scan of the dataset shown in Figure 1(a), the first transaction "1: bread, jam" containing two items is scanned in *L* order i.e., {*bread, jam*}, and the frequencies of the items *bread* and *jam* are updated by 1 in *MIS*-list. A branch of tree is constructed with two nodes, ⟨bread:1⟩ and ⟨jam:1⟩, where *bread* is linked as child of the root and *jam* is linked as child of *bread*. The updated *MIS*-tree after scanning first transaction is shown in Figure 2(b). The second transaction "2: bread, ball, pillow" containing three items is scanned in *L* order i.e., {*bread, ball, pillow*}, and the frequencies of the items *bread, ball* and *pillow* are updated by 1 in *MIS*-list. The sorted list of items in the second transaction will result in a branch where *bread* is linked to *root*, *ball* is linked to *bread* and *pillow* is linked to *ball*. However, this branch shares the common **prefix**, *bread*, with the existing path for the first transaction. Therefore, the count of *bread* node is incremented by 1 and new nodes are created for items *ball* and *pillow* such that *ball* is linked to *bread* and *jam* is linked to *ball*, and their nodes values are set to 1. The *MIS*-tree generated after scanning second transaction is shown in Figure 2(c). Similar process is repeated for the remaining transactions in the transactional dataset. A node link table is built for tree traversal. The constructed *MIS*-tree after scanning every transaction is shown in Figure 2(d).

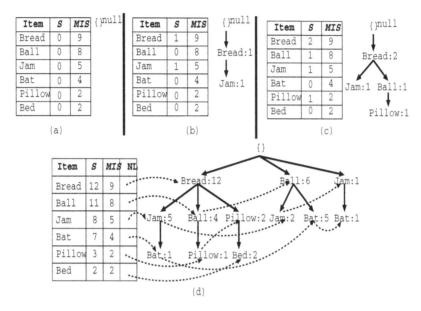

Fig. 2. *MIS*-tree for the transactional dataset shown in Fig. 1. (a) *MIS*-tree before scanning the transactional dataset. (b) *MIS*-tree after scanning first transaction. (c) *MIS*-tree after scanning second transaction. (d) *MIS*-tree after scanning entire transactional dataset.

Generating *compact MIS-tree*: The *MIS*-tree is constructed with every item in the transactional database. There may be items which do not generate any frequent pattern. So, using Lemma 1, we identify all those items which have support less than the lowest *MIS* value among all frequent items (or frequent 1-patterns) and prune them from the *MIS*-tree. In [9], it was observed that depending on the items' *MIS* values there exists scenarios where child nodes of a parent node can share a same item after pruning operation. So, tree-merging operation is performed on the pruned *MIS*-tree to merge such child nodes. The resultant *MIS*-tree is called compact *MIS*-tree. The algorithm and correctness for constructing *MIS*-tree and generating *compact MIS-tree* are discussed in [12].

Lemma 1. *In T, the items which have their support less than the lowest MIS value among all frequent items will not generate any frequent pattern.*

Proof. In the proposed model, one among the constraints for a pattern to be frequent is that it should satisfy the lowest *MIS* value among all its items. So items which have their supports less than the lowest *MIS* value among all the items cannot generate any frequent pattern. From the *sorted closure property*, we can say that in a frequent pattern, the item having lowest *MIS* value will be a frequent. Thus, we can prune all those items which have support less than the lowest *MIS* value among all the frequent items (or frequent 1-itemsets).

Continuing with the example, the frequent item having lowest *MIS* value is *bed*. Since, all items in the transactional dataset have their support values greater than or equal to $MIS(bed)$ no item is pruned from the *MIS*-tree. Thus, the existing *MIS*-tree is the *compact MIS-tree*.

Mining the *compact MIS-tree*:   Briefly, mining of frequent patterns from the *compact MIS-tree* is as follows. Choose each frequent length-1 pattern (or item) in the *compact MIS-tree* as the suffix-pattern. For this suffix-pattern construct its conditional pattern bases. From the conditional pattern bases, construct MIS-tree, called *conditional MIS-tree*, with all those prefix-subpaths that have satisfied the *MIS* value of the suffix-pattern and *mipd*. Finally, recursive mining on *conditional MIS-tree* results in generating all frequent patterns.

   The correctness of mining frequent patterns from the *compact MIS-tree* is based on the following Lemma.

Lemma 2. *Let α be a pattern in* compact *MIS-tree. Let MIS-minsup$_\alpha$ and mipd be the two constraints that α has to satisfy. Let B be α conditional pattern base, and β be an item in B. Let $S(\beta)$ and $S_B(\beta)$ be the support of β in the transactional database and in B respectively. Let $MIS(\beta)$ be the user-specified β's MIS value. If α is frequent and the support of β satisfies MIS-minsup$_\alpha$ and $S(\beta) - S_B(\beta) \leq mipd$, the pattern $< \alpha, \beta >$ is therefore also frequent.*

Proof. According to the definition of conditional pattern base and compact *MIS*-tree (or FP-tree), each subset in B occurs under the condition of the occurrence of α in the transactional database. If an item β appears in B for n times, it appearers with α in n times. Thus, from the definition of frequent pattern used in this model, if the $S_B(\beta) \geq MIS\text{-}minsup_\alpha$ and $S(\beta) - S_B(\beta) \leq mipd$ then $< \alpha, \beta >$ is a frequent pattern.

Mining the *compact MIS-tree* is shown in Table 1. Consider *bed*, which is the last item in the *MIS*-list. *Bed* occurs in one branch of the *compact MIS-tree* of Figure 2(d). The path formed is {*bread, pillow, bed*: 2}. Therefore, considering *bed* as a suffix, its corresponding prefix path ⟨bread, pillow: 2⟩, form its conditional pattern base. Its conditional *MIS*-tree contains only a single path, ⟨pillow: 2⟩, *bread* is not included because $S(bread) - S_{bed}(bread) > mipd$. The single path generates the frequent pattern {pillow, bed:2}. Similarly, by choosing every item in the *MIS*-list, the complete set of frequent patterns are generated. The frequent patterns generated in this model are {{bread}, {ball}, {jam}, {bat}, {pillow}, {bed}, {bread, jam}, {ball, bat}, {bed, pillow}}. It can be observed that this model has pruned all uninteresting patterns while finding frequent patterns consisting of frequent and rare items.

Table 1. Mining the *compact MIS-tree* by creating conditional pattern bases

Item	Conditional Pattern Base	Conditional *MIS*-tree	Frequent Patterns
bed	{*bread, pillow*: 2}	⟨*pillow*: 2⟩	{*pillow, bed*:2}
pillow	{*bread, ball*: 1} {*bread*: 2}	-	-
bat	{*bread, jam*: 1} {*ball*: 6}	⟨*ball*: 6⟩	{*ball, bat*: 6}
jam	{*bread*: 5} {*ball*:2}	⟨*bread*:5⟩	{*bread, jam*: 5}
ball	{*bread*: 5}	-	-

After generating frequent patterns, a procedure discussed in [1] can be used to generate association rules from the discovered frequent patterns.

3.3 Relation between the Frequent Patterns Generated in Different Models

Let F be the set of the frequent patterns generated when $minsup = x\%$. Let MCM be the set of frequent patterns generated in *minimum constraint model*, when items' MIS values are specified such that no items' MIS value is less than $x\%$. For the same items' MIS values and *mipd* value, let $MMCM$ be the set of frequent patterns generated in *minimum-maximum constraint model*. The relationship between these frequent patterns is $MMCM \subseteq MCM \subseteq F$.

4 Experimental Results

In this section, we present the performance comparison of the proposed model against *minimum constraint model* discussed in [7,9,11,12]. We also present the performance comparison of the proposed model with the model discussed in [13].

For experimental purposes we have chosen two kinds of datasets: (*i*) synthetic dataset and (*ii*) real-world dataset. The synthetic dataset is T10.I4.D100K dataset which is generated with the data generator [1]. This generator is widely used for evaluating association rule mining algorithms. It contains 1,00,000 number of transactions, 870 items, maximum number of items in a transaction is 29 and the average number of items in a transaction is 12. The real-world dataset is a retail dataset [14]. It contains 88,162 number of transactions, 16,470 items, maximum number of items in a transaction is 76 and the average number of items in each transaction is 5.8. Both of the datasets are available at Frequent Itemset MIning (FIMI) repository [15].

4.1 Experiment 1

In this experiment, we have compared the proposed model (*minimum-maximum constraint model*) against *minimum constraint model*. To specify MIS values for the items, we used the equation discussed in [11]. In this paper, we have described this equation in Equation 1 of Section 2. For both the datasets, we have chosen $LS = 0.1\%$ because we have considered that any pattern (or an item) having support less than 0.1% is uninteresting. Next, items' MIS values are specified by varying SD values at 0.25% and 1% respectively.

Both Fig. 3(a) and Fig. 3(b) show the number of frequent patterns generated in synthetic and real-world datasets at different items' *MIS* values and *mipd* values. The thick line in these figures represents the number of frequent patterns generated when *minsup* = 0.1%. The lines, titled "MCM(*SD* = *x*%, *LS* = *y*%)" and "MMCM(*SD* = *x*%, *LS* = *y*%)", represent the number of frequent patterns generated in *minimum constraint model* and *minimum-maximum constraint model* when items' *MIS* values are specified with *SD* = *x*% and *LS* = *y*% in Equation 1. The following observations can be drawn from these two figures.

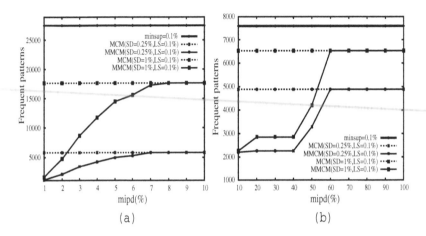

Fig. 3. Frequent patterns generated at different *mipd* values. (a) Synthetic (T10.I4.D100k) dataset and (b) Real-world (Retail) dataset.

First, increase in *SD* has increased the number of frequent patterns in both *minimum constraint* and *minimum-maximum constraint* models. The reason is as follows. Increase in *SD* has resulted in specifying low *MIS* values for the items. Low *MIS* values of the items facilitated patterns to satisfy low *minsup* values, thereby, increasing the number of frequent patterns.

Second, the *minimum constraint model* specifies *minsup* of a pattern by considering only the items' *MIS* values. Therefore, the number of frequent patterns discovered by this model are independent to *mipd* values.

Third, the *minimum-maximum constraint model* specifies *minsup* of a pattern by considering both items' *MIS* values and *ipd* values of a pattern. Therefore, the number of frequent patterns discovered by this model vary depending on the *mipd* value.

Fourth, it can be observed that for a fixed *SD* and *LS* values (or items' *MIS* values), the number of frequent patterns generated by *minimum-maximum constraint model* gets increased with increase in *mipd* value. The reason is that as *mipd* value increases, the patterns which have their support less than their respective maximal frequent item have generated as frequent patterns.

Analysis of the frequent patterns generated in retail dataset: In Figure 3(b) it can be observed that there is a sudden rise in the number of frequent patterns being generated when *mipd* values are varied from 40% to 60%. So, we have analyzed the frequent patterns generated at *mipd* = 40%, *mipd* = 50% and *mipd* = 60%. In our analysis, we have observed that when *mipd* > 40%, many of the frequent patterns generated contained *highly* frequent items and rare items appearing together with very less support values. Many of these patterns had their support value almost equivalent to 0.1%. The proposed approach is able to prune such uninteresting patterns when *mipd* < 40%.

4.2 Experiment 2

In this experiment, we compare the proposed model against the model discussed in [13]. In [13], the *minsup* used for discovering frequent 1-patterns is shown in Equation 4.

$$minsup = \frac{1}{2} \left[\sum_{i=1}^{n} TOTOCC_1 + \frac{MINS_1 + MAXS_1}{2} \right] \qquad (4)$$

where, $TOTOCC_l$, $MINS_1$ and $MAXS_1$ represent total support of items, minimum and maximum support of the items present in the transactional dataset.

Using Equation 4, the *minsup* derived for the synthetic and real-world datasets are 2.5% and 14.4% respectively. In synthetic dataset, at *minsup* = 2.5%, only 88 items out of 870 items have participated in generating frequent patterns. In real-world dataset, at *minsup* = 14.4%, only 14 items out of 16,470 items have participated in generating frequent patterns. The reason for deriving high *minsup* values in these datasets is due to the presence of items having high frequencies. We cannot remove these frequent items because we will miss some important interesting rules consisting of these items.

In the proposed model when $LS = 0.1\%$, 798 items out of 870 have participated in synthetic dataset, and 2118 items out of 16,470 items have participated in real-world dataset. Thus, the proposed model has facilitated more number of items to generate frequent patterns.

5   Conclusions and Future Work

The *minimum constraint model*, which uses "multiple minimum support framework" for finding rare association rules generates uninteresting frequent patterns if the items' frequencies in a dataset vary widely. In this paper, we explored "item-to-pattern difference" notion and extended it to the *minimum constraint model* so that the extended model can prune such patterns while mining rare association rules. For this extended model, an FP-growth-like approach is also presented. This FP-growth-like approach utilizes the prior knowledge provided by the user (items' *MIS* values) and discovers frequent patterns with a single scan on the dataset. We have evaluated the performance of the proposed model by conducting experiments on both synthetic and real-world datasets. The results show that as compared with *single minsup model* and *minimum constraint model*, the proposed model prunes more number of uninteresting rules while mining rare association rules.

As a part of future work, we are going to investigate an appropriate methodology for assigning confidence values in a dynamic manner to generate rare association rules.

References

1. Agrawal, R., Imielinski, T., Swami, A.: Mining association rules between sets of items in large databases. In: ACM SIGMOD International Conference on Management of Data, vol. 22, pp. 207–216. ACM Press, Washington (1993)
2. Hipp, J., Guntzer, U., Nakhaeizadeh, G.: Algorithms for Association Rule Mining - A General Survey and Comparision. ACM Special Interest Group on Knowledge Discovery and Data Mining 2(1), 58–64 (2000)
3. Melli, G., Osmar, R.Z., Kitts, B.: Introduction to the Special Issue on Successful Real-World Data Mining Applications. SIGKDD Explorations 8(1), 1–2 (2006)
4. Weiss, G.M.: Mining With Rarity: A Unifying Framework. SIGKDD Explorations 6(1), 7–19 (2004)
5. Mannila, H.: Methods and Problems in Data Mining. In: The International Conference on Database Theory, pp. 41–55 (1997)
6. Jiawei, H., Jian, P., Yiwen, Y., Runying, M.: Mining Frequent Patterns without Candidate Generation: A Frequent-Pattern Tree Approach. In: ACM SIGMOD Workshop on Research Issues in Data Mining and Knowledge Discovery, pp. 53–87 (2004)
7. Liu, B., Hsu, W., Ma, Y.: Mining Association Rules with Multiple Minimum Supports. In: ACM Special Interest Group on Knowledge Discovery and Data Mining Explorations, pp. 337–341 (1999)
8. Hahsler, M.: A Model-Based Frequency Constraint for Mining Associations from Transaction Data. In: Data Mining and Knowledge Discovery, pp. 137–166 (2006)
9. Hu, Y.-H., Chen, Y.-L.: Mining Association Rules with Multiple Minimum Supports: A New Algorithm and a Support Tuning Mechanism. Decision Support Systems 42(1), 1–24 (2006)
10. Zhou, L., Yau, S.: Association Rule and Quantitative Association Rule Mining among Infrequent Items. In: 8th International Workshop on Multimedia Data Mining, pp. 156–167 (2007)
11. Uday Kiran, R., Krishna Reddy, P.: An Improved Multiple Minimum Support Based Approach to Mine Rare Association Rules. In: IEEE Symposium on Computational Intelligence and Data Mining, pp. 340–347 (2009)
12. Uday Kiran, R., Krishna Reddy, P.: An Improved Frequent Pattern-growth Approach To Discover Rare Association rules. In: International Conference on Knowledge Discovery and Information Retrieval (2009)
13. Kanimonzhi Selvi, C.S., Tamilarasi, A.: Mining Association rules with Dynamic and Collective Support Thresholds. International Journal on Open Problems Computational Mathematics 2(3), 427–438 (2009)
14. Brijs, T., Swinnen, G., Vanhoof, K., Wets, G.: The use of association rules for product assortment decisions - a case study. In: Knowledge Discovery and Data Mining (1999)
15. Frequent Itemset MIning Repository, http://fimi.cs.helsinki.fi/data/

CAMLS: A Constraint-Based Apriori Algorithm for Mining Long Sequences*

Yaron Gonen, Nurit Gal-Oz, Ran Yahalom, and Ehud Gudes

Department of Computer Science, Ben Gurion University of the Negev, Israel
{yarongon,galoz,yahalomr,ehud}@cs.bgu.ac.il

Abstract. Mining sequential patterns is a key objective in the field of data mining due to its wide range of applications. Given a database of sequences, the challenge is to identify patterns which appear frequently in different sequences. Well known algorithms have proved to be efficient, however these algorithms do not perform well when mining databases that have long frequent sequences. We present CAMLS, Constraint-based Apriori Mining of Long Sequences, an efficient algorithm for mining long sequential patterns under constraints. CAMLS is based on the apriori property and consists of two phases, event-wise and sequence-wise, which employ an iterative process of candidate-generation followed by frequency-testing. The separation into these two phases allows us to: (i) introduce a novel candidate pruning strategy that increases the efficiency of the mining process and (ii) easily incorporate considerations of intra-event and inter-event constraints. Experiments on both synthetic and real datasets show that CAMLS outperforms previous algorithms when mining long sequences.

Keywords: data mining, sequential patterns, frequent sequences.

1 Introduction

The sequential pattern mining task has received much attention in the data mining field due to its broad spectrum of applications. Examples of such applications include analysis of web access, customers shopping patterns, stock markets trends, DNA chains and so on. This task was first introduced by Agrawal and Srikant in [4]: *Given a set of sequences, where each sequence consists of a list of elements and each element consists of a set of items, and given a user-specified min_support threshold, sequential pattern mining is to find all of the frequent subsequences, i.e. the subsequences whose occurrence frequency in the set of sequences is no less than min_support.* In recent years, many studies have contributed to the efficiency of sequential mining algorithms [2,14,4,8,9]. The two major approaches for sequence mining arising from these studies are: **apriori** and **sequence growth**.

---

* Supported by the IMG4 consortium under the MAGNET program of the Israel ministry of trade and industry; and the Lynn and William Frankel center for computer science.

H. Kitagawa et al. (Eds.): DASFAA 2010, Part I, LNCS 5981, pp. 63–77, 2010.

The **apriori** approach is based on the apriori property, as introduced in the context of association rules mining in [1]. This property states that if a pattern α is not frequent then any pattern β that contains α cannot be frequent. Two of the most successful algorithms that take this approach are GSP [14] and SPADE [2]. The major difference between the two is that GSP uses a horizontal data format while SPADE uses a vertical one.

The **sequence growth** approach does not require candidate generation, it gradually grows the frequent sequences. PrefixSpan [8], which originated from FP-growth [10], uses this approach as follows: first it finds the frequent single items, then it generates a set of projected databases, one database for each frequent item. Each of these databases is then mined recursively while concatenating the frequent single items into a frequent sequence. These algorithms perform well in databases consisting of short frequent sequences. However, when mining databases consisting of long frequent sequences, e.g. stocks values, DNA chains or machine monitoring data, their overall performance exacerbates by an order of magnitude.

Incorporating constraints in the process of mining sequential patterns, is a means to increase the efficiency of this process and to obviate ineffective and surplus output. cSPADE [3] is an extension of SPADE which efficiently considers a versatile set of syntactic constraints. These constraints are fully integrated inside the mining process, with no post-processing step. Pei et al. [7] also discuss the problem of pushing various constraints deep into sequential pattern mining. They identify the prefix-monotone property as the common property of constraints for sequential pattern mining and present a framework (Prefix-growth) that incorporates these constraints into the mining process. Prefix-growth leans on the sequence growth approach [8].

In this paper we introduce CAMLS, a constraint-based algorithm for mining long sequences, that adopts the apriori approach. The motivation for CAMLS emerged from the problem of aging equipment in the semiconductor industry. Statistics show that most semiconductor equipment suffer from significant unscheduled downtime in addition to downtime due to scheduled maintenance. This downtime amounts to a major loss of revenue. A key objective in this context is to extract patterns from monitored equipment data in order to predict its failure and reduce unnecessary downtime. Specifically, we investigated lamp behavior in terms of illumination intensity that was frequently sampled over a long period of time. This data yield a limited amount of possible items and potentially long sequences. Consequently, attempts to apply traditional algorithms, resulted in inadequate execution time.

CAMLS is designed for high performance on a class of domains characterized by long sequences in which each event is composed of a potentially large number of items, but the total number of frequent events is relatively small. CAMLS consists of two phases, event-wise and sequence-wise, which employ an iterative process of candidate-generation followed by frequency-testing. The event-wise phase discovers frequent events satisfying constraints within an event (e.g. two items that cannot reside within the same event). The sequence-wise phase constructs the frequent sequences and enforces constraints between events within these sequences (e.g. two events must occur one after another within a specified time interval). This separation allows the introduction of a novel pruning strategy which reduces the size of the search space considerably. We aim to utilize specific constraints that are relevant to the class of domains for which CAMLS is designed. Experimental results compare CAMLS to known algorithms and show that its advantage increases as the mined sequences get longer and the number of frequent patterns in them rises.

The major contributions of our algorithm are its novel pruning strategy and straightforward incorporation of constraints. This is what essentially gives CAMLS its high performance, despite of the large amount of frequent patterns that are very common in the domains for which it was designed.

The rest of the paper is organized as follows. Section 2 describes the class of domains for which CAMLS is designed and section 3 provides some necessary definitions. In section 4 we characterize the types of constraints handled by CAMLS and in section 5 we formally present CAMLS. Experimental results are presented in section 6. We conclude by discussing future research directions.

2   Characterization of Domain Class

The classic domain used to demonstrate sequential pattern mining, e.g. [4], is of a retail organization having a large database of customer transactions, where each transaction consists of customer-id, transaction time and the items bought in the transaction. In domains of this class there is no limitation on the total number of items, or the number of items in each transaction.

Consider a different domain such as the stock values domain, where every record consists of a stock id, a date and the value of the stock on closing the business that day. We know that a stock can have only a single value at the end of each day. In addition, since a stock value is numeric and needs to be discretized, the number of different values of a stock is limited by the number of the discretization bins. We also know that a sequence of stock values can have thousands of records, spreading over several years. We classify domains by this sort of properties. We may take advantage of prior knowledge we have on a class of domains, to make our algorithm more efficient. CAMLS aims at the class of domains characterized as follows:

- Large amount of frequent patterns.
- There is a relatively small number of frequent events.

Table 1 shows an example sequence database that will accompany us throughout this paper. It has three sequences. The first contains three events: $(acd), (bcd)$ and b in times 0, 5 and 10 respectively. The second contains three events: a, c and (db) in times 0, 4 and 8 respectively, and the third contains three events: $(de), e$ and (acd) in times 0, 7 and 11 respectively.

In section 6 we present another example concerning the behavior of a Quartz-Tungsten-Halogen lamp, which has similar characteristics and is used for experimental evaluation. Such lamps are used in the semiconductors industry for finding defects in a chip manufacturing process.

Table 1. Example sequence database. Every entry in the table is an event. The first column is the identifier of the sequence. The second column is the time difference from the beginning of the sequence to the occurrence of the event. The third column shows all of the items that constitute the event.

Sequence id (sid)	Event id (eid)	items		Sequence id (sid)	Event id (eid)	items
1	0	(acd)		3	0	(cde)
1	5	(bcd)		3	7	e
1	10	b		3	11	(acd)
2	0	a				
2	4	c				
2	8	(bd)				

3 Definitions

An *item* is a value assigned to an attribute in our domain. We denote an item as a letter of the alphabet: $a, b, ..., z$. Let $I = \{i_1, i_2, ..., i_m\}$ be the set of all possible items in our domain. An *event* is a nonempty set of items that occurred at the same time. We denote an event as $(i_1, i_2, ..., i_n)$, where $i_j \in I, 1 \leq j \leq n$. An event containing l items is called an l-event. For example, (bcd) is a 3-event. If every item of event e_1 is also an item of event e_2 then e_1 is said to be a *subevent* of e_2, denoted $e_1 \subseteq e_2$. Equivalently, we can say that e_2 is a *super-event* of e_1 or e_2 *contains* e_1. For simplicity, we denote 1-events without the parentheses. Without the loss of generality we assume that items in an event are ordered lexicographically, and that there is a radix order between different events. Notice that if an event e_1 is a proper superset of event e_2 then e_2 is radix ordered before e_1. For example, the event (bc) is radix ordered before the event (abc), and $(bc) \subseteq (abc)$.

A *sequence* is an ordered list of events, where the order of the events in the sequence is the order of their occurrences. We denote a sequence s as $\langle e_1, e_2, ..., e_k \rangle$ where e_j is an event, and e_{j-1} happened before e_j. Notice that an item can occur only once in an event but can occur multiple times in different events in the same sequence. A sequence containing k events is called a k-sequence, in contrast to the classic definition of k-sequence that refers to any sequence containing k items [14]. For example, $\langle (de)e(acd) \rangle$ is a 3-sequence. A sequence $s_1 = \langle e_1^1, e_2^1, ..., e_n^1 \rangle$ is a *subsequence* of sequence $s_2 = \langle e_1^2, e_2^2, ..., e_m^2 \rangle$, denoted $s_1 \subseteq s_2$, if there exists a series of integers $1 \leq j_1 < j_2 < ... < j_n \leq m$ such that $e_1^1 \subseteq e_{j_1}^2 \wedge e_2^1 \subseteq e_{j_2}^2 \wedge ... \wedge e_n^1 \subseteq e_{j_n}^2$. Equivalently we say that s_2 is a *super-sequence* of s_1 or s_2 *contains* s_1. For example $\langle ab \rangle$ and $\langle (bc)f \rangle$ are subsequence of $\langle a(bc)f \rangle$, however $\langle (ab)f \rangle$ is not. Notice that the subsequence relation is transitive, meaning that if $s_1 \subseteq s_2$ and $s_2 \subseteq s_3$ then $s_1 \subseteq s_3$. An *m-prefix* of an n-sequence s is any subsequence of s that contains the first m events of s where $m \leq n$. For example, $\langle a(bc) \rangle$ is a 2-prefix of $\langle a(bc)a(cf) \rangle$. An *m-suffix* of an n-sequence s is any subsequence of s that contains the last m events of s where $m \leq n$. For example, $\langle (cf) \rangle$ is a 1-suffix of $\langle a(bc)a(cf) \rangle$.

A *database of sequences* is an unordered set of sequences, where every sequence in the database has a unique identifier called *sid*. Each event in every sequence is associated with a timestamp which states the time duration that passed from the beginning of the sequence. This means that the first event in every sequence has a timestamp of 0. Since a timestamp is unique for each event in a sequence it is used as an event identifier and called *eid*. The *support* or *frequency* of a sequence s, denoted as $sup(s)$, in a sequence database D is the number of sequences in D that contain s. Given an input integer threshold, called *minimum support*, or *minSup*, we say that s is *frequent* if $sup(s) \geq minSup$. The *frequent patterns mining* task is to find all frequent subsequences in a sequence database for a given minimum support.

4   Constraints

A frequent patterns search may result in a huge number of patterns, most of which are of little interest or useless. Understanding the extent to which a pattern is considered interesting may help in both discarding "bad" patterns and reducing the search space which means faster execution time. Constraints are a means of defining the type of sequences one is looking for.

In their classical definition [4], frequent sequences are defined only by the number of times they appear in the dataset (i.e. frequency). When incorporating constraints, a sequence must also satisfy the constraints for it to be deemed frequent. As suggested in [7], this contributes to the sequence mining process in several aspects. We focus on the following two: **(i) Performance**. Frequent patterns search is a hard task, mainly due to the fact that the search space is extremely large. For example, with d items there are $O(2^{d^k})$ potentially frequent sequences of length k. Limiting the searched sequences via constraints may dramatically reduce the search space and therefore improve the performance. **(ii) Non-contributing patterns**. Usually, when one wishes to mine a sequence database, she is not interested in all frequent patterns, but only in those meeting certain criteria. For example, in a database that contains consecutive values of stocks, one might be interested only in patterns of very profitable stocks. In this case, patterns of unprofitable stocks are considered non-contributing patterns even if they are frequent. By applying constraints, we can disregard non-contributing patterns. We define two types of constraints: **intra-event constraints**, which refer to constraints that are not time related (such as values of attributes) and **inter-events constraints**, which relate to the temporal aspect of the data, i.e. values that can or cannot appear one after the other sequentially. For the purpose of the experiment conducted in this study and in accordance with our domain, we choose to incorporate two inter-event and two intra-events constraints. A formal definition follows.

Intra-event Constraints

– *Singletons* Let $A = \{A_1, A_2, ...A_n\}$ s.t. $A_i \subseteq I$, be the set of sets of items that cannot reside in the same event. Each A_i is called a singleton. For example,

the value of a stock is an item, however a stock cannot have more than one value at the same time, therefore, the set of all possible values for that stock is a singleton.
- *MaxEventLength* The maximum number of items in a single event.

Inter-events Constraints

- *MaxGap* The maximum amount of time allowed between consecutive events. A sequence containing two events that have a time gap which is grater than *MaxGap*, is considered uninteresting.
- *MaxSequenceLength* The maximum number of events in a sequence.

5   The Algorithm

We now present *CAMLS*, a Constraint-based Apriori algorithm for Mining Long Sequences, which is a combination of modified versions of the well known Apriori [4] and SPADE [2] algorithms. The algorithm has two phases, event-wise and sequences-wise, which are detailed in subsections 5.1 and 5.2, respectively. The distinction between the two phases corresponds to the difference between the two types of constraints. As explained below, this design enhances the efficiency of the algorithm and makes it readily extensible for accommodating different constraints. Pseudo code for CAMLS is presented in Algorithm 1.

Algorithm 1. CAMLS

Input *minSup*: minimum support for a frequent pattern.
maxGap: maximum time gap between consecutive events.
maxEventLength: maximum number of items in every event.
maxSeqLength: maximum number of events in a sequence.
D: data set. A: set of singletons.
Output F: the set of all frequent sequences.

---

procedure CAMLS($minSup, maxGap, maxEventLength,$
$maxSeqLength, D, A$)

1: {Event-wise phase}
2: $F_1 \leftarrow$
   allFrequentEvents($minSup, maxEventLength, A, D$)
3: $F \leftarrow F_1$
4: {Sequence-wise phase}
5: **for all** k such that $2 \leq k \leq maxSeqLength$ and $F_{k-1} \neq \phi$ **do**
6:     $C_k \leftarrow$ candidateGen($F_{k-1}, maxGap$)
7:     $F_k \leftarrow$ prune($F_{k-1}, C_k, minSup, maxGap$)
8:     $F \leftarrow F \cup F_k$
9: **end for**
10: **return** F

5.1 Event-Wise Phase

The input to this phase is the database itself and all of the intra-event constraints. During this phase, the algorithm partially disregards the sequential nature of the data, and treats the data as a set of events rather than a set of sequences. Since the data is not sequential, applying the intra-event constraints at this phase is very straightforward. The algorithm is similar to the Apriori algorithm for discovering frequent itemsets as presented in [4].

Similarly to Apriori, our algorithm utilizes an iterative approach where frequent $(k-1)$-events are used to discover frequent k-events. However, unlike Apriori, we calculate the support of an event by counting the number of sequences that it appears in rather than counting the number of events that contain it. This means that the appearance of an event in a sequence increases its support by one, regardless of the number of times that it appears in that sequence.

Denoting the set of frequent k-events by L_k (referred to as frequent k-itemsets in [4]), we begin this phase with a complete database scan in order to find L_1. Next, L_1 is used to find L_2, L_2 is used to find L_3 and so on, until the resulting set is empty, or we have reached the maximum number of items in an event as defined by *MaxEventLength*. Another difference between Apriori and this phase lies in the generation process of L_k from L_{k-1}. When we join two $(k-1)$-events to generate a k-event candidate we need to check whether the added item satisfies the rest of the intra-event constraints such as Singletons.

The output of this phase is a radix ordered set of all frequent events satisfying the intra-event constraints, where every event e_i is associated with an *occurrence index*. The occurrence index of the frequent event e_i is a compact representation of all occurrences of e_i in the database and is structured as follows: a list l_i of sids of all sequences in the dataset that contain e_i, where every sid is associated with the list of eids of events in this sequence that contain e_i. For example, Figure 1a shows the indices of events d and (cd) based on the example database of Table 1. We are able to keep this output in main memory, even for long sequences, due to the nature of our domain in which the number of frequent events is relatively small.

Since the support of e_i equals the number of elements in l_i, it is actually stored in e_i's occurrence index. Thus, support is obtained by querying the index instead of scanning the database. In fact, the database scan required to find L_1 is actually the only single scan of the database. Throughout the event-wise phase, additional scans are avoided by keeping the occurrence indices for each event. The *allFrequentEvents* procedure in line 2 of Algorithm 1 is responsible for executing the event-wise phase as described above.

Notice that for mining frequent events there are more efficient algorithms than Apriori, for example FP-growth [10], however, our tests show that the event-wise phase is negligible compared to the sequence-wise phase, so it has no real impact on the running time of the whole algorithm.

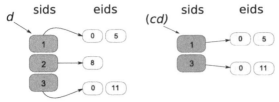

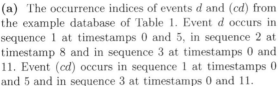

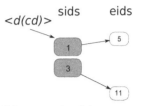

(a) The occurrence indices of events d and (cd) from the example database of Table 1. Event d occurs in sequence 1 at timestamps 0 and 5, in sequence 2 at timestamp 8 and in sequence 3 at timestamps 0 and 11. Event (cd) occurs in sequence 1 at timestamps 0 and 5 and in sequence 3 at timestamps 0 and 11.

(b) Example of the occurrence index for sequence $< d(cd) >$ generated from the intersection of the indices of sequences $< d >$ and $< (cd) >$.

Fig. 1. Example of occurrence index

5.2   Sequence-Wise Phase

The input to this phase is the output of the previous one. It is entirely temporal-based, therefore applying the inter-events constraints is straightforward. The output of this phase is a list of frequent sequences satisfying the inter-events constraints. Since it builds on frequent events that satisfied the intra-event constraints, this output amounts to the final list of sequences that satisfy the complete set of constraints.

Similarly to SPADE [2], this phase of the algorithm finds all frequent sequences by employing an iterative candidate generation method based on the apriori property. For the kth iteration, the algorithm outputs the set of frequent k-sequences, denoted as F_k, as follows: it starts by generating the set of all k-sequence candidates, denoted as C_k from F_{k-1} (found in the previous iteration). Then it prunes candidates that do not require support calculation in order to determine that they are non-frequent. Finally, it calculates the remaining candidates' support and removes the non-frequent ones.

The following subsections elaborate on each of the above steps. This phase of the algorithm is a modification of SPADE in which the candidate generation process is accelerated. Unlike SPADE, that does not have an event-wise phase and needs to generate the candidates at an item level (one item at a time), our algorithm generates the candidates at an event level, (one event at a time). This approach can be significantly faster because it allows us to use an efficient pruning method(section 5.2) that would otherwise not be possible.

Candidate Generation: We now describe the generation of C_k from F_{k-1}. For each pair of sequences $s_1, s_2 \in F_{k-1}$ that have a common $(k-2)$-prefix, we conditionally add two new k-sequence candidates to C_k as follows: (i) if s_1 is a generator (see section 5.2), we generate a new k-sequence candidate by concatenating the 1-suffix of s_2 to s_1; (ii) if s_2 is a generator, we generate a new k-sequence candidate by concatenating the 1-suffix of s_1 to s_2. It can be easily proven that if F_{k-1} is radix ordered then the generated C_k is also radix ordered. For example, consider the following two 3-sequences: $\langle (ab)c(bd)\rangle$ and $\langle (ab)cf\rangle$. Assume that these are frequent generator sequences from which we want to generate 4-sequence candidates. The common 2-prefix of both 3-sequences is $\langle (ab)c\rangle$ and their 1-suffixes are $\langle (bd)\rangle$ and $\langle f\rangle$. The two 2-sequences we get from the concatenation of the 1-suffixes are $\langle (bd)f\rangle$ and $\langle f(bd)\rangle$. Therefore, the resulting 4-sequence candidates are $\langle (ab)cf(bd)\rangle$ and $\langle (ab)c(bd)f\rangle$. Pseudo code for the candidate generation step is presented in Algorithm 2.

Algorithm 2. Candidate Generation

Input F_{k-1}: the set of frequent $(k-1)$-sequences.

Output C_k: the set of k-sequence candidates.

---

procedure candidateGen(F_{k-1})

1: **for all** sequence $s_1 \in F_{k-1}$ **do**
2:    **for all** sequence $s_2 \in F_{k-1}$ s.t. $s_2 \neq s_1$ **do**
3:       **if** $prefix(s_1) = prefix(s_2)$ **then**
4:          **if** $isGenerator(s_1)$ **then**
5:             $c \leftarrow concat(s_1, suffix(s_2))$
6:             $C_k \leftarrow C_k \cup \{c\}$
7:          **end if**
8:          **if** $isGenerator(s_2)$ **then**
9:             $c \leftarrow concat(s_2, suffix(s_1))$
10:           $C_k \leftarrow C_k \cup \{c\}$
11:          **end if**
12:       **end if**
13:    **end for**
14: **end for**
15: return C_k

Candidate Pruning: Candidate generation is followed by a pruning step. Pruned candidates are sequences identified as non-frequent based solely on the apriori property without calculating their support. Traditionally, [2,14], k-sequence candidates are only pruned if they have at least one $(k-1)$-subsequence that is not in F_{k-1}. However, our unique pruning strategy enables us to prune some of the k-sequence candidates for which this does not apply, due to the following observation: in the kth iteration of the candidate generation step, it is possible that one k-sequence candidate will turn out to be a super-sequence of another k-sequence. This happens when events of the subsequence are contained in the corresponding events of the super-sequence. More formally, if we have the two k-sequences $s_1 = \langle e_1^1, e_2^1, ..., e_k^1\rangle$ and $s_2 = \langle e_1^2, e_2^2, ..., e_k^2\rangle$ then $s_1 \subseteq s_2$ if $e_1^1 \subseteq e_1^2 \wedge e_2^1 \subseteq e_2^2 \wedge ... \wedge e_k^1 \subseteq e_k^2$. If s_1 was found to be non-frequent, s_2 can be pruned, thereby avoiding the calculation of its support. For example, if the candidate $\langle aac\rangle$ is not frequent, we can prune $\langle (ab)ac\rangle$, which was generated in the same iteration.

Our pruning algorithm is decribed as follows. We iterate over all candidates $c \in C_k$ in an ascending radix order (this does not require a radix sort of C_k because its members are generated in this order - see section 5.2). For each c, we check whether all of its $(k-1)$-subsequences are in F_{k-1}. If not, then c is not frequent and we add it to the set P_k which is a radix ordered list of pruned k-sequences that we maintain throughout the pruning step. Otherwise, we check whether a subsequence c_p of c exists in P_k. If so then again c is not frequent and we prune it. However, in this case, there is no need to add c to P_k because any candidate that would be pruned by c will also be pruned by c_p. Note that this check can be done efficiently since it only requires $O(log(|P_k|) \cdot k)$ comparisons of the events in the corresponding positions of c_p and c. Furthermore, if there are any k-subsequences of c in C_k that need to be pruned, the radix order of the process ensures that they have been already placed in P_k prior to the iteration in which c is checked. Finally, if c has not been pruned, we calculate its support. If c has passed the minSup threshold then it is frequent and we add it to F_k. Otherwise, it is not frequent, and we add it to P_k. Pseudocode for the candidate pruning is presented in Algorithm 3.

Support Calculation: In order to efficiently calculate the support of the k-sequence candidates we generate an occurrence index data structure for each candidate. This index is identical to the occurrence index of events, except that the list of eids represent candidates and not single events. A candidate is represented by the eid of the last event in it. The index for candidate $s_3 \in C_k$ is generated by intersecting the indices of $s_1, s_2 \in F_{k-1}$ from which s_3 is derived. Denoting the indices of s_1, s_2 and s_3 as inx_1, inx_2 and inx_3 respectively, the index intersection operation $inx_1 \odot inx_2 = inx_3$ is defined as follows: for each pair $sid_1 \in inx_1$ and $sid_2 \in inx_2$, we denote their associated eids list as $eidList(sid_1)$ and $eidList(sid_2)$, respectivly. For each $eid_1 \in eidList(sid_1)$ and $eid_2 \in eidList(sid_2)$, where $sid_1 = sid_2$ and $eid_1 < eid_2$, we add eid_2 to $eidList(sid_1)$ as new entry in inx_3. For example, consider the 1-sequences $s_1 = < d >$ and $s_2 = < (cd) >$ from the example database of Table 1. Their indices, inx_1 and inx_2, are described in Figure 1a. The index inx_3 for the 2-sequence $s_3 = < d(cd) >$ is generated as follows: (i) for $sid_1 = 1$ and $sid_2 = 1$, we have $eid_1 = 0$ and $eid_2 = 5$ which will cause $sid = 1$ to be added to inx_3 and $eid = 5$ to be added to $eidList(1)$ in inx_3; (ii) for $sid_1 = 3$ and $sid_2 = 3$, we have $eid_1 = 0$ and $eid_2 = 11$ which will cause $sid = 3$ to be added to inx_3 and $eid = 11$ to be added to $eidList(3)$ in inx_3. The resulting inx_3 is shown in Figure 1b. Notice that the support of s_3 can be obtained by counting the number of elements in the sid list of inx_3. Thus, the use of the occurrence index enables us to avoid any database scans which would otherwise be required for support calculation.

Handling the MaxGap Constraint: Consider the example database in Table 1, with $minSup = 0.5$ and $maxGap = 5$. Now, let us look at the following three 2-sequences: $\langle ab \rangle$, $\langle ac \rangle$, $\langle cb \rangle$, all in C_2 and have a support of 2 (see section 5.3 for more details). If we apply the $maxGap$ constraint, the sequence $\langle ab \rangle$ is no longer frequent and will not be added to F_2. This will prevent the algorithm from generating $\langle acb \rangle$, which is a frequent 3-sequence that does satisfy the $maxGap$ constraint. To overcome this problem, during the support calculation, we mark frequent sequences that satisfy the $maxGap$ constraint as *generators*. Sequences that do not satisfy the $maxGap$ constraint, but whose support is

Algorithm 3. Candidate Pruning

Input F_{k-1}: the set of frequent $(k-1)$-sequences.
C_k: the set of k-sequence candidates.
$minSup$: minimum support for a frequent pattern.
$maxGap$: maximum time gap between consecutive events.
Output F_k: the set of frequent k-sequence.

---

procedure prune($F_{k-1}, C_k, minSup, maxGap$)
1: $P_k \leftarrow \phi$
2: **for all** candidates $c \in C_k$ **do**
3:     $isGenerator(c) \leftarrow$ **true**;
4:     **if** $\exists s \subseteq c \wedge s$ is a $(k-1)$-sequence $\wedge s \notin F_{k-1}$ **then**
5:         $P_k.add(c)$
6:         continue
7:     **end if**
8:     **if** $\exists c_p \in P_k \wedge c_p \subset c$ **then**
9:         continue
10:     **end if**
11:     **if** $!(sup(c) \geq minSup)$ **then**
12:         $P_k.add(c)$
13:     **else**
14:         $F_k.add(c)$
15:         **if** $!(c$ satisfies $maxGap)$ **then**
16:             $isGenerator(c) \leftarrow$ **false**;
17:         **end if**
18:     **end if**
19: **end for**
20: **return** F_k

---

higher than $minSup$, are marked as *non-generators* and we refrain from pruning them. In the following iteration we generate new candidates only from frequent sequences that were marked as generators. The procedure *isGenerator* in Algorithm 3 returns the mark of a sequence (i.e., whether it is a generator or not).

5.3   Example

Consider the sequence database D given in Table 1 with $minSup$ set to 0.6 (i.e. 2 sequences), $maxGap$ set to 5 and the set $\{a, b\}$ is a Singleton.

Event-wise phase. Find all frequent events in D. They are: $\langle(a)\rangle : 3$, $\langle(b)\rangle : 2$, $\langle(c)\rangle : 3$, $\langle(d)\rangle : 3$, $\langle(ac)\rangle : 2$, $\langle(ad)\rangle : 2$, $\langle(bd)\rangle : 2$, $\langle(cd)\rangle : 2$ and $\langle(acd)\rangle : 2$, where $\langle(event)\rangle : support$ represents the frequent event and its support. Each of the frequent events is marked as a generator, and together they form the set F_1.

Sequence-wise phase. This phase iterates over the the candidates list until no more candidates are generated. *Iteration 1, step 1: Candidate generation.* F_1 performs a self join, and 81 candidates are generated: $\langle aa \rangle$, $\langle ab \rangle$, $\langle ac \rangle$, ..., $\langle a(acd) \rangle$, $\langle ba \rangle$, $\langle bb \rangle$, ..., $\langle(acd)(acd)\rangle$. Together they form C_2. *Iteration 1, step 2: Candidate pruning.* Let us consider the candidate $\langle aa \rangle$. All its 1-subsequences appear in F_1, so the F_{k-1} pruning passes. Next, P_2 still does not contain any sequences, so the P_k pruning passes as well. However, it does not pass the frequency test, since its support is 0, so $\langle aa \rangle$ is added to P_2. Now let us consider the candidate $\langle a(ac) \rangle$. It passes the F_{k-1} pruning, however, it does **not** pass the P_k pruning since a subsequence of it, $\langle aa \rangle$, appears in P_2. Finally, let us consider the candidate $\langle bc \rangle$. It passes both F_{k-1} and P_k pruning steps. Since it has a frequency of 2 it passes the frequency test, however, it does not satisfy the $maxGap$ constraint, so it is marked as a non-generator, but added to F_2. All the sequences marked as generators in F_2 are: $\langle ac \rangle : 2$, $\langle cb \rangle : 2$, $\langle cd \rangle : 2$ and $\langle c(bd) \rangle : 2$. The other sequences in F_2 are marked as non-generators and are: $\langle ab \rangle : 2$, $\langle ad \rangle : 2$, $\langle a(bd) \rangle : 2$, $\langle dc \rangle : 2$, $\langle dd \rangle : 2$ and $\langle d(cd) \rangle : 2$.

Iteration 2: At the end of this iteration, only one candidate passes all the pruning steps: $\langle acb \rangle : 2$, and since no candidates can be generated from one sequences, the process stops.

6   Experimental Results

In order to evaluate the performance of CAMLS, we implemented SPADE, PrefixSpan (and its constrained version, Prefix-growth) and CAMLS in Java 1.6 using the Weka [5] platform. We compared the run-time of the algorithms by applying them on both synthetic and real data sets. We conducted several runs with and without including constraints. Since cSPADE does not incorporate all of the constraints we have defined, we have excluded it from the latter runs. All tests were conducted on an AMD Athlon 64 processor box with 1GB of RAM and a SATA HD running Windows XP. The synthetic datasets mimic real-world behavior of a Quartz-Tungsten-Halogen lamp. Such lamps are used in the semiconductors industry for finding defects in a chip manufacturing process. Each dataset is organized in a table of synthetically generated illumination intensity values emitted by a lamp. A row in the table represents a specific day and a column represents a specific wave-length. The table is divided into blocks of rows where each block represents a lamp's life cycle, from the day it is

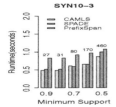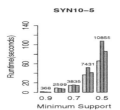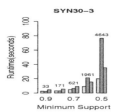

Fig. 2. Execution-time comparisons between CAMLS, SPADE and PrefixSpan on synthetic datasets for different values of minimum support, without constraints. The numbers appearing on top of each bar state the number of frequent patterns that exist for the corresponding minimum support.

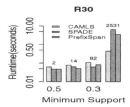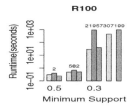

Fig. 3. Execution-time comparisons between CAMLS, SPADE and PrefixSpan on two real datasets for different values of minimum support, without constraints. The numbers appearing on top of each bar state the number of frequent patterns that exist for the corresponding minimum support.

first used to the last day it worked right before it burned out. To translate this data into events and sequences, the datasets were preprocessed as follows: first, the illumination intensity values were discretized into 50 bins using equal-width discretization [12]. Next, 5 items were generated: (i) the highest illumination intensity value out of all wave-lengths, (ii) the wave-length at which the highest illumination intensity value was received (iii) an indication whether or not the lamp burned out at the end of that day, (iv) the magnitude of the light intensity gradient between two consecutive measurements and (v) the direction of the light intensity gradient between two consecutive measurements. We then created two separate datasets. For each row in the original dataset, an event consisting of the first 3 items was formed for the first dataset and an event consisting of all 5 items was formed for the second one. Finally, in each dataset, a sequence was generated for every block of rows representing a lamp's life cycle from the events that correspond to these rows. We experimented with four such datasets each containing 1000 sequences and labeled SYNα-β where α stands for the sequence length and β stands for the event length. The real datasets, R30 and R100, were obtained from a repository of stock values [13]. The data consists of the values of 10 different stocks at the end of the business day, for a period of 30 or 100 days, respectively. The value of a stock for a given day corresponds to an event and the data for a given stock corresponds to a sequence, thus giving 10 sequences of either 30 (in R30) or 100 (in R100) events of length 1. As a preprocessing step, all numeric stock values were discretized into 50 bins using equal-frequency discretization [12].

Figure 2 compares CAMLS, SPADE and PrefixSpan on three synthetic datasets without using constraints. Each graph shows the change in execution-time as the minimum support descends from 0.9 to 0.5. The amount of frequent patterns found for each minimum support value is indicated by the number that appears above the respective bar. This comparison indicates that CAMLS has a slight advantage on datasets of short sequences with short events (SYN10-3). However, on datasets containing longer sequences (SYN30-3), the advantage of CAMLS becomes more pronounced as the amount of frequent patterns rises when decreasing the minimum support (around 5% faster than PrefixSpan and 35% faster than SPADE on the avarage). This is also true for datasets containing longer events (SYN10-5) despite the lag in the proccess (the advantage of CAMLS is gained only after lowering the minimum support below 0.7). This lag results from the increased length of events which causes the check for the containment relation in the pruning strategy (line 8 of 3) to take longer. Similar results can be seen in Figure 3 which compares CAMLS, SPADE and PrefixSpan on the two real datasets. In R30, SPADE and PrefixSpan has a slight advantage over CAMLS when using high minimum support values. We believe that this can be attributed to the event-wise phase that slows CAMLS down, compared to the other algorithms, when there are few frequent patterns. On the other hand, as the minimum support decreases, and the number of frequent patterns increases, the performance of CAMLS becomes better by an order of magnitude. In the R100 dataset, where sequences are especially long, CAMLS clearly outperforms both algorithms for all minimum support values tested. In the extreme case of the lowest value of minimum support, the execution of SPADE did not even end in a reasonable amount of time. Figure 4 compares CAMLS and Prefix-growth on SYN30-3, SYM30-5 and R100 with the usage of the *maxGap* and Singletons constraints. On all three datasets, CAMLS outperforms Prefix-growth.

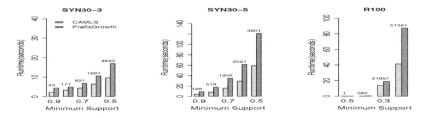

Fig. 4. Execution-time comparisons between CAMLS and Prefix-growth on SYN30-3 SYN30-5 and R100 with the *maxGap* and Singletons constraints for different values of minimum support. The numbers appearing on top of each bar state the number of frequent patterns that exist for the corresponding minimum support.

7   Discussion

In this paper we have presented CAMLS, a constraint-based algorithm for mining long sequences, that adopts the apriori approach. Many real-world domains require a substantial lowering of the minimum support in order to find any frequent patterns. This usually amounts to a large number of frequent patterns. Furthermore, some of these datasets may consist of many long sequences. Our motivation to develop CAMLS originated from realizing that well performing algorithms such as SPADE and PrefixSpan could not be applied on this class of domains. CAMLS consists of two phases reflecting a conceptual distinction between the treatment of temporal and non temporal data. Temporal aspects are only relevant during the sequence-wise phase while non temporal aspects are dealt with only in the event-wise phase. There are two primary advantages to this distinction. First, it allows us to apply a novel pruning strategy which accelerates the mining process. The accumulative effect of this strategy becomes especially apparent in the presence of many long frequent sequences. Second, the incorporation of inter-event and intra-event constraints, each in its associated phase, is straightforward and the algorithm can be easily extended to include other inter-events and intra-events constraints. We have shown that the advantage of CAMLS over state of the art algorithms such as SPADE, PrefixSpan and Prefixgrowth, increases as the mined sequences get longer and the number of frequent patterns in them rises.

We are currently extending our results to include different domains and compare CAMLS to other algorithms. In future work, we plan to improve the CAMLS algorithm to produce only closed sequences and to make our pruning strategy even more efficient.

References

1. Agrawal, R., Srikant, R.: Fast Algorithms for Mining Association Rules. In: 20th Int. Conf. Very Large Data Bases, VLDB, pp. 487–499. Morgan Kaufmann, San Francisco (1994)
2. Zaki, M.J.: SPADE: An efficient algorithm for mining frequent sequences. Machine Learning Journal, special issue on Unsupervised Learning, 31–60 (2001)
3. Zaki, M.J.: Sequence mining in categorical domains: incorporating constraints. In: 9th Int. Conf. on Information and knowledge management, pp. 422–429. ACM, New York (2000)
4. Agrawal, R., Srikant, R.: Mining Sequential Patterns. In: 11th Int. Conf. Data Engineering, pp. 3–14. IEEE Computer Society, Los Alamitos (1995)
5. Witten, I.H., Frank, E.: Data mining: practical machine learning tools and techniques with Java implementations. J. SIGMOD Rec. 31(1), 76–77 (2002)
6. Han, J., Kamber, M.: Data Mining: Concepts and Techniques. Morgan Kaufmann, San Francisco (2006)
7. Pei, J., Han, J., Wang, W.: Constraint-based sequential pattern mining: the pattern-growth methods. J. Intell. Inf. Syst. 28(2), 133–160 (2007)
8. Pei, J., Han, J., Mortazavi-Asl, B., Wang, J., Pinto, H., Chen, Q., Dayal, U., Hsu, M.: Mining sequential patterns by pattern-growth: The PrefixSpan approach. J. IEEE Transactions on Knowledge and Data Engineering 16 (2004)
9. Mannila, H., Toivonen, H., Verkamo, A.: Discovery of Frequent Episodes in Event Sequences. J. Data Min. Knowl. Discov. 1(3), 259–289 (1997)
10. Han, J., Pei, J., Yin, Y., Mao, R.: Mining Frequent Patterns without Candidate Generation A Frequent-Pattern Tree Approach. J. Data Min. Knowl. Discov. 8(1), 53–87 (2004)
11. Orlando, S., Perego, R., Silvestri, C.: A new algorithm for gap constrained sequence mining. In: The 2004 ACM symposium on Applied computing, pp. 540–547. ACM, New York (2004)
12. Pyle, D.: Data preparation for data mining. Morgan Kaufmann, San Francisco (1999)
13. Torgo, L.: Daily stock prices from January 1988 through October 1991, for ten aerospace companies,
http://www.liaad.up.pt/~ltorgo/Regression/DataSets.html
14. Srikant, R., Agrawal, R.: Mining sequential patterns: Generalizations and performance improvements. In: 5th International Conference on Extending Database Technology. Springer, Heidelberg (1996)

PGP-mc: Towards a Multicore Parallel Approach for Mining Gradual Patterns

Anne Laurent[1], Benjamin Negrevergne[2], Nicolas Sicard[3],
and Alexandre Termier[2]

[1] LIRMM - UM2- CNRS UMR 5506 - 161 rue Ada - 34392 Montpellier Cedex 5
laurent@lirmm.fr
http://www.lirmm.fr
[2] LIG - UJF-CNRS UMR 5217 - 681 rue de la Passerelle, B.P. 72, 38402 Saint
Martin d'Hères
Benjamin.Negrevergne@imag.fr, Alexandre.Termier@imag.fr
http://www.liglab.fr
[3] LRIE - EFREI - 30-32 av. de la république, 94 800 Villejuif
nicolas.sicard@efrei.fr
http://www.efrei.fr

Abstract. Gradual patterns highlight complex order correlations of the
form *"The more/less X, the more/less Y"*. Only recently algorithms have
appeared to mine efficiently gradual rules. However, due to the complexity of mining gradual rules, these algorithms cannot yet scale on huge
real world datasets. In this paper, we propose to exploit parallelism in
order to enhance the performances of the fastest existing one (GRITE).
Through a detailed experimental study, we show that our parallel algorithm scales very well with the number of cores available.

1 Introduction

Frequent pattern mining is a major domain of data mining. Its goal is to efficiently discover in data patterns having more occurrences than a pre-defined
threshold. This domain started with the analysis of transactional data (frequent
itemsets), and quickly expanded to the analysis of data having more complex
structures such as sequences, trees or graphs. Very recently, a new pattern mining problem appeared: mining frequent *gradual itemsets* (also known as *gradual
patterns*). This problem considers transactional databases where attributes can
have a numeric value. The goal is then to discover frequent co-variations between
attributes, such as: "The higher the age, the higher the salary". This problem
has numerous applications, as well for analyzing client databases for marketing
purposes as for analyzing patient databases in medical studies. Di Jorio et al. [1]
recently proposed GRITE, a first efficient algorithm for mining gradual itemsets
and gradual rules capable of handling databases with hundreds of attributes,
whereas previous algorithms where limited to six attributes [2]. However, as
gradual itemset mining is far more complex than traditionnal itemset mining,
GRITE cannot yet scale on large real databases, having millions of lines and
hundreds or thousands of attributes.

H. Kitagawa et al. (Eds.): DASFAA 2010, Part I, LNCS 5981, pp. 78–84, 2010.

One solution currently investigated by pattern mining researchers for reducing the mining time is to design algorithms dedicated for recent multi-core processors [3,4]. Analyzing their first results shows that the more complex the patterns to mine (trees, graphs), the better the scale-up results on multiple cores can be. This suggests that using multicore processors for mining gradual itemsets with the GRITE algorithm could give interesting results. We show in our experiments that indeed, there is a quasi-linear scale up with the number of cores for our multi-threaded algorithm.

The outline of this paper is as follows: In Section 2, we explain the notion of gradual itemsets. In Section 3, we present the related works on gradual patterns and parallel pattern mining. In Section 4, we present our parallel algorithm for mining frequent gradual itemsets, and Section 5 shows the results of our experimental evaluation. Last, we conclude and give some perspectives in Section 6.

2   Gradual Patterns

Gradual patterns refer to itemsets of the form *"The more/less X_1, ..., the more/less X_n"*. We assume here that we are given a database DB that consists of a single table whose tuples are defined on the attribute set $\mathcal{I}$. In this context, gradual patterns are defined to be subsets of $\mathcal{I}$ whose elements are associated with an ordering, meant to take into account increasing or decreasing variations. Note that $t[I]$ hereafter denotes the value of t over attribute I.

For instance, we consider the database given in Table 1 describing fruits and their characteristics.

Table 1. Fruit Characteristics

Id	Size (S)	Weight (W)	Sugar Rate (SR)
t_1	6	6	5.3
t_2	10	12	5.1
t_3	14	4	4.9
t_4	23	10	4.9
t_5	6	8	5.0
t_6	14	9	4.9
t_7	18	9	5.2

Definition 1. *(Gradual Itemset) Given a table DB over the attribute set $\mathcal{I}$, a gradual item is a pair (I, θ) where I is an attribute in $\mathcal{I}$ and θ a comparison operator in $\{\geq, \leq\}$.*

A gradual itemset $g = \{(I_1, \theta_1), ..., (I_k, \theta_k)\}$ is a set of gradual items of cardinality greater than or equal to 2.

For example, $(Size, \geq)$ is a gradual item, while $\{(Size, \geq), (Weight, \leq)\}$ is a gradual itemset.

The support of a gradual itemset in a database DB amounts to the extend to which a gradual pattern is present in a given database. Several support definitions have been proposed in the literature (see Section 3 below). In this paper, we consider the support as being defined as the number of tuples that can be ordered to support all item comparison operators:

Definition 2. *(Support of a Gradual Itemset) Let DB be a database and $g = \{(I_1, \theta_1), ..., (I_k, \theta_k)\}$ be a gradual itemset. The cardinality of g in DB, denoted by $\lambda(g, DB)$, is the length of the longest list $l = \langle t_1, \ldots, t_n \rangle$ of tuples in DB such that, for every $p = 1, \ldots, n - 1$ and every $j = 1, \ldots, k$, the comparison $t_p[I_j] \, \theta_j \, t_{p+1}[I_j]$ holds.*
The support of g in DB, denoted by $supp(g, DB)$, is the ratio of $\lambda(g, DB)$ over the cardinality of DB, which we denote by $|DB|$. That is, $supp(g, DB) = \frac{\lambda(g,DB)}{|DB|}$.

In the example database, for the gradual itemset $g = \{(S, \geq), (SR, \leq)\}$, we have $\lambda(g, DB) = 5$, with the list $l = \langle t_1, t_2, t_3, t_6, t_4 \rangle$. Hence $supp(g, DB) = \frac{5}{7}$.

3   Related Works

Gradual patterns and gradual rules have been studied for many years in the framework of control, command and recommendation. More recently, data mining algorithms have been studied in order to automatically mine such patterns [1,2,5,6,7,8].

The approach in [7] uses statistical analysis and linear regression in order to extract gradual rules. In [2], the authors formalize four kinds of gradual rules in the form *The more/less X is in A, then the more/less Y is in B*, and propose an Apriori-based algorithm to extract such rules. Despite a good theoretical study, the algorithm is limited to the extraction of gradual rules of length 3.

In [1] and [5], two methods to mine gradual patterns are proposed. The difference between these approaches lies in the computation of the support: whereas, in [5], a heuristic is used and an approximate support value is computed, in [1], the correct support value is computed.

In [8], the authors propose another way to compute the support, by using ranking such as the Kendall tau ranking correlation coefficient, which basically computes, instead of the length of the longest path, the number of pairs of lines that are correctly ordered (concordant and discordant pairs).

To the best of our knowledge, there are no existing *parallel* algorithms to mine gradual itemsets. The most advanced works in parallel pattern mining have been presented by [3] for parallel graph mining and [4] for parallel tree mining. These works have showed that one of the main limiting factor for scalable parallel perfomance was that the memory was shared among all the cores. So if all the cores request a lot of data simultaneously, the bus will be saturated and the performance will drop. The favorable case is to have compact data structures and complex patterns where a lot of computations have to be done for each chunk of data transfered from memory.

With its complex support computation and simple input data, gradual pattern mining is thus a favorable case for parallelization. The main difficulty will be to achieve a good load balance. We present our solution in the following section.

4 PGP-mc: Parallel Gradual Pattern Extraction

The sequential GRITE algorithm (see [1] for detailed algorithm) relies on a tree-based exploration, where every level $N + 1$ is built upon the previous level N. The first level of the tree is initialized with all attributes, which all become itemset siblings. Then, *itemsets* from the second level are computed by combining *frequent itemsets* siblings from the first level through what we call the $Join()$ procedure. Candidates which match a pre-defined threshold - they are considered as *frequent* - are retained in level $N + 1$.

In this approach, every level cannot be processed until the previous one has been completed, at least partially. So, we focused our efforts on the parallelization of each level construction where individual combinations of itemsets (through the $Join()$ procedure) are mostly independant tasks. The main problem is that the number of operations cannot be easily anticipated, at least for levels higher than 2. Moreover, the number of siblings may vary by a large margin depending of the considered itemsets. A simple parallel loop would lead to an irregular load distribution on several processing units.

In order to offset this irregularity, our approach dynamically attributes new tasks to a pool of threads on a "first come, first served" basis. At first, all frequent itemsets from the given level are marked unprocessed and queued in Q_i. A new frequent itemset i is dequeued and all its siblings are stored in a temporary queue Q_{si}. Each available thread then extracts the next unprocessed sibling j from Q_{si} and builds a new candidate k from i and j. The candidate is stored in level $N+1$ if it is considered frequent. When Q_{si} is empty, the next frequent itemset i is dequeued and Q_{si} is filled with its own siblings. The process is repeated until all itemsets i are processed (*e.g.*, Q_i is empty).

5 Experimental Results and Discussion

In this section we report experimental results from the execution of our program on two different workstations with up to 32 processing cores : COYOTE, with 8 AMD Opteron 852 processors (each with 4 cores), 64GB of RAM with Linux Centos 5.1, g++ 3.4.6 and IDKONN, with 4 Intel Xeon 7460 processors (each with 6 cores), 64GB of RAM with Linux Debian 5.0.2, g++ 4.3.2.

Most of the experiments are led on synthetic databases automatically generated by a tool based on an adapted version of IBM Synthetic Data Generation Code for Associations and Sequential Patterns[1]. This tool generates numeric databases depending on the following parameters: number of lines, number of attributes/columns and average number of distinct values per attribute.

---

[1] www.almaden.ibm.com/software/projects/hdb/resources.shtml

5.1 Scalability

The following figures illustrate how the proposed solution scales with both the increasing number of threads and the growing complexity of the problem. The complexity comes either from the number of lines or from the number of attributes in the database as the number of individual tasks is related to the number of attributes while the complexity of each individual task - itemsets joining - depends on the number of lines. In this paper, we report results for two sets of experiments.

The first experiment set involves databases with relatively few attributes but a significant number of lines. This kind of databases usually produces few frequent items with moderate to high thresholds. As a consequence the first two level computations represent the main part of the global execution time. Figures 1(a) and 1(b) show the evolution of execution time and speed-up respectively for 10000-line databases - ranging from 10 to 50 attributes - on COYOTE.

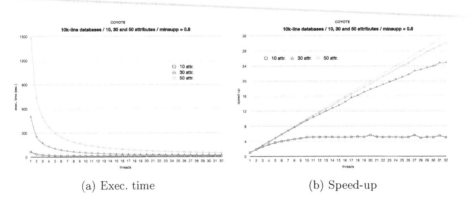

(a) Exec. time   (b) Speed-up

Fig. 1. Execution time and speed-up related to the number of threads. Test databases ranging from 10 to 50 attributes with 10k lines, on COYOTE.

As shown by Figure 1(a), speed-ups can reach very satisfying values in sufficiently complex situations. For example, speed-up is around 30 with 50 attributes where the theoretical maximum is 32. The upper limit for 10 and 20 attributes is not really surprising and can be explained by the lower number of individual tasks. As the number of tasks decreases and the complexity of each task increases, it becomes more and more difficult to reach an acceptable load balance. This phenomenon is especially tangible during the initial database loading phase (construction of the first level of the tree) where the number of tasks is exactly the number of attributes. For example, the sequential execution on the 10-attribute database takes around 64 seconds from which the database loading process takes 9 seconds. With 32 threads, the global execution time goes down to 13 seconds but more than 5.5 seconds are still used for the loading phase.

Experimental results on IDKONN are very similar to these figures as speed-ups go from a maximum of 4.8 with 24 threads on the 10-attribute database to a maximum of 22.3 with 24 threads on the 50-attribute database. Detailed results on IDKONN are available at `http://www.lirmm.fr/~laurent/DASFAA10`.

The second set of experiments reported in this article is about databases with growing complexity in term of attributes. Figures 2(a) and 2(b) show the evolution of execution time and speed-up respectively for 500-line databases with various number of attributes - ranging from 50 to 350 - on IDKONN.

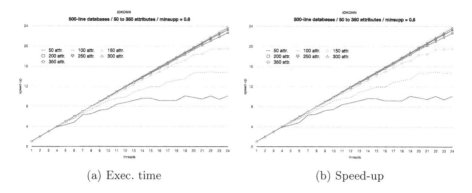

(a) Exec. time   (b) Speed-up

Fig. 2. Execution time and speed-up related to the number of threads. Test databases ranging from 50 to 350 attributes with 500 lines, on IDKONN.

As we can see, our solution is extremely efficient and scales very well for many attributes: we almost reach the theoretical maximum linear speed-up progression for 150 attributes or more. For example, the sequential processing of the 350 attributes database took more than five hours while it spend approximatively 13 minutes using 24 threads on IDKONN. Furthermore, speed-up results are particularly stable from one architecture to another[2], meaning that performances do not rely on very specific architectural features (caches, memory systems...).

With an execution time of less than 0.2 second with 16 threads, the 50-attribute database experiment illustrates how our approach can still achieve a very tangible acceleration on this particular case, which appears as crucial for real time or near real time data mining and applications (*e.g.*, intrusion/fraud detection).

6   Conclusion and Perspectives

In this paper, we propose an original parallel approach to mine large numeric databases for gradual patterns like *the oldest a people, the higher his/her salary*. Mining these rules is indeed very difficult as the algorithms must perform many time-consuming operations to get the frequent gradual patterns from the databases. In order to tackle this problem, our method intensively uses the multiple

---

[2] Complete experiments, detailed at *http://www.lirmm.fr/~laurent/DASFAA10*, show very similar results on COYOTE (with 32 threads).

processors and cores that are now available on recent computers. The experiments performed show the interest of our approach, by leading to quasi-linear speed-ups on problems that were previously very time-consuming or even impossible to manage, especially in the case of databases containing a lot of attributes.

This work opens many perspectives, not only based on technical improvements depending on ad-hoc architectures of the machines, but also based on other data mining paradigms. Hence we will consider closed gradual patterns in order to cut down the computation runtimes. We will also study the use of another parallel framework: clusters (including clusters of multi-core machines in order to benefit from both architectures).

Acknowledgements

The authors would like to acknowledge Lisa Di Jorio for providing the source code of the implementation of the GRITE algorithm [1].

References

1. Di Jorio, L., Laurent, A., Teisseire, M.: Mining frequent gradual itemsets from large databases. In: Adams, N.M., Robardet, C., Siebes, A., Boulicaut, J.-F. (eds.) IDA 2009. LNCS, vol. 5772, pp. 297–308. Springer, Heidelberg (2009)
2. Berzal, F., Cubero, J.C., Sanchez, D., Vila, M.A., Serrano, J.M.: An alternative approach to discover gradual dependencies. Int. Journal of Uncertainty, Fuzziness and Knowledge-Based Systems (IJUFKS) 15(5), 559–570 (2007)
3. Buehrer, G., Parthasarathy, S., Chen, Y.K.: Adaptive parallel graph mining for cmp architectures. In: ICDM, pp. 97–106 (2006)
4. Tatikonda, S., Parthasarathy, S.: Mining tree-structured data on multicore systems. In: VLDB 2009: Proceedings of the 35th international conference on Very large data bases (2009)
5. Di Jorio, L., Laurent, A., Teisseire, M.: Fast extraction of gradual association rules: A heuristic based method. In: IEEE/ACM Int. Conf. on Soft computing as Transdisciplinary Science and Technology, CSTST 2008 (2008)
6. Fiot, C., Masseglia, F., Laurent, A., Teisseire, M.: Gradual trends in fuzzy sequential patterns. In: Proc. of the Int. Conf. on Information Processing and Management of Uncertainty in Knowledge-based Systems, IPMU (2008)
7. Hüllermeier, E.: Association rules for expressing gradual dependencies. In: Elomaa, T., Mannila, H., Toivonen, H. (eds.) PKDD 2002. LNCS (LNAI), vol. 2431, pp. 200–211. Springer, Heidelberg (2002)
8. Laurent, A., Lesot, M.J., Rifqi, M.: Graank: Exploiting rank correlations for extracting gradual dependencies. In: Proc. of FQAS 2009 (2009)

Generalised Rule Mining

Florian Verhein

Ludwig-Maximilians-Universität, Munich, Germany
florian@verhein.com
www.florian.verhein.com

Abstract. Rules are an important pattern in data mining, but existing approaches are limited to conjunctions of binary literals, fixed measures and counting based algorithms. Rules can be much more diverse, useful and interesting! This work introduces and solves the *Generalised Rule Mining* (GRM) problem, which abstracts rule mining, removes restrictions on the semantics of rules and redefines rule mining by functions on vectors. This also lends to an interesting geometric interpretation for rule mining. The GRM framework and algorithm allow new methods that are not possible with existing algorithms, can speed up existing methods and separate rule semantics from algorithmic considerations. The GRM algorithm scales linearly in the number of rules found and provides orders of magnitude speed up over fast candidate generation type approaches (in cases where these can be applied).

1 Introduction

Rules are an important pattern in data mining due to their ease of interpretation and usefulness for prediction. They have been heavily explored as association patterns [2,3,8,5,9], "correlation" rules [3] and for associative classification [7,6,12]. These approaches consider only conjunctions of binary valued variables, use fixed measures of interestingness and counting based algorithms[1]. Furthermore, many rule mining algorithms are similar – the primary differences arise from the incorporation of the particular interestingness measures into existing algorithms.

The rule mining problem can be generalised by relaxing the restrictions on the semantics of the antecedent as well as the variable types, and supporting any interestingness measure on rules. A *generalised rule* $A' \to c$ describes a relationship between a set of variables in the antecedent $A' \subseteq A$ and a variable in the consequent $c \in C$, where A is the set of possible antecedent variables and C is the set of possible consequent variables. The goal in *Generalised Rule Mining* (GRM) is to find useful rules $A' \to c : A' \subseteq A \land c \in C \land c \notin A'$ given functions defining the measures and semantics of the variables and rules. Variables do not need to be binary valued. Semantics are not limited to conjunction.

This work introduces and solves the Generalised Rule Mining problem by proposing a vectorized framework and algorithm that are applicable to general-to-specific

---

[1] Such algorithms explicitly count instances/transactions that apply to a rule, typically through explicit counting, subset operations or tree/trie/graph traversals.

H. Kitagawa et al. (Eds.): DASFAA 2010, Part I, LNCS 5981, pp. 85–92, 2010.

methods[2]. By solving rule mining at the abstract level, the GRM framework clearly separates the semantics and measures of rules from the algorithm used to mine them. Hence, the algorithm can be exploited for any methodology mappable to the frameworks functions, such as [7,6,11,13]. In particular, complete contingency tables can be calculated, including columns for sub-rules as required by complex, statistically significant rule mining methods [11,13]. Perhaps more importantly, the framework allows and encourages novel methods (e.g. Section 4). Furthermore, this separation allows methods to automatically and immediately benefit from future advances in the GRM algorithm.

Rule mining is a challenging problem due to its exponential search space in $A \cup C$. At best, an algorithm's run time is linear in the number of interesting rules it finds. The GRM algorithm is optimal in this sense. It completely avoids "candidate generation" and does not build a compressed version of the dataset. Instead it operates directly on vectors. This also allows additional avenues for reducing the run time, such as automatic parallelisation of vector operations and exploitation of machine level operations. Finally, it can also exploit any mutual exclusion between variables, which is common in classification tasks.

2   Related Work

As far as the author is aware, there has been no previous attempt to solve the rule mining problem at the abstract level. Existing rule mining methods have concrete measures and semantics. Most are used for associative pattern mining or classification, only consider rules with conjunctive semantics of binary literals and use counting approaches in their algorithms. Association rule mining methods [2,5,8] typically mine item-sets before mining rules. Unlike GRM, they do not mine rules directly. Itemset mining methods are often extended to rule based classification [14,7,12]. The algorithms can be categorized into Apriori-like approaches [2] characterized by a breadth first search through the item lattice and multiple database scans, tree based approaches [5] characterized by a traversal over a pre-built compressed representation of the database, projection based approaches [12] characterised by depth first projection based searches and vertical approaches [8,4,10] that operate on columns. Vertical bit-map approaches [8,10] have received considerable interest due to their ability to outperform horizontal based techniques. Of existing work, the vectorized approach in GRM is most similar to the vertical approach. However, GRM mines rules directly and is not limited to bitmaps or support based techniques.

3   Generalised Rule Mining (GRM) Framework

Recall that elements of $A \cup C$ are called *variables* and the goal is to find useful rules $A' \rightarrow c : A' \subseteq A \wedge c \in C \wedge c \notin A'$. Each possible antecedent $A' \subseteq A$ and each possible consequent $c \in C$ are expressed as vectors, denoted by $x_{A'}$ and x_c respectively. As indicated in Figure 1, these vectors exist in the space X,

---

[2] That is, methods where a less specific rule $A'' \rightarrow c : A'' \subset A'$ is mined before the more specific one $A' \rightarrow c$.

the dimensions of which are *samples* $\{s_1, s_2, ...\}$ – depending on the application these may be instances, transactions, rows, etc. The database is the set of vectors corresponding to individual variables, $D = \{x_v : v \in A \cup C\}$.

Good rules have high prediction power. Geometrically, in such rules the antecedent vector is "close" to the consequent vector.

Definition 1. $m_R : X^2 \rightarrow \mathbb{R}$ *is a distance measure between the antecedent and consequent vectors* $x_{A'}$ *and* x_c. $m_R(x_{A'}, x_c)$ *evaluates the quality of the rule* $A' \rightarrow c$.

Fig. 1.

Any method based on counting is implemented using bit-vectors in the GRM framework. A bit is set in $x_{A'}$ (x_c) if the corresponding sample contains A' (c). Then, the number of samples containing A', $A' \rightarrow c$ and c are simply the number of set bits in $x_{A'}$, $x_{A'} \; AND \; x_c$ and x_c respectively. From these counts (n_{1+}, n_{11} and n_{+1} respectively) and the length of the vector, n, a complete contingency table can be constructed. Using this, m_R can evaluate a wide range of measures such as confidence, interest factor, lift, ϕ-coefficient and even statistical significance tests. Geometrically, in counting based applications m_R is the dot-product; $m_R(A' \rightarrow c) = x_{A'} \cdot x_c = n_{11}$.

Since x_c and all $x_a : a \in A$ corresponds to single variables they are available in the database. However, the $x_{A'} : A' \subset A \wedge |A'| > 1$ required for the evaluation of $m_R(\cdot)$ must be calculated. These are built incrementally from vectors $x_a \in D$ using the *aggregation function* $a_R(\cdot)$, which also defines the semantics of A':

Definition 2. $a_R : X^2 \rightarrow X$ *operates on vectors of the antecedent so that* $x_{A' \cup a} = a_R(x_{A'}, x_a)$ *where* $A' \subseteq A$ *and* $a \in (A - A')$.

In other words, $a_R(\cdot)$ combines the vector $x_{A'}$ for an *existing* antecedent $A' \subseteq A$ with the vector x_a for a new antecedent element $a \in A - A'$. The resulting vector $x_{A' \cup a}$ represents the larger antecedent $A' \cup a$. $a_R(\cdot)$ *also defines the semantics of the antecedent of the rule*: By defining how $x_{A' \cup a}$ is built, it must implicitly define the semantics between elements of $A' \cup a$. For rules with a conjunction of binary valued variables, vectors are represented as bit-vectors and hence $a_R(x_{A'}, x_a) = x_{A'} \; AND \; x_a$ defines the required semantics; Bits will be set in $x_{A' \cup a}$ corresponding to those samples that are matched by the conjunction $\wedge_{a_i \in A'} \wedge a$.

In some methods, a rule $A' \rightarrow c$ needs to be compared with its sub-rules $A'' \rightarrow c : A'' \subset A'$. This is particularly useful in order to find more specific rules that improve on their less specific sub-rules. Geometrically, this allows methods where the antecedent of a rule can be built by adding more variables in such a way that the corresponding vector $x_{A'}$ moves closer to x_c.

Definition 3. $M_R : \mathbb{R}^{|\mathcal{P}(A')|} \rightarrow \mathbb{R}$ *is a measure that evaluates a rule* $A' \rightarrow c$ *based on the value computed by* $m_R(\cdot)$ *for any sub-rule* $A'' \rightarrow c : A'' \subseteq A'$.

$M_R(\cdot)$ does not take vectors as arguments – it evaluates a rule based on m_R values that have already been calculated. If $M_R(\cdot)$ does not need access to m_R values of any sub-rules to perform its evaluation, it is called *trivial* since $m_R(\cdot)$ can perform the function instead. A *trivial* $M_R(\cdot)$ returns $m_R(\cdot)$ and has algorithmic advantages in terms of space and time complexity. For counting based approaches, $M_R(\cdot)$ can be used to evaluate measures on more complex contingency tables such as those in [13,11]. For example, to evaluate whether a rule significantly improves on its less specific sub-rules by using Fisher's Exact Test [13,11]. Together with I_R below, it can also be used to direct and prune the search space – even 'forcing' a measure implemented in m_R to be downward closed in order to implement a kind of greedy search.

Interesting rules are those that are a) desirable and should therefore be output and b) should be further improved in the sense that more specific rules should be mined by the GRM algorithm.

Definition 4. $I_R : \mathbb{R}^2 \to \{true, false\}$ *determines whether a rule* $A' \to c$ *is* interesting *based on the values produced by* $m_R(\cdot)$ *and* $M_R(\cdot)$.

Only interesting rules are further expanded and output. Hence, more specific rules (i.e. with more variables in the antecedent) will only be considered if $I_R(\cdot)$ returns true. Sometimes it is possible to preemptively determine that a rule is not interesting based purely on its antecedent, such as in support based methods.

Definition 5. $I_A : X \to \{true, false\}$. $I_A(x_A) = false$ *implies* $I_R(\cdot) = false$ *for all* $A' \to c : c \in C$.

4   Additional Motivational Examples

4.1   Probabilistic Association Rule Mining (PARM)

In a probabilistic database D, each row r_j contains a set of observations about variables $A \cup C$ together with their probabilities of being observed in r_j. Probabilistic databases arise when there is uncertainty or noise in the data and traditional methods cannot handle these. *Probabilistic Association Rules* can describe interesting patterns while taking into account the uncertainty of the data.

Problem Definition: find rules $A' \to c$ where the expected support $E(s(A' \to c)) = \frac{1}{n}\sum_{j=1}^n P(A' \to c \subseteq r_j)$ is above $minExpSup$. Under the assumption that the variables' existential probabilities in the rows are determined under independent observations, $P(A' \to c \subseteq r_j) = \Pi_{a \in A'}P(a \in r_j) \cdot P(c \in r_j)$.

In the *GRM* framework, each variable is represented by a vector x_i expressing the probabilities that the variable i exists in row j of the database: $x_i[j] = P(i \in r_j) : j \in \{1, n\}$. All the $\{P(A' \subseteq r_j) : r_j \in D\}$ can be calculated efficiently by incremental element-wise multiplication of individual vectors using the $a_R(\cdot)$ function: $a_R(x_{A'}, x_a)[j] = x_{A'}[j] * x_a[j]$. The GRM algorithm's use of $a_R(\cdot)$ ensures that there is no duplication of calculations throughout the mining process while keeping space usage to a minimum. $m_R(\cdot)$ is the expectation function and calculates the expected support: $m_R(A' \to c) = \frac{1}{n}\sum_{j=1}^n x_{A'}[j] * x_c[j]$. Geometrically, this is the scaled dot product of $x_{A'}$ and x_c: $m_R(A' \to c) = \frac{|x_{A'}||x_c|}{n}x_{A'} \cdot x_c$. $M_R(\cdot)$ is trivial. It can be shown that expected support is downwards closed (anti-monotonic): $E(s(A' \to c)) \leq E(s(A'' \to c)) : A'' \subseteq A'$, allowing pruning. Hence $I_R(A' \to c)$ returns *true* if and only if $m_R(A' \to c) \geq minExpSup$. $I_A(\cdot)$ can also be exploited to prune rules with $E(s(A')) < minExpSup$.

4.2 Various Correlated Rule Mining Methods

Other novel methods involve mining rules where the antecedent is highly correlated with the consequent and where the rule mining progresses to more specific rules if these reduce the angle between the antecedent and consequent vectors. Such methods (with various semantics, both real and binary vectors, and for descriptive or classification tasks) have been developed using the framework.

5   Generalised Rule Mining (GRM) Algorithm

The GRM algorithm efficiently solves any problem that can be expressed in the framework. The variant briefly described below assumes the measures used are anti-monotonic (or more specifically, at least order-anti-monotonic).

A Categorised Prefix Tree: (CPT) is a data structure where antecedents and rules are represented by single $PrefixNodes$ and variables in C can only be present in the leaves. Common prefixes are shared, so when $M_R(\cdot)$ is *non-trivial* the CPT efficiently stores rules in a compressed format. Otherwise, only the current path (rule) the algorithm is processing is in memory. Categories on sibling nodes provide a way to express mutual exclusion between variables and in turn allows the algorithm to exploit them. For instance, in classification tasks, attribute-values and classes are usually mutually exclusive.

The **GRM Algorithm**: (Figure 2) works by performing a strict depth first search (i.e. sibling nodes are not expanded

```
class PrefixNode {PrefixNode parent, String name,
    double value_m, double value_M}
//node: the PrefixNode (corresponding to A') to
// expand using the vectors in joinTo.
//x_A': the Vector corresponding to A'.
GRM(PrefixNode node, Vector x_A', List joinTo)
  List newJoinTo = new List();
  List currentCategory = new List();
  PrefixNode newNode = null;
  boolean addedConsequent = false;
  for each (x_v, v, lastInCategory) ∈ joinTo
    if (v ∈ C)
      double val_m = m_R(x_A', x_v);
      newNode = new PrefixNode(node, v, val_m, NaN);
      double val_M =evaluateAndSetM_R(newNode);
      if (I(val_m, val_M))
        if (M_R(·) non-trivial) store(newNode);
        outputRule(newNode);
        addedConsequent = true;
      else newNode = null;
    if (v ∈ A) //Note: possible that v ∈ A ∧ v ∈ C.
      Vector x_A'∪v = a_R(x_A', x_v);
      if (I_A(x_A'∪v))
        newNode = new PrefixNode(node, v, NaN, NaN);
      if (newNode ≠ null)
        GRM(newNode, newVector, newJoinTo);
        currentCategory.add(x_v, v, lastInCategory);
      else
        if (v ∈ C∧!addedConsequent)
          return; //prune early -- no super rules exist
      if (lastInCategory∧!currentCategory.isEmpty())
        currentCategory.last().lastInCategory = true;
      newJoinTo.addAll(currentCategory);
      currentCategory.clear();

main(File dataset)
  PrefixNode root = new PrefixNode(null, ε, NAN);
  Vector x_∞ = //initialise appropriately (e.g. ones)
  List joinTo = ... //read vectors from file**
  GRM(root, x_∞, joinTo);
```

Fig. 2. *Simplified* Generalized Rule Mining (GRM) algorithm. (**It is possible to implement the algorithm in the same runtime complexity without loading all vectors into memory at the same time.)

until absolutely necessary) and calculating vectors along the way. There is no "candidate-generation". The search is limited according to the interestingness function $I_R(\cdot)$ and $I_A(\cdot)$ and it progresses in depth by *"joining sibling nodes"* in the $CategorisedPrefixTree$, enabling auto-pruning. Vectors are calculated incrementally along a path in the search using $a_R(\cdot)$, avoiding any vector re-computations while maintaining optimal memory usage. The GRM algorithm *automatically* avoids considering rules that would violate any mutual exclusion constraints by carrying forward the categorization to the sibling lists ($joinTo$), and only joining siblings if they are from different categories. The search space is *automatically* pruned (i.e. without requiring explicit checking, thus saving vector calculations) by maintaining a list of siblings ($joinTo$), and only joining sibling nodes: Since only interesting rules as specified by $I_R(\cdot)$ are to be expanded, only interesting rules become siblings.

Theorem 1. *The run time complexity is $O(R \cdot |A| \cdot |C| \cdot (t(m_R) + t(M_R) + t(a_R) + t(I_R))$, where R is the number of rules mined by the algorithm and $t(X)$ is the time taken to compute function X from the framework.*

Hence the performance is linear in the number of *interesting* rules *found* by the algorithm (the number of rules for which $I_R(\cdot)$ returns *true*). It is therefore not possible to improve the algorithm other than by a constant factor. In most instantiations, $t(m_R) = t(a_R) = O(n)$ and $t(I_R) = t(I_A) = t(M_R) = O(1)$.

6   Experiments

GRM's runtime is evaluated here for conjunctive rules with bit-vectors (therefore covering counting based approaches) and PARM (as a representative of methods using real valued vectors). Due to the lack of existing algorithms capable of direct rule mining or of handling real valued data, a very efficient competing algorithm based on the commonly used Apriori ideas was developed, called *FastAprioriRules*[3], and in the evaluation it was given the advantage over GRM.

Existing rule mining methods require frequency counts. Hence, the algorithms are compared on the task of mining all conjunctive rules $A' \to c$ that are satisfied by (i.e. classify correctly) at least $minCount$ instances. By varying $minCount$, the number of interesting rules mined can be plotted against the run time. The UCI [1] Mushroom and the 2006 KDD Cup datasets were used as they are relatively large. Figure 3(a) clearly shows the linear relationship of Theorem 1

---

[3] *FastAprioriRules* is based on the common candidate generation and testing method-ology. However, it mines rules directly – that is, it does *not* mine sets first and then generate rules form these as this would be very inefficient. FastAprioriRules also exploits the mutual exclusion optimisation which greatly reduces the number of candidates generated, preemptively prunes by antecedents when possible and incor-porates the *pruneEarly* concept. Hence FastAprioriRules evaluates exactly the same number of rules as GRM so that the comparison is fair. To check if a rule matches an instance in the counting phase, a set based method is used, which proved to be much quicker than enumerating the sub-rules in an instance and using hashtree based lookup methods to find matching candidates. The dataset is kept in memory to avoid Apriori's downside of multiple passes, and unlike in the GRM experiments, I/O time is ignored.

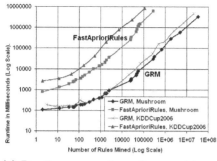 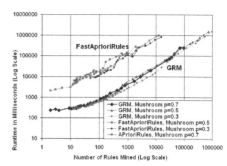

(a) Runtime comparison of support based conjunctive rules on the Mushroom dataset. GRM uses bitvectors.

(b) Runtime comparison of PARM on three probabilistic Mushroom data sets. GRM uses real vectors.

Fig. 3. Run-time is linear in the number of rules mined and orders of magnitude faster than FastAprioriRules, regardless of the method. **Log-log scale.**

for *over three orders of magnitude* before the experiments were stopped (this also holds in linear-linear scale). When few rules are found, setup factors dominate. More importantly, *GRM is consistently more than two orders of magnitude faster than FastAprioriRules.* It is also very insensitive to the dataset characteristics, unlike FastAprioriRules. Results on smaller datasets (UCI datasets Cleve and Heart) lead to identical conclusions and are omitted for clarity.

PARM is evaluated in a similar fashion by varying $minExpSup$. Here, the Mushroom dataset was converted to a probabilistic dataset by changing occurrences to a value chosen uniformly from $[0, 1)$ with a probability p. The resulting graph in Figure 3(b) shows the same linear relationship for the three values of $p \in \{0.3, 0.5, 0.7\}$. Here, GRM is at least one order of magnitude faster than the FastAprioriRules implementation.

References

1. Asuncion, D.N.A.: UCI machine learning repository (2007)
2. Agrawal, R., Srikant, R.: Fast algorithms for mining association rules. In: VLDB, pp. 487–499. Morgan Kaufmann, San Francisco (1994)
3. Brin, S., Motwani, R., Silverstein, C.: Beyond market baskets: generalizing association rules to correlations. In: ACM SIGMOD ICMD (1997)
4. Dunkel, B., Soparkar, N.: Data organization and access for efficient data mining. In: ICDE (1999)
5. Han, J., Pei, J., Yin, Y.: Mining frequent patterns without candidate generation. In: ACM SIGMOD ICMD, ACM Press, New York (2000)
6. Li, W., Han, J., Pei, J.: Cmar: Accurate and efficient classification based on multiple class-association rules. In: ICDM 2001. IEEE Computer Society, Los Alamitos (2001)
7. Liu, B., Hsu, W., Ma, Y.: Integrating classification and association rule mining. In: Knowledge Discovery and Data Mining, pp. 80–86 (1998)

8. Shenoy, P., Bhalotia, G., Haritsa, J.R., Bawa, M., Sudarshan, S., Shah, D.: Turbo-charging vertical mining of large databases. In: ACM SIGMOD ICMD (2000)

9. Srikant, R., Agrawal, R.: Mining quantitative association rules in large relational tables. In: ACM SIGMOD ICMD (1996)

10. Verhein, F., Chawla, S.: Geometrically inspired itemset mining. In: ICDM, pp. 655–666. IEEE Computer Society, Los Alamitos (2006)

11. Verhein, F., Chawla, S.: Using significant, positively and relatively class correlated rules for associative classification of imbalanced datasets. In: ICDM. IEEE Computer Society, Los Alamitos (2007)

12. Wang, J., Karypis, G.: Harmony: Efficiently mining the best rules for classification. In: SDM 2005 (2005)

13. Webb, G.I.: Discovering significant rules. In: ACM SIGKDD. ACM Press, New York (2006)

14. Yin, X., Han, J.: CPAR: Classification based on predictive association rules. In: SDM. SIAM, Philadelphia (2003)

An Effective Object-Level XML Keyword Search

Zhifeng Bao[1], Jiaheng Lu[2], Tok Wang Ling[1], Liang Xu[1], and Huayu Wu[1]

[1] School of Computing, National University of Singapore
{baozhife,lingtw,xuliang,wuhuayu}@comp.nus.edu.sg
[2] School of Information, Renmin University of China
jiahenglu@gmail.com

Abstract. Keyword search is widely recognized as a convenient way to retrieve information from XML data. In order to precisely meet users' search concerns, we study how to effectively return the targets that users intend to search for. We model XML document as a set of interconnected object-trees, where each object contains a subtree to represent a concept in the real world. Based on this model, we propose object-level matching semantics called *Interested Single Object* (ISO) and *Interested Related Object* (IRO) to capture single object and multiple objects as user's search targets respectively, and design a novel relevance oriented ranking framework for the matching results. We propose efficient algorithms to compute and rank the query results in one phase. Finally, comprehensive experiments show the efficiency and effectiveness of our approach, and an online demo of our system on DBLP data is available at **http://xmldb.ddns.comp.nus.edu.sg**.

1 Introduction

With the presence of clean and well organized knowledge domains such as Wikipedia, World Factbook, IMDB etc, the future search technology should appropriately help users precisely finding explicit objects of interest. For example, when people search DBLP by a query "Jim Gray database", they likely intend to find the *publications* object about "database" written by the *people* object "Jim Gray". As XML is becoming a standard in data exchange and representation in the internet, in order to achieve the goal of "finding only the *meaningful* and *relevant* data fragments corresponding to the interested objects (that users really concern on)", search techniques over XML document need to exploit the matching semantics at *object-level* due to the following two reasons.

First, the information in XML document can be recognized as a set of real world objects [16], each of which has attributes and interacts with other objects through relationships. E.g. Course and Lecturer can be recognized as objects in the XML data of Fig. 1. *Second*, whenever people issue a keyword query, they would like to find information about specific objects of interest, along with their relationships. E.g. when people search DBLP by a query "Codd relational model", they most likely intend to find the *publications* object about "relational model" written by "Codd". Therefore, it is desired that the search engine is able to find and extract the data fragments corresponding to the real world objects.

H. Kitagawa et al. (Eds.): DASFAA 2010, Part I, LNCS 5981, pp. 93–109, 2010.

1.1 Motivation

Early works on XML keyword search focus on LCA (which finds the Lowest Common Ancestor nodes that contain all keywords) or SLCA (smallest LCA) semantics, which solve the problem by examining the data set to find the smallest common ancestors [16,13,9,7,20]. This method, while pioneering, has the drawback that its result may not be meaningful in many cases. Ideally, a practical solution should satisfy two requirements: (1) it can return the *meaningful* results, meaning that the result subtree describes the information at *object-level*; and (2) the result is *relevant* to the query, meaning that it captures users' search concerns. Despite the bulk of XML keyword search literature (See Section 2), the existing solutions violate at least one of the above requirements.

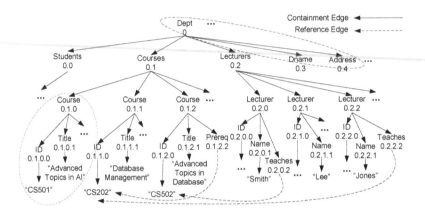

Fig. 1. Example XML data (with Dewey IDs)

Regarding to the *meaningfulness*, the query results should contain enough but non-overwhelming information. i.e. it should be of the same granularity as user's search concern. Unfortunately, the matching semantics proposed so far cannot achieve such goal. For example, for a query "Database" issued on Fig.1, both LCA and SLCA return title:0.1.1.1 and title:0.1.2.1 as results, while the desired result should be two subtrees rooted at Course:0.1.1 and Course:0.1.2, as they encapsulate enough information about a "Database" course. The recent competitors over SLCA include XSeek [18], CVLCA [13], MLCA [15] and MAXMATCH [17]. While those approaches propose some promising and improved matching semantics, the search target identification is still not clearly addressed. More importantly, their inability to exploit ID references in XML data causes some relevant results to be missed.

In order to complement the keyword search over tree model to find more relevant results, ID references in XML data are captured and matching semantics on *digraph data model* are designed. A widely adopted one is *reduced subtree*, which is the minimal subgraph containing all keywords. However, it suffers the same problem as those in tree model, as both of them exploit only the structure of XML data. Even worse, the problem of finding the results by increasing the sizes of reduced subtrees for keyword proximity is NP-hard [14], thus keyword search in digraph data model is heuristics-based and intrinsically expensive.

Regarding to the *relevance*, the query results should be relevant to user's search intention. However, the existing ranking strategies in both tree model [7] and digraph model [12,8] are built at XML node level, which do not meet user's search concern at object level. Moreover, ranking functions in digraph model even do not distinguish the containment edge and reference edge in XML data.

1.2 Our Approach

In this paper, we propose to model XML document as a set of object trees, where each real world object o (with its associated attributes) is encapsulated in an *object tree* whose root node is a representative node of o; two object trees are interconnected via a containment or reference edge in XML data. E.g. The part enclosed by a dotted circle in Fig. 1 shows an object tree for Dept and Course.

We propose our object-level matching semantics based on an analysis of user's search concern, namely ISO (*Interested Single Object*) and IRO (*Interested Related Object*). ISO is defined to capture user's concern on a single object that contains all keywords, while IRO is defined to capture user's concern on multiple objects. Compared to previous works, our *object-level* matching semantics have two main advantages. *First*, each object tree provides a more precise match with user's search concern, so that meaningless results (which even though contain all keywords) are filtered. *Second*, it captures the reference edges missed in tree model, and meanwhile achieves better efficiency than those solutions in digraph model by distinguishing the reference and containment edge in XML.

We design a customized ranking scheme for ISO and IRO results. The ranking function $ISORank$ designed for ISO result not only considers the *content* of result by extending the original TF*IDF [19] to object tree level, but also captures the keyword co-occurrence and specificity of the matching elements. The $IRORank$ designed for an IRO result considers both its self similarity score and the "bonus" score contributed from its interconnected objects. We design efficient algorithms and indices to dynamically compute and rank the matched ISO results and IRO results in one phase. Finally, we experimentally compare ISO and IRO algorithms to the best existing methods XSeek [16] and XReal [4] with real and synthetic data sets. The results reveal that our approach outperforms XReal by an order of magnitude in term of response time and is superior to XSeek in term of recall ratio, well confirming the advantage of our novel semantics and ranking strategies. A search engine prototype incorporating the above proposed techniques is implemented, and a demo of the system on DBLP data is available at http://xmldb.ddns.comp.nus.edu.sg [3].

2   Related Work

XML tree model: In tree data model, LCA is first proposed to find the lowest common ancestor containing all the keywords in their subtrees. SLCA [20] is proposed to find the smallest LCA that doesn't contain other LCA in its subtree. XSEarch [6] is a variation of LCA, which claims two nodes n_1 and n_2 are related if there is no two distinct nodes with same tag name on the paths from their LCA to n_1 and n_2. [15] incorporates SLCA into XQuery and proposes a Schema-Free XQuery where predicates in XQuery can be specified through SLCA concept. XSeek [16] studies how to infer the semantics of the search and identify the return nodes by recognizing possible entities and attributes inherently represented in XML data. The purpose of our research is also to maximize the possibility to understand user's search semantics, while we take a novel perspective by studying new semantics based on ID reference and designing effective ranking strategy.

XML graph model: The major matching semantics is to find a set of *reduced subtree* G' of database graph G, s.t. each G' is the smallest subgraph containing all keywords. However, the cost of finding all such G' ranked by size is intrinsically expensive due to its NP-hard nature[14]. Bidirectional expansion is proposed to find ranked reduced subtrees[12], but it requires the entire visited graph in memory, and suffers an inefficiency. BLINKS[8] improves it by designing a bi-level index for result pruning, with the tradeoffs in index size and maintenance cost. XKeyword[11] uses schema information to reduce search space, but its query evaluation is based on the method of DISCOVER [10] built on RDBMS, which cannot distinguish the containment and reference edges to further reduce search space. [5] builds a tree+IDRef model to capture ID references by avoiding the NP-hard complexity. However, this compromise may affect the results' meaningfulness and relevance, which are carefully investigated in this paper.

Results ranking: In IR field, TF*IDF similarity [19] is designed to measure the relevance of the keywords and the documents in keyword search over flat documents. XReal [4] addresses the keyword ambiguity problem by designing an XML TF*IDF on tree model, which takes the structural information of XML into account. XRANK [7] generalizes PageRank to XML element and rank among LCA results, where the rank of each element is computed statically in data preprocessing. In contrast, the ranking functions in this paper are designed to rank on the object trees and are computed dynamically during query processing.

3   Data Model

Definition 1 *(Object Tree). An object tree t in D is a subtree of the XML document, where its root node r is a representative node to denote a real world object o, and each attribute of o is represented as a child node of r.*

In an XML document D, a real-world object o is stored in form of a subtree due to its hierarchical inherency. How to identify the object trees is orthogonal to this paper; here, we adopt the inference rules in XSeek [16] to help identify the object trees, as clarified in Definition 1. As we can see from Fig. 1, there are 7 object trees (3 Course, 3 Lecturer and 1 Dept), and the part enclosed by a dotted circle is an object tree for Course:0.1.0 and Dept:0 respectively. Note that nodes Students, Courses and Lecturers of Dept:0 are *connection* nodes, which connect the object "Dept" and multiple objects "Student" ("Course" and "Lecturer").

Conceptual connection reflects the relationship among object trees, which is either a reference-connection or containment-connection defined as below.

Definition 2 *(Reference-connection).* *Two object trees u and v in an XML document D have a reference-connection (or are reference-connected) if there is an ID reference relationship between u and v in D.*

Definition 3 *(Containment-connection).* *Two object trees u and v in an XML document D have a containment-connection if there is a P-C relationship between the root node of u and v in D, regardless of the connection node.*

Definition 4 *(Interconnected object-trees model).* *Models an XML document D as a set of object trees, D=(T,C), where T is a set of object trees in D, and C is a set of conceptual connections between the object trees.*

In contrast to the model in XSeek [16], ID references in XML data is considered in our model to find more meaningful results. From Fig. 1, we can find Dept:0 and Course:0.1.0 are interconnected via a containment connection, and Lecturer:0.2.0 and Course:0.1.2 are reference-connected.

4   Object Level Matching Semantics

When a user issues a keyword query, his/her concern is either on a single object, or a pair (or group) of objects connected via somehow meaningful relationships. Therefore, we propose *Interested Single Object* (ISO) and *Interested Related Object* (IRO) to capture the above types of users' search concerns.

4.1   ISO Matching Semantics

Definition 5 *(ISO)* *Given a keyword query Q, an object tree o is the Interested Single Object (ISO) of Q, if o covers all keywords in Q.*

ISO can be viewed as an extension of LCA, which is designed to capture user's interest on a single object. E.g. for a query "database, management" issued on Fig. 1, LCA returns two subtrees rooted at Title:0.1.1.1 and Courses:0.1, neither of which is an object tree; while ISO returns an object tree rooted at Course:0.1.1.

4.2   IRO Matching Semantics

Consider a query "CS502, lecturer" issued on Fig. 1. ISO cannot find any qualified answer as there is no single object qualified while user's search concern is on multiple objects. However, there is a Lecturer:0.2.0 called "Smith" who teaches Course "CS502" (via a reference connection), which should be a relevant result. This motivates us to design IRO (*Interested Related Object*).

As a first step to define IRO pair and IRO group, we give a formal definition on the connections among these multiple objects.

Definition 6 *(n-hop-meaningful-connection). Two object trees u and v in an XML document have a n-hop-meaningful-connection (or are n-hop-meaningfully-connected) if there are $n - 1$ distinct intermediate object trees $t_1, ...t_{n-1}$, s.t.*

1. *there is either a reference connection or a containment connection between each pair of adjacent objects;*
2. *no two objects are connected via a common-ancestor relationship.*

Definition 7 *(IRO pair). For a given keyword query Q, two object trees u and v form an IRO pair w.r.t. Q if the following two properties hold:*

1. *Each of u and v covers some, and u and v together cover all keywords in Q.*
2. *u and v are n-hop-meaningfully-connected (with an upper limit L for n).*

IRO pair is designed to capture user's concern on two objects that have a direct or indirect conceptual connection. E.g. for query "Smith, Advanced, Database", two object trees Lecturer:0.2.0 and Course:0.1.2 form an IRO pair, as there is a *reference connection* between them. Intuitively, the larger the upper limit L is, more results can be found, but the relevance of those results decay accordingly. Lastly, *IRO group* is introduced to capture the relationships among three or more connected objects.

Definition 8 *(IRO group). For a given keyword query Q, a group G of object trees forms an IRO group if:*

1. *All the object trees in G collectively cover all keywords in Q.*
2. *There is an object tree $h \in G$ (playing a role of hub) connecting all other object trees in G by a n-hop-meaningful-connection (with an upper limit L' for n).*
3. *Each object tree in G is compulsory in the sense that, the removal of any object tree causes property (1) or (2) not to hold any more.*

As an example, for query "Jones, Smith, Database" issued on Fig. 1, four objects Course:0.1.1, Course:0.1.2, Lecturer:0.2.0 and Lecturer:0.2.2 form an IRO group (with $L' = 2$), where both Course:0.1.1 and Course:0.1.2 can be the hub. The connection is: Lecturer:0.2.2 "*Jones*" teaches Course:0.1.1, which is a pre-requisite of a "*Database*" Course:0.1.2 taught by Lecturer:0.2.0 "*Smith*".

An object involved in IRO semantics is called the *IRO object*; an ISO object o can form an IRO pair (or group) with an IRO object o', but o is not double counted as an IRO object.

4.3   Separation of ISO and IRO Results Display

As ISO and IRO correspond to different user search concerns, we separate the results of ISO and IRO in our online demo[1] [3], which is convenient for user to quickly recognize which category of results meet their search concern, thus a lot of user efforts are saved in result consumption.

---

[1] Note: in our previous demo, ISO was named as ICA, while IRO was named as IRA.

5   Relevance Oriented Result Ranking

As another equally important part of this paper, a relevance oriented ranking scheme is designed. Since ISO and IRO reflect different user search concerns, customized ranking functions are designed for ISO and IRO results respectively.

5.1   Ranking for ISO

In this section, we first outline the desired properties in ISO result ranking; then we design the corresponding ranking factors; lastly we present the *ISORank* formula which takes both the content and structure of the result into account.

Object-level TF*IOF similarity $(\rho(o, Q))$**:** Inspired by the extreme success of IR style keyword search over flat documents, we extend the traditional TF*IDF (Term frequency*Inverse document frequency) similarity [19] to our object-level XML data model, where flat document becomes the object tree. We call it as *TF*IOF* (Term frequency*Inverse object frequency) similarity. Such extension is adoptable since the object tree is an appropriate granularity for both query processing and result display in XML. Since TF*IDF only takes the *content* of results into account, but cannot capture XML's *hierarchical structure* we enforce the *structure* information for ranking in the following three factors.

F1. Weight of matching elements in object tree: The elements directly nested in an object may have different weights related to the object. So we provide an optional weight factor for advanced user to specify, where the default weight is 1. Thus, the *TF*IOF* similarity $\rho(o, Q)$ of object o to query Q is:

$$\rho(o, Q) = \frac{\sum_{\forall k \in o \cap Q} W_{Q,k} * W_{o,k}}{W_Q * W_o}, \quad W_{Q,k} = \frac{N}{1 + f_k}, \quad W_{o,k} = \sum_{\forall e \in attr(o,k)} tf_{e,k} * W_e \quad (1)$$

where $k \in o \cap Q$ means keyword k appears in both o and Q. $W_{Q,k}$ represents the weight of keyword k in query Q, playing a role of inverse object frequency (*IOF*); N is the total number of objects in xml document, and f_k is number of objects containing k. $W_{o,k}$ represents the weight of k in object o, counting the term frequency (*TF*) of k in o. $attr(o, k)$ denotes a set of attributes of o that directly contain k; $tf_{e,k}$ represents the frequency of k in attribute e, and W_e is the adjustable weight of matching element e in o, whose value is no less than 1, and W_e is set to 1 for all the experiments conducted in section 8.

Normalization factor of *TF*IOF* should be designed in the way that: on one hand the relevance of an object tree o containing the query-relevant child nodes should not be affected too much by other query-irrelevant child nodes; on the other hand, it should not favor the object tree of large size (as the larger the size of the object tree is, the larger chance that it contains more keywords). Therefore, in order to achieve such goals, two normalization factors W_o and W_Q are designed: W_o is set as the number of query-relevant child nodes of object o, i.e. $|attr(o, k)|$, and W_Q is set to be proportional to the size of Q, i.e. $|Q|$.

F2. Keyword co-occurrence $(c(o,Q))$**:** Intuitively, the less number of elements (nested in an object tree o) containing all keywords in Q is, o is likely to be more relevant, as keywords co-occur more closely. E.g. when finding papers in DBLP by a query "XML, database", a paper whose title contains all keywords should be ranked higher than another paper in "*database*" conference with title "*XML*".

Based on the above intuition, we present $c(o,Q)$ in Equation 2 (denominator part), which is modeled as *inversely proportional to the minimal number of attributes that are nested in o and together contain all keywords in Q*. Since this metric favors the single-keyword query, we put the number of query keywords (i.e. $|Q|$ in nominator part) as a normalization factor.

$$c(o,Q) = \frac{|Q|}{min(|\{E|E = attrSet(o) \ and \ (\forall k \in Q, \exists e \in E \ s.t. \ e.contain(k))\}|)} \tag{2}$$

F3. Specificity of matching elements $(s(o,Q))$**:** An attribute a of an object is fully (perfectly) specified by a keyword query Q if a only contains the keywords in Q (no matter whether all keywords are covered or not). Intuitively, *an object o with such fully specified attributes should be ranked higher; and the larger the number of such attribute is, the higher rank o is given.*

Example 1. When searching for a person by a query "David, Lee", a person p_1 with the exact name should be ranked higher than a person p_2 named "*David Lee Ming*", as p_1's name fully specifies the keywords in query, while p_2 doesn't.□

Thus, we model the specificity by measuring the *number of elements in the object tree that fully specify all query keywords*, namely $s(o,Q)$.

Note that $s(o,Q)$ is similar to TF*IDF at attribute level. However, we enforce the importance of full-specificity by modeling it as a boolean function; thus partial specificity is not considered, while it is considered in original TF*IDF.

So far, we have exploited both the *structure* (i.e. factors F1,F2,F3) and *content* (*TF*IOF* similarity) of an object tree o for our ranking design. Since there is no obvious comparability between structure score and content score, we use product instead of summation to combine them. Finally, the *ISORank(o,Q)* is:

$$ISORank(o,Q) = \rho(o,Q) * (c(o,Q) + s(o,Q)) \tag{3}$$

5.2 Ranking for IRO

IRO semantics is useful to find a pair or group of objects conceptually connected. As an IRO object does not contain all keywords, the relevance of an IRO object o, namely *IRORank*, should consist of two parts: its *self TF*IOF similarity* score, and the bonus score contributed from its IRO counterparts (i.e. the objects that form IRO pair/group with o). The overall formula is:

$$IRORank(o,Q) = \rho(o,Q) + Bonus(o,Q) \tag{4}$$

where $\rho(o,Q)$ is the *TF*IOF similarity* of object o to Q (Equation 1). *Bonus*(o,Q) is the extra contribution to o from all its IRO pair/group's counterparts for Q, which can be used as a *relative* relevance metric for IRO objects to Q, especially when they have a comparable *TF*IOF similarity* value. Regarding to the design of *Bonus* score to an IRO object o for Q, we present three guidelines first.

Guideline 1: IRO Connection Count. Intuitively, the more the IRO pair/ group that connect with an IRO object o is, the more likely that o is relevant to Q; and the closer the connections to o are, the more relevant o is.   ♣

For example, consider a query "interest, painting, sculpture" issued on XMark [2]. Suppose two persons Alice and Bob have interest in *"painting"*; Alice has conceptual connections to many persons about *"sculpture"* (indicated by attending the same auction), while Bob has connections to only a few of such auctions. Thus, Alice is most likely to be more relevant to the query than Bob.

Guideline 2: Distinction of different matching semantics. The IRO connection count contributed from the IRO objects under different matching semantics should be distinguished from each other.   ♣

Since IRO pair reflects a tighter relationship than IRO group, thus for a certain IRO object o, the connection count from its IRO pair's counterpart should have a larger importance than that from its IRO group's counterpart.

Example 2. Consider a query "XML, twig, query, processing" issued on DBLP. Suppose a paper p_0 contains "XML" and "twig"; p_1 contains "query" and "processing" and is cited by p_0; p_2 contains the same keywords as p_1; p_3 contains no keyword, but cites p_0 and p_2; p_4 contains "query" and p_5 contains "processing", and both cite p_0. By Definition 7-8, p_1 forms an IRO pair with p_0; p_2, p_3 and p_0 form an IRO group; p_0, p_4 and p_5 form an IRO group. Therefore, in computing the rank of p_0, the influence from p_1 should be greater than that of p_2 and p_3, and further greater than p_4 and p_5.   □

According to the above two guidelines, the *Bonus* score to an IRO object o is presented in Equation 5. *Bonus*(o,Q) consists of the weighted connection counts from its IRO pair and group respectively, which manifests Guideline 1. w_1 and w_2 are designed to reflect the weights of the counterparts of o's IRO pair and group respectively, where $w_1 > w_2$, which manifests Guideline 2.

$$Bonus(o,Q) = w_1 * BS_{IRO\_P}(o,Q) + w_2 * BS_{IRO\_G}(o,Q) \qquad (5)$$

Guideline 3: Distinction of different connected object types. The connection count coming from different conceptually related objects (under each matching semantics) should be distinguished from each other.   ♣

Example 3. Consider a query Q "XML, query, processing" issued on DBLP. The bonus score to a "query processing" paper from a related "XML" conference inproceedings should be distinguished from the bonus score coming from a related book whose title contains "XML", regardless of the self-similarity difference of this inproceedings and book.   □

Although the distinction of contributions from different object types under a certain matching semantics helps distinguish the $IRORank$ of an IRO object, it is preferable that we can distinguish the precise connection types to o to achieve a more exact *Bonus* score. However, it depends on a deeper analysis of the relationships among objects and more manual efforts. Therefore, in this paper we only enforce Guideline 1 and Guideline 2. As a result, the IRO bonus from the counterparts of o's IRO pair and IRO group is presented in Equation 6-7:

$$BS_{IRO\_P}(o,Q) = \sum_{\forall o'|(o,o') \in IROPair(Q,L)} \rho(o',Q) \tag{6}$$

$$BS_{IRO\_G}(o,Q) = \frac{\sum_{\forall g \in IROGroup(Q,L')|o \in g} BF(o,Q,g)}{|IRO\_Group(o,Q)|} \tag{7}$$

In Equation 6, $\rho(o',Q)$ is the *TF*IOF similarity* of o' w.r.t. Q, which is adopted as the contribution from o' to o. Such adoption is based on the intuition that, if an object tree o_1 connects to o'_1 s.t. o'_1 is closely relevant to Q, whereas object tree o_2 connects to o'_2 which is not as closely relevant to Q as o'_1, then it is likely that o_1 is more relevant to Q than o_2. In Equation 7, $BF(o,Q,g)$ can be set as the self similarity of the object in g containing the most number of keywords. As it is infeasible to design a one-fit-all bonus function, other alternatives may be adopted according to different application needs. L (in Equation 6) and L' (in Equation 7) is the upper limit of n in definition of *IRO pair* and *IRO group*.

6   Index Construction

As we model the XML document as the interconnected object-trees, the *first* index built is the *keyword inverted list*. An object tree o is in the corresponding list of a keyword k if o contains K. Each element in the list is in form of a tuple $(Oid, DL, w_{o,k})$, where Oid is the id of the object tree containing k (here we use the dewey label of the root node of object tree o as its oid, as it serves the purpose of unique identification) ; DL is a list of pairs containing the dewey labels of the exact locations of k and the associated attribute name; $w_{o,k}$ is the term frequency in o (see Equation 1). $c(o,Q)$ (in Equation 2) can be computed by investigating the list DL; $s(o,Q)$ is omitted in index building, algorithm design and experimental study later due to the high complexity to collect. Therefore, the $ISORank$ of an object tree can be efficiently computed. A B+ tree index is built on top of each inverted list to facilitate fast probing of an object in the list.

The *second* index built is *connection table CT*, where for each object c, it maintains a list of objects that have direct *conceptual connection* to c in document order. B+ tree is built on top of object id for efficient probes. Since it is similar to the adjacency list representation of graph, the task of finding the *n-hop-meaningfully-connected* objects of c (with an upper limit L for connection chain length) can be achieved through a depth limited (to L) search from c in CT. The worst case size is $O(|id|^2)$ if no restriction is enforced on L, where $|id|$ is number of object trees in database. However, we argue that in practice the size is much smaller as an object may not connect to every other object in database.

7   Algorithms

In this section, we present algorithms to compute and rank the ISO and IRO results.

Algorithm 1: KWSearch

Input: Keywords: $KW[m]$; Keyword Inverted List: $IL[m]$; Connection
Table: CT; upper limit: L, L' for IRO pair and group
Output: Ranked object list: RL

1 let $RL = ISO\_Result = IRO\_Result = \{\}$;
2 let HT be a hash table from object to its rank;
3 let IL_s be the shortest inverted list in $IL[m]$;
4 **for** *each object* $o \in IL_s$ **do**
5   let $K_o = $ getKeywords(IL, o);
6   **if** ($K_o == KW$)   /* o is an ISO object */
7   initRank(o, K_o, KW, HT); ISO_Result.add(o);
8   **else if** ($K_o \neq \emptyset$)
9          $IRO_Pair = $ getIROPairs(IL, o, o, CT, L)   /* Algorithm 2 */
9          $IRO_Group = $ getIROGroups(IL, o, o, CT, L', K_o) /* Algorithm 3 */
10 $RL = ISO\_Result \cup IRO\_Pair \cup IRO\_Group$;

Function initRank(o, K_o, KW, HT)
1 **if** (o not in HT)
2   HT.put(o.id, computeISORank(o,Q,KW));

Function computeRank($o, oList$)
1 **foreach** object $o' \in oList$
2      $K_{o'} = $ getKeywords(IL, o') ;
3   **if** ($K_{o'} == KW$)   /* o' is an ISO object */
4   initRank($o', K_{o'}, KW, HT$); ISO_Result.add(o');
5   **else if**($K_{o'} \neq \emptyset$ AND ($K_{o'} \cup K_o == KW$))   /* o' is IRO object */
6   initRank(o', K'_o, KW, HT);
7   IRO_Pair.add(o, o');
8   initRank(o, K_o, KW, HT);   /* o is an IRO object also */
9   updateIRORank($o, o', oList, HT$);

Function updateIRORank ($o, o', oList, HT$)
1 update the $IRORank$ of o based on Equation $5-7$;
2 put the updated ($o, IRORank$) into HT;

The backbone workflow is in Algorithm 1. Its main idea is to scan the shortest keyword inverted list IL_s, check the objects in the list and their connected objects, then compute and rank the ISO and IRO results. The details are: for each object tree o in IL_s, we find the keywords contained in o by calling function $getKeywords()$(line 5). If o contains all query keywords, then o is an ISO object, and we compute the $ISORank$ for o by calling $initRank()$, then store o together with its rank into hash table HT(line 6-7). If o contains some keywords, then o is an IRO object, and all its IRO pairs and groups are found by calling functions $getIROPairs()$ (Algorithm 2) and $getIROGroups$ (Algorithm 3) (line 8-10).

Function $computeRank()$ is used to compute/update the ranks of objects o' in $oList$, each forming an IRO pair with o. For each such o', it probes all inverted lists with o' to check three cases (line 1-2): (1) if o' is an ISO object containing all query keywords, then its $ISORank$ is computed and it is added into ISO_Result (line 3-4). (2) if both o and o' are IRO objects, their $TF*IOF$ similarity are initialized (if not yet), and their $IRORank$s are updated accordingly (line 5-9). Function $initRank()$ computes the $ISORank$ by Equation 3 if o is an ISO object, otherwise computes its TF*IOF similarity by Equation 1.

Algorithm 2 shows how to find all objects that form IRO pair with an IRO object src. It works in a recursive way, where input o is the current object visited, whose initial value is src. Since two objects are connected via either a reference or containment connection, line 2-3 deal with the counterparts of o via reference connection by calling $getConnectedList()$; line 4-7 deal with containment connection. Then it recursively finds such counterparts connecting to src indirectly in a depth limited search(line 8-10). $getIROGroups()$ in Algorithm 3 works in a similar way, the detail isn't shown due to space limit.

Algorithm 2: getIROPairs $(IL[m],\ src,\ o,$ $CT,\ L)$	**Algorithm 3:** getIROGroups $(IL[m],\ o,$ $CT,\ L',\ K_o)$
/* find all counterparts of o captured by IRO pair */	/* find all counterparts of o captured by IRO group */
1 **if** $L == 0$ **then** return ;	1 let $KS = \emptyset$; $count = 0$;
2 **let** oList = getConnectedList(o,CT) ;	2 cList = getConnectedList(o, CT, L');
3 computeRank(o, oList) ;	3 **for** $n=$ 1 to L' **do**
4 **let** ancList = getParent(o) ;	4 **foreach** $o' \in$ cList **do**
5 computeRank(o, ancList) ;	5 $KS =$ getKeywords(IL, o') $\cup\ KS$;
6 **let** desList = getChildren(o) ;	6 **if** $(KS \subset KW)$ **then**
7 computeRank(o, desList(o)) ;	7 $count$++; continue;
8 $L = L$ - 1 ;	8 **elseif** $(count>2)$ **then**
9 **foreach** $o' \in$ (oList $\cup$ ancList $\cup$ desList) s.t. o' is not IRO object yet	9 initialize group g containing such o and o';
10 getIROPairs(IL, src, o', CT, L) ;	10 IRO_Group.add(o,g);

The time complexity of $KWSearch$ algorithm is composed of three parts: (1) the cost of finding all IRO pairs is: $O(\sum_{o \in L_s} \sum_{i=1}^{L} |cList_i(o)| * \sum_{j=1}^{k} \log |L_j|)$, where L_s, o, $|cList_i(o)|$, k and $|L_j|$ represent the shortest inverted list of query keywords, an object ID in L_s, length of the list of objects forming an IRO pair with o with chain length $= i$ (limited to L), the number of query keywords, and the length of the jth keyword's inverted list respectively. (2) the cost of finding all IRO groups is: $O(\sum_{o \in L_s} \sum_{i=1}^{L'} |Q_{L'}| * \sum_{j=1}^{k} \log |L_j|)$, where the meaning of each parameter is same as part (1), and $|Q_{L'}|$ denotes the maximal number of object trees reached from o by depth limited search with chain length limit to L'. (3) the cost of finding all ISO objects is: $O(\sum_{o \in L_s} \sum_{j=1}^{k-1} \log |L_j|)$. The formation of each cost can be easily derived by tracing Algorithm 1-3.

8 Experimental Evaluation

Experiments run on a PC with Core2Duo 2.33GHz CPU and 3GB memory, and all codes are implemented in Java. Both real dataset DBLP(420 MB) and synthetic dataset XMark(115 MB) [2] are used in experiments. The inverted lists and connection table are created and stored in the disk with Berkeley DB [1] B+ trees. An online demo [3] of our system on DBLP, namely ICRA, is available at http://xmldb.ddns.comp.nus.edu.sg.

8.1 Effectiveness of ISO and IRO Matching Semantics

In order to evaluate the quality of our proposed ISO and IRO semantics, we investigate the overall recall of ISO, ISO+IRO with XSeek [16], XReal [4] and SLCA [20] on both DBLP and XMark. 20 queries are randomly generated for each dataset, and the result relevance is judged by five researchers in our database group. From the average recall shown in Table 1, we find: (1) ISO performs as well as XReal and XSeek, and is much better than SLCA. It is consistent with our conjecture that the search target of a user query is usually an object of interest, because the concept of object indeed is implicitly considered in the design of ISO, XReal and XSeek. (2) ISO+IRO has a higher recall than ISO alone, especially for queries on XMark, as there are more ID references in XMark that bring more relevant IRO results. In general, IRO semantics do help find more user-desired results while the other semantics designed for tree data model cannot.

Table 1. Recall Comparison

Data	SLCA	XSeek	XReal	ISO	ISO+IRO
DBLP	75%	82.5%	84.1%	84.1%	90.5%
XMark	55.6%	63.8%	60.4%	62.2%	80.7%

Table 2. Ranking Performance Comparison

Data	R-rank			MAP		
	XReal	ISO	IRO	XReal	ISO	IRO
DBLP	0.872	0.877	0.883	0.864	0.865	0.623
XMark	0.751	0.751	0.900	0.708	0.706	0.705

8.2 Efficiency and Scalability Test

Next, we compare the efficiency of our approach with SLCA and XReal [4] in tree model, and Bidirectional expansion [12] (Bidir for short) in digraph model. For each dataset, 40 random queries whose lengths vary from 2 to 5 words are generated, with 10 queries for each query size. The upper limit of connection chain length is set to 2 for IRO pair and 1 for IRO group, and accordingly we

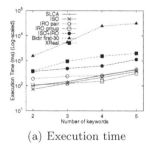

(a) Execution time   (b) Total result number

Fig. 2. Efficiency and scalability tests on DBLP

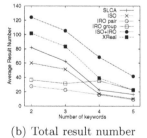

id	Query
Q1	David Giora
Q2	Dan Suciu semistructured
Q3	Jennifer Widom OLAP
Q4	Jim Gray transaction
Q5	VLDB Jim Gray
Q6	conceptual design relational database
Q7	join optimization parallel distributed environment

Fig. 3. Sample queries on DBLP

modify Bidir to not expand to a node of more than 2-hops away from a keyword node for a fair comparison. Besides, since Bidir searches as small portion of a graph as possible and generates the result during expansion, we only measure its time to find the first 30 results. The average response time on cold cache and the number of results returned by each approach are recorded in Fig. 2 and 4.

The log-scaled response time on DBLP is shown in Fig. 2(a), and we find: (1) Both SLCA and ISO+IRO are about one order of magnitude faster than XReal and Bidir for queries of all sizes. SLCA is twice faster than ISO+IRO, but considering the fact that ISO+IRO captures much more relevant results than SLCA (as evident from Table 1), such extra cost is worthwhile and ignorable. (2) ISO+IRO scales as well as SLCA w.r.t the number of query keywords, and ISO alone even has a better scalability than SLCA.

From Fig. 2(b), we find the result number of ISO is a bit smaller than that of SLCA, as ISO defines qualified result on (more restrictive) object level. Besides, ISO+IRO finds more results than SLCA and XReal, because many results that are connected by ID references can be identified by IRO. The result for XMark (see Fig. 4) is similar to DBLP, and the discussion is omitted due to space limit.

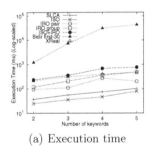

(a) Execution time

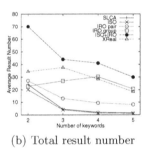

(b) Total result number

Table 3. sample query result number

Query	ISO result	IRO Result
Q1	16	42
Q2	14	58
Q3	1	56
Q4	14	230
Q5	8	1238
Q6	3	739
Q7	0	93

Fig. 4. Efficiency and scalability tests on XMark

8.3 Effectiveness of the Ranking Schemes

To evaluate the effectiveness of our ranking scheme on ISO and IRO results, we use two widely adopted metrics in IR: (1)Reciprocal rank (R-rank), which is 1 divided by the rank at which the first relevant result is returned. (2) Mean Average Precision (MAP). A precision is computed after each relevant one is identified when checking the ranked query results, and MAP is the average value of such precisions. R-Rank measures how good a search engine returns the first relevant result, while MAP measures the overall effectiveness for top-k results.

Here, we compute the R-rank and MAP for top-30 results returned by ISO, IRO and XReal, by issuing the same 20 random queries as describe in section 8.1 for each dataset. Specificity factor $s(o, Q)$ is ignored in computing $ISORank$; in computing the $IRORank$, $w_1 = 1$ and $w_2 = 0.7$ are chosen as the weights in Equation 5. The result is shown in Table 2. As ISO and XReal do not take into account the reference connection in XML data, it is fair to compare ISO with

XReal. We find ISO is as good as XReal in term of both R-rank and MAP, and even better on DBLP's testing. The ranking strategy for IRO result also works very well, whose average R-rank is over 0.88.

Besides the random queries, we choose 7 typical sample queries as shown in Fig. 3: $Q2$-$Q4$ intend to find publications on a certain topic by a certain author; $Q5$ intends to find publications of a particular author on a certain conference.

In particular, we compare our system [3] with some academic search engines such as Bidir in digraph model [12], XKSearch employing SLCA [20] in tree model, with commercial search engines, i.e. Google Scholar and Libra[2]. Since both Scholar and Libra can utilize abundant of web data to find more results than ours whose data source only comes from DBLP, it is infeasible and unfair to compare the total number of relevant results. Therefore, we only measure the number of top-k relevant results, where k=10, 20 and 30.

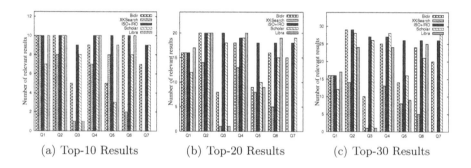

(a) Top-10 Results (b) Top-20 Results (c) Top-30 Results

Fig. 5. Result quality comparison

Since our system separates ISO results and IRO results (as mentioned in section 4.3), top-k results are collected in the way that, all ISO results are ordered before the IRO results. The total number of ISO results and IRO results are shown in Table 3, and the comparison for the top-30 results is shown in Fig. 5.

First, we compare ISO+IRO with Bidir and XKSearch. For queries that have both ISO and IRO results (e.g. $Q1$-$Q6$), our approach can find more relevant results, and rank them in most of the top-30 results. There is no ISO result for $Q7$, XKSearch also returns nothing; but 26 IRO results are actually relevant.

Second, we compare ISO+IRO with Libra and Scholar. From Fig. 5, we find our approach is comparable with Scholar and Libra for all sample queries. In particular, ISO+IRO is able to rank the most relevant ones in top-10 results for most queries, because its top-10 precision is nearly 100% for most queries, as evident in Figure 5(a). In addition, as Libra only supports keyword conjunction (similar to our ISO semantics), it does not work well for $Q3$ and $Q7$, as there is only 1 and 0 result containing all keywords for $Q3$ and $Q7$. As shown in Fig. 5(a), Scholar only finds 3 relevant results for $Q5$ in its top-10 answers, probably

---

[2] Google Scholar: http://scholar.google.com. Microsoft Libra: http://libra.msra.cn

because keywords "Jim" and "Gray" appear in many web pages causes many results that don't contain "VLDB" to still have a high rank, which is undesired.

Thirdly, as shown in Fig. 5, the average recall for each query generated by our ISO+IRO is above 80% at each of the three top-k levels, which confirms its advantage over any other approach.

9   Conclusion and Future Work

In this paper, we build a preliminary framework for object-level keyword search over XML data. In particular, we model XML data as the interconnected object-trees, based on which we propose two main matching semantics, namely ISO (*Interested Single Object*) and IRO (*Interested Related Object*), to capture different user search concerns. A customized ranking scheme is proposed by taking both the structure and content of the results into account. Efficient algorithms are designed to compute and rank the query results in one phase, and extensive experiments have been conducted to show the effectiveness and efficiency of our approach. In future, we plan to investigate how to distinguish the relationship types among objects and utilize them to define more precise matching semantics.

Acknowledgement

Jiaheng Lu was partially supported by 863 National High-Tech Research Plan of China (No: 2009AA01Z133, 2009AA01Z149), National Science Foundation of China (NSFC) (No.60903056, 60773217), Key Project in Ministry of Education (No: 109004) and SRFDP Fund for the Doctoral Program(No.20090004120002).

References

1. Berkeley DB, http://www.sleepycat.com/
2. http://www.xml-benchmark.org/
3. Bao, Z., Chen, B., Ling, T.W., Lu, J.: Demonstrating effective ranked XML keyword search with meaningful result display. In: Zhou, X., et al. (eds.) DASFAA 2009. LNCS, vol. 5463, pp. 750–754. Springer, Heidelberg (2009)
4. Bao, Z., Ling, T., Chen, B., Lu, J.: Effective XML keyword search with relevance oriented ranking. In: ICDE (2009)
5. Chen, B., Lu, J., Ling, T.: Exploiting id references for effective keyword search in xml documents. In: Haritsa, J.R., Kotagiri, R., Pudi, V. (eds.) DASFAA 2008. LNCS, vol. 4947, pp. 529–537. Springer, Heidelberg (2008)
6. Cohen, S., Mamou, J., Kanza, Y., Sagiv, Y.: XSEarch: A semantic search engine for XML. In: VLDB (2003)
7. Guo, L., Shao, F., Botev, C., Shanmugasundaram, J.: XRANK:ranked keyword search over XML documents. In: SIGMOD (2003)
8. He, H., Wang, H., Yang, J., Yu, P.S.: Blinks: ranked keyword searches on graphs. In: SIGMOD (2007)
9. Hristidis, V., Koudas, N., Papakonstantinou, Y., Srivastava, D.: Keyword proximity search in XML trees. In: TKDE, pp. 525–539 (2006)
10. Hristidis, V., Papakonstantinou, Y.: Discover: Keyword search in relational databases. In: VLDB (2002)

11. Hristidis, V., Papakonstantinou, Y., Balmin, A.: Keyword proximity search on XML graphs. In: ICDE (2003)
12. Kacholia, V., Pandit, S., Chakrabarti, S., Sudarshan, S., Desai, R.: Bidirectional expansion for keyword search on graph databases. In: VLDB (2005)
13. Li, G., Feng, J., Wang, J., Zhou, L.: Effective keyword search for valuable LCAS over XML documents. In: CIKM (2007)
14. Li, W., Candan, K., Vu, Q., Agrawal, D.: Retrieving and organizing web pages by information unit. In: WWW (2001)
15. Li, Y., Yu, C., Jagadish, H.V.: Schema-free xquery. In: VLDB (2004)
16. Liu, Z., Chen, Y.: Identifying meaningful return information for XML keyword search. In: SIGMOD (2007)
17. Liu, Z., Chen, Y.: Reasoning and identifying relevant matches for XML keyword search, vol. 1, pp. 921–932 (2008)
18. Liu, Z., Sun, P., Huang, Y., Cai, Y., Chen, Y.: Challenges, techniques and directions in building xseek: an XML search engine, vol. 32, pp. 36–43 (2009)
19. Salton, G., McGill, M.J.: Introduction to Modern Information Retrieval
20. Xu, Y., Papakonstantinou, Y.: Efficient keyword search for smallest LCAs in XML databases. In: SIGMOD (2005)

Effectively Inferring the Search-for Node Type in XML Keyword Search

Jiang Li and Junhu Wang

School of Information and Communication Technology
Griffith University, Gold Coast, Australia
Jiang.Li@student.griffith.edu.au, J.Wang@griffith.edu.au

Abstract. XML keyword search provides a simple and user-friendly way of retrieving data from XML databases, but the ambiguities of keywords make it difficult to *effectively* answer keyword queries. XReal [4] utilizes the statistics of underlying data to resolve keyword ambiguity problems. However, we found their proposed formula for inferring the search-for node type suffers from *inconsistency* and *abnormality* problems.

In this paper, we propose a *dynamic reduction factor* scheme as well as a novel algorithm DynamicInfer to resolve these two problems. Experimental results are provided to verify the effectiveness of our approach.

1 Introduction

Keyword search has long been used to retrieve information from collections of text documents. Recently, keyword search in XML databases re-attracted attention of the research community because of the convenience it brings to users - there is no need for users to know the underlying database schema or complicated query language.

Until now, a lot of research focuses on how to efficiently and meaningfully connect keyword match nodes (e.g., ELCA [6], XSEarch [5], MLCA [9], SLCA [12] and VLCA [8]) and generate meaningful, informative and compact results (e.g., GDMCT [7], XSeek [10] and MaxMatch [11]), but this only solves one side of the problem. The returned answers may be meaningful, but they may not be desired by users. Therefore, the other side of the problem is how to accurately acquire the users' search intention, which is a difficult task because keywords may have multiple meanings in an XML document, and keyword query lacks the ability to specify the meaning that is wanted. For example in Fig. 1, *11* appears as a text value of *volume* and *initPage* node, and *volume* exists as an XML tag name and a text value of *title* node, but which meaning is desired by the user is hard to determine just from the query {volume 11}.

In order to resolve the ambiguities of keywords, Bao et al [4] introduced the statistics of XML data into answering keyword queries, which provides an objective way of identifying users' *major* search intention. In their search engine XReal, they first use a formula, which is based on three guidelines (see Section 2 for details), to infer the *search-for node type* (SNT). Then they use an improved

H. Kitagawa et al. (Eds.): DASFAA 2010, Part I, LNCS 5981, pp. 110–124, 2010.

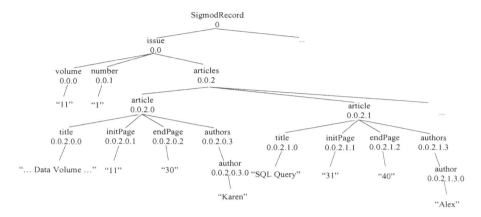

Fig. 1. An XML data tree of SigmodRecord

XML TF*IDF ranking strategy to rank individual matches of the identified SNT. For example, for the query {article Karen} over the data tree in Fig. 1, XReal may first infer *article* as the SNT. Then it ranks all of the subtrees rooted at *article* nodes that contain keywords "article" and "Karen".

In XReal's two-step approach, the accuracy of inferring SNT plays a crucial role in returning relevant final results. If the system selects an incorrect SNT, all of the final results would be irrelevant to the user. However, based on our experiments, their way of identifying the SNT suffers from *inconsistency* and *abnormality* problems even though each keyword in the query has only one meaning in an XML document. First, for the same query, XReal may return inconsistent SNTs when the data size changes (We will give examples to describe the details of these two problems in Section 3). In fact, XReal may infer different SNTs even though we simply replicate an XML document two times. This is an unreasonable behavior for an XML keyword search engine. Second, XReal may infer different SNTs when the keyword queries are similar queries. This is another type of inconsistency problem. For example, given the queries {article data} and {article SQL} over the data tree in Fig. 1, both keywords "data" and "SQL" appear as a text value of *title* node. Intuitively, these two queries should yield the same SNT, but XReal may infer different SNTs for them. Third, XReal may suggest unreasonable SNT when the frequency of keywords is low. However, users often submit keywords which appear as text values and have low frequencies, so this problem is serious for a search engine.

The two problems above show that the formula used by XReal can not effectively identify the SNT in some cases. In order to resolve these two problems, we propose a *dynamic reduction factor* scheme. Reduction factor is a *constant* value in the formula used by XReal to infer the SNT (see Section 2). In our solution, this factor is dynamic and changes on the fly. Its value is determined by a devised formula. We provide algorithm **DynamicInfer** which incorporates the dynamic reduction factor scheme to infer the SNT of a query. Extensive experiments verified the effectiveness of our approach to resolve the identified problems.

To summarize, we make the following contributions:

1. We identify two problems (i.e., inconsistency and abnormality problems) of XReal in inferring SNTs.
2. We propose a dynamic reduction factor scheme to resolve the identified problems. We provide algorithm DynamicInfer to incorporate this scheme to infer the SNT of a query.
3. We conducted an extensive experimental study which verified the effectiveness of our approach.

The rest of this paper is organized as follows. We briefly introduce the XML keyword search engine XReal and some other background information in Section 2. In Section 3, we illustrate and analyze the weaknesses of XReal with examples. The dynamic reduction factor scheme is presented in Section 4. Experimental studies are shown in Section 5. Section 6 presents related work followed by conclusion in Section 7.

2   Background

2.1   Notations

An XML document is modeled as an unordered tree, called the *data tree*. Each *internal node* (i.e., non-leaf node) has a label, and each *leaf node* has a value. The internal nodes represent elements or attributes, while the leaf nodes represent the values of elements or attributes. Each node v in the data tree has a unique Dewey code, which represents the position of that node in the data tree. With this coding scheme, ancestor-descendant relationship can be easily identified: for any two nodes v_1, v_2 in data tree t, v_1 is an ancestor of v_2 iff the dewey code of v_1 is a prefix of the dewey code of v_2. Fig. 1 shows an example data tree.

Keyword query: A keyword query is a finite set of keywords $K = \{k_1, ..., k_n\}$. Given a keyword k and a data tree t, the search of k in t will check both the labels of internal nodes and values of leaf nodes for possible occurrence of k.

Definition 1. *(Node Type) Let* n *be a node in data tree* t. *The node type of* n *is the* path *from the root to* n *if* n *is an internal node. If* n *is a leaf node, its node type is the node type of its parent.*

The type of a node actually represents the meaning of this node. In Fig. 1, the node type of *author* (0.0.2.0.3.0) is the path

(SigmodRecord.issue.articles.article.authors.author), and the node type of *title* (0.0.2.0.0) is the path

(SigmodRecord.issue.articles.article.title). The nodes *title* (0.0.2.0.0) and *title* (0.0.2.1.0) own the same node type because they share the same path. Node *volume* (0.0.0) and node *title* (0.0.2.0.0) have different node types.

Note: For simplicity, we will use the tag name instead of the path of a node to denote the node type throughout this paper if there is no confusion.

2.2   Overview of XReal

To make the paper self-contained, we review the XML keyword search engine XReal in this section. We will use a running example to briefly introduce its basic ideas.

Based on the fact that each query usually has only one desired node type to search for, keyword query processing in XReal is divided into two steps. First, three guidelines as well as a corresponding formula are used to identify the SNT. Second, an XML TF*IDF ranking mechanism is used to rank the matches of the identified SNT. We mainly discuss the first step because it is closely related to our work.

The three guidelines which are used to guide the identification of SNT are listed below. Given a keyword query q, XReal determines whether a node type T is the desired node type to search for based on the following three guidelines:

Guideline 1: T is intuitively related to every query keyword in q, i.e. for each keyword k, there should be some (if not many) T-typed nodes containing k in their subtrees.

Guideline 2: XML nodes of type T should be informative enough to contain enough relevant information.

Guideline 3: XML nodes of type T should not be overwhelming to contain too much irrelevant information.

To apply these guidelines, XReal uses the following formula to calculate the confidence score $C_{for}(T, q)$ of a node type T:

$$C_{for}(T, q) = log_e(1 + \prod_{k \in q} f_k^T) * r^{depth(T)} \tag{1}$$

where k represents a keyword in query q; f_k^T is the number of T-typed nodes that contain k as either values or tag names in their subtrees; r is some reduction factor with range $(0, 1]$ and normally chosen to be 0.8, and $depth(T)$ represents the depth of T-typed nodes in document.

In Formula (1), the first multiplier (i.e., $log_e(1 + \prod_{k \in q} f_k^T)$) enforces the first and third guidelines. The product of f_k^T ensures that the selected node type must be related to every keyword in the query, otherwise the score will be 0. For example, with the data in Fig. 1, the value of $\prod_{k \in \{volume, Karen\}} f_k^{author}$ is 0 because there is no subtree rooted at the node type $author$ that contains the keyword "volume". In addition, given a keyword k, the characteristics of tree structure determines that the node type T at lower levels has a greater chance to have larger values of f_k^T. For example, $f_{initPage}^{issue}$ is smaller than $f_{initPage}^{article}$ in Fig.1. Therefore, the first multiplier usually keeps the level of SNT low enough to make the result small. The second multiplier $r^{depth(T)}$ enforces the second guideline by making the level of SNT high enough to contain more information.

3   Analysis of XReal's Weaknesses

As we stated earlier, XReal uses Formula (1) to identify the search-for node type (SNT) of a query. However, there exist inconsistency and abnormality problems

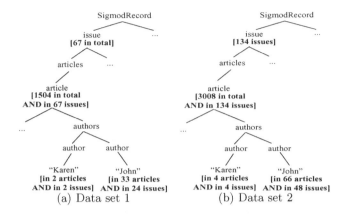

Fig. 2. Two Data Set

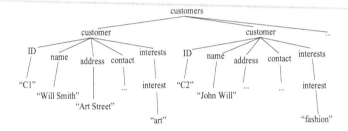

Fig. 3. An XML data tree of customers

when this formula is applied. In this section, we will use examples to explain the details of these problems, and discuss why these problems are serious to an XML keyword search engine. To facilitate our discussion, we first give a data set (i.e., Data set 1) in Fig. 2(a), which is identical to the SigmodRecord data set obtained from [1]. In the data set, we also provide the statistics of a certain meaning of the words that will be used in our examples. For example, the keyword "Karen" exists in two different articles as the text value of author node, and these two articles exist in two different issues. We replicated Data set 1 two times to get Data set 2 (Fig. 2(b)). All the examples in this section will be based on these two data sets.

[**Inconsistency problem 1**]: For the same query, Formula (1) may infer inconsistent SNTs when the size of a data set changes. To illustrate this problem, we simply replicate a data set two times to simulate the change of data size.

Example 1. Given the query {article, Karen} over Data set 1 and 2, we calculate and compare the confidence score $C_{for}(T, q)$ of the node types *article* and *issue* (Note: other node types, such as *SigmodRecord* and *articles*, are ignored here), and list the results in Table 1. Intuitively, *article* should be identified as the SNT no matter what the data size is, but the system infers inconsistent SNTs and selects an unreasonable node type *issue* when the data has a larger size.

Table 1. $C_{for}(T, q)$ for Inconsistency Problem (Query q: {article Karen})

Data set	T	$log_e(1 + \prod_{k \in \{article, Karen\}} f_k^T) * 0.8^{depth(T)}$	$C_{for}(T, q)$
Data set 1	issue	$log_e(1 + 67 * 2) * 0.8$	3.9242
	article	$log_e(1 + 1504 * 2) * 0.8^3$	4.1008
Data set 2	issue	$log_e(1 + 134 * 4) * 0.8$	5.0288
	article	$log_e(1 + 3008 * 4) * 0.8^3$	4.8104

Inconsistency problem is serious for an XML keyword search engine because the data in reality is not static. The data set may become larger or smaller when the data is inserted or deleted, but the precision of inferring SNT should not be affected by the scale of data.

[**Inconsistency Problem 2**]: For two similar queries, Formula (1) may infer inconsistent SNTs. Here by *similar queries* we mean their keywords have the same node types. Intuitively, similar queries are supposed to have the same SNT.

Example 2. Consider the queries {article John} and {article Karen} over Data set 1. These two queries are similar and the SNTs should be the same. However, as shown in Table 2, XReal returns *article* for {article Karen}, and returns *issue* for {article John}.

Table 2. $C_{for}(T, q)$ for Inconsistency Problem 2

q	T	$log_e(1 + \prod_{k \in q} f_k^T) * 0.8^{depth(T)}$	$C_{for}(T, q)$
{article Karen}	issue	$log_e(1 + 67 * 2) * 0.8$	3.9242
	article	$log_e(1 + 1504 * 2) * 0.8^3$	4.1008
{article John}	issue	$log_e(1 + 67 * 24) * 0.8$	5.9067
	article	$log_e(1 + 1504 * 33) * 0.8^3$	5.5360

[**Abnormality Problem**]: Formula (1) may infer unreasonable SNTs when the keywords in a query have low frequencies. We use the keywords that occur in text values to illustrate this problem.

Table 3. $C_{for}(T, q)$ for Abnormality Problem

q	T	$log_e(1 + \prod_{k \in \{Karen\}} f_k^T) * 0.8^{depth(T)}$	$C_{for}(T, q)$
{Karen}	issue	$log_e(1 + 2) * 0.8$	0.8789
	article	$log_e(1 + 2) * 0.8^3$	0.5625

Example 3. Consider the query {Karen} over the Data set 1. In this query, the keyword "Karen" has low frequency. We calculate $C_{for}(T, q)$ of the node type *article* and *issue*, and list the result in Table 3. From the result, it can be seen

that the node type *issue* will be selected as the SNT, but *article* is intuitively a better choice in this case. More seriously, if the user submits a keyword which has extremely low frequency, XReal even returns the node type *SigmodRecord* as the SNT.

The Abnormality Problem is also a serious problem for an XML keyword search engine, because it is very common for the user to submit keywords which appear as text values in a data set, and normally this kind of keywords have low frequencies.

Discussion: Inconsistency Problem 1 emerges because the logarithm function (i.e., $log_e(1 + \prod_{k \in q} f_k^T)$) in Formula (1) grows more and more slowly when its argument becomes larger due to the increased data size, but the exponential function (i.e., $r^{depth(T)}$) in Formula (1) decreases the value of the logarithm function more quickly when the depth increases. Therefore, the level of SNT tends to go up when the data size becomes larger. For similar queries, their argument values of the logarithm function may be very different but the reduction factor remains the same, so their SNTs are very likely to be different and lead to Inconsistency Problem 2. On the other hand, these two inconsistency problems are more likely to occur on *deep* tree structures than shallow ones. For example, the data tree in Fig. 3 is shallower than the SigmodRecord data tree in Fig. 1. It is very hard for the node type *customers* to be selected as the SNT because there is only one *customers* node in the data tree which results in its very small confidence score compared with the confidence score of *customer*. Although more *customer* nodes may be inserted into the data set, the number of *customers* node does not change.

The Abnormality Problem arises because for the node types at different levels, their values of the first multiplier in Formula (1) are so close when the frequency of keywords is low that the second multiplier can easily make the node types at higher levels to be the SNT. In other words, the first multiplier will become negligible when the keywords have low frequencies. For example in Data set 1, the value of $log_e f_{Karen}^{article}$ is the same with the value of $log_e f_{Karen}^{issue}$, so their confidence scores are mainly determined by the second multiplier in the formula. In addition, this problem is likely to happen on both deep and shallow data trees. For the data tree in Fig. 3, if the keyword has a very low frequency, the node type *customers* can also be inferred as the SNT, but this can not be accepted.

One may wonder whether these problems can be resolved by manually adjusting the reduction factor for each data set by the database administrator. The problem is the reduction factor is not closely related to the scale of the data but the argument value of the logarithm function in Formula (1). In other words, the reduction factor is closely related to the number of occurrences of each node type containing the keywords in a query. Therefore, it is impossible to choose a value which can fit for every query. The only way is to use a dynamic reduction factor (i.e., r) in the formula and setting its value on the fly. In this paper, we will explore how to achieve this.

4   Inferring Search-for Node

In this section, we introduce the statistics of underlying data utilized in our solution, and illustrate how to employ dynamic reduction factor scheme in improving the precision of SNT identification. Before discussing our solution, some preliminaries are introduced.

4.1   Preliminary Definitions

Entity nodes: Lots of XML documents in reality are well designed and conform to a pre-defined schema. Therefore, even though an XML document is modeled as a tree, it is actually a container of related entities in the real world. Consider the data tree in Fig. 1, which is a collection of *SigmodRecord, issue* and *article* entities. These entities are joined together through the ancestor-descendant relationship. The root of each entity is called *entity node*. We use an approach similar to that of [10] to identify entity nodes.

Definition 2. *Let t be a data tree. A node u in t is said to be a* simple node *if it is a leaf node, or has a single child which is a leaf node. A node u represents an* entity node *if: (1) it corresponds to a *-node in the DTD (if DTD exists), or has siblings with the same tag name as itself (if DTD does not exist), and (2) it is not a simple node.*

The entity type *of an entity node e refers to the node type (as defined in Definition 1) of e.*

Example 4. For the data tree in Fig. 1, nodes   issue (0.0), article (0.0.2.0) and article (0.0.2.1) are inferred as the entity nodes. The entity nodes article (0.0.2.0) and article (0.0.2.1) have the same entity-type, which is the node type SigmodRecord.issue.articles.article.

Definition 3. (Ancestor Node Type) *Given a node type $T \equiv l_0.l_1.\cdots.l_n$, we say the path $l_0.l_1.\cdots.l_i$ is an* ancestor node type *of T, for any $i \in [0, n-1]$.*

For example in Fig. 1, the node type Sigmodrecord.issue is an ancestor node type of SigmodRecord.issue.articles.article.

Definition 4. (The entity-type of a node type) *If a node type T is not the node type of some entity node, its entity-type is its ancestor node type T' such that (1) T' is the node type of some entity node, and (2) T' is the longest among all ancestor node types of T satisfying condition (1). If T is the node type of some entity node, its entity-type is itself. Node types that have the same entity-type are called* neighbor node types.

Note that every node type in the data tree owns one and only one entity-type.

Example 5. In Fig. 1, the entity-type of node type *initPage* is node type *article*. Node type *article*'s entity-type is itself (i.e., *article*). The node types *article*, *title*, *initPage*, *endPage*, *authors* and *author* are neighbor node types because they share the same entity-type *article*.

4.2   Dynamic Reduction Factor

We use a *dynamic reduction factor* scheme to adjust the reduction factor (i.e., r) in Formula (1) on the fly in order to resolve the inconsistency and abnormality problems.

Intuitively, the inconsistency and abnormality problems shown in Section 3 are all caused by an inappropriate reduction factor value and the inability of that value to adapt to the user query. The score of a node type becomes lower than that of some ancestor node types, causing the ancestor node type to be chosen as the search-for node type. In order to solve this problem, we propose the following guideline as the basis of our approach.

Guideline 4: Given a node type T which achieves the highest confidence score among its neighbor node types, the reduction factor should be such that it ensures that no ancestor node type T' of T achieves a higher confidence score than T if we ignore the occurrence of keywords in other parts of the data tree than subtrees rooted at T-typed nodes.

It should be noted that among a set of neighbor node types, the node type of the entity nodes normally has the highest confidence score according to Formula (1). For example, consider the query {SQL} over the data tree in Fig. 1. The confidence score of *article*, which is the node type of entity nodes, is larger than *title* (Note: Here we do not consider other neighbor node types, such as *initPage* and *endPage*, because their confidence scores are 0) because $f_{SQL}^{article}$ is equal to f_{SQL}^{title}, and *article*-typed nodes are higher than *title*-typed nodes in the data tree.

Example 6. The user submits query {article Karen} over the data tree in Fig. 1. After calculation, the node type *article* achieves the highest confidence score among all of the neighbor node types that have the entity-type *article*. According to Guideline 4 above, the new reduction factor should guarantee that the confidence score of *article* is larger than the confidence score of *articles*, *issue* and *SigmodRecord* if we ignore the occurrence of the keywords "article" and "Karen" in other parts of the data tree than the subtrees rooted at *article*-typed nodes.

To formalize Guideline 4, we propose the statistics $f_k^{T',T}$.

Definition 5. $f_k^{T',T}$ *is the number of T'-typed nodes that contain keyword k in the subtrees of their T-typed descendant nodes in the* XML *database.*

Example 7. Over the data tree in Fig. 1, $f_{data}^{issue,article}$ is the number of *issue*-typed nodes that contain keyword "data" in the subtrees rooted at *article*-typed nodes. $f_{data}^{issue,article}$ ignores the occurrence of keywords "data" in other parts of the data tree than the subtrees rooted at *article*-typed nodes.

Guideline 4 can be formally defined with the following formula:

$$\frac{log_e(1 + \prod_{k \in q} f_k^T) * r^{depth(T)}}{log_e(1 + \prod_{k \in q} f_k^{T',T}) * r^{depth(T')}} > 1, (T' \in Ancestors(T)) \qquad (2)$$

so the reduction factor r should satisfy the condition below.

$$\max\{ \sqrt[depth(T)-depth(T')]{\frac{log_e(1 + \prod_{k \in q} f_k^{T',T})}{log_e(1 + \prod_{k \in q} f_k^T)}}\} < r \leq 1, (T' \in Ancestors(T))$$

(3)

In the two formulas above, T is the node type which achieves the highest confidence score among all of its neighbor node types and is normally the node type of the entity nodes. T' is an ancestor node type of T. f_k^T is the number of T-typed nodes that contain k in their subtrees. $f_k^{T',T}$ is the number of T'-typed nodes that contain k in the subtrees of their T-typed descendant nodes.

Reservation Space: The value of reduction factor r should satisfy the condition in Formula (3). In practice, we need to determine the exact value of r, and we can add a small value to the max function. This small value is called *reservation space* (i.e., rs) in our approach, which can not be too small or too large. If rs is too small, the confidence score of T' is very easy to exceed the confidence score of T when there exist more T'-typed nodes that contain the keywords in the parts of the data tree excluding the subtrees rooted at T-typed nodes. If rs is too large, it is very difficult for T' to be selected as the SNT even though much more T'-typed nodes contain the keywords in the parts of the data tree excluding the subtrees rooted at T-typed nodes.

Based on our experiments, 0.05 is an appropriate value for rs. The formula for the reduction factor r is as follows, and the maximum value of r is 1.

$$r = \min\{\max\{ \sqrt[depth(T)-depth(T')]{\frac{log_e(1 + \prod_{k \in q} f_k^{T',T})}{log_e(1 + \prod_{k \in q} f_k^T)}}\} + 0.05, 1\}, (T' \in Ancestors(T))$$

(4)

4.3   Algorithm

XReal computes the confidence scores of all node types and selects the node type with the highest confidence score as the SNT. For our approach, in order to enforce Formula (4), we can not simply compute the confidence score of each node type, and need to design a new algorithm.

Our algorithm DynamicInfer for inferring the SNT is shown in Algorithm 1. Now we explain this algorithm. We first set the initial reduction factor as 0.8 (line 1). At line 2, we retrieve all of the leaf entity-types using procedure $GetLeafEntityTypes(NT)$. *Leaf entity-types* are the entity-types that do not have any descendant entity-types. For each leaf entity-type, we use a bottom-up strategy to infer the node type $T_{current}$ which achieves the highest confidence score in current bottom-up process. In each bottom-up process, we first use procedure $GetNeighborNodeTypes(NT, et)$ to collect all of the neighbor node types which have the entity-type et (line 7). Then we use procedure $GetNTWithHighestConfidenceScore$ to get the node type $T_{neighbor}$ that achieves the highest confidence score $C_{neighbor}$ among the neighbor node

Algorithm 1. DynamicInfer(NT, q)

Input: Node types $NT = \{nt_1, ..., nt_n\}$, a query q
Output: The search-for node type T_{for}

1: $r = 0.8$, $C_{for} = 0$, T_{for}=NULL//r: reduction factor; C_{for} is the confidence score of
$\qquad\qquad\qquad\qquad\qquad\qquad$ SNT T_{for}
2: $le = GetLeafEntityTypes(NT)$ //le is a list of leaf entity types
3: **for** $i = 1$ to $le.length()$ **do**
4: $\quad$ $C_{path} = 0$, T_{path}=NULL
5: $\quad$ $et = le[i]$
6: $\quad$ **while** $et \neq$ NULL **do**
7: $\quad\quad$ $nn = GetNeighborNodeTypes(NT, et)$; //$nn$ is a set of neighbor node types
$\qquad\qquad\qquad\qquad\qquad\qquad$ which have entity type et
8: $\quad\quad$ $(T_{neighbor}, C_{neighbor})$=$GetNTWithHighestConfidenceScore(nn)$
$\qquad\qquad\qquad\qquad\qquad$ //$T_{neighbor}$ is the node type which has
$\qquad\qquad\qquad\qquad\qquad$ the highest confidence score in nn, and its
$\qquad\qquad\qquad\qquad\qquad$ confidence score is $C_{neighbor}$
9: $\quad\quad$ **if** $C_{neighbor} > C_{current}$ **then** //$C_{current}$ is the confidence score of $T_{current}$
$\qquad\qquad\qquad\qquad\qquad$ which has the highest confidence score in
$\qquad\qquad\qquad\qquad\qquad$ current bottom-up process
10: $\quad\quad\quad$ $T_{current} = T_{neighbor}$
11: $\quad\quad\quad$ $r = AdjustReductionFactor(T_{neighbor}, et)$
12: $\quad\quad\quad$ $C_{neighbor} = CalculateConfidenceScore(T_{neighbor})$
13: $\quad\quad\quad$ $C_{current} = C_{neighbor}$
14: $\quad\quad$ $et = GetNextAncestorEntityType(et)$ //Get the closest ancestor entity
$\qquad\qquad\qquad\qquad\qquad$ type of et
15: $\quad$ $r = 0.8$
16: $\quad$ $C_{current} = CalculateConfidenceScore(T_{current})$
17: $\quad$ **if** $C_{for} < C_{current}$ **then**
18: $\quad\quad$ $T_{for} = T_{current}$
19: $\quad\quad$ $C_{for} = C_{current}$

20: **procedure** ADJUSTREDUCTIONFACTOR($T_{highest}, et$)
21: $\quad$ $A = GetAncestorNodeTypes(et)$ //A is a set of ancestor node types of et
22: $\quad$ $r = \min\{\max\{\ ^{depth(T_{highest})-depth(a)}\sqrt{\dfrac{log_e(1+\prod_{k \in q} f_k^{a,T_{highest}})}{log_e(1+\prod_{k \in q} f_k^{T_{highest}})}}\} + 0.05, 1\}, (a \in$
$\quad A)$
23: $\quad$ return r

---

types collected in the last step. If $C_{neighbor}$ is larger than $C_{current}$, we assign $T_{current}$ with $T_{neighbor}$ (line 10), and adjust the reduction factor with procedure *AdjustReductionFactor* (line 11). Next we need to recalculate the confidence score $C_{neighbor}$ of $T_{neighbor}$ by calling procedure *CalculateConfidenceScore* which uses Formula (1) (line 12), and set $C_{current}$ with $C_{neighbor}$ (line 13). We repeat the steps above until all of the ancestor entity-types of current leaf entity-type have been processed, and get the node type $T_{current}$ which has the highest confidence score in current bottom-up process. We reset r to 0.8 (line 15), and recalculate the confidence score of $T_{current}$ (line 16) because we want to compare the node types that achieve the highest confidence score in different bottom-up process using the same reduction factor. We start another bottom-up process if there are other leaf entity-types unprocessed. The initial value of reduction factor r in each bottom-up process is 0.8. Eventually, T_{for} is the node type which achieves the highest confidence score among all bottom-up processes.

We use the query {article John} to illustrate DynamicInfer.

Example 8. Consider the query {article John} over Data set 1. The algorithm first retrieves all of the leaf entity-types (line 2). In this data set, there is only one leaf entity-type *article*, so there is only one bottom-up process. The bottom-up process starts. We first collect all the neighbor node types that have the entity-type *article* (line 7) and apply Formula (1) to find the node type with the highest confidence score among these neighbor node types (line 8), which is *article* and its confidence score $C_{neighbor}$ is 5.5360 in this case. Because $C_{neighbor}$ is larger than $C_{current}$, we set $T_{current}$ with $T_{neighbor}$ (i.e., *article*) and adjust the reduction factor. After the adjustment (line 11), the new reduction factor is 0.8764. Then we recalculate the confidence score of *article* using Formula (1) (line 12). The new confidence score of *article* is 7.2783. Then we assign this value to $C_{current}$ (line 13). At line 14, we get the closest ancestor entity-type of *article*, which is *issue*. We also retrieve all of its neighbor node types and compute the confidence scores of these node types. We find that the node type *issue* has the highest confidence score 6.4708. Because 6.4708 is smaller than $C_{current}$ (i.e., 7.2783), we do not change $T_{current}$ and the reduction factor. Because *issue* does not have ancestor entity-type, the bottom-up process ends. At line 18, T_{for} is assigned with *article*. *article* is the only leaf entity-type in this case, so the algorithm ends. The node type *article* is identified as the SNT.

For the example above, XReal infers the node type *issue* as the SNT. For this case, *article* is a preferable SNT to *issue*.

5 Experiments

In this section, we present the experimental results on the accuracy of inferring the search-for node type (SNT) of our approach DynamicInfer against XReal [4]. We selected several queries which have specific search intentions, but XReal produces inconsistency and abnormality problems. We present the SNTs after applying our dynamic reduction factor scheme and the new reduction factor in the results.

5.1 Experimental Setup

The XML document parser we used is the XmlTextReader Interface of Libxml2 [3]. The keyword inverted list and statistics information are implemented in C++ and stored with Berkeley DB [2].

We implemented XReal and our approach in C++. All the experiments were performed on a 1.6GHz Intel Centrino Duo processor laptop with 1G RAM. The operating system is Windows XP. We used the data sets SigmodRecord and WSU obtained from [1] for evaluation.

5.2   Results of Inferring Search-for Node

Inconsistency Problem 1: We used SigmodRecord data set for the experiments on preventing Inconsistency Problem 1 because this data set has a deep structure. As we stated earlier, inconsistency problems are most likely to happen on deep data trees. We replicated the data set two times to simulate the data size changes. The selected queries (QI1-QI5), the SNTs inferred by XReal, the SNTs inferred by DynamicInfer and the new reduction factor are listed in Table 4. For these five queries, XReal infers unpreferable SNTs when the data set is double sized, so we only listed the new reduction factor for the queries over the double-sized data set in the table. It can be seen that our approach can resolve Inconsistency Problem 1 by applying the dynamic reduction factor scheme.

Inconsistency Problem 2: We also used SigmodRecord data set to do the experiments on preventing Inconsistency Problem 2. We selected three pairs of similar queries and listed the new reduction factor for the second query in each pair in Table 5. The results show that the dynamic reduction factor scheme can successfully solve Inconsistency Problem 2.

Abnormality Problem: Abnormality Problem is likely to happen on both deep and shallow tree structures, so we use SigmodRecord and WSU data set to do the experiments on preventing Abnormality Problem. WSU is a university courses data set which has a shallow structure. We selected three queries for each data set. From the results shown in Table 6, our approach can also resolve abnormality problems. It should be noted that the keywords in query QA3 have relatively high frequencies compared with other queries, but XReal also infers an unpreferable SNT.

6   Other Related Work

Most previously proposed XML keyword search systems used the concept of *lowest common ancestor* (LCA), or its variant, to connect the match nodes. XRank [6] proposed the excluding semantics to connect keyword matches, which connects

Table 4. Results on Resolving Inconsistency Problem 1

	Query	SNT of XReal		SNT of Our approach		r
		Original	Double-sized	Original	Double-sized	
QI1	{article Karen}	article	issue	article	article	0.8680
QI2	{article title SQL}	article	issue	article	article	0.8702
QI3	{article database}	article	issue	article	article	0.8675
QI4	{title data author}	article	issue	article	article	0.8595
QI5	{title web initPage}	article	issue	article	article	0.8614

Table 5. Results on Resolving Inconsistency Problem 2

	Query	SNT of XReal	SNT of Our approach	r
QI6	{author Karen}	article	article	
QI7	{author John}	issue	article	0.8764
QI8	{title database}	article	article	
QI9	{title query}	issue	article	0.8509
QI10	{article title database}	article	article	
QI11	{article title relational database}	issue	article	0.8568

Table 6. Results on Resolving Abnormality Problem

	Query	SNT of XReal	SNT of Our approach	r
QA1	{Karen}	issue	article	1
QA2	{XML}	SigmodRecord	article	1
QA3	{database system}	issue	article	0.9047
QA4	{crowe}	root	course	1
QA5	{models}	root	course	1
QA6	{labor}	root	course	1

keyword matches by the LCA nodes that contain at least one occurrence of all keywords after excluding the occurrences of keywords in their descendants that already contain all keywords. XKSearch [12] proposed the notion of Smallest LCA (SLCA) to connect keyword match nodes. A SLCA is the root of a subtree which contains all the keywords, and any subtree rooted at its descendants does not contain all the keywords. Li et al [9] proposed the concept of Meaningful LCA to connect keyword matches. A set of keyword matches are considered to be meaningfully related if every pair of the matches is meaningfully related. Two keyword matches are considered meaningfully related if they can be linked with a SLCA. The LCA based approaches have a common inherent problem, which is that they may link irrelevant XML nodes together and return large amount of useless information to the user. This problem is called false positive problem in [8]. In order to solve the false positive problem, Li et al [8] proposed the concept of valuable LCA (VLCA).

After keyword matches are meaningfully connected, it is important to determine how to output the results. Most LCA based approaches return the whole subtree rooted at the LCA or its variants (e.g., MLCA, SLCA, etc.). Sometimes, the return subtrees are too large for users to find needed information. Therefore, Hristidis et al. in [7] introduced minimum connecting trees to exclude the subtrees rooted at the LCA that do not contain keywords. This approach makes the results more compact. To make the answers more meaningful, XSeek [10] tried to recognize the possible entities and attributes in the data tree, distinguish between search predicates and return specifications in the keywords, and return nodes based on the analysis of both XML data structures and keyword match patterns.

7   Conclusion

In this paper, we identified the inconsistency and abnormality problems in the approach of inferring the search-for node type used by XReal. To resolve these problems, we propose a dynamic reduction factor scheme as well as algorithm DynamicInfer to apply this scheme.

We have implemented the proposed approach and the extensive experiments showed that our approach can resolve inconsistency and abnormality problems.

References

1. http://www.cs.washington.edu/research/xmldatasets
2. http://www.oracle.com/database/berkeley-db/
3. http://xmlsoft.org/
4. Bao, Z., Ling, T.W., Chen, B., Lu, J.: Effective XML keyword search with relevance oriented ranking. In: ICDE, pp. 517–528. IEEE, Los Alamitos (2009)
5. Cohen, S., Mamou, J., Kanza, Y., Sagiv, Y.: XSEarch: A semantic search engine for XML. In: VLDB, pp. 45–56 (2003)
6. Guo, L., Shao, F., Botev, C., Shanmugasundaram, J.: XRank: Ranked keyword search over XML documents. In: Halevy, A.Y., Ives, Z.G., Doan, A. (eds.) SIGMOD Conference, pp. 16–27. ACM, New York (2003)
7. Hristidis, V., Koudas, N., Papakonstantinou, Y., Srivastava, D.: Keyword proximity search in XML trees. IEEE Trans. Knowl. Data Eng. 18(4), 525–539 (2006)
8. Li, G., Feng, J., Wang, J., Zhou, L.: Effective keyword search for valuable LCAS over XML documents. In: Silva, M.J., Laender, A.H.F., Baeza-Yates, R.A., McGuinness, D.L., Olstad, B., Olsen, Ø.H., Falcão, A.O. (eds.) CIKM, pp. 31–40. ACM, New York (2007)
9. Li, Y., Yu, C., Jagadish, H.V.: Schema-free XQuery. In: Nascimento, M.A., Özsu, M.T., Kossmann, D., Miller, R.J., Blakeley, J.A., Schiefer, K.B. (eds.) VLDB, pp. 72–83. Morgan Kaufmann, San Francisco (2004)
10. Liu, Z., Chen, Y.: Identifying meaningful return information for XML keyword search. In: Chan, C.Y., Ooi, B.C., Zhou, A. (eds.) SIGMOD Conference, pp. 329–340. ACM, New York (2007)
11. Liu, Z., Chen, Y.: Reasoning and identifying relevant matches for XML keyword search. PVLDB 1(1), 921–932 (2008)
12. Xu, Y., Papakonstantinou, Y.: Efficient keyword search for smallest LCAs in XML databases. In: Özcan, F. (ed.) SIGMOD Conference, pp. 537–538. ACM, New York (2005)

Matching Top-k Answers of Twig Patterns in Probabilistic XML

Bo Ning[1,2], Chengfei Liu[1], Jeffrey Xu Yu[3], Guoren Wang[4], and Jianxin Li[1]

[1] Swinburne University of Technology, Melbourne, Australia
[2] Dalian Maritime University, Dalian, China
{BNing,CLiu,JianxinLi}@groupwise.swin.edu.au
[3] The Chinese University of Hongkong, Hongkong, China
yu@se.cuhk.edu.hk
[4] Northeastern University, Shenyang, China
wanggr@mail.neu.edu.cn

Abstract. The flexibility of XML data model allows a more natural representation of uncertain data compared with the relational model. The top-k matching of a twig pattern against probabilistic XML data is essential. Some classical twig pattern algorithms can be adjusted to process the probabilistic XML. However, as far as finding answers of the top-k probabilities is concerned, the existing algorithms suffer in performance, because many unnecessary intermediate path results, with small probabilities, need to be processed. To cope with this problem, we propose a new encoding scheme called *PEDewey* for probabilistic XML in this paper. Based on this encoding scheme, we then design two algorithms for finding answers of top-k probabilities for twig queries. One is called *ProTJFast*, to process probabilistic XML data based on element streams in document order, and the other is called *PTopKTwig*, based on the element streams ordered by the path probability values. Experiments have been conducted to study the performance of these algorithms.

1 Introduction

Uncertainty is inherent in many real applications, and uncertain data management is therefore becoming a critical issue. Unfortunately, current relational database technologies are not equipped to deal with this problem. Compared with the relational data model, the flexibility of XML data model allows a more natural representation of uncertain data. Many data models for probabilistic XML (PXML) have been studied in [4,5,6,7]. The queries on the probabilistic XML are often in the form of twig patterns. When querying probabilistic data, we have to compute the answers as well as the probability values of the answers. Three kinds of twig queries (B-Twig, C-Twig, and I-Twig) with different semantics were proposed, and their evaluations were studied in [8]. The paper [3] studied the query ranking in probabilistic XML by possible world model, and a dynamic programming approach was deployed that extends the dynamic programming approach of query ranking on uncertain relational data [9] to deal with the containment relationships in probabilistic XML.

H. Kitagawa et al. (Eds.): DASFAA 2010, Part I, LNCS 5981, pp. 125–139, 2010.

In this paper, we focus on the problem of efficiently finding twig answers with top-k probability values against probabilistic XML by using stream-based algorithms. Our data model for PXML is similar to PrXML$^{\{ind, mux\}}$ model in [7], in which the independent distribution and mutually-exclusive distribution are considered. We find that an effective encoding scheme for probabilistic XML requires new properties such as the path probability vision and ancestor probability vision. In addition, the encoding should also reflect the probability distribution information for handling, say, mutually-exclusive distribution. In this paper, we propose a new Dewey encoding scheme called *PEDewey* to meet these requirements.

Most of twig matching algorithms [12,2,13] for ordinary XML are based on the element streams ordered by the document order. They can be adjusted to process the probabilistic XML. However, for finding answers of the top-k probabilities, these algorithms suffer in performance, because many unnecessary computations are spent on elements and paths with small probabilities which may not contribute to answers. To improve the performance, we propose an algorithm called *PTopKTwig* which is based on the element streams ordered by the path probability values. For comparison purpose, we also propose an algorithm called *ProTJFast* based on document order. There are two definitions in ranking the top-k query results from uncertain relational databases. One definition [11] is ranking the results by the interplay between score and uncertainty. The work in [3] falls into this category. The other is to find the k most probable answers [10]. In this scenario, each answer has a probability instead of a score, which intuitively represents the confidence of its existence, and ranking is only based on probabilities. Our work falls into this category. As far as we know, there is no other work on ranking the top-k query results for PXML in this category.

2   Background and Problem Definition

2.1   Probabilistic XML Model

An XML document can be modeled as a rooted, ordered, and node-labeled tree, $T(V, E)$, where V represents a set of XML elements, and E represents a set of parent-child relationships (edges) between elements in XML. A probabilistic XML document T_P defines a probability distribution over an XML tree T and it can be regarded as a weighted XML tree $T_P(V_P, E_P)$. In T_P, $V_P = V_D \cup V$, where V is a set of ordinary elements that appear in T, and V_D is a set of distribution nodes, including independent nodes and mutually-exclusive nodes (*ind* and *mux* for short). An ordinary element, $u \in V_P$, may have different types of distribution nodes as its child elements in T_P that specify the probability distributions over its child elements in T. E_P is a set of edges, and an edge which starts from a distribution node can be associated a positive probability value as weight. Notice that, we can regard the probability of ordinary edges as 1. For example, in Figure 1, (a) is an ordinary XML document, and (b) is a probabilistic XML, which contains *ind* and *mux* nodes.

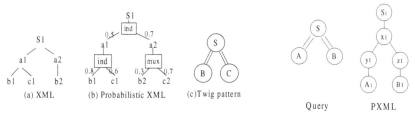

Fig. 1. Probabilistic XML   Fig. 2. Ancestor probability vision

2.2   Problem Statement

The answer of a twig query q with n nodes against an ordinary XML document is
a set of tuples. In each tuple, there are n elements from the XML document, and
those elements match the nodes in q and satisfy all the structural relationships
specified in q. Once the set of answers are obtained, we also need to evaluate
the probability associated with each answer. Given an answer expressed by a
tuple $t = (e_1, e_2, ..., e_n)$, there exist a subtree $T_s(V_s, E_s)$ of T_P, which contains
all those elements. The probability of t can be computed by the probability of
T_s using the probability model for independent events, as mutually-exclusive
distribution has been considered in determining the answer set. The probability
of T_s can be deduced by all the edges of T_s by the equation: $prob(t) = prob(T_s) = \Pi_{e_i \in E_s} prob(e_i)$. For example, in Figure 1, there are three answers of twig pattern
(c): $t_1(s_1, b_1, c_1)$, $t_2(s_1, b_1, c_2)$ and $t_3(s_1, b_2, c_1)$. The probability of answer t_1 is 0.24
(0.5*0.8*0.6). The tuple (s_1, b_2, c_2) is not an answer, because b_2 and c_2 are child
elements of a mux node.

We tackle the problem of finding the top-k matchings of a twig pattern against
a probabilistic XML document. This is defined by the top-k answers in terms of
their probability values. Given a twig query q, and probabilistic XML document
T_P, the answer set S_q contains all the matching results of q which satisfy both
structural relationships and the mutually-exclusive distribution specified in q.
The problem is to find the top-k answer set S_{TopK} which contains k tuples, and
for each tuple $t_i \in S_{TopK}$, its probability value is no less than that of any tuple
$t_j \in S_q \setminus S_{TopK}$.

For example, the probability values of the three answers of pattern (c) against
probability XML (b) in Figure 1 are $prob(t_1)(0.24)$, $prob(t_2)(0.196)$ and $prob(t_3)$
(0.063), respectively. Assume we find the top-2 answers, then t_1 and t_2 should
be returned.

3   Encoding Scheme for Probabilistic XML

3.1   Required Properties of Encoding for PXML

There are two kinds of encoding schemes for ordinary XML documents, region-
based encoding[1] and prefix encoding. Both encoding schemes support the structural relationships and keep the document order, and these two requirements are
essential for evaluating queries against ordinary XML documents.

As to the query evaluation against PXML, a new requirement of encoding appears, which is to record some probability values of elements. Depending on different kinds of processing, we may use the probability value of the current element which is under a distribution node in PXML (node-prob for short), or the probability value of the path from the root to the current element (path-prob for short). In fact, we may need both node-prob and path-prob in twig pattern matching against PXML. If only the node-prob value of current element is recorded, then during the calculation of the probability value of an answer, the node-prob values of the current element's ancestors are missing. Similarly, if only the path-prob value of the current element is used, it is easy to get the probability of a path, however, for the probability of a twig query answer, the path-prob value of the common prefix is also needed, but can not be found. For example, in Figure 2, the twig query is $S[//A]//B$, and the probability value of the answer $t_1:(S_1,A_1,B_1)$ can be calculated by the following formula, where the $pathProb(x_1)$ is calculated from the node-probs recorded in element A_1 or B_1.

$$prob(t_1) = pathProb(A_1) * pathProb(B_1)/pathProb(x_1);$$

The prefix encoding scheme owns the ideal property for supporting ancestor vision while a region-based encoding scheme does not. Therefore, it is better to encode PXML elements by a prefix encoding scheme. However, to match a twig pattern against a probabilistic XML document without accessing large number of ancestor elements, we also need to provide the ancestor probability vision in the encoding scheme.

3.2   PEDewey: Encoding PXML

Lu et al. proposed a prefix encoding scheme named extended Dewey [2]. In this paper, for the purpose of supporting twig pattern matching against probabilistic XML, we extend Lu's encoding scheme by adding the Properties of the *probability vision* and the *ancestor probability vision*, and propose a new encoding scheme called *PEDewey*.

Extended Dewey is a kind of Dewey encoding, which use the modulus operation to create a mapping from an integer to an element name, so that given a sequence of integers, it can be converted into the sequence of element names. Extended Dewey needs additional schema information about the child tag set for a certain tag in the DTD of an ordinary XML document. For example for the DTD in Figure 3 (a), ignoring the distribution nodes, tag A has $\{C,D,E\}$ as the child tag set. The child tag set of g is expressed as $CT(g)=g_0,g_1,...,g_{n-1}$. For any element e_i with tag name g_i, an integer x_i is assigned such that x_i mod $n = i$. Therefore, the tag name can be derived according to the value of x_i. By the depth-first traversal of the XML document, the encoding of each element can be generated. The extended Dewey encoding is an integer vector from the root to the current element, and by a Finite State Transducer, it can translate the encoding into a sequence of element names. The finite state transducer for the DTD in Figure 3(a) is shown in Figure 3(b).

Based on the extended Dewey, we propose a new encoding scheme named *PEDewey* for providing the probability vision and the ancestor probability vision. Given an element u, its encoding label(u) is defined as label$(s).x$, where s is the parent of u. (1) if u is a text value, then $x = -1$; (2) if u is an *ind* node, then $x = -2$; (3) if u is a *mux* node, then $x = -3$. (4)otherwise, assume that the tag name of u is the *k-th* tag in $CT(g_s)(k = 0, 1, ..., n\text{-}1)$, where g_s denotes the tag name of the parent element s. (4.1) if u is the first child of s, then $x = k$; (4.2) otherwise assume that the last component of the label of the left sibling of u is y, then

$$x = \begin{cases} \lfloor \frac{y}{n} \rfloor \cdot n + k & \text{if } (y \bmod n) < k; \\ \lceil \frac{y}{n} \rceil \cdot n + k & \text{otherwise.} \end{cases}$$

where n denotes the size of $CT(t_s)$.

PEDewey behaves the same as extended Dewey when judging an ancestor-descendant (or prefix) relationship between two elements by only checking whether the encoding of one element is the prefix of the other. However, *PEDewey* is different from extended Dewey when judging a parent-child (or tight prefix) relationship of two elements u and v. The condition label(u).*length* - label(v).*length* = 1 is checked by ignoring those components for distribution nodes in the *PEDewey* encodings.

In *PEDewey*, an additional float vector is assigned to each element compared with extended Dewey. The length of the vector is equal to that of a normal Dewey encoding, and each component holds the probability value of the element. From the encoding, the node-prob value of ancestors are recorded so the path-prob value of the current element and its ancestors can be easily obtained. The components for elements of ordinary, *ind* and *mux* are all assigned to 1.

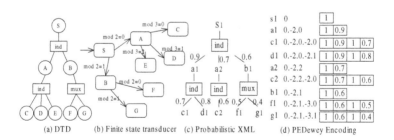

Fig. 3. Example of *PEDewey* encoding

We define some operations on the float vector. (1) Given element e, function *pathProb(e)* returns the path-prob of element e, which is calculated by multiplying the node-prob values of all ancestors of e in the float vector (2) Given element e and its ancestor e_a, function *ancPathProb(e, e_a)* returns the path-prob of e_a by multiplying those components from the root to e_a in e's float vector. (3) Given element e and its ancestor e_a, function *leafPathProb(e, e_a)* returns the path-prob of the path from e_a to e by multiply those components from e_a to e in

e's float vector. (4) Given elements e_i and e_j, function $twigProb(e_i, e_j)$, returns the probability of the twig whose leaves are e_i and e_j. Assume the e_i and e_j have common prefix e_c, and the probability of twig answer containing e_i and e_j is:

$$twigProb(e_i, e_j) = pathProb(e_i) * pathProb(e_j)/ancPathProb(e_i, e_c);$$

4   ProTJFast Algorithm Based on Document Order

4.1   Data Structures and Notations

For a twig query pattern q, a path pattern from the root to a leaf node f is denoted as p_f. We associate each leaf node f in q with a stream T_f, which contains the *PEDewey* encodings of elements with tag f. The elements are sorted by the ascending lexicography order, which is the document order of those elements in PXML document. The operations on the stream are *eof*, *advance*, and *get*.

Similar to *TJFast*, a set S_b is associated with each branching node b in query q in *ProTJFast*. In S_b, every two elements have an ancestor-descendant or a parent-child relationship. Each element cached in S_b may participate in the final query answers. Initially all the sets for branching nodes are empty. A list L_c for top-k candidates is associated with query q, and the function *lowerBound()* return the lowest probability value among those in L_c.

4.2   ProTJFast

We extend the twig matching algorithm *TJFast* to *ProTJFast*, which generates the twig answers with top-k probability values against probabilistic XML. The algorithm is presented in Algorithm 1. In *ProTJFast*, we need to generate twig answers from path answers as early as possible so that we can determine and then raise the lower bound for a top-k twig query. The point behind this is that we could effectively use the lower bound for filtering unnecessary computations.

In the main procedure of *ProTJFast* shown in Algorithm 1. Firstly, in each leaf stream, we find the first element whose encoding matches the individual root-leaf path pattern of q (Lines 2-3). Then we call function $getNext(q)$ to get the tag which is to be precessed next (Line 5). There are two tasks in function $getNext(q)$. The first task is to return the tag f_{act}, such that the head element e_{act} of $T_{f_{act}}$ has the minimal encoding in document order among the head elements of all leaf streams, and the second task is to add the element of branching node which is the ancestor of $get(T_{f_{act}})$ to the set S_b. The set S_b records the information of elements processed previously, therefore by the function $isAnswerOfTwig(q_{act})$, we can determine whether e_{act} can contribute to a twig answer with elements processed previously. If e_{act} is a part of twig answer (Line 6), we compute the twig answers which contains e_{act} by invoking function $mergeJoin(f_{act}, q)$, and store these twig answers in set S_{temp} (Line 7). Then we can update the lower bound by invoking function $updateLowerBound(S_{temp})$ (Line 8). Then we move the head element in stream $T_{f_{act}}$ to the next one whose

---

Algorithm 1. $ProTJFast(q)$

Data: Twig query q, and streams T_f of the leaf node in q.
Result: The matchings of twig pattern q with top-k probabilities.

1 **begin**
2   **foreach** $f \in leafNodes(q)$ **do**
3        $locateMatchedLabel(f)$;

4   **while** $\neg end(q)$ **do**
5        $f_{act} = getNext(q)$;
6   **if** $isAnswerOfTwig(f_{act})$ **then**
7           $S_{temp} = mergeJoin(f_{act}, q)$;
8           $updateLowerBound(S_{temp})$;
9           $locateMatchedLabel(f_{act})$;

10    $outputTopKSolutions()$;
11 **end**
12 **Procedure** $locateMatchedLabel(f)$
13 **begin**
14   **while** $get(T_f)$ *do not matchs pattern* p_f **do**
15       $advancebybound(T_f)$;

16 **end**
17 **Procedure** $advancebybound(T_f)$
18 **begin**
19   **while** $pathProb(get(T_f)) < lowerBound()$ **do**
20       $advance(T_f)$;

21 **end**
22 **Function** $end(q)$
23 **begin**
24   Return $\forall f \in leafNodes(q) \rightarrow eof(T_f)$;
25 **end**
26 **Function** $mergeJoin(f_{act}, q)$
27 **begin**
28    $e_{act} = get(T_{f_{act}})$;
29   **foreach** e_i *in the set of intermediate results* **do**
30       $e_{com} = commaonprefix(e_{act}, e_i)$;
31   **if** (e_{com} *matches all the branching nodes in* p_{act}) **then**
32   **if** (e_{com} *is not a mux node*) **then**
33   add the twig answers $[e_{act}, e_i]$ to temp set S_{temp};

34   return S_{temp};
35 **end**
36 **Procedure** $updateLowerBound(S_{temp})$
37 **begin**
38   **foreach** $m_i \in S_{temp}$ **do**
39   **if** $twigProb(m_i) > lowerbound$ **then**
40   Update the candidate list L_c to keep the present twig answers with top-k probabilities;

41   lowerbound=$\min(twigProb(c_i))$, $c_i \in L_c$;
42 **end**
43 **Function** $isAnswerOfTwig(f)$;
44 **begin**
45    $e = get(T_f)$;
46   return $\forall$ e_b which is prefix of e, and of of branching node$b$, $\rightarrow e \in S_b$;
47 **end**
48 **Function** $end(q)$
49 **begin**
50   Return $\forall f \in leafNodes(q) \rightarrow eof(T_f)$;
51 **end**

Algorithm 2. $getNext(n)$

---

1 **begin**
2   **if** *(isLeaf(n))* **then**
3   return n;
4   **else**
5   **if** *(n has only one child)* **then**
6   return $getNext(child(n))$ **else**
7   **for** $n_i \in children(n)$ **do**
8              $f_i = getNext(n_i)$;
9              $e_i = max\{p|p \in MatchedPrefixes(f_i, n)\}$;
10   max= $maxarg_i\{e_i\}$;
11   min= $minarg_i\{e_i\}$;
12   **forall** $e \in MatchedPrefixes(f_{min}, n)$ **do**
13   **if** $pathProb(e)\text{¿}lowerBound() \wedge e$ is a prefix of e_{max} **then**
14                 $moveToSet(S_n, e)$;
15   return n_{min};
16 **end**
17 **Function** $MatchedPrefixes(f, b)$
18 Return a set of element p that is an ancestor of get(T_f) such that p can match node b in the path solution of get(T_f) to path pattern p_f.
19 **Procedure** $moveToSet(S, e)$
20 Delete any element in S that has not ancestor-descendant (or parent-child) relationship with e;
21 Add e to set S_b;

---

encoding matches the individual root-leaf path pattern $p_{f_{act}}$ and the path probability is larger than the lower bound (Line 9). When the head element in any leaf stream reaches to the end, the answers with top-k probabilities are found.

Lines 32-33 in Algorithm 1 deal with the *mux* node in probabilistic XML. Firstly we get the common prefix of two path answers, and check whether the element of common prefix is a *mux* node (when the encoding of common prefix ends at -3). If so, these two path answers can not be merged into the twig pattern, because only one element among those elements under a *mux* node can appear.

For example, assume that given twig query q_1: S[//C]//D against probabilistic XML in Figure 4, and the answers with top-2 probabilities are required. Because Algorithm 1 is based on document order, firstly, two answers (c_1, d_1) and (c_2, d_1) are matched, and the initial lower bound is set to 0.512, which is the twig probability of (c_2, d_1). At this moment, (c_1, d_1) and (c_2, d_1) are the candidate answers in candidate list. The head elements in streams T_C and T_D are c_3 and d_2. Because the $pathProb(d_2)$ 0.49 is smaller than lower bound 0.512, d_2 can not contribute to the final answers definitely, and algorithm advance T_D to next element d_3 directly. There is no matching with c_3, therefore head element in T_C is advanced to c_4. c_4 and d_3 match twig answer and $twigProb(c_4, d_3)$ is larger than the current lower bound, so (c_4, d_3) is added to candidate list, and

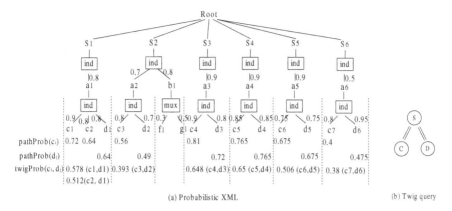

(a) Probabilistic XML   (b) Twig query

Fig. 4. Example of *ProTJFast*

the lower bound is updated to 0.578. When the head elements in streams are c_7 and d_6, the common ancestor is a_6, and the probability of a_6 is smaller than the lower bound, therefore the a_6 is not added to the set of branching node, and c_7 and d_6 can not contribute the final answers definitely. Once any stream reaches the end, the twig answers (c_4, d_3) and (c_5, d_4) with top-2 probabilities are returned. Notice that, if we query q_2: S[//F]//G, the elements pair (f_1, g_1) is not the answer, because their common prefix is *mux* node, and f_1 and g_1 can not appear in the XML document simultaneously.

5   PTopKTwig Algorithms Based on Probability Order

Intuitively the element with larger path probability value will more likely contribute to the twig answers with larger twig probability values. Keeping this idea in mind, we propose a new algorithm called *PTopKTwig* to deal with the top-k matching of twig queries against probabilistic XML based on the probability value order.

5.1   Data Structure and Notations

Similar to the data structure of algorithm *ProTJFast*, we also associate each leaf node f in a twig pattern q with a stream T_f, which contains *PEDewey* encoding of all elements that match the leaf node f. The elements in the stream are sorted by their path-prob values. It is very fast to sort those elements by using the float vector in *PEDewey* encodings. A list L_c for keeping top-k candidates is also allocated for q, and variable *lowerBound* records the lowest probability value of the twig answer among those in L_c. We maintain *cursorList*, a list pointing to the head elements of all leaf node streams. Using the function cursor(f), we can get the position of the head element in T_f.

5.2 Algorithm PTopKTwig

In the main algorithm of *PTopKTwig*(Algorithm 3), Lines 2-7 are used to find the initial k answers so that an initial lower bound can be obtained. In Lines 8-15, the rules of filtering by lower bound are used. When a new candidate is obtained, the lower bound is adjusted. Algorithm 3 proceeds in the probability order of all the leaf nodes in query q, by calling the function *getNextP()*. This function returns the tag name of the leaf node stream which has the biggest probability value in its head element among all leaf node streams. As such, each processed element will not be processed again, and the cursor list records the head elements to be processed next for all leaf node streams.

After function *getNextP()* returns a tag q_{act}, we may find new candidates which the head element in $T_{q_{act}}$ contributes to, by invoking function *matchTwig()*. Function *matchTwig()* has an argument $bFlag$, which determines whether the filtering rule based on the enhanced lower bound (see Section 5.3) needs to be applied. When we try to find the initial lower bound, there is not filtering rule used in function *matchTwig()*, so the argument is *"noBound"*.

During the process of finding other elements that contribute to the twig answers with $e_{q_{act}}$, there is no duplicated computation of comparing the prefixes, due to the order of probability values and the use of *cursorList*. The *cursorList* records the head elements in respective streams which is next to be processed. The elements before the head elements have been compared with elements in other streams, and the twig answers that these elements might contribute to have been considered. Therefore we only compare $e_{q_{act}}$ with the elements after the head elements in the related streams (Lines 3-4 in Algorithm 4).

5.3 Enhanced Lower Bounds

After getting the initial lower bound (Line 7 in Algorithm 3), we can get the lower bound for every stream, which is called the enhanced lower bound and is defined below:

Definition 1. *Enhanced Lower Bound of Stream*
Given a query q, leaf node stream T_f where $f \in leafNodes(q)$ and the lower-bound which is the probability value of the k-th twig answer, the enhanced lower bound of T_f is defined as $lowerBound_f$.

$$lowerBound_f = \frac{lowerBound}{\Pi max(predProb(f_i, f))}, (f_i \in leafNodes(q) \wedge f_i \neq f.)$$

We define a non-overlapping path of f_i relative to f as the path from the common ancestor of f_i and f to f_i which is denoted as $predPath(f_i,f)$. The function $predProb(f_i,f)$ returns a set of probability values of all instances of the non-overlapping path $predPath(f_i,f)$. In the above formula, the maximum value of the set $predProb(f_i,f)$ is selected for each f_i. The k-th probability value of L_c is the common lower bound for all the streams. Because $0 < \Pi$ $max(predProb(f_i,f)) \leq 1$, the enhanced lower bound is always larger than the

---

Algorithm 3. $PTopkTwig(q)$

Data: Twig query q, and streams T_f of the leaf node in q.
Result: The matchings of twig pattern q with top-k probabilities.

1 **begin**
2 $\quad$ **while** $length(CandidatesList) < k \land \neg\ end(q)$ **do**
3 $\quad\quad$ $q_{act}{=}getNextP(q)$;
4 $\quad\quad$ $tempTwigResults{=}matchTwig(q_{act}, q,\ "noBound")$;
5 $\quad\quad$ $add(L_c, tempTwigResults)$;
6 $\quad\quad$ $advanceCursor(cursor(q_{act}))$;

7 $\quad$ $lowerBound{=}twigProb(CandidateList[k])$;
8 $\quad$ **while** $\neg reachEnhancedBound(q)$ **do**
9 $\quad\quad$ $q_{act}{=}\text{getNextP}(q)$;
10 $\quad\quad$ $tempTwigResults{=}matchTwig(q_{act}, q,\ "withBound")$;
11 $\quad\quad$ $add(L_c, tempTwigResults)$;
12 $\quad\quad$ $lowerBound{=}twigProb(L_c[k])$;
13 $\quad\quad$ $advanceCursor(cursor(q_{act}))$;

14 $\quad$ Output k twig answers from candidates list L_c;
15 **end**
16 **Function** $end(q)$
17 **begin**
18 $\quad$ Return $\forall f \in leafNodes(q) \to eof(T_f)$;
19 **end**
20 **Function** $reachEnhancedBound(q)$
21 **begin**
22 $\quad$ $flag = true$;
23 $\quad$ **foreach** $q_i \in leafNodes(q)$ **do**
24 $\quad\quad$ $lowerbound_{q_i} = lowerbound/\Pi\ max(predProb(q_j, q_i));\ (q_j \in$
   $\quad\quad leafNodes(q) \land q_j \neq q_i)$;
25 $\quad\quad$ **if** $lowerbound_{q_i} > probpath(get(T_{q_i}))$ **then**
26 $\quad\quad\quad$ $flag = False$;

27 $\quad$ Return $flag$;
28 **end**
29 **Function** $getNextP(n)$
30 **begin**
31 $\quad$ **foreach** $q_i \in leafNodes(q)$ **do**
32 $\quad\quad$ $e_i = get(T_{q_i})$;

33 $\quad$ $max = maxarg_i(e_i)$;
34 $\quad$ return n_{max}
35 **end**

Algorithm 4. $matchTwig(q_{act}, q, bFlag)$

---

1 **begin**
2   **for** *any tags pair $[T_{q_a}, T_{q_b}]$ ($q_a, q_a \in leafNodes(q) \wedge q_a, q_b \neq q_{act}$)* **do**
3   Advance head element in T_{q_a} to the position of cursor(q_a);
4   Advance head element in T_{q_b} to the position of cursor(q_b);
5   **while** *(bFlag = "noBound" $\wedge \neg end(q)) \vee$ (bFlag = "withBound" $\wedge \neg$ reachEnhancedBound(q))* **do**
6   **if** *elements e_{q_a}, e_{q_a} match the common path pattern with $e_{q_{act}}$ in query q, and the common prefix of e_{q_a}, e_{q_a} match the common path pattern which is from the root to the branching node q_{bran} of q_a and q_b in query q, and the common prefix is not a element of mux node.* **then**
7   add e_{q_a}, e_{q_a} to the set of intermediate results.

8   return twig answers from the intermediate set.
9 **end**

---

common lower bound, i.e., the lower bound for T_f can be raised by considering non-overlapping paths from all other streams.

We apply the enhanced filtering rule based on the enhanced lower bound. Firstly, in Line 8 of the main algorithm (Algorithm 3), if the probability of any q_{act} is smaller than the enhanced lower bound in the corresponding streams, the algorithm stops. Secondly, during the process of matching twig answers by invoking $matchTwig()$ with $bFlag$ as "$withBound$", the elements with probabilities lower than the enhanced lower bound in the corresponding stream do not participate the comparison with the head element in $T_{q_{act}}$ in main algorithm.

During the process of calculating of the probability value of a twig answer that $e_{q_{act}}$ contributes to, for a leaf element e_{q_p} from the another stream T_{q_p}, if $leafPathProb(e_{q_p}, prefix(e_{q_p}, e_{q_{act}})) * pathProb(e_{q_{act}})$ is smaller than the lower bound, we can see that e_{q_p} can not contribute to a twig answer with top-k probabilities. Notice that, an enhanced lower bound in a leaf stream increases as the common lower bound increases.

For the same twig query q_1: S[//C]//D against probabilistic XML in Figure 4, assume again that the answers for top-2 probabilities are required. In Algorithm 3, streams T_C and T_D are scanned, and the elements in streams are sorted by path-prob values shown in Figure 5. The processing order of elements in streams are marked by dotted arrow line in Figure 5, which is obtained by invoking the $getNextP()$ function. Firstly, we find the initial lower bound. Tag C is returned by $getNextP()$ and c_4 is the head element in T_C, and is then used to find elements in T_D with which twig answers can be matched. Because the elements in T_D are sorted by path-prob values too, one answer (c_4, d_3) is found. We continue processing unprocessed elements with largest probability in all streams, until the initial two twig answers (c_5, d_4) and (c_4, d_3) are found, and the lower bound 0.648 is obtained, which is the probability of (c_5, d_4). From the lower bound, we can easily get the enhance lower bounds for T_C which is 0.72 and T_D which

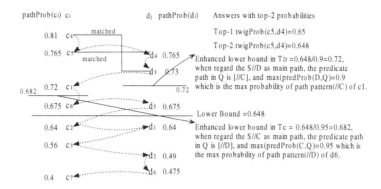

Fig. 5. Example of PTopKTwig

is 0.682. The enhanced lower bounds can be used to filter elements during the process of finding twig answers from a given element or to stop the process. For example, many elements (such as c_6, c_2, c_3, c_7, d_5, d_1, d_2, and d_6) need not to be processed. When the next element c_6 of T_C is to be processed, the path-prob value of c_6 is smaller than the enhanced lower bound of T_C, the algorithm stops, and outputs the top-2 answers (c_5, d_4) and (c_4, d_3) in the candidate list.

6   Experiments

6.1   Experimental Setup

We implemented Algorithms *ProTJFast* and *PTopKTwig* in JDK 1.4. All our experiments were performed on a PC with 1.86GHz Inter Pentium Dual processor and 2GB RAM running on Windows XP. We used both real-world data set DBLP and synthetic data set generated by IBM XML generator and a synthetic DTD. To generate the corresponding probabilistic XML documents, we inserted distribution nodes to the ordinary XML document and assigned probability distributions to the child elements of distribution nodes. The queries are listed in Table 1. To compare the performance between *ProTJFast* and *PTop-KTwig*, we used the metrics elapsed time and processed element rate $rate_{proc}$ $=num_{proc}/num_{all}$, where num_{proc} is the number of processed elements, and num_{all} is the number of all elements.

Table 1. Queries

ID	DBLP queries	ID	synthetic data queries
Q_1	dblp//article[//author]//title	Q_4	S//[//B][//C][//D]//A
Q_2	S//[//B]//A	Q_5	S//[//B][//C][//D][//E]//A
Q_3	S//[//B][//C]//A		

6.2   Performance Study

Influence of Number of Answers: We evaluated Q_1 against the DBLP data set of size 110MB by varying k from 10 to 50. Figures 6 and 7 show that when k increases, the elapsed time and the rate of processed elements of both algorithms increase as well. When k is small, the performance of $PTopKTwig$ is much better than $ProTJFast$. This is because that the smaller k is the better the enhanced lower bound is. When k becomes big, the enhanced lower bound degrades. From the figures, the elapsed time and the rate of processed elements of $PTopKTwig$ increases faster, though the performance is still better than $ProTJFast$. However, for a top-k query, most likely k keeps relatively small, so $PTopKTwig$ performs better than $ProTJFast$.

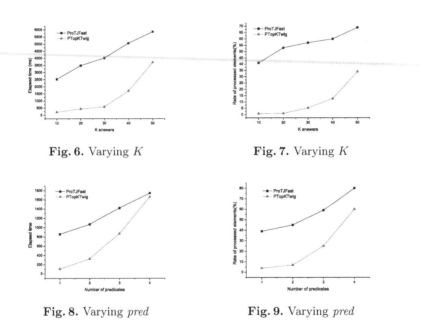

Fig. 6. Varying K **Fig. 7.** Varying K

Fig. 8. Varying $pred$ **Fig. 9.** Varying $pred$

Influence of Multiple Predicates: We evaluated the queries Q_2 to Q_5 on the synthetic data set, to test the influence of multiple predicates. The fan-out of these queries varies from 2 to 5. The results are shown in Figures 8 and 9. In Figure 8, the elapsed time of both algorithms increases when the fan-out increases. The situation is similar in Figure 9 when testing the rate of processed elements. As the number of predicates increases, the lower bound which is the value of k-th twig answer becomes smaller, therefore the elapsed time and the rate of processed elements of both algorithms increase. As to $PTopKTwig$, besides the reason of the small lower bound, another reason is that the matching of multiple leaf elements takes more time compared with streams in the document order. However, due to the enhanced lower bound, the rate of processed elements of $PTopKTwig$ is always smaller than that of $ProTJFast$.

7   Conclusions

In this paper, we studied how to find top-k matching of a twig pattern against probabilistic XML data. Firstly, we discussed the required properties for PXML encoding and proposed *PEDewey* - a new encoding scheme for PXML based on extended Dewey. Then we introduced two algorithms *ProTJFast* and *PTopK-Twig* which are based on *PEDewey*. The element streams in *ProTJFast* is by the document order, while the element streams in *PTopKTwig* is by the probability value order. Finally we presented and discussed experimental results on a range of real and synthetic data.

Acknowledgement. This research was supported by the Australian Research Council Discovery Project (Grant No. DP0878405), the National Natural Science Foundation of China (Grant No. 60873026,60773219, 60933001), the 863 Program (Grant No. 2007AA01Z192, 2009AA01Z131), and National Basic Research Program of China (Grant No. 2006CB303103).

References

1. Zhang, C., Naughton, J., DeWitt, D., Luo, Q., Lohman, G.: On Supporting Containment Queries in Relational Database Management Systems. In: Proceeding of SIGMOD, pp. 425–436 (2001)
2. Lu, J., Ling, T.W., Chan, C.-Y., Chen, T.: From region encoding to extended dewey: On efficient processing of XML twig pattern matching. In: Proceeding of VLDB, pp. 193–204 (2005)
3. Chang, L., Yu, J.X., Qin, L.: Query Ranking in Probabilistic XML Data. In: Proceeding of EDBT, pp. 156–167 (2009)
4. Hung, E., Getoor, L., Subrahmanian, V.S.: PXML: A probabilistic semistructured data model and algebra. In: Proceeding of ICDE, pp. 467–478 (2003)
5. Nierman, A., Jagasish, H.V.: ProTDB: Probabilistic data in XML. In: Proceeding of VLDB, pp. 646–657 (2002)
6. Senellart, P., Abiteboul, S.: On the complexity of managing probabilistic XML data. In: Proceeding of PODS, pp. 283–292 (2007)
7. Kimelfeld, B., Kosharovsky, Y., Sagiv, Y.: Query efficiency in probabilistic XML models. In: Proceeding of SIGMOD, pp. 701–714 (2008)
8. Kimelfeld, B., Sagiv, Y.: Matching twigs in probabilistic XML. In: Proceeding of VLDB, pp. 27–38 (2007)
9. Hua, M., Pei, J., Zhang, W., Lin, X.: Ranking queries on uncertain data: A probabilistic threshold approach. In: Proceeding of SIGMOD, pp. 673–686 (2008)
10. Re, C., Dalvi, N.N., Suciu, D.: Efficient top-k query evaluation on probabilistic data. In: Proc. of ICDE 2007, pp. 886–895 (2007)
11. Soliman, M.A., Ilyas, I.F., Chang, K.C.-C.: Top-k query processing in uncertain databases. In: Proc. of ICDE 2007, pp. 896–905 (2007)
12. Bruno, N., Srivastava, D., Koudas, N.: Holistic twig joins: Optimal XML pattern matching. In: Proceedings of SIGMOD, pp. 310–321 (2002)
13. Qin, L., Yu, J.X., Ding, B.: TwigList: Make Twig Pattern Matching Fast. In: Kotagiri, R., Radha Krishna, P., Mohania, M., Nantajeewarawat, E. (eds.) DASFAA 2007. LNCS, vol. 4443, pp. 850–862. Springer, Heidelberg (2007)

NOVA: A Novel and Efficient Framework for Finding Subgraph Isomorphism Mappings in Large Graphs

Ke Zhu, Ying Zhang, Xuemin Lin, Gaoping Zhu, and Wei Wang

School of Computer Science and Engineering,
University of New South Wales, Australia
{kez,yingz,lxue,gzhu,weiw}@cse.unsw.edu.au

Abstract. Considerable efforts have been spent in studying subgraph problem. Traditional subgraph containment query is to retrieve all database graphs which contain the query graph g. A variation to that is to find all occurrences of a particular pattern(the query) in a large database graph. We call it subgraph matching problem. The state of art solution to this problem is GADDI. In this paper, we will propose a more efficient index and algorithm to answer subgraph matching problem. The index is based on the label distribution of neighbourhood vertices and it is structured as a multi-dimensional vector signature. A novel algorithm is also proposed to further speed up the isomorphic enumeration process. This algorithm attempts to maximize the computational sharing. It also attempts to predict some enumeration state is impossible to lead to a final answer by eagerly pruning strategy. We have performed extensive experiments to demonstrate the efficiency and the effectiveness of our technique.

1 Introduction

In the real world, many complex objects are modelled as graphs. For instance, social networks, protein interaction networks, chemical compounds, World Wide Web, network design, work flows, and etc. It is a fundamental problem to find all data graphs which contain a specific pattern, the query graph q. This problem is well known to be subgraph containment query. There are already a considerable amount of existing work([1], [2], [3], [4], [5], [6], [7]) which study the subgraph containment query. In real life, it is often desirable to find the *occurrences* of a pattern instead of just finding which graphs contain it. For example, in protein-protein interaction(PPI) networks, biologists may want to recognize groups of proteins which match a particular pattern in a large PPI network. That pattern could be a interaction network among a number of protein types. Since each protein type could include a number of distinct proteins, we may find zero, one or more than one possible matches from the PPI network.

For another example, in order to prevent a privacy attack technique described in [8], we can use an efficient subgraph matching algorithm to test the data. In [8], the attacker is able to create a subgraph before a social network database is anonymized and released for research purpose. After the database is released, the attacker will need to find the occurrence of the subgraph he created so that this subgraph becomes the attacker's anchor point. Via this anchor point, he could locate the other vertices that he plans to attack.

The existing techniques proposed for subgraph containment query are focusing on answering whether the database graph ceontains the query pattern or not. GraphGrep[9]

H. Kitagawa et al. (Eds.): DASFAA 2010, Part I, LNCS 5981, pp. 140–154, 2010.

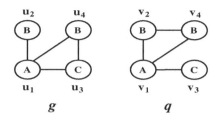

Fig. 1. Example database graph and query graph

could be extended to solve subgraph matching problem. However its construction cost is prohibitively large as shown in GADDI[10].

GADDI([10]), is the state of art technique to solve subgraph matching problem. Zhang *et. al.* proposed an index based on *Neighbourhood Discrimitive Substructures*. It counts the number of small substructures in induced intersection graph between the neighbourhood of two vertices. GADDI performs significantly better than GraphGrep[9] and Tale[11]. However, GADDI also has two major weakness. Firstly, since its index requires pairwise neighbourhood intersection and counting the number of small substructures in each intersection graph, its construction cost is very expensive. Secondly, since counting substructures are expensive, we could only afford to use very few of them to index the graph. In GADDI, only three unlabeled substructures are used. Without labels embedded in the substructures, they normally have limited filtering power. For example, in Fig 1, GADDI is not able to eliminate the possibility of u_1 being a potential match of v_1 because all the index could recognize is that the intersection graphs formed by v_1 with other vertices contains one unlabeled triangle.

In this paper, we propose a novel and efficient framework to solve subgraph matching problem. This framework utilizes a novel index called *nIndex* . In addition, we propose a novel algorithm to enumerate subgraph matchings. Our algorithm, named NOVA , will pre-order the query vertices in a way such that more computational cost could be shared. It will also employ an eagerly pruning strategy which could determine the current enumeration state is impossible to lead to a successful mapping, so that the enumeration process could exit early.

Our main contributions are summarized as follows:

1. We propose a flexible framework for finding subgraph matchings in large graph. This framework is based on a vector signature for each vertex in the graph.
2. We propose a novel vector domination based index(nIndex) which is efficient and effective in terms of index space, construction time and query response time. if a graph vertex is a match of a query vertex, it is a necessary(but not sufficient) condition for the vector of the graph vertex to dominate the vector of the query vertex. We propose a model to compress the index.
3. We propose a novel subgraph matching algorithm which attempts to maximize computational sharing. We will demonstrate a theortical cost model on which our strategy is based. This algorithm uses an eager pruning strategy to avoid expanding unnecessary enumeration state.
4. We perform extensive experiments on both real and synthetic data to show the effectiveness and the efficiency of our techniques.

The rest of the paper is organized as follows. Section 2 presents the problem definition. Section 3 presents our algorithm, namely NOVA . Section 4 introduces the nIndex and index compression. In Section 5, we will discuss how to extend our technique to boost its filtering power when number of labels decreases. We will report the experiment results in Section 6. Section 7 and 8 will discuss the related work and conclude the paper.

2   Definitions

In this section, we will introduce the common notations used in this paper. We will also present the fundamental definition and the problem statement in this section. In this paper, we assume the graphs are only vertex-labeled. However, it is straightforward to extend our techniques to edge-labeled graphs.

We use $V(g)$, $E(g)$, l_g to denote all vertices, all edges, and the labeling function for graph g respectively.

A subgraph matching from q to g is simply an injective relationship which maps a vertex $v \in q$ to a vertex $u \in g$.

Definition 1 (Subgraph Isomorphism Mapping). *Given two graphs* $G = (V(g), E(g), l_g)$ *and* $Q = (V(q), E(q), l_q)$, *an injective function* $f : q \to g$ *is a Subgraph Isomorphism Mapping if and only if:*

1. $\forall u \in V(q), f(u) \in V(g)$ *and* $l(u) = l(f(u))$
2. $\forall(u, v) \in E(q), (f(u), f(v)) \in E(g)$

Definition 2 (Subgraph Matching Query). *Given two graphs* g *and* q, *we need to find all possible subgraph isomorphism mappings.*

Generally, subgraph matching algorithms require prefiltering possible candidates for all vertices in q. In NOVA , we use a k-dimensional vector signature to choose possible candidates.

Definition 3 (Multi-dimensional Vector Signature). *For each* $u \in V(g)$, *we assign it a k-dimensional vector,* $sig(u) = < d_1(u), d_2(u), ..., d_k(u) >$, *as its signature. Each dimension* $sig(u)[d_i]$ *represents a particular characteristic of the vertex.*

Our framework allows any features to be placed in the multi-dimensional vector signature. For example, in the simplest case, we can use a 1-dimensional vector in which the only dimension is the labels. However, in our framework, we require the vector signatures to possess a specific property. We call this class of signatures *Admissible Multi-dimensional Vector Signature*. For simplicity, all multi-dimensional vector signatures appear in this paper are admissible unless explicitly specified otherwise.

Definition 4 (Admissible Multi-dimensional Vector Signature). *A signature is admissible if for any* $u \in V(g)$ *to be a match of* $v \in V(q)$, $sig(u)$ *dominates* $sig(v)$, *denoted by* $sig(u) \succeq sig(v)$. $sig(u)$ *dominates* $sig(v)$ *if: i)* $l(u) = l(v)$; *ii) For each* $d_i \in D$, $sig(u)[d_i] \succeq sig(v)[d_i]$, *where* $\succeq$ *is a partial order function defined on the specific attribute.*

Table 1. Notations

g, q	Database graph and query graph respectively
$V(g)$	Set of all vertices in g
$E(g)$	Set of all edges in g
u, v, w	A single vertex
(u, v)	A single edge with u and v being the end points
$L(G), l_g, l(u)$	the label domain of database G, the labeling function of g, the label of u, respectively
$sig(u)$	the multi-dimensional vector signature of u
$d_i(u)$	the value of ith dimension of $sig(u)$

In the paper, we will only use integers for all dimensions so that the $\succeq$ is naturally defined. It is straightforward to extend the definition to any customized partial order function.

3   Framework

Briefly, our framework is composed of following parts:

1. We preprocess the database graph g to construct a multi-dimensional vector signature for each vertex in the graph.
2. We will compress the original k-dimensional vector into a smaller m-dimensional vector by choosing the most selective non-overlapping dimensions. This step is optional.
3. For a given query graph q, we construct its signatures by using the same method. We test all vertices $u \in V(q)$ against vertices $v \in V(g)$ to obtain a candidate list for each u. A potential candidate v for u must satisfy two criteria: 1) $l(u) = l(v)$; 2) $sig(v) \succeq sig(u)$;
4. After obtaining a candidate list for all vertices in $V(q)$, we order the vertices in a way which could potentially maximize computational sharing.

4   Index

4.1   Index

In this section, we will propose an index scheme called nIndex based on label distribution. nIndex will be fully integrated into the vector domination model we proposed in Section 2. To index one vertex v, we need a user specified radius parameter r_{max}. For all integers r, where $0 < r \leq r_{max}$, we count the number of distinct length r simple paths who start from v and end at a vertex with label l.

Definition 5 (nIndex). *For a vertex u and a given value r_{max}, we define the nIndex signature of u to be:* $sig(v) = \{(r, l, count_{r,l}(v)) \,|\, 0 < r \leq r_{max} \wedge l \in l_g\}$. $count_{r,l}(v)$ *is the number of length r distinct simple paths with one end being v and the other end being a vertex of label l.*

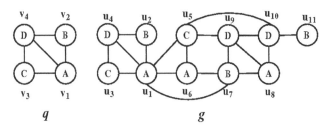

Fig. 2. A Running Example

Table 2. nIndex for figure 2 ($r = 2$)

(a) Index for q

Node	nIndex
v_1	(1, B, 1), (1, C, 1), (1, D, 1), (2, B, 1), (2, C, 1), (2, D, 2)
v_2	(1, A, 1), (1, D, 1), (2, A, 1), (2, C, 2), (2, D, 1)
v_3	(1, A, 1), (1, D, 1), (2, A, 1), (2, B, 2), (2, D, 1)
v_4	(1, A, 1), (1, B, 1), (1, C, 1), (2, A, 2), (2, B, 1), (2, C, 1)

(b) Index for g

Node	nIndex
u_1	(1, A, 1), (1, B, 2), (1, C, 2), (1, D, 1), (2, A, 3), (2, B, 2), (2, C, 2), (2, D, 5)
u_2	(1, A, 1), (1, D, 1), (2, A, 2), (2, B, 1), (2, C, 3), (2, D, 1)
u_3	(1, A, 1), (1, D, 1), (2, A, 2), (2, B, 3), (2, C, 1)
u_4	(1, A, 1), (1, B, 1), (1, C, 1), (2, A, 3), (2, B, 2), (2, C, 2)
u_5	(1, A, 2), (1, D, 2), (2, A, 4), (2, B, 4), (2, D, 3)
u_6	(1, A, 1), (1, B, 1), (1, C, 1), (2, A, 2), (2, B, 2), (2, C, 1), (2, D, 4)

Table 2 shows part of the index generated for Fig. 2. The rationale of using label distribution of distinct paths is because that it can partially reflect the neighbouring structrual characteristic. For example, adding an edge between u_2 and u_3 will have a significant effect on the signatures of q.

The space complexity for this index is $O(n) = kr_{max}n$ where $k = |L(G)|$, $n = |V(G)|$, r_{max} is the specified radius. However, for each vertex, there are many entries whose *count* is 0. These zero-entries are not stored. In next section, we will also demonstrate how to reduce the size of the index.

In NOVA , we found $r_{max} = 2$ is generally a good balance in terms of filtering power and construction efficiency.

Theorem 1. *In nIndex , $sig(u) \succeq sig(v)$ is a necessary condition if u is a matching of v in any isomorphism mapping.*

Proof. Let an arbitary isomorphism function be f, for any vertex $v \in V(q)$, we have a u where $u = f(v)$. For any simple path $p = \{v, v_1, ...v_r\}$, there must also exists a simple path p in datagraph g where $p = \{u, f(v_1), ...f(v_r)\}$ by the definition of subgraph isomorphism.

This could be easily proved by recalling Definition 1. For any dimension (r, l), we have:

$$sig_{r,l}(v) = count_{r,l}(v)$$
$$= |\{(v, v_1, ...v_r)|l(v_r) = l\}| \tag{1}$$
$$sig_{r,l}(u) = count_{r,l}(v)$$
$$= |\{(u, u_1, ...u_r)|l(u_r) = l\}| \tag{2}$$

Also, $\{(v, v_1, ...v_r)|l(v_r) = l\} \subset \{(u, u_1, ...u_r)|l(u_r) = l\}$ We could immediate conclude that $sig_{d,l}(u) \succeq sig_{d,l}(v)$ for any (d, l), therefore, $sig(u) \succeq sig(v)$.

4.2   Compression

In a k-dimension vector, not every dimension is equal in filtering power. Moreover, some dimensions are closely correlated and their filtering power is severely overshadowed by these correlated dimensions. Based on this observation, we can reduce a k-dimension vector into a m-dimension vector while preserving the filtering power as much as possible.

$$FP = \sum_{v_x \in Vg} selectiveness(sig(v_x))$$

$selectiveness(sig(v_x))$ denotes the selectivity of the signature of v_x.

$$selectiveness(sig(v_x)) = |\{v_y|v_y \in V(g) \land l(v_y) < l(v_x)\}|.$$

Let FP_k to denote the filtering power of a k-dimension vector, we want to select m out of k dimensions which results minimum loss of filtering power, that is to minimize $FP_k - FP_m$.

Our compression framework is composed of following steps:

1. From the k-dimension vector $\{d_1, d_2, ..., d_k\}$, we use a *selection function* to choose two dimension to merge.
2. For the chosen dimensions, d_x and d_y, we use a *merge function* to merge them.
3. Repeat the above steps until there are only m dimensions.

In this paper, we will choose the pair of dimensions whose correlation is maximum. Two dimensions are closely correlated if at most of the time, either they can both filter a candidate against a vertex or neither of them can. If two dimensions are closely correlated and one of them is unable to filter a vertex, then there will be little chance to filter this vertex with the filtering power of the other dimension. Intuitively, removal of one of these two correlated dimension shall cause limited loss of filtering power. Formally, correlation between two dimensions is defined as:

$$correlation(d_x, d_y) = |\{(v_i, v_j)|l(v_i) = l(v_j) \land sig(v_i)[d_x]$$
$$< sig(v_j)[d_x] \land sig(v_i)[d_y] < sig(v_j)[d_y]\}|$$

Having obtained the candidate merging pair, we will use the value of the one with higher filtering power. The filtering power of a single dimension is calculated as:

$$fp(d_x) = |\{(v_i, v_j)|l(v_i) = l(v_j) \land sig(v_i)[d_x] < sig(v_j)[d_x]\}|$$

The value of the merged dimension is $d_{xy} = max(fp(d_x), fp(d_y))$.

Algorithm 1. FindAllMatches(g, q, $index$)

Input: g: database graph, q: query graph, $index$: database index
Output: all subgraph isomorphism mapping
1 Build index for q;
2 **for each** $v \in V(q)$ **do**
3   **forall** $u \in V(g) \wedge l(u) = l(v)$ **do**
4   **if** $sig(u) \succeq sig(v)$ **then**
5   v.candidates.insert(u);
6   **end if**
7   **end for**
8 **end for**
9 Reorder all $v_i \in V(q)$;
10 Enumerate(0);

The algorithm to utilize the framework to compress the index is obvious and straight-forward. Due to space limitation, we will omit the compression algorithms.

5 Query

Our algorithm could be summarized into following steps:

1. We order vertices in q into a sequence. We say a ith vertex in the sequence is the *depth i* vertex. The ordering criteria will be explained later in this section.
2. For each vertices in q, we calculate the candidate list for it by using the labeling, degree, and the signature information.
3. Initially we start from depth 1, we will choose one candidate from the candidate list of depth 1 vertex. As the candidate is chosen, we will look up ahead to further filter the candidate lists of deeper vertices by using neighbouring information.
4. If current depth is equal to the number of vertices in q, we know the current matching is complete. We can backtrack to last depth and choose a different candidate. If there is no more candidates on the last depth, we need to backtrack further up and restore the changes made along the way.

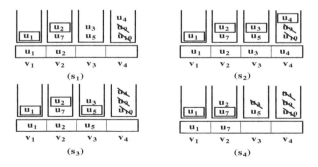

Fig. 3. Enumeration State

Algorithm 2. Enumerate($depth$)

Input: $depth$: the current enumeration depth

1 **if** $depth = |V(q)|$ **then**
2   **for each** $v_i \in V(q)$ **do**
3   output mapping $(v_i, v_i.\text{match})$;
4   **return**;
5   **end for**
6 **end if**
7 **for each** $u \in v_{depth}.candidates$ **do**
8       $v_{depth}.\text{match} = u$;
9   **forall** $i > depth$ **do**
10   remove u from v_i.candidates if it is a member;
11   **if** v_i *is a neighbour of* v_{depth} **then**
12   remove all non-neighbours of u from v_i.candidates;
13   **end if**
14   **if** $v_i.candidates.size() = 0$ **then**
15   add back removed candidates to their original list;
16   **end if**
17   **else**
18   Enumerate(depth + 1) ;
19   add back removed candidates to their original list;
20   **end if**
21   **end for**
22 **end for**

---

5. If there are any candidate lists have been reduced to empty list, we will immediate conclude the current matching is invalid. In this case, we will restore the changes made and choose a different candidate on this depth. If there is no more candidates on this depth, we need to backtrack further up and restore the changes made along the way.

We present our query algorithm in Alg 1 and Alg 2. We will use Fig 2 as a running example.

Example 1. The first step is to build up a candidate list for v_1 to v_4. We will discuss the filtering index in next Section. The next step is to order all vertices in q according to an ordering rule. This rule will be explained in next subsection. The resulting candidate lists and the enumeration order are shown in Fig 3. At enumeration depth 1, we have only one choice for v_1, which is u_1. The neighbours of v_1 are v_2, v_3 and v_4. Unfortunately, all candidates for v_2 and v_3 are neighbours of u_1, therefore we could not remove any vertices from these two candidate lists. However, we can prune u_9 and u_{10} because they are not neighbours of u_1. At the second depth, we choose all valid candidates for v_2 in order, the first one is u_2. Once we match u_2 to v_2, we will check the candidate lists of v_3 and v_4, who are v_2's neighbours. Keep going on and we will find the first correct mapping at depth 4 as shown in (s_2). After that we need to backtract to depth 3 and try to match u_5 to v_3. This time we find v_3 is connected to v_4 but the only candidate for

v_4, which is u_4, is not connected to u_5. We can conclude this matching is not correct. Having probed all possible vertices at depth 3, we will need to backtrack to depth 2. u_7 has not been probed at this depth. Again, v_2 is connected to v_4 but u_7 is not connected to u_4. As soon as we see the depth 4 vertex, v_4 has no more candidates, we can conclude this matching will not be correct.

5.1 Cost Model

The cost of calculating subgraph matchings greatly depends on a number of unpredictable factors, including the topology of the database graph, query graph, degrees etc. It is very difficult to give a precise estimation of the algorithm. Inspired by [7], we use the following expressions to approximate the overall cost. Let us suppose that Γ_i is the set of distinct submatchings at enumeration depth i. γ_i is one of the submatchings at depth i. To extend γ_i to next depth $i + 1$, there are $f_\gamma(|C_{i+1}|)$ ways of doing so, where C_{i+1} is the candidate sets of q_i and f_γ is an expansion function depending on γ. Formally, the number of distinct submatches at depth i is:

$$|\Gamma_i| = \begin{cases} |C_1| & \text{if } i = 1 \\ \sum_{\gamma \in \Gamma_{i-1}} f_\gamma(C_i) & \text{if } i > 1 \end{cases}$$

Everytime we extend an submapping γ we will incur an overhead cost δ_γ. The total approximated cost will be:

$$Cost_{total} = \sum_{i=1}^{|V(q)|} \sum_{\gamma \in \Gamma_{i-1}} f_\gamma(C_i) \times \delta_\gamma$$

It is not hard to see that the total cost is dominated by $\sum_{i=0}^{|V(|q|)} |\Gamma_i|$. Each $|\Gamma_i|$ is recursively affected by other Γ_k, where $k < i$, through a expansion function $f_\gamma(C_i)$. Although the precise and exact f_γ is difficult to determine, it is intuitive to expect $f_\gamma(C_i)$ is proportional to $|C_i|$. Since Γ_i is recursive, we expect the total cost to be minimized if we order Γs in non-descending order. This approach is essentially the same of as maximizing computational sharing in the enumeration process. Many subgraph isomorphism mappings may only differ to each other by a few vertices. We would like to reuse their shared parts as much as possible. In order to reuse as much as possible, we should move the shared parts to the earlier stage of enumeration process.

According to above analysis, we propose an ordering rule as such:

1. The v_0 will be the one with the least candidates.
2. The ith vertex must be a neighbour of at least one already ordered vertices. (without the loss of generality, we will assume q is a connected graph, so that you will always be able to find at least one such vertex.)
3. Among the ones satisfy the last criteria, we choose the one with the least candidates.

The rationale behind the first and the third rule is straightforward. The second rule is related to the eager verification strategy which we will detail in next subsection. Briefly, its rationale is to allow this vertex's candidate list to be pruned by previous vertices before reaching this enumeration depth.

5.2   Eager Verification

In traditional subgraph matching algorithms, we try to enumerate a submapping before testing whether the current mapping is subgraph isomorphic. We call them lazy verification strategy.

Instead of using the traditional approach, we propose an eager verification strategy in this paper. In this strategy, we try to reduce the number of candidates as early as possible. At ith depth of the enumeration process, if we match a u_x to the ordered query node v_i, we will check all candidate lists of v_j where $j > i$:

i) We remove u_x from v_j's candidate list.
ii) If v_j is a direct neighbour of v_i, we remove all vertices which are not direct neighbours of u_x from v_j's candidate list.

If any v_j's candidate list becomes empty as a result, we could immediately conclude u_x is not a correct match for v_i.

One of the most obvious advantage of eager verification is the ability of terminating the current enumeration early. For example, in Fig 3 state 4, we could terminate at the depth 2, whereas we have to look all the way through to depth 4 if lazy verification is used. Secondly, eager verification can save a lot of unnecessary isomorphism test. For example, in eager verification, a database graph vertex u_x was a candidate for query vertex v_j, however, it has already been pruned at depth i where $i < j$. u_x will not participate in the enumeration process at depth j. However, in the case of lazy verification, u_x will participate in the enumeration process every time when we want to expand a depth $j - 1$ submatching and then fail the isomorphism test because it will be exclusive with the vertex chosen at depth i.

Theorem 2. *In eager verification strategy, for any submatching formed at any depth i, it automatically satisfies the isomorphism rules, there is no isomorphism tests required for this submatching.*

Proof. Let us assume the theorem holds for depth $i - 1$, suppose there is a candidate vertex u_x for v_i and extend current submatching by u_x will violate the isomorphism rule. Let us recall Definition 1, if the current mapping functions is f, it is either:

i) $l(u_x) \neq l(v_i)$ or;
ii) $\exists v_k, k < i \wedge (v_k, v_i) \in E(q) \wedge (f(v_k), u_x) \notin E(g)$.

Obviously, the first condition is impossible because all candidates for v_i will have the same label as v_i. The second condition is also impossible because u_x was supposed to be removed from v_i's candidate list at depth k if eager verification strategy is used according to its pruning rules described above. This raises a contradiction.

Therefore, the theorem holds for i if it also holds for $i - 1$. Obviously the theorem holds for $i = 1$ therefore it will hold for all other i.

5.3   Correctness and Completeness

Correctness: All matching found by NOVA are correct.

Proof. Follows Theorem 2, the result is immediate.

Completeness: All correct matchings will be found by NOVA .

Proof. Let us denote a submatching of depth i with f_i. Suppose there is a missing matching f, there must exist a i such that f_i is a missing submatching but f_{i-1} is not. It is trivial to see that $i \neq 1$. For other $i > 0$, f_i could only become a missing submatching if $f(v_i)$ is removed from v_i's candidate list. According to the verification strategy of NOVA , it could only happen if there exists a k, such that $k < i$, $(v_k, v_i) \in E(q) \wedge (f(v_k), f(v_i)) \notin E(g)$. It is a contradiction against the assumption that f_i is a submatching according the definition of isomorphism mapping.

6   Experiment

We have performed extensive experiment to compare our techniques, NOVA (uncompressed version) and CNOVA(compressed version) with GADDI([10]), which is the only known competitor. We have also studied how the performance is affected by using the compressed index. In the experiments, we have used both real datasets and synthetic data to evaluate the performance. The query sets are generated by randomly choosing subgraphs from large graphs. We conducted all the experiments on a PC with a 2.4GHz quad-core processor, and 4GB main memory running Linux.

6.1   Real Data

HPRD([12]) is a human protein interaction network consisting of 9460 vertices and 37000 edges. We used its GO term description([13]) as its labels. NOVA spent 15 seconds to create an index of approximately 13MB while GADDI spent 512 and created an index of 92MB. Figure 4 shows the performance of these two techniques. We can see that when the degree of query increases, the performance of both techniques improve significantly. This is because vertices with higher degrees are generally more selective. It also shows that when the number of vertices in query increases, both technique respond slower. It is because more vertices mean more enumeration depth. We have shown NOVA is up to one magnitude faster than GADDI. CNOVA has an index size of 6MB and it performs roughly the same as NOVA.

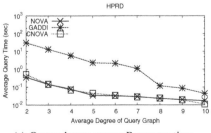
(a) Query degree versus Response time

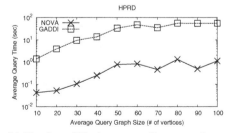
(b) Number of Vertices versus Response time

Fig. 4. HPRD Dataset

Polblog([14]) is a graph showing the links between politician blogs. There are 1490 vertices and each is representing a politician blog. There are more than 17000 edges and each is representing one hyperlink between them. We used origin of each blog as the labels and there are 46 of them. It is worth to mention that there are a few very powerful vertices closely linking to each other. Its label distribution is also very biased. For example, the top 2 labels representing 50% of all vertices. In this dataset, GADDI kept running for 36 hours and did not finish building the index and we have to terminate the process. This is because the intersection graph between the several powerful vertices are large. GADDI requires to count the occurrences of substructures which itself is a subgraph isomorphism enumeration process. While it is affordable when the intersection graph is small, the cost grows exponentially when the size grows. We did not use the compressing technique on this dataset because there is only a small number of labels so that further reducing its filtering power will not be meaningful. Figure 5 shows as the query degrees increases, the performance improves. It is also interesting to see that as the query size grows, NOVA initially runs faster then it slows down. This is because the label distribution is very biased. A small query graph means there is a high probability that all query vertices are very unselective. The resulted loss outweighs the gain from less enumeration depth. When the size of query reaches a certain number, selective vertices start to appear in queries and now the enumeration depth matters more.

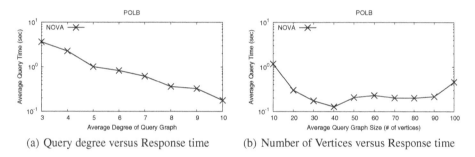

(a) Query degree versus Response time   (b) Number of Vertices versus Response time

Fig. 5. Politician Blog Dataset

6.2 Synthetic Dataset

We have generated synthetic datasets to evaluate the performance of our techniques as well as GADDI. The synthetic database has 5000 vertices, 40000 edges and 250 labels. The data graphs are generated by a social network generator([15]). The query sets have an average size of 25 vertices and average degree of 4. These are the standard settings and we will vary some of these parameters to show how they affect the performance.

The first graph in Figure 6 shows the response time increases as the size of the database graph increases for both NOVA and GADDI. In the second graph, we reduce the number of labels to only 10. Thus the vertices become very unselective. Under this condition, NOVA performs reasonably well. In both cases, NOVA is almost one magnitude faster than GADDI.

Figure 7 shows the index size of NOVA grows almost linearly. This is because our space complexity is $O(n) = krn$ where $k = |L(G)|$, $n = |V(G)|$, r is the specified

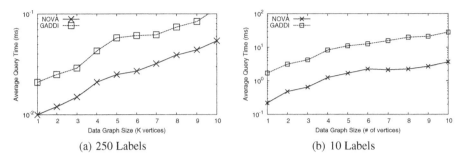

Fig. 6. Scalability against Database graph size

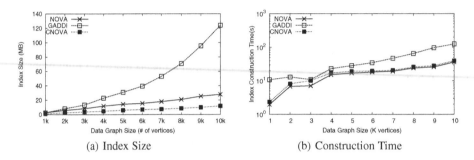

Fig. 7. Construction Cost

radius. In this case, kr is almost constant. We also evaluated NOVA with compression, which is denoted by *CNOVA*. We demonstrated that while *CNOVA* added a small amount of time in construction, the resulting index is 40% to 50% less than NOVA. Both of these techniques are more efficient and more effective than GADDI.

Figure 8 studies the scalability against query. As the degree increases, the response time improves for all of the three techniques. NOVA and CNOVA are still faster than GADDI. The gap narrows as the degree increases. This is because queries become more selective when degree increases. It is also interesting to see that CNOVA performs is slightly faster than NOVA when the degree is high. It is because when queries are more

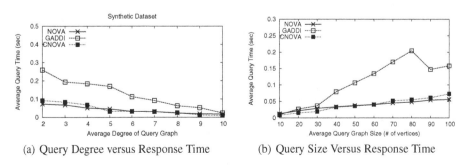

Fig. 8. Scalability against Query size

selective, NOVA does not have too much advantage over CNOVA because the filtering power is already strong. However, CNOVA's signature vectors are shorter, which means less overhead. Again, for the same reason, CNOVA is slightly faster than NOVA when the average size of query is small and slightly slower than NOVA when the size is large.

7   Related Work

There are already a large amount of existing work([1], [2], [3], [4], [5], [6], [7]), [16] related to subgraph containment in a database containing many small(less than 100 vertices) graphs. In this problem, given a query graph Q, the database needs to output all graphs which are supergraph isomorphic to Q. However, the isomorphism test is known to be NP-complete. In order to avoid expensive isomorphism test, the commonly used framework is as follow: i) Preprocess the database so that each graph in the database is indexed with a *signature*. ii) Calculate the signature of Q. iii) Verify all graphs in the database whose signatures are compatible with Q's. As the expensive part is performing isomorphism test, the above techniques are focusing on reducing verification time or reducing the total number of isomorphism tests to perform. The key differences between the subgraph containment problem and the problem studied in this paper are: i) In subgraph containment problem, there are multiple small graphs in the database. The above techniques are only suitable to index small graphs, whereas in our problem, we have a very large(thousands to tens of thousands vertices) single-graph database. ii) In subgraph containment problem, the verification stage only needs to find one match in each candidate to conclude Q is subgraph isomorphic to this candidate, whereas in our problem, we need to perform full isomorphism test to find all possible mappings.

   GraphGrep([9]) proposed to index all paths of length upto K and as well as an inverted index for them. By using the paths as a signature, the matching process will be much faster as they are very discriminative. However, in [10], Shijie Zhang *et. al.* has shown that GraphGrep does not scale well on large graphs.

   In [11], Tian *et. al.* proposed an efficient method to find approximate subgraph matchings in large database graphs. This technique is based on a neighbouring bitmap index. This method does not guarantee the connectiveness of result and it could only find one approximate subgraph matching for each database graph.

   GADDI([10]) is the state of art technique proposed for subgraph matching problem for large graph. It is based on a pair-wise index which records the number of tiny features contained in the induced intersection graph of neighbourhoods.

   Wang *et. al.*, in [16], also proposed a vector domination filtering technique based on neighbouring information to search patterns in a continuous graph stream. However, it has not studied how to effectively build indices for large graph databases based on the vector domination property.

   There are other interesting techniques to deal with data mining of large graphs. For example, [17] and [18] can be used to mine frequent graph patterns from a single large network. They propose how to define the support of frequent patterns in a large graph. Chen *et. al.*, in [19], propose a method to mine frequent patterns from a group of large graphs. In their techniques, frequent patterns are mined from the summarization graph which is much smaller than the original graph.

8 Conclusion

In this paper, we propose a novel index to solve the subgraph matching problem in large database graph. As shown in the experiment, the index has strong filtering power and is efficient in both construction and storage. We also proposed an efficient subgraph matching algorithm which attempts to maximize the computational sharing by pre-ordering the query vertices in the enumeration process. More importantly, the k-dimensional vector signature, which is the theme of our technique, could be used as a general framework for finding subgraph matchings in large database graph. Every dimension of the signature represents a specific feature of a vertex. The user can arbitrarily define the dominance function.

References

1. Cheng, J., Ke, Y., Ng, W., Lu, A.: Fg-index: towards verification-free query processing on graph databases. In: SIGMOD Conference (2007)
2. Jiang, H., Wang, H., Yu, P.S., Zhou, S.: Gstring: A novel approach for efficient search in graph databases. In: ICDE (2007)
3. Yan, X., Yu, P.S., Han, J.: Graph indexing: A frequent structure-based approach. In: SIGMOD Conference (2004)
4. Zhang, S., Hu, M., Yang, J.: Treepi: A novel graph indexing method. In: ICDE (2007)
5. Zhao, P., Yu, J.X., Yu, P.S.: Graph indexing: Tree + delta >= graph. In: VLDB (2007)
6. Zou, L., Chen 0002, L., Yu, J.X., Lu, Y.: A novel spectral coding in a large graph database. In: EDBT (2008)
7. Shang, H., Zhang, Y., Lin, X., Yu, J.X.: Taming verification hardness: an efficient algorithm for testing subgraph isomorphism. PVLDB 1(1) (2008)
8. Backstrom, L., Dwork, C., Kleinberg, J.M.: Wherefore art thou r3579x?: anonymized social networks, hidden patterns, and structural steganography. In: WWW (2007)
9. Giugno, R., Shasha, D.: Graphgrep: A fast and universal method for querying graphs. In: ICPR (2) (2002)
10. Zhang, S., Li, S., Yang, J.: Gaddi: distance index based subgraph matching in biological networks. In: EDBT (2009)
11. Tian, Y., Patel, J.M.: Tale: A tool for approximate large graph matching. In: ICDE (2008)
12. Human protein reference database, http://www.hprd.org/download
13. The gene ontology, http://www.geneontology.org
14. Adamic, L.A., Glance, N.: The political blogosphere and the 2004 us election. In: WWW 2005 Workshop on the Weblogging Ecosystem (2005)
15. Leskovec, J., Chakrabarti, D., Kleinberg, J., Faloutsos, C.: Realistic, mathematically tractable graph generation and evolution, using kronecker multiplication. In: Jorge, A.M., Torgo, L., Brazdil, P.B., Camacho, R., Gama, J. (eds.) PKDD 2005. LNCS (LNAI), vol. 3721, pp. 133–145. Springer, Heidelberg (2005)
16. Wang, C., Chen 0002, L.: Continuous subgraph pattern search over graph streams. In: ICDE (2009)
17. Chen, J., Hsu, W., Lee, M.L., Ng, S.K.: Nemofinder: Dissecting genome-wide protein-protein interactions with meso-scale network motifs. In: KDD (2006)
18. Kuramochi, M., Karypis, G.: Finding frequent patterns in a large sparse graph (2005)
19. Chen, C., Lin, C.X., Fredrikson, M., Christodorescu, M., Yan, X., Han, J.: Mining graph patterns efficiently via randomized summaries. In: VLDB (2009)

Efficiently Answering Probability Threshold-Based SP Queries over Uncertain Graphs

Ye Yuan[1], Lei Chen[2], and Guoren Wang[1]

[1] Northeastern University, Shenyang 110004, China
[2] Hong Kong University of Science and Technology, Hong Kong SAR, China
linuxyy@gmail.com

Abstract. Efficiently processing shortest path (SP) queries over stochastic networks attracted a lot of research attention as such queries are very popular in the emerging real world applications such as Intelligent Transportation Systems and communication networks whose edge weights can be modeled as a random variable. Some pervious works aim at finding the most likely SP (the path with largest probability to be SP), and others search the least-expected-weight path. In all these works, the definitions of the shortest path query are based on simple probabilistic models which can be converted into the multi-objective optimal issues on a weighted graph. However, these simple definitions miss important information about the internal structure of the probabilistic paths and the interplay among all the uncertain paths. Thus, in this paper, we propose a new SP definition based on the possible world semantics that has been widely adopted for probabilistic data management, and develop efficient methods to find threshold-based SP path queries over an uncertain graph. Extensive experiments based on real data sets verified the effectiveness of the proposed methods.

1 Introduction

In this paper, we study a novel problem, finding shortest paths (SPs) in an uncertain graph. Compared to its counterpart problem, finding shortest paths in a certain graph, the new problem has the following new characteristics: 1). We are working on an uncertain graph (UG). In a UG, an edge between any two vertices is associated with an existence probability[1] 2). A SP in a UG is associated with not only the weight of the path, but also a probability indicating the existence of the path. Similar to SP search over a certain graph, SP search over a UG has many applications as well. For example, due to the existence of uncertain links in wireless sensor networks or transportation networks, it is often essential to conduct a SP route query over a UG which is used to model these networks. In this work, we follow the widely used possible world data model [1,2] to describe uncertain graphs. Specifically, given a UG, each edge is associated with a weight and an existence probability. A possible world of UG is a graph having the exactly same set of vertices as UG and an instance of all the possible combination of edges. The probability of the possible world is the product of the probabilities of all the edges appeared in the possible world. With this possible world model, the SP search

---

[1] In this paper, we do not consider node (vertex) uncertainty and we will leave SP search over a node and edge uncertain graph as an interesting future work.

H. Kitagawa et al. (Eds.): DASFAA 2010, Part I, LNCS 5981, pp. 155–170, 2010.

over a UG is to find all SPs among all the possible worlds of the UG. The probability of a SP in each possible world is the possible world's probability, and the probability of the SP in the UG is the sum of the probabilities of all the possible worlds that the SP exists. In practice, it is often not useful to return all SPs over a UG since some SPs have very low probabilities. Therefore, in this paper, we study a probability threshold-based SP query over an uncertain graph. Specifically, given a probability threshold ϵ, a probabilistic threshold-based SP query over a UG returns a set of SPs, each SP has a probability greater or equal to ϵ. Note that T-SP query may return multiple SPs due to the existence of multiple possible worlds.

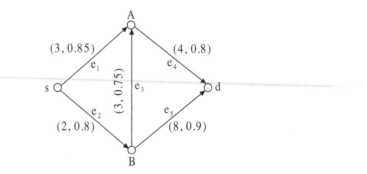

Fig. 1. Example of an uncertain graph with three paths (w, p)

Figure 1 shows an uncertain graph G and the parameters $(weight, probability)$ are shown as labels on the edges. Given a SP query of s and d on G shown in Figure 1, there are three possible paths, $(P_1 = e_1e_4, P_2 = e_2e_3e_4, P_1 = e_2e_5)$ from the source node (s) to the destination (d) as listed in Table 1 . The second row of Table 1 lists the probability of each possible world that a path (P_1, P_2, or P_3) is a SP. The third row lists the probability of a SP path in G, i.e., the sum of probabilities of all the possible worlds that the path is a SP. If the query probability threshold is 0.5, only P_1 is returned.

Clearly, a naive solution for SP search over a UG is to enumerate all the possible worlds and conduct SP search over each possible world, which is very inefficient. Therefore, in this paper, we first design a basic algorithm to avoid unfolding all possible worlds. The proposed algorithm can compute the exact probability for each candidate path by scanning the sorted list of paths (from s to d) only once. To further speed up the calculation of probability, we propose an advanced method which reduces the computation by combining isomorphic graphs. Finally, we also propose several pruning techniques to further reduce the search space.

The rest of the paper is organized as follows. Section 2 discusses the related works. The definitions of uncertain graph and probability threshold-based SP queries are given in Section 3. We present a basic SP probability computing method together with some pruning rules in Section 4. To speed up the query, Section 5 presents some improved algorithms and tighten probability bounds. Moreover, we discuss the results of the performance tests on real data sets in Section 6 and conclude in Section 7.

Table 1. Possible worlds and probability of each path (in Fig. 1) being shortest path

Path	$P_1 = e_1e_4$	$P_2 = e_2e_3e_4$	$P_3 = e_2e_5$
Possible worlds	$e_1e_2e_3e_4e_5$ (Pr=0.3672),	$\bar{e}_1e_2e_3e_4e_5$ (Pr=0.0648),	$e_1e_2e_3\bar{e}_4e_5$ (Pr=0.0918),
	$e_1\bar{e}_2e_3e_4e_5$ (Pr=0.0918),	$\bar{e}_1e_2e_3e_4\bar{e}_5$ (Pr=0.0072)	$\bar{e}_1e_2e_3\bar{e}_4e_5$ (Pr=0.0162),
	$e_1e_2\bar{e}_3e_4e_5$ (Pr=0.1224),		$e_1e_2\bar{e}_3\bar{e}_4e_5$ (Pr=0.0306),
	$e_1e_2e_3e_4\bar{e}_5$ (Pr=0.0408),		$\bar{e}_1e_2\bar{e}_3e_4e_5$ (Pr=0.0054)
	$e_1\bar{e}_2\bar{e}_3e_4e_5$ (Pr=0.0306),		$\bar{e}_1e_2\bar{e}_3e_4e_5$ (Pr=0.0216)
	$e_1e_2\bar{e}_3e_4\bar{e}_5$ (Pr=0.0136),		
	$e_1\bar{e}_2e_3e_4\bar{e}_5$ (Pr=0.0102),		
	$e_1\bar{e}_2\bar{e}_3e_4\bar{e}_5$ (Pr=0.0034)		
Sum of the probabilities	0.68	0.072	0.1656

2 Related Work

Key ideas in uncertain databases are presented in tutorials by Dalvi and Suciu [1], [2], and built on by systems such as Trio [4], MCDB [5] and MayBMS [3]. Initial research has focused on how to store and process uncertain data within database systems, and thus how to answer SQL-style queries. Subsequently, there has been a growing realization that in addition to storing and processing uncertain data such as as NN query [22], [26], range query [25], top-k query [23], [24] and skyline query [27]. There also exists serval advanced algorithms to analyze uncertain data, e.g., clustering uncertain data [9] and finding frequent items within uncertain data [10].

With respect to uncertain graphs, some works have considered the uncertain graph from their application fields. Papadimitriou [16], Chabini [17] and Fu [8] view the road networks as stochastic networks, and they consider the notion of shortest paths in expectation. Chabini et.al [17] focus on the expectation of travel times in addition to each edge and in this way the network can be captured by a measure of uncertainty, while Fu [8] studied the expected shortest paths in dynamic stochastic networks. Korzan B. [6] and [7] studied the shortest path issue in unreliable road networks in which arc reliability are random variables, and they compute the distribution of the SP to capture the uncertain. For the communication systems, Guerin and Orda investigated the problem of optimal routing when the state information is uncertain or inaccurate and expressed in some probabilistic manner [11], [12]. However, their probabilistic models are quite simple. These models miss important information about the internal structure of the probabilistic paths and the interplay among all the uncertain paths. In this paper, we study the SP search based on the possible world semantics, since the possible worlds model can reflect the intricate characteristics in an uncertain graph.

3   Problem Definition

In this section, we first formally introduce the uncertain graph (UG) model and define a probability threshold-based SP query over a UG, then, we highlight the key issue needed to be addressed to improve the efficiency of SP query processing.

Definition 1. *An uncertain graph G is denoted as $G = ((V, E), W, Pr)$, where (V, E) is a directed graph[2]; $W : E \rightarrow R$ is a weighted function, $Pr : E \rightarrow (0, 1]$ is a probability function denoting the existence of an edge in E.*

From the above definition, we can find that a certain graph is a special UG whose edge existent probability is 1, which can be denoted as $G = ((V, E), W, 1)$. As mentioned in Section 1, we use the possible world model to explain the semantics of UGs. Therefore, under the possible world model, a UG implicates a group of certain graphs (possible worlds). A certain graph $G' = ((V', E'), W)$ is implicated from a UG (denoted by $G \Rightarrow G'$) if $V' = V, E' \subseteq E$. Assume that the existences of different edges in a UG are independent to each other, we have,

$$Pr(G \Rightarrow G') = \prod_{e \in E'} Pr(e) \cdot \prod_{e \in (E \setminus E')} (1 - Pr(e)). \tag{1}$$

Let $Imp(G)$ denote the set of all certain graphs implicated by the UG G. Apparently the size of $Imp(G)$ is $2^{|E|}$ and $Pr(G \Rightarrow G') > 0$ for any certain graph G'. Moreover, we have $\sum_{G' \in Imp(G)} P(G \Rightarrow G') = 1$. For example in Fig. 1, there are total $2^5 = 32$ certain graphs implicated by G. Due to the space limit, Table 1 only lists 14 possible worlds together with their associated existence probability.

Definition 2. *Given an uncertain graph G, a probability threshold ϵ $(0 < \epsilon \leq 1)$, and two vertices s and d, a probabilistic threshold-based SP query returns a set of paths (from s to d), whose SP probability values are at least ϵ.*

We use $SP(G')$ to denote the SP returned by a SP query on two vertexes (s, d) of a certain graph G'. For a path $P \in G$, the SP probability of P is the probability that P is $SP(G')$ among all $G' \in Imp(G)$, that is,

$$Pr_{SP}(P) = \sum_{P = SP(G')} Pr(G \Rightarrow G'). \tag{2}$$

For any two vertexes s and d of G, the answer set to a probabilistic threshold-based SP query is a set of paths whose SP probability values are at least ϵ, that is,

$$Answer(SP, P, \epsilon) = \{P | P \in G, Pr_{SP}(P) \geq \epsilon\}. \tag{3}$$

To compute the answer set of a SP query on a UG, a naive solution is to enumerate all certain graphs, compute the SP probability of certain graphs, and select the paths satisfying the querying conditions (thresholds). Unfortunately, the naive method is inefficient since, as discussed before, there are exponential number possible worlds on a UG. Thus, our focus in this paper is to develop efficient algorithms.

---

[2] In this paper, we only consider directed graphs. For an undirected graph, an edge (u, v) can be replaced by two directed edge $u \rightarrow v, u \leftarrow v$.

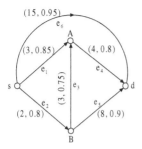 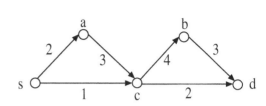

Fig. 2. Example of a graph (H) composed   **Fig. 3.** Example of a longer path in RH
of the pervious 4 optimal paths from s to d

Before illustrating the technical details of our efficient solution, we first point out
the key issues to implement the efficient computing and briefly summarize the key
techniques we proposed.

Definition 3. *Given vertices s and d over a UG G, assume that $P_1, ..., P_i$ is the list of
all paths (with increasing lengths from s to d) being investigated so far and $E_{P_1}, ..., E_{P_i}$
are the edge sets for i paths ($P_1, ..., P_i$), we use H to denote a graph composed by i
paths and associated nodes and $RH = H \backslash P_i$ to denote the graph reduced from H after
removing P_i.*

For an instance, Figure 2 gives an example of H composed of 4 paths. Compared to the
graph in Fig. 1, H has an additional edge e_6, which can be viewed as the fourth SP (P_4)
between s and d since its weight is larger than any of other three paths listed in Table 1.
As shown in Fig.2, if P_i is P_4, $RH = H \backslash P_4$ is just the graph in Fig. 1.

Now we define an important probabilistic event,

Definition 4. *Given RH, an event $Connect$ is defined as there exists at least one
shorter path (than P_i) in RH.*

Now we can compute the SP probability of path P_i as follows:

$$Pr_{SP}(P_i) = (1 - Pr(Connect)) \prod_{e \in E_{P_i}} Pr(e) \tag{4}$$

In this equation, $\prod_{e \in P_i} Pr(e)$ denotes the existent probability of P_i. The equation
indicates that if P_i is the SP, all the shorter paths (than P_i) in RH cannot be connected.
Otherwise P_i cannot be the shortest path. The **shorter paths** in the definition of event
$Connect$ is important, since RH might include paths longer than P_i. For example, there
are two paths $sacd$ (with length 7) and $scbd$ (with length 8) composed of RH shown in
Fig. 3, and the RH also includes a longer path ($sacbd$ with length 12) than P_i (assume
its length is 10).

The key issue of efficient computing $Pr_{SP}(P_i)$ is how to calculate the value of
$Pr(Connect)$ efficiently. Unfortunately, Valiant [19] pointed that the problem of computing probability that there is at least one path between given two vertexes in a
stochastic network is NP-Complete, which also means it is a hard problem to compute

$Pr(Connect)$. However, this is the case that users are interested in any SP over a UG with the probability greater than 0. In practice, users are only interested SP whose probability is above a predefined threshold ϵ. Thus, with this probability threshold ϵ, we propose several heuristics to remove paths from further evaluation if the probability upper bound to be SP for those paths are less than the threshold ϵ.

4 Probability Calculation of Shortest Path

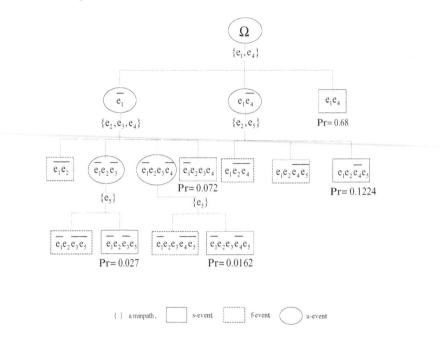

Fig. 4. Computing process for $Pr(Connect)$ of Fig.1 using the basic algorithm

4.1 Basic Calculation Algorithm

From the Equations (4) and (3), we can get,

$$Pr(Connect) \leq 1 - \frac{\epsilon}{\prod_{e \in E_{P_i}} Pr(e)}.$$

Let RHS of the formula be Q. Thus, the problem of determining whether a path P is a SP is converted into the following issue:

$$Pr(Connect) \leq Q. \tag{5}$$

In other words, for a path P, if its $Pr(Connect)$ satisfies Equation (5), it belongs to the answer set. Otherwise, P is marked as failed. According to the definition, to compute $Pr(Connect)$, we have to enumerate all the paths that connected s and d in RH, which is quite inefficient. Thus, in this section, we present an algorithm to avoid checking all the paths, called basic algorithm. The reason we call it basic algorithm because we will introduce some improvements later on to further improve the efficiency. Before giving the basic algorithm to compute $Pr(Connect)$, we first introduce some related concepts.

Definition 5. *Given RH containing two vertices s and d, an $s - d$ path is a sequence of edges that ensure the connection from s to d. If no proper subpath of an $s - d$ path is an $s - d$ path, it is called an $s - d$ minipath (denoted by $I_{min}(s - d)$).*

Definition 6. *Given RH, we define an edge set in RH as an event I, denoted by the existence of edges in the set. An edge i is* successful *under I if $i \in I$, otherwise it is* failed.

For example, given an event I, $I = \overline{i}j\overline{k}|_\varepsilon$, which stands for edge j is successful, edges i and k are failed, and other edges not appeared in I are *uncertain* (i.e., can be neither successful nor failed)[3]. Each I splits edge set E of RH into three disjoint subsets $E_s(I)$, $E_f(I)$ and $E_u(I)$, which record successful edges, failed edges and uncertain edges, respectively. An edge is uncertain if it does not appear in I. For $I = \overline{i}j\overline{k}|_\varepsilon$, $E_s(I) = \{j\}$, $E_f(I) = \{i, k\}$, $E_u(I) = E\backslash\{i, j, k\}$. That is to say $E_s(I)$ and $E_f(I)$ are the sets of successful and failed edges under event I correspondingly, $E_u(I)$ is the set of uncertain edges with respect to I, and $E_s(I) \bigcup E_f(I) \bigcup E_u(I) = E$. The probability of an event I is $Pr(I) = \prod_{e \in E_s} Pr(e) \prod_{e \in E_f} (1 - Pr(e))$.

Then, we define three special events.

Definition 7. *Given RH containing s and d, a successful event (denoted as s-event) is an event I if edges in $E_s(I)$ enable the connection from s to d. A failed event I (denoted as f-event) is an event I if edges in $E_f(I)$ disable the connection from s to d. If I is neither successful nor failed, we say I is an undetermined event (denoted by u-event).*

For example, in Fig. 1, $e_1e_4|_\varepsilon$ is an s-event since edges e_1e_4 connects s and d. $\overline{e_1}\overline{e_2}e_4|_\varepsilon$ is an f-event due to that the inexistences of e_1 and e_2 disable the connection from s to d. $e_2\overline{e_3}|_\varepsilon$ is a u-event, since it cannot be determined if connecting (or disconnecting) s and d by checking its $E_s(I) = e_2$ (or $E_f(I) = e_3$).

Theorem 1. *For each u-event I of RH, there is at least one $s - d$ path under I.*

Proof. Let $e_1...e_n = E_u(I)$. Then $e_1...e_n \bigcup E_s(I) = E \setminus E_f(I)$, where E is the edge set of RH. Since I is not failed, event $I \cdot e_1...e_n|_\varepsilon$ is an s-event. Since I is also not successful, $e_1...e_n$ is not empty. By definition, $e_1...e_n$ is an $s - d$ path under I. Q.E.D

Following the above example, for the u-event $e_2\overline{e_3}|_\varepsilon$, we can find an $I_{min}(s-d) = e_1e_4$ in its $E_u(I)$ such that $e_2\overline{e_3}|_\varepsilon \cdot e_1e_4|_\varepsilon$ is an s-event.

The theorem is the foundation of the basic algorithm. Initially, we do not know any successful events, and the entire event space $(E|_\varepsilon)$ of RH is a u-event. Theorem 1

---

[3] To distinguish an event consisting of edges and the edge set (path) consisting the same edges, we use $E|_\varepsilon$ to denote the event while E denotes the corresponding edge set.

guarantees that we can find an $I_{min}(s - d) = e_1...e_n$ under any u-event. Thus we can determine all the successful events in which there is a sub-event $I_{min}(s - d)|_\varepsilon$. The following equation guarantees producing disjoint events.

For a u-event I, from the probabilistic theory, we have:

$$I = I \cdot \Omega = I \cdot (\overline{e_1 e_2...e_n}|_\varepsilon + e_1 e_2...e_n|_\varepsilon)$$
$$= I \cdot \overline{e_1}|_\varepsilon + I \cdot e_1\overline{e_2}|_\varepsilon + \cdots + I \cdot e_1 e_2...\overline{e_n}|_\varepsilon + I \cdot e_1 e_2...e_n|_\varepsilon. \tag{6}$$

Let $e_1...e_n$ (shorter than P_i) be an $I_{min}(s - d)$, then $I \cdot e_1 e_2...e_n|_\varepsilon$ is an s-event. However, events $I^i = I \cdot e_1 e_2...\overline{e_i}|_\varepsilon$ $(1 \le i \le n)$ may be successful, failed or undetermined. Thus we adopt Equation (6) continually to produce successful events till no s-events can be produced. Then we sum up the probabilities of all successful events, which is $Pr(Connect)$.

The computing process of $Pr(Connect)$ can be denoted by a *solution tree* defined as follows,

Definition 8. *A solution tree is a tree structure, its root node denotes the universal event space Ω of RH and other nodes denote the events $I \cdot e_1...\overline{e_i}|_\varepsilon$ $(1 \le i \le n)$ or $I \cdot e_1...e_n|_\varepsilon$ given in Equation (6). Specially, a leaf node only denotes an s-event or an f-event, but an intermediate node can denote a u-event, an s-event or an f-event.*

The algorithm of computing $Pr(Connect)$ works as follows. Initially, the solution tree only contains the root node denoting the universal system space Ω (Ω is the initial u-event) where each edge is uncertain. The algorithm begins with Ω, and finds an $s - d$ minpath[4] to divide Ω into disjoint events according to Equation (6). These produced events consists of nodes in the first level of the *solution tree* [5]. In the first level, event $e_1...e_\Omega|_\varepsilon$ is successful, while events $e_1...\overline{e_i}|_\varepsilon$ for $1 \le i \le n_\Omega$ may classified be failed, successful and undetermined. Similarly, each u-event is further recursively divided into disjoint events until there is no u-event left. The *solution tree* grows from the u-events until only s-events and f-events are left, which compose of the leaf nodes of the *solution tree*. The value of $Pr(Connect)$ is the summation of the probabilities over all s-events.

For example, we want to compute the SP probability of $P_4 = e_6$ whose weights are larger than other three paths in Fig. 2. Thus Fig. 1 is just the graph RH. According to Equation (4), we only need to compute $Pr(Connect)$ of Fig. 1. Figure 4 gives the *solution tree* of the basic algorithm for Fig. 1. In the example, there is one root Ω at first, then the algorithm finds an $s - d$ minpath $e_1 e_4$ under Ω to produce events $\overline{e_1}|_\varepsilon$ (u-event), $e_1\overline{e_4}|_\varepsilon$ (u-event) and $e_1 e_4|_\varepsilon$ (s-event) in the first level of the tree. The u-events are continually decomposed until the tree only contains s/f-events that are shown as leaf nodes of Fig. 4. All the $s/f/u$-events have been listed in the figure, and the figure also shows the probabilities of all successful events. The sum of successful event probabilities is $Pr(Connect)$ that is $0.68 + 0.1224 + 0.072 + 0.0162 + 0.027 = 0.9176$.

Furthermore, let $Pr(I_s)$ and $Pr(I_f)$ be the current accumulated probability of s-events and f-events in the basic algorithm. For $Pr(Connect)$, we have $Pr(I_s) \le$

---

[4] There are many $s - d$ minpaths available, and we can choose any minpath since it might enumerate exponential number nodes of the solution tree by choosing any minpath.

[5] The root of a solution tree is defined as 0th level.

$Pr(Connect) \leq 1 - Pr(I_f)$, so we can conduct the following examinations. If $Pr(I_s)$ $> Q$, the algorithm stops and the path is marked failed. If $1 - Pr(I_f) < Q$, we can also stop the computing and return the path to the answer set. As the accumulations increase, we can get tighter bounds which help stopping the calculation as early as possible. We call the above bounding technique *s/f event bound method*.

4.2 Probabilistic Pruning Rules

So far, we implicitly have a requirement: all found paths satisfying the predicate in the SP query. However, a threshold probability SP query is interested in only those paths whose SP probabilities are higher than the probability threshold. Thus we develop some probabilistic pruning rules to remove paths before traversing all the paths.

Theorem 2. $Pr_{SP}(P) \leq Pr(P) \leq Pr_{subpath}(P)$.

The pruning rule means if the probability of a candidate path or its sub-path is smaller than the threshold probability, the path can be pruned. We generate optimal paths in a weight-increasing sequence. The shorter path that is marked failed may have influence on the later generated path. Theorem 3 gives this property.

Theorem 3. *Given two paths P and P', if the weight of P' is larger than that of P and $Pr(P') \leq Pr(P)$, $Pr(P) < \epsilon$, then $Pr(P') < \epsilon$.*

To use Theorem 3, we maintain the largest probability of the paths that have been marked failed. Any new checked path identified by the above pruning rule should be marked failed as well.

To determine the answer set, we need to enumerate optimal paths one by one. In fact, users only care about some shortest paths that also have high probability. Thus the answer set (also shown in the experiments) is very small, and we do not need to check many paths. However we need a stop condition that can mark all the unchecked paths failed. To achieve this, we provide a tight stopping condition given as follows,

Theorem 4. *For the ith SP path, if its $Pr(Connect) > 1 - \epsilon$, we can stop generating new optimal paths P_m ($m > i$), and all those paths are marked failed candidates.*

Proof: Let $I_1, ..., I_k$ be the events that shorter paths in RH of P_i are connected. Since

$$Pr(Connect) = Pr(I_1 \bigcup \cdots \bigcup I_k) = 1 - Pr(\overline{I_1 \bigcup \cdots \bigcup I_k}) = 1 - Pr(\overline{I}_1 \bigcap \cdots \bigcap \overline{I}_k)$$

and $Pr(Connect) > 1 - \epsilon$, we have $Pr(\overline{I}_1 \bigcap \cdots \bigcap \overline{I}_k) < \epsilon$. For $Pr(Connect)$ of P_m, there exists an integer $k' > k$, such that $Pr(Connect) = Pr(\overline{I}_1 \bigcap \cdots \bigcap \overline{I}_k \bigcap \cdots \bigcap \overline{I}_{k'})$. Since $\overline{I}_1 \bigcap \cdots \bigcap \overline{I}_k \bigcap \cdots \bigcap \overline{I}_{k'} \subset \overline{I}_1 \bigcap \cdots \bigcap \overline{I}_k$, $Pr(\overline{I}_1 \bigcap \cdots \bigcap \overline{I}_{k'}) < \epsilon$. This result leads to the conclusion.   Q.E.D

Theorem 4 provides an upper bound for paths that have not been seen yet. If the probability satisfies the condition, then the unseen paths do not need to be checked.

5   Optimization of Basic Calculation Algorithm

As noted in [19], it is a hard problem to compute the exact value of $Pr(Connect)$, thus the basic algorithm may need to enumerate huge number of events for a large graph before the algorithm satisfies the stop condition. In this section, we propose another optimization method, isomorphic graphs reduction, to stop the computing as early as possible.

5.1   Isomorphic Graphs Reduction

As noted in Equation (6), any event $I \cdot e_1 e_2 ... \overline{e_i}|_\varepsilon$ $(1 \leq i \leq n)$ may be u-event. Thus it is very likely to produce sequence u-events in low levels of a *solution tree* so that those u-events may produce large number of sub-u-events in exponential. That is to say, for a large RH, the width of the *solution tree* may increase in exponential as the tree height increases, which leads to a very large computing cost. However, we observe that there exists relation between a sequence u-events. This information can be used to greatly reduce the width of the tree. The method is proposed in this subsection.

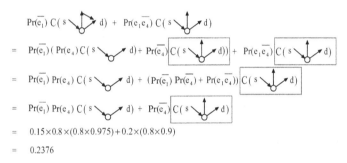

Fig. 5. Reduction of u-events in Fig. 4

For a u-event $E_u(I)|_\varepsilon$ of the *solution tree*, let $RH \backslash E_u(I)$ denote the subgraph of RH removing edge set $E_u(I)$ and $Pr(RH \backslash E_u(I))|_{con}$ denote the connected probability (from s to d) for $RH \backslash E_u(I)$. From the basic algorithm, we know that the contributed probability (to $Pr(Connect)$) of $E_u(I)|_\varepsilon$ is $Pr(E_u(I)) \cdot Pr(RH \backslash E_u(I))|_{con}$. For example, in Fig. 4, the contributed probability of $\overline{e}_1|_\varepsilon$ is $Pr(\overline{e}_1)Pr(RH \backslash e_1)|_{con}$, and the subtree rooted at $\overline{e}_1$ in Fig. 4 is the computing process of $Pr(RH \backslash e_1)|_{con}$.

For graph $S = RH \backslash E_u(I)$, let $S \backslash e$ denote that edge e is deleted from S and $S \diamond e$ denote that edge e is contracted (also called edge contraction) from S. *Edge contraction* is an operation which removes an edge from a graph while simultaneously merging together the two vertices that the edge used to connect. For S, there is an important property,

Theorem 5

$$Pr(S)|_{con} = Pr(e)Pr(S \diamond e)|_{con} + Pr(\overline{e})Pr(S \backslash e)|_{con}. \tag{7}$$

Proof: Let S_{con} denote the connected event from s to d in S. From the probabilistic theory, we have:

$$Pr(S)|_{con} = Pr(e)Pr(S_{con}|e) + Pr(\bar{e})Pr(S_{con}|\bar{e})$$

where $Pr(.|.)$ denotes the conditional probability.

Clearly, $Pr(S_{con}|\bar{e})$ equals $Pr(S\backslash e)|_{con}$. The probability of $Pr(S_{con}|e)$ denotes the potential impact on S_{con} given the condition that edge e is successful. Thus, we can contract e, and $Pr(S_{con}|e)$ equals $Pr(S \diamond e)|_{con}$.   Q.E.D

Let $I \cdot e_1...\overline{e_i}|_\varepsilon = I_i$ and $I \cdot e_1...\overline{e_{i+1}}|_\varepsilon = I_{i+1}$ be two neighbor u-events produced in the *solution tree*, and let F_i and F_{i+1} denote subgraphs $RH\backslash I_i$ and $RH\backslash I_{i+1}$ respectively.

Thus for two neighbor u-events, their contributed probability (to $Pr(Connect)$) is,

$$Pr(I_i)Pr(F_i)|_{con} + Pr(I_{i+1})Pr(F_{i+1})|_{con} = Pr(I_i)[Pr(e_{i+1})Pr(F_i \diamond e_{i+1})|_{con} + Pr(\bar{e}_{i+1}) \\ Pr(F_i\backslash e_{i+1})|_{con}] + Pr(I_{i+1})Pr(F_{i+1})|_{con}$$

For two neighbor u-events, F_i and F_{i+1} have same nodes, and F_i has an additional edge e_{i+1} than F_{i+1}. Hence $F_i\backslash e_{i+1}$ and F_{i+1} are same (isomorphic) graphs, then the above equation can be rewritten as:

$$Pr(I_i)Pr(e_{i+1})Pr(F_i \diamond e_{i+1})|_{con} + Pr(F_{i+1})|_{con}[Pr(I_i)Pr(\bar{e}_{i+1}) + Pr(I_{i+1})]$$

In the above deduction, F_i produces a graph that is isomorphic to F_{i+1}, and the two isomorphic graphs are combined into one. From the result, we know that two u-events produce two new events, and one of which has smaller graph size. Thus the combining process avoids producing many new u-events as the basic algorithm. For more than two neighbor u-events, we use the property given in Equation (7) to factor these u-events and combine isomorphic graphs. Finally we can get the same number of events as those u-events, while basic algorithm may produce exponential number events. Thus the width of the *solution tree* can be largely reduced. Figure 5 demonstrates the process of combining isomorphic graphs for two u-events ($\bar{e}_1$ and $e_1\bar{e}_4$) listed in Fig. 4. During the process, two isomorphic graphs (signed within the rectangle) are combined into one graph. From the figure, we know that original u-events produce two events with corresponding smaller graphs, and the produced events are both successful. Thus we can directly get the result. The final answer is 0.2376 which is the same as shown in Fig. 4 (0.027+0.0162+0.072+0.1224=0.2376). However in Fig. 4, two original u-events produce two u-events, again, the decomposition needs to be continued. To apply this improved algorithm, we reduce the sequence u-events continually until there are no neighbor u-events.

6   Performance Evaluation

We conduct an empirical study using real data sets on a PC with a 3.0 GHz Pentium4 CPU, 2.0 GB main memory, and a 160 GB hard disk. Our algorithms are implemented

Table 2. Data sets

	G_1	G_2	G_3	G_4	G_u
No. of nodes	1k	4k	8k	16k	21k
No. of edges	1.6k	6.4k	13k	25k	21.6k

in Microsoft Visual C++ 6.0. In the following experiments, *BA* denotes the basic algorithm, *ICA* denotes the isomorphic graphs combination method (*ICA*), *s,f-Bound* denotes the s/f event bound approach, and *Overall* denotes the combination of three techniques together.

As introduced in Section 1, the uncertain graph model is abstracted from the transportation network and communication fields. Thus we evaluate the proposed algorithms on the real road network data set: California Road Network with 21,047 nodes and 21,692 edges. The data set is extracted from the US Census Bureau TIGER/LineThere[6]. Note that the four algorithms can handle both undirected and directed graphs. In experiments, we represent the real database as a directed uncertain graph G_u. We further generate 4 subgraphs $G_1, ..., G_4$ from G_u with nodes varying from 1k to 16k. Each subgraph corresponds to a subarea of G_u. To test the sensitivity of the algorithms, we increase the density of G_4 by generating its edges to be 25k. The numbers of nodes and edges of $G_1, ..., G_4$ and G_u are listed in Table 2. Those graphs are all certain. To simulate an uncertain graph, we generate an existent probability for each edge following a Normal distribution $N(\mu, \sigma)$. The value of μ is set to the mean weight of all edges, i.e., $\mu = \sum_{i=1}^{|E|} w_i / |E|$, where w_i is the original weight of the corresponding edge in data sets. σ is also generated following Normal distribution $N(\mu_\sigma, \sigma_\sigma)$, where $\mu_\sigma = x\mu$, and μ is the mean weight. This simulation method follows the findings in studies on traffic simulations [14] and [15], which indicates that the travel time on paths in road networks follows the Normal distribution.

The path queries are generated as follows. The issued probability threshold is set from 0.2 to 0.7 and the default value is 0.4. The value of x is set to $x = 1\%$-5%, and the default value is 3%. For each parameter setting, we randomly run 50 path queries and report the average results.

Firstly, we test the pruning power of algorithms on a small data set (G_1) and a large data set (G_u). Figure 6 shows the number of traversed paths by queries and the answer size with different threshold probabilities and variances ($x\%$) for G_1. In Fig. 6(a), we know that the number of paths is decreasing as probabilities increase. This is because the small threshold probability leads to more answers (paths) that need to enumerate more paths to determine the final answers. Obviously *ICA* has a more pruning power than *BA*, since *ICA* reduces the number of traversed paths by combining isomorphic graphs. *s,f-Bound* beats *ICA*, and needs fewer traversed paths. Based on the whole improved methods, *Overall* prunes most paths, and the number of paths comes very close to the final result. As stated earlier, in the worst case, it may need to enumerate an exponential number of edges of RH to compute $Pr(Connect)$ in *BA*. But as shown in the result, our improved methods can avoid the case. The number of traversed paths for

---

[6] Topologically Integrated Geographic Encoding and Referencing system:
http://www.census.gov/geo/www/tiger/

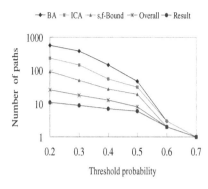
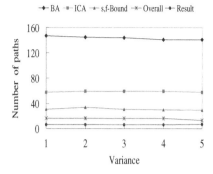

(a) Number of traversed paths vs. probability   (b) Number of traversed paths vs. variance
threshold

Fig. 6. Pruning power of the algorithms on G_1

different variances is given in Fig. 6(b). There is a vary small decreasing tendency for all algorithms and most of curves are stable, which indicates the pruning power is not sensitive to the varying probability distribution. For G_u, Figure 7 gives the results from which we know this data set needs much more paths than G_1 due to the fact that it leads to the largest uncertain graph. However as shown in Fig. 7(a), after adopting *overall*, we get a very small result that in the worst case (the threshold probability is 0.2) that we only need to traverse less than 70 paths to produce 23 results. Though G_u is a very large graph, the size of RH is small and hence the algorithms can answer queries efficiently. The results of above experiments with different threshold probabilities have a common feature that all curves of the four algorithms drop very quickly after the probability is 0.5. The numbers of paths are almost the same when the threshold probability is large, as shown in Fig. 6(a). The reason is that the answer set is very small when the threshold probability is large, and the small answer set leads to a small number of traversed paths.

We also evaluate the scalability of the algorithms. To evaluate the effeteness with different nodes, besides above two data sets, we further test the algorithms on graphs G_2, G_3 and G_4. Figure 8(a) shows the pruning power of the algorithms on the five data sets. All algorithms are scalable, and the improved algorithms that obliviously prune more paths than BA. The *s,f-Bound* have an efficient pruning power in the largest data set with 21.6k nodes, which is also indicated in Fig. 8(b) that shows the runtime on the five graphs. In this figure, *s,f-Bound* has a very short runtime of less than 10 seconds in the largest graph. If we apply *Overall*, the results can be much better and the runtime shown in Fig. 8(b) for all the graphs is less than 2 seconds. But as shown in Fig. 8, there is a sudden increase in G_4, then the curves decrease to G_u. The main reason is G_4 has more edges than G_u, which shows that both running efficiency and pruning power are sensitive to the number of edges of the uncertain graph. We also evaluate the scalability of pruning power and runtime with different edge sizes. To do this, we vary the density of G_2 by fixing the number of nodes as 4K while changing the number of edges. Five graphs are generated with 4K, 8K, 16K, 32K and 64K edges. We report the results in Fig. 9. The results confirm our finding that the number of traversed paths and runtime

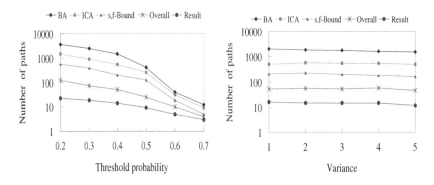

(a) Number of traversed paths vs. probability (b) Number of traversed paths vs. variance
threshold

Fig. 7. Pruning power of the algorithms on G_u

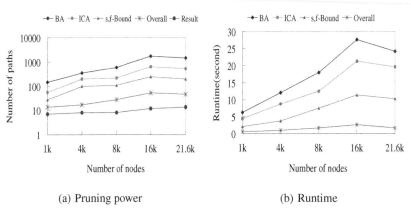

(a) Pruning power   (b) Runtime

Fig. 8. Scalability with different node sizes

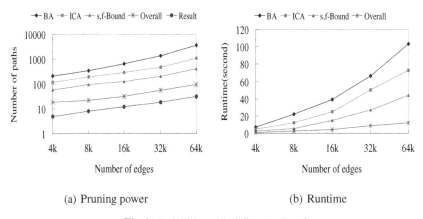

(a) Pruning power   (b) Runtime

Fig. 9. Scalability with different edge sizes

are sensitive to the number of edges. The results of runtime also indicate that, for each improved algorithm, the additional time of computing bounds or combining isomorphic graphs does not have an impact on the final runtime.

7   Conclusions

In this paper, the probability threshold-based SP path query is studied over a proposed uncertain graph model. An efficient basic probability computing method and several probabilistic pruning rules are proposed to avoid scanning all the possible worlds. An improved computing scheme based on combinations of isomorphic graphs can greatly reduce the probabilistic calculation cost. Also, a series of lower/upper bounds are given to stop the calculation as early as possible. Finally, we confirm our design through an extensive experimental study.

Acknowledgement. This research was supported by the National Natural Science Foundation of China (Grant No. 60873011,60773221, 60933001), the 863 Program (Grant No. 2007AA01Z192, 2009AA01Z131), and National Basic Research Program of China (Grant No. 2006CB303103). This research was also supported by National Natural Science Foundation of China (Grant No. 60803026) and Ph.D. Programs Foundation (Young Teacher) of Ministry of Education of China (Grant No. 20070145112). Lei Chen was supported by NSFC/RGC Joint Research Scheme under Project No. N HKUST602/08.

References

1. Suciu, D., Dalvi, N.N.: Foundations of probabilistic answers to queries. In: SIGMOD Conference (2005)
2. Dalvi, N.N., Suciu, D.: Management of probabilistic data: foundations and challenges. In: ACM PODS (2007)
3. Khoussainova, N., Balazinska, M., Suciu, D.: Towards correcting input data errors probabilistically using integrity constraints. In: ACM MobiDE Workshop (2006)
4. Benjelloun, O., Sarma, A.D., Hayworth, C., Widomn, J.: An introduction to ULDBs and the Trio system. IEEE Data Engineering Bulletin 29(1), 5–16 (2006)
5. Cormode, G., Garofalakis, M.: Sketching probabilistic data streams. In: ACM SIGMOD (2007)
6. Korzan, B.: Method of determining compromise paths in unreliable directed graphs. Bulletin of the Military University of Technology, Warsaw (1982)
7. Korzan, B.: Method of determining nondominated paths in unreliable directed graphs. Bulletin of the Military University of Technology, Warsaw (1983)
8. Sigal, C.E., Pritsker, A.A.B., Solberg, J.J.: The stochastic shortest route problem. Oper. Res. 28(5) (1980)
9. Cormode, G., McGregor, A.: Approximation algorithms for clustering uncertain data. In: ACM PODS (2008)
10. Zhang, Q., Li, F., Yi, K.: Finding frequent items in probabilistic data. In: ACM SIGMOD (2008)
11. Guerin, R.A., Orda, A.: QoS routing in networks with inaccurate information: Theory and algorithms. IEEE/ACM. Trans. (1999)

12. Lorenz, D.H., Orda, A.: QoS routing in networks with uncertain parameters. TON (1998)
13. Chen, S., Nahrstedt, K.: Distributed QoS routing in ad-hoc networks. In: IEEE JSAC (August 1999)
14. Fu, L., Rilett, L.R.: Expected shortest paths in dynamic and stochastic traffic networks. Transportation Research - part B (1998)
15. Turner, S.M., Brydia Robert, E., Liu, J.C.: ITS Data Management System: Year One Activities, Report No. FHWA/TX-98/1752-2. Texas Department of Transportation, Texas Transportation Institute (September 1997)
16. Papadimitriou, C.H., Yannakakis, M.: Shortest paths without a map. Theoretical Computer Science (1991)
17. Chabini, I.: Algorithms for k-shortest paths and other routing problems in time-dependent networks. Transportation Research Part B: Methodological (2002)
18. Silva, R., Craveirinha, J.: An Overview of routing models for MPLS Networks. In: Proc. of 1st Workshop on Multicriteria Modelling in Telecommunication Network Planning and Design (2004)
19. Valiant, L.G.: The Complexity of enumeration and reliability prblems. SIAM JL of Computing (August 1979)
20. Kerbache, L., Smith, J.: Multi-objective routing within large scale facilities using open finite queueing networks. Europ. J. Oper. Res. (2000)
21. Hershberger, J., Suri, S., Bhosle, A.: On the difficulty of some shortest path problems. In: Proc. of Sympos. Theoret. Aspects Comput. Sci. LNCS. Springer, Heidelberg (2003)
22. Ljosa, V., Singh, A.K.: APLA: indexing arbitrary probability distributions. In: Proc. of ICDE (2007)
23. Soliman, M.A., Ilyas, I.F., Chang, K.C.: Top-k query processing in uncertain databases. In: Proc. of ICDE (2007)
24. Re, C., Dalvi, N., Suciu, D.: Efficient top-k query evaluation on probabilistic data. In: Proc. of ICDE (2007)
25. Cheng, R., Xia, Y., et al.: Efficient indexing methods for probabilistic threshold queries over uncertain data. In: Proc. VLDB (2004)
26. Kriegel, H.P., Kunath, P., Renz, M.: Probabilistic nearest-neighbor query on uncertain objects. In: Kotagiri, R., Radha Krishna, P., Mohania, M., Nantajeewarawat, E. (eds.) DASFAA 2007. LNCS, vol. 4443, pp. 337–348. Springer, Heidelberg (2007)
27. Pei, J., Jiang, B., Lin, X., Yuan, Y.: Probabilistic skylines on uncertain data. In: Proc. of VLDB (2007)

Discovering Burst Areas in Fast Evolving Graphs

Zheng Liu and Jeffrey Xu Yu

Department Systems Engineering & Engineering Management,
The Chinese University of Hong Kong, Hong Kong
{zliu,yu}@se.cuhk.edu.hk

Abstract. Evolving graphs are used to model the relationship variations between objects in many application domains such as social networks, sensor networks, and telecommunication. In this paper, we study a new problem of discovering burst areas that exhibit dramatic changes during some periods in evolving graphs. We focus on finding the top-k results in a stream of fast graph evolutions. This problem is challenging because when the graph evolutions are coming in a high speed, the solution should be capable of handling a large amount of evolutions in short time and returning the top-k results as soon as possible. The experimental results on real data sets show that our proposed solution is very efficient and effective.

Keywords: Evolving Graphs, Burst Areas, Haar Wavelet.

1 Introduction

Graph patterns have the expressive ability to represent the complex structural relationships among objects in social networks, as well as in many other domains including Web analysis, sensor networks, and telecommunication. The popularity of social Web sites in recent years has attracted much attentions on social networks, hence the research interests on mining large graph data [6,2,9,5,4]. However, graphs are not static but evolving over time. Users in social Web sites participate in various activities such as writing blogs and commenting stories. These interactive activities happen all the time and cause the social networks changing rapidly and continuously.

In most social networks such as Digg [1], one of the common activities of users is to make comments on stories. Then a bipartite graph can be constructed by considering users and stories as vertices. There is an edge between a user and a story if the user submits a comment on the story. Let us assume Fig. 1(a) is a user-story graph at some time t. As time goes by, users submit more comments on stories and the graph evolves. Suppose at time $t + \delta t$, the user-story graph looks like one shown in Fig. 1(b). Since both the involvement of users and the popularity of stories are various, the degree of change may be different at each region in the graph. For example, as shown in the dotted line area, this region is much different from one at time t, while the remaining part looks similar, which means users in this region are more active and stories in it are more attractive.

H. Kitagawa et al. (Eds.): DASFAA 2010, Part I, LNCS 5981, pp. 171–185, 2010.

(a) Time t (b) Time $t + \delta t$

Fig. 1. An Evolving User-Story Graph

Inspiring by the motivation from Fig. 1, in this paper, we study a new problem of discovering the burst areas, that exhibit dramatic changes for a limited period, in fast graph evolutions. Intuitively, dramatic changes mean the total evolutions happened inside burst areas are much larger than one in other areas. There are several difficulties to solve this problem. First, evolving graphs in social networks are huge, which contain a large amount of vertices and edges. Second, sizes of burst areas could be various. And last, the duration of the burst period is difficult to predict, since a burst could last for minutes, hours, days or even weeks. All these difficulties make this problem challenging and interesting. A candidate solution must be efficient enough to deal with a great number of computations.

We focus on bipartite evolving graphs, since fast evolving graphs in social networks are mostly heterogeneous bipartite graphs. Similar to commenting stories, other possible activities of users could be either writing blogs, tagging photos, watching videos, or playing games. Each of such activities can be a fast evolving bipartite graph. In an evolving graph, there is a weight associated with each vertex or edge. The evolutions are in form of the change of weights of nodes/edges. The weight of a non-existing node/edge is zero, so it does not matter whether the coming nodes/edges are new to the graph.

The main contributions of this paper are summarized below.

– We formalize the problem of discovering burst areas in rapidly evolving graphs. The burst areas are ranked by the total evolutions happened inside and the top-k results are returned.
– Instead of calculating the total evolutions of every possible period, we propose to use Haar wavelet tree to maintain upper bounds of total evolutions for burst areas. We also develop an incremental algorithm to compute the burst areas of different sizes in order to minimize the memory usage.
– We present an evaluation of our proposed approach by using large real data sets demonstrating that our method is able to find burst areas efficiently.

The rest of this paper are organized as follows. Section 2 introduces the preliminary background knowledge and formalizes the problem of burst area discovery in an evolving graph. We present our computation approaches in Section 3. Experimental results are presented in Section 4, and Section 5 discusses the related work. Finally, Section 6 concludes the paper.

2   Problem Statement

We give an overview on how to model evolving graphs at first, before we come to any details about the problem and solution. There are mainly two approaches to model evolving graphs. One way is to represent an evolving graph as a sequence of graphs, $\mathcal{G} = (G_1, G_2, ...)$. Each graph G_i in the sequence is a snapshot of the evolving graph at time t_i. The advantage of this way is that it is convenient for users to study the characteristics of an evolving graph at a particular time stamp, as well as the differences between graphs of adjacent time stamps. One issue of this approach is the large storage cost in proportion to the size of the evolving graph and the time intervals between snapshots. The other method models an evolving graph as an initial graph, which is optional, and a stream of graph evolutions. This approach is more intuitive in most domains in the real world. For example, the interactive activities in social networks can be considered as an evolution stream. We model evolving graphs using the second approach in this paper since we are more interested in the burst areas of graph evolutions, not the graph characteristics at the current time.

An evolving graph $\mathcal{G} = (G, \Delta)$ consists of two parts, an initial graph G and a sequence of evolutions Δ. The initial graph G is a snapshot of the evolving graph at time t_0 with a set of vertices $V(G)$ and a set of edges $E(G)$. Let w_i denote the weight of vertex $v_i \in V$ and $w_{ij} \in E$ denote the weight of edge $e_{ij} = (v_i, v_j)$. Each item δ_t in the evolution stream Δ is a set of quantities indicating the weight changes of vertices or edges at the time t. There might be a number of evolutions at the same time. Let δ_t^i and δ_t^{ij} denote the weight change of vertex v_i and edge e_{ij} at time t, respectively. Without loss of generality, we assume the evolutions come periodically.

Given a large evolving graph $\mathcal{G} = (G, \Delta)$, we study the problem of finding burst areas. Since a burst region is actually a connected subgraph of the evolving graph, then any connected subgraph might be a possible burst area. Apparently, it is more likely that the total evolutions in a subgraph with many vertices/edges is greater than the one in a subgraph with fewer vertices/edges. Thus, it is insignificant to compare total evolutions among subgraphs with large differences in terms of vertex/edge quantity. Consequently, we introduce the r-radius subgraph, which is more meaningful and challenging.

For a given vertex v_i in a graph, the eccentricity EC of v_i is the maximum length of shortest paths between v_i and any other vertex in the graph. Based on the definition of eccentricity, the r-radius subgraph is defined as below.

Definition 1. *(r-Radius Subgraph)*
A subgraph $g = (V(g), E(g))$ in a graph G is an r-radius subgraph, if

$$\min_{v_i \in V(g)} EC(v_i) = r. \tag{1}$$

The r-radius subgraphs in a large graph may be overlapping and result in redundancy. To avoid this, we introduce the concept of maximum r-radius subgraph.

Definition 2. *(Maximal r-Radius Subgraph)*
An r-radius subgraph g^r is called maximal r-radius subgraph if there exists no other r-radius subgraph $g'^r \subseteq G$, which contains g^r.

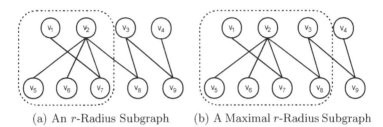

(a) An r-Radius Subgraph   (b) A Maximal r-Radius Subgraph

Fig. 2. An r-Radius Subgraph Example

Fig. 2 shows an example of r-radius subgraph and maximal r-radius subgraph. Suppose Fig. 2(a) and Fig. 2(b) demonstrate the same graph G. The subgraph in the dotted line area in Fig. 2(a) is a 2-radius subgraph. It is contained by the subgraph in the dotted line area in Fig. 2(b), which is a maximal r-radius subgraph.

It is a difficult task to identify all maximal r-radius subgraphs from a large evolving graph. We observe that a maximal r-radius subgraph is in fact a maximal r-hop neighborhood subgraph, which we define below. Let $N_{v_i}^r$ denote the set of vertices (except v_i) whose shortest path distance to v_i is less than or equal to r.

Definition 3. *(r-Hop Neighborhood Subgraph)*
An r-hop neighborhood subgraph $g_{v_i}^r$ in a graph G is defined as the subgraph of G containing all the vertices in $N_{v_i}^r$. v_i is called the center of $g_{v_i}^r$.

In the following paper, we might use r-hop subgraph for short.

We show the relationship between maximal r-radius subgraphs and r-hop neighborhood subgraphs in Theorem 1.

Theorem 1. *For each maximal r-radius subgraph $g^r \subseteq G$, there is a corresponding r-hop neighborhood subgraph $g_{v_i}^r \subseteq G$, and $g^r = g_{v_i}^r$.*

Proof Sketch: We will prove that (1) $\exists v_i \in g^r$, $g^r \subseteq g_{v_i}^r$, and (2) $\nexists v_j \in g_{v_i}^r$, $v_j \notin g^r$.

Let g^r be a maximal r-radius subgraph belonging to G. Recall that the eccentricity $EC(v_i)$ is the maximum length of shortest paths between v_i and any other vertex in g^r. Then based on Definition 1, there exists a vertex $v_i \in V(g^r)$ and $EC(v_i) = r$. Let $d(v_i, v_j)$ denote the length of the shortest path between v_i and v_j. Because $EC(v_i) = r$, so $\forall v_j \in V(g^r)$, we have $d(v_i, v_j) \leq r$. Let $g_{v_i}^r$ be an r-hop neighborhood subgraph based on Definition 3. Since $\forall v_j \in V(g^r), d(v_i, v_j) \leq r$, so $\forall v_j \in V(g^r)$, $v_j \in N_{v_i}^r$, where $N_{v_i}^r = V(g_{v_i}^r)$. Therefore, $\exists v_i \in g^r$, $g^r \subseteq g_{v_i}^r$.

Suppose $\exists v_k \in g_{v_i}^r$ and $v_k \notin g^r$, we have $d(v_i, v_k) \leq r$. Let g' denote the subgraph that $V(g') = V(g^r) \cup \{v_k\}$, then g' is also an r-radius subgraph, which is contradict to the condition that g^r is a maximal r-radius subgraph. Therefore, $\nexists v_j \in g_{v_i}^r, v_j \notin g^r$.   □

Theorem 1 indicates that a maximal r-radius subgraph must be an r-hop neighborhood subgraph. This is only a necessary condition, not sufficient. It is worth noting that an r-hop neighborhood subgraph might not be an r-radius subgraph. Take vertex v_4 in Fig. 2(a) as an example. Let us construct a 2-hop neighborhood subgraph $g_{v_4}^2$, which contains vertex v_3, v_4 and v_9. $g_{v_4}^2$ is not a maximal 2-radius subgraph, but a maximal 1-radius subgraph $g_{v_9}^1$, because $d(v_3, v_9) = d(v_4, v_9) = 1$. In this paper, we consider r-hop neighborhood subgraphs as the candidates of burst areas.

Recall δ_t^i and δ_t^{ij} denote the weight change of vertex v_i and edge e_{ij} at time t, respectively. Obviously, if $v_i \in N_{v_j}^r$, which means v_i is in the r-hop neighborhood subgraph $g_{v_j}^r$, then δ_t^i should be counted into the $g_{v_i}^r$. δ_t^{ij} belongs to an r-hop neighborhood subgraph when both v_i and v_j are in the subgraph. We define the burst score of an r-hop neighborhood subgraph as follows.

Definition 4. *(Burst Score)*
The vertex burst score of an r-hop neighborhood subgraph $g_{v_i}^r$ at time t is the total weights of the vertex evolutions happened inside.

$$BurstScore^V = \sum_{v_j \in g_{v_i}^r} \delta_t^j \qquad (2)$$

The edge burst score of an r-hop neighborhood subgraph $g_{v_i}^r$ at time t is the total weights of the edge evolutions happened inside.

$$BurstScore^E = \sum_{v_j, v_k \in g_{v_i}^r} \delta_t^{jk} \qquad (3)$$

So, the burst score of an r-hop neighborhood subgraph $g_{v_i}^r$ is sum of the vertex burst score and the edge burst score.

Now, we formally define the problem of discovering top-k burst areas in fast evolving graphs.

Problem 1. (Discovering Top-k Burst Areas)
For an evolving graph $\mathcal{G} = (G, \Delta)$, given a maximum hop size r_{max}, a burst window range (l_{min}, l_{max}), the top-k burst area discovery problem is that for each burst window size between l_{min} and l_{max} and hop size between 1 and r_{max}, finding the top-k r-hop neighborhood subgraphs with the highest burst scores in $\mathcal{G}$ at each time stamp continuously.

For conciseness, in the following, we focus on edge evolutions in heterogeneous bipartite evolving graphs. Our proposed solution can deal with vertex evolutions as well.

3   Discovering Burst Areas

A direct solution to discover top-k burst areas would be maintain total $(l_{max} - l_{min} + 1) \times r$ burst scores of each window size and hop size over sliding windows. At each time stamp t, these burst scores are updated based on the evolutions happened inside the corresponding r-hop neighborhood subgraphs. Then, for each window size and hop size, a list of top-k r-hop subgraphs based on burst scores is returned as the answer. Before we explain our proposed solution in details, which is much more efficient both in time complexity and memory consumption, we first introduce some background knowledge.

3.1   Haar Wavelet Decomposition

The Wavelet Decomposition is widely used in various domains, especially the signal processing. One of the conceptually simplest wavelet, Haar wavelet, is applied to compress the time series and speed up the similarity search in the time series database. The Haar wavelet decomposition is done by averaging two adjacent values on the time series repeatedly in multiple resolutions in a hierarchical structure, called Haar wavelet tree. The hierarchical structure can be constructed in $O(n)$ time. Fig. 3 illustrates how to construct the Haar wavelet tree[1] of a eight-value time series, which is at Level 0. Then at Level 1, there are four average value of adjacent values. The averaging process is repeated until there is only one average value left.

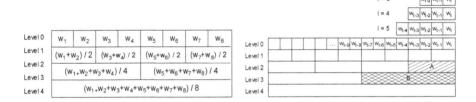

Fig. 3. Haar Wavelet Decomposition   **Fig. 4.** Upper Bounds of Burst Scores

3.2   Bounding Burst Scores of r-Hop Neighborhood Subgraphs

As defined in Definition 4, the burst score of an r-hop neighborhood subgraph is the total changed weights happened inside for a period of time. Given an evolving graph $\mathcal{G} = (G, \Delta)$ and a window size range (l_{min}, l_{max}), we introduce first how to bound r-hop burst scores for an r-hop subgraph.

Fig. 4 shows an example. w_t is the sum of all the changed weights happened in an r-hop neighborhood subgraph at time stamp t, Based on the Haar wavelet decomposition, we can construct the Haar wavelet tree as shown at the bottom in

---

[1] The Haar wavelet decomposition consists of both averages and differences. For conciseness, we ignore the difference coefficients which are not used in our solution.

Fig. 4. Suppose $l_{min} = 3$ and $l_{max} = 5$, the three corresponding burst windows are shown at the top. As we can see that, the burst windows of size 3 and 4 are contained in the window A at Level 2, while the burst window of size 5 is contained in windows B at Level 3. This leads to the following lemma.

Lemma 1. *A burst window of length l at time t is contained in the window at time t at Level $\lceil \log_2 l \rceil$ in the hierarchical Haar wavelet tree.*

Proof Sketch: Let $W = w_{t-l+1}, w_{t-l+2}, ..., w_t$ denote the burst window of length l. Based on the definition of the Haar wavelet decomposition, The length of window at time t at Level n is 2^n. Let $W' = w_t, w_{t-1}, ..., w'_{t-2^{(\lceil \log_2 l \rceil)+1}}$. Because $2^{(\lceil \log_2 l \rceil)} \geq l$, so $W \subseteq W'$. $\square$

Instead of average coefficients, we maintain sums of windows in the Haar wavelet tree. Since the changed weights are all positive, the sum in a window in the Haar wavelet tree is the upper bound of burst scores of all burst windows it contains.

Lemma 2. *The burst score of a length l burst window at time t is bounded by the sum coefficient of the window at time t at Level $\lceil \log_2 l \rceil$ in the Haar wavelet tree.*

We can use Lemma 2 to prune potential burst areas. If the burst score bound of an r-hop neighborhood subgraph for some window size is larger than the minimum score in the current top-k answers, then we perform a detailed search to check whether it is a true top-k answer. Otherwise, the r-hop subgraph is ignored. It is not necessary to build the whole Haar wavelet tree of all levels to compute burst score bounds of r-hop subgraphs. As we can see from Lemma 2, only the levels from $\lceil \log_2^{l_{min}} \rceil - 1$ to $\lceil \log_2^{l_{max}} \rceil$ are needed to compute the bound burst scores. Level $\lceil \log_2^{l_{min}} \rceil - 1$ is computed directly from Level 0.

Now, the problem is how to maintain the wavelet tree at each time stamp t, since graph evolutions come as a stream. There are mainly two approaches.

1. **Continuous Updating:** The entire Haar wavelet tree is updated at each time stamp t continuously. The approach ensures no delay in response time to return top-k answers.

2. **Lazy Updating:** Only windows at the lowest level are updated at each time stamp t. The sums maintained in the upper levels in the Haar wavelet tree are not computed until all data in the corresponding windows is available. For a burst window of size l, the response time delays at most $2^{\lceil \log_2 l \rceil}$.

In this paper, we propose to maintain Haar wavelet tree in a dynamic manner, which can achieve both low computation cost and no delay in response time. Fig. 5 presents a running example, which illustrates how the Haar wavelet tree changes as time goes by. Function $S(t, t')$ denotes the sum of weights in the window from time t to t'.

As shown in Fig. 5, suppose at time t, a Haar wavelet tree is built according to changed weights at Level 0. Then at time $t + 1$, instead of updating the entire Haar wavelet tree, we only shift each level left for one window and add the newly

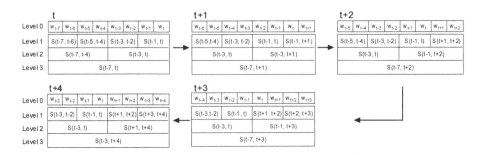

Fig. 5. Updating Haar Wavelet Tree

changed weight w_{t+1} to the last window at each level. Since changed weights are all positive, the sums of the last windows are still the upper bounds of burst scores of corresponding burst windows. At time $t+2$, since all weights at Level 0 used to compute the sum of the last window at Level 1 are available, we compute the actual sum of the last window at Level 1 based on Level 0. Then based on the weights at Level 1, the sum of the last window at Level 2 is recomputed. While the last window at Level 3 is not recomputed since the last two windows at Level 2 are overlapping. Instead, we add w_{t+1} to it. Time $t+3$ is similar to time $t+1$. At time $t+4$, last windows at all levels are recomputed, because the last two windows of lower levels are not overlapping. In general, last window at the lowest level (Level $\lceil \log_2^{l_{min}} \rceil - 1$) is computed every $2^{\lceil \log_2^{l_{min}} \rceil - 1}$ time stamps, while last windows at upper levels are recomputed once the last two windows at lower levels are not overlapping.

3.3   Incremental Computation of Multiple Hop Sizes

Suppose we need to monitor r-hop neighborhood subgraphs in multiple hop sizes, an easy solution is to maintain a Haar wavelet tree for each hop size of every r-hop subgraph. The total memory usage would be $O(rN)$, where N is the total number of vertices. Obviously, it is not efficient in the computation cost, and the memory assumption is high. In this section, we introduce our proposed algorithm to maintain burst score bounds of multiple hop sizes using at most $O(N)$ memory consumption. Our solution is to maintain Haar wavelet trees for 1-hop neighborhood subgraphs only. The burst score bounds of an r-hop subgraph is calculated from subgraphs of smaller hop size in an incremental manner.

Let first examine how the edge evolutions affect the burst scores of nearby r-hop neighborhood subgraphs using examples in Fig. 6. Fig. 6(a) shows an example for 1-hop neighborhood subgraph $g_{v_1}^1$. The dotted line is the edge evolution happened. It is apparent that if the edge evolution belongs to $g_{v_1}^1$, v_1 must be one of the vertices of the edge evolution. Fig. 6(b) shows an example for 2-hop neighborhood subgraph $g_{v_1}^2$. As we can see that if the edge evolution belongs to $g_{v_1}^2$, Either it is within $g_{v_1}^1$, or $N_{v_1}^2 \setminus \{v_1\}$. An edge evolution belongs $N_{v_1}^2 \setminus \{v_1\}$ means both vertices of the edge evolution belong to $N_{v_1}^2 \setminus \{v_1\}$. In this paper,

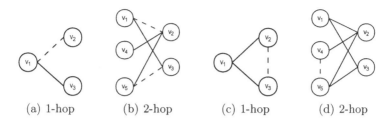

Fig. 6. Evolutions in 1-hop and 2-hop Subgraphs

we are focusing on heterogenous bipartite graph. Suppose v_1, v_4, v_5 and v_2, v_3 belong to the two sides of a bipartite graph, respectively. Then there are no such evolutions as shown by the dotted edges in Fig. 6(c) and 6(d).

Now we explain how to compute burst scores of r-hop neighborhood subgraphs incrementally from burst scores of 1-hop subgraphs. From the above obversion, we can see that the burst score of an r-hop subgraph is the sum of two parts. One is the burst score of $(r-2)$-hop subgraph, the other is the total evolutions within $N_{v_i}^r \setminus N_{v_i}^{r-2}$. While edge evolutions in $N_{v_i}^r \setminus N_{v_i}^{r-2}$ must be connected to one of the vertices in $N_{v_i}^{r-1} \setminus N_{v_i}^{r-2}$. Let $v_j \in N_{v_i}^{r-1} \setminus N_{v_i}^{r-2}$, then we have the following lemma.

Lemma 3. *The total edge evolutions in* $N_{v_i}^r \setminus N_{v_i}^{r-2}$ *equal to*

$$\sum_{v_j \in N_{v_i}^{r-1} \setminus N_{v_i}^{r-2}} BurstScore^1 v_j. \tag{4}$$

Based on Lemma 3, burst scores of r-hop neighborhood subgraphs are calculated incrementally by using the following equation.

$$BurstScore_{v_i}^r = BurstScore_{v_i}^{r-2} + \sum_{v_j \in N_{v_i}^{r-1} \setminus N_{v_i}^{r-2}} BurstSocre_{v_j}^1 \tag{5}$$

where $BurstScore_{v_i}^r$ denote the burst score of r-hop neighborhood subgraph $g_{v_i}^r$, and $BurstScore_{v_i}^0 = 0$. It is obvious that Eq. 5 is also correct if we substitute burst scores by their upper bounds.

3.4   Top-k Burst Area Discovery

The whole algorithm is presented in Algorithm 1. At each time t, the algorithm maintains Haar wavelet trees of all 1-hop neighborhood subgraphs at Line 3. If a vertex is saw for the first time, a new Haar wavelet tree is constructed. Otherwise, based on Section 3.2, Algorithm 1 updates all the Haar wavelet trees which have evolutions happened inside.

In each loop from Line 4 to Line 12, the algorithm discovers incrementally the top-k results from small hop size to large hop size. At Line 6, Algorithm 1 computes the upper bounds of burst scores according to Eq. 5. If the burst score

Algorithm 1. The Top-k Burst Area Discovery Algorithm

Input: An evolving graph $\mathcal{G} = (G, \Delta)$,
        a maximal hop size r_{max}, a window range (l_{min}, l_{max}), the value of k
Output: the top-k burst areas at each time t

1: **while** time $t \leq t_{max}$ **do**
2:    **for** $v_i \in V(\mathcal{G})$ **do**
3:        Update the Haar wavelet tree for 1-hop subgraph $g_{v_i}^1$;
4:    **for** $r = 1$ to r_{max} **do**
5:      **for** $v_i \in V(\mathcal{G})$ **do**
6:        Compute burst score bound $B_{v_i}^r$ of r-hop subgraphs $g_{v_i}^r$ using Eq. 5;
7:        **for** $l = l_{min}$ to l_{max} **do**
8:            $mink$ = the minimum burst score of the top-k list of hop size r and window length l;
9:            **if** $B_{v_i}^r > mink$ **then**
10:              Obtain $BurstScore_{v_i}^r$ by detailed search;
11:              **if** $BurstScore_{v_i}^r > mink$ **then**
12:                 remove the k-th vertex v_j in the corresponding top-k list;
13:                 add v_i to the corresponding top-k list.

bound of an r-hop neighborhood subgraph is larger than the minimum burst score $mink$ in the corresponding top-k list, Algorithm 1 performs a detailed search at Line 10 to verify whether it is a real top-k result. If the true burst score is larger than $mink$, it is added to the corresponding top-k list substituting the k-th item. To save memory space, instead of storing r-hop subgraphs, we only store centers of the r-hop subgraphs in the top-k list.

4   Experimental Evaluation

In this section, we report our experimental results on two real data sets to show both the effectiveness and the efficiency of our proposed algorithm.

4.1   Data Sets

The two real data sets are extracted from Digg [1]. Users can make their comments on stories in Digg. The vertices of the heterogenous bipartite evolving graph are users and stories. Graph evolutions are comments submitted by users.

The corpus of users' comments collected contains comments for around four month [7]. For better utilization, we split it into two two-month data sets, **Digg A** and **Digg B**. Comments in the data sets are categorized day by day and there are a large number of comments in each day. So, we further divide a day into four time stamps and randomly assigned the comments in the same day into one of the four time stamps. There are total 9583 users and 44005 articles. The total time stamps of both data sets is 232. The evolution characteristics of these two data sets are summarized in Fig. 7, which shows the total number of evolutions happened at each time stamp. There are periodic troughs, because users submit fewer comments during weekend.

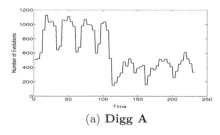 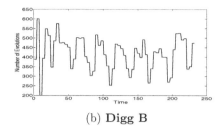

(a) **Digg A**   (b) **Digg B**

Fig. 7. Total Evolutions of The Evolving User-Story Graphs from Digg

4.2 Effectiveness

We demonstrate two examples discovered in Data Set **Digg A** in Fig. 8. The length of the burst window is 8. Among all the 10th burst areas of each time stamp, we select one with the highest burst score. Fig. 8(b) present the 1-hop neighborhood subgraph at time 35, while the corresponding 1-hop subgraph at time 27 is presented in Fig. 8(a).

The round vertices represent users, while the square vertices are stories. As we can see that the subgraph in Fig. 8(b) having more vertices and edges than one in Fig. 8(a), which indicates that the center vertex was the 10th active user during the burst period. Similar results could be found in Fig. 8(c) and Fig. 8(d), which show the 2-hop subgraphs with the highest burst score among all the 10th burst areas of each time stamp. The centers of these two subgraphs are shown as the central white vertices in the figures. These figures shows that the stories, which were commented by the user of the center vertex, also received many comments from other users during the burst period.

Fig. 9(a) and Fig. 9(b) present the center vertex ID of the top-1 1-hop and 2-hop burst areas from time 90 to time 140, respectively. At each time stamp, we plot the center vertex ID of top-1 burst area whose burst window length is 8, as well as the one of top-1 burst area whose burst window length is 16. The figure show that the top-1 burst area of large window length is not always the same as one of small window size, which explains why we need to find burst areas with different burst window lengths.

4.3 Efficiency

We perform our efficiency testing on Data Set **Digg A** and **Digg B**. Fig. 10(a) and Fig. 10(b) show the overall running time of the direct algorithm, which is discussed in the beginning in Section 2, as well as one of our proposed algorithm. The value of k in Fig. 10(a) and 10(b) is 10 and 20, respectively. As we can see, our proposed algorithm is much faster than the direct algorithm. The lower part of each bar of direct algorithm is the running time of updating all burst scores, and the lower part of each bar of our proposed algorithm is the running time of maintaining Haar wavelet trees. One advantage of our proposed algorithm is that the maintaining cost is less than 1/10 of one of the direct algorithm. This will

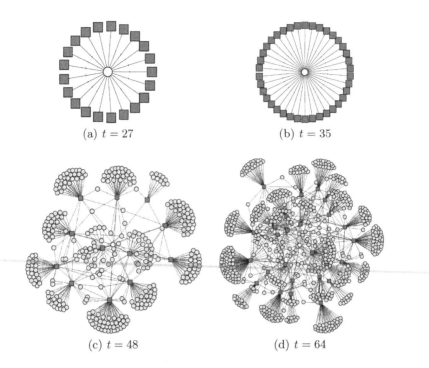

(a) $t = 27$ (b) $t = 35$

(c) $t = 48$ (d) $t = 64$

Fig. 8. Top-1 Burst Areas in **Digg A** ($l = 8$)

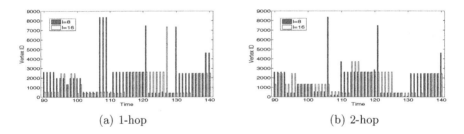

(a) 1-hop   (b) 2-hop

Fig. 9. Center Vertex ID of Top-1 Burst Areas

be useful when we do not monitor evolution streams continuously, but submit ad-hoc queries to find top-k burst areas at some interesting time stamps.

Fig. 10(c) and 10(d) show the overall running time for the direct algorithm and our proposed algorithm, when the value of k changes. The length l of burst window in Fig. 10(c) and 10(d) is 16 and 32, respectively. We can observe similar experimental results that our proposed algorithm uses much shorter time. Fig. 11 presents the corresponding results for Data Set **Digg B**, which prove again the efficiency of our proposed algorithm.

We report the pruning ability in Data Set **Digg A** and **Digg B** in Fig. 12. Fig. 12(a) and 12(b) shows the pruning ability of our proposed algorithm in

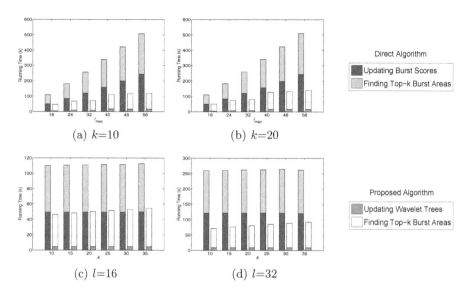

Fig. 10. Running Time of **Digg A**

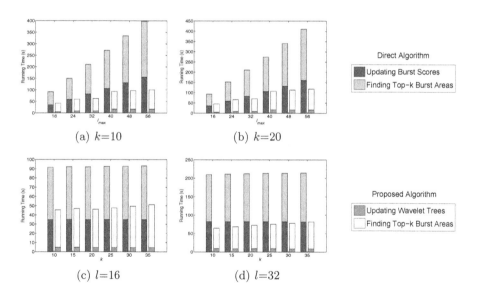

Fig. 11. Running Time of **Digg B**

Digg A and **Digg B** when the length of burst window varies. Fig. 12(c) and 12(d) shows the pruning ability of our proposed algorithm in **Digg A** and **Digg B** as the value of k changes. The results shows that our proposed algorithm is able to prune most of the detailed searches.

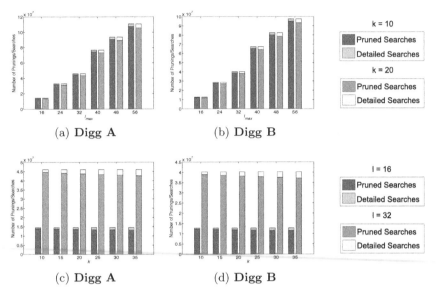

(a) **Digg A**   (b) **Digg B**

(c) **Digg A**   (d) **Digg B**

Fig. 12. Pruning Ability

5   Related Work

There are a few papers dealing with graph change. Liu et al. [8] proposed to spot significant changing subgraphs in evolving graphs. Significance is measured by the total change of similarities between vertex pairs inside subgraphs. After finding the most significant changing vertices, a clustering-manner algorithm is used to connect these vertices into subgraphs. Their problem is to find changing regions between two snapshots of a large evolving graph, while we concentrate on detecting burst areas over streams of fast graph evolutions in this paper.

The problem of identifying dense areas in large and sparse graphs has attracted considerable research efforts in literature. Such dense areas, especially in the domain of social science, are usually considered as communities. Most of the existing studies [3,5,4] only handle static graph data, while there are only a few studies [6,2,9] that deal with time-evolving graphs.

Kumar et al. [6] aimed to discover community bursts in a time-evolving blog graph. Their algorithm first extracts dense subgraphs from the blog graph to form potential communities. Then, the bursts within all the potential communities are identified by modeling the generation of events by an automaton. Bansal et al. [2] focused on seeking stable keyword clusters in a keyword graph, which evolves with additions of blog posts over time. All vertices in each maximum bi-connected subgraph are reported as a cluster. A cluster graph is further constructed using clusters as vertices and connecting the clusters in adjacent time stamps as edges. Finally, a path with the highest weight normalized by its length in the cluster graph is discovered and presented as the set of persistent keyword clusters. Sun et al. [9] proposed GraphScope that is able to discover communities in large and dynamic graphs. GraphScope then iteratively searches for the best community partition and time segmentation, which minimizes the encoding objective function based on Minimum Description Length.

In [10], Zhu et al. developed a framework to detect bursts in time series database. They used a shifted Haar wavelet tree, which is similar to haar wavelet tree with extra memory cost to improve the pruning ability. In our proposed solution, we could achieve similar pruning ability without extra memory cost by updating the Haar wavelet in a dynamic manner.

6   Conclusions

In this paper, we have studied the problem of finding top-k burst areas in fast graph evolutions. We proposed to update the Haar wavelet tree in a dynamic manner to avoid high computation complexity while keeping its high pruning ability. The top-k burst areas are computed incrementally from small hop size to large hop size in order to minimize memory consumption. Our experimental results on real data sets show our solution is very efficient and effective.

Acknowledgments. The work was supported by grants of the Research Grants Council of the Hong Kong SAR, China No. 419008 and 419109.

References

1. http://www.digg.com
2. Bansal, N., Chiang, F., Koudas, N., Tompa, F.W.: Seeking stable clusters in the blogosphere. In: VLDB, pp. 806–817 (2007)
3. Chakrabarti, D.: Autopart: parameter-free graph partitioning and outlier detection. In: Boulicaut, J.-F., Esposito, F., Giannotti, F., Pedreschi, D. (eds.) PKDD 2004. LNCS (LNAI), vol. 3202, pp. 112–124. Springer, Heidelberg (2004)
4. Dourisboure, Y., Geraci, F., Pellegrini, M.: Extraction and classification of dense communities in the web. In: WWW, pp. 461–470. ACM, New York (2007)
5. Gibson, D., Kumar, R., Tomkins, A.: Discovering large dense subgraphs in massive graphs. In: VLDB, pp. 721–732. VLDB Endowment (2005)
6. Kumar, R., Novak, J., Raghavan, P., Tomkins, A.: On the bursty evolution of blogspace. World Wide Web 8(2), 159–178 (2005)
7. Lin, Y.-R., Sun, J., Castro, P., Konuru, R.B., Sundaram, H., Kelliher, A.: Metafac: community discovery via relational hypergraph factorization. In: KDD, pp. 527–536 (2009)
8. Liu, Z., Yu, J.X., Ke, Y., Lin, X., Chen 0002, L.: Spotting significant changing subgraphs in evolving graphs. In: ICDM, pp. 917–922 (2008)
9. Sun, J., Faloutsos, C., Papadimitriou, S., Yu, P.S.: Graphscope: parameter-free mining of large time-evolving graphs. In: KDD, pp. 687–696. ACM, New York (2007)
10. Zhu, Y., Shasha, D.: Efficient elastic burst detection in data streams. In: KDD, pp. 336–345. ACM, New York (2003)

Answering Top-k Similar Region Queries*

Chang Sheng[1], Yu Zheng[2], Wynne Hsu[1], Mong Li Lee[1], and Xing Xie[2]

[1] School of Computing, National University of Singapore, Singapore
{shengcha,whsu,leeml}@comp.nus.edu.sg
[2] Microsoft Research Asia, Beijing, China
{yuzheng,Xing.Xie}@microsoft.com

Abstract. Advances in web technology have given rise to new information retrieval applications. In this paper, we present a model for geographical region search and call this class of query *similar region query*. Given a spatial map and a query region, a similar region search aims to find the top-k most similar regions to the query region on the spatial map. We design a quadtree based algorithm to access the spatial map at different resolution levels. The proposed search technique utilizes a filter-and-refine manner to prune regions that are not likely to be part of the top-k results, and refine the remaining regions. Experimental study based on a real world dataset verifies the effectiveness of the proposed region similarity measure and the efficiency of the algorithm.

1 Introduction

In the geo-spatial application, a *similar region query* happens when users want to find some similar regions to a query region on the map. The application scenarios include

- Similar region search. Due to the limitation of knowledge, people may only be familiar with some places where they visit frequently. For example, people go to the nearest entertainment region which include malls for shopping and the restaurants for dinner. Sometimes, people wish to know the alternative places as the options for both shopping and dinners. Base on their familiar entertainment region, similar region query retrieves the regions that have the similar functions to their familiar entertainment region.
- Disease surveillance. Similar region search query is also useful in identifying the potential high-risk areas that are prone to outbreak of diseases. Many infectious diseases thrive under the same geographical conditions. By querying regions that are similar in geographical characteristics, we can quickly highlight these high-risks areas.

The traditional IR model might be applied to answer similar region queries: A direct application is to partition the map into a set of disjoined regions, represent the region by a vector of PoI categories, and utilize the vector space model

---

* Part of this work was done when the first author worked as an intern in MSRA.

H. Kitagawa et al. (Eds.): DASFAA 2010, Part I, LNCS 5981, pp. 186–201, 2010.

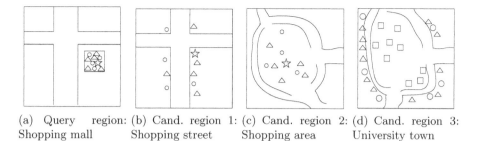

(a) Query region: Shopping mall (b) Cand. region 1: Shopping street (c) Cand. region 2: Shopping area (d) Cand. region 3: University town

Fig. 1. The first three plots show the distribution of five restaurants (triangles), four shops (circles) and one theater (star); The last plot gives a distribution of nine research institutes (rectangle) and many nearby restaurants (triangles) and shops (circles)

(VSM) [11] to evaluate the similarity of the regions. However, the traditional IR model is inadequate in supporting the "good" similar region queries due to two reasons. First, in traditional IR model, users are required to provide a set of keywords or terms to the search engine, and the search engine returns a list of texts which are relevant to the keywords. However, in similar region query, users only provide a query region instead of keywords. Second, similar region query searches the regions of similar region functionality, which is actually determined by the spatial objects (we call these spatial objects Point-of-Interests (PoIs) in the rest of paper) in the region and their spatial distribution in this region, i.e., *local distribution* or *distribution* in short. The traditional IR model does not take local distribution into account while computing the similarity.

For example, Figure 1 shows four local distributions, where the first three regions have the identical number and categories of PoIs, and the last region has different PoI categories from the first three regions. Given the query region shown in Figure 1(a), traditional IR model ranks Figure 1(b) and Figure 1(c) higher than Figure 1(d), because Figure 1(d) has different PoI categories. However, the traditional IR model could not distinguish the first three plots of Figure 1, which actually stand for three different region functionalities: Shopping malls are usually located in the communities as the entertainment centers; Shopping streets are located in the central business area for providing services to tourists; Shopping areas are located around the residential areas and the shops usually are groceries.

The above example highlights the importance of considering not only spatial objects categories but also their local distributions when answering similar region query. We present the problem for answering similar region query as follows.

Similar region query problem. Given a spatial map, a query region R_q, two coefficients to control the area of region μ_1 and μ_2, we aim to find the top-k most similar regions to R_q on the spatial map, such that 1) $\mu_1 \leq \frac{Area(R_i)}{Area(R_q)} \leq \mu_2$, R_i is a return region, and 2) any two return regions do not have large overlap[1].

---

[1] The degree of overlap is measured by the intersection ratio of two regions. In this paper, we set this ratio to be 0.8.

In this paper, we focus on two main issues in tackling the similar region search problem. The first issue is to provide a proper definition for region similarity. While there have been extensive researches into defining the document similarity [1], to the best of our knowledge, there is no existing similarity measure for regions. In this paper, we propose a reference distance feature which is consistent with the human routines to compare region similarity. Accordingly, we extend the VSM model to the Spatial Vector Space Model (SVSM) by using the reference distance feature to capture the local distributions of spatial object categories.

Second, the search space in the region search problem is a continuous spatial map. Exhaustive search on the continuous spatial map is too expensive to provide quick response to users. To solve this problem, we provide a quadtree based approximate region search approach. The basic idea follows the filter-and-refine approach which is described as follows. We maintain a top-k region set and the similarity threshold to be a top-k region. We extract the representative categories from the query region and filter the quadtree cells that do not contain the representative categories. We further prune those cells that are not likely to be the top-k most similar regions. The remaining cells are remained as seeds to expand gradually. We insert the expanded regions into top-k regions if their similarity values are greater than the similarity threshold, and accordingly update the similarity threshold.

The remainder of this paper is organized as follows. Section 2 discusses the related work. Section 3 gives preliminaries. Section 4 presents the spatial vector space model. Section 5 proposes the quadtree-based region search approach. Section 6 reports our experiment results. Finally, Section 7 concludes this paper.

2   Related Work

Text retrieval is one of the most related problems. Conventional text retrieval focuses on retrieving similar documents based on text contents, and a few of similarity models, such as vector space model [11] and latent semantic analysis model [5], are proposed to compare the similarity. Recently, location-aware text retrieval, which combines both location proximity and text contents in text retrieval, receives much attention. To perform efficient retrieval, both document locations and document contents are required to be indexed in the hybrid index structures, such as a combination of inverted file and R*-tree [14], a combination of signature files and R-tree [7], DIR tree [3]. Our work substantially differs from location-aware text retrieval queries because we consider the relative locations of spatial objects and the returned results are regions which are obtained by the space partition index Quadtree.

Image retrieval [4], particularly content based image retrieval (CBIR) [10], is another related problem. CBIR considers the color, texture, object shape, object topology and the other contents, and represent an image by a single feature vector or a bag of feature vectors for retrieval. CBIR is different from the similar region query problem because CBIR focuses on either the content of whole image or the relationship from one object to another object, while we search the similar regions based on the local distribution of one category to another category. In addition, the image retrieval system searches the similar images from an image database, while our algorithm finds the similar regions on one city map which need to be properly partitioned during retrieval.

There are two existing approaches to select features from spatial data. The first approach is based on spatial-related patterns, such as collocation patterns [8] and interaction patterns [13]. Both patterns are infeasible to be employed in the similar region queries because they capture the global distribution among different PoI types, not the local distribution. The second approach is the spatial statistical functions test, like cross K function test [2]. In spite of theoretic soundness, this approach need long training time so that it is impractical to provide efficient response to the query.

3   Preliminaries

Suppose $\mathcal{P}$ is a spatial map, and $\mathcal{T}$ is a set of PoI categories $\mathcal{T} = \{C_1, C_2, \ldots, C_K\}$. Each PoI may be labelled with multiple PoI categories. For example, a building is labelled both as "Cinema" and "Restaurant" if it houses a cinema and has at least one restaurant inside. The PoI database $\mathcal{D}$ contains a set of PoIs. Each PoI in $\mathcal{D}$ is presented by a tuple $o = \langle p_o; \mathcal{T}_o \rangle$, where $p_o = (x_o, y_o)$ denotes the location of o, and $\mathcal{T}_o$ is a set of o's PoI categories.

A *region* R is a spatial rectangle bounded by $[R_{x_{min}}, R_{x_{max}}] \times [R_{y_{min}}, R_{y_{max}}]$ which locates in map $\mathcal{P}$. A PoI $o = \langle p_o; \mathcal{T}_o \rangle$ is said to *occur* in region R if $p_o \in R$. We use $\mathcal{D}^R = \{o | o \in \mathcal{D} \wedge p_o \in R\}$ to denote all PoIs which occur in region R, and $\mathcal{D}^R_{C_i} = \{o | o \in \mathcal{D} \wedge p_o \in R \wedge C_i \in \mathcal{T}_o\}$ to denote a set of objects with category C_i which occur in region R.

By modifying the concepts of TF-IDF measure in VSM, we define the *CF-IRF* as follows. The *Category Frequency (CF)* of the category C_i in region R_j, denoted as $CF_{i,j}$, is the fraction of the number of PoIs with category C_i occurring in region R_j to the total number of PoIs in region R_j, that is,

$$CF_{i,j} = \frac{\mathcal{D}^{R_j}_{C_i}}{\mathcal{D}^{R_j}} \tag{1}$$

The importance of a category C_i depends on the distribution of PoIs with category C_i on the entire map. Suppose we impose a $g_x \times g_y$ grid on the map. The *Inverse Region Frequency (IRF)* of category C_i, denoted as IRF_i, is the logarithm of the fraction of the total number of grids to the number of grids that contain PoIs with category C_i.

$$IRF_i = \log \frac{g_x \times g_y}{|\{\mathcal{D}^{R_j}_{C_i} | \mathcal{D}^{R_j}_{C_i} \neq \emptyset\}|} \tag{2}$$

With CF and IRF, the significance of a category C_i in region R_j, denoted as CF-IRF$_{i,j}$, is defined as follows:

$$\text{CF-IRF}_{i,j} = CF_{i,j} \times IRF_i \tag{3}$$

The information content of a region R_j is denoted as a vector

$$\overrightarrow{R_j} = (f_{1,j}, f_{2,j}, \ldots, f_{K,j}) \tag{4}$$

where $f_{i,j}$ denotes the CF-IRF value of category C_i in region R_j. We use $|\overrightarrow{R_j}|$ denotes the Euclidean norm of vector $\overrightarrow{R_j}$.

$$|\overrightarrow{R_j}| = \sqrt{f_{1,j}^2 + \ldots + f_{K,j}^2} \tag{5}$$

The *information content similarity* of two regions R_i and R_j is the cosine similarity of the corresponding feature vectors of R_i and R_j.

$$Sim(R_i, R_j) = \cos(\overrightarrow{R_i}, \overrightarrow{R_j}) = \frac{\overrightarrow{R_i} \cdot \overrightarrow{R_j}}{|\overrightarrow{R_i}| \times |\overrightarrow{R_j}|} \tag{6}$$

4   Spatial Vector Space Model

A similarity measure is desirable to evaluate the similarity of two regions. To be consistent with the human routines to compare region similarity, we propose the intuitive two level evaluation criteria as follows.

1. Do the regions have a significant overlap in their representative categories? This is the basic gist when users compare the similarity of regions. For example, the regions shown in Figure 1(c) and Figure 1(b) share three common categories, and the regions shown in Figure 1(c) and Figure 1(d) share two common categories. Therefore, the region pair {Figure 1(c), Figure 1(b)} is considered to be more similar than the region pair {Figure 1(c), Figure 1(d)}.
2. If two regions share some common representative categories, do the PoIs of the common representative categories exhibit similar spatial distribution? We observe that Figure 1(a), Figure 1(b) and Figure 1(c) all share the same representative categories, however they are not considered similar as the distributions of the PoIs for each category are drastically different in the three figures. In other words, two regions are more similar if they have not only the common representative categories but also the similar spatial distribution of PoIs. Given a query region of shopping mall, Figure 1(a) is more similar to this query region than Figure 1(b) and Figure 1(c).

CF-IRF feature satisfies the first level evaluation criterion but does not satisfy the second level criterion because it ignores the local distribution of the PoIs. This motivates us to propose a distribution-aware spatial feature and Spatial Vector Space Model (SVSM). In SVSM, a region R_j is represented by a **spatial feature vector** of n entries, $\overrightarrow{R_j} = (f_{1,j}, f_{2,j}, \ldots, f_{n,j})$ where $f_{i,j}$ is the i-th **spatial feature entry** and n is total number of features or the dimension of the feature vector.

A desirable spatial feature will be insensitive to rotation variation and scale variation. In other words, if two regions are similar, rotating or magnifying one of

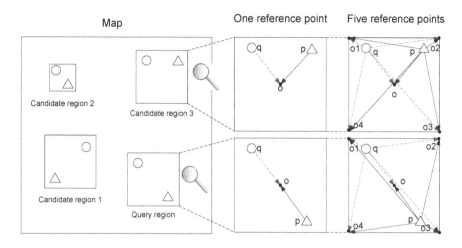

Fig. 2. An example of query region and its reference distance

the two regions will not affect their similarity. Figure 2 illustrates a query region and three candidate regions in a map. We consider candidate region 1 and 2 are similar to the query region because they are similar to query region after rotating by a scale variation (clockwise 270°) or magnifying by a scale variation, respectively. In contrast, candidate region 3 is not so similar as candidate region 1 and 2 to the query region. Motivated by the requirements to minimize the effects of scaling and allow for rotation invariant, we introduce the concept of reference distance to capture the spatial distributions.

Average nearest neighbor distance can be employed to measure the local distribution in statistics domain [6], but it is expensive ($O(n^2)$ if no spatial index is used, where n is the number of PoIs). Therefore, we propose the concept of reference point. The *reference points* are user-specified points in a region to capture the local distributions of the PoIs in this region. The intuition behind reference points is based on the observation that most users tend to use some reference points for determining region similarity. For example, while comparing two regions which have one cinema each, users tend to roughly estimate the average distances from the other categories to the cinema, and compare the estimated distances of two regions. Here, the cinema is a reference point.

It raises an issue to select the proper number of reference points and their locations. We consider two extreme cases as follows. On one hand, one or two reference points are not enough to capture the distribution. For example, Figure 2 shows that one reference point cannot distinguish the distribution of query region and candidate region 3. On the other hand, a larger number of reference points will give a more accurate picture of the spatial distributions among the PoIs, but at the expense of greater computational cost. In this paper, we seek the tradeoff between the two extreme cases. We propose that five reference points, including the center and four corners of the region, are proper to capture the local distribution. Figure 2 illustrates the five reference points, and the reference distances of two PoIs to the five reference points. The complexity to compute reference distance is $O(5 \cdot n)$, which is more efficient than nearest neighbor distance $O(n^2)$. Here, we do not claim that the selection of five point reference points is the best, but experiment results show that it is reasonable.

We now define the reference distance. Give a region R, a set of PoIs P, and five reference points $O=\{o_1, o_2, \ldots, o_5\}$. The distance of P to the i-th reference point $o_i \in O$ is

$$r(P, o_i) = \frac{1}{|P|} \sum_{p \in P} dist(p, o_i) \tag{7}$$

Assume region R has K different categories of PoIs. We use $r_{i,j}$ to denote the distance of PoIs with category C_i to the reference point o_j. The distance of K categories to the reference set O is a vector of five entries.

$$I = \{\overrightarrow{I_1}, \ldots, \overrightarrow{I_5}\} \tag{8}$$

where each entry is the distance of K categories to the reference point o_j, $\overrightarrow{I_i} = (r_{1,i}, r_{2,i}, \ldots, r_{K,i})$.

The similarity of two feature vector sets $I_{R_i} = \{\overrightarrow{I_{1,i}}, \ldots, \overrightarrow{I_{5,i}}\}$ and $I_{R_j} = \{\overrightarrow{I_{1,j}}, \ldots, \overrightarrow{I_{5,j}}\}$, is

$$Sim_r(I_{R_i}, I_{R_j}) = \frac{1}{5} \sum_{k=1}^{5} Sim(I_{k,i}, I_{k,j}) \tag{9}$$

We incorporate the rotation variation into similarity as follows. Given region R_j, we obtain four rotated regions R_{j1}, R_{j2}, R_{j3} and R_{j4} by rotating R_j 90 degree each time. The similarity of R_i and R_j is the similarity of R_i and the most similar rotated region of R_j, that is,

$$Sim_r(R_i, R_j) = arcmax\{Sim_r(I_{R_i}, I_{R_{jk}}), k = 1, 2, 3, 4\} \tag{10}$$

Lemma 1. *The reference distance feature is insensitive to rotation and scale variations.*

Proof: Based on Equation 10, we can derive that the reference distance feature is insensitive to rotation variation. Now we prove that the reference distance feature is insensitive to scale variation as follows. Assume R_j is obtained by scaling R_i by a factor σ. We have $\overrightarrow{I_{R_j}} = \sigma \overrightarrow{I_{R_i}}$ and $|\overrightarrow{I_{R_j}}| = \sigma |\overrightarrow{I_{R_i}}|$. Therefore,

$$Sim_r(R_i, R_j) = \cos(\overrightarrow{I_{R_i}}, \sigma \overrightarrow{I_{R_j}}) = \frac{\overrightarrow{I_{R_i}} \cdot \sigma \overrightarrow{I_{R_i}}}{|\overrightarrow{I_{R_i}}| \times \sigma |\overrightarrow{I_{R_i}}|} = 1. \qquad \square$$

5   Proposed Approach

Given a query region R_q and two coefficients to control the area of region returned, μ_1 and μ_2, the naive approach to answer similar region queries is to utilize a sliding window whose area is between $min\_area = \mu_1 \times area(R_q)$ and $max\_area = \mu_2 \times area(R_q)$. The sliding window is moved across the entire

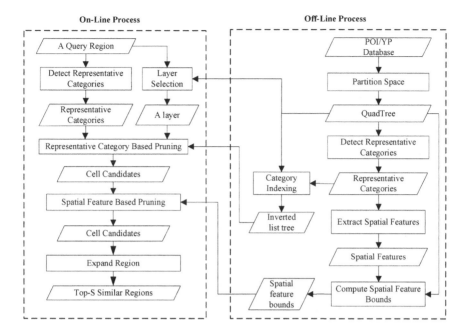

Fig. 3. Overview of system architecture

map and at each move, we compute the similarity between R_q and the sliding window. If we maintain a list of top-k regions having the k largest similarity values, then the time complexity of this naive approach in the worst case, given a $map_x \times map_y$ spatial map, is $O(k \times min\_area \times map_x \times map_y)$. This is because the total number of candidate regions is $min\_area \times map_x \times map_y$. For each candidate region we compute whether it overlaps with the existing top-k similar regions in $O(k)$ time complexity. Thus the overall time complexity of this algorithm is $O(k^2 \times min\_area \times map_x \times map_y)$, which is too expensive to provide a quick response to users.

To overcome the high complexity of the naive method, we propose a quadtree-based approximate approach. Figure 3 shows the system architecture overview of our approach. The architecture comprises an offline process and an online process. The offline process partitions the map into a hierarchical structure and builds a quadtree structure for quick retrieval of PoIs. The online process uses these index structures to perform region search queries efficiently. Given a query region, the system analyzes the shape and size of this region and determines the appropriate quadtree layer to initiate the similar region search process. At the same time, the system will compute the CF-IRF values to derive the representative categories of the query region. Once we know the starting level of the quadtree and the representative categories of the query region, we begin a filter-and-refine procedure to quickly reduce the search space that is unlikely to be in the top-k most similar regions.

5.1 Quadtree Structure

Given a PoI database and the map of this database, we partition the map and build a hierarchical quadtree structure [12] to facilitate the construction of multi-scale regions. In the quadtree, the root node indicates the whole map and each non-leaf node corresponds to one of the four partitioned cells from its parent's cell. At the lowest level, each leaf node corresponds to the partitioned cell with the smallest granularity. The depth of the quadtree depends on the smallest granularity requirement in applications. In our system, the leaf node is 100 meters by 100 meters, so the quadtree height is 10 for a city of 30 kilometers by 30 kilometers.

The quadtree structure enables an efficient handling of multi-granularity similar region queries. This is because we can adaptively select the different level of granularity by accessing the quadtree nodes at the appropriate level. For example, if the query region is the size of 200 meters by 200 meters and the parameter to control the minimal return region area $\mu_1 = 0.25$, we perform the search on the leaf node because the leaf node area is no less than μ_1 times of the query region area.

The quadtree allows the effective region pruning by storing the key statistical information at each node in the quadtree. Each node maintains the lower bound and upper bound of feature entries defined as follows.

Definition 1. *The lower bound feature vector of a node B, denoted as $\overrightarrow{B}_{lb}$, is $(f_{1,lb}, f_{2,lb}, \ldots, f_{n,lb})$, where $f_{i,lb}$ is the minimum i-th feature entry value of all descendant nodes of B.*

Definition 2. *The upper bound feature vector of a node B, denoted as $\overrightarrow{B}_{ub}$, is $(f_{1,ub}, f_{2,ub}, \ldots, f_{n,ub})$, where $f_{i,ub}$ is the maximum i-th feature value of all descendant nodes of B.*

Each quadtree node maintains the minimum/maximum CF-IRF vector and the minimum/maximum reference distance vector. These bounds are useful for pruning the candidate regions as stated in Lemma 2.

Lemma 2. *Let $\overrightarrow{R_q} = (f_{1,q}, f_{2,q}, \ldots, f_{n,q})$ to be the the feature vector of query region, δ to be the cosine similarity threshold of top-k regions. A node B can be pruned if for any feature entry $f_{i,q}$, we have $f_{i,ub} \cdot f_{i,q} \leq \frac{\delta}{n} \cdot |\overrightarrow{B}_{lb}| \cdot |\overrightarrow{R_q}|$.*

Proof: Let $f_{i,j}$ to be the i-th feature entry of region R_j where $R_j \in B$. Then $f_{i,lb} \leq f_{i,j} \leq f_{i,ub}$ and $|\overrightarrow{B}_{lb}| \leq |\overrightarrow{R_j}| \leq |\overrightarrow{B}_{ub}|$. Assume that $f_{i,ub} \cdot f_{i,q} \leq \frac{\delta}{n} \cdot |\overrightarrow{B}_{lb}| \cdot |\overrightarrow{R_q}|$. For the i-th feature entry $f_{i,j}$, we have $f_{i,j} \cdot f_{i,q} \leq f_{i,ub} \cdot f_{i,q} \leq \frac{\delta}{n} \cdot |\overrightarrow{B}_{lb}| \cdot |\overrightarrow{R_q}| \leq \frac{\delta}{n} \cdot |\overrightarrow{R_j}| \cdot |\overrightarrow{R_q}|$.

By summing up the inequalities, $\overrightarrow{R_j} \cdot \overrightarrow{R_q} = \sum_{p=1}^{n} f_{p,j} \cdot f_{p,q} \leq \delta \cdot |\overrightarrow{R_j}| \cdot |\overrightarrow{R_q}|$. So, we have $\cos(\overrightarrow{R_j}, \overrightarrow{R_q}) \leq \delta$, which means that any region R_j under B will not have a larger similarity than the top-k region similarity threshold. $\square$

With Lemma 2, we can prune all nodes B that have no chance of satisfying the similarity threshold δ. For example, suppose the quadtree node B has four child nodes B_1, B_2, B_3, B_4. Each feature vector of child node has five entries.

$$\overrightarrow{B_1} = (0.1, 0.3, 0.1, 0.8, 0.0), \ \overrightarrow{B_2} = (0.1, 0.7, 0.2, 0.7, 0.0)$$
$$\overrightarrow{B_3} = (0.0, 0.3, 0.1, 0.8, 0.2), \ \overrightarrow{B_4} = (0.2, 0.4, 0.2, 0.6, 0.1)$$

So we have $\overrightarrow{B}_{lb} = (0.0,\ 0.3,\ 0.1,\ 0.6,\ 0.0)$ and $\overrightarrow{B}_{ub} = (0.2,\ 0.7,\ 0.2,\ 0.8,\ 0.2)$.

Let the feature vector of query region is $\overrightarrow{R_q} = (0.9,\ 0.1,\ 0.9,\ 0.1,\ 0.8)$ and $\delta = 0.95$. We have $\frac{\delta}{n} \cdot |\overrightarrow{B}_{lb}| \cdot |\overrightarrow{R_q}| = 0.2468$. The node B can be pruned because each feature entry product of $\overrightarrow{R_q}$ and $\overrightarrow{B}_{ub}$ is less than 0.2468.

In addition, we also construct an inverted tree index on the representative categories to facilitate similar region search. The root node of the inverted tree has K entries, where each entry corresponds to a category. Each category, say C_i, of a non-leaf node is associated with a child node that has four entries. The entry value is 1 if the corresponding partitioned region has the C_i as a representative category; otherwise the entry value will be 0. This inverted list tree is recursively built until it reaches a leaf node of the quadtree structure or all four entries have value 0. Based on this inverted tree index, we can quickly identify the cells that have similar categories to the query region.

5.2 Region Search Algorithm

In this section, we present the search strategy based on the quadtree structure. The basic idea is to compute the proper search level in the quadtree in which the buckets of search level will be greater than the minimal area of returned regions, and on the search level we select a few bucket as seeds to gradually expand to larger regions of proper size and large similarity value to the query region.

Algorithm 1 gives a sketch of the region search process. Line 1 computes the search level based on the granularity of query region. Line 2 extracts the representative categories from the search region R_q. The function $ExtractCategory$ computes the CF-IRF values for each category on R_q and only maintains the top-m categories with the largest CF-IRF values. Line 3 adjusts the feature vector of R_q. The entries which correspond to the top-m representative categories remain and the other entries are set to be zero. Line 4 initializes the return region set to be an empty set and the similarity threshold δ to be 0. Line 5 calls procedure $SearchQTree$ to search the similar regions.

Procedure $SearchQTree$ recursively searches and prunes the candidate regions in quadtree. Line 8 is the validity checking for the top-k regions. A bucket is valid only if 1) it contains the CM representative categories, and 2) it cannot be pruned by Lemma 2. The inverted tree structure and the feature bounds of buckets facilitate the validity checking. If a bucket is valid and this bucket is higher level than l_{search} (Line 9), its child nodes need to be recursively detected further (Lines 10-11). Otherwise, Line 13 expands the valid buckets on l_{search} by calling the function $RegionExpansion$. Line 14 inserts the expanded region R to the top-k region set $\mathcal{R}$, if R has no big overlap with the existing top-k regions or R has overlap with one existing top-k regions but R has a larger similarity value. Line 15 updates the similarity threshold δ based on the k-th largest similarity value in $\mathcal{R}$ currently.

Algorithm 1. RegionSearch(R_q, T, k, m)

---

input   : Query region R_q; Quadtree T; Number of return regions k; Number of
           representative categories m.

output: Top-k similar regions

1 Compute the search level l_{search} on T based on R_q;

2 $CM = \texttt{ExtractCategory}(R_q, m)$;

3 $\texttt{Adjust}(\overrightarrow{R_q}, CM)$;

4 $\mathcal{R} = \emptyset$; $\delta = 0$;

5 $\texttt{SearchQTree}(\overrightarrow{R_q}, T.root, \delta, \mathcal{R})$;

6 **return** $\mathcal{R}$;

7 **Procedure SearchQTree**(R_q, B, δ, $\mathcal{R}$, CM)

8 **if** B has CM categories $\wedge$ B cannot be pruned by Lemma 2 **then**

9 | **if** $B.level < l_{search}$ **then**

10 | | **foreach** child node $B' \in B$ **do**

11 | | | $\texttt{SearchQTree}(R_q, B', \delta, \mathcal{R})$;

12 | **else**

13 | | $R = \texttt{RegionExpansion}(R_q, B')$;

14 | | $\mathcal{R} = \mathcal{R} \cup R$;

15 | | update δ;

16 **Function RegionExpansion**(R_q,R)

17 **repeat**

18 | **foreach** $dir \in \{LEFT, RIGHT, DOWN, UP\}$ **do**

19 | | $R'' = expand(R, dir)$;

20 | | $dir = arcmax(\texttt{Sim}(R_q, R''))$;

21 | $R' = expand(R, dir)$;

22 **until** $\texttt{Sim}(R_q, R) \leq \texttt{Sim}(R_q, R')$;

23 **return** R'

The *RegionExpansion* function (Lines 16-23) treats a region as a seed, performs the tentative expansion in four candidate directions, and selects the optimal expanded region which gives the largest similarity value. The step width of each expansion is the cell side of the quadtree leaf node in order to minimize the scope of expansion, which eventually approach the local most similar region. The expansion stops if there is no increase in the similarity value (Line 22).

Finally, Line 6 returns the top-k regions. If the number of regions in $\mathcal{R}$ is less than k, we decrease the value of m by 1 in Line 2, and search the cells which share exact $m-1$ common representative categories. We repeatedly decrease the m value by 1 till the number of return regions in $\mathcal{R}$ reaches k.

6   Experiment Studies

In this section, we present the results of our experiments to examine the performance of similar region search. We first describe the experiment settings and the evaluation approach. Then, we report the performance on the region queries.

6.1   Settings

In our experiments, we use the Beijing urban map, which ranges from latitude 39.77 to 40.036, and longitude 116.255 to 116.555. The spatial dataset consists of the real world yellow page data of Beijing city in China. This dataset has two parts. The first part contains the persistent stationery spatial objects, such as the large shopping malls, factories, gas stations, land-marks, etc. The second part is the set of short-term and spatial objects which are updated from time to time, such as small restaurants and individual groceries. The total number of PoIs are 687,773, and they are classified into 48 major categories by their properties and functions.

We construct a quadtree for the Beijing urban map. The quadtree height is 10, and the cell side of quadtree leaf node is about 100 meters and the number of leaf nodes is 512×512. For each node of quadtree, we compute the lower bound and upper bound for the two features, namely category frequency and reference distance. We set $\mu_1=0.25$ and $\mu_2=4$, which means the return region areas range from one quarter to four times of query region areas.

As we are not aware of any existing work that support top-k similar region queries, we only evaluate two variants of the RegionSearch algorithm as follows. 1) VSM: It is a baseline algorithm based on the CF-IRF vector space model, and 2) SVSM: It is a spatial vector space model based algorithm that measures region similarity by the reference distance feature vector.

Given a query region, VSM and SVSM return the top-5 most similar regions respectively. Five users who are familiar with Beijing city score the return regions from 0 to 3 according to the relevance of the query region and the return regions. The final score of a return region is the average scores of five users. Table 1 gives the meanings of each score level.

Table 1. Users' scores for the return region

Scores	Explanations
0	Totally irrelevant
1	A bit relevant, with at least one common functionality with the query region
2	Partially relevant, the functionality of return region cover that of query region
3	Identically relevant, the return and query regions have the same functionality

We employ DCG (discounted cumulative gain) to compare the ranking performance of VSM and SVSM. The criteria DCG is used to compute the relative-to-the-ideal performance of information retrieval techniques. For example, given $G =(2, 0, 2, 3, 1)$, we have $CG =(2, 2, 4, 7, 8)$ and $DCG =(2, 2, 3.59, 5.09, 5.52)$. The higher the scores computed by DCG, the more similar the return region. Please refer to [9] for the definitions of the cumulative gain (CG) and the discounted cumulative gain (DCG).

All the algorithms are implemented in C++ and the experiments are carried out on a server with dual Xeon 3GHZ processors and 4GB memory, running Windows server 2003.

6.2 Effectiveness Study

We select three typical types of query regions as the test queries.

- The shopping mall. The shopping mall is one of the commercial community whose spatial points are clustered in small regions.
- The commercial street. The commercial street is another commercial community whose spatial points distributed along the streets.
- The university. The spatial points has a star-like distribution where the institutes are located at the center and other facilities such as hotels and restaurants are located around the university.

Each type of query region is given three query regions, which are listed in Table 2. We evaluate the average DCG values for each type of query region.

Table 2. The type and size (meter × meter) of nine query regions

ID	Type	Query region	ID	Type	Query region	ID	Type	Query region
q_1	mall	150×150	q_4	street	150×600	q_7	university	1400×800
q_2	mall	100×300	q_5	street	200×500	q_8	university	1400×1100
q_3	mall	50×70	q_6	street	300×100	q_9	university	1200×1200

In order to find a proper number of representative categories, we run the two algorithms on different queries by varying the number of representative categories to be 3,5,10. We found that the average DCG curve of $m=5$ is better than the curves of $m=3$ and $m=10$. The result is consistent with our expectation because small m values are not enough to differentiate the region functionality, and large m values are likely to include some noise categories, both of which could affect the precision of return regions. We set $m=5$ in the rest experiments.

Figure 4 shows the average DCG curves for the three types of query regions. We observe that SVSM outperforms VSM for all of three query types, especially on shopping mall queries and street queries. This is expected because SVSM captures both the PoI categories and the local distribution in a region, which is consistent with human routines to evaluate the region similarity. In addition, SVSM did not have remarkable performance on the university queries. This is possibly because university query regions are larger than shopping mall queries and street queries, which results in the larger reference distances to decrease the contrast of spatial feature vector.

6.3 Efficiency Study

In this set of experiments, we study the efficiency of VSM and SVSM. Figure 5 gives the runtime for the three query types as m varies from 3 to 10. We see that both VSM and SVSM are scalable to m, but VSM is the faster than SVSM. This is because SVSM requires extra time to process the additional spatial features. We also observe that the runtime for SVSM decreases as m increases. This is because a larger number of representative categories lead to the pruning of more

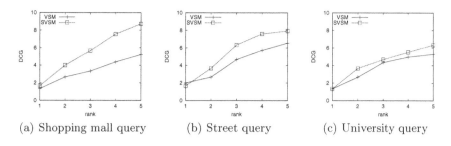

(a) Shopping mall query   (b) Street query   (c) University query

Fig. 4. Average DCG curves for three query types $(m = 5)$

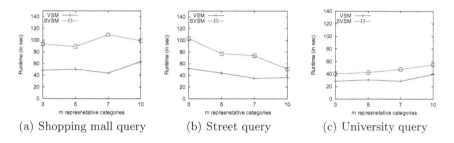

(a) Shopping mall query   (b) Street query   (c) University query

Fig. 5. Effect of m on runtime

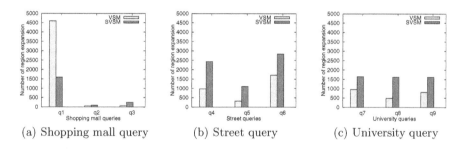

(a) Shopping mall query   (b) Street query   (c) University query

Fig. 6. Number of region expansion

candidate regions. In addition, the runtime for the university queries is smaller than the other two query types since the runtime is determined by the area constraint of return regions. For large query regions, the search starts at the higher levels of the quadtree which have small number of candidate regions.

Next, we check the performance of pruning strategy. Since only the candidate regions which pass the validity test are expanded, we evaluate the pruning strategy by counting the number of region expansion operations. Figure 6 shows the number of region expansion performed for the three query types. In this experiment, the shopping mall queries and street queries have around 260,000 candidate regions, and the university queries have around 16,000 candidate regions.

We observe that q_1 have more regions to be expanded than q_2 and q_3 (see Figure 6(a)). A closer look reveals that q_1 only contains three categories, hence many regions are considered as candidates. Figure 6(b) and Figure 6(c) show that less than 3,000 and 1,800 regions are expanded respectively, demonstrating the power of the pruning strategies.

7   Conclusion

In this paper, we introduce a similar region query problem in which spatial distribution is considered to measure region similarity. We propose the reference distance feature and spatial vector space model (SVSM) which extends the concept of vector space model to include reference distance features. We design a quadtree-based approximate search algorithm to filter and refine the search space by the lower and upper bounds of feature vectors. Experiments on the real world Beijing city map show that our approach is effective in retrieving similar regions, and the feature bounds are useful for pruning the search space. To the best of our knowledge, this is the first work on similar region search. We plan to investigate other types of spatial features for region similarity definition and hope to incorporate our techniques into the Microsoft Bing search engine.

References

1. Baeza-Yates, R., Ribeiro-Neto, B.: Modern Information Retrieval. Addison-Wesley Longman Publishing Co., Inc., Boston (1999)
2. Cassie, N.A.: Statistics for Spatial Data. Addison-Wesley Longman Publishing Co., Inc., Boston (1993)
3. Cong, G., Jensen, C.S., Wu, D.: Efficient retrieval of the top-k most relevant spatial web objects. PVLDB 2(1), 337–348 (2009)
4. Datta, R., Joshi, D., Li, J., Wang, J.Z.: Image retrieval: Ideas, influences, and trends of the new age. ACM Comput. Surv. 40(2), 1–60 (2008)
5. Deerwester, S., Dumais, S.T., Furnas, G.W., Landauer, T.K., Harshman, R.: Indexing by latent semantic analysis. Journal of the American Society for Information Science 41, 391–407 (1990)
6. Ebdon, D.: Statistics in geography. Wiley-Blackwell (1985)
7. Felipe, I.D., Hristidis, V., Rishe, N.: Keyword search on spatial databases. In: ICDE, pp. 656–665 (2008)
8. Huang, Y., Shekhar, S., Xiong, H.: Discovering colocation patterns from spatial data sets: a general approach. IEEE Transactions on Knowledge and Data Engineering 16(12), 1472–1485 (2004)
9. Jarvelin, K., Kekalainen, J.: Cumulated gain-based evaluation of ir techniques. ACM Transactions on Information Systems 20 (2002)
10. Lew, M.S., Sebe, N., Djeraba, C., Jain, R.: Content-based multimedia information retrieval: State of the art and challenges. ACM Trans. Multimedia Comput. Commun. Appl. 2(1), 1–19 (2006)

11. Salton, G., Wong, A., Yang, C.S.: A vector space model for automatic indexing. Commun. ACM 18(11), 613–620 (1975)
12. Samet, H.: The design and analysis of spatial data structures. Addison-Wesley Longman Publishing Co., Inc., Boston (1990)
13. Sheng, C., Hsu, W., Lee, M., Tung, A.K.H.: Discovering spatial interaction patterns. In: Haritsa, J.R., Kotagiri, R., Pudi, V. (eds.) DASFAA 2008. LNCS, vol. 4947, pp. 95–109. Springer, Heidelberg (2008)
14. Zhou, Y., Xie, X., Wang, C., Gong, Y., Ma, W.-Y.: Hybrid index structures for location-based web search. In: CIKM 2005, pp. 155–162. ACM, New York (2005)

Efficient Approximate Visibility Query in Large Dynamic Environments

Leyla Kazemi[1], Farnoush Banaei-Kashani[1], Cyrus Shahabi[1], and Ramesh Jain[2]

[1] InfoLab Computer Science Department
University of Southern California, Los Angeles, CA 90089-0781
{lkazemi,banaeika,shahabi}@usc.edu
[2] Bren School of Information and Computer Sciences
University of California, Irvine, CA 92697-3425
jain@ics.uci.edu

Abstract. Visibility query is fundamental to many analysis and decision-making tasks in virtual environments. Visibility computation is time complex and the complexity escalates in large and dynamic environments, where the visibility set (i.e., the set of visible objects) of any viewpoint is probe to change at any time. However, exact visibility query is rarely necessary. Besides, it is inefficient, if not infeasible, to obtain the exact result in a dynamic environment. In this paper, we formally define an *Approximate Visibility Query (AVQ)* as follows: given a viewpoint v, a distance ε and a probability p, the answer to an AVQ for the viewpoint v is an approximate visibility set such that its difference with the exact visibility set is guaranteed to be less than ε with confidence p. We propose an approach to correctly and efficiently answer AVQ in large and dynamic environments. Our extensive experiments verified the efficiency of our approach.

1 Introduction

Visibility computation, i.e., the process of deriving the set of visible objects with respect to some query viewpoint in an environment, is one of the main enabling operations with a majority of spatial analysis, decision-making, and visualization systems ranging from GIS and online mapping systems to computer games. Most recently the marriage of spatial queries and visibility queries to spatio-visual queries (e.g., k nearest visible-neighbor queries and nearest surrounder queries) has further motivated the study of visibility queries in the database community [4,11,12]. The main challenge with visibility queries is the time-complexity of visibility computation which renders naive on-the-fly computation of visibility impractical with most applications.

With some traditional applications (e.g., basic computer games with simple and unrealistic visualization), the virtual environment is simple and small in extent, and hence, it can be modeled merely by *memory-resident* data structures and/or synthetic data. With such applications, a combination of hardware solutions (e.g., high-end graphic cards with embedded visibility computation modules) and memory-based graphics software solutions are sufficient for visibility analysis. However, with emerging applications the virtual environment is

H. Kitagawa et al. (Eds.): DASFAA 2010, Part I, LNCS 5981, pp. 202–217, 2010.

becoming large and is modeled based on massive geo-realistic data (e.g., terrain models and complex 3D models) stored on disk (e.g., Google Earth, Second Life). With these applications, all proposed solutions [10,14,15] inevitably leverage pre-computation to answer visibility queries in real-time. While such solutions perform reasonably well, they are all rendered infeasible with dynamic environments, as they are not designed for efficient update of the pre-computed visibility information.

In this paper, for the first time we introduce *approximate visibility query (AVQ)* in large dynamic virtual environments. Given a viewpoint v, a distance ε, and a confidence probability p, the answer to AVQ with respect to the viewpoint v is an approximate visibility vector, which is guaranteed to be in a distance less than ε from the exact visibility vector of v with confidence p, where visibility vector of v is the ordered set of objects visible to v. The distance ε is defined in terms of the cosine similarity between the two visibility vectors. Approximation of the visibility is the proper approach for visibility computation with most applications because 1) exact answer is often unnecessary, 2) exact answer is sometimes infeasible to compute in real-time due to its time-complexity (particularly in large dynamic environment), and 3) a consistent approximation can always converge to the exact answer with arbitrary user-defined precision. To enable answering AVQs in large dynamic environments, we propose a pre-computation based method inspired by our observation that there is a strong *spatial auto-correlation* among visibility vectors of distinct viewpoints in an environment, i.e., the closer two viewpoints are in the environment, often the more similar are their visibility vectors. Consequently, one can approximate the visibility vector of a viewpoint v by the visibility vectors of its close neighbors. Towards this end, we propose an index structure, termed *Dynamic Visibility Tree (DV-tree* for short), with which we divide the space into disjoint partitions. For each partition we pick a representative point and pre-compute its visibility vector. The partitioning with DV-tree is such that the distance between the visibility vector of any point inside a partition and that of the representative point of the partition is less than ε with confidence p. Therefore, with DV-tree we can efficiently answer an AVQ for viewpoint v (with distance ε and confidence p) by returning the pre-computed visibility vector of the representative point of the partition in which v resides. Figure 1 depicts an example of AVQ answering, which we discuss in more detail in Section 3. Accordingly, Figure 1a shows the exact visibility for viewpoint v while Figure 1b depicts the approximate result returned by DV-tree (i.e., visibility of the representative point of the partition in which v resides).

DV-tree is particularly designed to be efficiently maintainable/updatable in support of visibility computation in *dynamic* environments. In a dynamic environment, at any time a set of moving objects are roaming throughout the space, and consequently, the visibility vectors of some viewpoints may change. Accordingly, the partitioning of the DV-tree must be updated to reflect the visibility changes. However, this process is costly as it requires computing the visibility vectors of *all* viewpoints within each (and *every*) partition, to be compared with the visibility vector of the representative point of the corresponding partition. We have devised a two-phase filtering technique that effectively reduces the overhead of the DV-tree partition maintenance. In the first phase, termed *viewpoint filtering*, we effectively filter out the viewpoints whose visibility remains unchanged despite the recent object movements in the environment. For the remaining viewpoints (i.e., those that are filtered in), we proceed with the second phase, termed *object filtering*. In this phase, before computing the visibility vector for each of the remaining viewpoints, we effectively filter out all objects in the environment whose visibility status with respect to the viewpoint remains unchanged despite the recent object movements in the environment. After the two-phase filtering process, we are left with a limited number of viewpoints whose visibility must be computed with respect to only a limited number of objects in each case. Therefore, we can efficiently compute their visibility and also revise the corresponding DV-tree partitions accordingly, if needed.

Finally, through extensive experiments, we show that our approach can efficiently answer AVQ in large dynamic environments. In particular, our experiments show that DV-tree result is more than 80% accurate in answering AVQ, while the update cost is tolerable in real scenarios. This validates our observation about the spatial auto-correlation in visibility vectors. Note that both the approximation error and the update cost can be interpreted as the visual *glitch* and frame rate delay, respectively, in visualization systems. In general, for a DV-tree with higher error-tolerance, larger partitions are generated. This results in more visual *glitches* as a viewpoint moves from one partition to another, since its visibility may encounter a noticeable change during this transition. However, less frame rate delay is expected, because the DV-tree update is less costly as compared to that of a less error-tolerant DV-tree. On the other hand, for a DV-tree with less error-tolerance, smaller partitions are generated, which results in less glitches, but higher frame rate delays. Thus, there is a trade-off between these two system faults. Our experiments also show that DV-tree significantly outperforms a competitive approach, HDoV-tree [15], in both query response time and update cost.

The rest of the paper is organized as follows. Section 2 reviews the related work. In Section 3, we formally define our problem, and successively in Section 4 we present an overview of our proposed solution. Thereafter, in Sections 5 and 6 we explain the processes of construction and update for our proposed index structure (DV-tree). Section 7 presents the experimental results. Finally, in Section 8 we conclude and discuss the future directions of this study.

2 Related Work

Visibility analysis is an active research topic in various fields, including computer graphics, computer vision, and most recently, databases. Below, we review the existing work on visibilty analysis in two categories: memory-based approaches for small environments and disk-based approaches for large environments.

2.1   Memory-Based Approaches

In [1,3,9,17], different approaches are proposed for fast and efficient rendering. The end goal of most of these studies is to develop efficient techniques to accelerate image generation for realistic walkthrough applications [6]. In addition, there are a few proposals [2,5,7,16] from the computer graphics community on visibility analysis in *dynamic* environments. However, the aforementioned work assume the data are memory-resident, and therefore, their main constraint is the computation time, rather than disk I/O. This is not a practical assumption considering the immense data size with today's emerging applications with large virtual environments.

2.2   Disk-Based Approaches

On the other hand, in database community, many spatial index structures (e.g., R-tree, quad-tree) are proposed for efficient access to large data. Here, the goal is to expedite the search and querying of *relevant* objects in databases (e.g., kNN queries) [13], where the relevance is defined in terms of spatial proximity rather than visibility. However, recently a number of approaches are introduced for efficient visibility analysis in *large* virtual environments that utilize such spatial index structures to maintain and retrieve the visibility data ([10,14,15]). In particular, [10,14] exploit spatial proximity to identify visible objects. However, there are two drawbacks with utilizing spatial proximity. First, the query might miss visible objects that are outside the query region (i.e., possible false negatives). Second, all non-visible objects inside the query region would also be retrieved (i.e., numerous false positives). Later, in [15], Shou et al. tackle these drawbacks by proposing a data structure, namely HDoV-tree, which precomputes visibility information and incorporates it into the spatial index structure. While these techniques facilitate answering visibility queries in large environments, they are not designed for dynamic environments where visibility might change at any time. They all employ a pre-computation of the environment that is intolerably expensive to maintain; hence, inefficient for visibility query answering in dynamic environments.

In this paper, we focus on answering visibility queries in virtual environments that are both large *and* dynamic. To the best of our knowledge, this problem has not been studied before.

3   Problem Definition

With visibility query, given a query point q the visibility vector of q (i.e., the set of objects visible to q) is returned. Correspondingly, with *approximate* visibility query for q the returned result is guaranteed to be within certain distance from the exact visibility vector of q, with a specified level of confidence. The distance (or alternatively, the similarity) between the approximate and exact visibility vectors is defined in terms of the cosine similarity between the two vectors.

In this section, first we define our terminology. Thereafter, we formally define *Approximate Visibility Query (AVQ)*.

Consider a virtual environment $\Omega \subset R^3$ comprising of a stationary environment φ as well as a set of moving objects μ (e.g., people and cars). The stationary environment includes the terrain as well as the static objects of the virtual environment (e.g., buildings). We assume the environment is represented by a TIN model, with which all objects and the terrain are modeled by a network of Delaunay triangles. We consider both static and moving objects of the environment for visibility computation. Also, we assume a query point (i.e., a viewpoint whose visibility vector must be computed) is always at height h (e.g., at eye level) above the stationary environment φ.

Definition 3.1. Given a viewpoint v, the 3D *shadow-set* of v with respect to an object $O \subset \Omega$, $S(v, O)$, is defined as follows:

$$S(v, O) = \{tr | tr \in \Omega,\ tr \notin O,\ \exists p \in tr\ \text{s.t.}\ \overline{vp} \cap O \neq \emptyset\} \tag{1}$$

i.e., $S(v, O)$ includes any triangle tr in Ω, for which a straight line $\overline{vp}$ exists that connects v to a point p on tr, and $\overline{vp}$ intersects with O.

Accordingly, we say a triangle t is visible to v, if t is not in the shadow-set of v with respect to any object in the environment. That is, t is visible to v, if we have:

$$t \in \{tr | tr \in \Omega - \bigcup_{\forall O \subset \Omega} S(v, O), dist(v, tr) \leq D\} \tag{2}$$

where $dist$ is defined as the distance between v and the farthest point from v on tr, and D is the *visibility range*, i.e., the maximum range visible from a viewpoint.

Note that without loss of generality, we assume boolean visibility for a triangle. Accordingly, we consider a visible triangle as the one which is only fully visible. However, triangle visibility can be defined differently (e.g., a triangle can be considered visible even if it is partially visible) and our proposed solutions remain valid.

Definition 3.2. Given a viewpoint v and an object $O \subset \Omega$, we define the *visibility value* of O with respect to v as follows:

$$vis_O^v = \frac{\sum_{tr \in T_v \wedge tr \in O} Area(tr)}{\sum_{tr \in T_v} Area(tr)} \times \frac{1}{Dist(v, O)} \tag{3}$$

where T_v is the set of all triangles visible to v, and $Dist(.,.)$ is the distance between a viewpoint and the farthest visible triangle of an object. In other words, the visibility value of an object O with respect to a viewpoint v is the fraction of visible triangles to v which belong to O, scaled by the distance between v and O. Intuitively, visibility value captures how visually significant an object is with respect to a viewpoint. Thus, according to our definition of visibility value, nearby and large objects are naturally more important than far away and small objects in terms of visibility. In general, there are many factors (some application-dependent) that can be considered in determining the visibility value of objects (e.g., size of the object, the view angle). Developing effective metrics to evaluate visibility value is orthogonal to the context of our study, and hence, beyond the scope of this paper.

Definition 3.3. For a viewpoint v, we define its *visibility vector* as follows:

$$VV_v = (vis^v_{O_1}, vis^v_{O_2}, ..., vis^v_{O_n}) \tag{4}$$

Visibility vector of v is the vector of visibility values for all the objects $O_i \in \Omega$ with respect to v.

Definition 3.4. Given two viewpoints, v_1 and v_2, the visibility similarity between the two viewpoints is defined as the *cosine similarity* between their visibility vectors as follows:

$$sim(v_1, v_2) = cosim(VV_{v_1}, VV_{v_1}) = \frac{VV_{v_1}.VV_{v_2}}{||VV_{v_1}||\ ||VV_{v_1}||} \tag{5}$$

We say the two viewpoints v_1 and v_2 (and correspondingly their visibility vectors) are $\alpha-similar$ if:

$$cosim(VV_{v_1}, VV_{v_1}) = \alpha \tag{6}$$

Equally, the *visibility distance* ε is defined based on the similarity α as $\varepsilon = 1 - \alpha$. Alternatively, we say the two viewpoints are ε-distant ($\varepsilon = 1 - \alpha$).

Definition 3.5. AVQ Problem
Given a query point q, a visibility distance ε, and a confidence probability p, the *Approximate Visibility Query* returns a vector $A \in \Omega$, such that the visibility vector VV_q of q and A have at least $(1 - \varepsilon)$-similarity with confidence p.

A visual example of AVQ query is shown in Figure 1. Given a query point q, $\varepsilon=30\%$, and $p=90\%$, Figure 1a depicts the exact visibility VV_q for q, whereas Figure 1b shows the AVQ result A, which approximates the visibility from viewpoint q with a user-defined approximation error.

4   Solution Overview

To answer AVQs, we develop an index structure, termed *Dynamic Visibility Tree* (*DV-tree* for short). Given a specific visibility distance ε and confidence p, a DV-tree is built to answer AVQs for any point q of the virtual environment Ω. The parameters ε and p are application-dependent and are defined at the system configuration time. Figure 2 depicts an example of DV-tree built for Ω. DV-tree is inspired by our observation that there exists a strong spatial auto-correlation among visibility vectors of the viewpoints. Accordingly, we utilize a spatial partitioning technique similar to quad-tree partitioning to divide the space into a set of disjoint partitions. However, unlike quad-tree that uses spatial distance between objects to decide on partitioning, with DV-tree we consider visibility distance between viewpoints to decide if a partition should be further partitioned into smaller regions. Particularly, we continue partitioning each region to four equal sub-regions until every viewpoint in each sub-region and the representative point of the sub-region (selected randomly) have at least $(1-\varepsilon)$-similarity in visibility with confidence p (see Figure 2). Once DV-tree is constructed based on such partitioning scheme, we also pre-compute and store the visibility vector of the representative point for every partition of the tree. Subsequently, once an AVQ for a query point q is received, it can be answered by first traversing DV-tree and locating the partition to which q belongs, and then returning the visibility vector of the partition's representative point as approximate visibility for q. For example, in Figure 2 the query point q is located at partition P_{31}. Thus, the answer to AVQ for the query point q is the visibility vector $VV_{r_{P_{31}}}$ of the representative point $r_{P_{31}}$ of the partition P_{31}.

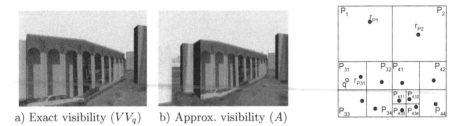

a) Exact visibility (VV_q)   b) Approx. visibility (A)

Fig. 1. Comparing approximate visibility (AVQ with $\varepsilon=30\%$ and $p=90\%$) with exact visibility for a viewpoint q

Fig. 2. DV-tree for Ω

However, in a dynamic environment the visibility (i.e., the visibility vectors of the viewpoints) may change as the objects move around. Accordingly, in order to guarantee correct AVQ answering, the pre-computed partitioning with DV-tree must be updated to maintain the distance between the representative point of the partition and the rest of the viewpoints within the partition. To enable efficient update, we propose a two-phase filtering technique which significantly reduces the update cost of DV-tree. In particular, the first phase prunes the unnecessary viewpoints whose pre-computed visibility data do not require update (Section 6.1), while the second phase prunes the set of objects that can be ignored while computing visibility for the remaining viewpoints (Section 6.2). Sections 5 and 6 discuss the construction and update processes for DV-tree, respectively.

5   Index Construction

In order to construct DV-tree with visibility distance ε and confidence p for a given environment, we iteratively divide the region covered by the environment into four smaller subregions, and for each partition we pick a random representative point, until with a confidence p all the viewpoints inside a partition reside in a visibility distance of less than ε, with the representative point of the partition (Figure 3). With this index construction process, at each step of the partitioning we need to compute the visibility vectors of all the viewpoints of a partition, and then find their visibility distance with the representative point. However, the computation cost for such operation is overwhelming. This stems from the fact that each partition should always satisfy the given ε and p for the approximation guarantee, and such task is hard to accomplish when all the viewpoints of a region are considered. More importantly, continuous maintenance of DV-tree requires repeated execution of the same operation. To address this problem, during each iteration of the DV-tree construction algorithm, instead of calculating the visibility distance for all the viewpoints of a partition, we calculate the visibility distance only for a chosen set of random sample points. In Figure 3, these sample points are marked in gray. Obviously, using samples would result in errors, because the samples are a subset of points from the entire partition. To achieve a correct answer for AVQ, we incorporate this *sampling error* into the approximation that in turn results in a probabilistic solution with confidence p.

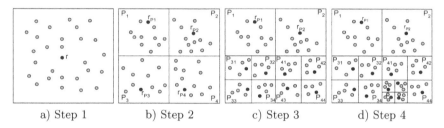

a) Step 1   b) Step 2   c) Step 3   d) Step 4

Fig. 3. Illustrating four steps of the DV-tree construction algorithm

In order to compute the sampling error, we first define a few notations. Given a partition with representative point r and assuming a visibility distance ε, we define the probability of success P for the partition as the probability that any random point q inside the partition has a visibility distance less than ε with r. In other words, we have:

$$Pr\{sim(r,q) \geqslant (1-\varepsilon)\} = P \qquad (7)$$

During the DV-tree construction, for a given visibility distance ε and confidence p, we say a partition satisfies the given ε and p if its probability of success P is equal to or larger than p (i.e., $P \geqslant p$).

Now, suppose we take n random sample points with replacement from the set of viewpoints inside a partition. We denote N as the number of sample points whose visibility distance to r is less than ε. Our goal is to find out the least value to expect for N, denoted by $\tilde{N}$, such that $P \geqslant p$ is guaranteed with high probability.

Lemma 1. *Given a partition with P as its probability of success, suppose we take n random sample points with replacement from the partition. Then, the random variable N follows a binomial distribution $B(n, P)$ where n is the number of trials and P is the probability of success.*

Proof. The proof is trivial and is therefore omitted due to lack of space. $\square$

Theorem 1. *Given a partition with P as its probability of success and n as the number of samples, the normal distribution $N(nP, nP(1 - P))$ is a good approximation for N, assuming large n and P not too close to 0 or 1.*

Proof. Since N has a binomial distribution (i.e., $B(n, P)$), according to the central limit theorem, for large n and P not too close to 0 or 1 (i.e., $nP(1 - P) > 10$) a good approximation to $B(n, P)$ is given by the normal distribution $N(nP, nP(1 - P))$. □

Consequently, we can approximate the value of $\tilde{N}$ with the normal distribution for n samples and a given *confidence interval* λ. Throughout the paper, we assume a fixed value for the confidence interval λ (i.e., $\lambda = 95\%$). Moreover, any value can be selected for n, as long as the constraint of Theorem 1 is satisfied. To illustrate, consider the following example, where for a given partition with $\varepsilon = 0.2$ and $p = 0.8$, we take 100 random sample points. Using the normal distribution $N(80, 16)$, with $\lambda = 95\%$, we have $\tilde{N} = 87$. This indicates if 87 out of 100 sample viewpoints have visibility distance of less than 0.2 with r, we are 95% confident that $P \geqslant p = 0.8$ holds for that partition.

Figure 4 illustrates the DV-tree that corresponds to the final partitioning depicted in Figure 3d. Each node P_I is of the form ($parent, P_{I1}, P_{I2}, P_{I3}, P_{I4},$ *Internal*), where P_{I1}, P_{I2}, P_{I3}, and P_{I4} are pointers to the node's children, *parent* is a pointer to the node's parent and *Internal* captures some information of the current node, which is required for the DV-tree maintenance as we explain in Section 6.

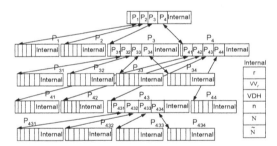

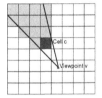

Fig. 4. DV-tree

Fig. 5. Potential occluded set of the cell c with respect to viewpoint v ($POS_c(v)$)

6   Index Maintenance

When an object moves from one location to another, not only the visibility vector of the representative point of a partition but also that of each viewpoint inside the partition might change. This affects the visibility distance of viewpoints inside the partition to the representative point, which consequently could invalidate the approximation guarantee. As a result, DV-tree may return incorrect answers for AVQ. Thus, in order to guarantee correct result, we must update the

DV-tree partitioning accordingly. Note that we assume a discrete-time model for the dynamics in the environment; i.e., an object located at location A at time t_0, may move to location B at time $t_0 + 1$. In this section, we propose a two-phase filtering technique which significantly reduces the cost of update: *viewpoint filtering* at phase 1 (Section 6.1) and *object filtering* at phase 2 (Section 6.2). After applying the two-step filtering, the visibility vectors of some of the viewpoints are updated. For each of these viewpoints, the visibility distance with the representative point of its partition might change. Consequently, the representative point might no longer remain as a *correct* representative of the viewpoints inside a partition. Accordingly, the partition must be revised such that a correct AVQ answer is guaranteed. This revision can be either by splitting or by merging the partitions (similar to quad-tree split and merge operations) to guarantee ε and p for the revised partitions. While splitting is necessary for correct query answering, merging only improves the efficiency of query answering. Therefore, to maintain DV-tree we split the partitions (when needed) immediately and merge the partitions in a lazy fashion (i.e., *lazy merge*). In the rest of this section, we explain our two-step filtering technique in more details.

6.1   Viewpoint Filtering

As discussed earlier, in order to efficiently maintain DV-tree, we exploit the fact that examining the visibility of all viewpoints is unnecessary for DV-tree maintenance. The viewpoints are categorized into the following two groups. The first group are those viewpoints that are not included in the sampling during DV-tree construction, and hence, any change in their visibility does not affect the DV-tree maintenance. These viewpoints are filtered out by sampling. This filtering step is performed only once during the DV-tree construction, and therefore we refer to it as *offline viewpoint filtering*. On the other hand, for the set of viewpoints that are sampled, maintaining visibility of *all* the samples with each object movement is unnecessary, because with each object movement only visibility of a subset of the sampled viewpoints changes. Thus, the second group of viewpoints are filtered out from the set of sampled viewpoints because their visibility cannot be affected by a particular object movement. We refer to this step as *online viewpoint filtering*. Below we elaborate on both of these viewpoint filtering steps.

Offline Viewpoint Filtering: By maintaining visibility of the sample viewpoints, we need to guarantee that our DV-tree is still properly maintained. Recall from Section 5 that N has a binomial distribution, if n samples are *randomly* selected. During an update of a partition, we need to guarantee that our sampled viewpoints which are stored in the visibility distance histogram VDH, remain valid random samples, and therefore, allow avoiding any resampling. This problem has been studied in [8], where given a large population of R tuples and a histogram of s samples from the tuples, to ensure that the histogram remains a valid representative of the set of tuples during tuple updates, one only needs to consider the updates for the sampled-set s. This guarantees that the histogram holds a valid random sample of size s from the current population. Accordingly, during object movements, when visibility vector of a viewpoint p inside the partition changes, if p is one of the sample points in VDH, its visibility vector is updated. Otherwise, the histogram remains unchanged. In both cases, the randomness of the sample points stored in VDH is guaranteed.

Online Viewpoint Filtering: As discussed earlier, once an object moves from location A to location B, only visibility vectors of a group of *relevant* viewpoints are affected. These are the viewpoints to which an object is visible when the object is either in location A or location B. With our online viewpoint filtering step, the idea is to pre-compute the set of relevant viewpoints for each point of the space. Accordingly, we propose *V-grid*. With V-grid, we partition the environment into a set of disjoint cells, and for each cell we maintain a list of moving objects that are currently within the cell, as well as a list of all leaf partitions in DV-tree (termed *RV-List*), that either their representative point or any of their sample points is a relevant viewpoint for that cell. When an object moves from one cell to another, we only need to update the visibility vectors of the relevant viewpoints of the two cells.

Our intuitive assumption for pre-computing the relevant viewpoints of a cell with V-grid is that moving objects of an environment (e.g., people, cars) have a limited height, while the static objects such as buildings can be of any height. We denote the maximum height of a moving object as H. In order to build V-grid, we impose a 3D grid on top of the stationary environment φ, where each cell of the grid is bounded with the height H on top of φ. Our observation is that any viewpoint that cannot see the surface of a cell, cannot see anything inside the cell either. Accordingly, all the viewpoints to which any point on the surface of a cell is visible, are considered as the relevant viewpoints of the cell.

6.2 Object Filtering

For each of the relevant viewpoints of the cell in which a movement occurs (i.e., when a moving object enters or leaves the cell), we need to recompute the visibility vector of the viewpoint. However, to compute the visibility vector for each relevant viewpoint, we only need to consider the visibility status of a subset of the environment objects that are potentially occluded by the cell in which the moving object resides before/after the movement. Below, we define how we can identify such objects.

Definition 6.2.1. Given a grid cell c, and a viewpoint v, let c_{xy} and v_{xy} be the 2D projection of c and v on the xy plane, respectively. We define the *potential occluded set* of c with respect to the viewpoint v (denoted by $POS_c(v)$) as the set of all cells, where for each cell $\widehat{c}$, there exists a line segment from v_{xy} to a point s in $\widehat{c}_{xy}$ which intersects c_{xy} in a point p such that the point p lies between the two points v_{xy} and s. The definition can be formulated as follows:

$$POS_c(v) = \{\widehat{c} | \exists s \in \widehat{c}_{xy}, \overline{sv_{xy}} \cap c_{xy} \neq \emptyset, \exists p \in \{\overline{sv_{xy}} \cap c_{xy}\}, |\overline{pv_{xy}}| < |\overline{sv_{xy}}|\} \quad (8)$$

where $|\overline{pv_{xy}}|$ and $|\overline{sv_{xy}}|$ are the lengths of line segments $\overline{pv_{xy}}$ and $\overline{sv_{xy}}$, respectively.

Figure 5 depicts an example of the potential occluded set of a viewpoint v with respect to a cell c as the gray area. We only need to recompute the visibility values of all the objects that reside in the POS of c with respect to v.

7   Performance Evaluation

We conducted extensive experiments to evaluate the performance of our solution in compare with the alternative work. Below, first we discuss our experimental methodology. Next, we present our experimental results.

7.1   Experimental Methodology

We performed three sets of experiments. With these experiments, we measured the accuracy and the response time of our proposed technique. With the first set of experiments, we evaluated the accuracy of DV-tree answers to AVQ. With the rest of the experiments, we compared the response time of DV-tree in both query and update costs with a competitive work. In comparing with a competitive work, since no work has been found on visibility queries in large dynamic virtual environments, we extended the HDoV approach [15] that answers visibility queries in large static virtual environments, to support dynamism as well. With HDoV-tree, the environment is divided into a set of disjoint cells, where for each cell the set of visible objects (i.e., union of all the objects which are visible to each viewpoint in the cell) are pre-computed, and incorporated into an R-tree-like structure. This spatial structure captures level-of-details (LoDs) (i.e., multi-resolution representations) of the objects in a hierarchical fashion, namely internal LoDs. Consequently, in a dynamic environment, not only the pre-computed visibility data of every cell should be updated, the internal LoDs should be updated as well. In this paper, we do not take into account the objects LoDs[1]. Conclusively, to perform a fair comparison with HDoV-tree, we only consider the cost of updating the pre-computed visibility data of every cell, while ignoring the cost of traversal and update of the tree hierarchy.

Because of the I/O bound nature of our experiments, we only report the response time in terms of I/O cost. Our DV-tree index structure is stored in memory, while the visibility data associated to each node of DV-tree (i.e., *Internal* of the node) is stored on disk. Thus, the traversal of DV-tree for point location query is memory-based, and the I/O cost considers only accessing the *Internal* of a node. Moreover, our V-grid is in memory as well, since RV-list of each cell holds only pointers to the relevant partitions.

We used a synthetic model built after a large area in the city of Los Angeles, CA as our data set with the size of 4GB, which contained numerous buildings. We also had a total number of 500 objects, with the maximum height of 1.5 meter (i.e., $H = 1.5m$) moving in the environment. For the movement model, we employed a network-based moving object generator, in which the objects move with a constant speed. Also, we picked 120 random samples during each

---

[1] Note that employing LoDs is orthogonal to our approach, and can be integrated into DV-tree; however, it is out of the scope of this paper.

iteration of DV-tree construction. For visibility computation, we set the visibility range D to 400 meters. Moreover, for visibility query, we ran 1000 queries using randomly selected query points, and reported the average of the results.

The experiments were performed on a DELL Precision 470 with Xeon 3.2 GHz processor and 3GB of RAM. Our disk page size was 32K bytes. Also, we employed an LRU buffer, which can hold 10% of the entire data.

7.2 Query Accuracy

With the first set of experiments, we evaluated the query accuracy of DV-tree. Given a DV-tree with visibility distance ε and a confidence p, we run n queries. Assuming N is the actual number of results satisfying ε, we expect $\frac{N}{n} \geq p$. We first varied the values for ε from 0.1 to 0.5 with p set to 70%. Figure 6a illustrates the percentage of queries satisfying ε for different values of ε. As Figure 6a depicts, for all cases this ratio is in the range of 92% to 95%, which is much higher than our expected ratio of 70% (shown with a dashed line in Figure 6a). Next, we varied the values for p from 50% to 90% with ε set to 0.4, and measured the percentage ratio of $\frac{N}{n}$ accordingly. As Figure 6b illustrates, for all cases this ratio is in the range of 82% to 97%. The dashed line in the figure shows the expected results. According to Figure 6b, the actual results are always above the dashed line. This demonstrates the guaranteed accuracy of our query result.

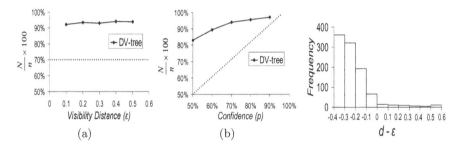

Fig. 6. Query accuracy   **Fig. 7.** Frequency distribution

The above results show the overall accuracy for a given ε, and p, but they do not represent the distribution of displacement of the query result's distance from the exact answer as compared with ε. For $\varepsilon = 0.4$, and $p = 70\%$, Figure 7 demonstrates the frequency distribution of this displacement (i.e., $d - \varepsilon$). The area under the curve represents the total number of queries, out of which we expect 70% to have a negative displacement. Figure 7 shows 94% query results satisfy this condition. The figure also shows that a large number of query results (68%) are only in distance of less than 0.2 from the exact result.

To summarize, this set of experiments confirms two issues. First, our use of sampling technique during DV-tree construction results in correct AVQ answering. Second, our initial observation about the spatial auto-correlation among the visibility vectors of the viewpoints is valid. This was also illustrated in Figure 1.

7.3   Query Response Time

With the second set of experiments, we compared the query response time of DV-tree with that of the extended HDoV-tree by varying the cell size in HDoV-tree. We set the values of ε and p to 0.4 and 70%, respectively. With DV-tree, the query response time is in terms of number of I/Os for retrieving visibility vector of the representative point of the partition to which the query point belongs, whereas with HDoV-tree the query response time is in terms of number of I/Os for retrieving union of the visibility vectors of all viewpoints inside the cell where query point is located. Thus, as the cell size grows, the query response time increases as well. Figure 8 illustrates such results, where the number of I/Os are shown in logarithmic scale. As the figure shows, in all cases DV-tree outperforms HDoV-tree during query answering, except for the case of HDoV-tree with the smallest cell size (i.e., one viewpoint per cell), where the query response time of HDoV-tree is identical to that of DV-tree.

Note that with DV-tree the query result is retrieved as the visibility vector of one viewpoint, which is restricted to the objects inside the viewpoint's visibility range. Conclusively, the query response time is independent of the values chosen for ε and p, as well as the data size.

7.4   Update Cost

The final set of experiments investigates the update cost for maintaining DV-tree. We set the number of moving objects to 50. First, we evaluate the DV-tree update cost by varying the values for ε, where p is fixed at 70%. As Figure 9a depicts, the update cost is in the range of 400 to 800 I/Os. The results show a decrease in I/O cost for large values of ε. The reason is that as the value of ε increases, larger partitions are generated, which leads to fewer updates. Thereafter, we fixed ε at 0.4, and evaluate the DV-tree update cost by varying the values for p. As Figure 9b illustrates, the number of I/Os varies between 500 to 650. Thus, the update cost slightly increases with the growing value of p. The reason is that higher value for p requires more number of viewpoints to satisfy ε in a partition (i.e., N), which results in maintaining more number of viewpoints.

Next, we compare the update cost of DV-tree with that of HDoV-tree. Similar to the previous experiments, we set ε to 0.4, and p to 70%, and vary the cell size for HDoV-tree. Figure 9c illustrates the number of I/Os for both data structures in a logarithmic scale. The significant difference between the two proves the effectiveness of our two-step filtering technique. For HDoV-tree with a small cell size, the number of cells whose visibility should get updated is large. As the cell size grows, the update cost slightly increases. The reason is that although visibility of less number of cells should be updated, retrieving the entire visible objects of a cell is still very costly.

Note that the accuracy evaluation in Section 7.2 is a good indicator that a DV-tree with a higher error tolerance allows for reasonable visualization of large-scale and dynamic environments with minor glitches. As Figure 9c depicts, each update of DV-tree with $\varepsilon = 0.4$ and $p = 70\%$ requires 500 I/Os on average.

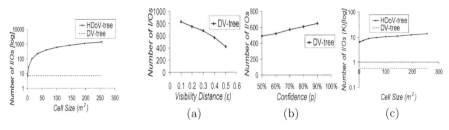

Fig. 8. Query response time   **Fig. 9.** Update cost

With every I/O taking 10ms on average, the update cost can be estimated as 5 seconds. Note that 5 second update cost is very practical as one can apply many techniques to hide this short delay. For instance, considering the case where trajectory of the moving objects are known a priori, location of the moving objects in the next 5 seconds can easily be predicted. This results in a smooth real-time rendering of the environment. Moreover, for the cases where the objects are constrained to move on predefined paths (e.g., road networks), one can prefetch from disk all the partitions whose visibility might change, and therefore, avoid the extra I/O cost for DV-tree update.

8   Conclusion and Future Work

In this paper, we introduced the novel concept of approximate visibility query in large dynamic environments. Accordingly, we proposed DV-tree for correct and efficient AVQ answering. To enable efficient maintenance for DV-tree, we also proposed a two-phase filtering technique that significantly reduces the update cost for DV-tree. With our experiments, we showed that our observation about the spatial auto-correlation of the visibility vectors is valid. We also demonstrated the overall superiority of our approach as compared to other approaches.

The focus of this paper has been on formal definition of the visibility problem and to propose a framework to address this problem through the use of approximation and DV-tree. For future work, we aim to explore optimization techniques which improve the query efficiency and maintenance cost of DV-tree. For example, while we chose to use a quad-tree-based approach to perform the spatial partitioning, other partitioning techniques exist that might result in a more optimal design for DV-tree. Furthermore, in this paper we make no assumption about movement of the objects. Another direction for our future work would be to use object models with known trajectory or restricted movement options (e.g., cars on roads), and exploit this knowledge to improve the DV-tree update cost.

Acknowledgement

This research has been funded in part by NSF grants IIS-0534761, and CNS-0831505 (CyberTrust), the NSF Center for Embedded Networked Sensing (CCR-0120778) and in part from the METRANS Transportation Center, under grants from USDOT and Caltrans. Any opinions, findings, and conclusions or recommendations expressed in this material are those of the author(s) and do not necessarily reflect the views of the National Science Foundation.

References

1. Airey, J.M., Rohlf, J.H., Brooks Jr., F.P.: Towards image realism with interactive update rates in complex virtual building environments. SIGGRAPH 24(2), 41–50 (1990)
2. Batagelo, H., Shin-Ting, W.: Dynamic scene occlusion culling using a regular grid. In: Computer Graphics and Image Processing, pp. 43–50 (2002)
3. Bittner, J., Havran, V., Slavk, P.: Hierarchical visibility culling with occlusion trees. In: CGI, pp. 207–219. IEEE, Los Alamitos (1998)
4. Bukauskas, L., Mark, L., Omiecinski, E., Böhlen, M.H.: itopn: incremental extraction of the n most visible objects. In: CIKM, pp. 461–468 (2003)
5. Chenney, S., Forsyth, D.: View-dependent culling of dynamic systems in virtual environments. In: SI3D, pp. 55–58 (1997)
6. Cohenor, D., Chrysanthou, Y., Silva, C.T., Durand, F.: A survey of visibility for walkthrough applications. IEEE Transactions on Visualization and Computer Graphics 9(3), 412–431 (2003)
7. Erikson, C., Manocha, D., Baxter III., V.: Hlods for faster display of large static and dynamic environments. In: I3D, pp. 111–120 (2001), doi:10.1145/364338.364376
8. Gibbons, P.B., Matias, Y., Poosala, V.: Fast incremental maintenance of approximate histograms. ACM Trans. Database Syst. 27(3), 261–298 (2002)
9. Greene, N., Kass, M., Miller, G.: Hierarchical z-buffer visibility. In: SIGGRAPH, pp. 231–238 (1993)
10. Kofler, M., Gervautz, M., Gruber, M.: R-trees for organizing and visualizing 3d gis databases. Journal of Visualization and Computer Animation 11(3), 129–143 (2000)
11. Lee, K.C.K., Lee, W.-C., Leong, H.V.: Nearest surrounder queries. In: ICDE, p. 85 (2006)
12. Nutanong, S., Tanin, E., Zhang, R.: Visible nearest neighbor queries. In: Kotagiri, R., Radha Krishna, P., Mohania, M., Nantajeewarawat, E. (eds.) DASFAA 2007. LNCS, vol. 4443, pp. 876–883. Springer, Heidelberg (2007)
13. Samet, H.: Foundations of Multidimensional and Metric Data Structures (2005)
14. Shou, L., Chionh, J., Huang, Z., Ruan, Y., Tan, K.-L.: Walking through a very large virtual environment in real-time. In: VLDB, pp. 401–410 (2001)
15. Shou, L., Huang, Z., Tan, K.-L.: Hdov-tree: The structure, the storage, the speed. In: ICDE, pp. 557–568 (2003)
16. Sudarsky, O., Gotsman, C.: Output-sensitive rendering and communication in dynamic virtual environments. In: VRST, pp. 217–223 (1997)
17. Teller, S.J., Séquin, C.H.: Visibility preprocessing for interactive walkthroughs. SIGGRAPH 25(4), 61–70 (1991)

The Objects Interaction Matrix for Modeling Cardinal Directions in Spatial Databases*

Tao Chen, Markus Schneider, Ganesh Viswanathan, and Wenjie Yuan

Department of Computer & Information Science & Engineering
University of Florida
Gainesville, FL 32611, USA
{tachen,mschneid,gv1,wyuan}@cise.ufl.edu

Abstract. Besides topological relations and approximate relations, *cardinal directions* have turned out to be an important class of qualitative spatial relations. In spatial databases and GIS they are frequently used as selection criteria in spatial queries. But the available models of cardinal relations suffer from a number of problems like the unequal treatment of the two spatial objects as arguments of a cardinal direction relation, the use of too coarse approximations of the two spatial operand objects in terms of single representative points or minimum bounding rectangles, the lacking property of converseness of the cardinal directions computed, the partial restriction and limited applicability to simple spatial objects only, and the computation of incorrect results in some cases. This paper proposes a novel two-phase method that solves these problems and consists of a tiling phase and an interpretation phase. In the first phase, a tiling strategy first determines the zones belonging to the nine cardinal directions of *each* spatial object and then intersects them. The result leads to a bounded grid called *objects interaction grid*. For each grid cell the information about the spatial objects that intersect it is stored in an *objects interaction matrix*. In the second phase, an interpretation method is applied to such a matrix and determines the cardinal direction. These results are integrated into spatial queries using directional predicates.

1   Introduction

Research on *cardinal directions* has had a long tradition in spatial databases, Geographic Information Systems (GIS), and other disciplines like cognitive science, robotics, artificial intelligence, and qualitative spatial reasoning. Cardinal directions are an important qualitative spatial concept and form a special kind of *directional relationships*. They represent *absolute* directional relationships like *north* and *southwest* with respect to a given reference system in contrast to *relative* directional relationships like *front* and *left*; thus, cardinal directions describe an order in space. In spatial databases they are, in particular, relevant as selection and join conditions in spatial queries. Hence, the determination of

---

* This work was partially supported by the National Science Foundation under grant number NSF-CAREER-IIS-0347574.

H. Kitagawa et al. (Eds.): DASFAA 2010, Part I, LNCS 5981, pp. 218–232, 2010.

and reasoning with cardinal directions between spatial objects is an important research issue.

In the past, several approaches have been proposed to model cardinal directions. They all suffer from at least one of four main problems. First, some models use quite coarse approximations of the two spatial operand objects of a cardinal direction relation in terms of single representative points or minimum bounding boxes. This can lead to inaccurate results. Second, some models assume that the two spatial objects for which we intend to determine the cardinal direction have different roles. They create a scenario in which a *target object* A is placed relative to a dominating *reference object* B that is considered as the center of reference. This is counterintuitive and does not correspond to our cognitive understanding. Third, some models do not support *inverse cardinal directions*. This means that once the direction $dir(A, B)$ between two objects A and B is computed, the reverse direction $inv(dir(A, B))$ from B to A should be deducible, i.e., $inv(dir(A, B)) = dir(B, A)$. For example, if A is northwest and north of B, then the inverse should directly yield that B is to the southeast and south of A. Fourth, some models only work well if the spatial objects involved in direction computations have a *simple* structure. This is in contrast to the common consensus in the spatial database community that *complex* spatial objects are needed in spatial applications. As a consequence of these problems, some models can yield wrong or counterintuitive results for certain spatial scenarios.

The goal of this paper is to propose and design a computation model for cardinal directions that overcomes the aforementioned problems by taking better into account the shape of spatial operand objects, treating both spatial operand objects as equal partners, ensuring the property of converseness of cardinal directions ($A \; p \; B \Leftrightarrow B \; inv(p) \; A$), accepting complex spatial objects as arguments, and avoiding the wrong results computed by some approaches.

Our solution consists in a novel two-phase method that includes a *tiling phase* followed by an *interpretation phase*. In the first phase, we apply a tiling strategy that first determines the zones belonging to the nine cardinal directions of *each* spatial object and then intersects them. The result leads to a closed grid that we call *objects interaction grid* (*OIG*). For each grid cell we derive the information about the spatial objects that intersect it and store this information in a so-called *objects interaction matrix* (*OIM*). In the second phase, we apply an interpretation method to such a matrix and determine the cardinal direction.

Section 2 discusses related work and summarizes the available approaches to compute cardinal directions. In Section 3, the objects interaction matrix model is introduced in detail. Section 4 compares the OIM model to past approaches. Section 5 defines directional predicates for integrating cardinal directions into spatial queries. Finally, Section 6 draws conclusions and depicts future work.

2   Related Work

Several models have been proposed to capture cardinal direction relations between spatial objects (like *point*, *line*, and *region* objects) as instances of *spatial*

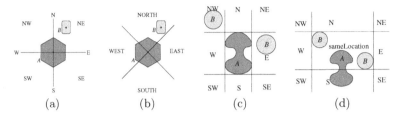

Fig. 1. Projection-based (a) and cone-shaped (b) models, and the Direction-Relation Matrix model with A as reference object (c) and with B as reference object (d)

data types [1]. These models can be classified into *tiling-based* models and *minimum bounding rectangle-based* (*MBR-based*) models.

Tiling-based models define cardinal direction relations by using partitioning lines that subdivide the plane into tiles. They can be further classified into *projection-based* models and *cone-shaped* models, both of which assign different roles to the two spatial objects involved. The first object represents the *target object* that is put into relationship to the second, dominant object called the *reference object*.

The *projection-based* models define direction relations by using partitioning lines parallel to the coordinate axes. The early approach in [2] first generalizes the reference (target) object by a reference (target) point (commonly the centroid of the object). Then it partitions the embedding space according to the reference point into four non-overlapping zones and uses the composition of two basic cardinal directions, that is, *north*, *west*, *south*, and *east*, in each zone to assign one of the four pairwise opposite directions *northwest*, *northeast*, *southeast*, and *southwest* to it (Figure 1a). The direction is then determined by the zone in which the target point falls. A problem of this approach is that the intermediate generalization step completely ignores the shape and extent of the spatial operand objects and thus leads to easy to use but rather inaccurate models. The *Direction-Relation Matrix* model [3,4] presents a major improvement of this approach by better capturing the influence of the objects' shapes (Figure 1c). In this model, the partitioning lines are given by the infinite extensions of the minimum bounding rectangle segments of the reference object. This leads to a tiling with the nine zones of *north*, *west*, *east*, *south*, *northwest*, *northeast*, *southwest*, *southeast*, as well as a *central zone* named *sameLocation* and given by the minimum bounding rectangle of the reference object. The target object contributes with its exact shape, and a direction-relation matrix stores for each tile whether it is intersected by the target object. Thus this model suffers from the problem of unequal treatment of objects leading to incorrect and counterintuitive determinations of cardinal directions. For example, the Figure 5a shows a map of the two countries China and Mongolia. If China is used as the reference object, Mongolia is located in the minimum bounding rectangle of China, and thus the model only yields *sameLocation* as a result. This is not what we would intuitively expect. A further problem of this model is that it does not enable us to directly imply the inverse relation. For example, Figure 1c and Figure 1d show the same spatial configuration. If A is the reference object (Figure 1c), the model derives that parts of B are northwest and east of A. We would now expect that then A is southeast and west of B. But the model determines *sameLocation* and south as cardinal directions (Figure 1d).

The *cone-shaped* models define direction relations by using angular zones. The early approach in [5] first also generalizes a reference object and a target object by point objects. Two axis-parallel partitioning lines through the reference point are then rotated by 45 degrees and span four zones with the cardinal directions *north*, *west*, *east*, and *south* (Figure 1b). Due to the generalization step, this model can produce incorrect results. The *Cone-Based Directional Relations* concept [6] is an improvement of the early approach and uses the minimum bounding rectangle of the reference object to subdivide the space around it with partitioning lines emanating from the corners of the rectangle with different angles. This model has the problems of an unequal treatment of the operand objects and the lack of inverse cardinal relations.

MBR-based models approximate both spatial operand objects of a directional relationship through minimum bounding rectangles and bring the sides of these rectangles into relation with each other by means of Allen's interval relations [7]. By using these interval relations, the *2D-string* model [8] constructs a direction-specifying 2D string as a pair of 1D strings, each representing the symbolic projection of the spatial objects on the x- and y-axis respectively. The 2D String model may not provide correct inverse direction relations. Another weakness of this model (and its extensions) is the lack of the ability to uniquely define directional relations between spatial objects since they are based on the projection of the objects along both standard axes.

The *Minimum Bounding Rectangle (MBR)* model [9,10] also makes use of the minimum bounding rectangles of both operand objects and applies Allen's 13 interval relations to the rectangle projections on the x- and y-axes respectively. 169 different relations are obtained [11] that are expressive enough to cover all possible directional relation configurations of two rectangles. A weakness of this model is that it can give misleading directional relations when objects are overlapping, intertwined, or horseshoe shaped. A comparison with the Direction-Relation Matrix model reveals that spatial configurations exist whose cardinal direction is better captured by either model.

3   The Objects Interaction Matrix Model

The Objects Interaction Matrix (OIM) model belongs to the tiling-based models, especially to the projection-based models. Figure 2 shows the two-phase strategy of our model for calculating the cardinal direction relations between two objects A and B. We assume that A and B are non-empty values of the complex spatial data type *region* [1]. The *tiling phase* in Section 3.1 details our novel tiling strategy that produces *objects interaction grids*, and shows how they are represented by *objects interaction matrices*. The *interpretation phase* in Section 3.2 leverages the objects interaction matrix to derive the directional relationship between two spatial region objects.

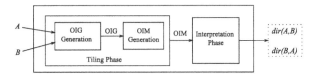

Fig. 2. Overview of the two phases of the Objects Interaction Matrix (OIM) model

3.1 The Tiling Phase: Representing Interactions of Objects with the Objects Interaction Grid and Matrix

In this section, we focus on the *tiling phase* as the first phase of our OIM model. The general idea of our tiling strategy is to superimpose a grid called *objects interaction grid* (*OIG*) on a configuration of two spatial objects (regions). Such a grid is determined by the two vertical and two horizontal *partitioning lines* of *each* object. The two vertical (two horizontal) partitioning lines of an object are given as infinite extensions of the two vertical (two horizontal) segments of the object's minimum bounding rectangle. The four partitioning lines of an object create a partition of the Euclidean plane consisting of nine mutually exclusive, directional *tiles* or *zones* from which one is bounded and eight are unbounded (Figures 1c and 1d). Further, these lines partition an object into non-overlapping components where each component is located in a different tile. This essentially describes the tiling strategy of the Direction-Relation Matrix model (Section 2).

However, our fundamental difference and improvement is that we apply this tiling strategy to *both* spatial operand objects, thus obtain two separate grid partitions (Figures 1c and 1d), and then overlay both partitions (Figure 3a). This leads to an entirely novel cardinal direction model. The overlay achieves a co-equal interaction and symmetric treatment of both objects. In the most general case, all partitioning lines are different from each other, and we obtain an overlay partition that shows 9 central, bounded tiles and 16 peripheral, unbounded tiles (indicated by the dashed segments in Figure 3a). The unbounded tiles are irrelevant for our further considerations since they cannot interact with both objects. Therefore, we exclude them and obtain a grid space that is a bounded proper subset of $\mathbb{R}^2$, as Definition 1 states. This is in contrast to the partitions of all other tiling-based models that are unbounded and equal to $\mathbb{R}^2$.

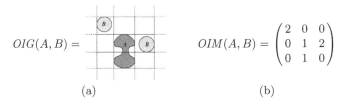

$$OIG(A, B) = \qquad\qquad OIM(A, B) = \begin{pmatrix} 2 & 0 & 0 \\ 0 & 1 & 2 \\ 0 & 1 & 0 \end{pmatrix}$$

(a)   (b)

Fig. 3. The objects interaction grid $OIG(A, B)$ for the two region objects A and B in Figures 1c and 1d (a) and the derived objects interaction matrix $OIM(A, B)$ (b)

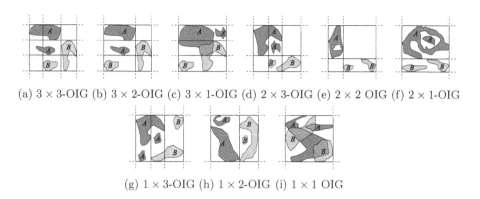

(a) 3×3-OIG (b) 3×2-OIG (c) 3×1-OIG (d) 2×3-OIG (e) 2×2 OIG (f) 2×1-OIG

(g) 1×3-OIG (h) 1×2-OIG (i) 1×1 OIG

Fig. 4. Examples of the nine possible sizes of objects interaction grids

Definition 1. *Let* $A, B \in$ *region with* $A \neq \varnothing$ *and* $B \neq \varnothing$, *and let* $min_x^r = \min\{x \mid (x, y) \in r\}$, $max_x^r = \max\{x \mid (x, y) \in r\}$, $min_y^r = \min\{y \mid (x, y) \in r\}$, *and* $max_y^r = \max\{y \mid (x, y) \in r\}$ *for* $r \in \{A, B\}$. *Then the* objects interaction grid space (OIGS) *of* A *and* B *is given as*

$$\text{OIGS}(A, B) = \{(x, y) \in \mathbb{R}^2 \mid \min(min_x^A, min_x^B) \leq x \leq \max(max_x^A, max_x^B) \wedge \\ \min(min_y^A, min_y^B) \leq y \leq \max(max_y^A, max_y^B)\}$$

Definition 2 determines the bounded grid formed as a part of the partitioning lines and superimposed on $OIGS(A, B)$.

Definition 2. *Let* seg *be a function that constructs a segment between any two given points* $p, q \in \mathbb{R}^2$, *i.e.,* $seg(p, q) = \{t \mid t = (1 - \lambda)p + \lambda q, 0 \leq \lambda \leq 1\}$. *Let* $H_r = \{seg((min_x^r, min_y^r), (max_x^r, min_y^r)), seg((min_x^r, max_y^r), (max_x^r, max_y^r))\}$ *and* $V_r = \{seg((min_x^r, min_y^r), (min_x^r, max_y^r)), seg((max_x^r, min_y^r), (max_x^r, max_y^r))\}$ *for* $r \in \{A, B\}$. *We call the elements of* H_A, H_B, V_A, *and* V_B objects interaction grid segments. *Then the* objects interaction grid (OIG) *for* A *and* B *is given as*

$$OIG(A, B) = H_A \cup V_A \cup H_B \cup V_B.$$

This definition comprises the description of all grids that can arise. In the most general case, if $H_A \cap H_B = \varnothing$ and $V_A \cap V_B = \varnothing$, we obtain a bounded 3×3-grid. Special cases arise if $H_A \cap H_B \neq \varnothing$ and/or $V_A \cap V_B \neq \varnothing$. Then equal grid segments coincide in the union of all grid segments. As a result, depending on the relative position of two objects to each other, objects interaction grids can be of different size. However, due to the non-empty property of a region object, not all grid segments can coincide. This means that at least two horizontal grid segments and at least two vertical grid segments must be maintained. Figure 4 shows examples for all nine possible sizes of objects interaction grids, and Definition 3 gives a corresponding formal characterization.

Definition 3. *An objects interaction grid* $OIG(A, B)$ *is of size* $m \times n$, *with* $m, n \in \{1, 2, 3\}$, *if* $|H_A \cap H_B| = 3 - m$ *and* $|V_A \cap V_B| = 3 - n$.

The objects interaction grid partitions the objects interaction grid space into *objects interaction grid tiles* (*zones, cells*). Definition 4 provides their definition.

Definition 4. *Given $A, B \in$ region with $A \neq \varnothing$ and $B \neq \varnothing$, $\text{OIGS}(A, B)$, and $OIG(A, B)$, we define $c_H = |H_A \cup H_B| = |H_A| + |H_B| - |H_A \cap H_B|$ and c_V correspondingly. Let $H_{AB} = H_A \cup H_B = \{h_1, \ldots, h_{c_H}\}$ such that (i) $\forall 1 \leq i \leq c_H : h_i = seg((x_i^1, y_i), (x_i^2, y_i))$ with $x_i^1 < x_i^2$, and (ii) $\forall 1 \leq i < j \leq c_H : h_i < h_j$ (we say that $h_i < h_j :\Leftrightarrow y_j < y_i$). Further, let $V_{AB} = V_A \cup V_B = \{v_1, \ldots, v_{c_V}\}$ such that (i) $\forall 1 \leq i \leq c_V : v_i = seg((x_i, y_i^1), (x_i, y_i^2))$ with $y_i^1 < y_i^2$, and (ii) $\forall 1 \leq i < j \leq c_V : v_i < v_j$ (we say that $v_i < v_j :\Leftrightarrow x_i < x_j$).*

Next, we define four auxiliary predicates that check the position of a point (x, y) with respect to a grid segment:

$$
\begin{aligned}
below((x, y), h_i) &\Leftrightarrow x_i^1 \leq x \leq x_i^2 \wedge y \leq y_i \\
above((x, y), h_i) &\Leftrightarrow x_i^1 \leq x \leq x_i^2 \wedge y \geq y_i \\
right\_of((x, y), v_i) &\Leftrightarrow y_i^1 \leq y \leq y_i^2 \wedge x \geq x_i \\
left\_of((x, y), v_i) &\Leftrightarrow y_i^1 \leq y \leq y_i^2 \wedge x \leq x_i
\end{aligned}
$$

An objects interaction grid tile $t_{i,j}$ *with* $1 \leq i < c_H$ *and* $1 \leq j < c_V$ *is then defined as*

$$
t_{i,j} = \{(x, y) \in \text{OIGS}(A, B) \mid below((x, y), h_i) \wedge above((x, y), h_{i+1}) \wedge \\
right\_of((x, y), v_j) \wedge left\_of((x, y), v_{j+1})\}
$$

The definition indicates that all tiles are bounded and that two adjacent tiles share their common boundary. Let $OIGT(A, B)$ be the set of all tiles $t_{i,j}$ imposed by $OIG(A, B)$ on $OIGS(A, B)$. An $m \times n$-grid contains $m \cdot n$ bounded tiles.

By applying our tiling strategy, an objects interaction grid can be generated for any two region objects A and B. It provides us with the valuable information which region object intersects which tile. Definition 5 provides us with a definition of the *interaction* of A and B with a tile.

Definition 5. *Given $A, B \in$ region with $A \neq \varnothing$ and $B \neq \varnothing$ and $\text{OIGT}(A, B)$, let ι be a function that encodes the* interaction *of A and B with a tile $t_{i,j}$, and checks whether no region, A only, B only, or both regions intersect a tile. We define this function as*

$$
\iota(A, B, t_{i,j}) = \begin{cases}
0 \text{ if} & A° \cap t_{i,j}° = \varnothing \wedge B° \cap t_{i,j}° = \varnothing \\
1 \text{ if} & A° \cap t_{i,j}° \neq \varnothing \wedge B° \cap t_{i,j}° = \varnothing \\
2 \text{ if} & A° \cap t_{i,j}° = \varnothing \wedge B° \cap t_{i,j}° \neq \varnothing \\
3 \text{ if} & A° \cap t_{i,j}° \neq \varnothing \wedge B° \cap t_{i,j}° \neq \varnothing
\end{cases}
$$

The operator $°$ denotes the point-set topological *interior* operator and yields a region without its boundary. For each grid cell $t_{i,j}$ in the ith row and jth column of an $m \times n$-grid with $1 \leq i \leq m$ and $1 \leq j \leq n$, we store the coded information in an *objects interaction matrix* (*OIM*) in cell $OIM(A, B)_{i,j}$. Since directional relationships have a qualitative and not a quantitative or metric nature, we abstract from the geometry of the *objects interaction grid space* and only keep the information which region intersects which tile. The OIM for $m = n = 3$ is shown below, and Figure 3b gives an example.

$$OIM(A, B) = \begin{pmatrix} \iota(A, B, t_{1,1}) & \iota(A, B, t_{1,2}) & \iota(A, B, t_{1,3}) \\ \iota(A, B, t_{2,1}) & \iota(A, B, t_{2,2}) & \iota(A, B, t_{2,3}) \\ \iota(A, B, t_{3,1}) & \iota(A, B, t_{3,2}) & \iota(A, B, t_{3,3}) \end{pmatrix}$$

3.2 The Interpretation Phase: Assigning Semantics to the Objects Interaction Matrix

The second phase of the OIM model is the *interpretation phase*. This phase takes an objects interaction matrix obtained as the result of the tiling phase as input and uses it to generate a set of cardinal directions as output. This is achieved by separately identifying the locations of both objects in the objects interaction matrix and by pairwise interpreting these locations in terms of cardinal directions. The union of all these cardinal directions is the result.

In a first step, we define a function *loc* (see Definition 6) that acts on one of the region objects A or B and their common objects interaction matrix and determines all locations of components of each object in the matrix. Let $I_{m,n} = \{(i, j) \mid 1 \leq i \leq m, 1 \leq j \leq n\}$. We use an index pair $(i, j) \in I_{m,n}$ to represent the location of the element $M_{i,j} \in \{0, 1, 2, 3\}$ and thus the location of an object component from A or B in an $m \times n$ objects interaction matrix.

Definition 6. *Let M be the $m \times n$-objects interaction matrix of two region objects A and B. Then the function loc is defined as:*

$$loc(A, M) = \{(i, j) \mid 1 \leq i \leq m, 1 \leq j \leq n, M_{i,j} = 1 \ \vee \ M_{i,j} = 3\}$$
$$loc(B, M) = \{(i, j) \mid 1 \leq i \leq m, 1 \leq j \leq n, M_{i,j} = 2 \ \vee \ M_{i,j} = 3\}$$

For example, in Figure 3b, object A occupies the locations (2,2) and (3,2), and object B occupies the locations (1,1) and (2,3) in the objects interaction matrix $OIM(A, B)$. Therefore, we obtain $loc(A, OIM(A, B)) = \{(2, 2), (3, 2)\}$ and $loc(B, OIM(A, B)) = \{(1, 1), (2, 3)\}$.

In a second step, we define an *interpretation function* ψ to determine the cardinal direction between any two object components of A and B on the basis of their locations in the objects interaction matrix. We use a popular model with the nine cardinal directions *north* (N), *northwest* (NW), *west* (W), *southwest* (SW), *south* (S), *southeast* (SE), *east* (E), *northeast* (NE), and *origin* (O) to symbolize the possible cardinal directions between *object components*. In summary, we obtain the set $CD = \{N, NW, W, SW, S, SE, E, NE, O\}$ of *basic cardinal directions*. A different set of basic cardinal directions would lead to a different interpretation function and hence to a different interpretation of index pairs. Definition 7 provides the interpretation function ψ with the signature $\psi : I_{m,n} \times I_{m,n} \to CD$.

Table 1. Interpretation table for the interpretation function ψ

(i,j) $\diagdown$ (i',j')	(1,1)	(1,2)	(1,3)	(2,1)	(2,2)	(2,3)	(3,1)	(3,2)	(3,3)
(1,1)	O	W	W	N	NW	NW	N	NW	NW
(1,2)	E	O	W	NE	N	NW	NE	N	NW
(1,3)	E	E	O	NE	NE	N	NE	NE	N
(2,1)	S	SW	SW	O	W	W	N	NW	NW
(2,2)	SE	S	SW	E	O	W	NE	N	NW
(2,3)	SE	SE	S	E	E	O	NE	NE	N
(3,1)	S	SW	SW	S	SW	SW	O	W	W
(3,2)	SE	S	SW	SE	S	SW	E	O	W
(3,3)	SE	SE	S	SE	SE	S	E	E	O

Definition 7. *Given* $(i,j),(i',j') \in I_{m,n}$, *the interpretation function* ψ *on the basis of the set* $CD = \{N, NW, W, SW, S, SE, E, NE, O\}$ *of* basic cardinal directions *is defined as*

$$\psi((i,j),(i',j')) = \begin{cases} N & \text{if } i < i' \wedge j = j' \\ NW & \text{if } i < i' \wedge j < j' \\ W & \text{if } i = i' \wedge j < j' \\ SW & \text{if } i > i' \wedge j < j' \\ S & \text{if } i > i' \wedge j = j' \\ SE & \text{if } i > i' \wedge j > j' \\ E & \text{if } i = i' \wedge j > j' \\ NE & \text{if } i < i' \wedge j > j' \\ O & \text{if } i = i' \wedge j = j' \end{cases}$$

For example, in Figure 3b, we obtain that $\psi((3,2),(1,1)) = SE$ and $\psi((2,2), (2,3)) = W$ where holds that $(2,2),(3,2) \in loc(A, OIM(A,B))$ and $(1,1), (2,3) \in loc(B, OIM(A,B))$. Table 1 called *interpretation table* shows the possible results of the interpretation function for all index pairs.

In a third and final step, we specify the *cardinal direction function* named *dir* which determines the *composite cardinal direction* for two region objects A and B. This function has the signature $dir : region \times region \to 2^{CD}$ and yields a set of basic cardinal directions as its result. In order to compute the function *dir*, we first generalize the signature of our interpretation function ψ to $\psi : 2^{I_{m,n}} \times 2^{I_{m,n}} \to 2^{CD}$ such that for any two sets $X, Y \subseteq I_{m,n}$ holds: $\psi(X,Y) = \{\psi((i,j),(i',j')) \mid (i,j) \in X, (i',j') \in Y\}$. We are now able to specify the cardinal direction function *dir* in Definition 8.

Definition 8. *Let* $A, B \in region$. *Then the* cardinal direction function dir *is defined as*

$$dir(A, B) = \psi(loc(A, OIM(A,B)), loc(B, OIM(A,B)))$$

We apply this definition to our example in Figure 3. With $loc(A, OIM(A, B)) = \{(2,2),(3,2)\}$ and $loc(B, OIM(A,B)) = \{(1,1),(2,3)\}$ we obtain

$$dir(A, B) = \psi(\{(2,2),(3,2)\}, \{(1,1),(2,3)\})$$
$$= \{\psi((2,2),(1,1)), \psi((2,2),(2,3)), \psi((3,2),(1,1)), \psi((3,2),(2,3))\}$$
$$= \{SE, W, SW\}$$

Similarly, we obtain the inverse cardinal direction as:

$$dir(B, A) = \psi(\{(1,1),(2,3)\}, \{(2,2),(3,2)\})$$
$$= \{\psi((1,1),(2,2)), \psi((1,1),(3,2)), \psi((2,3),(2,2)), \psi((2,3),(3,2))\}$$
$$= \{NW, E, NE\}$$

Syntactically function dir yields a set of basic cardinal directions. The question is what the exact meaning of such a set is. We give the intended semantics of the function result in Lemma 1.

Lemma 1. Let $A, B \in region$. Then $dir(A, B) = \{d_1, \ldots, d_k\}$ if the following conditions hold:

(i) $1 \le k \le 9$
(ii) $\forall 1 \le i \le k : d_i \in CD$
(iii) $\exists r_{11}, \ldots, r_{1k}, r_{21}, \ldots, r_{2k} \in region :$
  (a) $\forall 1 \le i \le k : r_{1i} \subseteq A, r_{2i} \subseteq B$
  (b) $dir(r_{11}, r_{21}) = d_1 \wedge \ldots \wedge dir(r_{1k}, r_{2k}) = d_k$

Several r_{1i} from A as well as several r_{2i} from B might be equal. Thus at most k parts from A and at most k parts from B are needed to produce the k basic cardinal directions of the result. There can be further parts from A and B but their cardinal direction is not a new contribution to the result.

Finally we can say regarding Figure 3 that "Object A is *partly southeast*, *partly west*, and *partly southwest* of object B" and that "Object B is *partly northwest*, *partly east*, and *partly northeast* of object A", which is consistent.

4   Comparison to Past Approaches

We now review the problems raised in the Introduction and show how our OIM model overcomes them. The first problem is the *coarse approximation problem* that leads to imprecise results. Models that capture directions between region objects have evolved from reducing these objects to points, to the use of minimum bounding rectangles to approximate their extent, and ultimately to the final goal of considering their shapes. From this perspective, the *Directional-Relation Matrix* (DRM) model is superior to the MBR model due to the fact that it captures the shape of the target object. However, it only represents an intermediate step between the MBR model and the final goal because only the shape of one object is considered and the shape of the other object does not contribute at all. The OIM model that we propose in this paper is the first model that considers the shapes of both region objects.

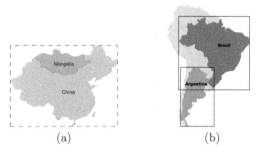

(a)　　　　　　　(b)

Fig. 5. Determining the cardinal direction

Table 2. Cardinal directions between Argentina (A) and Brazil (B) in Figure 5b from different models

Model α	$dir_\alpha(A, B)$	$dir_\alpha(B, A)$
MBR	$\{weak\_bounded\_south\}$	$\{weak\_north\}$
DRM	$\{sL^\dagger, S\}$	$\{sL^\dagger, NW, N, NE, E\}$
2D-String	$\{S\}$	$\{NW, N, NE\}$
OIM	$\{S, W, SW, O, SE\}$	$\{N, E, NE, O, NW\}$

† sL means *sameLocation*

Unlike the MBR model and the 2D string model, where the region objects play the same role, the DRM model suffers from the *unequal treatment problem*. A target object is tested for intersection against the tiles created around a reference object. As a result, components of the target object inside different tiles contribute to the final direction relations while the reference object contributes as a whole object. This unequal treatment causes imprecision. Let $dir_{DRM}(A, B)$ be a function that determines the cardinal directions for two simple regions A and B in the DRM model where A is the target object and B is the reference object. Then, in Figure 5b, the DRM model determines the cardinal direction between Argentina (A) and Brazil (B) as $dir_{DRM}(A, B) = \{sameLocation, S\}$. This is not precise because for the major part of Brazil, Argentina lies to the *southwest*, and it also lies to the *west* of some part of Brazil. In our OIM model, objects are treated equally, and thus both contribute to the final cardinal direction. Our model yields the result $dir(A, B) = \{SE, S, SW, W, O\}$, which captures the cardinal directions precisely.

The *converse problem* is a common problem shared by most models. It means that models generate inconsistent results when swapping their operand objects. Table 2 shows the different cardinal directions between Argentina and Brazil in Figure 5b as they are obtained by different cardinal direction models. The results show that the MBR model, the DRM model, and the 2D-String model do not maintain the converseness property, i.e., $dir_\alpha(A, B) \neq inv(dir_\alpha(B, A))$ for $\alpha \in \{MBR, DRM, 2D\text{-}String\}$. Only the OIM model supports the inverse operation, i.e., $dir(A, B) = inv(dir(B, A))$ holds. Therefore, by applying the OIM model, we obtain consistent results corresponding to human intuition.

Further, the MBR model, the DRM model, and the 2D string model have originally been designed for simple regions only. Since all these models are based on the moving bounding rectangle approximation of at least one object, an extension to complex regions and their minimum bounding rectangles is feasible without difficulty. However, this procedure usually generates rather poor results. For example, in Figure 1d, if we take the minimum bounding rectangle of the entire object B, then object A is to the *weak_bounded_south* of object B according to the MBR model, and object A is to the *sameLocation* and *south* of object B according to the DRM model. Both results are imprecise since the western direction of A to one component of B is not captured. Although variants exist for the models to handle complex objects more precisely, considerable efforts are required. Our model natively supports complex objects and is able to yield much more precise results. For the same example in Figure 3a, our model generates the result $dir(A, B) = \{SE, W, SW\}$, which describes object A to be partly *southeast*, partly *west*, and partly *southwest* of object B.

As a summary, we show in Table 3 the evaluation of the four major models based on the four criteria of shape capturing, equal treatment of operand objects, support for the inverse operation, and support for complex objects.

5   Defining Directional Predicates within Databases

Based on the OIM model and the interpretation mechanism described in the previous sections, we can identify the cardinal directions between any given two complex region objects. To integrate the cardinal directions into spatial databases as selection and join conditions in spatial queries, binary *directional predicates* need to be formally defined. For example, a query like "Find all states that are strictly north of Florida" requires a directional predicate like *strict_north* as a selection condition of a spatial join. Assuming a relation *states* with attributes *sname* of type string and *loc* of type *region*, we can express the last query in an SQL-like style as follows:

```
SELECT s1.sname FROM states s1, states s2
WHERE  s2.sname='Florida' and strict_north(s1.loc,s2.loc);
```

The *dir* function, which produces the final cardinal directions between two complex region objects A and B, yields a subset of the set $CD = \{N, NW, W, SW, S, SE, E, NE, O\}$ of *basic cardinal directions*. As a result, a total number of

Table 3. Comparison of the OIM model with other cardinal direction models

Models	Shape Capturing	Equal Treatment	Inverse Operation	Complex Objects
MBR	no	yes	no	ps[†]
DRM	partially	no	no	ps[†]
2D-String	no	yes	no	ps[†]
OIM	yes	yes	yes	yes

[†] "ps" means "poorly supported"

$2^9 = 512$ cardinal directions can be identified. Therefore, at most 512 directional predicates can be defined to provide an *exclusive* and *complete* coverage of all possible directional relationships. We can assume that users will not be interested in such a large, overwhelming collection of detailed predicates since they will find it difficult to distinguish, remember and handle them. Instead, a reduced and manageable set is preferred. Such a set should be user-defined and/or application specific. It should be application specific since different applications may have different criteria for the distinction of directional relationships. For example, one application could require a clear distinction between the cardinal direction *north* and *northwest*, whereas another application could perhaps accept no distinction between the two and regard them both as *northern*. Thus it should also offer user the flexibility of defining their own set of predicates.

As a first step, in Definition 9, we propose nine *existential directional predicates* that ensure the existence of a particular basic cardinal direction between parts of two region objects A and B.

Definition 9. *Let $A, B \in region$. Then the existential directional predicates are defined as*

$$\begin{aligned}
exists\_north(A, B) &\equiv (N \in dir(A, B)) \\
exists\_south(A, B) &\equiv (S \in dir(A, B)) \\
exists\_east(A, B) &\equiv (E \in dir(A, B)) \\
exists\_west(A, B) &\equiv (W \in dir(A, B)) \\
exists\_origin(A, B) &\equiv (O \in dir(A, B)) \\
exists\_northeast(A, B) &\equiv (NE \in dir(A, B)) \\
exists\_southeast(A, B) &\equiv (SE \in dir(A, B)) \\
exists\_northwest(A, B) &\equiv (NW \in dir(A, B)) \\
exists\_southwest(A, B) &\equiv (SW \in dir(A, B))
\end{aligned}$$

For example, $exists\_north(A, B)$ returns *true* if a part of A is located to the north of B; this does not exclude the existence of other cardinal directions. Later, by using this set of existential directional predicates together with $\neg$, $\vee$ and $\wedge$ operators, the user will be able to define any set of composite directional predicates for their own applications.

The following Lemma 2 shows that by using the existential directional predicates and the logical operators $\neg$, $\vee$ and $\wedge$, we can obtain a complete coverage and distinction of all possible 512 basic and composite cardinal directions from the OIM model based on the CD set.

Lemma 2. *Let the list $\langle d_1, d_2, d_3, d_4, d_5, d_6, d_7, d_8, d_9 \rangle$ denote the cardinal direction list $\langle N, S, E, W, O, NE, SE, NW, SW \rangle$ and let the list $\langle p_1, p_2, p_3, p_4, p_5, p_6, p_7, p_8, p_9 \rangle$ denote the basic directional predicates list $\langle exists\_north, exists\_south, exists\_east, exists\_west, exists\_origin, exists\_northeast, exists\_southeast, exists\_northwest, exists\_southwest \rangle$. Let $A, B \in region$ and $1 \leq i, j \leq 9$. Then for any basic or composite cardinal direction provided by $dir(A, B)$, the following logical expression returns true:*

$$\bigwedge_{d_i \in dir(A,B)} p_i \;\wedge\; \bigwedge_{d_j \notin dir(A,B)} \neg p_j$$

The existential predicates provide an interface for the user to define their own *derived directional predicates*. We give two examples.

The first set is designed to handle *strict directional predicates* between two region objects. *Strict* means that two region objects are in exactly one basic cardinal direction to each other. Definition 10 shows an example of *strict_north* by using the existential predicates.

Definition 10. *Let* $A, B \in$ *region. Then* strict_north *is defined as:*

$$
\begin{aligned}
strict\_north(A, B) = \; & exists\_north(A, B) \wedge \neg exists\_south(A, B) \\
& \wedge \neg exists\_west(A, B) \wedge \neg exists\_east(A, B) \\
& \wedge \neg exists\_northwest(A, B) \wedge \neg exists\_northeast(A, B) \\
& \wedge \neg exists\_southwest(A, B) \wedge \neg exists\_southeast(A, B) \\
& \wedge \neg exists\_origin(A, B)
\end{aligned}
$$

The other strict directional predicates *strict_south*, *strict_east*, *strict_west*, *strict_origin*, *strict_northeast*, *strict_northwest*, *strict_southeast*, *strict_southwest* are defined in a similar way.

The second set of predicates is designed to handle *similarly oriented directional predicates* between two regions. *Similarly oriented* means that several cardinal directions facing the same general orientation belong to the same group. Definition 11 shows an example of *northern* by using the existential predicates.

Definition 11. *Let* $A, B \in$ *region. Then* northern *is defined as:*

$$
\begin{aligned}
northern(A, B) = \; & (exists\_north(A, B) \vee exists\_northwest(A, B) \\
& \vee exists\_northeast(A, B)) \\
& \wedge \neg exists\_east(A, B) \wedge \neg exists\_west(A, B) \\
& \wedge \neg exists\_south(A, B) \wedge \neg exists\_southwest(A, B) \\
& \wedge \neg exists\_southeast(A, B) \wedge \neg exists\_origin(A, B)
\end{aligned}
$$

The other similarly oriented directional predicates *southern*, *eastern*, and *western* are defined in a similar way. From Definition 11, we can see that due to the disjunction of three existential directional predicates each similarly oriented directional predicate represents multiple directional relationships between two objects. For example, if A is in the northern part of B, then $dir(A, B) \in \{\{N\}, \{NW\}, \{NE\}, \{N, NW\}, \{N, NE\}, \{NW, NE\}, \{N, NW, NE\}\}$.

We can now employ predicates like *strict_north*, *northern* and *exists_north* in queries. For example, assuming we are given the two relations:

```
states(sname:string, area:region)
national_parks(pname:string, area:region)
```

We can pose the following three queries: *Determine the national park names where the national park is located* (a) *strictly to the north of Florida,* (b) *to the northern of Florida,* and (c) *partially to the north of Florida.* The corresponding SQL queries are as follows:

```
(a) SELECT P.pname FROM national_park P, states S
    WHERE  S.sname='Florida' and strict_north(P.area, S.area)

(b) SELECT P.pname FROM national_park P, states S
    WHERE  S.sname='Florida' and northern(P.area, S.area)

(c) SELECT P.pname FROM national_park P, states S
    WHERE  S.sname='Florida' and exists_north(P.area, S.area)
```

6   Conclusions and Future Work

In this paper, we have laid the foundation of a novel concept, called *Objects Interaction Matrix* (*OIM*), for determining cardinal directions between region objects. We have shown how different kinds of directional predicates can be derived from the cardinal directions and how these predicates can be employed in spatial queries. In the future, we plan to extend our approach to handle two complex point objects, two complex line objects, and all mixed combinations of spatial data types. Other research issues refer to the efficient implementation and the design of spatial reasoning techniques based on the OIM model.

References

1. Schneider, M. (ed.): Spatial Data Types for Database Systems. LNCS, vol. 1288. Springer, Heidelberg (1997)
2. Frank, A.: Qualitative Spatial Reasoning: Cardinal Directions as an Example. International Journal of Geographical Information Science 10(3), 269–290 (1996)
3. Goyal, R., Egenhofer, M.: Cardinal Directions between Extended Spatial Objects (2000) (unpublished manuscript)
4. Skiadopoulos, S., Koubarakis, M.: Composing Cardinal Direction Relations. Artificial Intelligence 152(2), 143–171 (2004)
5. Haar, R.: Computational Models of Spatial Relations. Technical Report: TR-478, MSC-72-03610 (1976)
6. Skiadopoulos, S., Sarkas, N., Sellis, T., Koubarakis, M.: A Family of Directional Relation Models for Extended Objects. IEEE Trans. on Knowledge and Data Engineering (TKDE) 19(8), 1116–1130 (2007)
7. Allen, J.F.: Maintaining Knowledge about Temporal Intervals. Journal of the Association for Computing Machinery 26(11), 832–843 (1983)
8. Chang, S.K.: Principles of Pictorial Information Systems Design. Prentice-Hall, Englewood Cliffs (1989)
9. Papadias, D., Egenhofer, M.: Algorithms for Hierarchical Spatial Reasoning. GeoInformatica 1(3), 251–273 (1997)
10. Papadias, D., Theodoridis, Y., Sellis, T.: The Retrieval of Direction Relations Using R-trees. In: Karagiannis, D. (ed.) DEXA 1994. LNCS, vol. 856, pp. 173–182. Springer, Heidelberg (1994)
11. Papadias, D., Egenhofer, M., Sharma, J.: Hierarchical Reasoning about Direction Relations. In: 4th ACM workshop on Advances in Geographic Information Systems, pp. 105–112 (1996)

Efficient Algorithms to Monitor Continuous Constrained k Nearest Neighbor Queries

Mahady Hasan[1], Muhammad Aamir Cheema[1], Wenyu Qu[2], and Xuemin Lin[1]

[1] The University of New South Wales, Australia
{mahadyh,macheema,lxue}@cse.unsw.edu.au
[2] College of Information Science and Technology,
Dalian Maritime University, China
quwenyu.dl@gmail.com

Abstract. Continuous monitoring of spatial queries has received significant research attention in the past few years. In this paper, we propose two efficient algorithms for the continuous monitoring of the constrained k nearest neighbor (kNN) queries. In contrast to the conventional k nearest neighbors (kNN) queries, a constrained kNN query considers only the objects that lie within a region specified by some user defined constraints (e.g., a polygon). Similar to the previous works, we also use grid-based data structure and propose two novel grid access methods. Our proposed algorithms are based on these access methods and guarantee that the number of cells that are accessed to compute the constrained kNNs is minimal. Extensive experiments demonstrate that our algorithms are several times faster than the previous algorithm and use considerably less memory.

1 Introduction

With the availability of inexpensive position locators and mobile devices, continuous monitoring of spatial queries has gained significant research attention. For this reason, several algorithms have been proposed to continuously monitor the k nearest neighbor (kNN) queries [1, 2, 3], range queries [13, 4] and reverse nearest neighbor queries [5, 6] etc.

A k nearest neighbors (kNN) query retrieves k objects closest to the query. A continuous kNN query is to update the kNNs continuously in real-time when the underlying data issues updates. Continuous monitoring of kNN queries has many applications such as fleet management, geo-social networking (also called location-based networking), traffic monitoring, enhanced 911 services, location-based games and strategic planning etc. Consider the example of a fleet management company. A driver might issue a kNN query to monitor their k closest vehicles and may contact them from time to time to seek or provide assistance. Consider another example of the location based reality game BotFighter in which the players are rewarded for shooting the other nearby players. To be able to earn more points, the players might issue a continuous kNN query to monitor their k closest players.

H. Kitagawa et al. (Eds.): DASFAA 2010, Part I, LNCS 5981, pp. 233–249, 2010.

We are often required to focus on the objects within some specific region. For example, a user might be interested in finding the k closest gas stations in North-East from his location. Constrained kNN queries [7] consider only the objects that lie within a specified region (also called constrained region). We formally define the constrained kNN queries in Section 2.1. In this paper, we study the problem of continuous monitoring of constrained kNN queries.

The applications of the continuous constrained kNN queries are similar to the applications of kNN queries. Consider the example of the fleet management company where a driver is heading towards the downtown area. The driver might only be interested in k closest vehicles that are within the downtown area. Consider the example of BotFighter game, the players might only be interested in the k closest players within their colleges so that they could eliminate their fellow students. Continuous constrained kNN queries are also used to continuously monitor reverse kNN queries. For example, six continuous constrained kNN queries are issued in [5, 8] to monitor the set of candidate objects. Similarly, constrained NNs are retrieved in [6] to prune the search space.

Although previous algorithms can be extended to continuously monitor constrained kNN queries, they are not very efficient because no previous algorithm has been specifically designed for monitoring constrained kNN queries. In this paper, we design two simple and efficient algorithms for continuous monitoring of constrained kNN queries. Our algorithms significantly reduce the computation time as well as the memory usage. The algorithms are applicable to any arbitrary shape of constrained region as long as a function is provided that checks whether a point or a rectangle intersects the constrained region or not. Our contributions in this paper are as follows;

- We introduce two novel grid access methods named Conceptual Grid-tree and ArcTrip. The proposed access methods can be used to return the grid cells that lie within any constrained region in order (ascending or descending) of their proximity to the query point.
- We propose two efficient algorithms to continuously monitor constrained kNN queries based on the above mentioned grid access methods. It can be proved that both the algorithms visit minimum number of cells to monitor the constrained kNN queries. Our algorithms significantly reduce the computational time and the memory consumption.
- Our extensive experiments demonstrate significant improvement over previous algorithms in terms of computation time and memory usage.

2   Background Information

2.1   Preliminaries

Definition 1. *Let O be a set of objects, q be a query point and R be a constrained region. Let $O_R \subseteq O$ be a set of objects that lie within the constrained region R, a constrained kNN query returns an answer set $A \subseteq O_R$ that contains k objects such that for any $o \in A$ and any $o' \in (O_R - A)$, $dist(o, q) \leq dist(o', q)$ where dist is a distance metric assumed Euclidean in this paper.*

Please note that a conventional kNN query is a special case of the constrained kNN queries where the constrained region is the whole data space.

In dynamic environment, the objects and queries issue updates frequently. The problem of continuous monitoring of constrained kNN queries is to continuously update the constrained kNNs of the query.

Like many existing algorithms, we use time stamp model. In time stamp model, the objects and queries report their locations at every time stamp (i.e., after every t time units) and the server updates the results and reports to the client who issued the query. Our algorithm consists of two phases: 1) In *initial computation*, the initial results of the queries are computed; 2) In *continuous monitoring*, the results of the queries are updated continuously at each time stamp.

Grid data structure is preferred [1] for the dynamic data sets because it can be efficiently updated in contrast to the more complex data structures (e.g., R-trees, Quad-trees etc). For this reason, we use an in-memory grid data structure where entire space is partitioned into equal sized cells. The cell width in any direction is denoted by δ. A cell $c[i,j]$ denotes the cell at column i and row j. Clearly, an object o lies into the cell $c[\lfloor o.x/\delta \rfloor, \lfloor o.y/\delta \rfloor]$ where $o.x$ and $o.y$ represent x and y co-ordinate values of the object location.

Let q be a query, R be the constrained region and rec be a rectangle. Below, we define minimum and maximum constrained distances.

Definition 2. *Minimum constrained distance $MinConstDist(rec, q)$ is the minimum distance of q to the part of the rectangle rec that lies in the constrained region R. If rec completely lies outside the constrained region R then the minimum constrained distance is infinity. The maximum constrained distance $Max ConstDist(rec, q)$ is defined in a similar way.*

Please note that if the constrained region is a complex shape, computing the minimum (maximum) constrained distance might be expensive or not possible. In such cases, we use $mindist(rec, q)$ and $maxdist(rec, q)$ which denote minimum and maximum distances of q from the rectangle rec, respectively. Fig. 1 shows examples of minimum and maximum constrained distances for two rectangles $rec1$ and $rec2$ where the constrained region is a rectangular region and

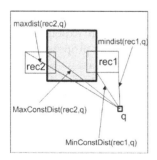 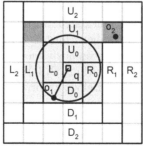 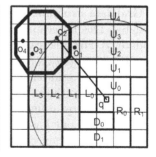

Fig. 1. Constrained distances   **Fig. 2.** Illustration of CPM   **Fig. 3.** CPM for constrained kNN queries

is shown shaded. We use these distances to avoid visiting un-necessary rectangles. Please note that minimum and maximum constrained distances give better bounds compared to minimum and maximum distances and hence we prefer to use constrained distances if available.

Table 1 defines the notations used throughout this paper.

Table 1. Notations

Notation	Definition
o, q	an object, a query
$o.x, o.y, q.x, q.y$	the coordinates (x-axis, y-axis) of o and q
$c, c[i, j]$	a cell c (at i^{th} column and j^{th} row)
$dist(x, y)$	the distance between two points x and y
$q.CkNN$	the set of constrained k nearest neighbors of q
δ	the side length of a cell
R	the constrained region
$q.dist_k$	the distance between the k^{th}NN and the query q
$mindist(c, q), maxdist(c, q)$	minimum, maximum distance between q and the cell c
$MinConstDist(c, q),$ $MaxConstDist(c, q)$	minimum, maximum distance between q and the part of cell c that lies in the constrained region

2.2   Related Work

Ferhatosmanoglu et al. [7] are first to introduce the constrained kNN queries. They solve the problem for static data objects and static queries. Their proposed solution traverses R-tree [9] in best-first [10] manner and prune the intermediate entries by using several interesting pruning rules. They show that their technique is optimal in terms of I/O. Gao et. al [11] studied the problem of finding k-nearest trajectories in a constrained region.

Now, we focus on the related work on continuous nearest neighbor queries [12, 2, 3, 1, 5] where the queries and/or objects change their locations frequently and the results are to be updated continuously. Voronoi diagram based approaches (e.g., [13]) have also been proposed for the conventional kNN queries but they are mainly designed for the case when only the queries are moving.

Grid data structures are preferred when the underlying datasets issue frequent updates. This is because more complex structures (e.g., R-tree) are expensive to update [1]. For this reason, several algorithms [2, 3, 1, 5] have been proposed that use grid-based data structure to continuously monitor kNN queries. .

Most of the grid-based kNN algorithms [2, 3, 1, 14] iteratively access the cells that are close to the query location. Below, we briefly introduce CPM [1] because it is a well-known algorithm for continuously monitoring kNN queries. Also, to the best of our knowledge, this is the only work for which an extension to continuous constrained kNN queries has been presented.

CPM [1] organizes the cells into conceptual rectangles and assigns each rectangle a direction (right, down, left, up) and a level number (the number of cells in between the rectangle and q as shown in Fig. 2). CPM first initializes an empty min-heap H. It inserts the query cell c_q and the level zero rectangles (R_0, D_0, L_0, U_0) with the keys set to minimum distances between the query and the rectangles/cell into H. The entries are de-heaped iteratively. If a de-heaped entry e is a cell then it checks all the objects inside the cell and updates $q.kNN$ (the set of kNNs) and $q.dist_k$ (the distance of current k^{th}NN from q). If e is a rectangle, it inserts all the cells inside the rectangle and the next level rectangle in the same direction into the heap. The algorithm stops when the heap becomes empty or when e has minimum distance from query not less than $q.dist_k$. The Fig. 2 shows 1NN query where the NN is o_1. The algorithm accesses the shaded cells. For more details, please see [1].

CPM can also be used to answer continuous constrained kNN queries by making a small change. More specifically, only the rectangles and cells that intersect the constrained region are inserted in the heap. Fig. 3 shows an example where the constrained region is a polygon. The constrained NN is o_2 and the rectangles shown shaded are inserted into the heap.

2.3   Motivation

At the end of Section 2.2, we briefly introduced how CPM can be used to answer continuous constrained kNN queries. Fig. 3 shows the computation of a constrained kNN query and the rectangles that were inserted into the heap are shown shaded. Recall that whenever CPM de-heaps a rectangle, it inserts all the cells into the heap. In the case of a constrained kNN queries, it inserts only the cells that intersect the constrained region. Please note that it may require to check a large number of cells to see if they intersect the constrained region or not. In the example of Fig. 3, for every shaded cell, CPM checks whether it intersects the constrained region or not.

The problem mentioned above motivates us to find a more natural grid access method. In this paper, we present two novel access methods called Conceptual Grid-tree and ArcTrip. Then, we introduce our algorithms based on these access methods which significantly perform better than CPM.

3   Grid-Tree Based Algorithm

In this section, first we revisit the Conceptual Grid-tree we briefly introduced in [6] to address a different problem. Then, we present Grid-tree based algorithm to continuously monitor the constrained kNN queries.

3.1   The Conceptual Grid-Tree

Consider a grid that consists of $2^n \times 2^n$ cells (Fig. 4 shows an example of a 4×4 grid). The grid is treated as a conceptual tree where the root contains $2^n \times 2^n$ grid cells[1]. Each entry e (and root) is recursively divided into four children of equal sized rectangles such that each child of an entry e contains $x/4$ cells where

---

[1] If the grid size is not $2^n \times 2^n$, it can be divided into several smaller grids such that each grid is $2^i \times 2^i$ for $i > 0$. For example, a 8×10 grid can be divided into 5 smaller grids (i.e., one 8×8 grid and four 2×2 grids).

x is the number of cells contained in e. The leaf level entries contain four cells each (the root, intermediate entries and the grid cells are shown in Fig. 4).

Please note that the Grid-tree is just a *conceptual visualization* of the grid and *it does not exist physically* (i.e., we do not need pointers to store entries and its children). More specifically, the root is a rectangle with each side of length 1 (we assume unit space). To retrieve the children of an entry (or root), we divide its rectangle into four equal sized rectangles such that each child has side length $l/2$ where l is the side length of its parent. A rectangle with side length equal to δ (the width of a gird cell) refers to a cell $c[i, j]$ of the grid. The cell $c[i, j]$ can be identified by the coordinates of the rectangle. More specifically, let a be the center of the rectangle, then the cell $c[i, j]$ is $c[\lfloor a.x/\delta \rfloor, \lfloor a.y/\delta \rfloor]$.

3.2 Initial Computation

Algorithm 1 presents the technique to compute the initial results of a constrained kNN query using the Conceptual Grid-tree. The basic idea is similar to that of applying BFS search [10] on R-tree based data structure. More specifically, the algorithm starts by inserting the root of the Grid-tree into a min-heap H (root is a rectangle with side length 1). The algorithm iteratively de-heaps the entries. If a de-heaped entry e is a grid cell then it looks in this cell and update $q.CkNN$ and $q.dist_k$ where $q.CkNN$ is the set of constrained kNNs and $q.dist_k$ is the distance of kth nearest neighbor from q (lines 7 and 8). If $q.CkNN$ contains less than k objects, then $q.dist_k$ is set to infinity. Recall that to check whether an entry e is a grid cell or not, the algorithm only needs to check if its side width is δ.

Algorithm 1. Grid-based Initial Computation

Input: q: query point; k: an integer
Output: $q.CkNN$
1: $q.dist_k = \infty$; $q.CkNN = \phi$; $H = \phi$
2: Initialize the H with root entry of Grid-Tree
3: **while** $H \neq \phi$ **do**
4:    de-heap an entry e
5:    **if** $MinConstDist(e, q) \geq q.dist_k$ **then**
6:       **return** $q.CkNN$
7:    **if** e is a cell in the grid **then**
8:       update $q.CkNN$ and $q.dist_k$ by the objects in e
9:    **else**
10:       **for** each of the four children c **do**
11:          **if** c intersects the constrained region **then**
12:             insert c into H with key $MinConstDist(c, q)$
13: **return** $q.CkNN$

If the de-heaped entry e is not a grid cell, then the algorithm inserts its children into the heap H according to their minimum constrained distances[2] from q. A

---
[2] Recall that we use minimum distance in case the constrained region is a complex shape such that minimum constrained distance computation is either complicated or not possible.

child c that does not intersect the constrained region is not inserted (lines 10 to 13). The algorithm terminates when the heap becomes empty or when a de-heaped entry e has $MinConstDist(e, q) \geq q.dist_k$ (line 5). This is because any cell c for which $MinConstDist(c, q) \geq q.dist_k$ cannot contain an object that lies in the constrained region and is closer than the kth nearest neighbor. Since the de-heaped entry e has $MinConstdist(e, q) \geq q.dist_k$, every remaining entry e' has $MinConstDist(e', q) \geq q.dist_k$ because the entries are accessed in ascending order of their minimum constrained distances.

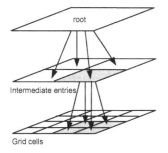 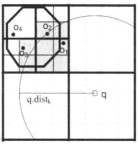 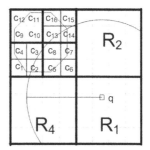

Fig. 4. The Conceptual Grid-tree

Fig. 5. Illustration of Algorithm 1

Fig. 6. Illustration of the pruned entries

Example 1. Fig. 5 shows an example of a constrained kNN ($k = 1$) query q and the constrained region is a polygon (we assume that the function to compute minimum constrained distance is not available, so we use minimum distance). To illustrate the working of our algorithm, the entries of the Grid-tree are shown in Fig. 6. An entry $C_{[i \to j]}$ refers to the rectangle that encloses the cells $c_i, c_{i+1}, ..., c_j$. For example, $C_{[9 \to 12]}$ refers to the top-left small rectangle that contains the cells c_9, c_{10}, c_{11} and c_{12}. To further improve the illustration, we show the steps of the execution in Table 1. Please refer to Fig. 5, 6 and Table 1 for rest of the example. Below, we explain the execution of the algorithm for some of the steps.

1. The root of the tree is inserted in the heap. The set of $q.CkNN$ is set to empty and $q.dist_k$ is set to infinity.

2. Root is de-heaped. Its children $R1$, $R2$ and $R4$ are not inserted into the heap because they do not intersect the constrained region. The only child that is inserted is $C_{[1 \to 16]}$.

3. $C_{[1 \to 16]}$ is de-heaped and all four children are inserted in the heap because they intersect the constrained region.

4. $C_{[5 \to 8]}$ is de-heaped and its children (cells c_5, c_6, c_7 and c_8) are inserted in the heap.

5-8. The cells c_6, c_5, c_7 and c_8 are de-heaped in this order and the algorithm looks for the objects that lie inside it. Only one object o_1 is found (in cell c_7) but it lies outside the constrained region so it is ignored.

The algorithm continues in this way.

Table 2. Grid-tree access

Step	Deheaped Entries	Heap content	$q.CkNN$	$q.dist_k$
1	ϕ	$root$	ϕ	∞
2	$root$	$C_{[1\rightarrow16]}$	ϕ	∞
3	$C_{[1\rightarrow16]}$	$C_{[5\rightarrow8]}, C_{[1\rightarrow4]}, C_{[13\rightarrow16]}, C_{[9\rightarrow12]}$	ϕ	∞
4	$C_{[5\rightarrow8]}$	$c_6, c_5, c_7, c_8, C_{[1\rightarrow4]}, C_{[13\rightarrow16]}, C_{[9\rightarrow12]}$	ϕ	∞
5-8	$c_6, c_5, c_7, c_8,$	$C_{[1\rightarrow4]}, C_{[13\rightarrow16]}, C_{[9\rightarrow12]}$	ϕ	∞
9	$C_{[1\rightarrow4]}$	$c_2, C_{[13\rightarrow16]}, c_3, c_1, C_{[9\rightarrow12]}, c_4$	ϕ	∞
10	c_2	$C_{[13\rightarrow16]}, c_3, c_1, C_{[9\rightarrow12]}, c_4$	ϕ	∞
11	$C_{[13\rightarrow16]}$	$c_{14}, c_3, c_{13}, c_{15}, c_1, C_{[9\rightarrow12]}, c_{16}, c_4$	ϕ	∞
12	c_{14}	$c_3, c_{13}, c_{15}, c_1, C_{[9\rightarrow12]}, c_{16}, c_4$	ϕ	∞
13	c_3	$c_{13}, c_{15}, c_1, C_{9-12}, c_{16}, c_4$	o_3	$dist(o_3,q)$
14	c_{13}	$c_{15}, c_1, C_{[9\rightarrow12]}, c_{16}, c_4$	o_2	$dist(o_2,q)$
15	c_{15}	$c_1, C_{[9\rightarrow12]}, c_{16}, c_4$	o_2	$dist(o_2,q)$

13. At step 13, the cell c_3 is de-heaped and an object o_3 is found that lies in the constrained region. $q.CkNN$ is updated to o_3 and $q.dist_k$ is set to $dist(o_3,q)$.
14. c_{13} is de-heaped and an object o_2 is found. Since o_2 is closer to q than o_3, o_3 is deleted from $q.CkNN$ and o_2 is inserted. $q.dist_k$ is set to $dist(o_2,q)$.
15. The next de-heaped cell c_{15} has $mindist(c_{15},q) \geq dist(o_2,q)$ so the algorithm terminates and o_2 is returned as the answer.

3.3 Continuous Monitoring

Data Structure: The system stores a query table and an object table to record the information about the queries and the objects. More specifically, an object table stores the object id and location of every object. The query table stores the query id, query location and the set of its constrained kNNs.

Each cell of the grid stores two lists namely *object list* and *influence list*. The object list of a cell c contains the object id of every object that lies in c. The influence list of a cell c contains the id of every query q that has visited c (by visiting c we mean that it has considered the objects that lie inside it (line 8 of Algorithm 1)). The influence list is used to quickly identify the queries that might have been affected by the object movement in a cell c.

Handling a single update: In the timestamp model, the objects report their locations at every timestamp (i.e., after every t time units). Assume that an object o reports a location update and o_{old} and o_{new} correspond to its old and new locations, respectively. The object update can affect the results of a query q in the following three ways;

1. *internal update:* $dist(o_{old},q) \leq q.dist_k$ and $dist(o_{new},q) \leq q.dist_k$; clearly, only the order of the constrained kNNs may have been affected, so we update $q.CkNN$ accordingly.

2. *incoming update:* $dist(o_{old}, q) > q.dist_k$ and $dist(o_{new}, q) \leq q.dist_k$; o is inserted in $q.CkNN$
3. *outgoing update:* $dist(o_{old}, q) \leq q.dist_k$ and $dist(o_{new}, q) > q.dist_k$; o is not a constrained kNN anymore, so we delete it from $q.CkNN$.

It is important to note that $dist(o, q)$ is considered infinity if o lies outside the constrained region. Now, we present our complete update handling module.

The complete update handling module: The update handling module consists of two phases. In first phase, we receive query and object updates and reflect their effect on the results. In the second phase, we compute the final results. Algorithm 2 presents the details.

Algorithm 2. Continuous Monitoring

Input: location updates
Output: $q.CkNN$
Phase 1: *receive updates*
 1: **for** each query update q **do**
 2:     insert q in Q_{moved}
 3: **for** each object update o **do**
 4:     $Q_{affected} = c_{o_{old}}.Influence\_list \cup c_{o_{new}}.Influence\_list$
 5:     **for** each query q in $(Q_{affected} - Q_{moved})$ **do**
 6:         if internal update; update the order of $q.CkNN$
 7:         if incoming update; insert o in $q.CkNN$
 8:         if outgoing update; remove o from $q.CkNN$
Phase 2: *update results*
 9: **for** each query q **do**
10:     if $q \in Q_{moved}$; call initial computation module
11:     if $|q.CkNN| \geq k$; keep top k objects in $q.CkNN$ and update $q.dist_k$
12:     if $|q.CkNN| < k$; expand $q.CkNN$

Phase 1: First, we receive the query updates and mark all the queries that have moved (line 1 to 2). For such queries, we will compute the results from scratch (similar to CPM). Then, for each object update, we identify the queries that might have been affected by this update. It can be immediately verified that only the queries in the influence lists of c_{old} and c_{new} may have been affected where c_{old} and c_{new} denote the old and new cells of the object, respectively. For each affected query q, the update is handled (lines 5 to 8) as mentioned previously (e.g., internal update, incoming update or outgoing update).

Phase 2: After all the updates are received, the results of the queries are updated as follows; If a query is marked as moved, its results are computed by calling the initial computation algorithm. If $q.CkNN$ contains more than k objects in it (more incoming updates than the outgoing updates), the results are updated by keeping only the top k objects. Otherwise, if $q.CkNN$ contains less than k objects, we expand the $q.CkNN$ so that it contains k objects.

The expansion is similar to the initial computation algorithm except the following change. The cells that have $MaxConstDist(c, q) \leq q.dist_k$ are not inserted into the heap. This is because such cells are already visited.

3.4   Remarks

A cell c is called visited if the algorithm retrieves the objects that lie inside it. The number of visited cells has direct impact on the performance of the algorithm. Our algorithm is optimal[3] in the sense that it visits minimum number of cells (i.e., if any of these cells are not visited, the algorithm may report incorrect results). Moreover, the correctness of the algorithm follows from the fact that it visits all such cells. Due to space limitations, we omit the proof of correctness and the optimality. However, the proof is very similar to the proof for a slightly different problem (please see Chapter 4.5 of [15]).

We would like to remark that the Conceptual Grid-tree provides a robust access method that can be used to access cells in order of any preference function. For example, it can be naturally extended to access cells in decreasing order of their minimum L_1 distances from the query point. As another example, in [4] we use grid-tree to access the cells in order of their minimum distances to the boundary of a given circle.

4   ArcTrip Based Algorithm

In this section, we first present a grid access method called ArcTrip. Then, we present the algorithm to continuously monitor the constrained kNN queries based on the ArcTrip.

4.1   ArcTrip

ArcTrip is a more general case of our previous work CircularTrip [16]. Given a query point q and a radius r, the CircularTrip returns the cells that intersect the circle centered at query location q and has radius r. More specifically, it returns every cell c for which $mindist(c, q) \leq r$ and $maxdist(c, q) > r$. Fig. 7 shows the CircularTrip where the shaded cells are returned by the algorithm.

The algorithm maintains two directions called D_{cur} and D_{next} (Fig. 7 shows the directions for the cells in different quadrants based on the location of q). The main observation is that if a cell c intersects the circle then at least one of the cells in either direction D_{cur} or D_{next} also intersects the circle. The algorithm starts with any cell that intersects the circle. It always checks the cell in the direction D_{cur} and returns the cell if it intersects the circle. Otherwise, the cell in the direction D_{next} is returned. The algorithm stops when it reaches the cell from where it had started the CircularTrip.

Given a query point q, radius r and angle range$\langle \theta_{st}, \theta_{end} \rangle$, ArcTrip returns every cell c that i) intersects the circle of radius r with center at q and ii) lies within the angle range $\langle \theta_{st}, \theta_{end} \rangle$. Note that when the angle range is $\langle 0, 2\pi \rangle$,

---

[3] The proof assumes that the functions to compute minimum and maximum constrained distances are available. Moreover, the case when the query changes its location is exception to the claim of optimality (we choose to compute the results from scratch when the query moves).

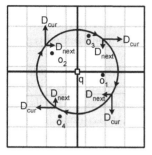
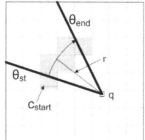
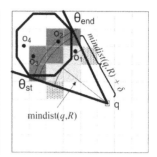

Fig. 7. CircularTrip with radius r

Fig. 8. ArcTrip with radius r

Fig. 9. Illustration of Algorithm 3

ArcTrip is same as the CircularTrip. In Fig. 8, $ArcTrip(q, r, \langle\theta_{st}, \theta_{end}\rangle)$ returns the shaded cells. ArcTrip works similar to the CircularTrip except that it starts with a cell c_{start} that intersects the circle at angle θ_{st} and stops when the next cell to be returned is outside the angle range.

4.2   Initial Computation

Let $\langle\theta_{st}, \theta_{end}\rangle$ be the angle range that covers the constrained region and the minimum distance of the constrained region from the query q is $mindist(q, R)$ (as shown in Fig. 9). The basic idea of the ArcTrip based algorithm is to call ArcTrip with the angle range $\langle\theta_{st}, \theta_{end}\rangle$ and radius r set to $mindist(q, R)$. The radius is iteratively increased by δ (the cell width) and the returned cells are visited in ascending order of their minimum distances from the query unless k constrained NNs are found. It can be guaranteed that the algorithm does not miss any cell if the radius is iteratively increased by δ [16].

Algorithm 3. ArcTrip Based Initial Computation

Input: q: query point; k: an integer
Output: $q.CkNN$

1: $q.dist_k = \infty$; $q.CkNN = \phi$; $H = \phi$
2: compute $mindist(q, R)$ and $\langle\theta_{st}, \theta_{start}\rangle$
3: $r = mindist(q, R)$
4: **for** each cell c returned by ArcTrip$(q, r, \langle\theta_{st}, \theta_{end}\rangle)$ **do**
5:    insert c in H with key $MinConstDist(c, q)$ if it intersects the constrained region
6: **while** $H \neq \phi$ **do**
7:    de-heap an entry e
8:    **If** $MinConstDist(e, q) \geq q.dist_k$; **return** $q.CkNN$
9:    update $q.CkNN$ and $q.dist_k$ by the objects in e
10:   **if** $H = \phi$ **then**
11:      $r = min\{r + \delta, q.dist_k\}$
12:      **for** each cell c returned by ArcTrip$(q, r, \langle\theta_{st}, \theta_{end}\rangle)$ **do**
13:         insert c into H with key $MinConstDist(c, q)$ if the cell is not visited before and intersects the constrained region
14: **return** $q.CkNN$

Algorithm 3 shows the details of the initial computation. The radius r of the ArcTrip is set to $mindist(q, R)$. The algorithm inserts the cells returned by the ArcTrip$(q, r, \langle \theta_{st}, \theta_{end} \rangle)$ into a min-heap if they intersect the constrained region (lines 4 and 5).

The cells are de-heaped iteratively and $q.CkNN$ and $q.dist_k$ are updated accordingly (line 9). When the heap becomes empty, the algorithm calls ArcTrip by increasing the radius (i.e., $r = min\{r + \delta, q.dist_k\}$). The returned cells are again inserted in the heap (lines 10 to 13). Note that the ArcTrip may returns some cells that were visited before, so such cells are not inserted in the heap (line 13).

The algorithm stops when the heap becomes empty or when the next de-heaped entry has $MinConstDist(e, q) \geq q.dist_k$ (line 8). The proof of correctness and the proof that the algorithm visits minimum possible cells is similar to Theorem 1 in [16].

Example 2. Fig. 9 shows the computation of a constrained NN query. Initially, the ArcTrip is called with radius r set to $mindist(q, R)$ and the light shaded cells are returned. The dotted cells are not inserted in the heap because they do not intersect the constrained region. Other light shaded cells are visited in ascending order but no valid object is found. ArcTrip is now called with the radius increased by δ and the dark shaded cells are returned. Upon visiting these cells, the object o_2 and o_3 are found. Since o_2 is closer, it is kept in $q.CkNN$ and $q.dist_k$ is set to $dist(o_2, q)$. Finally, ArcTrip with radius $q.dist_k$ is called to guarantee the correctness. No cell is inserted in the heap because all the cells returned by ArcTrip have been visited. The algorithm terminates and reports o_2 as the result.

4.3   Continuous Monitoring

The continuous monitoring algorithm (and the data structure) is similar to the continuous monitoring of Grid-based algorithm (Algorithm 2) except the way $q.CkNN$ is expanded at line 12. The set of constrained kNNs is expanded in a similar way to the initial computation module described above except that the starting radius of ArcTrip is set to $r = q.dist_k$.

4.4   Remarks

Similar to the Grid-tree based algorithm, ArcTrip based algorithm is optimal in number of visited cells. The proof of optimality and correctness is also similar (due to space limits, we do not present the proofs and refer the readers to [15]).

ArcTrip is expected to check lesser number of cells that intersect the constrained region as compared to the Grid-tree based access method. However, retrieving the cells that intersect the circle is more complex than the Grid-tree based access method. In our experiments, we found that both the algorithms have similar overall performance. Similar to the grid-based access method, Arc-Trip can be used to access cells in increasing or decreasing order of minimum or maximum Euclidean distance of the cells from q.

We remark that although we observed in our experiments that the grid-tree based algorithm and ArcTrip-based algorithm demonstrate very similar performance, they are two substantially different grid access methods. The proposed grid access methods can be applied to several other types of queries (e.g., furthest neighbor queries). It would be interesting to compare the performance of both proposed access methods for different types of queries and we leave it as our future work.

5 Experiments

In this section, we compare our algorithms GTree (Grid-tree algorithm) and ARC (ArcTrip algorithm) with CPM [1] which is the only known algorithm for continuous monitoring of the constrained kNN queries. In accordance with the experiment settings in [1], we use Brinkhoff data generator [17] to generate objects moving on the road network of Oldenburg, a German city. The agility of object data sets corresponds to the percentage of objects that reports location updates at a given time stamp. The default speeds of generator (slow, medium and fast) are used to generate the data sets. If the data universe is unit, the objects with slow speed travel the unit distance in 250 time stamps. The medium and fast speeds are 5 and 25 times faster, respectively. The queries are generated similarly. Each query is monitored for 100 time stamps and the total time is reported in the experiments. In accordance with [7], for each query, a random constrained region is generated with random selectivity (i.e., a rectangle at a random location with randomly selected length and width). Table 3 shows the parameters used in our experiments and the default values are shown in bold.

Table 3. System Parameters

Parameter	Range
Grid Size	$16^2, 32^2, 64^2, \mathbf{128^2}, 256^2, 512^2$
Number of objects ($\times 1000$)	20, 40, 60, 80, **100**
Number of queries	100, 200, 500, **1000**, 2500, 5000
Value of k	2, 4, 8, **16**, 32, 64, 128
Object/query Speed	slow, **medium**, fast
Object/query agility (in %)	10, 30, **50**, 70, 90

Effect of grid size: Since we use grid structure, we first study the effect of grid cardinality in Fig. 10. Fig. 10(a) shows the performance of each algorithm on different grid sizes with other parameters set to default values. In accordance with previous work that use grid based approach, the performance degrades if the grid size is too small or too large. More specifically, if the grid has too low cardinality, the cost of constrained kNN queries increase because each cell contains larger number of objects. On the other hand, if the grid cardinality is too high then many of the cells are empty and the number of visited cells is increased.

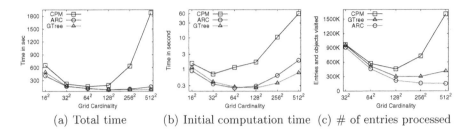

(a) Total time	(b) Initial computation time	(c) # of entries processed

Fig. 10. Effect of Grid Size

We compare the initial computation costs of the three algorithms in Fig. 10(b). The initial computation costs of our algorithms are several times better than CPM.

In Section 2.3, we showed an example that CPM may process a large number of entries (rectangles, cells and objects) to see if they intersect with the constrained region. Although several other factors contribute to the query execution cost, the number of entries for which the intersection is checked is one of the major factors that affect the query cost.

Fig. 10(c) shows the number of entries (rectangles, cells and objects) for which the intersection with the constrained region is checked. As expected, the number is large if the cells are too large or too small. If the cells are large, the number of objects for which the intersection is checked is large. On the other hand, if the cells are small, the number of cells (and conceptual rectangles) for which the intersection is checked is large. When grid cardinality is low, all three algorithms process similar number of entries. This is because most of the entries are objects inside the cells. Since each algorithm visits similar number of cells when cardinality is low, all the objects within each cell are checked for the intersection.

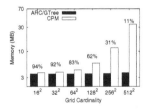

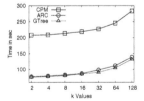

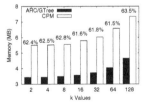

| **Fig. 11.** Grid vs memory | **Fig. 12.** k vs time | **Fig. 13.** k vs memory |

Fig. 11 compares the memory usage of the three algorithms. GTree and Arc-Trip based algorithm both store the same data structure and hence have same memory usage. To efficiently update the results, for each query, the CPM stores the heap and a *visit list* (the visit list contains the cells that have been visited by the query). The percentage on top of the bars represents the ratio of the memory usage of the algorithms (e.g., 62% means that our algorithms require 62% of the total memory used by CPM).

Effect of k values: Fig. 12 studies the effect of k on all the algorithms. Clearly, our algorithms outperform CPM for all k values. Interestingly, both of our algorithms show very similar performances and trends for most of the data settings. We carefully conducted the experiments and observed that the initial computation cost and the cost for expanding $q.CkNN$ (line 12 of Algorithm 2) of both algorithms is similar. The way queries are updated is also similar. Hence, for most of the data settings, they have similar performances and trends.

We observe that all three algorithms are less sensitive for small values of k. This is because each cell contains around 30 objects on average for the default grid size. For small k values, a small number of cells are visited to compute the results. Hence, the main cost for small k values is identifying the cells that lie in the constrained region.

Fig. 13 shows the memory usage of each algorithm for different k values. The memory usage is increased with k. As explained earlier, the change is less significant for small k values because the number of cells visited is almost same when k is small.

Effect of data size: We study the performance of each algorithm for different object and query data sets. More specifically, Fig. 14 shows the total query execution time for data sets with different number of objects. The cost of all algorithms increase because the algorithms need to handle more location updates for larger data sets.

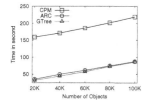
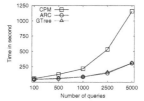
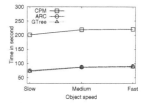

Fig. 14. # of objects   **Fig. 15.** # of queries   **Fig. 16.** Object speed

Fig. 15 shows the time for each algorithm for the data sets with different number of queries. Both of our algorithms show similar performance and scale better than CPM. CPM is up to around 4 times slower than our algorithms.

Effect of speed: In this Section, we study the effect of object and query speed on the computation time. Fig. 16 and 17 show the effect of object and query speed, respectively. As noted in [1] for kNN queries, we observe that the speed does not affect any of the three constrained kNN algorithms.

Effect of agility: As described earlier, agility corresponds to the percentage of objects (or queries) that issues location updates at a given time stamp. Fig. 18 studies the effect of data agility. As expected, the costs of all algorithms increase. This is because the algorithms need to handle more object updates as the agility increases.

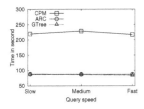   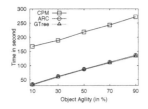   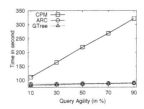

Fig. 17. Query speed   **Fig. 18.** Object agility   **Fig. 19.** Query agility

Fig. 19 shows the effect of query agility. The cost of CPM increases with increase in query agility because whenever a query changes the location the results are computed from scratch. Interestingly, the query agility does not have a significant effect on our algorithms. This is mainly because the initial computation cost (the case when a query moves) is not significantly higher than the update cost.

6   Conclusion

We propose two continuous constrained kNN algorithms based on two novel grid access methods. The proposed algorithms are optimal in the sense that they visit minimum number of cells to monitor the queries. Moreover, they use significantly less memory compared to the previous algorithm. Extensive experiments demonstrate that our algorithms are several times faster than the previous algorithm.

Acknowledgments. Research of Wenyu Qu was supported by National Natural Science Foundation of China under the grants numbered 90818002 and 60973115. Xuemin Lin was supported by the ARC Discovery Grants (DP0987557, DP0881035, DP0987273 and DP0666428), Google Research Award and NICTA.

References

1. Mouratidis, K., Hadjieleftheriou, M., Papadias, D.: Conceptual partitioning: An efficient method for continuous nearest neighbor monitoring. In: SIGMOD Conference, pp. 634–645 (2005)
2. Yu, X., Pu, K.Q., Koudas, N.: Monitoring k-nearest neighbor queries over moving objects. In: ICDE, pp. 631–642 (2005)
3. Xiong, X., Mokbel, M.F., Aref, W.G.: Sea-cnn: Scalable processing of continuous k-nearest neighbor queries in spatio-temporal databases. In: ICDE, pp. 643–654 (2005)
4. Cheema, M.A., Brankovic, L., Lin, X., Zhang, W., Wang, W.: Multi-guarded safe zone: An effective technique to monitor moving circular range queries. To appear in ICDE (2010)
5. Xia, T., Zhang, D.: Continuous reverse nearest neighbor monitoring. In: ICDE, p. 77 (2006)

6. Cheema, M.A., Lin, X., Zhang, Y., Wang, W., Zhang, W.: Lazy updates: An efficient technique to continuously monitoring reverse knn. VLDB 2(1), 1138–1149 (2009)
7. Ferhatosmanoglu, H., Stanoi, I., Agrawal, D., Abbadi, A.E.: Constrained nearest neighbor queries. In: SSTD, pp. 257–278 (2001)
8. Wu, W., Yang, F., Chan, C.Y., Tan, K.L.: Continuous reverse k-nearest-neighbor monitoring. In: MDM, pp. 132–139 (2008)
9. Guttman, A.: R-trees: A dynamic index structure for spatial searching. In: SIGMOD Conference, pp. 47–57 (1984)
10. Hjaltason, G.R., Samet, H.: Ranking in spatial databases. In: SSD, pp. 83–95 (1995)
11. Gao, Y., Chen, G., Li, Q., Li, C., Chen, C.: Constrained k-nearest neighbor query processing over moving object trajectories. In: Haritsa, J.R., Kotagiri, R., Pudi, V. (eds.) DASFAA 2008. LNCS, vol. 4947, pp. 635–643. Springer, Heidelberg (2008)
12. Mokbel, M.F., Xiong, X., Aref, W.G.: Sina: Scalable incremental processing of continuous queries in spatio-temporal databases. In: SIGMOD Conference, pp. 623–634 (2004)
13. Zhang, J., Zhu, M., Papadias, D., Tao, Y., Lee, D.L.: Location-based spatial queries. In: SIGMOD Conference, pp. 443–454 (2003)
14. Wu, W., Tan, K.L.: isee: Efficient continuous k-nearest-neighbor monitoring over moving objects. In: SSDBM, p. 36 (2007)
15. Cheema, M.A.: Circulartrip and arctrip: effective grid access methods for continuous spatial queries. UNSW Masters Thesis (2007),
http://handle.unsw.edu.au/1959.4/40512
16. Cheema, M.A., Yuan, Y., Lin, X.: Circulartrip: An effective algorithm for continuous nn queries. In: Kotagiri, R., Radha Krishna, P., Mohania, M., Nantajeewarawat, E. (eds.) DASFAA 2007. LNCS, vol. 4443, pp. 863–869. Springer, Heidelberg (2007)
17. Brinkhoff, T.: A framework for generating network-based moving objects. GeoInformatica 6(2), 153–180 (2002)

Chasing Tree Patterns under Recursive DTDs

Junhu Wang[1] and Jeffrey Xu Yu[2]

[1] Griffith University, Gold Coast, Australia
J.Wang@griffith.edu.au
[2] The Chinese University of Hong Kong, China
yu@se.cuhk.edu.hk

Abstract. Finding a homomorphism between tree patterns is an important technique for testing tree pattern containment, and it is the main technique behind algorithms for rewriting tree pattern queries using views. Recent work has shown that for tree patterns P and Q that involve parent-child (/) edges, ancestor-descendant (//) edges, and branching ([]) only, under a non-disjunctive, non-recursive DTD G, testing whether P is contained in Q can be done by chasing P into P' using five types of constraints derivable from G, and then testing whether P' is contained in Q without G, which in turn can be done by finding a homomorphism from Q to P'. We extend this work to non-disjunctive, recursive DTDs. We identify three new types of constraints that may be implied by a non-disjunctive recursive DTD, and show that together with the previous five types of constraints, they are necessary, and sufficient in some important cases, to consider for testing containment of tree patterns involving /, //, and [] under G. We present two sets of chase rules to chase a tree pattern repeatedly, and compare the advantages of these chase rules.

1 Introduction

XPath plays a central role in all XML query languages. A major fragment of XPath can be represented as tree patterns [5]. Finding homomorphism between tree patterns is an important technique for efficiently testing tree pattern containment, and for finding contained/equivalent rewritings of a tree pattern using a view [8,4,3]. It is shown in [5] that, when P and Q belong to several classes of tree patterns, P is contained in Q if and only if there is a homomorphism from Q to P. Unfortunately, when a DTD is present, the existence of a homomorphism from Q to P is no longer a necessary condition for P to be contained in Q. It is shown in [7], however, that if the DTD is *duplicate-free* and the tree patterns involve / and [] only, then testing whether tree pattern, P, is contained in another pattern, Q, under the DTD can be reduced to testing whether P is contained in Q under two types of constraints implied by the DTD. This result was extended in [3] to tree patterns involving /,// and [] (known as the class $P^{\{/,//,[]\}}$ [5]), under non-recursive and non-disjunctive DTDs. It is shown that in this case, testing whether P is contained in Q under the DTD can be done by chasing P to P' using five types of constraints implied by the DTD, and then testing whether P' is contained in P without the DTD.

H. Kitagawa et al. (Eds.): DASFAA 2010, Part I, LNCS 5981, pp. 250–261, 2010.
© Springer-Verlag Berlin Heidelberg 2010

In this work we extend the work [3] to non-disjunctive, recursive DTDs. This is motivated by the fact that the majority of real-world DTDs allow recursion[2]. We focus on tree patterns in $P^{\{/,//,[]\}}$ under DTDs that can be represented as (possibly cyclic) schema graphs [3]. Our focus is the transformation of a tree pattern P into P' under a DTD G so that the containment of P in any other pattern Q under DTD G can be tested by identifying a homomorphism from Q to P'. This is of practical application in tree pattern rewriting using views since the main approaches of rewriting algorithms are all based on the testing containment using homomorphism [8,4,3]. We are not concerned with the completeness, efficiency, and theoretical complexity of tree pattern containment under recursive DTDs, which have been studied already in [6].

Our main contributions are:

- We identify three new types of constraints derivable from a recursive DTD, and provide an efficient algorithm to extract all such constraints.
- For tree patterns in $P^{\{/,//,[]\}}$, we show the three new types of constraints together with the constraints identified in [3] are sufficient to catch the structural restrictions imposed by a recursive DTD for the purpose of containment test.
- We present two sets of chase rules, Chase1 and Chase2, with respect to the new constraints. Chase1 chases $P \in P^{\{/,//,[]\}}$ to a set S of tree patterns in $P^{\{/,//,[]\}}$ such that, if the chase terminates, then $P \subseteq_G Q$ iff every pattern in S is contained in Q without DTD. Chase1 is inefficient and it may not even terminate. Chase2 chases $P \in P^{\{/,//,[]\}}$ to a pattern that involves an additional type of edges - the *descendant-or-self* axis. Chase2 is more efficient than Chase1 in many cases.
- As required by Chase2, we define tree patterns involving $/, //, []$ and the *descendant-or-self* axis, and show that for such tree patterns P and Q, P is contained in Q if but not only if there is a homomorphism from Q to P. We also identify subclasses of such tree patterns for which the existence of homomorphism is both sufficient and necessary for $P \subseteq Q$.

The rest of the paper is organized as follows. Section 2 provides the preliminaries. We define the new constraints in and provide the algorithm to find all such constraints implied by a DTD in Section 3. The chase rules are presented in Section 4. Finally, Section 5 concludes the paper.

2   Preliminaries

2.1   DTD, XTree and Tree Patterns

Let Σ be an infinite set of tags. We model an XML document as a tree (called an XTree) with every node labeled with some tag in Σ, and model a DTD as a connected directed graph G satisfying the following conditions: (1) Each node is labeled with a distinct tag in Σ. (2) Each edge is labeled with one of 1, ?, +, and *, which indicate "exactly one", "one or zero", "one or many", and "zero or many", respectively. Here, the default edge label is 1. (3) There is a unique node, called the root, which *may* have an incoming degree of zero, and all other nodes have incoming degrees greater than 0. The set of tags occurring in G is denoted Σ_G. Because a node in a DTD G has a unique label, we also refer to a node by its label. A DTD is said to be *recursive* if it has a cycle. To ensure that all conforming XML trees (see below) are finite, we require that in every cycle, there is at least one edge marked with ? or $*$. A recursive DTD example is shown in Fig. 1 (a).

Let v be a node in an XTree t, the label of v is denoted $label(v)$. Let $N(t)$ (resp. $N(G)$) denote the set of all nodes in XTree t (resp. DTD G), and $rt(t)$ (resp. $rt(G)$) denote the root of t (resp. G). A tree t is said to conform to DTD G if (1) for every node $v \in N(t)$, $label(v) \in \Sigma_G$, (2) $label(rt(t)) = label(rt(G))$, (3) for every edge (u, v) in t, there is a corresponding edge $(label(u), label(v))$ in G, and (4) for every node $v \in N(t)$, the number of children of v labeled with x is constrained by the label of the edge $(label(v), x)$ in G. The set of all XTrees conforming to G is denoted T_G.

A *tree pattern* (TP) in $P^{\{/,//,[]\}}$ is a tree P with a unique *output node* (denoted O_P), with every node labeled with a tag in Σ, and every edge labeled with either $/$ or $//$. The path from the root to the output node is called the *output path*. A TP corresponds to an XPath expression. Figures 1 (b), (c) and (d) show three TPs, P, P' and Q. They correspond to the XPath expressions $a[//c]//d$ and $a[//c]/x//d$, and $a/x//d$ respectively. Here, single and double lines represent $/$-edges and $//$-edges respectively, a branch represents a condition ([]) in an XPath expression, and a circle indicates the *output node*.

Let $N(P)$ (resp. $rt(P)$) denote the set of all nodes in a TP P (resp. the root of P). An *embedding* of a TP P in an XTree t is a mapping δ from $N(P)$ to $N(t)$ which is (1) *label-preserving*, i.e., $\forall v \in N(P)$, $label(v) = label(\delta(v))$, (2) *root-preserving*, i.e., $\delta(rt(P)) = rt(t)$, and (3) *structure-preserving*, i.e., for every edge (x, y) in P, if it is a $/$-edge, then $\delta(y)$ is a child of $\delta(x)$; if it is a $//$-edge, then $\delta(y)$ is a descendant of $\delta(x)$, i.e, there is a path from $\delta(x)$ to $\delta(y)$. Each embedding δ produces a node $\delta(O_P)$, which is known as an *answer* to the TP. We use $P(t)$ to denote the *answer set* (i.e., set of all answers) of P on t.

A TP P is said to be *satisfiable* under DTD G if there exists $t \in T_G$ such that $P(t)$ is not empty. In this paper we implicitly assume all TPs are satisfiable under the DTDs in discussion. We will also use the following terms and notations. An *x-node* means a node labeled x. An *x-child* (resp. *x-parent*, *x-descendant*) means a child (resp. parent, descendant) labeled x. A */-child* (resp. *//-child*) means a child connected to the parent via a $/$-edge (resp. $//$-edge). A *(/,x)-child* means a $/$-child labeled x. An *x//y-edge* (resp. *x/y-edge*) means a $//$-edge (resp. $/$-edge) from an x-node to a y-node.

2.2   TP Containment and Boolean Patterns

A TP P is said to be *contained* in another TP Q, denoted $P \subseteq Q$, if for every XTree t, $P(t) \subseteq Q(t)$. When a DTD G is present, P is said to be *contained in Q under G*, denoted $P \subseteq_G Q$, if for every XTree $t \in T_G$, $P(t) \subseteq Q(t)$.

It is noted in [5] that the containment problem of tree patterns can be reduced to the containment problem of boolean patterns. A *boolean pattern* is a tree pattern without any output node. Given a boolean pattern P and an XML tree t, $P(t)$ returns TRUE iff there is an embedding of P in t. Given two boolean tree patterns P and Q, P is said to be contained in Q (under G) if for all XTree t ($\in T_G$), $P(t) \rightarrow Q(t)$. By similar argument, when a DTD is present, the containment problem of tree patterns can be reduced to that of boolean patterns. In the rest of the paper, all of the tree patterns in discussion are assumed to be boolean tree patterns.

Given $P, Q \in P^{\{/,//,[]\}}$, $P \subseteq Q$ iff there is a homomorphism from Q to P [1]. Recall: let $P, Q \in P^{\{/,//,[]\}}$. A *homomorphism* from Q to P is a mapping δ from $N(Q)$ to $N(P)$ that is label-preserving, root-preserving, structure-preserving as discussed in the last section.

Lakshmanan et al. showed in [3] that, if P and Q are both in $P^{\{/,//,[]\}}$ and the DTD G is acyclic, then $P \subseteq Q$ can be reduced to TP containment under a set Δ of constraints (referred to as the LWZ constraints hereafter) implied by G. To check containment under Δ, Lakshmanan et al. used some chase rules to chase P repeatedly until no more change can be made, resulting a new TP P' (the chased pattern), and showed that $P \subseteq_G G$ iff $P' \subseteq Q$.

The LWZ constraints are listed below.

(1) **Parent-Child Constraints (PC)**, denoted $a \Downarrow^1 x$, which means that whenever an x-node is the descendant of an a-node, it must be the child of the a-node.

(2) **Sibling Constraints (SC)**, denoted $a{:}S{\downarrow}y$, where $S = \{x\}$ or $S = \emptyset$, and a, x, y are labels in Σ. In the first case the constraint means that for every a-node, if it has an x-child, then it also has a y-child. In the second case, it means that every a-node must have an y-child. Note: for our DTDs, the only case of SC is of the form $a : \emptyset \downarrow y$, which we will abbreviate as $a :\downarrow y$.

(3) **Cousin Constraints (CC)**, denoted $a : x \Downarrow y$. The constraint means that for every a-node, if it has an x-descendant, then it also has a y-descendant.

(4) **Intermediate Node Constraints (IC)**, denoted $a \xrightarrow{x} y$, which means that every path from an a-node to a y-node must pass through an x-node.

(5) **Functional Constraints (FC)**, denoted $a \twoheadrightarrow x$, which means that every a-node has at most one x-child.

The LWZ rules are as follows.

PC-rule: If $G \vDash a \Downarrow^1 x$, then change any $a//x$-edge to a/x-edge.

SC-rule: If $G \vDash a :\downarrow y$, then add a $(/, y)$-child to any a-node, if the a-node does not have such a y-child already.

CC-rule: If $G \vDash a : x \Downarrow y$, and an a-node, a^0, has a $(//, x)$-child, then add a $(//, y)$-child to a^0 if a^0 does not have a y-descendant already.

IC-rule: If $G \vDash a \xrightarrow{x} y$, then replace any $a//y$-edge $a^0//y^0$ with an $a//x$-edge $a^0//x^0$ and the $x//y$-edge $x^0//y^0$.

FC-rule: If $G \vDash a \twoheadrightarrow x$, and there are two or more $(/, b)$-children of an a-node, then merge these b-children.

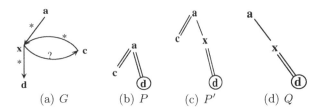

(a) G (b) P (c) P' (d) Q

Fig. 1. Recursive DTD G and some TPs conforming to G

3   New Constraints Derivable from a Recursive DTD

With a recursive DTD the LWZ constraints are no longer sufficient to capture the structural restrictions imposed by the DTD that may affect containment between two TPs, as shown in the following examples.

Example 1. Consider the recursive DTD in Fig. 1 (a) and the queries $P = a//d$ and $Q = a/x//d$. Using the LWZ constraints we can chase P to $a//x//d$ but no further. There is no homomorphism from Q to the chased pattern. Hence we cannot find $P \subseteq_G Q$.

Observe in Example 1 that, in any tree conforming to the DTD, every path from an a-node to a d-node passes through an x-node (i.e., $a \xrightarrow{x} d$ is implied by the DTD). Moreover, the node immediately following the a-node on the path is an x-node. This information enables us to transform the query $a//d$ to $a/x//d$, and thus to establish the equivalence of P and Q in the example.

Example 2. Consider the TPs P and Q and DTDs G_1 and G_2 in Fig. 2. It is easy to see $P \subseteq_{G_1} Q$ and $P \subseteq_{G_2} Q$. But using the LWZ constraints we cannot transform P to any other form under either G_1 or G_2, and there is no homomorphism from Q to P. Therefore we cannot detect $P \subseteq_{G_i} Q$ for $i = 1, 2$.

Unlike Example 1, in the DTDs of Example 2, the paths from a to b do not have a fixed immediate following node of a or a fixed immediate preceding node of b. However, all paths from a to b must pass through the edge (a, b). Thus $a//b$ represents the disjunction of a/b, $a//a/b$, $a/b//b$ and $a//a/b//b$. Therefore, P is equivalent to the union of the TPs $a/a/b/b$, $a/a//a/b/b$, $a/a/b//b/b$ and $a/a//a/b//b/b$ under the DTDs. Since every TP in the union is contained in Q, we know P is contained in Q under the DTDs.

    Next we formally define three new types of constraints. The first two types catch the useful information shown in Example 1, the third constraint captures the useful information shown in Example 2.

Definition 1. *A* child of first node constraint *(CFN) is of the form* $x \xrightarrow{/b} y$, *where* $x, y, b \in \Sigma_G$. *It means that on every path from an x-node to a y-node, the node immediately following the x-node must be a b-node.*

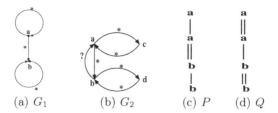

Fig. 2. Recursive DTDs and TPs

A parent of last node constraint (PLN) *is of the form* $x \xrightarrow{b/} y$, *which means that for every path from an x-node to a y-node, the node immediately preceding the y-node is a b-node.*

An essential edge constraint (EE) *is of the form* $x \xrightarrow{a/b} y$, *it means that every path from an x-node to a y-node must contain an edge from an a-node to a b-node.*

Note the following special forms of CFN, PLN, and EE constraints:

$$x \xrightarrow{/y} y, \; x \xrightarrow{x/} y \text{ and } x \xrightarrow{x/y} y \text{ (abbreviated as } x \xrightarrow{/} y\text{)}.$$

These special forms have the following properties:

Proposition 1

$$G \vDash x \xrightarrow{b} y, \; x \xrightarrow{/b} b \; \Leftrightarrow G \vDash x \xrightarrow{/b} y$$
$$G \vDash x \xrightarrow{b} y, \; b \xrightarrow{b/} y \; \Leftrightarrow G \vDash x \xrightarrow{b/} y$$
$$G \vDash x \xrightarrow{a} y, \; a \xrightarrow{b} y, \; a \xrightarrow{/} b \; \Leftrightarrow G \vDash x \xrightarrow{a/b} y$$

Therefore, in the following, we will focus only on the above special cases of constraints, and when we say CFN, PLN, EE constraints, we mean the above special forms. Also, when $x = y = b$, the above three constraints become $x \xrightarrow{/x} x$, $x \xrightarrow{x/} x$, and $x \xrightarrow{/} x$. They are equivalent in the sense that one implies the others. They mean that on any path from one x-node to another there can be only x-nodes. We will refer to the LWZ constraints and the CFD, PLN, and EE constraints simply as the *constraints*, and use Δ_G to denote the set of all such constraints implied by DTD G. These constraints represent part of the restrictions imposed by G on the structure of XML trees conforming to G. For example, the DTD in Fig. 1 implies the CFNs $a \xrightarrow{/x} x$, $a \xrightarrow{/x} d$ and the PLNs $x \xrightarrow{x/} d$, $a \xrightarrow{x/} d$, while either of the DTDs in Fig. 2 implies the EE $a \xrightarrow{/} b$ (among others).

3.1   Finding the New Constraints Implied by DTD

A *trivial* constraint is one which can be derived from every DTD over Σ. For example, $x : y \Downarrow y$. A trivial constraint is useless. There are other constraints

---

Algorithm 1. Finding new constraints implied by DTD G

1: Initialize S_{CFN}, S_{PLN} and S_{EE} as empty set.
2: **for** every edge (x, y) in G (possibly $x = y$) **do**
3:     **if** there is no path from x to y in $G - \{(x, y)\}$ **then**
4:         add $x \xrightarrow{/} y$ to S_{EE}
5:         **if** for all $u \in child(x) - \{y\}$, there is no path from u to y **then**
6:             add $x \xrightarrow{/y} y$ to S_{CFN}
7:         **if** for all $v \in parent(y) - \{x\}$, there is no path from x to v **then**
8:             add $x \xrightarrow{x/} y$ to S_{PLN}

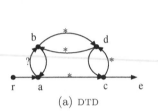

start	end	constraints
b	a	$b \xrightarrow{b/} a$
c	a	$c \xrightarrow{/d,b/} a$
c	b	$c \xrightarrow{/d} b$
c	d	$c \xrightarrow{/d} d$
d	a	$d \xrightarrow{b/} a$
d	b	$d \xrightarrow{/} b$

(a) DTD   (b) New constraints

Fig. 3. Recursive DTD and the new constraints found using Algorithm 1

that are useless. For example, when the DTD does not include the tags x or y, or there is no path from x to y in the DTD, the constraints $x : y \Downarrow z$, $x \rightarrow y$, $x \Downarrow^1 y$, $x \xrightarrow{z} y$, $x \xrightarrow{/z} y$ and $x \xrightarrow{z/} y$ will be true but useless. We will call such constraints *vacuous*. We are only interested in non-trivial and non-vacuous constraints.

Algorithms for finding all LWZ constraints have already been studied in [3]. Therefore we focus on the new constraints. Let us use S_{CFN}, S_{PLN}, and S_{EE} to denote the sets of all CFN, PLN, and EE constraints implied by G. Let

$child(x) = \{y \,|(x,y) \text{ is an edge in } G\}$, and
$parent(y) = \{x \,|(x,y) \text{ is an edge in } G\}$.

Algorithm 1 shows an algorithm for finding all EE, CFN, and PLN constraints implied by G. The algorithm checks each edge (x, y) in G, to see whether there is a path from x to y that does not pass through (x, y). If no such path is found, then $x \xrightarrow{/} y$ is implied by G (line 3-4). Then it further checks whether there is a child u of x other than y such that there is a path from u to y, if no such path can be found, then $x \xrightarrow{/y} y$ is implied by G (line 5-6). Similarly, if there is no path from any parent of y (except x) to y, then $x \xrightarrow{x/} y$ is implied by G (line 7-8). Checking whether there is a path from one node to another can be done in $O(|E(G)|)$, and there are no more than $|N(G)|$ children (parents) of any node, thus the algorithm takes time $O(|E(G)|^2 \times |N(G)|)$. The correctness of the algorithm is straightforward.

Example 3. Consider the DTD in Fig. 3 (a). The new constraints found by Algorithm 1 are listed in Fig. 3 (b). Note that since the IC constraints $a \xrightarrow{b} a$ and

$c \xrightarrow{d} c$ are also implied by G, we can combine them with $b \xrightarrow{b/} a$ and $c \xrightarrow{/d} d$ to obtain $a \xrightarrow{b/} a$ and $c \xrightarrow{/d} c$.

4   Chasing TPs with Constraints

To use the new constraints to test TP containment, we need to define some chase rules. For easy description of the rules, we use superscripted characters to indicate nodes in a TP that are labeled with that character. For example, $x^0, x^1, \ldots$ represent nodes that are labeled x. We also use $x^0 // y^0$ (resp. x^0 / y^0) to represent the //-edge (resp. /-edge) (x^0, y^0). We will define two sets of chase rules/algorithms, referred to as *Chase1* and *Chase2*. Each set of rules has its own advantages. In what follows, Δ_G denotes the set of all non-trivial, non-vacuous constraints implied by DTD G.

4.1   Chase1

Given DTD G and TP $P \in P^{\{/,//,[]\}}$, Chase1 transforms P using the LWZ rules and the following rules:

1. If $G \nvDash x \Downarrow^1 y$, but $G \vDash x \xrightarrow{/y} y$, and P contains the edge $x^0 // y^0$, then split P into two TPs P_1 and P_2, such that $x^0 // y^0$ is replaced with x^0 / y^0 and $x^0 / y^1 // y^0$, respectively, in P_1 and P_2.

2. If $G \nvDash x \Downarrow^1 y$, but $G \vDash x \xrightarrow{x/} y$, and P contains the edge $x^0 // y^0$, then split P into two TPs P_1 and P_2, such that $x^0 // y^0$ is replaced with x^0 / y^0 and $x^0 // x^1 / y^0$, respectively, in P_1 and P_2.

3. If $G \nvDash x \xrightarrow{x/} y$, $G \nvDash x \xrightarrow{/y} y$, but $G \vDash x \xrightarrow{/} y$, and P contains the edge $x^0 // y^0$, then split P into four TPs P_1, P_2, P_3, P_4, such that $x^0 // y^0$ is replaced with x^0 / y^0, $x^0 // x^1 / y^0$, $x^0 / y^1 // y^0$, and $x^0 // x^1 / y^1 // y^0$, respectively, in P_1, P_2, P_3 and P_4.

4. If $G \vDash x \xrightarrow{/} x$, and P contains the edge $x^0 // x^1$, then split P into three TPs P_1, P_2 and P_3, such that $x^0 // x^1$ is replaced with x^0 / x^1, $x^0 / x^2 / x^1$, and $x^0 / x^2 // x^3 / x^1$, respectively, in P_1, P_2 and P_3.

Let G be a DTD. Let $P, Q \in P^{\{/,//,[]\}}$ be satisfiable under G. To test whether $P \subseteq_G Q$, we repeatedly chase P with Δ_G, using the rules in Chase1 and the LWZ rules, into a set S of TPs: Initially, $S = \{P\}$. After each application of Chase1, S is updated. It is easy to see that, under G, P is equivalent to the union of the TPs in S. Therefore, we have

Proposition 2. *If at some stage of chasing P with Δ_G using the rules in Chase1, there is a homomorphism from Q to every TP in S, then $P \subseteq_G Q$.*

Example 4. Consider the DTD G and the TPs P, Q in Figures 4 (a), (b) and (e). P and Q are satisfiable under G. Δ_G contains $x \xrightarrow{x/} y$. Thus P can be chased to P_1 and P_2 as shown in Figures 4 (c) and (d). There is a homomorphism from Q to P_1, and a homomorphism from Q to P_2. Therefore, we know $P \subseteq_G Q$.

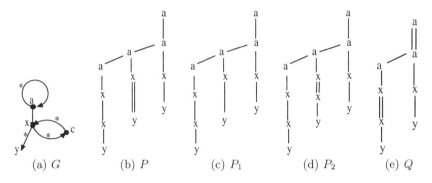

(a) G (b) P (c) P_1 (d) P_2 (e) Q

Fig. 4. $P \subseteq_G Q$ because P can be chased to $P_1 \cup P_2$ under G, and $P_1, P_2 \subseteq Q$

The following is an important property of Chase1.

Theorem 1. *Let $P, Q \in P^{\{/,//,[]\}}$ be satisfiable under G. If P cannot be chased using the rules in Chase1, then $P \subseteq_G Q$ iff there is a homomorphism from Q to P.*

The proof of the above theorem is in the full version of this paper.

The above result is important because it implies that for testing containment of TPs in $P^{\{/,//,[]\}}$ (that terminate with Chase1) under non-disjunctive recursive DTDs, we only need to consider their containment under the corresponding CFN, PLN, EEC and LWZ constraints. In other words, these constraints are sufficient to catch the essential structural restrictions (imposed by the DTD) that may affect containment of TPs in $P^{\{/,//,[]\}}$.

Observe that Chase1 may not terminate, that is, it can go on infinitely. For example, when $G \models x \xrightarrow{/} x$, the TP $x//x$ is chased into three TPs, and one of them still has $x//x$. Also, the number of TPs we will obtain may grow exponentially. Despite these problems, Chase1 is still useful in several cases. In particular, Chase1 can detect some cases of containment which cannot be detected using Chase2. For instance, the $P \subseteq_G Q$ in Example 4 cannot be detected using Chase2.

Next we present Chase2, which is more efficient than Chase1 in many cases.

4.2 Chase2

Chase2 makes use of the *descendant-or-self* axis, denoted $\sim$, of XPath to represent the union of several TPs in $P^{\{/,//,[]\}}$ as a single pattern in $P^{\{/,//,\sim,[]\}}$ (the set of all patterns involving $/, //, []$ and $\sim$). A tree pattern in $P^{\{/,//,\sim,[]\}}$ is like a TP in $P^{\{/,//,[]\}}$ except that an edge may be labeled with $\sim$ (in addition to $/$ and $//$). Such edges will be referred to as $\sim$-edges. To eliminate impossible cases, we require that the nodes on both sides of any $\sim$-edge have identical labels. Observe that if there are several consecutive $\sim$ edges, they can be replaced with a single one. For example, $x \sim x \sim x$ can be replaced with $x \sim x$. When drawing TPs in $P^{\{/,//,\sim,[]\}}$, we use double dotted lines to represent $\sim$-edges (refer to Fig. 5 (a)).

Given TPs $P \in P^{\{/,//,\sim,[]\}}$ and an XML tree t, an *embedding* of P in t is a mapping from $N(P)$ to $N(t)$ that is root-preserving and label-preserving as defined before, and structure-preserving which now includes the condition that, if (x,y) is a $\sim$-edge in P, then $\delta(y)$ is $\delta(x)$ or a descendant of $\delta(x)$. Note when P does not have $\sim$-edges, the embedding reduces to an embedding of TPs $\in P^{\{/,//,[]\}}$ in t. Similarly, given $Q \in P^{\{/,//,\sim,[]\}}$, a *homomorphism* from Q to P is defined in the same way as for the case where both P and Q are in $P^{\{/,//,[]\}}$, except that structure-preserving now means if (x,y) is a /-edge in Q, then $(h(x), h(y))$ is a /-edge in P; if (x,y) is a //-edge, then there is a path from $h(x)$ to $h(y)$ *which contains at least one /-edge or //-edge*; if (x,y) is a $\sim$-edge, then either $h(x) = h(y)$ or there is a path from $h(x)$ to $h(y)$. With these definitions, we have the following result.

Theorem 2. *For any TPs $P \in P^{\{/,//,\sim,[]\}}$ and $Q \in P^{\{/,//,[]\}}$, $P \subseteq Q$ if, but not only if, there is a homomorphism from Q to P.*

Proof. The "if" is straightforward, the "not only if" is shown by Example 5.

Example 5. For the TPs Q and P in Fig. 5, $P \subseteq Q$ but there is no homomorphism from Q to P.

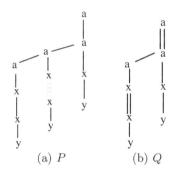

(a) P (b) Q

Fig. 5. $P \subseteq Q$, but no homomorphism exists from Q to P

There are some special cases where the existence of a homomorphism is both necessary and sufficient for $P \subseteq Q$, where $P, Q \in P^{\{/,//,\sim,[]\}}$.

Theorem 3. *$P \subseteq Q$ iff there is a homomorphism from Q to P in the following cases:*

1. *$P \in P^{\{/,//,\sim,[]\}}$ and $Q \in P^{\{/,//,[]\}}$, and for every $\sim$-edge between two x-nodes in P, there does not exist a //-edge between two x-nodes in Q.*
2. *P and Q are both in $P^{\{/,//,\sim\}}$.*
3. *$P \in P^{\{/,//,\sim,[]\}}$ and $Q \in P^{\{/,//\}}$.*
4. *$P \in P^{\{/,//,\sim\}}$ and $Q \in P^{\{/,//,[]\}}$.*

The proof of the theorem is in the full version of this paper.

Rules in Chase2

Chase 2 consists of the LWZ rules and the following rules.

1. If $G \nvDash x \Downarrow^1 y$, but $G \vDash x \xrightarrow{/y} y$, then replace any $x//y$-edge, $x^0//y^0$, with $x^0/y^1 \sim y^0$.

2. If $G \nvDash x \Downarrow^1 y$, but $G \vDash x \xrightarrow{x/} y$, then replace any $x//y$-edge, $x^0//y^0$, with $x^0 \sim x^1/y^0$.

3. If $G \nvDash x \xrightarrow{/y} y$, $G \nvDash x \xrightarrow{x/} y$, but $G \vDash x \xrightarrow{/} y$, then replace any $x//y$-edge, $x^0//y^0$, with $x^0 \sim x^1/y^1 \sim y^0$.

4. If $G \vDash x \xrightarrow{/} x$, then replace any $x//x$-edge, $x^0//x^1$, with $x^0/x^2 \sim x^1$ or with $x^0 \sim x^2/x^1$.

Chase2 can be used to chase a TP repeatedly until no more change can be obtained. The following proposition is straightforward.

Proposition 3. *Let G be a* DTD, *Δ be a subset of constraints implied by G, and $P, Q \in P^{\{/,//,[]\}}$ be* TP*s. Let P' be the* TP *obtained by applying the Chase2 rules to P using the constraints in Δ. If there is a homomorphism from Q to P', then $P \subseteq_G Q$.*

The example below demonstrates the advantage of Chase2 over Chase1, that is, the better efficiency in some cases.

Example 6. Consider the TPs P, Q and the DTD G_1 in Fig. 2. G_1 implies $a \xrightarrow{/} b$, but not $a \xrightarrow{/b} b$ or $a \xrightarrow{a/} b$. Therefore P can be chased into $P' = a/a \sim a/b \sim b/b$ using Chase2. Clearly there is a homomorphism from Q to P'. Therefore we can conclude $P \subseteq_{G_1} Q$ and $P \subseteq_{G_2} Q$.

However, generally $P \subseteq_G Q$ does not imply the existence of a homomorphism from Q to $\Delta_G(P)$. One reason for this deficiency of Chase2 is because of Theorem 2. Another reason is that the rules in Chase2 are not sufficient such that, in some cases, not all useful constraints can be applied in the chase.

We point out that the Chase2 rules can be extended with additional rules to make the chase "complete" for some special cases, that is, $P \subseteq_G Q$ can be tested by testing the existence of a homomorphism from Q to P', where P' is the TP obtained by applying Chase2 and the additional rules. The details can be found in the full version of this paper.

5   Conclusion

We identified three new types of constraints that may be implied by a recursive DTD G, and presented an algorithm for finding them. These constraints are used to transform a tree pattern in $P^{\{/,//,[]\}}$ in order to test whether $P \subseteq_G Q$ using homomorphism. We provided two sets of chase rules for this purpose. As a by-product, we showed that the existence of a homomorphism is sufficient but not necessary for the containment of tree patterns with self-or-descendant edges, and identified special cases where the existence of homomorphism is necessary.

Acknowledgement. This work is supported by Griffith University New Researcher's Grant No. GUNRG36621 and the grant of the Research Grants Council of the Hong Kong SAR, China No. 418206.

References

1. Amer-Yahia, S., Cho, S., Lakshmanan, L.V.S., Srivastava, D.: Minimization of tree pattern queries. In: SIGMOD, pp. 497–508 (2001)
2. Choi, B.: What are real dtds like? In: WebDB, pp. 43–48 (2002)
3. Lakshmanan, L.V.S., Wang, H., Zhao, Z.J.: Answering tree pattern queries using views. In: VLDB, pp. 571–582 (2006)
4. Mandhani, B., Suciu, D.: Query caching and view selection for XML databases. In: VLDB, pp. 469–480 (2005)
5. Miklau, G., Suciu, D.: Containment and equivalence for an XPath fragment. In: PODS, pp. 65–76 (2002)
6. Neven, F., Schwentick, T.: On the complexity of XPath containment in the presence of disjunction, DTDs, and variables. Logical Methods in Computer Science 2(3) (2006)
7. Wood, P.T.: Containment for XPath fragments under DTD constraints. In: Calvanese, D., Lenzerini, M., Motwani, R. (eds.) ICDT 2003. LNCS, vol. 2572, pp. 300–314. Springer, Heidelberg (2002)
8. Xu, W., Özsoyoglu, Z.M.: Rewriting XPath queries using materialized views. In: VLDB, pp. 121–132 (2005)

Efficient Label Encoding for Range-Based Dynamic XML Labeling Schemes

Liang Xu, Tok Wang Ling, Zhifeng Bao, and Huayu Wu

School of Computing, National University of Singapore
{xuliang,lingtw,baozhife,wuhuayu}@comp.nus.edu.sg

Abstract. Designing dynamic labeling schemes to support order-sensitive queries for XML documents has been recognized as an important research problem. In this work, we consider the problem of making range-based XML labeling schemes dynamic through the process of encoding. We point out the problems of existing encoding algorithms which include computational and memory inefficiencies. We introduce a novel Search Tree-based (ST) encoding technique to overcome these problems. We show that ST encoding is widely applicable to different dynamic labels and prove the optimality of our results. In addition, when combining with encoding table compression, ST encoding provides high flexibility of memory usage. Experimental results confirm the benefits of our encoding techniques over the previous encoding algorithms.

1 Introduction

XML is becoming an increasingly important standard for data exchange and representation on the Web and elsewhere. To query XML data that conforms to an *ordered tree-structured* data model, XML labeling schemes have attracted a lot of research and industrial attention for their effectiveness and efficiency. XML Labeling schemes assign the nodes in the XML tree unique labels from which their structural relationships such as ancestor/descendant, parent/child can be established efficiently.

Range-based labeling schemes[6,11,12] are popular in many XML database management systems. Compared with prefix labeling schemes[7,2,13], a key advantage of range-based labeling schemes is that their label size as well as query performance are not affected by the structure (depth, fan-out, etc) of the XML documents, which may be unknown in advance. Range-based labeling schemes are preferred for XML documents that are deep and complex, in which case prefix labeling schemes perform poorly because the lengths of prefix labels increase linearly with their depths. However, prefix labeling schemes appear to be inherently more robust than range-based labeling schemes. If negative numbers are allowed for local orders, prefix labeling schemes require re-labeling only if a new node is inserted between two consecutive siblings. Such insertions can be processed without re-labeling based on existing solutions[14,9]. On the other hand, any insertion can trigger the re-labeling of other nodes with range-based labeling schemes.

H. Kitagawa et al. (Eds.): DASFAA 2010, Part I, LNCS 5981, pp. 262–276, 2010.

The state-of-the-art approach to design dynamic range-based labeling schemes is based on the notion of *encoding*. It is also the only approach that has been proposed which can *completely* avoid re-labeling. By applying an encoding scheme to a range-based labeling scheme, the original labels are transformed to some dynamic format which can efficiently process updates without re-labeling. Existing encoding schemes include CDBS[4], QED[3,5] and Vector[8] encoding schemes which transform the original labels to binary strings, quaternary strings and vector codes respectively. The following example illustrates the applications of QED encoding scheme to containment labeling scheme, which is the representative of range-based labeling schemes.

Example 1. In Figure 1 (a), every node in the XML tree is labeled with a containment label of the form: *start*, *end* and *level*. When QED encoding scheme is applied, the *start* and *end* values are transformed into QED codes based on the encoding table in (b). We refer to the resulting labels as QED-Containment labels which are shown in (c). QED-Containment labels not only preserve the property of containment labels, but also allows dynamic insertions with respect to lexicographical order[3].

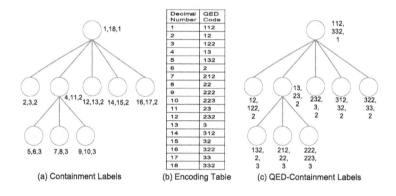

Fig. 1. Applying QED encoding scheme to containment labeling scheme

Formally speaking, we consider an encoding scheme as a mapping f from the original labels to the target labels. Let X and Y denote the set of order-sensitive codes in the original labels and target labels respectively, f maps each element x in X to an element $y = f(x)$ in Y. For the mapping to be both correct and effective, f should satisfy the following properties:

1. **Order Preserving**: The target labels must preserve the order of the original labels, i.e. $f(x_i) < f(x_j)$ if and only if $x_i < x_j$ for any x_i, $x_j \in X$.
2. **Optimal Size**: To reduce the storage cost and optimize query performance, the target labels should be of optimal size, i.e. the total size of $f(x_i)$ should be be minimized for a given range. To satisfy this property, f has to take the range to be encoded into consideration. The mappings may be different for different ranges.

The following example illustrates how this mapping in Figure 1 (b) is derived based on QED encoding scheme.

Example 2. To create the encoding table in Figure 1 (b), QED encoding scheme first extends the encoding range to (0, 19) and assigns two empty QED codes to positions 0 and 19. Next, the $(1/3)^{th}$ (6=round(0+(19-0)/3)) and $(2/3)^{th}$ (13=round(0+(19-0)×2/3)) positions are encoded by applying an insertion algorithm with the QED codes of positions 0 and 19 as input. The QED insertion algorithm takes two QED codes as input and computes two QED codes that are lexicographically between them which are as short as possible (Such insertions are always possible because QED codes are dynamic). The output QED codes are assigned to the $(1/3)^{th}$ and $(2/3)^{th}$ positions which are then used to partition range (0, 19) into three sub-ranges. This process is recursively applied for each of the three sub-ranges until all the positions are assigned QED codes. CDBS and Vector encoding schemes adopt similar algorithms.

We classify these algorithms i.e. CDBS, QED and Vector, as *insertion-based* algorithms since they make use of the property that the target labels allow dynamic insertions. However, a drawback of the insertion-based approach is that by assuming the entire encoding table fits into memory, it may fail to process large XML documents due to memory constraint. Since the size of the encoding table can be prohibitively large for large XML documents and main memory remains the limiting resource, it is desirable to have a memory efficient encoding algorithm. Moreover, the insertion-based approach requires costly table creation for every range, which is computationally inefficient for encoding multiple ranges of multiple documents.

In this paper, we show that only a single encoding table is needed for the encoding of multiple ranges. As a result, encoding a range can be translated into indexing mapping of the encoding table which is not only very efficient, but also has an adjustable memory usage. The main contributions of this paper include:

- We propose a novel Search Tree-based (ST) encoding technique which has a wide application domain. We illustrate how ST encoding technique can be applied to binary string, quaternary string and vector code and prove the optimality of our results.
- We introduce encoding table compression which can be seamlessly integrated into our ST encoding techniques to adapt to the amount of memory available.
- We propose Tree Partitioning (TP) technique as an optimization to further enhance the performance of ST encoding for multiple documents.
- Experimental results demonstrate the high efficiency and scalability of our ST encoding techniques.

2   Preliminary

2.1   Range-Based Labeling Schemes

In containment labeling scheme, every label is of the form (*start, end, level*) where *start* and *end* define an interval and *level* refers to the level in the

XML document tree. Assume node n has label (s_1, e_1, l_1) and node m has label (s_2, e_2, l_2), n is an ancestor of m if and only if $s_1 < s_2 < e_2 < e_1$. i.e. interval (s_1, e_1) contains interval (s_2, e_2). n is the parent of m if and only if n is an *ancestor* of m and $l_1 = l_2 - 1$. Other range-based labeling schemes[11,12] have similar properties.

Example 3. In Figure 1 (a), node(1,18,1) is an ancestor of node(7,8,3) because 1<7<8<18. Node(4,11,2) is the parent of node(5,6,3) because 4<5<6<11 and 2=3-1.

Although range-based labeling schemes work well for *static* XML documents, insertions of new nodes may lead to costly re-labeling. Leaving gaps[12] only allows limited number of insertions before re-labeling is required. Floating point numbers have been suggested to be used[1]. However, the precision of floating point number is limited by the fixed number of bits in its mantissa. As a result, re-labeling is still necessary when the number of insertions exceeds certain limits.

2.2   Dynamic Formats

Dynamic formats proposed in the literature include binary strings that end with 1[4], quaternary strings that end with 2 or 3[3] and vector codes[8]. They are dynamic in the sense that arbitrary insertions can be made between two consecutive codes without affecting other codes. We use binary strings to illustrate the property of dynamic formats. We include the descriptions of quaternary strings and vector codes in the extended version of this paper[10].

Definition 1. *(Binary String) Given a set of binary numbers $A = \{0, 1\}$ where each number is stored with 1 bit. A binary string is a sequence of elements in A.*

Binary strings are compared based on lexicographical order. The following theorem formalizes the dynamic property of binary strings that end with 1.

Theorem 1. *Given two binary strings C_l and C_r which both end with 1 such that C_l precedes C_r in lexicographical order (denoted as $C_l \prec C_r$), we can always find C_m which also ends with 1 and $C_l \prec C_m \prec C_r$.*

Theorem 1 can be proved based on Algorithm 1.

Example 4. Given three binary strings 01, 11 and 111, it follows from lexicographical order that $01 \prec 11 \prec 111$. Insertion between 01 and 11 will produce 011, since length(01) $\geq$ length(11) ($01 \oplus 1$, Algorithm 1 line 2). And insertion between 11 and 111 gives 1101, since length(11) < length(111) (111 with the last 1 change to 01, Algorithm 1 line 4).

3   ST Encoding Technique

In this section, we present the details of our ST encoding technique which can be applied to binary string, quaternary string and vector codes, and are called STB, STQ and STV encoding schemes respectively.

Algorithm 1. InsertBinaryString(C_l, C_r)

Data: C_l and C_r which are both binary strings that end with 1 and $C_l \prec C_r$
Result: C_m which ends with 1 and $C_l \prec C_m \prec C_r$
1 **if** $length(C_l) \geq length(C_r)$ **then**
2     $C_m = C_l \oplus 1$ /* $\oplus$ means concatenation */;
3 **end**
4 **else** $C_m = C_r$ with the last number 1 change to 01;
5 **return** C_m;

3.1 ST-Binary (STB)

Data structure: Our STB encoding is based by the data structure we call STB tree. An **STB tree** is a complete binary tree where each node is associated with a binary string that ends with 1, which we refer to as an STB code. The STB code of the root is 1.

Given a node n in the STB tree, the STB code of its left child lc and right child rc can be derived as follows:

- $C_{lc} = C_n$ with the last 1 replaced with 01
- $C_{rc} = C_n \oplus 1$ ($\oplus$ means concatenation)

Two STB trees with 6 and 12 nodes are shown in Figure 2 (b) and (c).

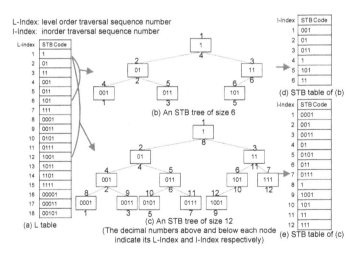

Fig. 2. STB encoding of two ranges 6 and 12

Lemma 1. *The left subtree of a node n contains only STB codes lexicographically less than C_n; The right subtree of n contains only STB codes lexicographically greater than C_n.*

Proof. [**Sketch**] Given any STB code n which is a binary string that ends with 1, we denote C_n as "$S1$" where "S" is a binary string or an empty string. It follows that C_{lc}="$S01$" and similarly, $C_{lc.lc}$="$S001$" and $C_{lc.rc}$="$S011$". Now it is easy to see that all the STB codes in the left subtree have "$S0$" as their prefix. Since "$S0$" precedes "$S1$" in lexicographical order, all the STB codes in the left subtree are lexicographically less than C_n. The rest of the lemma follows similarly.

Theorem 2. *An STB tree is a binary search tree based on lexicographical order.*

Proof. Theorem 2 follows directly from Lemma 1.

An **L table** stores the STB codes of an STB tree in order of *level order traversal*. We denote the index of an L table as **L-Index** and use L to denote the set of decimal numbers in L-Index. An important observation about L table is that it can be shared by STB trees of different sizes: the first m rows of the L table represents an STB tree of size m in level order. An **STB table** stores the STB codes of an STB tree in order of *inorder traversal*. We denote the index of an STB table as **I-Index** and use I to denote the set of decimal numbers in I-Index.

Example 5. Consider the STB tree of size 6 in Figure 2 (b). If we order its STB codes according to level order traversal sequence, they match the first 6 rows of the L table in (a). Ordering the codes in order of inorder traversal sequence would produce the STB table in (d). Similar observation can be made for the STB tree in (c).

Algorithms. To encode a range m with STB encoding is to realize the mappings represented by an STB table of size m. Intuitively, this can be achieved by traversing the STB tree of size m in inorder.

Formally speaking, STB encoding defines a mapping $f : I \rightarrow B$ where B denotes the set of STB codes. More specifically, f is established through two levels of mappings: $f(i) = h(g(i))$ where $g : I \rightarrow L$ and $h : L \rightarrow B$. Deriving h is straight forward from the L table. Depending on the range to be encoded, the size of L table can be extended dynamically. How g can be established is shown in Algorithm 2 which is based on inorder traversal of a binary tree. First a stack *path* is initialized to store the L-Indices of a root-to-leaf path(line 1). Then we proceed to call Function **PushLeftPath** which pushes the L-Index of the leftmost path (starting from the root) into *path* (line 2). For each $i \in I$, we map i to the top element in *path* (Recall that during an inorder traversal, the leftmost element is always visited first). Then the L-Index of the leftmost path that starts from the right child of the top element is pushed into *path* (line 3 to 6).

Next we show that STB encoding is order preserving and of optimal size.

Theorem 3. *Given a range m and any two numbers j and k such that $1 \leq j < k \leq m$, it follows that $C_j \prec C_k$ where C_j and C_k denote the STB codes transformed from j and k based on STB encoding.*

Algorithm 2. ItoLMapping(m)

Data: m which is the range to be encoded.
Result: The mapping from I-Index to L-Index stored in an array $ItoL[1 \ldots m]$.
1 Initialize Stack $path$;
2 PushLeftPath($path$, 1, m);
3 **for** $i{=}1$ **to** m **do**
4 $\quad$ $l{=}path.\textbf{Pop}()$;
5 $\quad$ $ItoL[i] = l$;
6 $\quad$ PushLeftPath($path$, $2 \times l + 1$, m)$\qquad$ /* $2 \times l + 1 \longrightarrow$ right child */
7 **end**

Function PushLeftPath($path$, l, m)

---

while $l \leq m$ **do**
$\quad$ $path.\textbf{Push}(l)$;
$\quad$ $l = 2 \times l$ $\qquad\qquad\qquad\qquad\qquad\qquad$ /* $2 \times l \longrightarrow$ left child */
end

Proof. Since an STB tree is a binary search tree (Theorem 2), an inorder traversal of the STB tree visits the STB codes in increasing lexicographical order. In other words, STB encoding is order preserving.

Lemma 2. *Level i of an STB tree has 2^{i-1} STB codes (except possibly the last level) of length i. (Assume the root is of level 1).*

Lemma 2 easily follows from the properties of STB trees.

Since an STB code is a binary string that ends with 1, there are 2^{i-1} possible STB codes of length i. From Lemma 2, we can see that an STB tree has all the possible STB codes of length i at level i (except possibly the lowest level). The fact that an STB tree is a complete binary tree implies that STB codes with length i are always used up before STB codes with length $i + 1$ are used. Therefore STB encoding produces labels with optimal size.

3.2   ST-Quaternary (STQ)

We illustrate our STQ encoding scheme using the data structure we call STQ tree. An **STQ tree** is a complete ternary tree. Each node of the STQ tree is associated with two STQ codes: left code (L) and right code (R) where $R = L$ with the last number 2 change to 3. L and R of the root are 2 and 3 respectively.

Given a node n in the STQ tree, the left code of its left child (lc), middle child (mc) and right child (rc) can be derived as follows:

- $L_{lc} = L_n$ with the last number 2 change to 12;
- $L_{mc} = L_n \oplus 2$ ($\oplus$ means concatenation);
- $L_{rc} = R_n \oplus 2$.

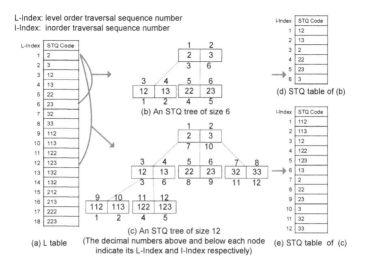

Fig. 3. STQ Encoding of two ranges 6 and 12

For every node, we have $R = L$ with the last number 2 change to 3.

Two STQ trees with 6 and 12 codes are shown in Figure 3 (b) and (c).

Lemma 3. *The left subtree of a node n contains only STQ codes lexicographically less than L_n; The middle subtree of n contains only STQ codes lexicographically between L_n and R_n; The right subtree of n contains only STQ codes lexicographically greater than R_n.*

The proof is similar to that of Lemma 1, so we omit it here. Given Lemma 3, an STQ tree can be seen as a search tree if we define the inorder traversal sequence to be in order of: (1) Traverse the left subtree; (2) Visit L of the root; (3) Traverse the middle subtree; (4) Visit R of the root and (5) Traverse the right subtree. In this way, we can define **I-Index**, **L-Index**, **STQ table** and **L table** similar to those of STB tree.

STQ encoding defines the mapping from I-Index to STQ codes which is achieved through two levels of mappings: from I-Index to L-Index and from L-Index to STQ codes. As shown in Figure 3 (a), the mappings from L-Index to STQ codes are stored a single L table which can be shared by multiple ranges. The mappings from I-Index to L-Index can be derived from Algorithm 4 which performs an inorder traversal of the STQ tree.

The correctness of our STQ encoding algorithm follows from the fact that an inorder traversal visits the STQ codes in increasing lexicographical order. The size of the encoded labels is also optimal as, intuitively, our algorithm favors STQ codes with smaller lengths.

Algorithm 4. ItoLMapping(m)

   Data: m which is range to be encoded.
   Result: The mapping from I-Index to L-Index stored in an array $ItoL[1 \ldots m]$.
1 Initialize Stack *path*;
2 PushLeftPath(*path, 1, m*);
3 **for** $i=1$ **to** m **do**
4 | l=*path*.**Pop**();
5 | $ItoL[i] = l$;
6 | **if** $l \ mod \ 2 =1$ **then**   /* $l \longrightarrow$ lcode */
7 | | PushLeftPath(*path*, $3 \times l + 2$, *m*) /* $3 \times l + 2 \longrightarrow$ middle child */
8 | **else**   /* $l \longrightarrow$ rcode */
9 | | PushLeftPath(*path*, $3 \times l + 1$, *m*) /* $3 \times l + 1 \longrightarrow$ right child */
10 | **end**
11 **end**

Function PushLeftPath(*path, l, m*)

---
   while $l \leq m$ **do**
     | *path*.**Push**($l + 1$);
     | *path*.**Push**(l);
     | $l = 3 \times l$ /* $3 \times l \longrightarrow$ left child */
   end

3.3   ST-Vector (STV)

Our STV encoding scheme is based on the data structure we call STV tree. It is a complete binary tree where each node is associated with a vector code: C. The vector codes of the root, its left child and right child are (1,1), (2,1) and (1,2) respectively.

Given a node n and its parent p in the STV tree, the vector codes of its left child (lc) and right child (rc) can be derived as follows: If n is the left child of p, $C_{lc}=2 \times C_n - C_p$; $C_{rc}=C_n + C_p$; Else, $C_{lc}=C_n + C_p$; $C_{rc}=2 \times C_n - C_p$. An example of STV tree is shown in Figure 4.

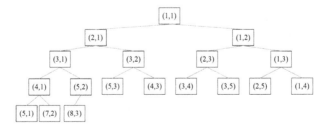

Fig. 4. STV tree

Theorem 4. *An STV tree is a binary search tree based on vector order.*

The proof is based on mathematical induction, we omit it here. Given the STV tree, we can define L table similar to that of STB encoding which stores the mapping from L index to Vector codes. Moreover, since STV tree is a binary search, Algorithm 2 can be directly applied to derive the mapping from I to L index. We ignore the details of STV encoding since it is similar to STB encoding.

3.4 Comparison with Insertion-Based Approach

Compared with the insertion-based approach, our design of ST encoding as a two level mapping has the following advantages: (1) Since $h : L \rightarrow STB/STQ/STVcode$ remains the same for different ranges, the cost of encoding a new range is only to compute $g : I \rightarrow L$. By sharing h for different ranges, we avoid costly table creation for every range; (2) Compression technique can be conveniently applied to L table to provide high flexibility of memory usage (Section 4). The compression technique is easily incorporable because compressing L table only affects h while h and g are independent of each other; (3) By exploiting the common mappings of different ranges, we can further speed up the encoding of multiple ranges (Section 5).

4   Encoding Table Compression

The L table of STB is shown in Figure 5 (a). Considering its STB codes with indices from 2 onwards, we can see that every STB code at index $2i + 1$ can be deduced from the STB code at index $2i$ by changing the second last number to 1. Therefore we can compress this L table to half by only retaining the rows with even indices ((b)). Thus, the mapping from L-Index to STB codes for becomes:

$$h(l) \rightarrow \begin{cases} LTable[l/2] & , when\ l\ mod\ 2 = 0 \\ LTable[\lfloor l/2 \rfloor]with\ the\ sec\text{-} \\ ond\ last\ number\ change\ to\ 1 & , when\ l\ mod\ 2 = 1 \end{cases} \quad (1)$$

The table in (b) can be further compressed by a factor of 2 if we consider the STB codes with indices from 2 onwards. We exclude the STB codes with odd indices since they can be derived from the STB codes with even indices by changing the *third* last number to 1 ((c)). In this way, we can compress the L table of STB by factors of $2, 4, 8 \dots 2^C$ and we denote C as the compression factor.

By analyzing the L table of STQ in Figure 5 (d), the straight forward compression is to exclude the STQ codes with even indices since they can be derived from the STQ codes with odd indices by changing the last 2 to 3 ((b)). Therefore the mapping from L-Index to STQ codes becomes:

$$h(l) \rightarrow \begin{cases} LTable[\lceil l/2 \rceil] & , when\ l\ mod\ 2 = 1 \\ LTable[l/2]\ with\ the \\ last\ number\ change\ to\ 3 & , when\ l\ mod\ 2 = 0 \end{cases} \quad (2)$$

L	STB Code
1	1
2	01
3	11
4	001
5	011
6	101
7	111
8	0001
9	0011
10	0101
11	0111
12	1001
13	1011
14	1101
15	1111
16	00001
17	00011
18	00101

(a) The original L table of STB

L	STB Code
1	01
2	001
3	101
4	0001
5	0101
6	1001
7	1101
8	00001
9	00101

(b) Compressed L table with C=1

L	STB Code
1	001
2	0001
3	1001
4	00001

(c) Compressed L table with C=2

L	STQ Code
1	2
2	3
3	12
4	13
5	22
6	23
7	32
8	33
9	112
10	113
11	122
12	123
13	132
14	133
15	212
16	213
17	222
18	223

(d) The original L table of STQ

L	STQ Code
1	2
2	12
3	22
4	32
5	112
6	122
7	132
8	212
9	222

(e) Compressed L table with C=0

L	STQ Code
1	12
2	112
3	212

(f) Compressed L table with C=1

Fig. 5. Compress L tables of STB and STQ by factors of 2^C and 2×3^C respectively

Consider the table in Figure 5 (e), it can be further compressed by a factor of 3 if we consider the STQ codes from index 2 onwards. The STQ codes at indices $3i$ and $3i + 1$ can be derived from the STQ code at index $3i - 1$ by changing the second last number to 2 and 3. Therefore we exclude the STQ codes at indices $3i$ and $3i + 1$ and the resulting table is shown in (f). In summary the L table of STQ can be compressed by factors of $2, 6, 18 \ldots 2 \times 3^C$.

The L table of STV can be compressed by a factor of 2 based on the bilateral symmetry we observe in the STV tree (Figure 4). Further compression is possible based on the symmetry at lower levels. Overall we can achieve compression factors of 2^C.

5   Tree Partitioning (TP)

We introduce Tree Partitioning (TP) as an optimization to further enhance the performance of ST encoding technique. We use STB tree to illustrate the idea of TP. Our optimization technique can be easily adapted for STQ and STV trees.

STB encoding technique, as we have shown, is a mapping $f(i) = h(g(i))$ where $g : I \rightarrow L$ and $h : L \rightarrow B$. Since h remains the same for different ranges, the cost of encoding a range is dominated by g. The motivation for TP optimization is that, given multiple ranges to be encoded, the computational cost of g can be reduced if we can exploit the common mappings for ranges that are close to some extent.

Suppose there are two STB trees T of size s_1 and T' of size s_2 (without loss of generality, we assume $s_1 < s_2$), we analyze the common mapping of the two trees when they have the same height, say k, i.e. $2^k \le s_1 < s_2 < 2^{k+1}$.

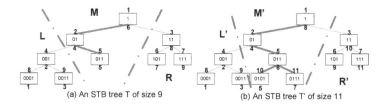

Fig. 6. TP Optimization

Our TP algorithm divides T' into three partitions:

L partition: All the nodes on the left of the path from the root to the node with L-Index=$s_1 + 1$.

R partition: All the nodes on the right of the path from the root to the node with L-Index=s_2

M partition: The rest of the nodes in the STB tree

T is also divided into three partitions: L', R' and M'. L' and L partitions have the same L-Index and so do R' and R partitions. And the rest of the nodes fall into M'. g in L and L' partitions are the same as the two partitions overlap and are visited first during inorder traversal. If we increase all the I-Index in R by $s_2 - s_1$, g in R and R' also coincide.

Example 6. Two STB trees T and T' in Figure 6 (a) and (b) are partitioned based on our TP algorithm. In the resulting partitions, g in L and L' are the same. g in region R can be derived from that in R' if we increase the L-Index in R by $11 - 9 = 2$.

Since both M and M' bounded by two root-to-leaf paths, Algorithm 2 can be easily modified to compute the mappings in them (an intermediate state can be calculated based on direct calculation which is available in [10]). By partitioning the range to be encoded, we can re-use some of the previously-computed mappings and avoid re-computing g for the whole range.

6   Experiments and Results

In this section, we experimentally evaluate and compare the various encoding techniques developed in this paper against the insertion-based encoding schemes including CDBS, QED and Vector. The comparison of CDBS, QED and Vector with the previous labeling scheme are beyond the scope of this paper and can be found in [5,8].

We used data sets from XMark benchmark, Treebank, SwissProt and DBLP datasets for our experiments. The characteristic of these data sets are shown in Table 1. We used JAVA for our implementation and our experiments are performed on Pentium IV 3 GHz with 1G of RAM running on windows XP.

Table 1. Test data sets

Data set	Max/average fan-out	Max/average depth	No. of nodes
XMark	25500/3242	12/6	179689
Treebank	56384/1623	36/8	1666315
SwissProt	50000/301	5/3	2437666
DBLP	328858/65930	6/3	3332130

6.1   Encoding Time

First we evaluate the encoding time of these encoding schemes using containment labels of the XMark data set. We randomly generated 80 XMark documents whose sizes range from 1 MB to 90 MB. In Figure 7, we observe clear time difference between ST encodings and insertion-based encodings: our STB and STV encodings are both approximately 3 times faster than CDBS and Vector encoding; Moreover, our STQ encoding is approximately 7 times faster than QED encoding. The reason is clear from the comparison of algorithms: insertion-based encodings need to create an encoding table for every range, which is significantly slower than our ST encodings that perform index mapping of a single table. The advantages of ST encoding are more significant when we apply TP optimization which exploits common mappings of encoding multiple ranges. Overall ST encodings with TP are by a factor of 5-11 times faster than insertion-based encodings for containment labels. The results confirm that our ST encoding techniques are highly efficient for encoding multiple ranges and substantially surpass the insertion-based encodings.

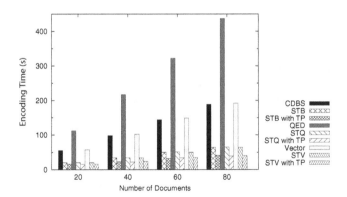

Fig. 7. Encoding containment labels of multiple documents

6.2   Memory Usage and Encoding Table Compression

We compare the memory usage of different algorithms which is dominated by the size of the encoding tables and the results are shown in Figure 8. Without any compression, the table size of STB and CDBS are the same, and so are their table creation times. However, unlike CDBS whose table size is fixed, our STB encoding can adjust its table size by varying the compression factor C. A larger C yields a smaller table size and less table creation time. Similar observation can be made in Figure 8 (c) and (d) for quaternary strings. The table creation time of STQ is less than that of QED due to the complexity of the QED insertion algorithms. By adjusting the compression factor, our ST encoding can process large XML data sets with limited memory available.

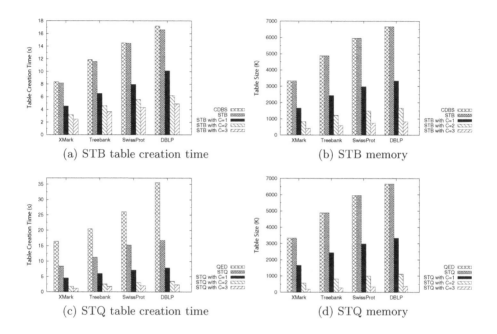

(a) STB table creation time   (b) STB memory

(c) STQ table creation time   (d) STQ memory

Fig. 8. Encoding table compression

6.3   Label Size and Query Performance

We empirically evaluate the label size and query performance of different labeling schemes. We have proved that both STB and STQ encodings produce labels of optimal sizes. The labels of vector and STV encoding schemes are stored as UTF8 strings. From our experimental results, their label sizes may differ by a small amount which is overall negligible, so we ignore the diagrams here. Moreover, since the labels produced by ST encoding and its insertion-based counterpart are of the same format, their query performance is also the same. In summary, the labels produced by our ST encoding techniques are of optimal quality.

7   Conclusion

In this paper, we take the initiative to address the problem of efficient label encoding. We propose ST encoding technique which can be applied to range-based labeling schemes to produce dynamic labels. We show that ST encoding technique is highly efficient and has a wide application domain. Compared with insertion-based encodings which are main memory-based and have fixed memory requirements, our ST encoding technique has an adjustable memory usage and is therefore able to process very large XML documents with limited memory available. An interesting future research direction is to explore more dynamic formats and study how the application scope of ST encoding could be extended to these formats.

References

1. Amagasa, T., Yoshikawa, M., Uemura, S.: QRS: A Robust Numbering Scheme for XML Documents. In: ICDE (2003)
2. Cohen, E., Kaplan, H., Milo, T.: Labeling Dynamic XML Trees. In: SPDS (2002)
3. Li, C., Ling, T.W.: QED: A Novel Quaternary Encoding to Completely Avoid Re-labeling in XML Updates. In: CIKM (2005)
4. Li, C., Ling, T.W., Hu, M.: Efficient Processing of Updates in Dynamic XML Data. In: ICDE (2006)
5. Li, C., Ling, T.W., Hu, M.: Efficient Updates in Dynamic XML Data: from Binary String to Quaternary String. In: VLDB J. (2008)
6. Zhang, C., Naughton, J.F., DeWitt, D.J., Luo, Q., Lohman, G.M.: On Supporting Containment Queries in Relational Database Management Systems. In: SIGMOD (2001)
7. Tatarinov, I., Viglas, S., Beyer, K.S., Shanmugasundaram, J., Shekita, E.J., Zhang, C.: Storing and Querying Ordered XML Using a Relational Database System. In: SIGMOD (2002)
8. Xu, L., Bao, Z., Ling, T.W.: A Dynamic Labeling Scheme Using Vectors. In: Wagner, R., Revell, N., Pernul, G. (eds.) DEXA 2007. LNCS, vol. 4653, pp. 130–140. Springer, Heidelberg (2007)
9. Xu, L., Ling, T.W., Wu, H., Bao, Z.: DDE: from dewey to a fully dynamic XML labeling scheme. In: SIGMOD (2009)
10. Xu, L., Ling, T.W., Bao, Z.: Efficient Label Encoding for Range-based Dynamic XML Labeling Schemes (Extended),
www.comp.nus.edu.sg/~xuliang/encodingextend.pdf
11. Dietz, P.F.: Maintaining order in a linked list. In: Annual ACM Symposium on Theory of Computing (1982)
12. Li, Q., Moon, B.: Indexing and Querying XML Data for Regular Path Expressions. In: VLDB (2001)
13. Abiteboul, S., Alstrup, S., Kaplan, H., Milo, T., Rauhe, T.: Compact Labeling Scheme for Ancestor Queries. SIAM J. Comput. (2006)
14. O'Neil, P., O'Neil, E., Pal, S., Cseri, I., Schaller, G., Westbury, N.: ORDPATHs: Insert-friendly XML Node Labels. In: SIGMOD (2004)

An Efficient Parallel PathStack Algorithm for Processing XML Twig Queries on Multi-core Systems

Jianhua Feng, Le Liu, Guoliang Li, Jianhui Li, and Yuanhao Sun

Tsinghua National Laboratory for Information Science and Technology, Department of
Computer Science and Technology, Tsinghua University, Beijing 100084, China
{fengjh@,le-liu02@mails,liguoliang@}tsinghua.edu.cn,
{jian.hui.li,yuanhao.sun}@intel.com

Abstract. Multi-cores are more and more popular recently and have being altered the course of computing. Traditional XPath query evaluation algorithms cannot take full advantages of multi-cores, and it is not straightforward to adapt such algorithms on multi-cores. In this paper, we propose an efficient parallel PathStack algorithm, named P-PathStack, for processing XML twig queries. The algorithm first efficiently partitions input element lists into multiple buckets, and then processes data in each bucket in parallel. With efficient partitioning method, our proposed algorithm can avoid many useless elements and achieve very good speedup ratio. We have implemented the algorithm and experimental results show that it achieves high performance and speedup ratio.

Keywords: Multi-core, Partition, Parallel, PathStack.

1 Introduction

As XML has become the *de facto* standard of data representation and exchange over the Internet, it plays an essential role in many modern computer and business systems. It has become one of the hottest topics to store and query XML documents for database researchers. Some XML query languages, such as XPath[1], XQuery[2], XML-QL[3], have been standardized and implemented. One key technique in these query languages is to use a path expression to express and search particular structure patterns. To efficiently evaluate path expressions, XML documents can be labeled with numbers [4], and by incorporating numbers to labels. One can quickly determine the relationships of parent-child or ancestor-descendant between element nodes and attribute nodes using the labeled numbers, without traversing the whole original XML document.

Many algorithms have been proposed for processing XPath queries recently. AI-Khalifa et al. [5] proposed the structural join algorithm, which solved XPath queries with linear complexity. But structural join algorithm will generate large intermediate results. Then holistic twig join algorithms were proposed for processing XPath twigs which can avoid large intermediate results, such as TwigStack [6], TSGeneric [7], TJFast [8], iTwigJoin [9].

H. Kitagawa et al. (Eds.): DASFAA 2010, Part I, LNCS 5981, pp. 277–291, 2010.
© Springer-Verlag Berlin Heidelberg 2010

In addition, Multi-cores are more and more popular recently and have being altered the course of computing. All the above algorithms have a common characteristic: they are proposed for single-core CPUs. They cannot take fully advantages of multi-core CPUs. To take advantage of multi-cores, efficient parallel algorithms are desirable for evaluating XPath queries. In our previous work [31], we proposed a parallel structural join algorithm (PSJ) for processing XPath in parallel. However it cannot handle XML twigs in holistic and thus is not efficient for XML twig queries.

In this paper, we parallelize PathStack algorithm [6] and enhance it on two aspects: one is the data partition, and the other is the task partition. It is very critical to evenly partition input ordered lists for parallel XPath query processing. Guoliang Li et al. [10] proposed an even partition based method, which partitions the input XML element lists into buckets evenly and may skip many ancestor or descendant elements. We borrow the idea of even partition from [10], and adapt the even partition approach to our problem, and take full use of the excellence of skipping ancestor or descendant nodes. We use region number instead of BBTC (Blocked Binary-Tree Coding scheme) [11], as region encoding is simple and useful while BBTC is strong but complicated.

A parallel XPath query algorithm is the other key issue for parallel algorithms. This paper proposes an efficient parallel algorithm to process XPath twig queries in parallel. Although the PSJ algorithm has good performance and good speedup ratio against traditional structural join algorithm, it has the inherent shortages of binary structural joins. It processes a couple of two nodes at a time and produces large immediate results. Also, it is less efficient than algorithms which at least process a root-to-leaf path at a time, such as PathStack/TwigStack[6], TSGeneric[7], and etc. To overcome these shortcomings, we use PathStack [6] as the baseline algorithm, and devise a parallel PathStack algorithm, named P-PathStack. The algorithm P-PathStack is more efficient and produces less immediate results.

Similar to parallel structural join algorithm, P-PathStack algorithm consists of two steps. First it evenly partitions XML data into multiple buckets, then evaluates root-to-leaf path in each bucket in parallel, and finally merges the results of all paths. Algorithm P-PathStack avoids large immediate results. The experimental results prove that P-PathStack has good speed up ratio and outperforms the parallel structural join algorithm significantly.

Our main contributions are summarized as follows:

1)   We adapt even partition approach from [10] to our problem, take full use of the excellence of skipping ancestor or descendant nodes and partition XML elements more evenly.

2)   We propose a parallel PathStack algorithm P-PathStack, which evaluates twig-XPath in parallel. The algorithm achieves high efficiency by using our optimization.

The rest of the paper is organized as follows. Section 2 gives some previous work on XML query processing in parallel. We give the preliminary of P-PathStack algorithm in Section 3. Section 4 presents the parallel PathStack. In Section 5, we give experimental results of the parallel algorithm and analyze the algorithm. Finally we conclude in Section 6.

2 Related Work

XML Query Processing: Many algorithms have been proposed for processing XML queries. Stack-tree-Desc/Anc [5] was the first stack-based algorithm, which has linear complexity. Zhang et al. [19] proposed a multi-predicate merge join (MPMGJN) algorithm based on <start, end, level> labeling of XML elements for binary structural join. Li et al. [20] proposed EE/EA Join, which decomposed the structure join into element-element join and element-attribute join. To scan elements much faster, index-based approaches are also proposed [21], [22], [23], in which the indices of B+-tree, R-tree and XR-tree are examined to improve the efficiency of XML queries processing. The later work [24] studied the problem of binary join order selection for complex queries on a cost model, which took into consideration factors such as selectivity and intermediate results size. Although structure join algorithm is more efficient than navigation, it will involve large intermediate results.

To solve the problem of huge intermediate results, holistic twig join algorithm is proposed. Bruno et al. [6] proposed a holistic twig join algorithm TwigStack to avoid large intermediate results. With a chain of linked stacks to compactly represent partial results of individual query root-to-leaf paths, TwigStack merged the sorted lists of participating element sets altogether, without involving large intermediate results. TwigStack had been proved to be optimal in terms of input and output sizes for twigs with only A-D (Ancestor-Descendant) edges [31].

In [7] Jiang et al. studied holistic twig join on all/partly indexed XML documents. The algorithms used indices to efficiently skip the elements that did not contribute to final answers, but it could not reduce the size of intermediate results. Lu et al. [8] proposed a novel algorithm, TJFast, on extended Dewey that only used leaf nodes' streams and saved I/O consumption. Lu et al. [26] proposed the algorithm Twig-StackList, which was better than any of previous work in term of the size of intermediate results for matching XML twig patterns with both P-C and A-D edges. Chen et al. [27] proposed an algorithm iTwigJoin, which was still based on region encoding, but worked with different data partition strategies (e.g. Tag+Level and Prefix Path Streaming). Tag+Level streaming can be optimal for both A-D and P-C only twig patterns whereas PPS streaming could be optimal for A-D only, P-C only and one branch node only twig patterns assuming there was no repetitive tag in the twig patterns [31]. Our prior work TJEssential algorithm [30] combined root-to-leaf with leaf-to-root way to improve the performance of XML query processing.

In addition, Wei Lu et al. [12] proposed a parallel approach for XML parsing, which first uses an initial pass to determine the logical tree structure of an XML document and then divides the document between the chunks occur at well-defined points. Wei Lu et al. [13] introduced the concept of work stealing into the field of XML processing. In multi-threads environment, when a thread is idle, the thread will choose a busy thread, and steal a half of work from the busy thread. When all work has been done, all threads will exit. According to common prefix of XQuery queries, Xiaogang Li et al. [14] distributed XML data into different machines, and then evaluated the partitioned data on each machine, finally merged the distributed results to generate final results.

In our prior work [31] we proposed a parallel structural join algorithm (PSJ) to execute XPath queries which contains only one root-to-leaf in parallel. However, as structural join algorithm, PSJ will produce large immediate results. So PSJ is not very fit for processing twig-XPath queries. As a result, we improve the classic PathStack algorithm [6] and make it execute twig-XPath queries in parallel to take full advantages of computing resources on multi-core processors.

Multi-Core: A multi-core CPU includes two or more independent cores, which are integrated onto a single integrated circuit die. In this paper, we study how to use multi-core CPU to improve the performance of XML query processing.

3 Preliminaries

Li et al. [10, 31] proposed even partition based method to improve the performance of processing XML queries. The proposed even partition based approach divides AList (the input list of ancestor elements) and DList (the input list of descendant list) into different buckets, $AList_i$ (the *i-th* bucket of AList) and $DList_i$ (the *i-th* bucket of DList) respectively, and only structure joins of suited buckets are useful to the join results. The even partition based approach can guarantee the following equation [10, 31],

$$AList \bullet DList = \bigcup_{i=1}^{n_b} (AList_i \bullet DList_i), \qquad (1)$$

where $AList_i$ and $DList_i$ respectively denote the element sets of AList and DList in the i-th bucket after partition, and n_b denotes the number of buckets. In other words, only $AList_i \cdot DList_i$ is useful to the final results and $AList_i \cdot DList_j$ $(i \neq j)$ will not.

As stated in [10], we first partition DList into different buckets $DList_i$ ($i=0,1,\ldots,$ n_b), and the size of each bucket (except the last one) is constant, denoted as b_s. We then partition AList into the buckets $AList_i$ ($i=0,1,\ldots,$ n_b) based on DList. For any element e contained in AList, the element e belongs to $AList_i$ if and only if e has one or more descendants in $DList_i$. In most cases, this partition approach can assure that all elements which are in different buckets do not have the ancestor-descendant or parent-child relationships, that is, $AList_i \cdot DList_j = \varnothing$ $(i \neq j)$. Even if $AList_i \cdot DList_j \neq \varnothing$ in some cases, the partitioning conditions below assure that $AList_i \cdot DList_j \subseteq AList_j \cdot DList_j$, so we only need to evaluate $AList_i \cdot DList_i$ ($i=0,1,\ldots,$ n_b) and merge their results to get the finally result.

Besides the equation (1) above, the partitioned buckets also satisfy the following three conditions [10]:

$$DList = \bigcup_{i=1}^{n_b} DList_i \text{ and } DList_i \cap DList_j = \varnothing (i \neq j) \qquad (2)$$

$$\forall i, 1 \leq i < n_b = \left\lceil |DList|/b_s \right\rceil, |DList_i| = b_s \text{ and } |DList_{n_b}| = |DList| - b_s * (n_b - 1) \qquad (3)$$

$$\bigcup_{n=1}^{n_b} AList_i \subseteq AList \qquad (4)$$

To partition AList and DList, we first partition descendant elements (i.e. DList) into several buckets orderly. And then determine which ancestor elements belong to the bucket using a binary search based method. We will introduce the details in Section 4.

For ease of presentation, we give an example. Suppose AList={all the elements whose local name is A in Fig. 1(a)} and DList={all the elements whose local name is D in Fig. 1(a)}. Then AList and DList can be partitioned into different buckets as shown in Fig. 1(a), 1(b). In Fig. 1(a), each bucket contains two four D elements, and Fig. 1(b) two D elements. In Fig. 1(b) the element $A2$ will be put into both $AList_1$ and $AList_2$. Although $AList_1 \cdot DList_2 = \{D_3\}$, $AList_2 \cdot DList_2 = \{D_3, D_4\}$, that is, $AList_1 \cdot DList_2 \subseteq AList_2 \cdot DList_2$, so we can safely ignore the operation $AList_1 \cdot DList_2$. The same occurs for $AList_2$ and $DList_1$.

Even partition can also skip ancestor or descendant elements. In Fig. 1(a), the element $A1$ will be skipped, because $A1$'s region encode <2,3,1> doesn't intersect with any D element's region encode. Also the elements $D7$ and $D8$ will be skipped.

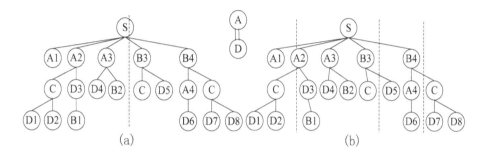

Fig. 1. An XML tree and corresponding partition results

4   P-PathStack: Parallel PathStack Algorithm

In this section, we first introduce even partition based approach, and then describe the algorithm P-PathStack in detail.

4.1   Even Partition

In this paper, we employ region encoding <start, end, level> to encode XML documents. Using region encoding, we can quickly determine the relationship between two nodes, such as parent-child or ancestor-descendant relationship. As the size of XML document increases, the space cost by region encoding increases linearly. Suppose AList and DList denote the ancestor element list and the descendant element list respectively, and they are in document order. We will partition AList and DList into different buckets $bucket_i$ ($i=0,1,..., n_b$), where $bucket_i$ contains both $AList_i$ and $DList_i$.

Now we are ready to introduce two Rules [10, 31] to partition DList and AList into different buckets.

Rule 1: Partition DList

```
for i = 0 … (n_b-1)
        bucket_i.dstartpos = i* b_s
        if i<b_s-1
            bucket_i.dendpos = (i+1)* b_s - 1;
        else
            bucket_i.dendpos = |DList| - 1
end for
```

Rule 1 means that DList is partitioned into n_b buckets, and each one (except the last one) contains b_s descendant elements and the last contains the remaining elements. The variable *dstartpos* denotes the start position of the bucket, while *dendpos* denotes the end position.

Rule 2: Partition AList

```
for i = 0 … (n_b-1)
     bucket_i.astartpos = min {p|a_p.end > bucket_i.minstart,
   0≤p<|AList|, a_p∈AList}
        bucket_i.aendpos = max {p|a_p.start < bucket_i.maxend,
   0≤p<|AList|, a_p∈AList}
end for
```

`bucket_i.minstart` means the minimal start value of region encode of descendant element in `bucket_i`, while `bucket_i.maxend` means the maximal end value of region encode <start, end,level>.

Rule 2 means that if one or more ancestor elements have descendants in `bucket_i`, there is a start position and end position in AList. The elements between `astartpos` and `aendpos` are contained by `bucket_i`. If `aendpos` < `astartpos`, there are not ancestor elements in `bucket_i`, then the result of PathStack algorithm in `bucket_i` will be empty.

For an XPath root-to-leaf path A0//A1//A2//A3, we first partition leaf node A3 according to Rule 1, and partition A3's parent A2 according to Rule 2; then based on the partitioning results of A2, partition its parent node A1 according to Rule 2. Repeat these steps, until the root node A0.

In fact, the partition process can be reversed. We can firstly partition the root-node elements (A0), and then partition its child A1, repeat these steps until the leaf node A3. Though the details are not completely the same with Rule 2 when partitioning from the root to the leaf, the basic idea is the same.

In this paper, we suppose there are not nested elements in an XML document, i.e. an element doesn't contain sub-elements which have the same name with the parent. This assumption is very nature for most of XML documents. Based on this condition, we can use binary search to find the first ancestor element whose end value of region encode is larger than `minstart` value of this bucket. We can also use binary search to find the last ancestor element whose start value of region encode is smaller than maxend value of this bucket. Using binary search, the time cost in partition period is little, less than 5% of total elapsed time. The experimental results prove it in Section 5.

4.2 Work Balance

The purpose of even partition is to evaluate XPath in parallel. Thus we should make each bucket contain the same number of elements as much as possible. It cannot achieve high speed up ratio of parallel algorithm for unbalanced partition.

We can see that Rule 1 in Section 4.1 makes the number of the leaf-node element in each bucket almost the same. The two rules cannot assure the total number of elements from root to leaf in each bucket is the same, but for those XML documents with elements distributed evenly, the total number in each bucket is close to each other. Because the number of leaf-node elements in each bucket is almost the same, and according to our partition rules, the number of its parent node elements assigned in each bucket will be close if the XML document with elements distributed evenly.

On the other hand, we should also consider the balance between threads in parallel. The number of buckets assigned to each thread should be the same. So we should consider the number of threads used to determine the value of n_b (number of buckets) and b_s (the number of leaf-node elements in each bucket).

We determine the values of n_b and b_s as below:

```
// thread_number: number of threads used
nb = thread_number*16;
bs = |DList|/nb;
/* sizehigh: the upper limit of number of leaf-node elements
   in each bucket*/
while bs > sizehigh
    nb *= 2;
    bs = |DList|/nb;
end while
/* sizelow: the lower limit of number of leaf-node elements
   in each bucket */
while bs < sizelow
    if(nb == thread_number)
        break;
    else
        nb /= 2;
    bs = |DList|/nb;
end while
```

From the above pseudo codes, we partition DList into n_b buckets, and n_b is 16 times of number of threads used (because it achieves higher performance for n_b =16), then the same number of buckets will be assigned to each thread. Each thread is assigned at least one bucket. And we set upper limit and lower limit for b_s. If b_s is too large, it will make against work balance, as the total number of elements contained in each bucket will differ more; if b_s is too small, it will make against the exertion of parallel predominance, as parallel scheduling needs extra cost and partitioning data costs more time. Similarly, we can easily get the pseudo codes of partitioning from the root node to the leaf node, see Section 4.1.

4.3 Elements Skip

We have mentioned the skipping function of even partition below. The excellence of skipping ancestor or descendant elements is very important in our parallel algorithm.

We can see that when partitioning AList according to Rule 2 in Section 4.1, we only need to find the first ancestor element es whose end value of the region encode is larger than the minstart value of this bucket and the last one el whose start value of region encode is smaller than maxend value. In other words, elements located before es or after el will be skipped.

The detailed procedure consists of two steps. Firstly partition from the leaf node to the root node. We partition DList into one bucket, and determine which elements of its parent node belong to this bucket according to Rule 2. Repeat these steps until the root node. Secondly partition from the root node to the leaf node. We put the remaining elements of the root node into one bucket, and determine which elements of the child node belong to this bucket, as described in Section 4.2. After this procedure, many useless elements will be skipped. This approach can skip many useless elements at the head and at the end of the element list.

For example, in Fig. 1(a), the first step will skip the element A1, and the second step will skip the elements D7 and D8. The remaining elements are {A2,A3,A4} and {D1, D2, D3, D4, D5,D6}.

4.4 P-PathStack: Parallel PathStack Algorithm

In this section, we describe the parallel PathStack algorithm, which is the key part of this paper.

We propose Algorithm P-PathStack to compute answers of a query twig pattern. Fig. 2 illustrates the algorithm. We first partition the data into difference buckets, and then for the data in each bucket, we use algorithm PathStack [6].

```
Algorithm P-PathStack(q)
1. Determine  the  value  of  n_b  and  b_s  as  described  in
   Section 4.2
2. Partition each root-to-leaf path into multiple buckets
   as in Section 4.1
3. For each root-to-leaf path, skip elements at the head
   and at the end of element list, as in Section 4.3
4. Call  PathStack[6]  algorithm  for  all  buckets  in
   parallel using OpenMP [15]
5. MergeAllPathSolutions()
```

Fig. 2. Algorithm P-PathStack

From Fig. 2, we can see that the key idea of the parallel algorithm is data partition and parallel execution. Line 1 to Line 3 partition data into several buckets, and the complexity is $O(n_b \,(\log|A_1List|+ \log|A_2List|+...+ \log|A_nList|))$. $|A_iList|$ denotes the number of elements with the name A_i ($i=1,2,...,n$) and A_i is a node contained by the root-to-leaf path. The complexity is much cheaper than the PathStack algorithm's linear complexity.

In Line 4 we call the standard PathStack algorithm to calculate XPath result in each bucket. Note that we put all the buckets of all paths into thread pool to execute in parallel. As a result, not only execute in parallel between different paths, also between buckets which belong to the same path. We make bucket as the parallel unit, because we can reduce the granularity of parallel and make work load between threads more balanced, then we can enhance the efficiency of parallel algorithm. In this paper, we use OpenMP [15] to implement parallel execution. OpenMP uses thread pool technology; we can set the number of threads used in program, and OpenMP will assign buckets evenly to the threads and assign the threads to different CPU cores on multi-core systems.

For example, consider the XPath A[//B][//D]. There are two root-to-leaf paths, A//D and A//B. For path A//D, after Line 1 in Fig. 2 we get {A2, A3, A4; D1, D2, D3, D4, D5, D6}. Suppose we partition them into two buckets, bucket1 {A2; D1, D2, D3} and bucket2 {A3, A4; D4, D5, D6}. For path A//B, after Line 1 in Fig. 2 we get {A2, A3; B1, B2}; we partition them into two buckets, bucket3 {A2; B1} and bucket4 {A3; B2}. Now we get all four buckets and we use OpenMP to evaluate XPath result in the four buckets in parallel. The results of the four buckets respectively are {A2, A2, A2}, {A3}, {A2} and {A3}. Combine the result which belong to the same root-to-leaf path, we get elements {A2, A2, A2, A3} and {A2, A3}, then merge them, and finally we get the final result {A2, A3}.

5   Experimental Study

We conducted a set of extensive experiments to study the performance of P-PathStack algorithm in this section. We tested 6 XPath queries on two datasets, XMark [16] and DBLP [17]. All the experiments are carried out on an Intel Xeon E5310 CPU (64 bits, 8 cores), 4G main memory, Red Hat 5 operating system, with Linux kernel version 2.6.18.5. We used C++ language and Intel C++ Compiler 11 (icc), which supports OpenMP 3.0 by default.

Table 1. XPath queries tested on datasets

Dataset	XPath NO	XPath
XMark	Q1	item//mailbox//mail/text//keyword
	Q2	site[//category//description//keyword]//mailbox//mail//text
	Q3	site[//person//name][//open_auction//increase]//mailbox//text
DBLP	Q4	dblp//article[@year='2005']
	Q5	dblp[//article[@mdate='2002-01-03']]//proceedings//number/text()='1'
	Q6	article[@year='2005'][@key.contains("sigmod")]//cite//label

The dataset DBLP is a set of real bibliography files, and the size of the raw text files is about 380MB. We generated the XMark data with scale factor = 4 and the raw text file is about 450MB. Table 1 lists the XPath queries tested on the two datasets. There are three query patterns for the 6 queries. Q1 and Q4 belong to the patterns of root-to-leaf path, Q2 and Q5 are twig patterns with two paths, and Q3 and Q6 are pattern with three paths.

5.1 Performance of P-PathStack

In this experiment we study the performance and speed up ratio of P-PathStack. We tested the six XPath queries listed in Table 1.

In this paper, we define the speedup ratio of P-PathStack below:

$$speedup\_ratio = time\_cost\_by\_PathStack / time\_cost\_by\_n\text{-}t \qquad (5)$$

where `time_cost_by_PathStack` denotes the elapsed time cost by traditional PathStack algorithm and `time_cost_by_n-t` denotes the elapsed time cost by P-PathStack algorithm with n threads.

In order to scale the efficiency of parallel, we define Relative Parallel Efficiency (RPE),

$$RPE(Q) = speedup\_ratio / n \qquad (6)$$

where `speedup_ratio` denotes the speedup ratio of P-PathStack algorithm with n threads, and n is the number of threads used.

Fig. 3 shows the experimental results on the two datasets. We can see that the speed up ratio is very significant; even it increases faster than number of threads used. We can see that in most cases the RPE value is larger than 1. Only for a few cases RPE is smaller than 1. For example, RPE equals 72.7% for Q6 with 8 threads and 78.4% with 7 threads. In other words, the speed up ratio is larger than the number of threads used.

With one thread, the smallest RPE equals 136% for Q3, and up to 201% for the query Q5. It seems something is wrong that RPE is larger than 1, but nothing is wrong. That's mainly because even partition approach we use for partitioning data can skip many useless ancestor and descendant elements. If enough elements are skipped, P-PathStack algorithm with one thread may cost less time than the traditional PathStack algorithm and it's probable that RPE exceeds 1. Fig. 4 shows the number of elements P-PathStack algorithm and the traditional PathStack algorithm read. The elements P-PathStack read is much less than that PathStack reads. The percentage for P-PathStack to PathStack is 65.9% at most for Q2 and only 47.2% at least for Q5. The average percentage is about 57%. As a result, the RPE value exceeds 1 even up to 2. With this excellence of even partition, our parallel algorithm achieves excellent performance and high speed up ratio.

We also define another variable to scale the parallel efficiency of the parallel algorithm, Average Relative Parallel Efficiency (ARPE),

$$ARPE(Q) = average(RPE \text{ for all threads}) \qquad (7)$$

With the formula (6) we calculate the ARPE value easily. For Q1, ARPE is 137.6%, Q2 134.7%, Q3 107.6% and Q4 154.7%, Q5 164.4%, Q6 115.8%.

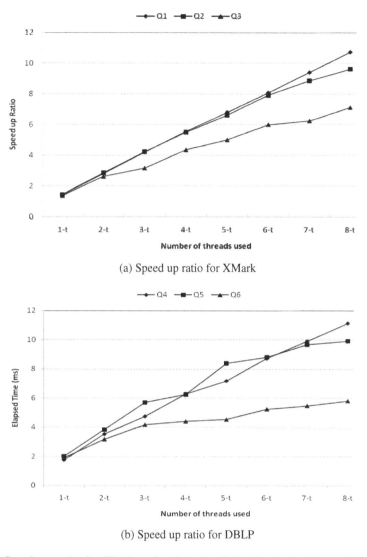

(a) Speed up ratio for XMark

(b) Speed up ratio for DBLP

Fig. 3. Speed up ratio for XPath quries (n-t: the P-PathStack algorithm with n threads, n=1,2,...,8)

In Fig. 3, the speed up ratio does not increase linearly as number of threads used increases. There are several factors which impact efficiency of parallel.

The more threads used, the more cost parallel scheduling needs. Operating system needs extra cost for parallel scheduling. In order to reduce the extra cost, the number of threads used is always smaller than number of cores in a CPU. As a result, we can assign all threads to different cores, and then reduce even remove switch of threads on CPU cores.

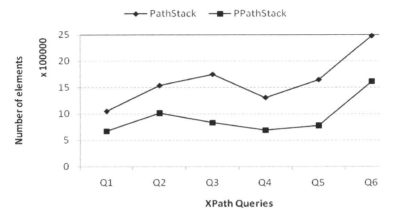

Fig. 4. Number of elements read

The more threads used, the more data accessing conflicts occur. Our algorithm runs on memory-shared multi-core system. All the cores share the main memory, and also share L2 cache at most cases. In the algorithm, there needs much write/read operations. In the experiments, there are at most 8 threads evaluating XPath queries in parallel. When threads access shared resources like main memory and L2 cache, resource competition occurs. Operating system must cost much time to solve the resource competition problem. In fact, in our experiments, we assign separated memory blocks to each thread for reducing accessing conflicts.

The more threads used, the less time cost by parallel execution. But partitioning data and merging results run in serially, so the time cost by them will account for a greater percentage of total running time. This will also lower the speed up ratio. Fig. 5 shows percentage partitioning data accounts of total elapsed time. Basically the percentage increases as number of threads used increase.

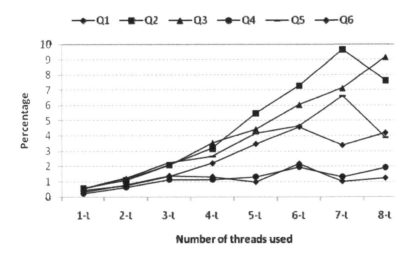

Fig. 5. Percentage accounted by even partition

5.2 Even Partition

Data partition is the key technology in parallel algorithm. In our algorithm, we use even partition as the data partition method and it contributes much to the parallel algorithm. So in this section we will discuss the experimental results about even partition.

Fig. 4 shows the number of elements read by PathStack and P-PathStack. We find that even partition skip about a half of elements, which is very helpful to improve the efficiency of P-PathStack algorithm.

Fig. 5 shows percentage even partition accounts of total elapsed time. In most cases the percentage is smaller than 5%, only in few cases it exceeds 5%, and the peak occurs in Q2 with 7 threads, whose value is 9.66%. The average percentage of all queries and all threads is 2.83%. It makes high speed up ratio of parallel execution that even partition accounts a very low percentage of total elapsed time.

In section 4.2, we mention that size of bucket will impact work balance and parallel efficiency. Experimental results show that it's true. Give Q5 with 4 threads for an example, see Table 2.

Q5 consists of two paths, path1: dblp//article[@mdate='2002-01-03'], path2: dblp //proceedings//number/text()='1'. In the columns of path1 and path2 in Table 2, the first number denotes the number of buckets (n_b, n_b can't be smaller than number of threads used, 4 in here), and the second denotes the number of leaf-node elements (b_s) in this path. The columns of sizelow and sizehigh denote the lower limit and the upper limit of b_s. We can see that when we set the lower limit the value 1000 or 3000 and the upper limit 6-times of the lower limit, the speed up ratio is about 1.5 times of others in the table. Generally speaking, small buckets are more helpful for work balance but need more partitioning time. In our experiments, we find that lower limit 1000 for DBLP and lower limit 3000 for XMark are the best.

Table 2. Size of buckets for Q5 with 4 threads

sizelow	sizehigh	Speedup Ratio	path1	path2
1000	6000	6.13	16*1895	32*1944
3000	18000	6.03	8*3791	16*3888
5000	30000	4.19	4*7583	8*7777
7000	42000	4.19	4*7583	8*7777
9000	54000	4.20	4*7583	4*15555
10000	60000	4.20	4*7583	4*15555

6 Conclusion and Future Work

As multi-core CPUs become more and more popular, parallel algorithms for XPath and XQuery processing become more critical by utilizing the computing resources of multi-core CPUs. We have made a good attempt in this paper for processing XML twig queries in parallel. In our prior work [31], we proposed a parallel structural join algorithm and in this paper we proposed a parallel PathStack algorithm P-PathStack to process XML twig queries in parallel, which can skip many more unnecessary elements and avoid huge intermediate results. Evidently P-PathStack is more efficient for memory-shared multi-core systems.

For future work, we want to devise more effective data partition algorithms which can skip many more unnecessary elements and achieve better work balance, and devise more efficient parallel computing model for XPath/XQuery processing.

Acknowledgement

This work is partly supported by the National Natural Science Foundation of China under Grant No. 60873065, the National High Technology Development 863 Program of China under Grant No.2007AA01Z152 and 2009AA011906.

References

1. Jamex, C., et al.: XML path language (XPath), http://www.w3.org/TR/xpath
2. Boag, S., et al.: XQuery: An XML Query Language,
   http://www.w3.org/TR/xquery/
3. Deutsch, A., Fernandez, M., Florescu, D., et al.: A query language for XML. WWW (1999)
4. Florescu, D., Kossman, D.: Sorting and Querying XML Data using an RDBMS. IEEE Data Engineering (1999)
5. AI-Khalifa, S., Jagadish, H.V., et al.: Structural Joins: A Primitive for Efficient XML Query Pattern Matching. In: Proceedings of the 18th International Conference on Data Engineering, San Jose, California, USA, pp. 141–152 (2002)
6. Bruno, N., Koudas, N., Srivastava, D.: Holistic Twig Joins:Optimal XML Pattern Matching. In: SIGMOD, pp. 310–321 (2002)
7. Jiang, H., et al.: Holistic Twig Joins on Indexed XML Documents. In: VLDB (2003)
8. Lu, J., Ling, T.W., Chan, C.-Y., Chen, T.: From Region Encoding To Extended Dewey: On Efficient Processing of XML Twig Pattern Matching. In: VLDB, pp. 193–204 (2005)
9. Chen, T., Lu, J., Ling, T.W.: On Boosting Holism In XML Twig Pattern Matching Using Structural Indexing Techniques. In: SIGMOD, pp. 455–466 (2005)
10. Li, G., Feng, J., et al.: Exploiting Even Partition to Accelerate Structure Join. In: IEEE XWICT (2006)
11. Feng, J.H., Li, G., Zhou, L., et al.: BBTC: A New Update-supporting Coding Scheme for XML Documents. In: Fan, W., Wu, Z., Yang, J. (eds.) WAIM 2005. LNCS, vol. 3739, pp. 32–44. Springer, Heidelberg (2005)
12. Lu, W., Chiu, K., Pan, Y.: A Parallel Approach to XML Parsing. In: 7th IEEE/ACM International Conference on Grid Computing, pp. 223–230 (2006)
13. Lu, W., Gannon, D.: Parallel XML Processing by Work Stealing. In: SOCP (2007)
14. Li, X.: Efficient and Parallel Evaluation of XQuery. Doctor Dissertation. The Ohio State University (2006)
15. OpenMP: Simple, Portable, Scalable SMP Programming,
    http://www.openmp.org/drupal/
16. http://www.xml-benchmark.org
17. http://dblp.uni-trier.de/xml/
18. Miklau, G.: UW XML Repository,
    http://www.cs.washington.edu/research/xmldatasets

19. Zhang, C., Naughton, J.F., DeWitt, D.J., Luo, Q., Lohman, G.M.: On Supporting Containment Queries in Relational Database Management Systems. In: SIGMOD, pp. 425–436 (2001)
20. Li, Q., Moon, B.: Indexing and Querying XML Data for Regular Path Expressions. In: VLDB, pp. 361–370 (2001)
21. Chien, S.Y., Vagena, Z., Zhang, D., Tsotras, V.J., Zaniolo, C.: Efficient Structural Joins on Indexed XML Documents. In: VLDB, pp. 263–274 (2002)
22. Grust, T.: Accelerating XPath Location Steps. In: SIGMOD, pp. 109–120 (2002)
23. Jiang, H.F., Lu, H.J., Ooi, B.C., Wang, W.: XR-Tree:Indexing XML Data for Efficient Structural Joins. In: ICDE (2003)
24. Wu, Y., Patel, J., Jagadish, H.: Structural join order selection for XML query optimization. In: ICDE, pp. 443–454 (2003)
25. Choi, B., Mahoui, M., Wood, D.: On the Optimality of Holistic Algorithms for Twig Queries. In: Mařík, V., Štěpánková, O., Retschitzegger, W. (eds.) DEXA 2003. LNCS, vol. 2736, pp. 28–37. Springer, Heidelberg (2003)
26. Lu, J., Chen, T., Ling, T.W.: Efficient Processing of XML Twig Patterns with Parent Child Edges: A Look-ahead Approach. In: CIKM, pp. 533–542 (2004)
27. Chen, T., Lu, J., Ling, T.W.: On Boosting Holism In XML Twig Pattern Matching Using Structural Indexing Techniques. In: SIGMO, pp. 455–466 (2005)
28. Mathis, C., Harder, T., Haustein, M.P.: Locking-aware structural join operators for XML query processing. In: SIGMOD, pp. 467–478 (2006)
29. Chen, S., Li, H.-G., Tatemura, J., Hsiung, W.-P., Agrawal, D., Candan, K.S.: Twig^{2}Stack: Bottom-up Processing of Generalized Tree Pattern Queries over XML Documents. In: VLDB (2006)
30. Li, G., Feng, J., Zhou, L.: Efficient holistic twig joins in leaf-to-root combining with root-to-leaf way. In: Kotagiri, R., Radha Krishna, P., Mohania, M., Nantajeewarawat, E. (eds.) DASFAA 2007. LNCS, vol. 4443, pp. 834–849. Springer, Heidelberg (2007)
31. Liu, L., Feng, J., Li, G., Qian, Q., Li, J.: Parallel Structural Join Algorithm on Memory-shared Multi-core Systems. In: WAIM, Zhang Jiajie, China (2008)

Keyword Search on Hybrid XML-Relational Databases Using XRjoin

Liru Zhang, Tadashi Ohmori, and Mamoru Hoshi

Graduate School of Information Systems, The University of Electro-Communications
1-5-1, Chofugaoka, Chofu City, Tokyo, Japan
{zhangliru, omori}@hol.is.uec.ac.jp

Abstract. Keyword search on databases has been popularized since it enables users to get information under databases without any knowledge of the schema or query languages. There are a number of keyword-search techniques on both RDB and XML DB, but those may still miss appropriate answers when some substructures of XMLs are related by relational linkage information. To overcome this problem, we consider integration of XML and relational data and this paper proposes a new method of keyword search on hybrid XML-Relational databases. As a new concept, a new join operator, named *XRjoin*, is designed and utilized to join XML with relational data. Experiments on DB2 v9.5 show effects of our proposal.

1 Introduction

Currently, a variety of semi-structured data as well as relational data has been stored in structural databases (e.g. IBM DB2 v9.5 [5]), and there is much increasing need for users to retrieve both XML and relational data by keyword search [4]. A relational database containing XML without any change of its format is termed a *hybrid XML-Relational database (XML-RDB)*. Motivated by this background, this paper proposes a new method of keyword search on this hybrid database system. Applying the keyword search on a XML-RDB is a challenging task, because keyword-search techniques on RDB widely differ from those on XML DB. We are motivated by the fact that the keyword-search results from the same data set under RDB and XML DB are different not only on their ranks but also on their contents.

To explain our motivation, we use a simple DBLP-style data set of Fig. 1. In Fig. 1 Table *Conference* has one XML column *XML1*, which includes the "Conf-Session-Paper" hierarchy. Table *Authors* has one XML column *XML2*, which includes the "Author-Paper" hierarchy. Because of storing XML data, the two tables in the XML-RDB are called **hybrid entities**. In contrast to a hybrid entity, Table *Paper* is a relational entity without any XML. There are *m:n* relationships in Table *Paper-Author* between the tuples of *Authors* and those of *Paper*. The *1:n* relationships between tuples of *Conference* and those of *Paper* are presented in *Paper* by columns *CID* and *PID*. Table *Citation* contains the reference information between papers. Note that the "paper" elements in XMLs are represented by the attribute *pid* only, and exclude any detail of papers. The *pid* is related by the tuple id *PID* of *Paper*.

H. Kitagawa et al. (Eds.): DASFAA 2010, Part I, LNCS 5981, pp. 292–298, 2010.

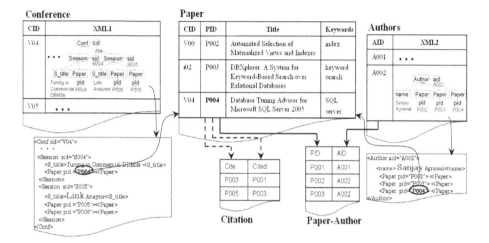

Fig. 1. A fragment of our experimental XML-RDB

Fig.1 shows the case where the keyword query **{link, sanjay}** is given to this XML-RDB. Here we should find two answers, where two XML-subtrees including the keywords are connected by some linkage information. That is, the first answer is the one where the conference(V04)'s subtree satisfying "link" has the same paper (P004) as the author(A002)'s subtree satisfying "sanjay" has. This answer tells that the two XML-subtrees have the same "paper" element. The second answer is the one where the conference(V04)'s subtree satisfying "link" has a paper P005, which cites another paper P003 owned by the author(A002)'s subtree satisfying "sanjay". This answer uses the citation relationship which links the minimal subtrees of the XMLs.

The ability for existing studies to retrieve the above appropriate answers is very limited. For example, XRANK [2] of XML DB extracts subtrees of LCA (least common ancestor) from the nodes hit by keywords, and ignores linkage information; thus it cannot find the above answers made of XML-subtrees linked by relational linkages. When normalizing Fig. 1 into relational tables, DBXplorer [1] of RDB does not consider XMLs; thus it cannot get an appropriate subtree of XML, where, for example, two sessions satisfying the keywords belong to the same conference. DISCOVER [3] of RDB can find such an answer, but its maximum Candidate Network's size T limits its ability to retrieve subtrees of XMLs having any depth or heterogeneity.

To overcome problems of these methods, **our objective** is to provide a keyword-search method on a XML-RDB to get substructures of XMLs with any depth and heterogeneity that are related by relational linkages. A join operator to do join between XML and relational data is imperative. Thus we develop a new operator named *XRjoin* to accomplish it.

Section 2 draws a hybrid XML-RDB system and describes our approach of keyword search on this system. In Section 3, we propose the new join operator *XRjoin* and demonstrate how it works. And the concluding remarks are given in Section 4.

2   Keyword Search on a Hybrid XML-RDB System

2.1   Data Model

Based on the ER model of a RDB, we firstly design a new schema for a hybrid XML-RDB. Fig. 2 shows the two data models of a RDB and an XML-RDB where the corresponding data of a subset DBLP are stored.

Fig. 2(a) shows an ER model of the RDB. The RDB has four entities *Conference, Session, Paper, Authors*, and four relationships *Conf-Sess, Sess-Paper, Paper-Author, Citation*. The "Conf-Session-Paper" and "Author-Paper" hierarchies have been decomposed into several tables.

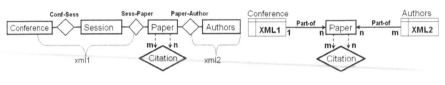

(a) The ER model of a RDB   (b) The schema of a hybrid XML-RDB

Fig. 2. Data model

Fig. 2(b) shows the schema of an XML-RDB. This schema includes three entities *Conference, Paper, Authors*, two m:n relationships *Paper-Authors* (omitted in Fig. 2(b)), *Citation* and two **part-of**[1] relationships between *Conference.XML1* and *PID*, *Authors.XML2* and *PID*. The instances of XMLs are expressed in Fig. 1. *XML1* makes four tables (*Conference, Conf-Sess, Session, Sess-Paper*) of the RDB into one hybrid entity of the XML-RDB. And *XML2* contains the information between one author and one or more papers in one XML. When the value of the attribute *pid* in one *XML1* including a keyword is equal to the value of *PID* of the tuple including another keyword, it is possible to do our proposed *XRjoin* between *XML1* and *Paper*.

2.2   Our Approach of Keyword Search

The processes of our keyword search on the hybrid XML-RDB system are composed of the following four steps: (1) identify entities including each keyword, (2) enumerate **join-trees**[2], (3) generate SQL/XML query statements, (4) execute these statements and obtain results.

Keyword entities: The entities including each keyword are first identified by using auxiliary tables. In this XML-RDB, three auxiliary tables are made as follows.

Table 1. Symtbl_relation (value, tuple_id, column_name, relation_name)
Table 2. Symtbl_XML (value, DeweyID, hybridEntity_name)
Table 3. Symtbl_link (tuple_id, DeweyID)

---

[1] *Part-of* is a relationship where one tuple id "belongs" to (is a part or member of) another object (element of XML in a hybrid entity).

[2] A *join-tree* is composed of the entities connected by foreign-keys where each keyword appears at least one. The join-tree decides how to join between these entities.

The Table 1 *Symtbl_relation* stores the value, the tuple id, the column name and the relation name for each tuple of all relational entities. For each text-node for all XMLs of hybrid entities, the Table 2 *Symtbl_XML* stores its value, DeweyID (an effective labeling method for XML [2]) and its hybrid relation name. We do the full-text search on *Symtbl_relation* and *Symtbl_XML* to identify where keywords are. The Table 3 *Symtbl_link* contains the tuple id (*Paper.PID*) of the relational entity (same as the attribute *pid* of a "paper" element) and the *DeweyID* of the "paper" element. For simplicity, each DeweyID is unique because of containing the tuple id of XML in hybrid entities. It is utilized to compute the LCA (least common ancestor) when XRjoin is executed.

Join-trees: Based on those entities, the system enumerates the join-trees, the minimum-cost Steiner-trees in a schema-graph, where each keyword appears at least once. After enumeration, all join-trees are listed and ordered by a score that is proportionate to the number of entities in a join-tree. Users can select any join-tree to do next step to get detailed information.

As an example, Fig. 3(a) shows a case where two keywords K1 and K2 hit entities of the schema of Fig. 2(b). Fig. 3(b) shows two join-trees for this case. Because there are hybrid entities in two join-trees, join between XML and relational data are necessary. Thus, we design the *XRjoin* (described in Section 3 in detail) to do it. Join-tree1 in Fig. 3(b) just does XRjoin once between a hybrid entity *Conference* and a relational entity *Paper*. Join-tree2 has two hybrid entities, so we must do XRjoins not only between *Conference* and *Paper*, but also between *Authors* and *Paper*. A natural join by foreign-key will be done between the two temporary resulting tables of XRjoins to get final results. Our system is also enhanced to enumerate join-trees including citation relationship.

After finding join-trees, one join-tree is selected by a user. Then, according to the join-tree, the SQL/XML query statement is generated and executed automatically to finish the join operations among the entities.

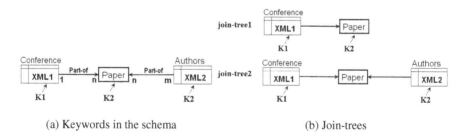

(a) Keywords in the schema   (b) Join-trees

Fig. 3. An example of join-tree enumeration

3   XRjoin

3.1   Definition of XRjoin

XRjoin is an operator to join XML data with relational data, which is written formally as *XRjoin ((X, K1), (E, K2))*, shown in Fig. 4(a). In Fig. 4(a), *X* is a hybrid entity and

E is a relational entity, and *K1* and *K2* are keywords in the two tables respectively. The result of the XRjoin is the set of all combinations of tuples in *X* and *E* including keywords (*K1* and *K2*) respectively, where the *part-of* relationship is satisfied.

The algorithm of the notation *XRjoin ((X, K1), (E, K2))* are shown in Fig. 4(b). At line 1 of Fig. 4(b), σ_{K2} (**E**) is a set of tuples including *K2* in *E*, and each tuple id of *e* is represented by an attribute *e_id* in *X*. Line 2 shows *x* is a tuple in *X*, whose XML part includes *K1*. Only one XML column is permitted in *X*. At line 3, *XRjoin ((X, K1), (E, K2))* is defined as a set of new tuples [**e, LCA-T (e_id, K1) on x**]. **LCA-T (e_id, K1) on x** is a (non-empty) subtree whose root element is the LCA for the element including *e_id* and an element satisfying *K1* in the XML of *x*. *T* is an output hybrid entity that contains these new hybrid tuples. *XRjoin ((X, K1), (E, K2))* retrieves the XML information satisfying *K1* (in *X*) which further contains a relational data-item satisfying *K2* (in *E*).

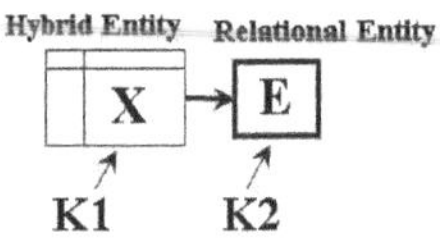

```
Hybrid Entity   Relational Entity      Input: X, K1, E, K2
                                        Output: T
   ┌─┐    ┌─┐                           1:  for each e ∈ σ_K2 (E)
   │X│ ─▶ │E│                           2:    for each x ∈ X where x contains K1
   └─┘    └─┘                           3:      insert a tuple [e, LCA-T (e_id, K1) on x] into T
    ↗      ↗                            4:    endfor
   K1     K2                            5:  endfor
                                        6:  return T
```

(a) the schema graph   (b) algorithm

Fig. 4. The *XRjoin ((X, K1), (E, K2))*

When multiple keywords hit, the algorithm of XRjoin mentioned above is advanced for general usage. Generalized XRjoin permits ***K1*** on *X* to be a set of *{K_1, K_2, ..., K_n}(n>0)*, and ***K2*** on *E* to be null or one keyword. (i.e., if ***K2*** on *E* has multiple keywords, they are logically-ANDed into one.) When multiple keywords hit the hybrid entity *X*, **LCA-T (*e_id, K1*) on x** at line 3 of Fig. 4(b) is changed into **LCA-T (*e_id, K_1, K_2, ..., K_n*) on x,** which means to extract the subtree from LCA including all elements of keywords *{K_1, K_2, ..., K_n}* and the element *e_id*. If no keyword hits the relational entity *E*, the algorithm of XRjoin changes line 1 of Fig. 4(b), `for each e ∈ σ_K2 (E),` into `for each e ∈ E`, which means all tuples in *E* are handled.

3.2 Behavior of XRjoin

According to the algorithm mentioned above, we present an illustration of concrete contents to explain how XRjoin works. Fig. 5 shows some tuples of *Paper*, and an instance of *XML1* in hybrid entity *Conference*. K1 is the keyword "tuning". K2 is the keyword "base". K1 exists in the session title (*S_title*) of *XML1*, and K2 hits the *title* of three tuples **P004, P005, P006** in *Paper*. This XRjoin is presented as *XRjoin ((Conference, "tuning"), (Paper, "base"))*. Fig. 6 shows the resulting hybrid entity *T*. XRjoin extracts each LCA between the element including "tuning" and one of the paper elements whose attributes are **P004, P005, P006**. The subtree from the LCA is inserted as a new XML column named *SUBTREE* in *T*. Besides *SUBTREE*, related relational data (*PID, Title, Keywords*) in *Paper* are also inserted into *T*.

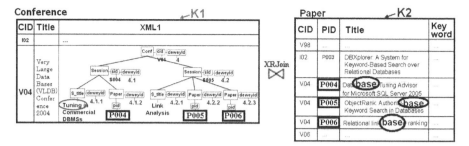

Fig. 5. The instances of hybrid entity *Conference* and relational entity *Paper*

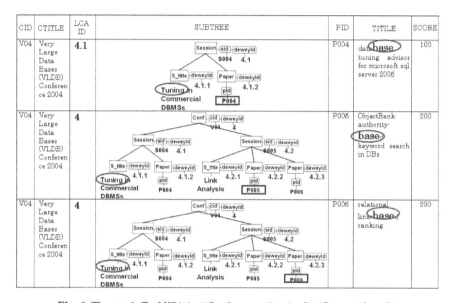

Fig. 6. The result *T* of *XRjoin ((Conference, "tuning"), (Paper, "base"))*

Score: When XRjoin extracts the LCA, its *Score* is calculated according to the difference of the lengths between *LCAID* and the one of *K1nodeid* and *K2nodeid* which has a longer DeweyID. It is directly proportional to the depth of the subtree. We list the hybrid tuples in ascending order of the *Score* (see Fig. 6).

$$\text{Score} = 100 \times (\text{Length}[\ K1nodeid \mid K2nodeid\ ] - \text{Length}[\ LCAID\ ])$$
where Length [] is a function to calculate the length of a DeweyID.

As another example, Fig. 7(a) shows a case of join-tree2 (of Fig. 3(b)). The join-tree2 decides the steps of join as follows, when there are two keywords *K1*(link) and *K2*(sanjay). The first step is to do XRjoin between *Conference* and *Paper* as the notation *XRjoin ((Conference, K1), (Paper))*. The second step is to do XRjoin between *Authors* and *Paper* as the notation *XRjoin ((Authors, K2), (Paper))*. The last step is to do natural join by a SQL statement between the two XRjoin resulting tables. The result is shown in Fig. 7(b), which means that an author named "sanjay" has written a paper which has been published in conference VLDB 2004 and the conference has a

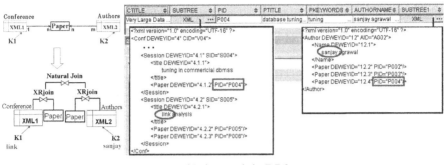

(a) the join-tree2   (b) the result in DB2

Fig. 7. Query {link, sanjay}

session titled "link analysis". The answer cannot be obtained by existing techniques [1], [2], [3] (when T=2).

4 Concluding Remarks

In this paper, we firstly stated our motivation about this study, and briefly analyzed the existing keyword-search methods [1], [2], [3] on RDB and XML DB, and gave the examples of the results that they may miss. To get appropriate and ordered XML-relational hybrid information, we proposed a new method of keyword search on a hybrid XML-Relational database. We defined a new join operator XRjoin to join XML data with relational data, and implemented the XRjoin in our system that is a keyword search engine system on a subset of DBLP. Experiments showed that our system built on DB2 v9.5 can work well in practice and it can find the answers that we want to get, and we will demonstrate this system at DEMO session.

Our proposal is successful in extracting substructures of XMLs linked by relational linkages. In the near future, we will improve this system to generate Candidate networks to do join that DBXplorer does not do, and enhance XRjoin accordingly.

References

1. Agrawal, S., Chaudhuri, S., Das, G.: DBXplorer: A System for Keyword-Based Search over Relational Databases. In: 18th International Conference on Data Engineering, pp. 5–16. IEEE Computer Society, San Jose (2002)
2. Guo, L., Shao, F., Botev, C., Shanmugasundaram, J.: XRANK: Ranked Keyword Search over XML Documents. In: ACM SIGMOD International Conference on Management of Data, pp. 16–27. ACM, California (2003)
3. Hristidis, V., Papakonstantinou, Y.: DISCOVER: Keyword search in relational databases. In: 28th International Conference on Very Large Data Bases, pp. 670–681. Morgan Kaufmann, Hong Kong (2002)
4. Yi, C., Wei, W., Ziyang, L., Xuemin, L.: Keyword search on structured and semi-structured data. In: ACM SIGMOD International Conference on Management of Data, pp. 1005–1010. ACM, Rhode Island (2009)
5. IBM DB2 Database Information Center,
   http://publib.boulder.ibm.com/infocenter/db2luw/

Efficient Database-Driven Evaluation of Security Clearance for Federated Access Control of Dynamic XML Documents

Erwin Leonardi[1], Sourav S. Bhowmick[1], and Mizuho Iwaihara[2]

[1] School of Computer Engineering, Nanyang Technological University, Singapore
[2] Graduate School of Information, Production and System, Waseda University, Japan
assourav@ntu.edu.sg, iwaihara@aoni.waseda.jp

Abstract. Achieving data security over cooperating web services is becoming a reality, but existing XML access control architectures do not consider this federated service computing. In this paper, we consider a federated access control model, in which *Data Provider* and *Policy Enforcers* are separated into different organizations; the *Data Provider* is responsible for evaluating criticality of requested XML documents based on co-occurrence of security objects, and issuing security clearances. The *Policy Enforcers* enforce access control rules reflecting their organization-specific policies. A user's query is sent to the *Data Provider* and she needs to obtain a permission from the *Policy Enforcer* in her organization to read the results of her query. The *Data Provider* evaluates the query and also evaluate *criticality* of the query, where evaluation of sensitiveness is carried out by using *clearance rules*. In this setting, we present a novel approach, called the DIFF approach, to evaluate security clearance by the *Data Provider*. Our technique is build on top of relational framework and utilizes pre-evaluated clearances by taking the differences (or deltas) between query results.

1 Introduction

Increasingly, data and services over the Web are becoming decentralized in nature. For example, the architecture of web services is becoming more decentralized; a number of servers stretching over different locations/organizations are orchestrating together to provide a unified service, sometimes referred to as *cloud computing* [5]. In this setting, access control for protecting sensitive data in XML format should also be cross-organizational, where a user, an access requester, and the *Data Provider* holding sensitive data, belong to different organizations. Figure 1 shows a conceptual depiction of such a federated access control model. The *Data Providers* and *Policy Enforcers* are separated into different organizations; each *Data Provider* is responsible for evaluating criticality of requested XML documents based on co-occurrence of security objects, and issuing security clearances. The *Policy Enforcers* enforce access control rules which reflect their own organization-specific policies. We assume that these organizations have agreed on *global* security policies for information exchange.

Let us illustrate the architecture with an example depicted in Figure 2(a) containing a part of a travel plan produced by a travel agency. We assume that the travel agency respond to request from clients and users using XML documents. These documents may

H. Kitagawa et al. (Eds.): DASFAA 2010, Part I, LNCS 5981, pp. 299–306, 2010.

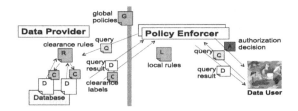

Fig. 1. Overview of federated access control

(a) A Published Document   (b) Clearance Rules

Fig. 2. Example

contain sensitive information. Suppose that an XML document containing relevant results is requested by a user in an airline company providing flights for the tour. The user needs to obtain a permission from the *Policy Enforcer* in his/her organization to read the document. The user query is sent to the *Data Provider* (in this case, the travel agency). The *Data Provider* evaluates the query and also evaluate criticality of the query, where evaluation of sensitiveness is carried out by using *clearance rules R*. A *clearance rule* $r \in R$ is a 2-tuple $[O, L]$, where O is a set of objects existing in XML documents and L is a *clearance label* that defines necessary security clearance the user should have. In this paper, we limit the scope of the objects to be text nodes of an XML tree. Note that these objects can be results of a set of XPath queries. A rule r raises a security caution defined by L iff $O \subseteq B$ where $B = \{b_1, b_2, \ldots, b_n\}$ is a bag of objects in a query result q. Figure 2(b) illustrates a sample of clearance rules represented as a table. If we apply the clearance rules to the document shown in Figure 2(a), we obtain the clearance labels $L1, L2$, and $L3$. For instance, the objects Alice and Nagoya appear in the document and matches the rule (Alice, Nagoya, $L1$). Likewise, Jane and Tokyo appear in the document and matches the rule (Jane, Tokyo, $L2$). The co-occurrence of Tom, Kyoto, Diabetic meal raises the label $L3$. A partial order between labels (such as $L1 < L2 < L3$) can be introduced, where '$A < B$' means that B is superior or more cautious than A. A query may raise a set of clearance labels, but if a priority order between labels is defined, a label that is dominated by another superior label can be ignored.

Finally, the *Policy Enforcer* receives the clearance labels C, and decides whether the user is eligible for the clearance by mapping the labels C to its local roles, and checking whether the user is assigned to one of these roles. Observe that by issuing clearance, the *Data Provider* can export the task of access authorization to the *Policy Enforcer*, thus realizing federated access control.

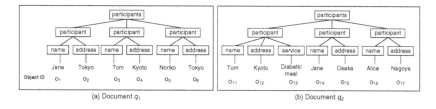

(a) Document q_1 (b) Document q_2

Fig. 3. Example of results evaluated by the *Data Provider*

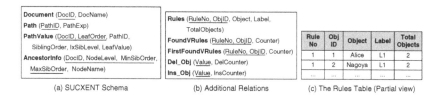

(a) SUCXENT Schema   (b) Additional Relations   (c) The Rules Table (Partial view)

Fig. 4. Relational schemas

Motivation and overview. There has been a number of efforts to realize federated access control [1,2,3,6,8]. However, to the best of our knowledge, none of these efforts have undertaken a systematic study of the security clearance technique supported by the *Data Provider* in a **dynamic** environment where the underlying XML documents may evolve with time. In this paper, we propose *a database-driven, diff-based strategy to address this issue*. Our proposed technique compliments existing research on federated access control strategies for XML documents.

Since the access control policies are realized through integration of the clearance rules at the *Data Provider* and the local rules at *Policy Enforcers*, at first glance, it may seem that we could take the strategy of pre-evaluating clearance rules at the *Data Provider* as much as possible and cache obtained clearance labels. The advantage of this strategy is that clearance evaluation for repeated queries can be avoided. However, this approach is not a feasible strategy as illustrated by the following example. Consider the documents q_1 and q_2 sent by the *Data Provider* to a *Policy Enforcer* in response to a user's queries at times t_1 and t_2, respectively, where $t_1 < t_2$. We assume that the *Data Provider* represents the query results in XML format and the set of clearance rules in Figure 2(b) must be satisfied by the documents. It is quite possible for q_1 and q_2 to share some data objects due to the following reasons: (a) q_1 and q_2 are results of the *same* query that is issued at times t_1 and t_2. The results may not be identical as the underlying data have evolved during this time period; (b) q_1 and q_2 are results of two different queries. However, some fragments of the underlying data may satisfy both the queries. Consequently, only the second rule in Figure 2(b) is valid for q_1. However, in q_2 this rule does not hold anymore. On the contrary, now the first and third rules are valid for q_2. In other words, updates to the underlying data invalidates caching of the clearance rules of q_1.

In this paper, we take a novel approach for evaluating security clearance by exploiting the overlapping nature of query results. Specifically, we investigate taking differences (deltas) of XML representations of the query results, so that valid clearance labels can be detected and reused. We compute the clearance labels of the first result (q_1) by scanning the entire result. Subsequently, labels of subsequent results are computed efficiently by analyzing the differences between the results. We refer to this strategy as the DIFF approach. Since we store the clearance rules and XML results in a RDBMS, the DIFF approach detects differences in the query results and clearance rules using a series of SQL statements. In the next section, we elaborate on this approach. Note that due to space constraints, the naïve approach of scanning the entire resultset for *every* request (referred to as the SCAN approach) is discussed in [7].

2   The DIFF Approach

Consider a set of query results $Q = \{q_1, q_2, \ldots, q_n\}$ in XML format. We refer to these results as *versions* in the sequel. Assume that the clearance labels for q_1 are cached, but no cache entry exists for the remaining results q_i where $i > 1$. How can the *Data Provider* evaluate clearance labels for the remaining $(n-1)$ versions efficiently? In the DIFF approach, we take advantage of the significant overlaps between q_1 and remaining results by reusing cached clearance labels whenever possible, and re-evaluate the clearance rules that are only affected by the changes (deltas) to the results. Note that often the size of the deltas are typically smaller than the size of q_i.

2.1   Relational Schema

We first present the relational schema that we use for storing results and clearance rules in the database for both SCAN and DIFF approaches. As the results requested by a *Policy Enforcer* are represented in an XML format, we can use any existing techniques for XML storage built on top of a RDBMS [4] to store these results. We use the SUCXENT schema [9] depicted in Figure 4(a) for storing the request results in a RDBMS. SUCXENT is a tree-unaware approach for storing and querying XML documents in relational databases. Particularly, in this paper, only the LeafValue attribute of the PathValue table is used for security clearance evaluation. The PathValue table stores the textual content of the leaf nodes of an XML tree in the LeafValue column. Hence, we do not elaborate on the remaining attributes and tables in Figure 4(a).

The clearance rules are stored in the Rules table (Figure 4(b)). The RuleNo attribute is used as an unique identifier of a rule. The TotalObjects attribute maintains the total number of sensitive objects in a rule r whose co-occurrences raise security cautions. The level of security caution is stored in the Label attribute. The ObjID and Object attributes store the identifier and value of the text objects in the query results, respectively. For example, Figure 4(c) depicts how the first rule (Alice, Nagoya, L1) in Figure 2(b) is stored in the Rules table. The FoundVRules and FirstFoundVRules tables are used to keep track of the number of sensitive objects that appeared in the requested query results. The number of occurrences of k-th sensitive object of a rule r is stored in the Counter attribute. The remaining tables shall be elaborated in Section 2.3.

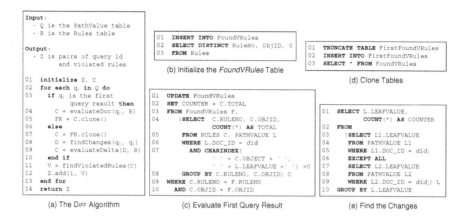

Fig. 5. The DIFF algorithm and SQL queries

2.2 Effects of the Changes

Suppose we have two query results, namely q_1 and q_2, and a set of clearance rules $C = \{c_1, c_2, \ldots, c_n\}$. After evaluating q_1, let $R = \{r_1, r_2, \ldots, r_n\}$ be the set of clearance rules that match with q_1 where $R \subseteq C$. Let O_{q1} and O_{q2} be the bags of objects in q_1 and q_2, respectively.

Let us now discuss the effects of the changes to the query results on the clearance rules. In this paper, we focus on two types of change operations to the query results: *deletion* and *insertion* of text objects. Note that the *update* of a text object can be represented as a sequence of delete and insert operations. An object o_{del} is a **deleted** object iff $o_{del} \in O_{q1}$ and $o_{del} \notin O_{q2}$. Similarly, an object o_{ins} is an **inserted** object iff $o_{ins} \notin O_{q1}$ and $o_{ins} \in O_{q2}$.

Property 1. A deletion of an object o_{del} will cause the removal of clearance rule $r \in R$ iff co-occurrence o_{del} with $o_k \in O_{q1}$ forms the clearance rule $r \in R$, and there does not exist $o_{k'} \in O_{q1}$ such that $value(o_{k'}) = value(o_{del})$ where $value(o)$ is the text value of object o.

Property 2. An insertion of an object o_{ins_1} will cause an addition of clearance rule r into R if o-occurrence of o_{ins_1} with $o_j \in O_{q1}$ forms a clearance rule $r \in C$, or co-occurrence of o_{ins_1} with another inserted object o_{ins_2} forms a clearance rule $r \in C$.

2.3 The DIFF Algorithm

The DIFF algorithm is depicted in Figure 5(a). The input to the algorithm are two relational tables, namely the `PathValue` table (denoted as Q in Figure 5(a)) and the `Rules` table (denoted as R in Figure 5(a)). Note that the requested results are stored in the `PathValue` table. The first step is to initialize the `FoundVRules` table (denoted as C in Figure 5(a)) by invoking an SQL query depicted in Figure 5(b) and a list Z. For each query result, the algorithm will do the followings (Lines 02–13). If the current

```
01  UPDATE FoundVRules SET COUNTER =  C.TOTAL
02  FROM  FoundVRules F,
03     (SELECT C.RULENO, C.OBJID,
           F.COUNTER - COUNT(*) AS TOTAL
04     FROM DEL_OBJ D, RULES C, FirstFoundVRules F
05     WHERE CHARINDEX('  ' + C.OBJECT + '  ',
                       '  ' + D.VALUE  +'  ') >0
06        AND C.RULENO = F.RULENO
07        AND C.OBJID = F.OBJID
08     GROUP BY C.RULENO, C.OBJID, F.COUNTER) C
09  WHERE C.RULENO = F.RULENO  AND  C.OBJID = F.OBJID
```

```
01  SELECT DISTINCT C.RULENO, C.CLABEL
02  FROM RULES C,
03     (SELECT F.RULENO,
             COUNT(F.OBJID) AS VOBJ
04     FROM FoundVRules F
05     WHERE F.COUNTER >0
06     GROUP BY F.RULENO) F
07  WHERE F.RULENO = C.RULENO
08     AND F.VOBJ = C.TOTALOBJECT
```

(a) Analyze the Changes   (b) Find Violated Rules

Fig. 6. SQL queries used in DIFF approach

query result is the first one (q_1), then it evaluates the occurrences of sensitive objects in q_1 (Lines 03–05). The evaluation is done by executing the SQL query depicted in Figure 5(c). The objective of this query is to update the value of Counter attribute of the FoundVRules tables to the number of occurrences of a sensitive object in a particular rule (Lines 04-08, Figure 5(c)). Next, the algorithm clones the FoundVRules table into the FirstFoundVRules table. The FirstFoundVRules table stores the results generated by evaluation of q_1.

If the current requested query results is *not* the first one (denoted as q_i where $i > 1$), then the algorithm will do the followings (Lines 06–10). First, it clones the FirstFoundVRules table into the FoundVRules table (Line 07) using the SQL query depicted in Figure 5(d). This step is important as we want to evaluate clearance for q_i using the clearance of q_1. Next, the algorithm determines the differences between q_1 and q_i by executing two SQL queries. The first SQL query is used to find the deleted objects (Figure 5(e)). Note that did_1 and did_2 will be replaced by the ids of q_1 and q_i, respectively. The result of this SQL query is stored in the Del_Obj table (Figure 4(b)). The second SQL query is used to detect the inserted objects. We use the same SQL query as shown in Figure 5(e); however, $did1$ and $did2$ will be replaced by the ids of q_i and q_1, respectively. The results of this SQL query is kept in the Ins_Obj table (Figure 4(b)).

Having found the differences between q_1 and q_i, the algorithm analyzes the deleted and inserted objects based on the Property 1 and Property 2, respectively, in order to determine the clearance of q_i. The SQL query depicted in Figure 6(a) is executed to analyze the set of deleted objects. Line 3 is used to decrease the number of appearances of sensitive objects if the sensitive objects are deleted. Similarly, this query is slightly modified to analyze the inserted objects. The modifications are as following. The "-" in Line 3 is replaced by "+". In Line 4, we replace "DEL_OBJ" with "INS_OBJ".

The last step in evaluating each requested result q_i is to find the rules that raise security cautions by querying the FoundVRules table (denoted by V). Figure 6(b) presents the SQL query for determining such rules. Then, we add a pair of request ids i and V into Z. Finally, the algorithm returns Z which may be analyzed further in order to determine which requested results are safe for publication.

3   Experimental Results

In this section, we present the experiments conducted to evaluate the performance of our proposed approach and report some of the results obtained. A more detailed results is

Dataset	Number of Leaf Nodes	Filesize (KB)
1	258	13
2	427	21
3	703	34
4	1,437	70
5	2,151	104
6	3,734	180

N	R	Total Objects
2	50	100
4	50	200
6	50	300
8	50	400
10	50	500

N	R	Total Objects
2	500	1,000
4	500	2,000
6	500	3,000
8	500	4,000
10	500	5,000

N	R	Total Objects
2	5,000	10,000
4	5,000	20,000
6	5,000	30,000
8	5,000	40,000
10	5,000	50,000

(a) Data Set   (b) Clearance Rules Characteristics

Fig. 7. Dataset and clearance rules characteristics

Data Set	N=2								N=10							
	1	2 A	2 B	2 Total	3 A	3 B	3 Total	Avg	1	2 A	2 B	2 Total	3 A	3 B	3 Total	Avg
1	1.31	0.02	0.06	0.08	0.02	0.04	0.06	0.48	7.79	0.05	0.12	0.17	0.05	0.10	0.15	2.70
2	2.05	0.02	0.06	0.08	0.02	0.05	0.06	0.73	12.53	0.05	0.12	0.17	0.05	0.09	0.14	4.28
3	3.25	0.03	0.05	0.07	0.02	0.05	0.07	1.13	20.39	0.06	0.12	0.17	0.06	0.10	0.16	6.91
4	8.39	0.03	0.06	0.08	0.03	0.05	0.08	2.85	41.36	0.07	0.12	0.19	0.07	0.09	0.17	13.91
5	12.48	0.05	0.06	0.11	0.05	0.05	0.09	4.23	81.71	0.08	0.12	0.20	0.08	0.09	0.18	20.69
6	21.52	0.07	0.05	0.13	0.07	0.05	0.12	7.25	106.72	0.10	0.12	0.22	0.10	0.09	0.20	35.71

(a) R=500

Data Set	N=2								N=10							
	1	2 A	2 B	2 Total	3 A	3 B	3 Total	Avg	1	2 A	2 B	2 Total	3 A	3 B	3 Total	Avg
1	15.32	0.08	0.17	0.25	0.08	0.13	0.20	5.26	75.86	0.30	0.55	0.86	0.30	0.33	0.63	25.78
2	24.99	0.08	0.17	0.25	0.08	0.11	0.20	8.48	123.75	0.30	0.56	0.86	0.31	0.33	0.64	41.75
3	40.64	0.08	0.17	0.25	0.09	0.12	0.21	13.70	202.21	0.31	0.56	0.87	0.30	0.34	0.64	67.91
4	82.63	0.10	0.16	0.26	0.10	0.12	0.23	27.70	412.95	0.36	0.56	0.92	0.37	0.34	0.70	138.19
5	123.42	0.12	0.17	0.29	0.12	0.12	0.23	41.31	618.45	0.38	0.57	0.94	0.38	0.33	0.72	206.70
6	219.92	0.13	0.17	0.29	0.14	0.12	0.26	73.49	1069.9	0.40	0.57	0.97	0.40	0.33	0.74	357.20

(b) R=5000

Fig. 8. Experimental results: The DIFF approach (in seconds)

available in [7]. The experiments were conducted on a computer with Pentium 4 3GHz processor and 1GB RAM. The operating system was Windows XP Professional. All the approaches were implemented using Java JDK 1.6. We use Microsoft SQL Server 2005 Developer Edition as our backend database system.

We use synthetic XML documents that are generated based on the DTD of SIGMOD Record XML. We assume that these documents represent results requested by the *Policy Enforcers*. Each data set has three different versions. Figure 7(a) depicts the characteristics of our data sets. The clearance rules are generated by randomly choosing the objects that co-occur together. The numbers of clearance rules (denoted as R) are between 50 and 5,000 rules, and the number of objects in each rule (denoted as N) are between 2 and 10. Hence, the total number of sensitive objects in the clearance rules is between 100 and 50,000 (Figure 7(b)).

The performance of the DIFF approach for $R = \{500, 5000\}$ and $N = \{2, 10\}$ is depicted in Figure 8. The "A" and "B" columns denote the execution times of finding the changes and of analyzing the changes, respectively. Observe that as the values of N and R increase, the performance becomes slower. The performance of analyzing the first document version is slower compared to the subsequent versions as the whole document is analyzed for the clearance rules. The performance of analyzing the subsequent versions is significantly faster as much lesser number of objects are evaluated.

4   Conclusions and Future Work

In this paper, we have presented a novel and sophisticated approach for automatically evaluating sensitiveness of publishing a batch of XML documents in a federated XML access control environment, and giving security clearance based on the sensitiveness. We use the differences between requested query results for clearance evaluation in our model. Our experimental results show that the proposed diff-based approach is efficient in determining security clearance. As part of future work, we would like to extend our framework to support clearance of security objects that are semantically related.

References

1. Liberty Alliance Project Homepage, http://www.projectliberty.org/
2. OpenID Foundation, http://openid.net/
3. Cohen, E., Thomas, R.K., Winsborough, W., Shands, D.: Models for Coalition-based Access Control. In: ACM SACMAT (2002)
4. Gou, G., Chirkova, R.: Efficiently Querying Large XML Data Repositories: A Survey. In: IEEE TKDE, vol. 19(10) (2007)
5. Hayes, B.: Cloud Computing. Communications of the ACM 51(7), 9–11 (2008)
6. Kern, A., Kuhlmann, M., Schaad, A., Moffett, J.: Observations on the Role Life-Cycle in the Context of Enterprise Security Management. In: ACM SACMAT (2002)
7. Leonardi, E., Bhowmick, S.S., Iwaihara, M.: Efficient Database-Driven Evaluation of Security Clearance for Federated Access Control of Dynamic XML Documents. Technical Report (2009),
   www.cais.ntu.edu.sg/~assourav/TechReports/XMLSecurity-TR.pdf
8. Lin, D., Rao, P., Bertino, E., et al.: Policy Decomposition for Collaborative Access Control. In: SACMAT (2008)
9. Prakash, S., Bhowmick, S.S., Madria, S.K.: SUCXENT: An Efficient Path–Based Approach to Store and Query XML Documents. In: Galindo, F., Takizawa, M., Traunmüller, R. (eds.) DEXA 2004. LNCS, vol. 3180, pp. 285–295. Springer, Heidelberg (2004)

Speeding Up Complex Video Copy Detection Queries

Ira Assent[1], Hardy Kremer[2], and Thomas Seidl[2]

[1] Department of Computer Science, Aalborg University, Denmark
`ira@cs.aau.dk`
[2] Data management and exploration group, RWTH Aachen University, Germany
`{kremer,seidl}@cs.rwth-aachen.de`

Abstract. Massive amounts of video data from digital tv channels, online video communities, peer-to-peer networks, and video blogs require automated techniques for copyright enforcement and usage tracking. Effective video copy distortion models usually incur high computational cost. We propose an index supported multistep filter-and-refine algorithm for a complex copy detection model. We characterize a class of filters for which we prove completeness of the result, and provide further runtime improvement by a novel tight approximation. In thorough experiments, we demonstrate that our algorithm substantially improves processing times.

1 Introduction

Videos are increasingly abundant due to widespread use of online video communities (e.g. YouTube), peer-to-peer networks for video sharing, video blogs, and digital tv channels. With easy distribution and many available tools for recording, editing, altering, storing and re-distributing video content, copyright protection and usage tracking face enormous challenges.

Video copy detection algorithms aim at automatic identification of video content that is identical to the query or represents an altered version of the original video [9,24,14,3]. As opposed to content-based similarity search in video databases [10,15,16], the aim is not searching for similar topics or otherwise related content in video material, but to discover videos that have undergone technical or manual changes, such as change in contrast or editing of the order of scenes in the video [14]. Content-based copy detection (CBCD) is based on the video content alone, i.e. 'the media itself is the watermark' [9]. Each video is characterized by a signature computed using a feature extraction algorithm. CBCD techniques can also be used to complement watermarking, i.e. embedding of information for identification [7].

As the computational task of video copy detection is inherently complex, some video copy approaches also contain speed-up techniques and index support. Window-based approaches compare only subsequences of the videos of fixed length [9,25,13]. Key-frame-based methods represent videos via fewer key frames for detected shots [11,7,23,5,6]. Some approaches use very compact fingerprints

H. Kitagawa et al. (Eds.): DASFAA 2010, Part I, LNCS 5981, pp. 307–321, 2010.

of the videos [11]. Indexing of features for faster comparison has been studied e.g. in [5,6], where an extended R-tree stores video frame features for retrieval.

In our prior work, a model for robust adaptable video copy detection ($RAVC$) was introduced [3]. It effectively detects video copies without prior compression or key frame extraction. The identification of videos is based on matching approaches in both the image and time domain. This $RAVC$ approach has been shown to successfully and reliably identify video copies that have been subjected to alterations that originate in a number of benchmark scenarios. Query processing under this model is, however, a computational expensive task.

In this paper, we propose an index-supported algorithm for speeding up copy detection under the existing $RAVC$ model. For efficient query processing, we propose a multistep filter-and-refine algorithm. Our novel generic filter exploits the properties of the underlying EMD and DTW distance functions. We prove that this generic filter is a lower bound of the $RAVC$. In our multistep filter-and-refine algorithms, the lower bounding property guarantees no false dismissals, and therefore ensures that our algorithm will return all copies as defined by $RAVC$. Moreover, we show how this general filter can be easily combined with other filter approaches, thus directly benefiting from ongoing active research on the EMD and DTW distance functions in a flexible "plug-and-play" fashion.

To support the VA-file-index [22] for our algorithm, we introduce the so-called *dual MinDist* function. This function allows computing the filter distance between query lower bounds directly for the indexed, quantized, features.

Optionally, we propose a novel approximation of the $RAVC$ distance function. We demonstrate in the experiments that this approximation provides substantial further runtime improvements with remarkably little degradation in the copy detection accuracy. Our main contributions include

- dual *MinDist* function (VA-file indexing of multidimensional video features)
- generic flexible filter with proof of lower bounding property (completeness)
- very tight novel approximation of the $RAVC$ (optional further speed-up)

2   Video Copy Detection

We start by summarizing the robust adaptable video copy detection scheme RAVC [3] illustrated in Figure 2. The general idea is to compare two videos by matching their video histograms. A video histogram $V = (v_1, \ldots, v_n)$ is a series of (frame) image histograms v_i (e.g. color histograms) in their chronological order. As video copies are hardly ever exact copies, i.e. identical digital representations, RAVC accounts for two main types of video alterations: image (e.g. changes in contrast or screen ratio) and time (e.g. re-sampling of frames for different video encoding standards or re-ordering of scenes). For time, RAVC uses an extension of Dynamic Time Warping (DTW) [19] to multidimensional time series. Figure 1 illustrates the idea behind DTW: Given two time series, Euclidean Distance (left) compares only corresponding time points, whereas the DTW distance (center) computes the best matching by stretching and scaling (warping) along the time axis. To avoid degenerate matchings, warping is usually

a)   b)   c)

Fig. 1. a) Euclidean: only corresponding time points are compared (vertical lines); b) DTW: the bursty pattern is aligned, but one point in the top blue time series is matched against several in the lower series; c) DTW with k-band: warping is restricted

restricted to at most k positions (right). Each point in time is not associated with a single value, but with a multidimensional vector, i.e. the image histogram of a frame. The comparison of each frame, the ground distance GD, uses the Earth Mover's Distance (EMD) [17]. RAVC is defined as the minimum matching of recursively shorter video histograms.

Definition 1. Robust Adaptable Video Copy distance. *The Robust Adaptable Video Copy distance between two video histograms $X = (x_1, \ldots, x_n)$ and $Y = (y_1, \ldots, y_n)$ with respect to a ground distance given by a cost matrix $C = [c_{ij}]$ and with respect to a band constraint k is defined as follows:*

$$RAVC(X, Y) = GD_{EMD}(X, Y) + min \begin{cases} RAVC(start(X), start(Y)) \\ RAVC(X, start(Y)) \\ RAVC(start(X), Y) \end{cases}$$

$$RAVC(\emptyset, \emptyset) = 0, RAVC(X, \emptyset) = RAVC(\emptyset, Y) = \infty$$

with

$$GD_{EMD}\left((x_1, \ldots, x_u), (y_1, \ldots, y_v)\right) = \begin{cases} EMD(x_u, y_v) & |u - v| \leq k \\ \infty & else \end{cases}$$

$RAVC$ is recursively defined like DTW for univariate time series. The main difference in $RAVC$ is that EMD is used to match frames. EMD reflects the discordance in the best match (necessary "flow" to transform one into the other) based on a ground distance in feature space (e.g. similarity between colors).

Definition 2. Earth Mover's Distance. *The Earth Mover's Distance between two normalized frame histograms $u = (u_1, \ldots, u_d)$ and $v = (v_1, \ldots, v_d)$ with respect to a ground distance given by a cost matrix $C = [c_{ij}]$ is:*

$$EMD_C(u, v) = \min_{\mathbf{F}} \left\{ \sum_{i=1}^{d} \sum_{j=1}^{d} c_{ij} f_{ij} \mid Pos \wedge SumUp \right\}$$

$$Pos : \forall_{1 \leq i,j \leq d} \ f_{ij} \geq 0 \quad SumUp : \forall_{1 \leq i \leq d} \sum_{j=1}^{d} f_{ij} = u_i \ \wedge \ \forall_{1 \leq j \leq d} \sum_{i=1}^{d} f_{ij} = v_j$$

where $\mathbf{F}$ denotes the set of possible flow matrices. Thus, the minimum flow between the histograms is computed, under the constraints that the flows be positive, and sum up to the histogram bin values.

Fig. 2. RAVC model: videos are matched according to Dynamic Time Warping alignment (in the time domain) of the Earth Mover's Distance values (in the image domain)

As the straightforward calculation of a matching in both the image and the time domain is a computationally costly task (dynamic programming algorithms for DTW are of quadratic complexity, EMD is of worst case exponential complexity, yet in practice quadratic or cubic runtimes are observed [17]), we propose a novel algorithm for efficient query processing.

3 Query Processing

In this section, we present our algorithm for efficient video copy detection under the $RAVC$ model. Our approach follows the multistep filter-and-refine paradigm, and provides index support based on the VA-file as illustrated in Figure 3: In the multistep filter-and-refine algorithm, we speed up query processing by using an efficiently computable filter function to generate potential video copy candidates. Only these candidates undergo further filtering and, if needed, refinement with the full $RAVC$ model. Note that this multistep filter-and-refine algorithm uses a feedback loop to minimize the number of candidates that have to be processed as suggested in KNOP [20]. In order to achieve efficient video copy detection without degradation of quality of the $RAVC$ model, the $ICES$ filter criteria should be met [1]: the filter should be *indexable*, i.e. compatible with an index for faster access to relevant videos; it should be *complete*, i.e. no true result should be falsely dismissed from the answer set; it should be *efficient*, i.e. remarkably faster than the exact $RAVC$ model to ensure substantial runtime gains; and finally, it should be *selective*, i.e. the set of candidates should be small to avoid unnecessary calls to the costly exact $RAVC$. For indexing, we use the VA-file, an index structure that was developed specifically for high-dimensional data [22], and has been successfully used for indexing of music time series data in [18]. To process a query, a first filter that uses VA-file indexing on quantized features is used to fill a priority queue of candidate copies for the next filter step (cf. Fig. 3): The VA-file quantizes the data space and assigns compact bit codes for quick sequential reading of the compressed data. For a video histogram $V = (v_1, \ldots, v_n)$, we denote its quantized bit code representation as $V^I = (v_1^I, \ldots, v_n^I)$. Processing of queries of the VA-file is based on the minimal distance between the query and objects with a certain bit code, the so-called *MinDist* (we introduce it for filtering of quantized video histograms in Section 3.2).

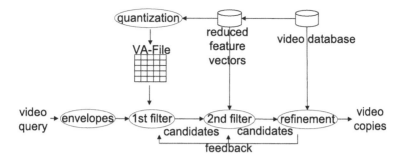

Fig. 3. Video originals are enveloped (LB_{Keogh} lower bound of DTW) for the 1st filter, where candidates are extracted from quantized videos in the VA-file. A 2nd filter (exact DTW with IM lower bound of EMD) is followed by $RAVC$ refinement using exact video features. A feedback loop ensures that the number of candidates is minimized.

3.1   Generic Filter

In the following, we propose a generic filter for the $RAVC$ model that fulfills the ICES criteria. As $RAVC$ is based on the DTW and EMD distances, we build on work that has been done in this very active research area. The generic filter is based on filters either in the image or in the time domain and can be flexibly instantiated using existing approaches for DTW or EMD. However, it is important to note that existing filters assume simpler settings, i.e. univariate time series for DTW, or non-quantized features for the EMD. We therefore introduce methods for handling complex video histograms.

LB_{Keogh}, a DTW filter technique for univariate time series, computes an *envelope* around the time series that reflects the maximal and minimal values within the k-band constraint: $Env_k(x) = ((x_1^L, x_1^U), \ldots, (x_n^L, x_n^U))$ with $x_i^L = min_{-k \leq j \leq k}(x_{i+j})$, $x_i^U = max_{-k \leq j \leq k}(x_{i+j})$ [12,26]. The distance to values above or below the envelope lower bounds the actual DTW distance. We take dimension-wise envelopes of video histograms $V = (v_1, \ldots, v_n)$ of length n and dimensionality d: $ENV_k(V) = (Env_k(V_1), \ldots, Env_k(V_d)) = V^E$. We use this lower bound with a recent speed-up technique [4].

Note that instead of constructing the envelope around the query, this could also be done around all videos indexed in the database. This poses no difficulty for our algorithm, yet constructing of the envelope and subsequent quantization (or vice versa), leads to very large hyperrectangles and consequently to poor pruning power. The relatively weaker performance of quantized envelopes has also been validated in preliminary experiments, and an example of this effect is given in the experiments in Section 4.

Through constraint relaxation, it becomes possible to compute a EMD lower bound in a dimensionwise fashion, thereby reducing the complexity of the filter to a linear problem only [1]. This IM (independent minimization) filter enlarges the search space for the minimum through constraint relaxation.

Our generic filter (*1st filter* in Fig. 3) consists of computing the *dual MinDist* (DMD) between query envelopes and quantized video features.

Definition 3. *Generic Filter (GF)*
For a query video histogram $X = (x_1, \ldots, x_n)$ and indexed video histogram $Y = (y_1, \ldots, y_n)$, the generic filter for RAVC is defined using the envelope of X, X^E, and quantization of Y, Y^I:

$$GF(X^E, Y^I) = \sum_{i=1}^{n} DMD_{EMD}(x_i^E, y_i^I)$$

The generic filter requires computing the *MinDist*, i.e. the smallest distance between the query envelope and quantized video. Both the envelope and the quantized video describe ranges in the feature space that can be visualized as hyperrectangles. However, in the literature, the *EMD MinDist* is only available for comparison between a hyperrectangle and a vector. We introduce the *dual MinDist* for hyperrectangles in this work.

3.2 Dual MinDist

For a multidimensional vector and a hyperrectangle, the *EMD MinDist* has been derived in [2]. It adapts constraints in the minimization by replacing the *SumUp* criterion with a range R that reflects the upper and lower bounds of the hyperrectangle.

For a normalized frame histogram $u = (u_1, \ldots, u_d)$ and a hyperrectangle $v^I = ((v_1^L, v_1^U), \ldots, (v_d^L, v_d^U))$ the *MinDist* of the *EMD* with respect to a ground distance given by a cost matrix $C = [c_{ij}]$ is defined as:

$$MinDist_{EMD_C}(u, v^I) = \min_{\mathbf{F}} \left\{ \sum_{i=1}^{d} \sum_{j=1}^{d} c_{ij} f_{ij} \mid Pos \wedge R \right\}$$

$$R : \forall_{1 \leq i \leq d} \sum_{j=1}^{d} f_{ij} = u_i \quad \wedge \quad \forall_{1 \leq j \leq d} \; v_j^L \leq \sum_{i=1}^{d} f_{ij} \leq v_j^U$$

As mentioned above, comparison of quantized video histograms with video histograms represented via envelopes corresponds to comparison of two series of hyperrectangles. Instead of the *MinDist* that computes the smallest distance between any point and a hyperrectangle, we thus need the *Dual MinDist* that computes the smallest distance between two hyperrectangles.

We define this *EMD Dual MinDist* by adapting the R constraint such that the matching takes the upper and lower boundaries of both hyperrectangles into account. However, simply relaxing the constraint for both hyperrectangles would not constitute the *Dual MinDist*. The minimization would not be forced to actually find a valid matching between vectors in both hyperrectangles. The reason for this is that neither sum would have to equal the sum of *weights* in any histogram. Instead, matching parts of histogram bin entries on either side would be falsely determined as the minimum. We therefore introduce an additional constraint W that ensures correctness of this *Dual MinDist* by requiring that the sum of all flows equals one, i.e. the sum of entries in normalized histograms.

Definition 4. *Dual MinDist for* EMD

For hyperrectangles $u^I = ((u_1^L, u_1^U), \ldots, (u_d^L, u_d^U))$ and $v^I = ((v_1^L, v_1^U), \ldots, (v_d^L, v_d^U))$ the Dual MinDist for the EMD with respect to a ground distance given by a cost matrix $C = [c_{ij}]$ is defined as:

$$DMD_{EMD_C}(u^I, v^I) = \min_{\mathbf{F}} \left\{ \sum_{i=1}^{d} \sum_{j=1}^{d} c_{ij} f_{ij} \mid Pos \ \wedge W \ \wedge \ R' \right\}$$

$$R' : \forall_{1 \le i \le d} \ u_i^L \le \sum_{j=1}^{d} f_{ij} \le u_i^U \ \wedge \ \forall_{1 \le j \le d} \ v_j^L \le \sum_{i=1}^{d} f_{ij} \le v_j^U \quad W : \sum_{i=1}^{d} \sum_{j=1}^{d} f_{ij} = 1$$

The constraint R' ensures that the range of the minimization problem is indeed within the corresponding hyperrectangles, whereas W restricts the minimization to histograms of normalized weight. Our novel *Dual MinDist* provides the means for actually using a multistep filter-and-refine algorithm as outlined before. We now benefit from the lower bounding for DTW in envelope-based hyperrectangles of the query and from the indexing in the VA-file which corresponds to the quantization-based hyperrectangles of the database videos. As a further benefit, we may employ lower bounds for the $EMD \ MinDist$ as well. This is based on the observation that the $MinDist$ is a special case of the *Dual MinDist*. This can be easily verified by choosing a hyperrectangle with identical upper and lower delimiters, i.e. a point. The important observation that we make here, is that the above mentioned IM lower bound can be modified to fit our new *Dual MinDist* by similar constraint relaxation. We therefore use it as a lower bound in our query processing algorithm. In Figure 3 this is depicted as the *2nd filter*.

3.3 Completeness

It has been shown that for multistep algorithms in the GEMINI or KNOP frameworks, proving that the filter function is a *lower bound* to the original distance suffices to show completeness [8,20]. We prove completeness of the generic filter by showing that it or any any lower bound thereof indeed lower bounds the $RAVC$ scheme:

Theorem 3.1. *The generic filter GF (Def. 3) lower bounds the $RAVC$, i.e. for two video histograms $X = (x_1, \ldots, x_n)$ and $Y = (y_1, \ldots, y_n)$ holds:*

$$GF(X^E, Y^I) \le RAVC(X, Y)$$

Proof. From $GF(X^E, Y^I) \le RAVC(X, Y)$ we have by definition of GF and $RAVC$: $\sum_{t=1}^{n} DMD_{EMD}(x_t^E, y_t^I) \le DTW_{EMD}(X, Y)$. Let the optimal DTW alignment be described as a series of positions $P = (p_1^X, p_1^Y), \ldots, (p_R^X, p_R^Y)$, then these are exactly the images compared using EMD:

$$\sum_{t=1}^{n} DMD_{EMD}(x_t^E, y_t^I) \le \sum_{r=1}^{R} EMD\left(x_{(p_r^X)}, y_{(p_r^Y)}\right)$$

Using a similar argument as for the derivation of LB_{Keogh}, namely that we can compute for each of the n positions the minimal distance possible under band constraint k:

$$\sum_{t=1}^{n} DMD_{EMD}(x_t^E, y_t^I) \leq \sum_{t=1}^{n} \min_{z \in \{x_{t-k},...,x_{t+k}\}} \{EMD(z, y_t)\}$$

It suffices to show that the above statement holds for each individual t:

$$DMD_{EMD}(x_t^E, y_t^I) \leq \min_{z \in \{x_{t-k},...,x_{t+k}\}} \{EMD(z, y_t)\}$$

By EMD definition

$$\leq \min_{z \in \{x_{t-k},...,x_{t+k}\}} \left\{ \min_{\mathbf{F}} \left\{ \sum_{i=1}^{d} \sum_{j=1}^{d} c_{ij} f_{ij} \mid Pos \wedge SumUp \right\} \right\}$$

with: $Pos : \forall_{1 \leq i,j \leq d} f_{ij} \geq 0 \ SumUp : \forall_{1 \leq i \leq d} \sum_{j=1}^{d} f_{ij} = z_i \wedge \forall_{1 \leq j \leq d} \sum_{i=1}^{d} f_{ij} = y_{(t,\ j)}$. The outer minimum can be pulled inside, since the constraints are independent. By weaking the constraints, the search space for the minimization increases, yielding a smaller or at most equal result:

$$DMD_{EMD}(x_t^E, y_t^I) \leq \min_{\mathbf{F}} \left\{ \sum_{i=1}^{d} \sum_{j=1}^{d} c_{ij} f_{ij} \mid Pos \wedge M \right\}$$

$$M : \forall_{1 \leq j \leq d} \sum_{i=1}^{d} f_{ij} = v_j \wedge \forall_{1 \leq i \leq d} \sum_{j=1}^{d} f_{ij} \geq \min_{z_{t-k,i}:z_{t+k,i}} \wedge \forall_{1 \leq i \leq d} \sum_{j=1}^{d} f_{ij} \leq \max_{z_{t-k,i}:z_{t+k,i}}$$

Now the right side is the $MinDist$: $DMD_{EMD}(x_t^E, y_t^I) \leq MinDist_{EMD}(x_t^E, y_t) \diamond$

This concludes the proof of the lower bounding property for the generic filters, thus we guarantee completeness in multistep filter-and-refine algorithms as desired. The proof of the general filter allows for easy plugging in of other lower bounds of DTW and EMD with immediate guarantees on completeness.

3.4 Speedup Using Approximations

In the $RAVC$ model, the exact computation of the EMD is computationally expensive. For more efficient query processing, we propose an additional new approximation of the EMD that is of only linear complexity. Starting from the existing IM lower bound of the EMD that has linear complexity, we propose a new upper bound of the EMD of the same complexity. Our novel approximation of the EMD is the average of these two lower and upper bounds. Both of these bounds are very tight, as we show in the experiments, and their average is very close to the exact EMD value. Consequently, trading off very little accuracy results in substantial efficiency gains.

We construct our new upper bound OM (ordered minimization) based on the nature of the EMD constraints. As opposed to constraint relaxation for the lower bound IM, the upper bound OM is based on introduction of an additional constraint. We construct this constraint such that linear computation is achieved.

To linearize the computation for an upper bound of the EMD, we introduce a new constraint to the minimization problem that requires assigning flows between the histograms in the order of their cost entries. Then, the upper bound OM can be computed sequentially on an ordered list of dimensions.

Definition 5. *Ordered Minimization (OM)*
For two histograms $u = (u_1, \ldots, u_d)$ and $v = (v_1, \ldots, v_d)$ the Ordered Minimization with respect to a ground distance given by cost matrix $C = [c_{ij}]$ is defined as:

$$OM_C(u, v) = \min_{\mathbf{F}} \left\{ \sum_{i=1}^{d} \sum_{j=1}^{d} c_{ij} f_{ij} \mid Pos \wedge SumUp \wedge O \right\}$$

$$O : \forall_{1 \leq i,j \leq d} \ with \ f_{ij} > 0 \ \forall_{1 \leq x,y \leq d} \ with \ c_{xy} < c_{ij} \quad \sum_{i=1}^{d} f_{xi} = u_x \quad \vee \quad \sum_{i=1}^{d} f_{iy} = v_y$$

The new constraint O ensures that constraints on the histogram dimensions are satisfied in the order given by the cost matrix entries. This means that in a manner similar to a greedy approach, the cheapest cost values are preferred in the matching, regardless of possible increase in the overall value of the solution. This allows linearization of the optimization problem. By simply ordering the dimensions with respect to their cost, we can efficiently compute OM. As this solution satisfies additional constraints, the minimum detected is an upper bound of the one found under the weaker EMD constraints. We formally state that OM is indeed an upper bound and prove this theorem.

Theorem 3.2. *For two histograms u and v and a cost matrix $C = [c_{ij}]$ the EMD is bounded by the OM in Definition 5:*

$$EMD_C(u, v) \leq OM_{EMD_C}(u, v)$$

Proof
We show that the solution space S' of the optimization problem given by the constraints of the OM is fully included in the solution space S described by the constraints of the EMD. Then, as a consequence, the minimum determined in this smaller solution space cannot be smaller than the minimum in the original space: for any subset S' of S: $S' \subseteq S \Rightarrow min(S) = min(min(S \backslash S'), min(S')) \leq min(S')$.

This inclusion property is straightforward from the definition of the solution problem via the constraint sets of EMD and OM, respectively: $Pos \wedge SumUp \supseteq Pos \wedge SumUp \wedge O$. Consequently, the minimum found by EMD cannot be larger than the one described by OM, and the upper bounding property holds.   ◇

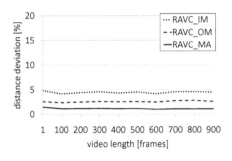 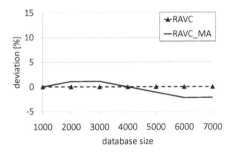

Fig. 4. Distance deviation w.r.t. length   **Fig. 5.** Recall deviation w.r.t. db size

We now have both a lower and an upper bound of the EMD. We propose using their average MA (minimization average), i.e. the average of the Independent Minimization and the Ordered Minimization, as a close approximation of the costly EMD:

Definition 6. *Minimization Average (MA)*
For two histograms $u = (u_1, \ldots, u_d)$ and $v = (v_1, \ldots, v_d)$ EMD approximation is defined as

$$MA(u, v) = \frac{IM(u,v) + OM(u,v)}{2}$$

We demonstrate in the experiments in Section 4 that this approximation is indeed a very close one. $RAVC$ can thus be used without approximation for full correctness of the result, or with this approximation for more efficient query processing times.

4   Experiments

We demonstrate in these experimental evaluations that video copy detection under the robust adaptable video copy detection scheme, $RAVC$, is substantially speeded up by our multistep filter-and-refine approach.

We implemented $RAVC$ and our multistep filter-and-refine algorithm based on the VA-file (Sect. 3) in Java, using lower bounds for DTW and EMD as discussed in Section 3.1. We evaluate both the exact $RAVC$ scheme as proposed in Section 3.3 and the approximate speed-up in Section 3.4. Experiments were conducted on 2.33GHz Intel XEON CPU machines running Windows Server 2008. The video data sets are the TRECVid[1] benchmark data of about 43 hours of videos [21]. The videos have an aspect ratio of 320x200. We use query videos as in the original $RAVC$ paper [3], which are generated using benchmark scenarios described in [14]: changes in contrast, black bars as a result of screen ratio changes, gauss filters, and changes in the temporal order. All results are averaged over 40 queries. Where not described differently, we set the band constraint in

---

[1] http://www-nlpir.nist.gov/projects/trecvid/

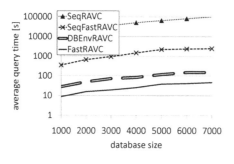
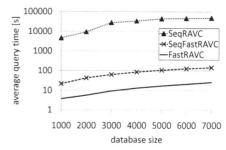

Fig. 6. DB size scalability (20d)   **Fig. 7.** DB size scalability (4d)

DTW to $k = 80$, the dataset size to the full TRECVid database, the video length to 512 frames. The color histograms were computed in extended HLS space with dimensionalities of 4 or 20.

As a first experiment, we show that the approximation through the Minimization Average (MA), introduced in Section 3, is indeed a very close approximation of the full distance computation. Figure 4 shows the distance value deviation of the three distance functions IM, OM, and MA to the full $RAVC$ model. As we can see, both the upper (OM) and lower (IM) bound tightly enclose the $RAVC$ model. By taking their average (MA), the distance value deviation is always below 1.5%, making it a remarkably close approximation.

Additionally, we evaluate the corresponding accuracy of the query results for $RAVC\_MA$ compared to $RAVC$. The accuracy is measured as the recall of the closest match found in the database, i.e. how many times the top result is indeed the original to the altered query copy. Results are depicted in Figure 5. As we can see, the overall deviation (in percent of the number of queries) to the exact values is very low. These changes are within a two percent point range, thus do not indicate any major differences.

We now study the performance of our approach in terms of runtimes. We compare sequential computation of the $RAVC$ model, sequential computation of our proposed approximation, denoted as $SeqFastRAVC$, as well as our index-supported multistep filter-and-refine algorithm for the approximation, referred to as $FastRAVC$. We evaluate the performance with respect to different scalability issues: we vary the size of the video database, the length of the videos, the band of the DTW distance, as well as the result size of the copy detection scheme. In our first efficiency experiment, we study the scalability in terms of database size for video histograms with a resolution of 20 dimensions. Figure 6 shows the results. Please note that the y-axis uses a logarithmic scale. As we can see, sequential computation of the $RAVC$ model is prohibitively slow. For a database of about 7000 videos, the runtime is more than 100,000 seconds which corresponds to over 28 hours. This is in stark contrast to the $FastRAVC$ which takes only less than 47 seconds. This is a tremendous speed-up of more than three orders of magnitude. Sequential computation of the approximation, $SeqFastRAVC$ results in runtimes of slightly less than 2500 seconds, i.e. more

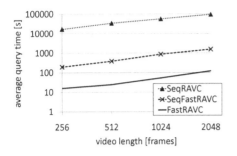 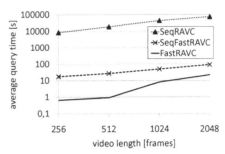

Fig. 8. Video length scalability (20d)   **Fig. 9.** Video length scalability (4d)

than 40 minutes. Thus, our approximation set-up contributes substantially to the runtime improvements. The index-supported filtering provides another order of magnitude. Clearly, our approach reduces video copy detection times that are practically infeasible to runtimes that are much more realistic for practical applications. This is observed across all database sizes. As an additional comparison, we have also included an example of moving the envelope, i.e. the lower bound for DTW from the query to the database ($DBEnvRAVC$). As discussed in Subsection 3.1, performance of this lower bounding approach is consistently poorer than our proposed technique. Its computation takes more than 150 seconds, which is almost four times that of $FastRAVC$. If the envelope is put around the videos in the index, this means that the hyperrectangles that originate from the VA-file induced quantization grow substantially due to the lower bounding envelopes. As a consequence, the lower bound is a lot less tight, and pruning is not as effective. We have observed this relationship consistently in all our experiments. In the following, we do not include it in the diagrams.

We validated these results for different histogram resolutions. Figure 7 shows this exemplarily for 4-dimensional frame histograms. As we can see, even though the dimensionality of the video histograms is very different, the general behavior of the different computation methods is similar. The difference between $FastRAVC$ and $SeqFastRAVC$ is slightly less pronounced, but the overall achieved speed-up remains at more than three orders of magnitude. Our next performance evaluation studies the scalability with respect to the video length, i.e. the number of frames in each video. Figure 8 shows the results for 20-dimensional video histograms. Also in this experiment, we observe speed-up rates around three orders of magnitude, with a slightly larger difference between the $SeqFastRAVC$ and the $FastRAVC$ for video databases of shorter length. This might be due to the fact that shorter video lengths correspond to a lower dimensionality for the index. Consequently, runtimes benefit more from the index support. Still, this results only in minor differences.

The same setup, but for video features with a resolution of only 4 dimensions per frame histogram, clearly supports this hypothesis. We show the runtime results in Figure 9. Once again, there is a huge efficiency gain for our method, but there is a noticeable jump from the shorter videos, where we observe a larger

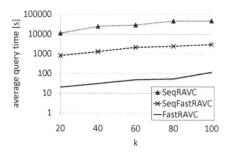

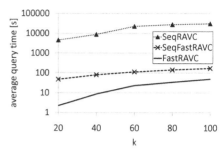

Fig. 10. k-band scalability (20d)   **Fig. 11.** k-band scalability (4d)

gap between *SeqFastRAVC* and its index-supported counterpart *FastRAVC*, to the longer videos, where the gap decreases. Still, the index support provides speed-ups of at least four times even in the closest case. The overall gains are consistently more than three orders of magnitude. An important parameter in the RAVC model is the amount of stretching along the time axis that is permitted in *DTW*. This k-band constraint regulates the amount of editing in the temporal domain that is considered to still be a video copy (cf. Page 309). As this band constraint is also known to play an important role with respect to *DTW* runtime performance (where larger bandwidths usually correspond to longer runtimes), we study the effect of this parameter on our proposed algorithm. Figure 10 illustrates the results for 20-dimensional video histograms. As we can see, our approach is remarkably robust with respect to the bandwidth parameter. The general tendencies for *RAVC*, *SeqFastRAVC* and *FastRAVC* are the same as in previous experiments, and runtime gains are remarkable even for wider bands.

Once again, we contrast our findings with the same setup on a low-resolution histogram database. Figure 11 shows the performance for different bandwidths for 4-dimensional video histograms. For these features, we see that there is a change in the relationship between *FastRAVC* and *SeqFastRAVC*. As we can see, the influence of the bandwidth constraint is more pronounced for the index-based vs. sequential speed-up algorithm. This might be due to the fact that with a low resolution of histograms and a wide area of potential matchings, the quantized values in the index might not differentiate enough between the different videos in the database. Still, *FastRAVC* remains much more efficient. As a final study, we vary the result size for video copy detection. This is specified as the number of potential copies retrieved by the algorithms. Usually, a video is assumed to be a copy of at most one element of the database. Slightly increasing the result size parameter leads to more false alarms, but reduces misses. Figure 12 illustrates the results for the 20-dimensional video histograms for one to five potential copies in the result. As we can see, the results are quite stable with respect to changes in the result size, and we observe only slightly less speed-up in the index-based *FastRAVC* for the low-dimensional features. In this case, as depicted in Figure 13, the gain from the index decreases for increasing result size. As mentioned before, this might be due to the fact that lower resolution video histograms that are quantized in the index are not as differentiable. As

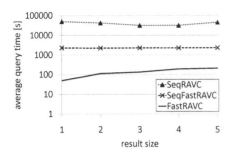
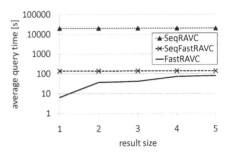

Fig. 12. Result size scalability (20d)   **Fig. 13.** Result size scalability (4d)

the tolerance threshold is increased implicitly by requesting a larger result size, this probably leads to the effect that the very good speed-up ratios for higher resolution histograms are not matched. Still, the index-based solution is always the best performing across all experiments. Especially for setups that are very relevant for video copy detection, i.e. small result size, high resolution features, large databases, we have seen very good performance of our algorithm.

5   Conclusion

We propose an efficient multistep filter-and-refine algorithm for the $RAVC$ model for video copy detection. It includes a novel generic filter that builds on the distance measures used in $RAVC$. Moreover, we integrate VA-file based indexing for which we devised the $MinDist$ function that allows comparing indexed videos with lower bounding approximations of queries. Optionally, we provide a novel tight approximation of the $RAVC$ for further speed up. Our evaluation on real benchmark video data demonstrates that our algorithm achieves substantial runtime improvements of several orders of magnitude. With these speed-ups, reliable $RAVC$ copy detection becomes feasible for practical applications.

Acknowledgments

This work was partially funded by DFG grant SE 1039/1-3.

References

1. Assent, I., Wenning, A., Seidl, T.: Approximation techniques for indexing the Earth Mover's Distance in multimedia databases. In: ICDE (2006)
2. Assent, I., Wichterich, M., Meisen, T., Seidl, T.: Efficient similarity search using the earth mover's distance for large multimedia databases. In: ICDE (2008)
3. Assent, I., Kremer, H.: Robust adaptable video copy detection. In: Mamoulis, N., Seidl, T., Pedersen, T.B., Torp, K., Assent, I. (eds.) Advances in Spatial and Temporal Databases. LNCS, vol. 5644, pp. 380–385. Springer, Heidelberg (2009)

4. Assent, I., Wichterich, M., Krieger, R., Kremer, H., Seidl, T.: Anticipatory dtw for efficient similarity search in time series databases. PVLDB 2(1), 826–837 (2009)
5. Böhm, C., Gruber, M., Kunath, P., Pryakhin, A., Schubert, M.: ProVeR: Probabilistic Video Retrieval using the Gauss-Tree. In: ICDE, pp. 1521–1522 (2007)
6. Böhm, C., Kunath, P., Pryakhin, A., Schubert, M.: Effective and efficient indexing for large video databases. In: BTW, pp. 132–151 (2007)
7. Chiu, C.Y., Li, C.H., Wang, H.A., Chen, C.S., Chien, L.F.: A time warping based approach for video copy detection. In: ICPR, pp. 228–231 (2006)
8. Faloutsos, C.: Searching Multimedia Databases by Content. Kluwer, Norwell (1996)
9. Hampapur, A., Hyun, K., Bolle, R.: Comparison of sequence matching techniques for video copy detection. In: Conf. on Storage and Retrieval for Media Databases (2002)
10. Hanjalic, A.: Content-based Analysis of Digital Video. Kluwer, Dordrecht (2004)
11. Joly, A., Frelicot, C., Buisson, O.: Robust Content-Based Video Copy Identification in a Large Reference Database. In: CIVR (2002)
12. Keogh, E.: Exact indexing of dynamic time warping. In: Proc. Int. Conf. on Very Large Data Bases (VLDB), pp. 406–417 (2002)
13. Kim, C., Vasudev, B.: Spatiotemporal sequence matching for efficient video copy detection. Trans. on Circuits and Systems for Video Technology 15(1) (2005)
14. Law-To, J., Chen, L., Joly, A., Laptev, I., Buisson, O., Gouet-Brunet, V., Boujemaa, N., Stentiford, F.: Video copy detection: a comparative study. In: CIVR (2007)
15. Lee, J., Oh, J., Hwang, S.: STRG-Index: spatio-temporal region graph indexing for large video databases. In: SIGMOD, pp. 718–729 (2005)
16. Lu, H., Xue, X., Tan, Y.: Content-Based Image and Video Indexing and Retrieval. In: Lu, R., Siekmann, J.H., Ullrich, C. (eds.) Joint Chinese German Workshops. LNCS (LNAI), vol. 4429, p. 118. Springer, Heidelberg (2007)
17. Rubner, Y., Tomasi, C.: Perceptual Metrics for Image Database Navigation. Kluwer, Dordrecht (2001)
18. Ruxanda, M.M., Jensen, C.S.: Efficient similarity retrieval in music databases. In: Proc. Int. Conf. on Management of Data, pp. 56–67 (2006)
19. Sakoe, H., Chiba, S.: Dynamic programming algorithm optimization for spoken word recognition. Trans. Acoustics, Speech, Signal Processing 26(1), 43–49 (1978)
20. Seidl, T., Kriegel, H.P.: Optimal multi-step k-nearest neighbor search. In: Proc. ACM Int. Conf. on Management of Data (SIGMOD), pp. 154–165 (1998)
21. Smeaton, A.F., Over, P., Kraaij, W.: Evaluation campaigns and trecvid. In: Int. Workshop on Multimedia Information Retrieval (MIR), pp. 321–330 (2006)
22. Weber, R., Schek, H.J., Blott, S.: A quantitative analysis and performance study for similarity-search methods in high-dimensional spaces. In: VLDB (1998)
23. Yan, Y., Ooi, B., Zhou, A.: Continuous Content-Based Copy Detection over Streaming Videos. In: ICDE, pp. 853–862 (2008)
24. Yang, X., Sun, Q., Tian, Q.: Content-based video identification: a survey. In: Proc. Int. Conf. on Information Technology: Research and Education (2003)
25. Yuan, J., Tian, Q., Ranganath, S.: Fast and robust search method for short video clips from large video collection. In: ICPR, pp. 866–869 (2004)
26. Zhu, Y., Shasha, D.: Warping indexes with envelope transforms for query by humming. In: SIGMOD, pp. 181–192 (2003)

Efficient Skyline Maintenance for Streaming Data with Partially-Ordered Domains*

Yuan Fang[1] and Chee-Yong Chan[2]

[1] Institute for Infocomm Research, Singapore
`yfang@i2r.a-star.edu.sg`
[2] National University of Singapore
`chancy@comp.nus.edu.sg`

Abstract. We address the problem of skyline query processing for a count-based window of continuous streaming data that involves both totally- and partially-ordered attribute domains. In this problem, a fixed-size buffer of the N most recent tuples is dynamically maintained and the key challenge is how to efficiently maintain the skyline of the sliding window of N tuples as new tuples arrive and old tuples expire. We identify the limitations of the state-of-the-art approach STARS, and propose two new approaches, STARS$^+$ and SkyGrid, to address its drawbacks. STARS$^+$ is an enhancement of STARS with three new optimization techniques, while SkyGrid is a simplification STARS that eliminates a key data structure used in STARS. While both new approaches outperform STARS significantly, the surprising result is that the best approach turns out to be the simplest approach, SkyGrid.

1 Introduction

Due to the usefulness of skyline queries in identifying interesting data points and its conceptual simplicity, there is a lot of research attention on how to efficiently process skyline queries. Given a set of tuples S, a skyline query (with respect to a collection A of attributes of interest) returns the subset of S (the so called "*skyline*") that are dominating with respect to A. Specifically, a tuple t_x **dominates** another tuple t_y iff t_x is better than or equal to t_y in every attribute in A, and is strictly better in at least one such attribute. Thus, the **skyline** of S, which consists of all tuples in S that are not dominated by any tuple in S, represents the subset of the most interesting points (with respect to A).

Using the popular example of a tourist who is looking for a hotel that is both cheap as well as close to the city, the "skyline" hotels that are of interest to the tourist are the hotels not dominated by any other hotel, where a hotel h_x dominates another hotel h_y if it satisfies the following conditions: (1) $h_x.price \leq h_y.price$, (2) $h_x.distance \leq h_y.distance$, and (3) at least one of the inequalities in (1) and (2) is strict.

---

* Part of this work was done when the author was a student at National University of Singapore.

H. Kitagawa et al. (Eds.): DASFAA 2010, Part I, LNCS 5981, pp. 322–336, 2010.
© Springer-Verlag Berlin Heidelberg 2010

Much of the early work on skyline queries are in the context of attributes with *totally-ordered* domains (as illustrated by the skyline hotel example), and focuses on query processing in *offline* environment where a skyline result is computed in response to a query on a disk-resident dataset (e.g., [5,8,6]). Recent research effort has shifted towards query processing in *online* environment, where a skyline result is dynamically maintained for a long-standing skyline query over continuous streaming data [10,12].

The key challenge for the streaming data environment is how to efficiently update the skyline for a *sliding window* of tuples. There are two models for the sliding window length N in streaming data applications. In the *time-based window* model, N represents the lifespan of each tuple in some number of time units. Each arriving tuple t_i has an arrival time-stamp s_i and expires after $s_i + N$ time units. Thus, the skyline is computed over all non-expired tuples and is updated whenever a new tuple arrives or an existing tuple expires [12]. In the *count-based window* model, the skyline is maintained for the most recent N tuples [10]. Thus, the skyline is updated whenever a new tuple arrives, and the arrival of the new tuple may also cause the oldest existing tuple to expire if there are already N tuples before the new arrival.

Several recent work on skyline queries have broadened the scope to include categorical attributes with *partially-ordered* domains in both offline [2,9] as well as online [10] environment. Categorical attributes are more general than numerical attributes as the dominance relationships among the domain values for a categorical attribute are based on a partial ordering instead of a total ordering.

In this paper, we address the problem of skyline query processing for a count-based window of continuous streaming data that involves both totally- and partially-ordered attribute domains. In this problem, a fixed-size buffer of the N most recent tuples is dynamically maintained and the key challenge is how to efficiently maintain the skyline of the sliding window of N tuples as new tuples arrive and old tuples expire. The state-of-the-art approach for this skyline problem is the STARS method [10], which is based on two key data structures: a multi-dimensional grid to organize the tuples in the buffer, and a geometric arrangement structure to organize the skyline tuples.

There are many interesting applications that require dynamic skyline maintenance of streaming objects in our setting. Consider an Internet search webservice that continuously accepts search requests from users, where each search request is associated with various categorical attributes of interest such as the search language, geographical region of the request, and the browser software and operating system used. The webservice can define a partial order over the attributes indicating its preferences of the search requests it wants to track. The system will then filter out and maintain a subset of recent interesting search requests, which can be exploited to study trends for better search results. Another application in news services is illustrated in [10].

In this paper, we make the following contributions. We identify the limitations of the STARS method and propose two new approaches, STARS$^+$ and SkyGrid, to address its drawbacks. STARS$^+$ is an enhancement of STARS with three new optimization techniques, while SkyGrid is a simplification STARS that completely eliminates a key data structure used in STARS. While both new approaches outperform STARS significantly, the surprising result is that the best approach turns out to be the simplest approach, SkyGrid, which outperforms both STARS and STARS$^+$ by up to a factor of 3 and 2.1, respectively.

The rest of this paper is organized as follows. In Section 2 we review the STARS algorithm. We present two new approaches, STARS$^+$ and SkyGrid, in Sections 3 and 4, respectively. In Section 5, we present an experimental evaluation of the proposed algorithms. Finally, Section 6 concludes the paper. Due to space constraint, all proofs are omitted.

2   Overview of **STARS** Approach

In this section, we give an overview of the STARS approach [10], which is the state-of-the-art algorithm for the skyline problem that we are addressing in this paper and the basis of our STARS$^+$ approach.

The STARS approach, which is based on the count-based sliding window model, maintains a fixed-size buffer of N tuples and updates the skyline of the N most recent tuples as new tuples arrive and old tuples expire. Whenever a new tuple t_{in} arrives and the buffer is already full with N tuples, the oldest tuple t_{out} in the buffer is expired and t_{in} becomes part of the skyline if it is not dominated by any other tuples in the buffer. Moreover, if t_{out} was a skyline tuple, then it is possible for some of the non-skyline tuples in the buffer to be *promoted* to become skyline tuples. Specifically, for each non-skyline tuple t in the buffer, if t is exclusively dominated by t_{out} (i.e., t_{out} is the only skyline tuple that dominates t), then t is promoted to a skyline tuple.

To avoid unnecessary dominance comparisons, STARS minimizes the set of tuples in the buffer by discarding *irrelevant tuples* from the buffer. A tuple t in the buffer is classified as irrelevant if it is dominated by a younger tuple t' (i.e., t' arrives later than t). The reason is that since t' will only expire after t, t is guaranteed to be dominated by at least one tuple throughout its remaining lifespan in the buffer which means that t can never be promoted to a skyline tuple. Thus, any tuple t that is dominated by a newly arrived tuple t_{in} is irrelevant and can be immediately discarded from the buffer. STARS refers to the buffer containing only relevant tuples as the **skybuffer**.

The skyline maintenance algorithm in STARS, which is invoked whenever a new tuple arrives, is shown in Fig. 1. The algorithm has the following inputs: SB is the buffer of relevant tuples (skybuffer), $S \subseteq SB$ is the set of skyline tuples, t_{in} is the newly arrived tuple, and t_{out} is the oldest tuple to be expired.

The key operations in the maintenance algorithm can be classified into the following three types of queries. (1) **D-query**: Given a tuple t, return the set of buffer tuples that are dominated by t; (2) **S-query**: Given a tuple t, determine whether t is dominated by any skyline tuple; and (3) **P-query**: Given a skyline tuple t that expires, return the set of buffer tuples that are promoted to skyline tuples due to the expiry of t. In Fig. 1, a D-query is used in steps 2 and 4,

(a) DAG for $\mathcal{D}$ (b) Skybuffer grid

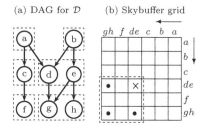

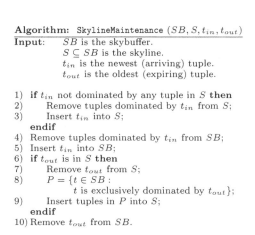

Algorithm: SkylineMaintenance (SB, S, t_{in}, t_{out})

Input: SB is the skybuffer.
            $S \subseteq SB$ is the skyline.
            t_{in} is the newest (arriving) tuple.
            t_{out} is the oldest (expiring) tuple.

1) **if** t_{in} not dominated by any tuple in S **then**
2)   Remove tuples dominated by t_{in} from S;
3)   Insert t_{in} into S;
   endif
4) Remove tuples dominated by t_{in} from SB;
5) Insert t_{in} into SB;
6) **if** t_{out} is in S **then**
7)   Remove t_{out} from S;
8) $P = \{t \in SB :$
          t is exclusively dominated by $t_{out}\}$;
9)   Insert tuples in P into S;
   endif
10) Remove t_{out} from SB.

Fig. 2. Skybuffer organization

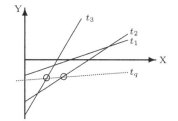

Fig. 1. Skyline maintenance framework of STARS

Fig. 3. Skyline organization

an S-query is used in step 1, and a P-query is used in step 8. Note that a P-query can be evaluated in terms of a D-query and multiple S-queries: given an expiring skyline tuple t, a D-query is first used to find the set of tuples T that are dominated by t, and then for each tuple t' in T, an S-query is used to check if t' is exclusively dominated by t.

To efficiently support the core operations (i.e. D-queries and S-queries) for skyline maintenance, the STARS approach organizes the buffer tuples and skyline tuples using two key data structures.

Multi-dimensional Grid. Suppose the skyline is computed wrt d attributes, $A_1, \cdots A_d$. STARS organizes the tuples in the buffer using a d-dimensional grid, where the ith dimension corresponds to attribute A_i. The objective is to map and store each tuple into a grid cell to support efficient D-queries.

To enable this mapping, the partially-ordered domain of each categorical attribute is linearized into a total ordering by a topological sort of the attribute domain's partial order. More specifically, the partially-ordered domain of a categorical attribute is represented by a directed acyclic graph (DAG), where each vertex in the DAG represents a domain value, and each edge represents the dominance relationship between two attribute values that cannot be be inferred by transitivity such that a value v is better than v' iff there exists a directed path from v's vertex to v''s vertex in the DAG. Let $r(v)$ denote the rank of the vertex corresponding to value v in a topological sort of the DAG. It follows that if $r(v) > r(v')$, then v cannot dominate v'. However, if $r(v) < r(v')$, then either v dominates v' or the two values are incomparable.

In this way, the scales of the grid on each dimension is bucketized into as many buckets as the number of domain values. Thus, each tuple $t = (a_1, \cdots, a_d)$

is mapped into the cell given by $\langle r(a_1), \cdots, r(a_d) \rangle$. To find the set of buffer tuples that are dominated by t (i.e., evaluate a D-query), STARS only needs to consider tuples $t' = (a'_1, \cdots, a'_d)$ that are located in the cells satisfying the following range query wrt t: $r(a_1) \leq r(a'_1), \cdots$, and $r(a_d) \leq r(a'_d)$. Additionally, many cells satisfying the range query are false positives and can be pruned as well. Thus, the grid organization enables STARS to eliminate many unnecessary dominance comparisons against tuples that cannot be dominated by t. To make the method scale to a large number of attributes or large attribute domains, STARS introduces techniques to control grid granularity by grouping multiple values into the same bucket. We use Fig. 2 (from [10]) to illustrate this.

Example 1. Consider the domain $\mathcal{D}$ of a categorical attribute consisting of the values $\{a, \cdots, h\}$ that are organized into the partial order depicted in Fig. 2(a). A possible topological sort is $a, \cdots, h$, which can be grouped into six buckets, each indicated by a dotted box in Fig. 2(a). A grid to organize a 2D dataset on $\mathcal{D} \times \mathcal{D}$ is depicted in Fig. 2(b). Consider a tuple t that is mapped to the cell marked $\times$ in Fig. 2(b). The dotted region in Fig. 2(b), which corresponds to the range query wrt t, represents the set of cells that could contain tuples dominated by t. Note that among the nine cells in the region, only the three cells marked $\bullet$ are candidate cells; the remaining six cells are false positives that cannot contain tuples dominated by t, which can be eliminated as well.   □

Geometric Arrangements. To efficiently support S-queries, STARS organizes the skyline tuples using a geometric arrangement of lines that maps skyline tuples onto a 2D plane. For this mapping, STARS needs to first choose two of the attributes (say A_i and A_j) among the attributes of interest for the skyline computation. Then each skyline tuple $t = (a_1, \cdots, a_d)$ is represented by a line $y = r(a_i) \cdot x - r(a_j)$ in the 2D plane, where a_i and a_j are t's values for attributes A_i and A_j, respectively. Based on this geometric line arrangement, two tuples t and t' are incomparable if the intersection point (x_I, y_I) of their line representations has $x_I < 0$. STARS uses the doubly-connected-edge-list (DCEL) data structure [4] to represent the positive half (wrt the x-axis) of the line representation of each skyline tuple. Using DCEL, STARS is able to efficiently evaluate an S-query wrt a tuple t by retrieving the lines that intersect with t's line in the positive half of the x-axis. In this way, many skyline tuples incomparable to t are pruned. Moreover, evaluating an S-query is progressive as it can terminate once an intersecting skyline tuple is found. We use the example (from [10]) shown in Fig. 3 to illustrate this concept.

Example 2. Suppose there are three skyline tuples t_1, t_2, and t_3. Their line representations are shown as labelled in Fig. 3. Consider an S-query wrt a tuple t_q, which is represented by the line as labelled in Fig. 3. Using DCEL, t_3 is the first line found to intersect with t_q, and a dominance comparison is performed to check if t_q is dominated by t_3. If so, then the the evaluation of the S-query completes; otherwise, the next line that intersects t_q is t_2 and a dominance comparison between t_2 and t_q is performed and so on. Observe that t_1 is pruned as it intersects with t_q at the point (x_I, y_I) where $x_I < 0$.   □

3   STARS$^+$ Approach

In this section, we present three optimization techniques to improve the performance of STARS. We refer to this optimized variant as STARS$^+$. The first technique reduces the number of S-queries required for evaluating P-queries. The second technique introduces auxiliary structures to improve the evaluation of D-queries. The third technique optimizes the line arrangement technique to improve the evaluation of S-queries. Our experimental results show that STARS$^+$ significantly outperforms the unoptimized STARS.

3.1   Dominating Tuple (DT) Optimization

In the STARS approach, each P-query (step 6 in `SkylineMaintenance` algorithm) to find the tuples that are exclusively dominated by an expiring skyline tuple is evaluated in terms of one D-query and multiple S-queries, which incurs a rather high computation overhead. One way to speed up a P-query evaluation is to reduce the number of S-query evaluations.

One approach to reduce the the number of S-query evaluations is to keep track of the number of tuples that dominate each tuple. This idea is referred to as the "eager approach" [12] in contrast to the non-optimized "lazy approach". Specifically, each tuple t is associated with a counter, denoted by $t.counter$, which represents the number of skyline tuples that dominate t. When t first arrives, $t.counter$ is initialized to the number of skyline tuples that dominate t. Subsequently, whenever a skyline tuple t_{out} expires, for each tuple t dominated by t_{out}, $t.counter$ is decremented by one to indicate that t is dominated by one fewer tuple due to the expiry of t_{out}. Clearly, if $t.counter > 0$, we can conclude that t is not exclusively dominated by t_{out} without requiring a S-query evaluation. While the advantage of the eager approach is that a P-query can be evaluated with significantly fewer S-queries, the drawback is that the initialization of $t.counter$ requires the entire skyline to be scanned when t arrives. In fact, the performance of the eager approach was shown to be worse than the lazy approach [12].

To avoid the overhead of the eager approach, we adopt a "semi-eager" approach for STARS$^+$ that simply associates each tuple t with a single skyline tuple, denoted by $t.dt$, that dominates t. The knowledge about this dominating tuple $t.dt$ is available virtually "for free" as part of the S-query issued to check if t is a skyline tuple when t arrives; thus, only a minor modification to the S-query evaluation procedure is needed to return a skyline tuple that dominates t when t is not a skyline tuple. Subsequently, whenever a skyline tuple t_{out} expires, for each tuple t dominated by t_{out}, if $t.dt$ is not equal to t_{out}, then t is not exclusively dominated by t_{out} and we save the cost of an S-query to determine this. However, if $t.dt$ is equal to t_{out}, an S-query is invoked to check if there is another skyline tuple (besides t_{out}) that dominates t. If there is indeed another tuple t' that dominates t, then we update $t.dt$ to be t' and conclude that t is not exclusively dominated by t_{out}.

The following result shows that our proposed DT optimization can significantly reduce the number of S-queries to be evaluated for a P-query.

Theorem 1. *Suppose that a newly arrived tuple t in the skybuffer is dominated by all the existing skyline tuples. Let s denote the number of the skyline tuples. If no new skyline tuple is encountered, then with the* DT *optimization, the total expected number of S-queries that are executed (to check if t should be promoted) due to the expiry of the s skyline tuples is bounded by $\Theta(\ln s)$. In contrast, the total number of such S-query evaluations for the lazy approach is s.*

3.2   Empty Cell (EC) Optimization

Recall that in STARS, a D-query wrt a tuple t is evaluated by examining all tuples in each candidate cell within the region specified by a range query wrt t. We observe that many candidate cells are empty, particularly for high-dimensional data as the number of grid cells grows exponentially with data dimensionality. Consequently, a large overhead is incurred in examining empty cells.

To get an idea of the sparsity of the grid cells, let us consider a d-dimensional dataset and a buffer size of N tuples. Assuming the attribute values are independent, the average number of tuples in the skybuffer is given by $O(\ln^d N)$ [7]. Let the granularity of each grid dimension be g (i.e., each dimension scale has g buckets). Then, the number of tuples per grid cell is given by $\rho = \frac{\ln^d N}{g^d}$. For high-dimensional data, ρ is often very small (e.g., $\rho = 0.022$ when $N = 10^5, d = 4$ and $g = 30$).

To reduce the overhead of examining empty grid cells when evaluating D-queries in a d-dimensional grid, the Empty Cell (EC) optimization technique maintains $d - 1$ additional structures, termed *index grids*, to keep track of the number of tuples in the grid. Each index grid C_i ($1 \le i \le d-1$) is i-dimensional, having the same scales as the first i dimensions of the original grid. All the cells in C_i have an initial value of 0. When a tuple is added to or removed from the buffer, EC-Indexing in Fig. 4(a) is invoked to update the index grids. During the evaluation of a D-query, the candidate cells are examined by enumerating the cell coordinates in a systematic manner: for each prefix of the d-length enumeration, STARS$^+$ invokes EC-Checking in Fig. 4(b) to check if the enumeration for the current prefix can be terminated due to an empty region. If a *true* value is returned, STARS$^+$ terminates further enumeration for the current prefix and backtracks.

Example 3. For a 3D grid, two index grids C_1 and C_2 are maintained. All of their cells are initialized to 0. Suppose a tuple is added to the buffer at $\langle 2, 5, 3 \rangle$. Then $C_1 \langle 2 \rangle$ and $C_2 \langle 2, 5 \rangle$ are updated to 1. A D-query evaluation starts enumerating the candidate cells to be examined with the enumeration prefix $\langle 1 \rangle$. Since $C_1 \langle 1 \rangle = 0$, STARS$^+$ terminates further enumeration with this prefix, and backtracks to the next prefix $\langle 2 \rangle$. Since $C_1 \langle 2 \rangle \ne 0$, STARS$^+$ continues the enumeration with the next dimension to consider $\langle 2, 1 \rangle$. Since $C_2 \langle 2, 1 \rangle = 0$, STARS$^+$ terminates further enumeration with $\langle 2, 1 \rangle$ and backtracks to $\langle 2, 2 \rangle$. Since $C_2 \langle 2, 2 \rangle = 0$, STARS$^+$ continues backtracking until $\langle 2, 5 \rangle$.   □

When a D-query evaluation is enumerating a prefix with i dimensions, there is a probability of $p_i = k^{g^{d-i}}$ that the enumeration will backtrack, where k is

Alg. (a): EC-Indexing $(\langle k_1, k_2 \ldots k_d \rangle, e)$

Input: $\langle k_1, k_2 \ldots k_d \rangle$ are the coordinates
in the skybuffer grid, where a tuple
is added or removed.
e indicates an add or remove event.

1) **for** $i = 1$ **to** $d - 1$ **do**
2)   **if** e is "add" **then**
3)   Increase $C_i \langle k_1, k_2 \ldots k_i \rangle$ by 1;
4)   **else**
        /* e is "remove" */
5)   Decrease $C_i \langle k_1, k_2 \ldots k_i \rangle$ by 1;
    endif
endfor

Alg. (b): EC-Checking $(\langle k_1, k_2 \ldots k_m \rangle)$

Input: $\langle k_1, k_2 \ldots k_m \rangle$ $(1 \leq m < d)$ is the
coordinates prefix in the skybuffer
grid, enumerated in a D-query.
Output: a boolean indicating if all cells with
coordinates prefix $\langle k_1, k_2 \ldots k_m \rangle$
are empty.

1) **for** $i = 1$ **to** m **do**
2)   **if** $C_i \langle k_1, k_2 \ldots k_i \rangle$ is 0 **then**
3)   **return** *true*
    endif
endfor
4) **return** *false*

Fig. 4. Empty Cell (EC) optimization

the average probability that a cell is empty. Therefore, a D-query evaluation *with* EC is expected to examine a fraction λ of the candidate cells, where $\lambda = \prod_{i=1}^{d-1}(1 - p_i) = \prod_{i=1}^{d-1}(1 - k^{g^{d-i}}) \leq 1 - k^g$. Suppose $k = 0.99$ and $g = 30$, then $\lambda < 0.26$. As d or g increases, k approaches 1, and so EC becomes more effective.

The overhead incurred by EC is low. The cost to update the index grids when a tuple is added or removed is $O(d)$ which is negligible since d is usually small. The space overhead for each index grid C_i is $O(g^i)$; thus, the total space requirement of $O(g^{d-1})$ is insignificant relative to the $O(g^d)$ space requirement of the original grid.

3.3 Geometric Arrangement (Minmax) Optimization

Our third optimization concerns the geometric arrangement technique for evaluating S-queries. In STARS, the two attributes used for line mapping are selected arbitrarily when data dimensionality is higher than two. To assess the performance impact of the choice of the attribute pair, we conducted an experiment to compare the performance of skyline maintenance for every possible attribute pair and found that the performance gap between the best and worst pair can exceed 20%. Thus, the choice of the attribute pair for the mapping is important but there is no clear heuristic that can be used to optimize this selection. Another drawback of STARS is that it utilizes only two attributes for the mapping. Intuitively, using more attributes is likely to provide better pruning power as more information about the data is being exploited.

We present an enhanced variant of the line mapping, termed Minmax, that utilizes all attributes. Consider a d-tuple $t = (a_1, \cdots, a_d)$. Minmax maps t to the line $y = C \cdot x - D$, where $C = \max(r(t.a_1), \cdots, r(t.a_d))$ and $D = \min(r(t.a_1), \cdots, r(t.a_d))$. The following result establishes its correctness.

Theorem 2. *Let l_1 and l_2 represent the two lines mapped from two d-tuples t_1 and t_2 based on Minmax, respectively. If l_1 and l_2 intersect at the point (x_I, y_I) where $x_I < 0$, then t_1 and t_2 are incomparable.*

4 SkyGrid Approach

In this section, we present a more extreme approach to optimize STARS by actually eliminating the use of the geometric arrangement technique for S-queries. Instead, all skyline maintenance operations are performed using only the grid data structure. To distinguish between the skyline and non-skyline tuples in the buffer, each tuple is associated with a single bit that is set to *true* iff the tuple is a skyline tuple. We refer to this new approach as SkyGrid.

The simplified skyline maintenance framework for SkyGrid is shown in Fig. 5. Clearly, using only a single data structure in SkyGrid simplfies the skyline maintanance operations. Recall that for both STARS and STARS$^+$ (refer to Fig. 1), if t_{in} is a skyline tuple, we need to update two structures with the following operations: (1) remove the line representations of any skyline tuples that are dominated by t_{in} (step 2); (2) insert the line representation of t_{in} (step 3); (3) remove the tuples in the buffer that are dominated by t_{in} (step 4); and (4) insert the t_{in} into the buffer (step 5). In contrast, for SkyGrid, if t_{in} is a skyline tuple, only the grid structure needs to be updated with the following operations (refer to Fig. 5): (1) insert t_{in} into the buffer (step 3); and (2) remove the tuples in the buffer that are dominated by t_{in} (step 4).

The simpler skyline maintenance operations in SkyGrid results in better performance. In STARS$^+$, the cost of inserting or removing the line representation of a skyline tuple in the geometric arrangement is $O(s)$ using DCEL, where s is the number of skyline tuples [4]. In contrast, for SkyGrid, the cost for promoting a tuple into the skyline is only $O(1)$ (by marking a skyline status bit).

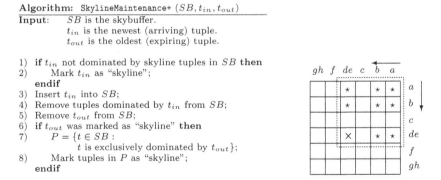

Algorithm: SkylineMaintenance+ (SB, t_{in}, t_{out})

Input: SB is the skybuffer.
              t_{in} is the newest (arriving) tuple.
              t_{out} is the oldest (expiring) tuple.

1) **if** t_{in} not dominated by skyline tuples in SB **then**
2)   Mark t_{in} as "skyline";
   endif
3) Insert t_{in} into SB;
4) Remove tuples dominated by t_{in} from SB;
5) Remove t_{out} from SB;
6) **if** t_{out} was marked as "skyline" **then**
7) $P = \{t \in SB :$
            t is exclusively dominated by $t_{out}\}$;
8)   Mark tuples in P as "skyline";
   endif

Fig. 5. Simplified skyline maintenance framework   **Fig. 6.** S-query in SkyGrid

To support S-queries, SkyGrid can simply find the candidate cells in a similar way as in D-queries, but in the opposite direction. Specifically, SkyGrid only needs to consider d-tuples $t' = (a'_1, \cdots, a'_d)$ that are located in the cells satisfying the following range query wrt $t = (a_1, \cdots, a_d)$: $r(a_1) \geq r(a'_1), \cdots$, and $r(a_d) \geq r(a'_d)$. The following example illustrates this idea.

Example 4. Consider the domain $\mathcal{D}$ depicted in Fig. 2(a). A grid to organize a 2D dataset on $\mathcal{D} \times \mathcal{D}$ is depicted in Fig. 6. Consider a tuple t that is mapped to the cell marked $\times$ in Fig. 6. The dotted region in Fig. 6, which corresponds to the range query wrt t, represents the set of cells that could contain tuples dominating t. The actual candidate cells for the S-query are marked by $\star$.   □

Next, we compare the pruning potential of STARS$^+$ and SkyGrid. As S-query is progressive in both, we ignore the progressiveness. Assuming independent attribute values, the following result states the expected pruning ratio of STARS$^+$.

Theorem 3. *STARS$^+$ (utilizing the Minmax mapping) is expected to prune fewer than half of the number of skyline tuples in an S-query evaluation.*

On the other hand, we expect SkyGrid to be able to prune more skyline tuples. While a formal computation is difficult as it depends on data domains, we can obtain an estimation. The evaluation of an S-query wrt to a tuple t examines a fraction $\prod_{i=1}^{d} k_i$ of all the cells as candidates, where $k_i \in [0, 1]$ is the fraction of buckets dominating t on each dimension. Hence $(1 - \prod_{i=1}^{d} k_i)$ of the cells are pruned. For high dimensional data with a reasonable value of k_i (e.g., $d > 2$ and $k_i < 0.7$), the estimated number of cells (and hence tuples) that are pruned is more than half.

To further improve performance, SkyGrid also incorporates both the DT and EC optimizations of STARS$^+$. Note that we can use two sets of EC index grids for the buffer and skyline, respectively. In this way, when evaluating an S-query, SkyGrid identifies candidate cells by utilizing only the index grids for the skyline. This avoids the need to examine most candidate cells that contain no skyline tuples, thereby reducing the number of skyline status bits that have to be checked.

In terms of space requirement, the cost for STARS$^+$ is $O(s^2)$ using DCEL to organize s skyline tuples as a geometric arrangement [4]. In contrast, since SkyGrid organizes the skyline tuples as part of the skybuffer, no additional space is required. However, each tuple in SkyGrid requires a skyline status bit, so an extra $O(s_b)$ space is needed, where s_b is the size of the skybuffer. When s is reasonably large, STARS$^+$ incurs a higher space overhead than SkyGrid.

5   Experimental Evaluation

5.1   Experiment Settings

In our experiments, we generated synthetic partially-ordered domains following the approach in [10]. Each domain is modeled as a DAG and is characterized by the parameters (m, h, c, f), where m is the number of vertices, h is the height of the DAG, $c \in (0, 1]$ is the fraction of the vertices at the next level that are connected to a vertex, and f refers to the type of DAG which is either "t" for tree-like or "w" for wall-like DAG. We refer to an attribute domain by these parameters; for instance, $(500, 8, 0.3, t)$.

We generated four 4-dimensional datasets shown in Table 1, where each column corresponds to one dataset and the ith row corresponds to the domain for the ith attribute; d-dimensional datasets, where $d \in \{2,3\}$, are generated from Table 1 by simply considering only the first d rows of the table. For each algorithm being evaluated, we ran it on each of the four datasets, and report the average performance over the four datasets (unless stated otherwise).

For each data domain, we also considered three different distributions: (1) *independent*, where the attribute values of the tuples follow a uniform distribution; (2) *correlated*, where a tuple that is good in one attribute also tends to be good in other attributes; (3) *anti-correlated*, where a tuple that is good in one attribute tends to be bad in at least one other attribute [1,5,11].

Table 1. Synthesized sets of data domains

Dataset I	Dataset II	Dataset III	Dataset IV
(250, 7, 0.3, t)	(120, 7, 0.2, t)	(100, 10, 0.1, w)	(500, 8, 0.3, t)
(180, 6, 0.6, t)	(120, 7, 0.2, t)	(100, 10, 0.2, w)	(500, 8, 0.3, t)
(180, 20, 0.3, w)	(120, 5, 0.2, t)	(100, 10, 0.4, w)	(500, 8, 0.3, t)
(90, 4, 0.2, t)	(120, 5, 0.2, t)	(100, 10, 0.8, w)	(500, 8, 0.3, t)

Table 2. Skyline sizes

Dim	Corr	Indep	Anti
$d = 2$	240	25	45
$d = 3$	395	480	418
$d = 4$	636	3779	4444
$d = 5$	1298	12363	15875

The number of data dimensions, denoted by d, was varied from 2 to 4. Table 2 shows the skyline sizes for datasets with domain $(500, 8, 0.3, t)$ on each attribute, using a 100K buffer with different data distributions. Note that datasets with partially-ordered domains have much more skylines than totally-ordered datasets since two tuples are more likely to be incomparable [1,11]. For independent or anti-correlated datasets, the size of the skylines becomes very large once $d \geq 5$; therefore, finding conventional skylines for $d \geq 5$ becomes less interesting [3].

Furthermore, we varied buffer sizes from 10K to 1M. Lastly, we chose $g = 20$ as the default grid granularity if it is not stated.

All the algorithms were implemented using Java. The experiments were conducted on a 3.0GHz PC with 3GB of main memory running Windows OS.

5.2 Evaluating STARS$^+$ Optimizations

In this subsection, we evaluate the effectiveness of each of the three optimizations DT, EC and Minmax that are introduced for STARS$^+$.

Dominating Tuple. STARS$^+$ utilizes DT to improve the performance of P-queries. Figure 7 shows the average time per P-query evaluation *without* DT (normalized wrt *with* DT)[1]. DT clearly improves the evaluation of P-queries, up to 2.3 times faster. The speed-up is greater when the skyline is larger, which is often caused by a larger buffer, particularly when $d = 4$. For $d \in \{2,3\}$, we notice a non-monotonous speed-up wrt buffer size. On lower dimensional datasets, the skyline sizes are much smaller. Their skylines soon become "saturated" (i.e., no longer growing and maybe shrinking due to randomness in data) when the buffers become larger. Hence, when the buffer increases beyond the saturation point, their skyline sizes become non-monotonous, resulting in the non-monotonicity of the performance speed-up by DT. When $d = 4$, the saturation point is well beyond 1M, so we only observe an improving speed-up.

---

[1] To be fair, both used the same attribute pair for line mapping.

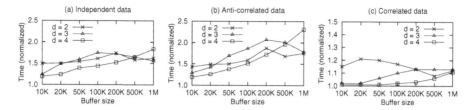

Fig. 7. Effectiveness of DT on P-query evaluation

Empty Cell. STARS$^+$ utilizes EC to improve the performance of D-query evaluation. The average time per D-query *without* EC (normalized wrt *with* DT)[1] is shown in Fig. 8. There is negligible improvement when $d = 2$, as the number of cells is small. However, when $d > 2$, EC becomes very effective. This is especially so for correlated data, where the tuples distribute unevenly in the grid resulting in more empty cells.

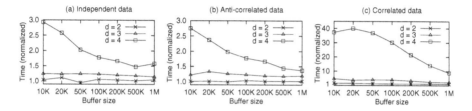

Fig. 8. Effectiveness of EC on D-query evaluation

Figure 9 compares the average evaluation time per D-query under varying grid granularity. When the granularity is initially increased from a small value, the performance of D-query evaluation both *with* and *without* EC improve due to a finer grid. However, as the granularity increases beyond 20, the performance without EC quickly deteriorates due to the rapid growth of the number of empty cells. In contrast, with EC the performance degradation is less pronounced, as most of the empty cells are pruned.

Pruning Efficiency of Minmax. STARS$^+$ utilizes Minmax to improve the skyline organization by pruning more skyline tuples in an S-query evaluation.

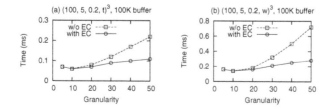

Fig. 9. Effect of grid granularity on D-query evaluation

Following [10], we define the pruning efficiency (PE) of an S-query as the fraction of skyline tuples that require dominance comparison (i.e., that are not pruned). Thus, smaller PE values are better. Figure 10 compares the PE of S-queries for STARS, STARS$^+$ and SkyGrid, where $d = 4$. For the performance results of STARS, instead of arbitrarily choosing two attributes for line mapping, we evaluated STARS with *all* possible attribute pairs, and present the performance results corresponding to the best pair (STARS-Best) as well as the worst pair (STARS-Worst) for comparison.

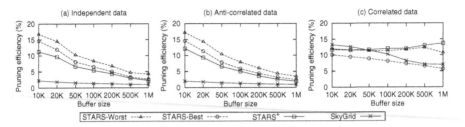

Fig. 10. Pruning efficiency of S-query

The results reveal that there could be a performance gap between STARS-Best and STARS-Worst. In Figs. 10(a) and (b), STARS$^+$ not only closes the gap, but also maintains a lead over STARS-Best. In addition, SkyGrid is much better than STARS$^+$ in terms of PE.

However, in Fig. 10(c) on correlated data, the PE of STARS$^+$ is generally on par with STARS-Worst. The reason is that the Minmax optimization in STARS$^+$ is not effective on correlated data. Consider two tuples with correlated attribute values that map to the lines $l_1 : y = C_1 \cdot x - D_1$ and $l_2 : y = C_2 \cdot x - D_2$, respectively. If $C_1 > C_2$, it is likely that $D_1 > D_2$; therefore, it is also likely that $\frac{D_1-D_2}{C_1-C_2} > 0$, the x-coordinate where l_1 and l_2 intersect. By Theorem 2, the two tuples are unlikely to be pruned. On the other hand, SkyGrid outperforms both Minmax and STARS-Worst, but loses marginally to STARS-Best on buffers larger than 50K. The reason is that tuples with correlated attribute values distribute unevenly in the grid, resulting in less efficient S-queries. However, SkyGrid still achieves the best overall performance despite this (see Section 5.3).

5.3   Evaluating Overall Performance

In this subsection, we compare the overall performance of STARS (both STARS-Best and STARS-Worst), STARS$^+$ and SkyGrid. To be fair to STARS and SkyGrid, we also implemented in them the two optimizations of STARS$^+$, DT and EC. Due to space constraints, we only present the results for $d = 4$; similar trends are observed for $d \in \{2, 3\}$.

Tuple update time. We measure the average time per tuple update, which corresponds to the time for one invocation of the `SkylineMaintenance` algorithm. The results are presented in Fig. 11.

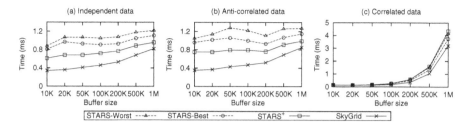

Fig. 11. Comparison of tuple update time with DT and EC

The results for independent and anti-correlated data in Figs. 11(a) and (b) reveal that SkyGrid achieves the best overall performance, followed by STARS$^+$, and lastly STARS. However, with large buffers, the performance gap between SkyGrid and STARS$^+$ narrows. The reason is that although the performance of SkyGrid for P-queries is much better than STARS$^+$, P-queries occur less frequently with large buffers due to the decreased probability for an expiring tuple to be a skyline tuple [10]. However, despite this, SkyGrid still performs better than STARS$^+$ by a clear margin with buffers as large as 1M. Note that SkyGrid is still preferable in time-critical applications, where the cost of each individual update is more important than the amortized cost. SkyGrid greatly improves the otherwise very expensive tuple updates that involve a P-query.

On correlated data, as shown in Fig. 11(c), STARS$^+$ is only marginally outperformed by STARS, although the former performs poorly in PE. Tuples with correlated attributes tend to distribute unevenly in the grid, resulting in less efficient D-queries. Thus, the performance of P-queries becomes a less dominating factor in the overall performance. Also, skylines on correlated data are generally smaller, resulting in a lower frequency of P-queries. So the PE of S-queries matters less to overall performance. This also explains why SkyGrid has the best overall performance even though it is not so in terms of PE.

Figure 12 studies the effect of grid granularity on the average time per tuple update of the three approaches with a 100K buffer. Observe that SkyGrid remains the best approach under different grid granularities.

Space requirement. The memory usage of the three approaches is shown in Table 3, with a 100K buffer on Dataset IV ($d = 4$). Clearly, SkyGrid uses the least

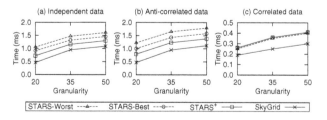

Fig. 12. Effects of granularity on tuple update time

Table 3. Memory usage

	Corr	Indep	Anti
Skyline	636	3779	4444
	Memory (MB)		
STARS-Worst	55	574	731
STARS-Best	56	505	693
STARS$^+$	55	442	589
SkyGrid	54	55	55

memory, as it requires little extra space for the skyline representation. On the other hand, STARS$^+$ and STARS require comparable amount of memory, since they both use a geometric arrangement to organize the skyline. Also note that the differences are insignificant on correlated data because of a smaller skyline.

6 Conclusion

In this paper, we have presented two new approaches, STARS$^+$ and SkyGrid, to compute skylines for streaming data that involves partially-ordered attribute domains. Our experimental results show that both STARS$^+$ and SkyGrid outperform the state-of-the-art STARS approach, with the surprisingly result that the simplest approach, SkyGrid is the best approach.

References

1. Börzsönyi, S., Kossmann, D., Stocker, K.: The skyline operator. In: ICDE, pp. 421–430 (2001)
2. Chan, C.-Y., Eng, P.-K., Tan, K.-L.: Stratified computation of skylines with partially-ordered domains. In: SIGMOD, pp. 203–214 (2005)
3. Chan, C.-Y., Jagadish, H.V., Tan, K.-L., Tung, A.K.H., Zhang, Z.: Finding k-dominant skylines in high dimensional space. In: SIGMOD, pp. 503–514 (2006)
4. de Berg, M., van Kreveld, M., Overmars, M., Schwarzkopf, O.: Computational Geometry: Algorithms and Applications, 2nd edn. Springer, Heidelberg (2000)
5. Kossmann, D., Ramsak, F., Rost, S.: Shooting stars in the sky: an online algorithm for skyline queries. In: VLDB, pp. 275–286 (2002)
6. Lee, K.C., Zheng, B., Li, H., Lee, W.-C.: Approaching the skyline in Z order. In: VLDB, pp. 279–290 (2007)
7. Lin, X., Yuan, Y., Wang, W., Lu, H.: Stabbing the sky: Efficient skyline computation over sliding windows. In: ICDE, pp. 502–513 (2005)
8. Papadias, D., Tao, Y., Fu, G., Seeger, B.: An optimal and progressive algorithm for skyline queries. In: SIGMOD, pp. 467–478 (2003)
9. Sacharidis, D., Papadopoulos, S., Papadias, D.: Topologically-sorted skyline for partially-ordered domains. In: ICDE (2009)
10. Sarkas, N., Das, G., Koudas, N., Tung, A.K.: Categorical skylines for streaming data. In: SIGMOD, pp. 239–250 (2008)
11. Tan, K.-L., Eng, P.-K., Ooi, B.C.: Efficient progressive skyline computation. In: VLDB, pp. 301–310 (2001)
12. Tao, Y., Papadias, D.: Maintaining sliding window skylines on data streams. IEEE TKDE 18(3), 377–391 (2006)

A Simple, Yet Effective and Efficient, Sliding Window Sampling Algorithm*

Xuesong Lu[1], Wee Hyong Tok[2,**], Chedy Raissi[3], and Stéphane Bressan[4]

[1] School of Computing, National University of Singapore
`xuesong@nus.edu.sg`
[2] Microsoft Corporation
`weetok@microsoft.com`
[3] School of Computing, National University of Singapore
`raissi@lirmm.fr`
[4] School of Computing, National University of Singapore
`steph@nus.edu.sg`

Abstract. Sampling streams of continuous data with limited memory, or reservoir sampling, is a utility algorithm. Standard reservoir sampling maintains a random sample of the entire stream as it has arrived so far. This restriction does not meet the requirement of many applications that need to give preference to recent data. The simplest algorithm for maintaining a random sample of a sliding window reproduces periodically the same sample design. This is also undesirable for many applications. Other existing algorithms are using variable size memory, variable size samples or maintain biased samples and allow expired data in the sample.

We propose an effective algorithm, which is very simple and therefore efficient, for maintaining a near random fixed size sample of a sliding window. Indeed our algorithm maintains a biased sample that may contain expired data. Yet it is a good approximation of a random sample with expired data being present with low probability. We analytically explain why and under which parameter settings the algorithm is effective. We empirically evaluate its performance (effectiveness) and compare it with the performance of existing representatives of random sampling over sliding windows and biased sampling algorithm.

1 Introduction

Sampling is a utility task used in diverse applications such as data mining [7,9,12], query processing [2], and sensor data management [4,11].

There are many different algorithms dedicated to effectively building a random sample of a small fixed dataset whose efficiency varies and depends on how data is stored and accessed. It is however less obvious to incrementally build the

---

* This work was partially funded under research grant number R-252-000-301-646 from NUS.
** Wee Hyong Tok contributed to this work when he was a graduate student at National University of Singapore.

H. Kitagawa et al. (Eds.): DASFAA 2010, Part I, LNCS 5981, pp. 337–351, 2010.

sample in a single pass or, more generally, to maintain a random sample when the dataset is updated. This is for instance the case when dealing with data streams. Reservoir sampling is a family of algorithms, first introduced by McLeod et al. in [8] and revisited by Vitter in [13], precisely able to incrementally sample a data set in one pass or, similarly, data stream. In reservoir sampling the first n points in the data set are stored at the initialization step. The $(t + 1)^{th}$ data replaces a randomly selected data in the reservoir with probability $\frac{n}{t+1}$. As the data set length increases, the probability of the insertion reduces to guarantee that every sample is equiprobable. The fact that reservoir sampling produces samples of the entire data set as opposed to samples of data in a sliding window is a clear disadvantage for data stream applications where users need to consider the most recent information.

A generalization of the problem considers both insertions and deletions. The algorithms in [5], *random pairing* and *resizing samples*, cater for such updates. Random pairing considers each insertion as a compensation of a previous deletion. In both algorithms samples are resized to maintain their randomness.

Users may also be concerned by the freshness of data. They may need to focus on most recent data or consider that data is expiring. The algorithm in [1], *biased sampling*, allows a bias towards recent data. We show here that this approach cannot be parameterized to become a good approximation of a sliding window.

The algorithms in [3], *simple window sampling* and *chain window sampling*, sample from a sliding window. Data outside the window, older data, expire. The simple algorithm is efficient but periodic. It lazily replaces data in the sample when they expire and therefore reproduces the same sampling design for each tumbling window. The chain sample algorithm does not suffer from periodicity. However, the needed memory (the size of the chains) is only statistically bound.

In this paper, we consider the problem of maintaining an approximate uniform random sample of a fixed specified size n over a data stream based on a sequence based *sliding window model*. The algorithm presented in this work belongs to the class of reservoir sampling. It is called FIFO (First In First Out) sliding window sampling, or FIFO for short. It maintains a sample of size n and requires a memory of size n. The algorithm is biased towards recent data. The algorithm however does not produce true random samples of the sliding window and may contain expired data. However, as we argue, it can be parameterized to produce almost random samples. It relies on a simple queue data structure. We present some analytical results and empirically and comparatively evaluate its effectiveness. We compare it with the main reservoir and sliding window reservoir algorithms. We conclude that FIFO is both efficient and effective.

The remainder of the paper is organized as follows. In Section 2 we present the main existing sampling algorithms in detail. In Section 3 we propose our novel algorithm FIFO and study it analytically. In Section 4 we empirically evaluate its performance (effectiveness) and compare it to that of the main algorithms. Finally, we draw our conclusions in Section 5.

2   Background and Related Work

In this section, we introduce a variety of existing sampling algorithms in detail.

2.1   Reservoir Algorithm

We introduce algorithm R and the updated version algorithm Z [13] in this section.

Algorithm R. Algorithm R works as follows: t denotes the number of the data in the dataset that have been processed. If $t+1 \leq n$, the $(t+1)^{th}$ data is directly inserted into the reservoir. Otherwise, the data is made a candidate and replaces one of the old candidates in the reservoir with probability $n/(t+1)$. The replaced data is uniformly selected from the reservoir.

Algorithm Z. The basic idea of Algorithm Z is to skip data that are not going to be selected, and rather select the index of next data. A random variable $\varphi(n,t)$ is defined to be the number of data that are skipped over before the next data is chosen for the reservoir, where n is the size of the sample and t is the number of data items processed so far. This technique reduces the number of data items that need to be processed and thus the number of calls to $RANDOM$ ($RANDOM$ is a function to generate a uniform random variable between 0 and 1).

2.2   Sampling with Updates

As we discussed above, the reservoir algorithm can only produce samples of the insertion-only dataset. In [5], two algorithms *"Random Pairing"* (RP) and *"Resizing Samples"* (RS) are proposed to cater for deletions.

Random Pairing. The basic idea behind random pairing is to avoid accessing the base data set by considering the new insertion as a *compensation* for the previous deletion. In the long term, every deletion from the data set is eventually compensated by a corresponding insertion. The algorithm maintains two counters c_1 and c_2, which respectively denote the numbers of uncompensated deletions in the sample S and in the base data set R. Initially c_1 and c_2 are both set to 0. If $c_1 + c_2 = 0$, the reservoir algorithm is applied. If $c_1 + c_2 \neq 0$, the new data has a probability $c_1/(c_1 + c_2)$ to be chosen for S; otherwise, it is excluded. Then c_1 or c_2 are modified accordingly. When the transaction consists of a sequence deletion of the wth before last element immediately compensated by an insertion, this is the case of a sliding window of size w, random pairing degenerates into the simple algorithm of [3].

Resizing Samples. The general idea of any resizing algorithm is to generate a sample S of size at most n from the initial dataset R and after some finite transactions of insertions and deletions, produce a sample S' of size n' from the new base dataset R', where $n < n' < |R|$. The proposed algorithm follows this general idea by using a random variable based on the binomial distribution.

2.3 Sampling Sliding Window

In [3], two types of sliding window are defined: (*i*) the *sequence-based window* and (*ii*) the *timestamp-based window*. In this paper, we focus on algorithms of the former type, which generate a sample of size n from a window of size w.

Simple Algorithm. The *simple algorithm* first generates a sample of size n from the first w data using the reservoir algorithm. Then, the window moves. The sample is maintained until a new coming data causes an old data in the sample to expire. The new data is then inserted into the sample and the expired data is discarded. This algorithm can efficiently maintain a uniform random sample of the sliding window. However, the sample design is reproduced for every tumbling window. If the ith data is in the sample for the current window, the $(i + cw)^{th}$ data is guaranteed to be included into the sample sometime in future, where c is an arbitrary integer constant.

Chain-sample Algorithm. The *chain-sample algorithm* generates a sample of size 1 for each chain. So in order to get our sample of size n, n chains need to be maintained. When the ith data enters the window, it is selected to be the sample with probability $\frac{Min(i,w)}{w}$. If the data is selected, the index of the data that replaces it when it expires is uniformly chosen from $i + 1$ to $i + w$. When the data with the selected index arrives, the algorithm puts it into the sample and calculates the new replacement index, etc. Thus, a chain of elements that can replace the outdated data is built.

2.4 Biased Reservoir Sampling

The authors of [1] propose a biased reservoir sampling algorithm. The bias is defined a priori by bias function that gives more recent data a higher probability to be put in the sample. The probability of the rth data included in the reservoir at the arrival of the tth data is proportional to the bias $f(r, t)$:

$$f(r, t) = e^{-\lambda(t-r)}, \tag{1}$$

where λ is the bias rate lying between 0 and 1.

The algorithm first maintains an empty reservoir of capacity $n = \lceil 1/\lambda \rceil$. Assume that at the arrival of the tth data, the fraction of the reservoir filled is $F(t)$. The $(t + 1)^{th}$ data is deterministically inserted into the reservoir and replaces one randomly selected old data with probability $F(t)$. Otherwise, no deletion occurs and the reservoir size increases by 1.

3 FIFO Sampling Algorithm

The goal of our work is to always extract a uniform random sample from the current window. Suppose the sample size is n and the window size is w. In this scenario, after the tth (we assume that $t > w$) data is processed, all the expired data which arrive before the $(t-w+1)^{th}$ point should have null probability to be

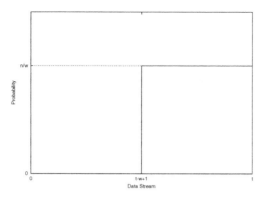

Fig. 1. The probability distribution of uniform sampling of the window. t is the number of processed data in the stream.

included in the sample, while all the data in the window should have a probability n/w to be included. We plot the ideal probability distribution in Figure 1. Below we compare the probability distributions of different stream sampling algorithms and show that FIFO best approximates it without reproducing its sample design.

The main idea of FIFO is that whenever a new data in the stream arrives, we insert it into the sample with a fixed probability p and simply discard the oldest data in the sample. Below we see that if we appropriately select the value of p, the algorithm can approximate a random sample. The complete algorithm is given in Algorithm 1.

Algorithm 1. FIFO Sampling Algorithm

Data: n : sample size, p : inclusion probability, DS : data stream
Result: The sample S

1 $S \leftarrow \{\}$;
2 *Insert sequentially the first n data of DS into S;*
3 **while** *NOT EndOFStream(DS)* **do**
4   *Randomly generate a number φ in the interval $[0, 1)$;*
5   **if** $\varphi < p$ **then**
6   *Insert the next data into S;*
7   *Discard the oldest data in S;*
8
9 **end**

3.1   Probability Analysis

In this section, we analyze the probability distribution of FIFO. We start by giving the probability formula of each data to be included in the sample. Obviously, a data is contained in the sample at a certain point if and only if (a)

the data is selected into the sample when it is processed and (*b*) it has not been discarded. Because FIFO always discards the oldest data in the sample, a newly inserted data is not replaced out until n of its subsequent data have entered the sample, that is, if the rth data is inserted into the sample and less than n data are selected into the sample from the $(r+1)^{th}$ data to the tth data, the rth data is in the sample after the tth data is processed. Denote by n the sample size and by $P_{(r,t)}$ the probability that the rth data is in the sample after t data have been processed, where $t \geq r$. Without loss of generality, we assume that $t \gg n$. We divide the stream into three intervals: (*i*)[1, n]; (*ii*)[n+1, t-n]; (*iii*)[t-n+1, t]. Note that the probability distributions in these three intervals are different.

When $1 \leq r \leq n$, the case is slightly different. Initially the first n data are sequentially inserted into the sample. So for the first data, $n-1$ of its successors have been inserted into the sample. It is discarded once there is one more data to be selected into the sample from the $(n+1)^{th}$ data to the tth data. For the second data, two more insertions after the first n insertions cause it to be discarded, etc. Thus we can derive the probability formula for the first n data, which is:

$$P_{(r,t)} = \sum_{k=0}^{r-1} Binomial(k; t-n, p), \qquad (2)$$

where the function $Binomial()$ represents the Binomial distribution [10]. A binomial function $Binomial(k; m, p_0) = \binom{m}{k} p_0^k (1-p_0)^{m-k}$ calculates the probability that an event happens for k times in m tests, with a probability of p_0 each time. In the above formula, $t-n$ is the total number of remaining data in the stream, k is the number of data selected for the sample and p is the probability.

When $n+1 \leq r \leq t-n$, the situation is similar to the case that we discussed above except that each data initially enters the sample with probability p. So the formula is:

$$P_{(r,t)} = p \times \sum_{k=0}^{n-1} Binomial(k; t-r, p). \qquad (3)$$

The last n data are not replaced out once they are selected into the sample, as each of them has less than n successors. Thus the formula is trivial:

$$P_{(r,t)} = p. \qquad (4)$$

The complete probability formulae are as follows:

$$P_{(r,t)} = \begin{cases} \sum_{k=0}^{r-1} B(k; t-n, p) & \text{for} \quad 1 \leq r \leq n \\ p \sum_{k=0}^{n-1} B(k; t-r, p) & \text{for} \quad n+1 \leq r \leq t-n \\ p & \text{for} \quad t-n+1 \leq r \leq t, \end{cases} \qquad (5)$$

where $B()$ is the abbreviation of $Binomial()$.

Figure 2 shows the probability distributions obtained using the above formulae. In the figure, we vary the value of p and plot the probability for each data

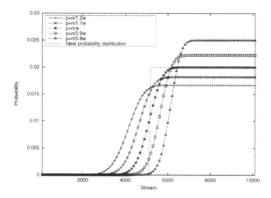

Fig. 2. Probability of each data to be included at $t = 10,000$ for varying p. $n = 100$, $w = 5,000$.

to be finally included in the sample at a given time t. The X-axis represents the variable r and the Y-axis is the probability $P_{(r,t)}$. The number of processed data in the figure is 10,000, the window size is 5,000 and the sample size is 100. We also plot the corresponding ideal distribution.

In the figure, the line-point graphs show the probability distributions for different values of p according to our analysis, and the dashed graph shows the ideal probability distribution as discussed above. From the graphs, we can see that the probability distribution of $p = n/w$ seems best approximate the ideal distribution. Expired data have lower probabilities to be included in the sample as their age increases while most data in the window have near equi-probability n/w to be selected. The figure suggests that the optimum of this situation is obtained near $p = n/w$. Although equi-probability is only a necessary condition for a sampling algorithm to generate random samples, no further dependency being imposed, it is clear that it is a sufficient condition for FIFO.

In Figure 3, we show, for given n, w and p ($p = n/w$), the probability distributions for selected varying values of t. The graphs being parallel indicates that the same effectiveness is maintained for the successive windows.

3.2   Optimal Selection Probability

One important question is the choice of the optimal value of the probability p. That is the value of p that yields the best approximation of a random sampling of the sliding window. We estimate this value in two different ways: analytically and empirically. On Figure 2, we see the probability distributions for different values of p. $p = n/w$ seems to be the optimal value. Let us confirm this.

We can estimate the optimal value of the probability p by comparing the difference D between the distribution of our algorithm for various values of p and the ideal distribution:

$$D = \sum_{r=1}^{t} |P_{(r,t)} - P_{(r)}|, \tag{6}$$

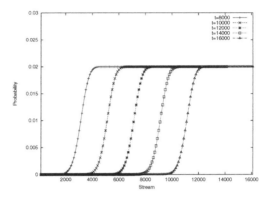

Fig. 3. Probability distributions for varying values of t

where $P_{(r,t)}$ can be calculated by Equation 5 and $P_{(r)}$ is the probability in theory defined as follows:

$$P_{(r)} = \begin{cases} 0 & \text{if } r \leq t - w \\ n/w & \text{else.} \end{cases} \tag{7}$$

Empirically, the optimal value of the probability p should coincide with the smallest value D.

By experiments, we find that for a given pair of w and n, D always reaches the minimum value when the inclusion probability is set to be n/w. Thus we believe that the optimal probability should be $\frac{n}{w} \pm d$, where d is a very small real number for adjustment. Figure 4 and Figure 5 show the results on two sets of parameters.

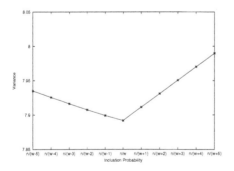

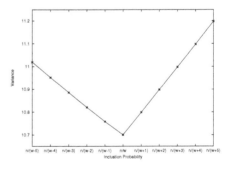

Fig. 4. Sample 100 data with window size of 5,000 from a 10,000 data stream

Fig. 5. Sample 200 data with window size of 2,000 from a 20,000 data stream

We can also calculate the probability p from another point of view. Recall that in Figure 2 and 3, that the probability for a data to be included in the sample always firstly increases fast as t gets larger. After a certain point, the increase ratio becomes smaller and smaller and eventually the probability stays at p. This point is called an inflexion point. For a fixed pair of n and w, we believe that the best approximation of a random sample is obtained when the inflexion point coincides with $t - w + 1$.

Because the distribution is discrete, we cannot calculate the inflexion point by a classical derivation. However, we can use the following approach. Let $d_{(r,t)}$ be the difference between probabilities of the rth data and the $(r + 1)^{th}$ data in the sample after the processing of the tth data, that is, $d_{(r,t)} = P_{(r+1,t)} - P_{(r,t)}$. We also define $\Delta d_{(r,t)} = d_{(r,t)} - d_{(r-1,t)}$. If r is an inflexion point, we have that $\Delta d_{(r-1,t)} \geq 0$ and $\Delta d_{(r,t)} \leq 0$. By replacing with the formulae in Section 3 (we assume that $n + 1 \leq r \leq t - n$ for the general case), we get:

$$d_{(r,t)} = p \sum_{k=0}^{n-1} [B(k; t - r - 1, p) - B(k; t - r, p)]. \tag{8}$$

A result in the case of a *Binomial Distribution* [10] states that if $X \sim B(x, p)$ and $Y \sim B(y, p)$ then $X + Y \sim B(x + y, p)$. Thus:

$$B(k; t - r, p) = B(k; t - r - 1, p)B(0; 1, p) + B(k - 1; t - r - 1, p)B(1; 1, p)$$
$$= (1 - p)B(k; t - r - 1, p) + pB(k - 1; t - r - 1, p). \tag{9}$$

By taking Equation 9 into Equation 8, we have:

$$d_{(r,t)} = p^2 \sum_{k=1}^{n-1} [B(k; t - r - 1, p) - B(k - 1; t - r - 1, p)]$$
$$+ p[B(0; t - r - 1, p) - B(0; t - r, p)]$$
$$= p^2 B(n - 1; t - r - 1, p). \tag{10}$$

Thus,

$$\Delta d_{(r,t)} = p^2 [B(n - 1; t - r - 1, p) - B(n - 1; t - r, p)]. \tag{11}$$

We can use Poisson distribution, $Poisson(k_0, \lambda) = \lambda^{k_0} e^{-\lambda}/k_0!$ to approximate the Binomial distribution $Binomial(k_0; m, p_0)$, if m is sufficiently large and p_0 is sufficiently small, where $\lambda = mp_0$ [10]. Our formulae satisfy the constraint, so we get:

$$\Delta d_{(r,t)} = p^2 \{ \frac{[(t - r - 1)p]^{n-1} e^{-(t-r-1)p}}{(n - 1)!} - \frac{[(t - r)p]^{n-1} e^{-(t-r)p}}{(n - 1)!} \}$$
$$= \frac{p^{(n+1)} e^{-(t-r-1)p}}{(n - 1)!} [(t - r - 1)^{n-1} - \frac{(t - r)^{n-1}}{e^p}]. \tag{12}$$

Let $\Delta d'_{(r,t)} = (t - r - 1)^{n-1} - \frac{(t-r)^{n-1}}{e^p}$. Obviously, $\Delta d_{(r,t)}$ and $\Delta d'_{(r,t)}$ are of the same positive and negative shape. So if r is an inflexion point, we have

$\Delta d'_{(r-1,t)} \geq 0$ and $\Delta d'_{(r,t)} \leq 0$. Because we empirically claim that the inflexion point coinciding with $t - w + 1$ is the optimum, we replace r by $t - w + 1$, so we get:

$$\begin{cases} (w-1)^{n-1} - \frac{w^{n-1}}{e^p} & \geq & 0 \\ (w-2)^{n-1} - \frac{(w-1)^{n-1}}{e^p} & \leq & 0. \end{cases} \tag{13}$$

Thus we get a bound of p, which is,

$$(n-1)\ln\frac{w}{w-1} \leq p \leq (n-1)\ln\frac{w-1}{w-2}. \tag{14}$$

Table 1 illustrates the previous result. We can find that this bound for the

Table 1. Some bounds calculated by specified n and w

n	w	Lower Bound	Upper Bound	$\frac{n}{w}$
500	2000	0.249562	0.249687	0.25
100	2000	0.0495124	0.0495372	0.05
1000	5000	0.19982	0.19986	0.2
1000	12000	0.0832535	0.0832604	0.0833333

optimal probability is close to n/w, which coincides with the results of the first approach. Thus, we believe that the probability n/w is a good approximation to the optimal probability.

4   Performance Evaluation

In this section, we compare FIFO's performance with that of the various sampling algorithms discussed. We compare the analytical bias function and empirically compare the distribution divergence.

The results confirm the inappropriateness of the reservoir, resizing sample and biased sampling algorithm for the problem of sliding windows. They also show that, in practice FIFO performs as effectively as the simple sliding window sampling (and random pairing which degenerates into simple random sampling).

4.1   Comparison of Analytical Bias Functions

In this section, we analytically compare the bias function of FIFO with the optimal probability p with the bias functions of the simple algorithm and the biased reservoir sampling algorithm, respectively. We show that FIFO approximates a random sample without a predictable sample design.

In the simple algorithm, expired data are discarded and all the data in the sample come from the current window. As the first w data are processed using the reservoir sampling algorithm, each of the data have a probability n/w to be included. Thus, data in the current window also have the optimal inclusion probability. However, the simple algorithm is periodical.

In the biased sampling algorithm [1], the probability of the rth data in the stream being included in the sample after the processing of the tth data is $P_{(r,t)} = e^{-(t-r)/n}$, where n is the sample size. Figure 6 and Figure 7 show the results based on two sets of parameters. In Figure 6, the case is generating a sample of size 200 from a set of 5,000 data with window size 1,000. Figure 7 shows the result of sampling from a dataset of 10,000 data with window size being set to 5,000 and sample size being set to 500. In both figures, we also plot the probability distribution of reservoir sampling. The results demonstrate that the inclusion probability of biased sampling algorithm suddenly increases to 1 when the stream is close to the end. Data with a little distant history in the window has a very small probability to be included. Obviously, FIFO approximates the ideal distribution better than the biased reservoir sampling algorithm does.

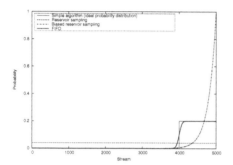
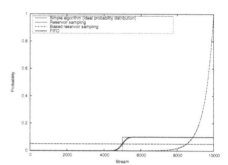

Fig. 6. Probability distributions of different sampling algorithms. $n = 200$, $w = 1,000$, $p = n/w$.

Fig. 7. Probability distributions of different sampling algorithms. $n = 500$, $w = 5,000$, $p = n/w$.

4.2 Empirical Performance Evaluation: Setup

We compare our implementation of algorithm R, algorithm Z, Random pairing (RP), the simple algorithm, the chain-sample algorithm, the biased reservoir algorithm and FIFO. We consider a sliding window on the stream.

The empirical performance evaluation uses the *Jensen-Shannon Divergence* to quantify the difference between the distributions of the data in the sliding window and in the sample. A small Jensen-Shannon divergence indicates similar distributions. For two distributions $P = \{p_1, p_2 \ldots p_n\}$ and $Q = \{q_1, q_2 \ldots q_n\}$, Jensen-Shannon divergence measures their similarity as follows:

$$D_{JS}(P\|Q) = \frac{1}{2}D_{KL}(P\|M) + \frac{1}{2}D_{KL}(Q\|M), \tag{15}$$

where $M = \frac{1}{2}(P+Q)$, D_{KL} is the *Kullback-Leibler Divergence*, which is defined as follows:

$$D_{KL}(P\|Q) = \sum_{i=1}^{n} p_i \log(p_i/q_i). \tag{16}$$

We use Jensen-Shannon divergence to measure the similarity of distributions of the successive samples with the distribution in the successive sliding windows.

We empirically evaluate the performance of the algorithms with both synthetic and real datasets.

4.3 Empirical Performance Evaluation: Synthetic Data

The synthetic dataset that we used is a set of 1,000,000 integers with values ranging from 1 to 10 chosen from a **Zipfian** distribution. In order to create changes in the distribution, we shuffle the distribution every 100,000 data. The sample size is fixed to 1,000 and the window size is 50,000. We plot the graph of the Jensen-Shannon divergence for each algorithm.

The results are shown in Figure 8. As we discussed above, algorithm R and Z produce successive samples of the entire dataset, but not of the sliding windows. As expected, the Jensen-Shannon divergence increases. Indeed the sample in R and Z is representative of the entire stream so far and therefore diverges from the distribution in the window that contains only the most recent data. The reshuffling corresponding to the changes in distribution are clearly visible on the plot. The chain-sample algorithm produces low Jensen-Shannon divergence values. The biased reservoir sampling algorithm generates very high peaks at the interfaces of two intervals with different distributions. It is too sensitive to the changes, as expected as well. The simple algorithm performs the best in the above algorithms. The RP algorithm degenerates into the simple algorithm. Our FIFO algorithm performs similarly to the simple algorithm, which can be seen clearly in Figure 9.

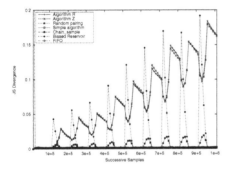

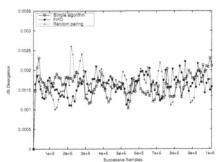

Fig. 8. The Jensen-Shannon divergence values of successive samples and sliding windows

Fig. 9. Comparison of FIFO, RP and simple algorithm, 10 datasets of value range from 1 to 10

We then extend our experiment to further compare the performances of FIFO with that of the simple algorithm. Figure 10 and Figure 11 show the results of the experiments on 100 datasets containing integers with values ranging from 1 to 100 and 100 datasets containing integers with values ranging from

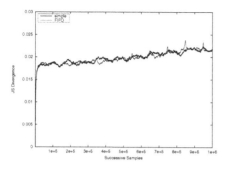
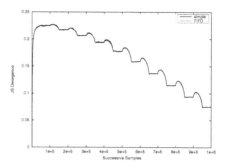

Fig. 10. Comparison of FIFO and simple algorithm, 100 datasets of value range from 1 to 100.

Fig. 11. Comparison of FIFO and simple algorithm, 100 datasets of value range from 1 to 1,000.

1 to 1,000 respectively. The reader remembers that, although the two algorithms perform equally well, the simple algorithm sample design is periodical.

We also confirm the optimal value for p by evaluating FIFO's performance by setting different inclusion probabilities. Figure 12 and Figure 13 show the results on the 10 datasets of value from 1 to 10 and 100 datasets of value from 1 to 100 respectively. From the figure, we can see that the optimal probability should near n/w, which coincides with the results in Figure 2.

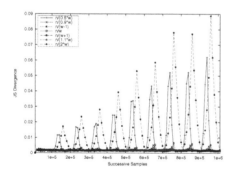
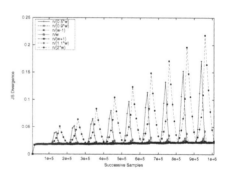

Fig. 12. Comparison of FIFO with different inclusion probabilities, 10 datasets of value range from 1 to 10, $w = 50,000$, $n = 1,000$.

Fig. 13. Comparison of FIFO with different inclusion probabilities, 100 datasets of value range from 1 to 100, $w = 50,000$, $n = 1,000$.

4.4   Empirical Performance Evaluation: Real Data

The real life dataset that we used in the experiments is the weather data collected at [6] which records the surface synoptic weather information for the entire globe from December 1981 to November 1991. The reports come from land stations and ships located in particular positions of the earth. We use the reports from land stations in 1990, and we are interested in the current weather attribute which has a domain of 48 integer values. The data is sorted in chronological order. The total number of the records is $1,344,024$. We set the sample size to be 1,000 and the window size to be 50,000. We still calculate successive Jensen-Shannon divergence values as in the synthetic dataset experiment to evaluate the performances of the algorithms. The results are shown as Figure 14. We can see that algorithm R, algorithm Z and the biased reservoir sampling algorithm produce higher Jensen-Shannon divergence values. The other four algorithms perform relatively better. However, RP degenerates into the simple algorithm. In Figure 15, we can see their performances more clearly. The chain-sample algorithm produces a slightly higher Jensen-Shannon divergence than FIFO and the simple algorithm. FIFO and the simple algorithm are still the best.

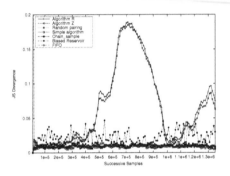 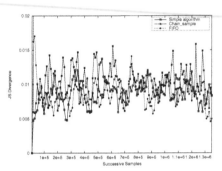

Fig. 14. Performance evaluation on real life dataset

Fig. 15. Performance evaluation on real life dataset

5   Conclusions

In this paper, we propose a new sampling algorithm, called FIFO sliding window sampling or FIFO, for short, for sampling sliding windows over data stream. We compare its performance to that of existing stream and sliding window sampling algorithms. We analyze the properties of FIFO analytically and empirically to show that our new algorithm is effective: it can maintain a near random sample of a sliding window with fixed memory and without reproducing its sample design. FIFO is also very efficient as it only maintains a queue and can be straightforwardly further improved by skipping data, as in algorithm Z. It is therefore able to process high arrival rate data streams. Although we have shown empirically and argued analytically that FIFO is most effective for p near n/w, we are now trying to obtain an exact analytical formula for this optimum.

References

1. Aggarwal, C.C.: On Biased Reservoir Sampling in the Presence of Stream Evolution. In: VLDB 2006, pp. 607–618 (2006)
2. Babcock, B., Chaudhuri, S., Das, G.: Dynamic Sample Selection for Approximate Query Processing. In: SIGMOD Conference 2003, pp. 539–550 (2003)
3. Babcock, B., Datar, M., Motwani, R.: Sampling from a moving window over streaming data. In: SODA 2002, pp. 633–634 (2002)
4. Considine, J., Li, F., Kollios, G., Byers, J.W.: Approximate Aggregation Techniques for Sensor Databases. In: ICDE 2004, pp. 449–460 (2004)
5. Gemulla, R., Lehner, W., Haas, P.J.: A Dip in the Reservoir: Maintaining Sample Synopses of Evolving Datasets. In: VLDB 2006, pp. 595–606 (2006)
6. Hagn, C.J., Warren, S.G., London, J.: Edited synoptic cloud reports from ships and land stations over the globe, 1982-1991 (1996),
   http://cdiac.esd.ornl.gov/ftp/ndp026b
7. John, G.H., Langley, P.: Static Versus Dynamic Sampling for Data Mining. In: KDD 1996, pp. 367–370 (1996)
8. McLeod, A.I., Bellhouse, D.R.: A convenient algorithm for drawing a simple random sample. Appl, Stat., pp. 182–184 (1983)
9. Raissi, C., Poncelet, P.: Sampling for Sequential Pattern Mining: From Static Databases to Data Streams. In: ICDM 2007, pp. 631–636 (2007)
10. Ross, S.: A first course in probability, 5th edn. (1997)
11. Silberstein, A., Braynard, R., Ellis, C.S., Munagala, K., Yang, J.: A Sampling-Based Approach to Optimizing Top-k Queries in Sensor Networks. In: ICDE 2006, p. 68 (2006)
12. Toivonen, H.: Sampling Large Databases for Association Rules. In: VLDB 1996, pp. 134–145 (1996)
13. Vitter, J.S.: Random sampling with a reservoir. ACM Transactions on Mathematical Software 11(1), 37–57 (1985)

Detecting Leaders from Correlated Time Series

Di Wu[1], Yiping Ke[1], Jeffrey Xu Yu[1], Philip S. Yu[2], and Lei Chen[3]

[1] The Chinese University of Hong Kong
{dwu,ypke,yu}@se.cuhk.edu.hk
[2] University of Illinois at Chicago
psyu@cs.uic.edu
[3] The Hong Kong University of Science and Technology
leichen@cse.ust.hk

Abstract. Analyzing the relationships of time series is an important problem for many applications, including climate monitoring, stock investment, traffic control, etc. Existing research mainly focuses on studying the relationship between a pair of time series. In this paper, we study the problem of discovering leaders among a set of time series by analyzing lead-lag relations. A time series is considered to be one of the leaders if its rise or fall impacts the behavior of many other time series. At each time point, we compute the lagged correlation between each pair of time series and model them in a graph. Then, the leadership rank is computed from the graph, which brings order to time series. Based on the leadership ranking, the leaders of time series are extracted. However, the problem poses great challenges as time goes by, since the dynamic nature of time series results in highly evolving relationships between time series. We propose an efficient algorithm which is able to track the lagged correlation and compute the leaders incrementally, while still achieving good accuracy. Our experiments on real climate science data and stock data show that our algorithm is able to compute time series leaders efficiently in a real-time manner and the detected leaders demonstrate high predictive power on the event of general time series entities, which can enlighten both climate monitoring and financial risk control.

1 Introduction

In the literature, the lagged correlation between two streams has been well studied in empirical research [5,1,12] and efficient algorithms to discover lagged correlations have also been developed [13]. However, the study on summarizing the relationships across multiple data streams is still lacking. The comprehensive relationships among multiple data streams are very helpful in many applications to monitor and control the overall movement of the entity where the data streams are generated. Two application examples are given as follows.

Earth Science: In climate teleconnection network, each stream represents the weather observations (e.g., temperature, pressure and precipitation) [16] of a specific point on the latitude-longitude spherical grids. The lagged correlation between two streams indicates that the weather change in one location can affect the weather in another location with some time delay. By analyzing lead-lag on observations in multiple locations, the earth scientists can understand better from which location the climate phenomena originates and how it evolves.

H. Kitagawa et al. (Eds.): DASFAA 2010, Part I, LNCS 5981, pp. 352–367, 2010.

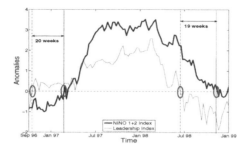

Fig. 1. Leadership Index VS. NINO 1+2 Index from 1996-1999

Finance: The stock market can be modeled as a financial network, in which each stream represents the price of a stock. The lead-lag effect between two streams implies that the price change of one stock influences that of another [12]. In finance crisis, when the market goes down dramatically and the government plans to launch finance bailout, the regulators desire to know the subset of stocks which poses risks (influences) on others and triggers the movement of the whole market. They can then apply a program to these market leaders and control the overall systemic risk.

In this paper, we study the problem of discovering leaders among a set of time series by analyzing lead-lag relations. We target to extract leaders from multiple time series in a real-time manner. Here, we demonstrate the significance of the problem and the usefulness of the discovered leaders on a real climate dataset. We analyze the streams of the sea surface temperature (SST) on the Pacific ocean ($30°S - 30°N$, $55°E - 80°W$) where the famous Nino phenomena occurs irregularly every 4-5 years. We study a period from 1996-1999. In Fig. 1, the bold blue line shows the weekly NINO 1+2 index which is a standard climate index developed by earth scientists to study SST anomalies in a Nino region off the coast of Peru. A positive value of the index indicates significant anomalies. As shown in the figure, the NINO 1+2 index begins to increase in January 1997 and goes above 0 in March 1997. Later, it begins to drop and eventually falls below 0 in November 1998. On the other hand, we sample 125 streams of SST time series from that region and extract weekly leaders from them. We then form a leadership index using the extracted leaders weighted by their normalized leadership scores. The red line in Fig. 1 gives the leadership index which exhibits a similar but earlier trend to NINO 1+2 index. It begins to increase in September 1996 and rises above 0 in October 1996, which is 20 weeks earlier than NINO 1+2 index. Later, it falls below 0 in July 1998, which is 19 weeks earlier. To further confirm the relationship between the two indices, we conduct a Granger-causality analysis [7] by performing F-test on the lagged value of both indices. After selecting the optimal lagged value for the regression model (lag = 2 for NINO 1+2 index and 1 for leadership index), the result suggests that the leadership index Granger-causes NINO index (the F-Statistics is 6.64) while NINO index does not Granger-cause leadership index (the F-Statistics is statistically insignificant).

Through this example and many other experimental results, we find that the discovered leaders are able to bring enlightening information. First, leaders are good representatives of the whole entity. An event usually introduces some changes to leaders, whose effect then propagates to related time series. As a result, analysts only need to monitor and analyze leaders in order to evaluate the overall entity movement triggered by events. Second, since the leadership is defined by the lagged correlation, leaders have the predictive power within the computed lag as shown in Fig. 1. Therefore, analyzing leaders can detect the trend of an event at an early stage. In climate observation and control, this predictive power is very helpful in giving the scientists an early alert on the climate phenomena and allowing them to do better preventions for the coming disasters.

The problem of finding the leaders among multiple time series poses great challenges. First, the observations of time series (e.g., temperature, intra-day stock price) usually change rapidly over time, which implies that the leaderships among them may also change from time to time. Therefore, the lagged correlations between pairs of time series, which are used for leadership identification, must be re-computed for every new time tick, while the correlation computation at each time tick is already costly. This high computational complexity makes the design of an efficient solution difficult. Second, after computing the lagged correlation between each pair of streams, how to define and extract useful leaders out of the whole set of time series is also a big challenge.

In this paper, we propose an efficient streaming algorithm to address the problem. The main contributions of the paper are summarized as follows. First, we formalize a new problem of discovering the leadership among multiple time series, which well captures the overall co-movements of time series. Second, we devise an efficient solution that discovers the leaders in a real-time manner. Our solution utilizes an effective update strategy, which significantly reduces the computational complexity in a stream environment. Third, we justify the efficiency of our solution, the effectiveness of our update strategy, as well as the usefulness of the discovered leaders by conducting extensive experiments over the real climate data and financial data.

The rest of the paper is organized as follows. Section 2 gives the preliminaries. Section 3 defines the problem of leadership discovery and discusses the main idea of our solution. Section 4 presents the incremental correlation update strategy. Section 5 reports the performance evaluation. Finally, Section 6 reviews some related work and Section 7 concludes the paper.

2   Preliminaries

We consider a set of N synchronized time series $\{S^1, S^2, \ldots, S^N\}$, where each time series $S^j = (s_1^j, \ldots, s_t^j)$ is a sequence of discrete observations over time, and s_t^j is the value of S^j at the most recent time point t. Given a length w and a time point t, a sliding window for time series S^j, denoted as $s_{t,w}^j$, is the subsequence $(s_{t-w+1}^j, \ldots, s_t^j)$. And the lagged correlation between two sliding windows $s_{t,w}^i$ and $s_{t,w}^j$ of two time series S^i and S^j at lag l, denoted as $\rho_{t,w}^{ij}(l)$, is computed by considering the common parts of the shifted sequences:

$$\rho_{t,w}^{ij}(l) = \begin{cases} \dfrac{\sum_{\tau=t-w+1}^{t-l} (s_{\tau+l}^i - \overline{s_{t,w-l}^i})(s_\tau^j - \overline{s_{t-l,w-l}^j})}{\sigma_{t,w-l}^i \sigma_{t-l,w-l}^j}, & l \geq 0; \\ \rho_{t,w}^{ji}(-l), & l < 0, \end{cases} \qquad (1)$$

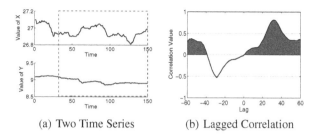

(a) Two Time Series   (b) Lagged Correlation

Fig. 2. Two Time Series and the Lagged Correlation Plot over their Local Sliding Windows

where $\overline{s_{t,w-l}^i}$ and $\overline{s_{t-l,w-l}^j}$ are the mean values in the shifted sliding windows $s_{t,w-l}^i$ and $s_{t-l,w-l}^j$, and $\sigma_{t,w-l}^i$ and $\sigma_{t-l,w-l}^j$ are the standard deviations. In particular, $\rho_{t,w}^{ij}(0)$ is the correlation with zero lag (known as the local Pearson's correlation [11]). When $l > 0$, $\rho_{t,w}^{ij}(l)$ denotes the correlation between the sliding windows $s_{t,w}^i$ and $s_{t,w}^j$ by delaying S^i with a lag l. The case when $l < 0$ can be easily handled symmetrically. Since $\rho_{t,w}^{ij}(l)$ is computed on the common parts of two windows, l is less than the window length w, and in practice $|l| \leq w/2$ as suggested in [2]. In a stream context, it is not desirable to compute $\rho_{t,w}^{ij}(l)$ from scratch at each time point t. As shown in [18,13], the lagged correlation can be computed efficiently by tracking the following statistics: the inner product, the sum of squares and the sum of the shifted windows $s_{t,w-l}^i$ and $s_{t-l,w-l}^j$.

3   Leadership Discovery

In this section, we first define the problem of leadership discovery.

Problem Definition. The problem of leadership discovery is to find the leaders among N synchronized time series, $S^1, S^2, \ldots, S^N$, that exhibit significant lead-lag relations over the set of time series in a real-time manner, where the lead-lag relation is measured by the concept of lagged correlation.

Solution Overview. Our solution to the problem of leadership discovery has three main steps: (1) compute the lagged correlation between each pair of time series; (2) construct an edge-weighted directed graph based on lagged correlations to analyze the lead-lag relation among the set of time series; (3) detect the leaders by analyzing the leadership transmission in the graph. We now discuss each step in detail.

3.1   Lagged Correlation Computation

The first step is to compute the lagged correlation between each pair of time series. Existing work [13] on computing lagged correlations cannot be directly applied to our problem, since i) it tries to capture lag correlation in the whole history of streams while our objective is to obtain the local lags in the current sliding window, and ii) the approximation in their updating algorithm has accuracy preference to the points with small lags and may generate a large error for large lags, which is not desirable for our problem.

Therefore, we propose to aggregate the effects of various lags and define an *aggregated lagged correlation*. Without loss of generality, we focus on positive correlation, while negative correlation can be handled similarly. We explain how to compute the aggregated lagged correlation by the following example. Fig. 2(a) shows two time series X (top) and Y (bottom) with a length of 150. The window length is set to be 120 and we consider the window marked by the dotted rectangle. Fig. 2(b) shows the lagged correlation at each lag l computed by Eq. (1) over the two windows. The maximum lag $m = 60$, i.e., $|l| \leq 60$. When $l < 0$ (i.e., Y is delayed), the positive correlation only exists for $l \in [-60, -39]$ (the shadowed area). When $l \geq 0$ (i.e., X is delayed), starting from $l = 1$, we observe a strong increase in positive correlation and it achieves a peak value of 0.81 at $l = 32$. In order to identify the leadership (X leads Y or Y leads X), we need to aggregate all the observed correlation values over the entire lag span and take the expected correlation value given the two cases of l. The aggregated lagged correlation between two time series S^i and S^j, denoted as $E^{ij}(\rho)$, is then defined as the larger expected correlation value:

$$E^{ij}(\rho) = \max(E^{ij}(\rho|l \geq 0), E^{ij}(\rho|l < 0)). \qquad (2)$$

We say that S^i *leads* S^j if $E^{ij}(\rho) = E^{ij}(\rho|l < 0)$, *and* S^i *is led by* S^j *otherwise if* $E^{ij}(\rho) = E^{ij}(\rho|l \geq 0)$. Such leadership ($S^i$ leads S^j or vice versa) is also called the *lead-lag* relation between S^i and S^j. The value of $E^{ij}(\rho|l \geq 0)$ is computed as

$$E^{ij}(\rho|l \geq 0) = \sum_{l=0}^{m} \max(\rho^{ij}(l), 0) \cdot p(l|l \geq 0), \qquad (3)$$

where $\max(\rho^{ij}(l), 0)$ takes only positive correlations and $p(l|l \geq 0)$ takes the value of $1/(m+1)$ since the contribution of each lag is equal. $E^{ij}(\rho|l < 0)$ can be computed symmetrically. In Fig. 2, by Eq. (3), $E^{XY}(\rho|l < 0) = 0.1056$ and $E^{XY}(\rho|l \geq 0) = 0.4017$. Thus, $E^{XY}(\rho) = \max(0.1056, 0.4017) = 0.4017$ indicating X is led by Y.

3.2   Graph Construction

In order to model the leadership relationships among a set of time series, we construct a simple edge-weighted directed graph, $G(\mathcal{V}, \mathcal{E})$, where the set of nodes $\mathcal{V} = \{S^1, S^2, \ldots, S^N\}$ represents N time series, and the set of directed edges $\mathcal{E}$ represents lead-lag relations between time series. An edge (S^i, S^j) indicates that S^i is led by S^j and its weight is set as $E^{ij}(\rho)$. Since we are interested in significant lead-lag relations, we set a *correlation threshold* γ such that only those pairs S^i and S^j with $E^{ij}(\rho) > \gamma$ have edges in G. It is important to note that, when the window slides, the edges and their weights in G will change dynamically.

3.3   Leader Extraction

Given the graph G, we now extract leaders from it. Since a good leader needs to capture both direct and indirect leaderships, we first analyze the leadership transmission in G. Suppose that each time series has a leadership score, based on which a ranking among time series can be obtained. We now discuss how to assign a good leadership score.

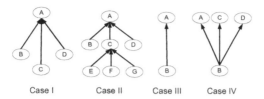

Fig. 3. Comparison of Leadership Score on Different Graph Structures

Consider the leadership score of A under different graphs as shown in Fig. 3. In case I and II, A directly leads 3 time series, B, C, and D. In case I, all of the three have zero in-degree. In case II, C has an in-degree of 3, which implies that A indirectly leads the three that are led by C as well as the three directly led by A itself. It indicates that the leadership score of A in case II should be larger than that in case I. On the other hand, consider case III and case IV. In case III, B is exclusively led by A, whereas in case IV, B is led by A as well as the other two, C and D. The leadership score of A in case III should be larger than that in case IV. Therefore, we define leadership score as

$$score^j = \sum_{S^i \in L_{S^j}} \frac{score^i E^{ij}(\rho)}{d_{out}(S^i)}, \qquad (4)$$

where L_{S^j} is the set of time series that are led by S^j, $score^i$ is the leadership score of S^i and $d_{out}(S^i)$ is the summation of out edge weights of S^i. This leadership score defined above is similar to that defined for the Web Graph on which PageRank score is computed to represent the popularity of web pages. In this paper, we adopt PageRank [4] as the leadership score of a time series to quantify its importance in the graph G.

Finally, based on the structure of G and the PageRank values of time series, we extract the leaders by eliminating redundant leaderships. The basic idea is to first sort the time series by the descending order of their PageRank values and then to remove iteratively the time series that is led either by previously found leaders or by the descendant of previously found leaders.

3.4   The Overall Algorithm

Our solution is presented in Algorithm 1. Given the latest values in time series at time point t, the algorithm first updates the statistics needed in computing lagged correlations as stated in Section 2. It then computes pairwise aggregated correlations (Lines 2-5). Graph G is then constructed (Line 6) and the power method computes the PageRank vector π (Line 7). Finally, the *ExtractLeaders* procedure (Algorithm 2) identifies leaders. In *ExtractLeaders*, time series are first sorted by the descending order of the rank π. Then starting from the time series with the highest rank, it checks the time series led by it and removes them as well as their descendants from the list. The procedure *RemoveDescendant* repeats the process recursively until all descendants of the current leader are removed. The remaining time series on the list are returned as leaders.

We now analyze the complexity of Algorithm 1. Correlation computation in Lines 2-5 needs to compute $(2m+1)N^2$ correlation values, which involves complex mathematical calculation. PageRank computation and the *ExtractLeaders* procedure take $O(kN^2)$ and $O(N)$ time, respectively, where k is the number of power method iterations. Thus, the most time-consuming steps in Algorithm 1 are in computing correlations and PageRank. The space complexity of the algorithm is $O(mN^2)$ for storing the correlation statistics and $O(N^2)$ for storing the values in power method.

Algorithm 1. DiscoverLeaders	**Algorithm 2.** ExtractLeaders
INPUT: N time series, $S^1,\ldots,S^N$, up to current time t, sliding window length w, maximum lag m, correlation threshold γ OUTPUT: *leaders* 1: Update statistics needed for correlation computation; 2: **for** every pair of time series S^i and S^j **do** 3: Compute correlation $\rho_{t,w}^{ij}(l)$, for $|l| \le m$; 4: Compute aggregated lagged correlation $E^{ij}(\rho)$ by Eq. (2); 5: **end for** 6: Construct graph G with respect to γ; 7: Compute PageRank vector π on G; 8: $L \leftarrow ExtractLeaders(G,\pi)$; 9: **return** L;	INPUT: graph G, rank vector π OUTPUT: *leaders* 1: $L \leftarrow$ Sort time series in descending order by π; 2: **for** each time series S^j in L **do** 3: *RemoveDescendant*(L, G, S^j); 4: **end for** 5: **return** L; 6: **Procedure** *RemoveDescendant*(L, G, S^j) 7: **for** each time series S^i in L after S^j **do** 8: **if** (S^i, S^j) is an edge in G **then** 9: *RemoveDescendant*(L, G, S^i); 10: Remove S^i from L; 11: **end if** 12: **end for**

In a stream environment, correlation computation becomes the bottleneck of Algorithm 1 since the implementation of PageRank is fast when the graph is small enough to store in the main memory (e.g., $N = 500$). Too many correlation values need to be computed at each time point and there are endless time points coming into the stream. In order to accomplish prompt leadership detection, we further propose an effective update approach that is able to reduce the number of correlation computations and meanwhile retaining high accuracy, which is described in the following section.

4   Real-Time Correlation Update

In order to speed up the computation of the aggregated lagged correlation for a pair of time series, we propose an efficient update approach by investigating the evolutionary characteristics of lagged correlations. Recall that in Eq. (3), all positive lagged correlation values are aggregated, i.e., we compute the area with positive correlations. Therefore, compared with the exact correlation value at each lag, the area formed by these positive correlations is more crucial to determine the lead-lag relation. We call this area the *interesting area*. The basic idea of our update approach is to track the interesting area. More specifically, at an initial time point, we compute the exact correlation value at each lag and record the interesting area. Then at the subsequent time point, we track

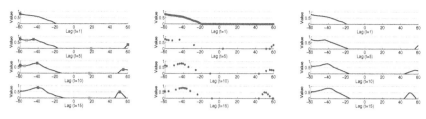

(a) Exact Interesting Area   (b) Probed Correlation Values   (c) Approx. Interesting Area

Fig. 4. Tracking the Interesting Area

and update this interesting area by computing the correlation for only a small number of lags. We then use this interesting area to approximate the aggregated lagged correlation.

We now discuss how to track and update the interesting area. Fig. 4(a) gives an example of the evolutionary shapes of the interesting area between two time series. The lagged correlation is computed at each lag $l \in [-60, 60]$. At time $t = 1$, the interesting area spans from $l = -60$ to $l = -20$ and the corresponding correlation value decreases gradually from 0.8 to 0. We call such continuous area a *wave*. When $t = 5$, we note that there are two waves of the interesting area. The first one spans from $l = -60$ to $l = -17$, which is obviously an evolution from the previous wave. Compared with the wave at $t = 1$, the boundary of this wave enlarges from $l = -20$ to $l = -17$. Hereafter, we call this type of wave an *existing wave*. The second wave spans from $l = 55$ to $l = 60$. Since this wave does not exist at $t = 1$, we call this type of wave a *new wave*. When $t = 10$ and $t = 15$, the existing wave changes slowly, while this new wave enhances its effect.

The above example shows that, in order to keep track of the interesting area, we need to capture the evolutionary pattern of two types of waves, existing waves and new waves. Our solution is based on two observations.

Observation 1. *An existing wave at time t is relatively stable at subsequent time points after t.*

Observation 1 can be explained as follows. For a specific lag l, the correlation $\rho_{t,w}^{ij}(l)$ at time t is computed on two shifted windows $s_{t,w-l}^{i}$ and $s_{t-l,w-l}^{j}$. When the time moves to $t + 1$, correlation $\rho_{t+1,w}^{ij}(l)$ is computed on $s_{t+1,w-l}^{i}$ and $s_{t-l+1,w-l}^{j}$. Notice that there is a large overlap in these two sets of windows. Specifically, the difference between $s_{t,w-l}^{i}$ and $s_{t+1,w-l}^{i}$ (also between the other two windows) is only one point. As a result, the two correlations $\rho_{t,w}^{ij}(l)$ and $\rho_{t+1,w}^{ij}(l)$ cannot differ a lot. Therefore, we have the above observation of an existing wave.

Using Observation 1, we can track an existing wave as follows. The most important features of a wave are its magnitude and width. The magnitude of a wave can be characterized by its maximum points, while the width can be characterized by the minimum points. Therefore, we propose to approximate the area of an existing wave by tracking its peak points. Specifically, after we compute the exact correlation value for each lag at the initial time point, we record the peak points for the existing wave. Then, at the subsequent time point, we only compute the exact correlation value for the lag of each maximum peak point and conduct a geometric progression probing to both sides of the lag until the probe reaches the boundary. The boundary can be either the adjacent minimum peak point, the maximum lag $\pm m$ or the point with a negative correlation value. Then, we conduct a linear interpolation over the computed correlation points to approximate the area of the wave. Finally, the peak points are updated according to the probed correlation values so that they can be used for the subsequent time point.

Fig. 4(b) shows the points, at which we compute (probe) correlation values. Suppose that $t = 1$ is an initial time point. We compute all the lagged correlation values for $l \in [-60, 60]$ and record a maximum peak point at $l = -60$. When $t = 5$, we probe from the maximum peak point $l = -60$ until reaching the boundary, where we detect a negative correlation. In this process, the probing step is increased exponentially so that the approximated wave has higher accuracy around the peak point. There are altogether 7 correlation values computed in the probing process. Then, as shown in Fig. 4(c), linear interpolation is applied to these 7 points to form the approximated existing wave. As further shown in $t = 10$ and $t = 15$, this existing wave can be well tracked.

Now, the remaining problem is to track a new wave. As there is no existent evidence of a new wave at the initial time point, we are not able to record its peaks for tracking purpose. Fortunately, we have the following observation of new waves.

Observation 2. *A new wave at t only emerges at maximum lag values of $\pm m$.*

Observation 2 can be explained as follows. We first consider the case when $0 \le l \le m$. At a specific time t, the correlation $\rho_{t,w}^{ij}(l)$ is computed on two windows of length $(w - l)$. Therefore, with the increase of l from 0 to m, the window length, on which $\rho_{t,w}^{ij}(l)$ is computed, decreases. On the other hand, compared with the previous time point $t - 1$, each time series evolves by adding a new data point to and deleting an old data point from the sliding window. This causes the value of $\rho_{t,w}^{ij}(l)$ to be different from $\rho_{t-1,w}^{ij}(l)$. However, the effect of the time series evolvement on the value of $\rho_{t,w}^{ij}(l)$ is different for different lag l. With the increase of l, the windows, on which $\rho_{t,w}^{ij}(l)$ is computed, becomes smaller and thus the effect of the evolvement becomes larger, which results in larger difference of $\rho_{t,w}^{ij}(l)$ and $\rho_{t-1,w}^{ij}(l)$. This explains why a new wave may emerge at the largest lag $l = m$. Similarly, a new wave is also likely to emerge at $l = -m$.

According to Observation 2, we can track new waves by monitoring the correlation values at $l = \pm m$. As shown in Fig. 4(b), although there is no sign of a new wave at $l = 60$ when $t = 1$, we also compute its correlation at $t = 5$. This strategy successfully detects a positive correlation value at $l = 60$. Then, we take it as an existing wave and track it using the approach we have discussed above. In summary, at $t = 5$, we use 11 points to track the whole interesting area, saving 91% of correlation computation.

Our update approach, *UpdateCorrelation*, is presented in Algorithm 3. It first checks the correlation values at the two maximum lag points to detect potential new waves (Line 2). If there exists a new wave, the algorithm treats it as an existing wave (Lines 3-5). Then, the algorithm approximates each existing wave by two procedures *Probe* and *Interpo* (Lines 7-11). Procedure *Probe* is shown in Algorithm 4. After computing the correlation value at the maximum peak point, it probes the points on its two sides in

---

Algorithm 3. UpdateCorrelation

INPUT: new value at t for two time series S^i and S^j, sliding window length w, maximum lag m, the set of peak points $peak_{t-1}^{ij}$ at time $t-1$

OUTPUT: the lead-lag relation of S^i and S^j

1: **if** there is no existing wave at $l = \pm m$ **then**
2:    Compute $\rho_{t,w}^{ij}(m)$ and $\rho_{t,w}^{ij}(-m)$ to detect potential new waves;
3:    **if** there exists new waves **then**
4:       Add the corresponding l to $peak_{t-1}^{ij}$;
5:    **end if**
6: **end if**
7: **for** each maximum peak point $ptMax$ in $peak_{t-1}^{ij}$ **do**
8:    $sampleWavePointSet = Probe(ptMax)$;
9:    $wavePointSet = Interpo(sampleWavePointSet)$;
10:    Add $wavePointSet$ to corresponding $\rho_{t,w}^{ij}(l)$;
11: **end for**
12: $peak_t^{ij} = detectPeak(\rho_{t,w}^{ij}(l))$;
13: Compute aggregated lagged correlation $E^{ij}(\rho)$ by Eq. (2);
14: Decide the lead-lag relation of S^i and S^j;

Algorithm 4. Probe

INPUT: a peak point $ptMax$

OUTPUT: $sampleWavePointSet$

1: $sampleWavePointSet \leftarrow$ Compute $\rho_{t,w}^{ij}(ptMax)$;
2: $step = 1$;
3: $index = ptMax \mp step$;   // + for right side probe
4: **while** $index$ is not a left(right) boundary point **do**
5:    $sampleWavePointSet \leftarrow$ Compute $\rho_{t,w}^{ij}(index)$;
6:    $step = step \times 2$;
7:    $index = ptMax \mp step$;   // + for right side probe
8: **end while**

---

a geometric progression style. The probing stops when the boundary is met, which we have discussed above. As for the procedure *Interpo*, we use the linear interpolation [10] to connect the probed values and form the approximated interesting area. We then detect and update peak points according to the probed correlation values (Line 12), which can be implemented by an existing peak detection algorithm [3]. Finally, we decide the lead-lag relation based on the approximated interesting area (Lines 13-14).

The *UpdateCorrelation* algorithm enables us to track the interesting area using only $O(\log m)$ correlation computations instead of $O(m)$ that a brute-force approach requires. Moreover, since we start probing from the maximum peak points and stop probing when detecting the boundary, the actual number of correlation computations is much smaller. We further study the efficiency improvement of *UpdateCorrelation* in Section 5.

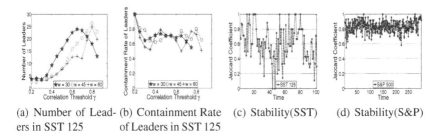

(a) Number of Lead- (b) Containment Rate   (c) Stability(SST)   (d) Stability(S&P)
ers in SST 125   of Leaders in SST 125

Fig. 5. Parameter Sensitivity and Leaders Stability

5   Experimental Results

In this section, we design a set of experiments to answer the following questions:

(1) What are the effects of the parameters (e.g., the sliding window length, the correlation threshold) on the performance of our algorithm in terms of discovered leaders?
(2) How does the set of discovered leaders evolve as the sliding window moves forward? Does the set of leaders remain stable or evolve a lot with time?
(3) Are detected leaders interesting and useful? How can we use them appropriately?
(4) How effective is *UpdateCorrelation*? How good is its approximation accuracy? Does the accuracy degrade over time?

We perform our experiments on a PC with a Pentium IV 3.4GHz CPU and 2GB RAM and the algorithm is implemented with Matlab. We test by using two real datasets.

- **SST 125.** It contains 125 streams of weekly sea surface temperature on the Pacific ocean from 1990-present[1]. Each stream is normalized using Z-Score [14].
- **S&P 500.** It contains 500 streams of high-frequency stock transaction data which we retrieve from the NYSE Trade and Quote (TAQ) database. We extract the tick data of stock prices by computing the Volume Weighted Average Price (VWAP) for transactions at each tick as $VWAP = \frac{Number of Share Bought \times Share Price}{Total Share Bought}$.

Sensitivity of Parameters: There are three parameters in our algorithm: the window length w, the correlation threshold γ and the maximum lag m. As suggested in [2], m is set to be $w/2$. Therefore, we only test two paremeters γ and w. We test on 100 consecutive time ticks in SST 125 and vary γ from 0.2 to 0.85 with a step of 0.05. We also test three values of $w = 30, 45, 60$. Fig. 5(a) presents the number of leaders detected at each γ. For all w, we find a clear rise in the number of leaders when γ increases from 0.2 to 0.6. This is because the number of edges in G decreases with the increase in γ. As G becomes sparser, the locations are less likely to be covered by the same leader, which results in more leaders. For $w = 30$, when γ exceeds 0.7, there is a drop in the number of leaders. This is because when γ is set too high, many locations become isolated and are not led by any others. Therefore, the number of leaders decreases when

---

[1] http://www.cdc.noaa.gov/data/gridded/

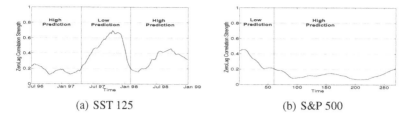

(a) SST 125   (b) S&P 500

Fig. 6. Zero-Lag Strength of Leaders

γ is high and becomes 0 when γ is set as 1, i.e., no edge in $\mathcal{G}$. We also observe similar phenomena for other values of w but with different turning points. In order to study the evolution of leaders when varying γ, we compute the containment rate of leaders between two consecutive γ as $\frac{|Leaders(\gamma_i) \cap Leaders(\gamma_{i-1})|}{|Leaders(\gamma_{i-1})|}$. As shown in Fig. 5(b), for all w, the containment rate at different γ remains high (averagely 0.7). This indicates that most of the leaders found at a low γ can also be found at a high γ. This gives us a hint in choosing γ. Normally, γ can be set around 0.3 since it tends to select a small number of leaders. If users want to be more confident with the lead-lag relation, γ can be set higher and a higher γ also covers most of the results that are produced by lower ones.

Stability of Leaders Over Time: A user may raise the following question: since the leaders are updated at every time tick, can I trust the current detected leaders? We now study the stability of leaders over time. We adopt the Jaccard coefficient [15] to measure the similarity between the leaders extracted at two consecutive time ticks, which is computed as $\frac{|Leaders(t_i) \cap Leaders(t_{i-1})|}{|Leaders(t_i) \cup Leaders(t_{i-1})|}$. For SST 125, we set $w = 30$, $\gamma = 0.3$ and extract leaders at 104 consecutive time ticks in 1997-1998. As shown in Fig. 5(c), the stability generally remains high (the average similarity is 0.61). The average leader duration (i.e., the time length in which a stock continues to be a leader) is 5.3 ticks (one and a half months) and the maximum duration is 12 ticks (three months). The result suggests that the detected leaders have a certain degree of stability although the interval between two consecutive time ticks is as long as 1 week. Nevertheless, there is a drop of stability in the middle of the Nino phenomena(around $t = 55$). This is because all locations have high anomaly scores as shown in Fig. 1 at that time. Therefore, the lead-lag effect is not significant and the leaders vary from time to time, which results in relatively low leadership stability. For S&P 500, we set $w = 120$, $\gamma = 0.3$ and extract leaders at 270 consecutive time ticks in an entire trading day. In Fig. 5(d), we find that the average similarity is high as 0.82 and is quite stable. This is because its graph $\mathcal{G}$ is large and a small number of altered edges are not likely to affect the stocks' PageRank. In summary, the results indicate a certain degree of stability for the evolution of the leaders.

Predictive Power: We now demonstrate the usefulness of detected leaders by constructing a Leadership Index, where the weight β_i of each leader in the index portfolio is determined by its relative PageRank value, i.e., $\beta_i = \frac{\pi_i}{\sum_{j \in Leaders} \pi_j}$. Fig. 7 presents the Leadership Index on S&P 500. We extract 1-minute interval data and set $w = 60$, $\gamma = 0.3$. Among the 500 stocks, we extract an average of 10.8 leaders in a trading day. Compared with the market index formed of S&P 500, we find there are five phases

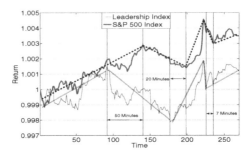

Fig. 7. Leadership Index VS. S&P 500 Market Index

in both indices with the upward/downward trend. In the first phase, these two indices rise together with some minor delay in S&P 500 Index. Then, at $t = 95$, the Leadership Index begins to go down first while S&P 500 Index keeps rising until meets its first turning point at $t = 145$, which is delayed by 50 minutes. After that, Leadership Index rebounds at $t = 177$ with a first steady rising trend followed by a steep burst at $t = 209$. In contrast, S&P 500 Index starts the rising trend at $t = 197$ and meets the burst point at $t = 214$, which are both delayed with Leadership Index. The final turning point of S&P 500 Index is at $t = 233$, which is delayed with Leadership Index by 7 minutes. In summary, in the first phase, Leadership Index leads S&P 500 Index with very small lags; while in other phases, Leadership Index leads S&P index with larger lags and the lag decreases from 50 minutes at the beginning to 7 minutes at the end. We conduct Granger-causality analysis over these two indices and the result suggests that Leadership Index Granger-causes S&P 500 index where the optimal lagged value is 1 for both indices with a significant F-Statistics of 9.65. We find similar results in SST 125 dataset. Recall that in Fig. 1, at the beginning and the ending of Nino phenomena, Leadership Index leads Nino 1+2 index with large lags, whilst in the middle phase of the phenomena, the lead-lag effect is not so significant with small lags.

The above findings indicate that the leadership index indeed exhibits a predictive ability. However, its predictive power has different strengths at different time. Then, how can we know the predictive strength of the Leadership Index at a specific point of time? We study again the shape of the *interesting area* and differentiate two types of waves, the zero-lag wave and the non-zero-lag wave. The zero-lag wave is centered around the lag value of 0. Two time series having a zero-lag wave tend to have a low predictive power due to the small lag. On the other hand, a non-zero-lag wave indicates a large time lag, which is the cause of the high predictive power. We define the strength of zero-lag correlations as the fraction of the edges in G that have zero-lag waves. The strength indicates the extent that the graph G is contributed by zero-lag correlations. Therefore, a low zero-lag correlation strength indicates a high predictive power and vice versa. Fig. 5 presents the zero-lag correlation strength over time on the two datasets SST 125 and S&P500. We find that the strength for SST 125 is low at the beginning when the Nino phenomena starts to emerge. After the Nino phenomena develops fully, all the locations tend to have synchronized anomalies and the strength becomes high as 0.7. Finally, when the phenomena begins to diminish, some locations lead others to drop

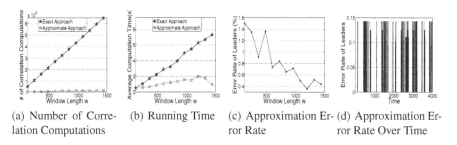

(a) Number of Corre-   (b) Running Time   (c) Approximation Er-   (d) Approximation Er-
lation Computations   ror Rate   ror Rate Over Time

Fig. 8. Performance of Correlation Update

and the strength falls down again, which results in the increase of predictive power. On the other hand, for S&P 500, we observe a high but decreasing strength curve starting from $t = 1$ and it reaches 0.1 at $t = 95$ (matching with the end of the first rising phase of Leadership Index in Fig. 7). It then stays very low below 0.2 until the end of the trading day. Therefore, the evolution pattern of the zero-lag strength coincides with the change of the predictive power of Leadership Index.

Correlation Update: We now study the effectiveness of the *UpdateCorrelation* algorithm. In order to have a longer and consistent time series to test, we extract 30 stocks with tick frequency of 5 seconds and vary w from 120 to 1440. For each w, we move forward the sliding window over that trading day and compare our approximate approach with the exact approach. Fig. 8(a) reports the number of correlation computations. When $w = 120$, the exact approach needs around 54,000 correlation computations, while our approximate approach only needs 7571 computations. The number of correlation computations for the exact approach increases linearly with w, while our approximate approach grows very slowly with w. When $w = 1440$, our approximate approach needs to compute 20,767 correlation values, which is over 30 times less than 648,000 computations of exact approach. Fig. 8(b) presents the average running time for the two approaches, which shares a similar trend with the correlation computations in Fig. 8(a). When $w = 1440$, the running time for approximate approach is 0.94s, which is an order of magnitude faster than 9.3s of the exact approach. Fig. 8(c) shows the accuracy of the approximation. The error rate is computed as the Jaccard distance between the two sets of leaders detected by the two approaches. And the average error rate is less than 1.5% and decreases when w increases. Fig. 8(d) also presents the approximation error rate over time when we move forward the sliding window by setting $w = 360$, $\gamma = 0.3$. It shows that the error is always lower than 0.15 as time goes far away from the initial time tick. This justifies our approximate approach refines peak values and can achieve good approximation accuracy.

6   Related Work

There are several existing studies on multiple time series stream mining. Spiros et al. [11] tracked local correlations by comparing the local auto-covariance matrices of each time series. Zhu and Shasha [18] monitored thousands of time series data but focused on finding high cross-correlation pairs of them. Tan et al. [14] analyzed the linear correlation of multiple climate time series and attemptedX to construct climate index using clustering. Sakurai et al. [13] proposed an algorithm named BRAID to detect arbitrary lag correlations among time series. BRAID uses a geometric probing strategy and sequence smoothing to approximate the lag value wave. Since BRAID always starts probing from lag $l = 0$, the approximation generates larger error when l becomes larger. In our work, on the contrary, we track features of each interesting area, i.e., the peaks and boundaries, and probe from each local maximum peaks. This gives a good approximation accuracy for the wave at large l. To the best of our knowledge, our work is the first to discover the leadership among multiple time series. We are also aware of a stream of work [6,17,8,9] that constructs a weighted graph on time series in order to discover different interesting patterns. Dorr and Denton [6] proposed to construct a hierarchic graph by analyzing similar subsequence of time series to discover timing patterns (e.g., a subsequence of one time series "begins earlier", "ends later", or is "longer" than another). Idé and Kashima [8] proposed an anomaly detection method by analyzing the eigenspace of the dependency matrix. Later, Idé et al. [9] computed the anomaly score of a time series by investigating its k-neighborhood time series. Instead, our work discovers leaders by constructing a graph based on the lead-lag relations of time series.

7   Conclusions

In this paper, we formalize a novel problem of discovering leaders from multiple time series based on lagged correlation. A time series is identified as a leader if its movement triggers the co-movement of many other time series. We develop an efficient algorithm to detect leaders in a real-time manner. The experiments on real climate science data and financial data show that the discovered leaders demonstrate high predictive power on the event of general time series entities and the approximate correlation update approach is up to an order of magnitude faster than the exact approach at a relative low error rate.

Acknowledgment. The work was supported by grants of the Research Grants Council of the Hong Kong SAR, China No. 419008 and 419109.

References

1. Bhuyan, R.: Information, alternative markets, and security price processes: A survey of literature. Finance 0211002, EconWPA (2002)
2. Box, G., Jenkins, G.M., Reinsel, G.: Time Series Analysis: Forecasting and Control. Prentice Hall, Englewood Cliffs (1994)
3. Brent, R.P.: Algorithms for Minimization Without Derivatives. Dover Publications, New York (2002)
4. Brin, S., Page, L.: The anatomy of a large-scale hypertextual Web search engine. Comput. Netw. ISDN Syst. 30(1-7), 107–117 (1998)
5. Chan, K.: A further analysis of the lead-lag relationship between the cash market and stock index futures market. Review of Financial Studies 5(1), 123–152 (1992)
6. Dorr, D.H., Denton, A.M.: Establishing relationships among patterns in stock market data. In: Data & Knowledge Engineering (2008)

7. Granger, C.W.J.: Investigating causal relations by econometric models and cross-spectral methods. Econometrica 37(3), 424–438 (1969)
8. Idé, T., Kashima, H.: Eigenspace-based anomaly detection in computer systems. In: KDD, pp. 440–449 (2004)
9. Idé, T., Papadimitriou, S., Vlachos, M.: Computing correlation anomaly scores using stochastic nearest neighbors. In: ICDM, pp. 523–528 (2007)
10. Meijering, E.: Chronology of interpolation: From ancient astronomy to modern signal and image processing. In: Proc. of the IEEE, pp. 319–342 (2002)
11. Papadimitriou, S., Sun, J., Yu, P.S.: Local correlation tracking in time series. In: ICDM, pp. 456–465 (2006)
12. Säfvenblad, P.: Lead-lag effects when prices reveal cross-security information. Working Paper Series in Economics and Finance 189, Stockholm School of Economics (September 1997)
13. Sakurai, Y., Papadimitriou, S., Faloutsos, C.: Braid: Stream mining through group lag correlations. In: SIGMOD, pp. 599–610 (2005)
14. Steinbach, M., Tan, P.-N., Kumar, V., Klooster, S.A., Potter, C.: Discovery of climate indices using clustering. In: KDD, pp. 446–455 (2003)
15. Tan, P.-N., Steinbach, M., Kumar, V.: Introduction to Data Mining. Addison-Wesley, Reading (2006)
16. von Storch, H., Zwiers, F.W.: Statistical Analysis in Climate Research. Cambridge University Press, Cambridge (2002)
17. Wichard, J.D., Merkwirth, C., Ogorzałek, M.: Detecting correlation in stock market. Physica A: Statistical Mechanics and its Applications 344(1-2), 308–311 (2004)
18. Zhu, Y., Shasha, D.: Statstream: Statistical monitoring of thousands of data streams in real time. In: VLDB, pp. 358–369 (2002)

Mining Outliers with Ensemble of Heterogeneous Detectors on Random Subspaces

Hoang Vu Nguyen, Hock Hee Ang, and Vivekanand Gopalkrishnan

Nanyang Technological University, Singapore

Abstract. Outlier detection has many practical applications, especially in domains that have scope for abnormal behavior. Despite the importance of detecting outliers, defining outliers in fact is a nontrivial task which is normally application-dependent. On the other hand, detection techniques are constructed around the chosen definitions. As a consequence, available detection techniques vary significantly in terms of accuracy, performance and issues of the detection problem which they address. In this paper, we propose a unified framework for combining different outlier detection algorithms. Unlike existing work, our approach combines non-compatible techniques of different types to improve the outlier detection accuracy compared to other ensemble and individual approaches. Through extensive empirical studies, our framework is shown to be very effective in detecting outliers in the real-world context.

1 Introduction

The problem of detecting abnormal events, also called outliers, has been widely studied in recent years [1–3]. Researchers have developed several techniques to mine outliers in static databases and also recently in data streams. Existing outlier detection methods can be classified as distance-based [3–5], density-based [1, 2] and evolutionary-based [6]. There are many ways in practice to define what outliers exactly are, e.g., r-neighborhood Distance-based Outlier [3], k^{th} Nearest Neighbor Distance-based Outlier [5] (a.k.a. k-NN) and Cumulative Neighborhood [4]. Since detection methods are usually constructed around specific outlier notions, their detection qualities vary significantly among datasets. For example, a recent study [7] shows that the Nearest-Neighbor (NN) method performs well when outliers are located in sparse regions whereas LOF [1] performs well when outliers are located in dense regions of normal data. Existing techniques usually compute distances (*in full feature space*) of every data sample to its neighborhood to determine whether it is an outlier or not [1–3, 6]. This causes two side-effects. First, for high-dimensional datasets the concept of locality as well as neighbors becomes less meaningful [8]. Second, not all features are relevant for outlier mining. More specifically, popular distance functions like Euclidean and Mahalanobis are extremely sensitive to noisy features [7]. Despite the presence of the curse of dimensionality, it is difficult in practice to choose a relevant subset of features for the learning purpose [6, 9, 10].

H. Kitagawa et al. (Eds.): DASFAA 2010, Part I, LNCS 5981, pp. 368–383, 2010.

While the nature of data is unpredictable, there is a need for an efficient technique to combine different outlier detection techniques to overcome the drawback of each single method and yield higher detection accuracy. The motivation here is similar to the advent of ensemble classifiers in the machine learning area [9, 11]. With the feasibility of ensemble learning and subspace mining demonstrated, the natural progression would be to combine them both. Lazarevic and Kumar [10] propose the first solution for *semi-supervised* ensemble outlier detection in feature subspace. That work assumes the existence of outlier scores where a *combine* function can be applied directly. However, this is not practically true since different detection methods can produce outlier scores of different *scales*. For example, it can be recognized that the scores produced using k^{th} Nearest Neighbor Distance-based Outlier [5] are smaller in scale than those using Cumulative Neighborhood [4]. Furthermore, as pointed out in Section 3.3, different detection techniques also produce different *types* of score vectors. In particular, some vectors are real-valued while others are binary-valued. This leads to the need of a unified notion of outlier score and an efficient technique to specifically deal with scores' heterogeneity. The availability of such notion would facilitate the task of combination.

Problem Statement. Consider a dataset DS with N data samples in dim dimensions. While most of the data samples in DS are normal, some are outliers, and our task is to detect these outliers. While few outliers can be found when all dimensions are taken into account, most of them can only be identified when looking at some subsets of features. In addition, some features of DS are noisy, and cause the full distance computation to be inaccurate if they are included. Given a set of base outlier detection technique(s), our goal is to build an efficient method to combine the results obtained from them while overcoming their individual drawbacks when applying on DS. The ensemble framework should: (a) alleviate of the curse of dimensionality and noisy features, (b) efficiently combine outlier score vectors of base techniques having different *scales* and different *characteristics*, and (c) provide higher detection quality than each individual base technique used in the ensemble (when applied on full feature space). In order to address this problem, we present the _H_eterogeneous _D_etector _E_nsemble on _R_andom _S_ubspaces (HeDES) framework. The advantage of using HeDES lies in its ability to incorporate various heuristics for combining different types of score vectors. The main contributions of this work can be summarized as follows:

– We introduce a unified notion of outlier score function and show how existing outlier definitions can be represented using it. We demonstrate how to identify different types of outlier scores in literature by using this new notion of outlier score function.
– We propose a generalized framework for ensemble outlier detection in feature subspaces - HeDES. Unlike the existing simple framework [10], HeDES is able to combine different techniques producing outlier scores of different scales or even different types of scores (e.g., real-valued v/s. binary-valued).

Through extensive empirical studies, we demonstrate that the HeDES framework can outperform state-of-the-art detection techniques and is therefore suitable for outlier detection in real-world applications. The rest of this paper is organized as follows. Related work and background knowledge are presented in the next section. Details of our approach are provided in Section 3 and empirical comparison with other current-best approaches is discussed in Section 4. Finally, the paper is summarized in Section 5 with directions for future work.

2   Literature Review

Distance-based outlier detection techniques in general exploit the distance of a data sample to its neighborhood to determine whether it is outlier or not. Distances can be computed either using only one neighbor [5] or using k nearest neighbors [4]. The notion of distance-based outlier was first introduced by Knorr and Ng [3] and then refined in [5]. Breunig *et al.* [1] propose the first notion of density-based outliers. The outlier score used, called Local Outlier Factor (LOF), is a measure of difference in neighborhood density of a data sample p and that of data samples in its local neighborhood. LOF for data samples belonging to a cluster is approximately equal to 1, while that for outliers should be much higher. Experimental results from [7] show that LOF outperforms other detection techniques in most cases. Papadimitriou *et al.* [2] introduce a new definition of density-based outliers. Instead of using the k nearest neighbors of a data sample p in computing its outlier score, they employ the r-neighborhood of p. The outlier score of each data sample, called MDEF, is used to compare against the normalized deviation of its neighborhood's scores and standard-deviation is employed in the outlier flagging decision. This removes the need of using any static cutoff or score ranking.

Both distance-based and density-based techniques involve the computation of distances from each data sample to its neighborhood. However, for high-dimensional datasets the concept of locality as well as neighbors becomes less meaningful [8]. This limitation is addressed by an evolutionary-based technique introduced by Aggarwal and Yu [6]. The method first performs a grid *discretization* of the data by dividing each data attribute into $\emptyset$ equi-depth ranges. Then, a genetic approach is employed to mine subspaces whose densities are in the top smallest values. Nevertheless, it suffers the intrinsic problems of evolutionary approaches - its accuracy is unstable and varies depending on the selection of initial population size as well as the crossover and mutation probabilities. The problem of mining in subspaces has also been studied in supervised learning [8, 9]. Ho [9] point out that constructing different classifiers by using randomized initial conditions or data perturbations cannot ensure high classification accuracy. Instead, randomly sampling subsets of feature space (i.e., feature subspaces) for different classifiers seems to be a very promising solution. Likewise, Lazarevic and Kumar [10] tackled the outlier detection problem using an ensemble of outlier techniques built on the problem subspaces. By assuming that information about normal behavior in the underlying dataset is known, they reported findings similar to that of [9]. Their technique, called Feature Bagging, consists of two variants of combine functions: *Breadth First* and *Cumulative Sum*. The Breadth First combine method (a) first sorts all outlier score vectors, (b) then takes the data samples with highest outlier score from all outlier detection algorithms, and (c) finally appends their indices at the end of the final index vector (and so on). On the other hand, Cumulative Sum simply sums up all the score vectors and returns the result as the final outcome. Nevertheless, Feature Bagging does not specify clearly how to integrate outlier scores with different scales and different characteristics (e.g. *real-valued* vector vs. *binary* vector). Furthermore, Breadth First is reported to be sensitive to the order of detection algorithms applied [10]. Another notion of ensemble outlier mining is presented in [12]. However, like Feature Bagging, no consideration is given to the heterogeneity of outlier scores produced by different techniques. Furthermore, it lacks of details on how to process the score vectors to make its proposed combine functions be applicable whereas a direct application is impossible (c.f. Section 3). In addition, several aspects of the ensemble outlier detection problem (as mentioned in Section 1) are not discussed.

Abe *et al.* [13] propose an approach for constructing an ensemble of dichotomizers for mining outliers using artificially generated data. Their approach, called Active Outlier, first reduces the problem of outlier detection to classification. Active learning (a form of data sub-sampling) is used to construct a set of dichotomizers, combined results of which are used to identify outliers. Active Outlier is indeed a type of ensemble learning using data sub-sampling. As mentioned in [9, 10], building ensembles using data perturbation cannot enrich the homogeneity or de-correlate the relationship among learners in the ensemble as efficiently as feature sub-sampling. Our empirical studies on real-life datasets (c.f., Section 4) support this claim.

3   Methodology

The HeDES framework is a *generalized* framework for mining outliers in subspaces using ensemble of outlier detection techniques (henceforth termed detectors). In the following, we present the details of constructing the ensemble and explain how it is applied in HeDES.

3.1   Ensemble Construction

The process of constructing the ensemble of detectors is displayed in Algorithm 1. In each of the total R rounds, we first sample a detector T from the pool of techniques considered $(\mathcal{T})$ on a *round-robin* basis. Practically, R should be chosen as a multiple of the pool size. Next, we form a subspace S where T will operate by randomly choosing N_f features from the full feature space. Here, N_f is sampled from the uniformly distributed range $[\lfloor dim/2 \rfloor, dim-1]$. The pair (T, S) is then added to the ensemble. By sampling N_f from the range $[\lfloor dim/2 \rfloor, dim-1]$ instead of fixing it to $\lfloor dim/2 \rfloor$ like in [9], we increase the possibility of generating different subsets of features for each detector in the ensemble. Since the detection capability of each detector relies on its own notion of dissimilarity measure, this increases the chance that they generalize their prediction in ways different to each other. Hence, the above process of constructing the ensemble takes advantage of high-dimensional feature space and weakens the curse of dimensionality.

After identifying all the detectors to be used in the ensemble, we adjust their weights by running the ensemble against an *unlabeled* training set. The intuition behind this weight-adjust is that some detection techniques are more powerful than others on some certain types of data. For example, recent study by Lazarevic *et al.* [7] shows that the Nearest-Neighbor (*NN*) method outperforms LOF when outliers are located in sparse regions whereas LOF [1] yields higher performance than NN when outliers are located in dense regions of normal data. Even though the detectors in the ensemble are applied on the same dataset during testing, the subspaces where they operate are homogeneous. Furthermore, subspace distributions are different whereas detectors' prediction performance is dependent on their respective subspace. Thus, our argument on detectors' superiority over the others in some certain data still holds in our ensemble learning. Since the nature of subspaces is unpredictable, assigning fixed weights for detectors is not a good solution. Intuitively, had we known which detectors would work better, we would give higher weights to them. In the absence of this knowledge, a possible strategy is to use the result of detectors on a separate validation dataset, or even their performance on the training dataset, as an estimate of their future performance.

Algorithm 1. CONSTRUCTING HeDES

---
1 **for** $i = 1$ **to** R **do**
2 | Choose a detector $T_i \in \mathcal{T}$
3 | Randomly sample N_f from $[\lfloor dim/2 \rfloor, dim - 1]$
4 | Randomly sample a subset of features S_i of size N_f from the feature set of DS
5 | Add (T_i, S_i) into the ensemble
6 Apply the ensemble to the synthetic training dataset
7 Adjust the weight of each detector in the ensemble

This paper, similar to AdaBoost [14], employs the latter strategy. However, since the training set is unlabeled, a direct weight-adjust is not straightforward. To overcome this problem, we construct a labeled synthetic training dataset from the original (unlabeled) one by applying the technique presented in [13]. In brief, the synthetic set is comprised of normal data drawn from the original one and artificially generated outliers. The artificial outliers here are created by using a uniform distribution U that is defined within a bounded subspace whose minimum and maximum are limited to be 10% beyond the observed minimum and maximum, respectively. Let the original training set be S_{tr}, we construct the set of artificial outliers S_{out} of size $|S_{tr}|$ according to U on the bounded domain.

Algorithm 2. MINING OUTLIERS WITH HEDES

1 Normalize DS
2 **foreach** *detector type j* **do**
3 $\quad$ $TVS_j = \emptyset$
4 **for** $i = 1$ **to** R **do**
5 $\quad$ Choose the detector (T_i, S_i) from the ensemble
6 $\quad$ $j = $ type of T_i
7 $\quad$ $RVS_i = $ apply T_i to DS projected on S_i
8 $\quad$ $TVS_j = TVS_j \cup \{RVS_i\}$
9 **foreach** *detector type j* **do**
10 $\quad$ $VS_j = SUBCOMBINE(TVS_j)$
11 $VS_{FINAL} = COMBINE(VS_1, VS_2, \ldots)$

The synthetic training set is then set to be $S_{tr} \cup S_{out}$. More details are given in [13]. The use of this set helps us estimate the performance of each detector in the ensemble and adjust its weight correspondingly despite the lack of knowledge on anomalous behavior. Since outlier detectors in the ensemble are unsupervised, they are less susceptible to the overfitting problem. In other words, the weights trained are loosely coupled with the synthetic training set. Furthermore, this artificial data generation has been shown to be successful in training highly accurate classifiers [13]. Thus, the weights obtained in the training phase are likely to have very high generalization capability on unseen test data. By using the weight-adjusted scheme, the effect of detection techniques that are not as relevant as the others can be reduced. This becomes even more critical when irrelevant techniques may lead to a significantly wrong assignment of outlier score (c.f., Section 4).

3.2 HeDES Framework

Our proposed approach, HeDES, is described in Algorithm 2, and functions as follows. The testing dataset is passed through the ensemble. For every pair (T, S) in the ensemble, we apply T to DS projected on subspace S and obtain a raw vector score. This raw vector score is stored together with other vector scores generated by the same detector type j in TVS_j. After finishing R rounds, each set of vector scores (vectors in the same set are of the same type) are combined separately using SUBCOMBINE function to yield a vector score VS_j. Finally, the COMBINE function is invoked using all the VS's obtained to produce the final vector score VS_{FINAL}. The interpretation (combination) of VS and VS_{FINAL} depends on the specific combine functions utilized which are explored in detail in Section 3.4. Note that the two most important components in this framework are: (a) the outlier score function, and (b) the (SUB)COMBINE functions. The main difference between the simple subspace ensemble framework in [10] and our generalized framework lies in the multi-staged combine function which allows much more flexible integration among the heterogeneous types of outlier detection techniques. It is highlighted that similar to other ensemble classifiers [9, 14], ensemble outlier detection method is a *parallel learning* algorithm [10]. Since each round of running is independent of the other, a parallel implementation can be employed for faster learning.

3.3   Outlier Score Function

Assume a metric distance function D exists on DS, using which we can measure the dissimilarity between two arbitrary data samples in any arbitrary subspace. A general approach that has been used by most of the existing outlier detection methods [1, 3, 6] is to assign an *outlier score* (based on the distance function) to each individual data point, and then design the detection process based on this score. The use of the outlier score is analogous to the mapping of multi-dimensional datasets to $\mathbb{R}$ space (the set of real numbers). In other words, we can define the outlier score function (F_{out}) which maps each data sample in DS to a unique value in $\mathbb{R}$. Intuitively, to create an outlier score function, we first identify a set of measurements based on some specified criteria, then define a mechanism g for combining them, and finally generate a function (F_{out}) based on g. Most the existing techniques utilize only a single measurement, i.e., g becomes a *uni-variable* function that is related directly to the only measurement taken into account. With reference to the k-NN [5], let the measurement considered be the distance from a data pattern p to its kth nearest neighbor (D^k), then a possible choice of F_{out} is $F_{out} = g(D^k) = D^k$.

Outlier score function classification. Among existing approaches to outlier detection problem, we can classify F_{out} into *global* and *local* score functions. An outlier score function is called *global* when the value it assigns to a data sample $p \in DS$ can be used to compare globally with other data samples. More specifically, for two arbitrary data samples p_1 and p_2 in DS, $F_{out}(p_1)$ and $F_{out}(p_2)$ can be compared with each other, and if $F_{out}(p_1) > F_{out}(p_2)$, p_1 has a larger possibility than p_2 to be an outlier. The definitions proposed by Angiulli *et al.* [4], Breunig *et al.* [1], and Ramaswamy *et al.* [5] straightforwardly adhere to this category. On the other hand, the definition of Knorr and Ng [3] can be converted to this category by taking the inverse of the number of neighbors within distance r of each data point. In contrast, a *local* outlier score function assigns to each data sample p, a score that can only be used to compare within some local neighborhood. Example of such a function is proposed in [2], where the local comparison space is the set of data samples lying within the *circle* centered by p and the *radius* is user-defined. The choice of a global or local outlier score function clearly affects later stages of the algorithm design process.

A classification of detection techniques using F_{out}. Using the notion of F_{out} defined above, existing outlier detection techniques can be classified into two types: (a) Threshold-based where a local F_{out} is usually used, and (b) Ranking-based where a global F_{out} is employed, (c.f., Definitions 1 and 2, respectively). According to this classification, the methods proposed in [4–6] using global score functions are classified as Ranking-based. On the other hand, LOCI [2] with local score function is classified as Threshold-based. Although the technique in [3] utilizes a global F_{out}, it is classified as Threshold-based by letting $F_{out}(p) = \frac{1}{|S(p)|}$ and choosing $t = \frac{1}{1-P}$. In this case, a data sample $p \in DS$ is an outlier if $F_{out}(p) > t$, i.e., $F_{out}(p) > \frac{1}{1-P}$. Note that the threshold t in LOCI [2] is dynamic, whereas that of [3] is static (dependant on the pre-defined variable P).

Definition 1. [THRESHOLD-BASED] *Given a (dynamic or static) threshold t, a data sample p is an outlier of DS if $F_{out}(p) > t$.*

Definition 2. [RANKING-BASED OR TOP-n-OUTLIER] *Given a positive integer n, a data sample p is an nth outlier of DS if no more than $n - 1$ other points in DS have a higher value of F_{out} than p. An algorithm based on this definition outputs the top n outliers.*

When F_{out} is global, a Ranking-based technique is normally preferred since the assigned score values of data samples can be compared globally to produce the top points with largest scores. The resultant score vector is then real-valued and identical to the values that F_{out} assigns to data samples. On the other hand, if F_{out} is a local one, a Threshold-based approach becomes a reasonable choice. As a consequence, the score vector obtained contains only binary values (0 for non-outliers and 1 for outliers) since the scores produced by F_{out} are already *discretized* through a threshold-based test. Therefore, score vectors produced by different detection techniques are heterogeneous and need to be processed carefully to facilitate the COMBINE process.

Issue of converting F_{out} to the posterior probabilities. Assume by applying an outlier detector T with outlier score function F_{out} onto DS, we obtain the score vector: $RVS = \{F_{out}(p_1), F_{out}(p_2), \ldots, F_{out}(p_N)\}$. The problem of outlier detection is equivalent to a binary classification problem with two classes: O (outlier class) and M (normal class). One important question which has not been addressed well by the research community is how to compute the posterior probability $P(O|F_{out}(p_i))$ using the knowledge on RVS. Gao and Tan [15] propose two methods attempting to solve this problem. The first method bases on the assumption that the posterior probabilities follow a logistic sigmoid function and the normal and anomalous samples have similar forms of outlier score distribution (same covariance matrix). It then tries to learn the function's parameters using RVS. The second learner on the other hand models the likelihood probability distributions $P(F_{out}(p_i)|O)$ and $P(F_{out}(p_i)|M)$ as a Gaussian and an exponential distribution, respectively. The posterior probabilities are then computed using Bayes theorem. Among the two methods, mixture modeling is more suitable for ensemble learning as demonstrated in [15].

The main intuition leading to this mixture model is derived from the empirical studies using k-NN [5] as the score function. However, the argument used in [15] does not hold for density-based approaches, such as LOF, where density of a data sample is compared (divided) to that of its neighbors. Because of limited space, we omit the demonstration here. Our empirical studies (c.f., Section 4) point out that processing the outlier scores directly (like in HeDES and Feature Bagging) instead of converting to posterior probabilities will yield better detection results.

3.4 COMBINE Functions

As discussed in Section 2, Lazarevic and Kumar [10] introduce two combine functions (Cumulative Sum and Breadth First) which have been successfully used in ensemble-based outlier mining. Here, we present three novel combine functions which are *Weighted Sum, Weighted Majority Voting* and *OR Voting*. Unlike Breadth First, these functions are invariant to the order of the detectors. Since accuracy is the most critical factor in ensemble learning, this property becomes an advantage of our approach. Among them, the first two functions are shown to be very efficient in ensemble classification and have been widely employed in many practical applications [14, 16]. The intuition for utilizing weighted combine functions were also discussed in details above. Weighted Majority Voting is known to excel in combining class labels assigned by different classifiers in the ensemble. On the other hand, Weighted Sum in classification is normally applied on posterior probabilities [9]. Conversely, in HeDES, it is used to combine *normalized* outlier scores produced by different detectors of the ensemble. Finally, Or Voting is a natural combine function for integrating heterogeneous types of output scores as demonstrated later. It is important to note here that exploring all possible combine functions is not a focus in this paper. Nevertheless, our chosen combine functions are still able to encompass almost all available types of outlier scores in the field.

Although HeDES provides an easy extension to score vectors of various types (depending on the purpose of learners), in this paper score vectors are either real-valued or binary-valued. An natural approach (Ensemble Voting) to combine different types of score vectors is to simply normalize and *discretize* the real-valued score vectors (convert all score vectors to the same type), and thereafter integrate all the binary-valued score vectors (inclusive of the discretized real-valued score vectors) using Weighted Majority Voting. However, such a natural approach is not sufficient and does not produce good results (c.f., Section 4). The set of input score vectors to the (SUB)COMBINE function is classified into two groups in which the first group contains score vectors (TVS_R) resulting from applying Ranking-based techniques, whereas the second group contains score vectors (TVS_T) of Threshold-based ones. Our strategy is to apply some combine function on TVS_R and TVS_T separately to obtain VS_R and VS_T. Finally, a special combine function is used to integrate VS_R and VS_T to produce the final score vector VS_{FINAL}. It is noted that the problem of combining results of Ranking-based and Threshold-based techniques here is very similar to the problem of combining detection results of categorical and continuous features in mixed-attribute datasets as addressed in [17]. In both cases, we process real values and binary/categorical values separately. Eventually, a heuristic is used to integrate the results obtained. This is the base intuition for our Or Voting combine function.

Processing outlier score vectors. Because of the different nature between Ranking-based and Threshold-based techniques, outlier score vectors produced by them need different treatments. Assume the data samples in DS are $p_1, p_2, \ldots, p_N$. A detection technique T using a specific score function F_{out} is applied to identify outliers in DS. We denote T's resultant score vector as $RVS = \{F_{out}(p_1), F_{out}(p_2), \ldots, F_{out}(p_N)\}$.

If T is a Ranking-based technique: Vectors of different Ranking-based techniques may have different scales [5]. Hence, to apply combine functions, real-valued vectors need to have equivalent scale. In other words, normalization is necessary. In HeDES, RVS is normalized using the standardization technique. One of the most important characteristics of this normalization technique is its ability to maintain the detectability of extreme values after performing normalization [18]. As argued in [10], this facilitates combining real-valued vectors since a data sample receiving a high score value by one detector, after summing up its score with those produced by other detectors, may still have large values and be flagged as outliers. We define the *normalized* value of $F_{out}(p_j)$ in RVS as: $Score_{norm}(p_j) = \frac{F_{out}(p_j) - m}{s}$ where $m = \frac{1}{N}(\sum_{i=1}^{N} F_{out}(p_i))$ and $s = \frac{1}{N}(\sum_{i=1}^{N} |F_{out}(p_i) - m|)$. By applying normalization, the range of outlier score becomes independent of the technique used. Since all normalized vectors score have comparable scale, it is feasible to integrate them.

If T is a Threshold-based technique: We preserve RVS as it is. This is because each individual element in RVS already indicates the posterior probability of being outlier for data points. Thus, if an ensemble employs techniques from both Ranking-based and Threshold-based, we need a special combine function. Since *Cumulative Sum* and *Breadth First* functions ignore the score vectors' heterogeneity, they are not suitable for use.

Weighted Sum. This function is used for vectors in TVS_R. Let us denote the *weight* of the detector $T_i \in \mathcal{T}$ at round i with score vector RVS_i as W_i. The final score vector of all vectors in TVS_R is defined as: $VS_R = \sum_i W_i \times RVS_i$. Weighted Sum is in fact a modified version of Cumulative Sum proposed in [10]. However, the weight-based strategy helps boost the performance of more efficient detectors. This cannot be obtained in equi-weight schemes.

Weighted Majority Voting. This combine function is used for processing vectors in TVS_T. Although similar to most of the existing ensemble classifiers [9, 14], the problem here is much simpler since we are only interested in two classes of data: normal (class M) and outlier (class O). Since all vectors in TVS_T only contain binary values, they are suitable for Weighted Majority Voting. As in the case of Weighted Sum, the weight of each vector is determined by the performance of the corresponding detection technique on training datasets.

OR Voting. This function is used for combining VS_R and VS_T. However, its input vectors must contain only binary values. Therefore, we perform a *discretization* process on VS_R where its top values are converted to 1, and the rest are converted to 0. Under this scheme, we have: $VS_{FINAL} = VS_R \vee VS_T$ where "$\vee$" is the usual Boolean operator.

Interpretation of VS_{FINAL}. If the pool of detection techniques $\mathcal{T}$ contains only Ranking-based techniques, we then flag those data samples having highest scores in VS_{FINAL} as outliers. In case $\mathcal{T}$ contains only Threshold-based techniques, outliers are those points having score in VS_{FINAL} equal to 1. Finally, if $\mathcal{T}$ contains both Ranking-based and Threshold-based methods, outliers are those whose scores equal to 1 in VS_{FINAL}. Thus, the flagging mechanism for "mixed" $\mathcal{T}$ is similar to that of an ensemble containing only Threshold-based methods. This is because by applying the OR function, the real-valued vector VS_R is already converted to a binary-valued one. Similar to [10], the number of outliers to flag for Ranking-based methods depends on the specific dataset used.

4   Experimental Evaluation

To verify the effectiveness of the proposed combination framework, we conducted the experiments on several real datasets which are taken from UCI Machine Repository [1]. These datasets are used widely in outlier detection as well as in rare class mining [10], and are summarized in Table 1. The setup procedure (converting datasets into binary-class sets, etc.) employed here follows exactly that of Feature Bagging. In the field of outlier detection, ROC curve (as well as AUC) is an important metric used to evaluate detection quality. Similar to [4, 7, 10, 15], AUC (area under the ROC curve) was chosen as performance benchmark in this paper because of its proved relevance for outlier detection [7, 10]. In each experiment, due to space limitation, we only report how AUC changes when the number of rounds R is varied for KDD Cup 1999 dataset. This dataset is chosen as it has the largest number of instances as well as attributes among all the datasets considered, and hence is a good representative. For other sets, the results are similar and average AUC with $R = 10$ is presented (setting R to 10 was suggested in [10, 15]). For every dataset, each reported result is a 95% confidence interval of the AUC obtained by averaging the outcomes of running the algorithms 10 times on each of its generated binary-class sets. In our empirical studies, two different base detectors are considered: LOF [1] and LOCI [2], and are tested using full feature space. The former is known to be one of the best Ranking-based techniques [7] while the latter is a well-known Threshold-based technique [2]. By choosing these high quality base detectors, we are able to highlight the improvement of HeDES in detection accuracy. For LOF, the parameter $MinPts$ was set to 20. For LOCI, we chose $n_{min} = 20$, $n_{max} = 50$, $\alpha = 1/2$, and $k_\alpha = 3$. Those values were derived from the corresponding papers [1, 2]. Apart from the two base techniques, we compared our approach with other ensemble approaches including: Feature Bagging [10], Active Outlier [13], Mixture Model [15], and Ensemble Voting (c.f., Section 3.4). Feature Bagging uses two combine functions: Cumulative Sum and Breadth First. For each dataset under consideration, we choose to display the highest AUC value among the two for Feature Bagging. Active Outlier constructs an ensemble after t rounds of training, i.e. the ensemble contains t detectors. Here, t was set to R for fair comparison. Since Active Outlier does not use any base detector, its performance remains the same regardless of which base detector is chosen for other ensemble techniques.

---

[1] http://www.ics.uci.edu/ mlearn/MLRepository.html

Table 1. Characteristics of datasets used for measuring accuracy of techniques

Dataset	Classes	Attributes	Instances	Outlier v/s. Normal
Ann-thyroid 1	3	21	3428	class 1 v/s. 3
Ann-thyroid 2	3	21	3428	class 2 v/s. 3
Lymphography	4	18	148	merged class 2 & 4 v/s. rest
Satimage	7	36	6435	smallest class v/s. rest
Shuttle	7	9	14500	class 2, 3, 5, 6, 7 v/s. 1
KDD Cup 1999	2	42	60839	class U2R v/s. normal
Breast Cancer	2	32	569	class 2 v/s. 1
Segment	7	19	2310	each class v/s. rest
Letter	26	16	6238	each class v/s. rest

Experiment on Ranking-based technique. This experiment aims to investigate the performance of the our proposed combine function, Weighted Sum, when applied to the Ranking-based technique. We compared our method against LOF, Feature Bagging (FB), Mixture Model (MM), and Active Outlier (AO). The results are shown in Figure 1 and Table 2. It can be observed that Weighted Sum strategy yields very good results in all test cases. Even in the case where the base technique, LOF, performs no better than random guessing due to high dimensionality of the dataset (Satimage), our approach is still able to bring very good improvement. The results also indicate that using full feature space in outlier detection may yield low accuracy, especially when the number of features is large and it is likely that some features are noisy. The performance of Mixture Model over the datasets used is worse than Active Outlier and Feature Bagging. This agrees with our argument about the applicability of Mixture Model on other notions of outliers. In particular, the outlier score proposed in LOF is density-based whereas k-NN is distance-based. Extensive studies in the field have pointed out the significant differences between these two notions. These in addition to the results obtained show that the assumption made in Mixture Model is not flexible enough to encompass the scores produced by LOF. For all ensemble techniques considered (including our approach), AUC value increases as the number of detectors included in the ensemble increases. However, Weighted Sum and Feature Bagging tend to work better than AO. This can be attributed to the fact that ensemble learning by subspace sampling produces more efficient learners than data sub-sampling one [9].

Experiment on Threshold-based technique. In this experiment, we study the effect of our proposed combine function, Weighted Majority Voting (WMV), for Threshold-based techniques. Thus, LOCI is selected as the base detector. Our approach's performance is assessed against LOCI, Feature Bagging (FB, also utilizes LOCI), and Active Outlier (AO). Mixture Model is omitted here since the posterior probabilities can be derived directly from the binary-valued scores. In fact, the results achieved by Mixture Model under this setting are the same as that of Feature Bagging. From Figure 1 and Table 3, it can be seen that Weighted Majority Voting yields the best or nearly best results in

Table 2. Ranking-based technique: AUC values of LOF, Feature Bagging, Mixture Model, Active Outlier, and Weighted Sum ($R = 10$)

Dataset	LOF	FB	MM	AO	WS
Ann-thyroid 1	0.869	0.869 ± 0.015	0.855 ± 0.021	0.856 ± 0.023	**0.892 ± 0.005**
Ann-thyroid 2	0.761	0.769 ± 0.003	0.759 ± 0.007	0.753 ± 0.009	**0.798 ± 0.008**
Lymphography	0.924	0.967 ± 0.009	0.921 ± 0.001	0.843 ± 0.041	**0.984 ± 0.004**
Satimage	0.510	0.558 ± 0.031	0.562 ± 0.025	0.646 ± 0.024	**0.703 ± 0.022**
Shuttle	0.825	0.839 ± 0.004	0.724 ± 0.017	0.843 ± 0.006	**0.861 ± 0.002**
Breast Cancer	0.805	0.825 ± 0.022	0.758 ± 0.012	0.822 ± 0.015	**0.866 ± 0.017**
Segment	0.820	0.847 ± 0.017	0.798 ± 0.005	0.836 ± 0.002	**0.882 ± 0.003**
Letter	0.816	0.821 ± 0.003	0.722 ± 0.014	0.824 ± 0.002	**0.848 ± 0.001**

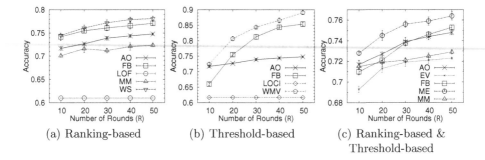

(a) Ranking-based   (b) Threshold-based   (c) Ranking-based & Threshold-based

Fig. 1. AUC values of all competing approaches on the KDD Cup 1999 dataset

all cases (the margin with respect to the best one is negligible). For Feature Bagging, neither Cumulative Sum nor Breadth First works well in combining vectors of Threshold-based techniques. This indicates that specialized schemes are required. With the results achieved in this test, Weighted Majority Voting is shown to be a promising candidate.

Overall, we can observe that ensemble outlier detection (Feature Bagging, Weighted Majority Voting, Active Outlier) results in good improvements over the base technique. We again observe the same pattern as in the previous experiment: the accuracy of ensemble techniques grows as the number of detectors increases and that of Active Outlier is dominated by our approach's and Feature Bagging's.

Experiment on Ranking-based & Threshold-based techniques. So far in our empirical studies, the ensemble contains either only Ranking-based (LOF) or only Threshold-based (LOCI) detection techniques. We now investigate our last proposed combine strategy, the OR Voting, in an ensemble where both types of techniques are considered. Therefore, in this experiment, both LOF (Ranking-based) and LOCI (Threshold-based) are employed. We call our method under this setting Mixed Ensemble (ME). More specifically, we use Weighted Sum for Ranking-based technique whereas with Threshold-based technique, we apply Weighted Majority Voting. The results from each group are combined using the OR Voting. Our proposed approach is compared against Feature Bagging

Table 3. Threshold-based technique: AUC values of LOCI, Feature Bagging, Active Outlier, and Weighted Majority Voting ($R = 10$)

Dataset	LOCI	FB	AO	WMV
Ann-thyroid 1	0.871	**0.873 ± 0.003**	0.856 ± 0.023	0.872 ± 0.021
Ann-thyroid 2	0.747	0.754 ± 0.026	0.753 ± 0.009	**0.812 ± 0.015**
Lymphography	0.892	0.932 ± 0.007	0.843 ± 0.041	**0.987 ± 0.003**
Satimage	0.529	0.535 ± 0.022	0.646 ± 0.024	**0.654 ± 0.024**
Shuttle	0.822	0.856 ± 0.011	0.843 ± 0.006	**0.873 ± 0.004**
Breast Cancer	0.801	0.827 ± 0.002	0.822 ± 0.015	**0.842 ± 0.001**
Segment	0.835	**0.852 ± 0.002**	0.836 ± 0.002	0.850 ± 0.014
Letter	0.811	0.834 ± 0.016	0.824 ± 0.002	**0.872 ± 0.004**

Table 4. Ranking-based & Threshold-based techniques: AUC values of Feature Bagging, Mixture Model, Active Outlier, Ensemble Voting, and Mixed Ensemble ($R = 10$)

Dataset	FB	MM	AO	EV	ME
Ann-thyroid 1	0.870 ± 0.015	0.813 ± 0.013	0.856 ± 0.023	0.832 ± 0.012	**0.883 ± 0.020**
Ann-thyroid 2	0.768 ± 0.031	0.684 ± 0.001	0.753 ± 0.009	0.754 ± 0.012	**0.792 ± 0.004**
Lymphography	**0.955 ± 0.033**	0.735 ± 0.002	0.843 ± 0.041	0.901 ± 0.235	0.952 ± 0.014
Satimage	0.531 ± 0.003	0.517 ± 0.043	0.646 ± 0.024	0.544 ± 0.007	**0.780 ± 0.005**
Shuttle	0.853 ± 0.028	0.729 ± 0.013	0.843 ± 0.006	0.827 ± 0.024	**0.871 ± 0.016**
Breast Cancer	0.824 ± 0.013	0.755 ± 0.023	0.822 ± 0.015	0.837 ± 0.017	**0.864 ± 0.015**
Segment	0.845 ± 0.007	0.792 ± 0.016	0.836 ± 0.002	0.840 ± 0.004	**0.852 ± 0.006**
Letter	0.841 ± 0.004	0.785 ± 0.011	0.824 ± 0.002	0.836 ± 0.003	**0.877 ± 0.018**

(FB), Mixture Model (MM), Active Outlier (AO) and the natural combination approach (Ensemble Voting, a.k.a. EV). Ensemble Voting, similar to ensemble classifier using weighted majority voting (e.g., AdaBoost), is shown to yield very high accuracy in the classification problem [14]. However, through this experiment we point out that it is not very applicable for ensemble outlier detection. For Cumulative Sum of Feature Bagging, we simply sum up all score vectors after performing normalization. The AUC values of all methods are presented in Figure 1 and Table 4. Our approach (Mixed Ensemble) once again performs very well compared to other techniques. The results also show that when an ensemble contains both Ranking-based and Threshold-based techniques, natural sum-up scheme of Cumulative Sum as well as usual ensemble learning based on Weighted Majority Voting does not help much. Instead, we need special combine functions to deal specifically with different types of score vectors.

5   Conclusions and Future Work

In this paper, the problem of ensemble outlier detection in high-dimensional datasets were studied in detail. A formal notion of outlier score which helps to identify different types of outlier score vectors was introduced. Using the new notion, we presented a heterogeneous detector ensemble on random subspaces (HeDES) framework using different relevant combine functions to tackle the problem of heterogeneity of techniques. Extensive empirical studies on several popular real-life datasets show that our approach can outperform contemporary techniques in the field. In future work, we are considering a systematic extension to test all possible combine functions. Furthermore, we intend to expand the scope of our empirical studies by performing experiments on more large and high-dimensional datasets with different base outlier detection techniques. These will bring us a better understanding about the benefit of ensemble outlier detection for real-world applications. Last but not least, we would like to investigate how the selection of subspaces and base outlier detection techniques affects the detection accuracy of the ensemble.

References

1. Breunig, M.M., Kriegel, H.P., Ng, R.T., Sander, J.: LOF: Identifying density-based local outliers. In: SIGMOD, pp. 93–104 (2000)
2. Papadimitriou, S., Kitagawa, H., Gibbons, P.B., Faloutsos, C.: LOCI: Fast outlier detection using the local correlation integral. In: ICDE, pp. 315–324 (2003)
3. Knorr, E.M., Ng, R.T.: Algorithms for mining distance-based outliers in large datasets. In: VLDB, pp. 392–403 (1998)
4. Angiulli, F., Basta, S., Pizzuti, C.: Distance-based detection and prediction of outliers. IEEE Transactions on Knowledge and Data Engineering 18(2), 145–160 (2006)
5. Ramaswamy, S., Rastogi, R., Shim, K.: Efficient algorithms for mining outliers from large data sets. In: SIGMOD, pp. 427–438 (2000)
6. Aggarwal, C.C., Yu, P.S.: An effective and efficient algorithm for high-dimensional outlier detection. VLDB J. 14(2), 211–221 (2005)
7. Lazarevic, A., Ertöz, L., Kumar, V., Ozgur, A., Srivastava, J.: A comparative study of anomaly detection schemes in network intrusion detection. In: SDM (2003)
8. Beyer, K.S., Goldstein, J., Ramakrishnan, R., Shaft, U.: When is "nearest neighbor" meaningful? In: ICDT, pp. 217–235 (1999)
9. Ho, T.K.: The random subspace method for constructing decision forests. IEEE Transactions on Pattern Analysis and Machine Intelligence 20(8), 832–844 (1998)
10. Lazarevic, A., Kumar, V.: Feature bagging for outlier detection. In: KDD, pp. 157–166 (2005)
11. Kong, E.B., Dietterich, T.G.: Error-correcting output coding corrects bias and variance. In: ICML, pp. 313–321 (1995)
12. He, Z., Deng, S., Xu, X.: A unified subspace outlier ensemble framework for outlier detection. In: Fan, W., Wu, Z., Yang, J. (eds.) WAIM 2005. LNCS, vol. 3739, pp. 632–637. Springer, Heidelberg (2005)
13. Abe, N., Zadrozny, B., Langford, J.: Outlier detection by active learning. In: KDD, pp. 504–509 (2006)
14. Freund, Y., Schapire, R.E.: A decision-theoretic generalization of on-line learning and an application to boosting. Journal of Computer and System Sciences 55(1), 119–139 (1997)

15. Gao, J., Tan, P.N.: Converting output scores from outlier detection algorithms into probability estimates. In: ICDM, pp. 212–221 (2006)
16. Strehl, A., Ghosh, J.: Cluster ensembles - a knowledge reuse framework for combining multiple partitions. Journal of Machine Learning Research 3, 583–617 (2003)
17. Otey, M.E., Ghoting, A., Parthasarathy, S.: Fast distributed outlier detection in mixed-attribute data sets. Data Mining and Knowledge Discovery 12(2-3), 203–228 (2006)
18. Hawkins, D.M.: Identification of Outliers. Chapman and Hall, London (1980)

Mining Diversity on Networks

Lu Liu[1], Feida Zhu[3], Chen Chen[2], Xifeng Yan[4],
Jiawei Han[2], Philip Yu[5], and Shiqiang Yang[1]

[1] Tsinghua University
[2] University of Illinois at Urbana-Champaign
[3] Singapore Management University
[4] University of California at Santa Barbara
[5] University of Illinois at Chicago

Abstract. Despite the recent emergence of many large-scale networks in different application domains, an important measure that captures a participant's *diversity* in the network has been largely neglected in previous studies. Namely, diversity characterizes how diverse a given node connects with its peers. In this paper, we give a comprehensive study of this concept. We first lay out two criteria that capture the semantic meaning of diversity, and then propose a compliant definition which is simple enough to embed the idea. An efficient top-k diversity ranking algorithm is developed for computation on dynamic networks. Experiments on both synthetic and real datasets give interesting results, where individual nodes identified with high diversities are intuitive.

1 Introduction

Mining diversity is an important problem in various areas and finds many applications in real-life scenarios. For example, in information retrieval, people use information entropy to measure the diversity based on a certain distribution, e.g., one person's research interests diversity[12]. In social literature, diversity, which has been proposed under other terminologies like *bridging social capital*, proves its importance in many social phenomena. Putnam found that bridging social capital benefits societies, governments, individuals and communities[11]. In particular, bridging social capital helps reduce an individual's chance of catching certain diseases and the chance of dying, e.g., joining an organization cuts in half an individual's chance of dying within the next year, leading to the conclusion that "Network diversity is a predictor of lower mortality".

Mining diversity on network data is also critical for network analysis as network data emerge in abundance in many of today's real world applications. For example, advertisers may be very interested in the most diverse users in social network because they connect with users of many different types, which means "word of mouth" marketing on these users could reach potential customers of a much wider spectrum of varied tastes and budgets. In a research collaboration network of computer scientists, the diversity of a node could indicate the corresponding researcher's working style. A highly diverse researcher collaborates with colleagues from a wide range of institutions and communities, while a less diverse one might only work with a small group of people, e.g., his/her students. As such, an interesting query on such a network could be "Who

H. Kitagawa et al. (Eds.): DASFAA 2010, Part I, LNCS 5981, pp. 384–398, 2010.

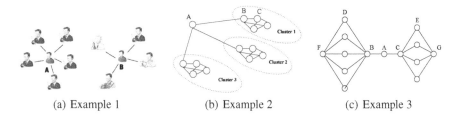

| (a) Example 1 | (b) Example 2 | (c) Example 3 |

Fig. 1. Three Examples

are the top ten diversely-collaborating researchers in the data mining community?". To illustrate the intuition of diversity on networks, let us look at an example.

Example 1. Consider a social network example in which nodes represent people and edges represent social connections between corresponding parties. Suppose we examine two nodes A and B in Fig.1(a) where A connects to 5 neighbors and B connects to 4 neighbors. However, the 5 neighbors of A are all from the same profession and the same community, while the 4 neighbors of B are from 4 different professions and/or communities. Here, although the neighborhood of B is smaller than that of A, it is obvious that B connects to a more diverse group of people, which could have important implications regarding the role he/she may play in the network, e.g., the profitability and impact if we are to choose a node to launch a marketing campaign.

Example 1 demonstrates that the diversity of a node on network is determined by the characteristics of its neighborhood. Greater difference between the neighbors translates into greater diversity of the node. In Example 1, the attributes or the labels are used to distinguish the neighbors. Then how can we measure the diversity if no attribute information is given? Example 2 illustrates another way to mine diversity which is based on the topological structure of the network.

Example 2. In Fig.1(b), comparing nodes A and C with the same degree of 3, it is easy to observe significant difference between the diversities of their neighborhoods. A connects to three neighbors, each of which belongs to a distinct community, while C connects to three closely connected neighbors that form a cohort. In many applications, A might be more interesting, because of its role of joining different persons together.

The two examples above give two different ways to measure diversity on networks. However regardless of using either neighborhood attributes or topology, certain common principles conveying the semantic meaning of diversity underlie any particular kind of computation or definition of diversity. In fact, it is our observation that there are two basic factors impacting the diversity measure on a network.

- *All else being equal, the greater the size of the neighborhood, the greater the diversity.*

  When all the neighbors are the same, in terms of both associated labels and neighborhood topology, more neighbors lead to a greater diversity.

- *The greater the differences among the neighbors, the greater the diversity.*

The neighbors can be distinguished either by their attributes and labels or by the topological information of the neighborhood. Whichever way, a larger difference should translate into a greater diversity.

The above two factors can also been treated as two criteria taken as the basis for proposing a reasonable definition for measuring diversity. In this paper, we focus on mining the diversity on network based on the topological structure. As pointed out in Section 2, existing measures like centrality can not accurately capture the notion of diversity in general, although certain degree of correlation between them can be observed for some data sets.

Our contributions can be summarized as follows.

- As far as we know, there has been no research work to investigate diversity on network structure data based on network characteristics. We are the first to propose the diversity concept on network and give two criteria that capture the semantic meaning of diversity.

- We investigate mining diversity based on topological information of a network, find a function which is simple enough to embed the two criteria and propose an efficient algorithms to obtain top-k diverse nodes on dynamic networks.

- Extensive experiment studies are conducted on synthetic and real data sets including DBLP. The results are interesting, where individual nodes identified with great diversities are highly intuitive.

The remaining of this paper is organized as follows. In Section 2, the related work is introduced and compared with our work. In Section 3, we propose a diversity definition based on topological information of network and develop an efficient top-k diversity ranking algorithm for dynamic networks in Section 4. The experiment results are reported in Section 5. Other kinds of diversity definition are discussed in Section 6. Section 7 concludes this study.

2   Related Work

As network data emerge in abundance in many of today's real world applications, many research work has been done on network analysis in recent literatures. Properties reflecting the overall characteristics of network, such as density, small world, hierarchical modularity and power law [15,5,2,10], have been observed for a long time. Compared to these, many measures that focus on individual components, e.g., degree, betweenness, closeness centrality, clustering coefficient, authority and etc, have also been proposed to distinguish the roles of nodes in network [13,9,14,7]. Besides, some other types of patterns, e.g., frequent subgraphs that focus more on local topologies [8,16], can be mined from the network.

However, all these measures are different from diversity and thus could not accurately capture the idea behind. Degree centrality, which is defined as the number of links for a given node, does not consider whether the neighbors are similar. Betweenness centrality assigns higher value to nodes appearing on the shortest paths of more node pairs. As we shall observe in the experiments, it might be correlated with diversity to some extent in particular data scenarios, but it is not a direct modeling of diverseness and thus would not satisfy the two criteria we have proposed in general. Closeness centrality, which measures the average shortest-path length from a node to all other nodes in the network, has similar problems. Moreover, such shortest-path based measures require the global computation of all-pair shortest paths, which leads to the time-consuming measure calculations on a large network. The clustering coefficient value of a node corresponds to the number of edges among its neighbors normalized by the maximum number of such edges; intuitively, with higher clustering coefficient, the neighbors have more connections among them and thus are more similar to each other, which leads to lower diversity. However, clustering coefficient does not consider the scale of the neighborhood and only counts number of edges as the sole parameter, which is inevitably restricted. Interestingly, it can be treated as a degenerated version of our diversity definition when the latter is confined to a very special setting.

3 Diversity Definition

In this section, we will propose concrete diversity definitions based on nodes' neighborhood topology. First, a simple definition is given out and the calculation results on Example 2 illustrate that it matches our intuition of diversity. Then we will propose a general definition and show its calculation results on more examples, in which we analyze its parameters and compare it with centrality.

3.1 Terminology and Representation

Let an undirected unweighted network be $G = \{(V, E) \mid V$ is a set of nodes and E is a set of edges, $E \in V \times V$, an edge $e = (i, j)$ connects two nodes i and $j, i, j \in V, e \in E\}$. $N(v)$ denotes the set of v's neighbors. $|N(v)|$ denotes the cardinality of $N(v)$, i.e., the number of neighbors. r is the radius of the neighborhood. If it is set to be 1, $N(v)$ is the set of directly connected nodes and $|N(v)|$ equals to the degree of node v. $N_{-u}(v)$ denotes the set of v's neighbors which excludes the nodes that become v's neighbors through u. For example, when $r = 1$, $N_{-u}(v)$ is the set of the direct neighbors of v except u itself; when $r = 2$, $N_{-u}(v) = N(v) - \{x|$there is only one shortest path from v to x which is through $u\}$. $L(i, j)$ denotes the length of shortest path from node i to node j.

3.2 A Simple Diversity Example

To illustrate the diversity measure, we first use a simple definition as below, which can get the intuitive results of Example 2 in Fig.1(b).

Definition 1. *Given a network G and a node $v \in V(G)$, the diversity $D(v)$ is defined as*

$$D(v) = \sum_{u \in N(v)} \left(1 - \frac{|N(v) \bigcap N(u)|}{|N(u)|}\right) \tag{1}$$

The underlying intuition of the definition is that, for a target node v, if a neighbor u has fewer connections with other neighbors of v, u is considered to contribute more to the diversity of v. Therefore the diversity of v is defined as the aggregation of every neighboring node u's contribution which equals to the probability of leaving the direct neighborhood of v through u [7].

Based on this definition, we can get that the diversity values of A,B,C in Example 2 are 3, 2, 1.167 respectively. The relative values match our intuition of diversity ranking on this network.

3.3   Diversity: General Definition

While the previous definition based on direct common neighborhood is simple and intuitive in some cases, we need more flexibility and generality in the diversity definition for most applications to capture the measure more accurately. As we discussed above, the diversity in general grows in proportion with the size of the neighborhood. With this notion of each neighbor contributing to the diversity of the central node, we propose the general definition of diversity in an aggregate form as follows.

Definition 2 *[Diversity]. The diversity of a node v is defined as an aggregation of each neighbor u's contribution to v's diversity.*

$$D(v) = \sum_{u \in N(v)} w_v(u) * F(u, v) \tag{2}$$

where $F(u, v)$ is a function measuring the diversity introduced by u. $w_v(u)$ is u's weight in the aggregation.

According to our guiding principles, if a neighbor u is less similar to other neighbors of v, u would contribute more to v's diversity. Thus $F(u, v)$ is a function evaluating the dissimilarity between u and other neighbors of v in the set radius r, i.e., the set $N_{-u}(v)$. In general, $F(u, v)$ can be defined as a linear function of the similarity between u and $N_{-u}(v)$ as

$$F(u, v) = 1 - \alpha * S(u, N_{-u}(v)) \tag{3}$$

$S(u, N_{-u}(v))$ is a function measuring the similarity between u and $N_{-u}(v)$ up to a normalization. α indicates its weight, which can be set empirically. We define $S(u, N_{-u}(v))$ as the average similarity between u and each node x of $N_{-u}(v)$. There are various ways to measure the similarity between two nodes u and x, e.g., shortest path is a reasonable choice for many real-world scenario. However, computing shortest paths on a global scale is inefficient. Fortunately, since diversity is a local property defined on a neighborhood with a set radius, we can use the following definition based on local shortest path computation.

Definition 3 *[Similarity Between Node Pair]. The similarity between two nodes u and x is defined as:*

$$S(u, x) = \begin{cases} \delta^{(l-1)}, 0 < \delta < 1 & \text{if } L(u, x) = l \leq r \\ 0 & \text{otherwise} \end{cases}$$

Table 1. Computation Results for Example 2

Node	DC	BC	Diversity ($\alpha = 0.8 \; \delta = 0.8$)			
			r=1	r=2	r=3	r=4
A	3	48	3	5.208	5.208	5.208
B	4	27	1.6	2.763	4.147	4.245
C	3	0	0.867	1.767	2.962	4.489

If two nodes are too far apart, in the sense that their distance is larger than the neighborhood radius r of our interest, their similarity is considered to be zero; Otherwise, their similarity is inversely proportional to their distance. δ is a damping factor to reflect the notion that nodes farther apart share less similarity. The effect of δ is further explored in Section 3.4. With the similarity between a pair of nodes defined, we can give the definition of similarity between a node and a set of nodes.

Definition 4 *[Similarity Between Node and Node Set]. The similarity between a node u and a set of nodes $N_{-u}(v)$ is defined as*

$$S(u, N_{-u}(v)) = \frac{\sum_{x \in N_{-u}(v) \cap N_{-v}(u)} (w_v(x) * S(u, x))}{\sum_{x \in N_{-v}(u)} S(u, x)} \qquad (4)$$

where $w_v(x)$ is the weight of x in v's neighborhood.

The purpose of setting weight, e.g., $w_v(u)$ and $w_v(x)$, is to prioritize all the nodes in v's neighborhood. There are more than one possible ways to define the weights. In this paper, we define $w_v(x) = S(v, x)$ based on the argument that distance-based similarity is an appropriate way to evaluate the priority of a node in v's neighborhood when a radius larger than 1 is needed. Putting it together, we have

$$S(u, N_{-u}(v)) = \frac{\sum_{x \in N_{-u}(v) \cap N_{-v}(u)} (S(v, x) * S(u, x))}{\sum_{x \in N_{-u}(v)} S(u, x)} \qquad (5)$$

It is easy to notice that the definition in Section 3.2 is a special case of this general definition.

3.4   Examples and Analysis

To illustrate the intuition of the diversity measure above and analyze the impact of its parameters, we get the computation results for Example 2 and 3 in Fig.1(b)(c) with changing parameters and show them in Table 1 and 2, where the computation results of degree and betweenness centrality are also listed[1].

Comparison with Degree and Betweenness. Example 2 demonstrates that diversity does not equal to degree. E.g., A and C are with the same degree but their diversities differ a lot. In Example 3, as the neighbors of all the nodes are not directly connected with

---

[1] DC and BC denote degree and betweenness centrality for short respectively in this paper.

Table 2. Computation Results for Example 3

Node	DC	BC	Diversity ($\alpha = 0.8$, $\delta = 0.5$)						Diversity ($\alpha = 0.8$, $\delta = 0.8$)					
			r=1	r=2	r=3	r=4	r=5	r=6	r=1	r=2	r=3	r=4	r=5	r=6
A	2	42	2	4.70	4.74	4.74	4.74	4.74	2	5.31	4.97	4.97	4.97	4.97
B	6	47	6	3.19	3.92	3.99	3.99	3.99	6	3.04	4.37	4.39	4.39	4.39
C	5	43	5	2.98	3.90	3.96	3.96	3.96	5	2.85	4.50	4.51	4.51	4.51
D	2	1.6	2	2.39	2.69	3.19	3.24	3.24	2	2.33	2.96	4.25	4.38	4.37
E	2	2.25	2	2.16	2.48	3.10	3.15	3.15	2	2.14	2.82	4.41	4.51	4.51
F	5	5	5	2.34	2.73	3.15	3.39	3.41	5	2.13	3.01	4.11	5.06	5.18
G	4	3	4	2.08	2.47	2.90	3.19	3.21	4	1.92	2.83	3.94	5.13	5.25

each other, the value of diversity equals to degree when $r = 1$. But when r increases from 1 to 2, the diversity ranking changes. Example 3 demonstrates that diversity does not equal to betweenness centrality either. E.g., betweenness centrality of A and C in Fig.1(c) are roughly the same, but their diversities are obviously different.

Radius of Neighborhood. Table 1 and 2 show all the calculation results when r changes from 1 to the possible maximal value (it means that the neighborhood would no longer change when r increases more). It is found that a larger radius may lead to counter-intuitive ranking results. However, it is our belief and definition that diversity should measure an aspect of a node's interaction with its local neighborhood. To judge a node's diversity on a global scale (e.g., considering all the nodes as neighbors of the center node) is semantically controversial. On the other hand, it is discovered that *"small world"* phenomenon applies to a wide range of networks such as the Internet, the social networks like Facebook and the bio-gene networks, which means most nodes in these networks are found to be within a small number of hops from each other. In particular, the theory of "six degrees of separation" indicates that in social network most people can reach any other individuals through six persons. It follows that when r increases beyond a small number, a node's diversity would be aggregated by nearly all the nodes' contributions in the network, which deviates away from what diversity is meant to capture based on our previous discussion. Therefore, a small radius should be chosen in the computation. Furthermore, the results show that the top-k results in the diversity ranking become stable when $r = 2$ or $r = 3$ in most cases.

Damping Factor. The damping factor δ controls a neighbor's impact on the diversity measure in relation to its distance to the central node. Intuitively, neighbors far away should have smaller impact on the central node's diversity. As we discussed above, diversity is influenced mainly by two factors: the size of the neighborhood and the difference among the neighbors. On real data sets, as the radius increases, the number of neighbors increases enormously, which makes the size of neighborhood be a dominating factor of diversity computation. This imbalance would sometimes distort the ranking result. Therefore an appropriate damping factor can be chosen to balance the two factors, e.g., $\delta = 0.5$ in Table 2 .

4   Top-K Diversity Ranking Algorithm

In real applications, top-k diversity ranking for query-based dynamic networks is often required in data scenarios. Still take the DBLP example. Suppose the original input network is the entire DBLP co-authorship network G generated by including papers from all the eligible conferences. If a user poses a query "Who are the most diverse researcher in Database community?", it would result in the dropping of edges which correspond to papers published in non-database conferences. Diversity ranking is then computed on the resulting sub-network. The challenge for computing measures on dynamic networks is that it is no longer possible to compute once for all and answer all the queries by retrieving saved results. As such, the task is to develop efficient algorithms for top-k diversity measure on dynamic networks generated by user queries.

Our strategy is to find ways to quickly estimate an upper-bound of $D(v)$ for each node v in the new sub-network. Meanwhile we store the smallest diversity value of top k candidates which is denoted as l_bound. If the upper-bound of v is smaller than l_bound, it can be tossed away to save computation. Otherwise we perform more costly computation to get the accurate measure value of $D(v)$ and update l_bound.

We obtain the upper-bound based on two scenarios. First, the diversity of a node should be smaller than the cardinality of its neighborhood. When all the neighbors have no connections, the diversity reaches the maximal value. On the other hand, as the query-based dynamic network is a subgraph of original network, one node's neighborhood should be the sub-set of its original neighborhood. Thus two nodes' similarity should be smaller than their similarity on the original network. By using the monotonicity property, we obtain the upper-bounds and propose an efficient top-k diversity ranking algorithm.

For any quantity W computed on a network G, we use W' to represent the same quantity computed on a sub-network $G' \subseteq G$. We use $N_u(v)$ to denote the set of nodes in v's r-neighborhood which can only be reached by shortest paths passing through u, i.e., $N_u(v) = N(v) \setminus N_{-u}(v)$.

Lemma 1. *For a network G and a node $v \in V(G)$, $D(v) \leq \sum_{u \in N(v)} w_v(u)$.*

Lemma 1 is due to the fact that $F(u, v) \leq 1$ by definition and $F(u, v) = 1$ only when all the neighbors of v have no connections.

Lemma 2. *For a network G and a sub-network $G' \subseteq G$, for any two nodes $u, v \in V(G)$, $0 \leq S'(u, v) \leq S(u, v) \leq 1$.*

Lemma 2 is due to the fact that the length of the shortest path $L(u, v)$ for any two nodes u and v in G increases monotonically in sub-network G'.

We define some notations to simplify the formulas. We set $C(v) = \sum_{u \in N(v)} w_v(u)$. According to Lemma 1, $C(v)$ is an upper bound of $D(v)$. Since in this paper we define $w_v(u) = S(u, v)$, we also have $C(v) = \sum_{u \in N(v)} S(u, v)$. Hence, for any sub-network $G' \subseteq G$, $C'(v) = \sum_{u \in N'(v)} S'(u, v)$. We denote $S = \sum_{x \in N_{-u}(v) \cap N_{-v}(u)} (S(v, x) * S(u, x))$ for short.

Input: Sub-network G' and K
Output: A set T of K nodes with top diversity
1: $Q \leftarrow$ Queue of $V(G')$, sorted by $C'(v)$
2: $l\_bound \leftarrow 0; T \leftarrow \emptyset;$
3: Pop out the top node v in Q
4:      **if** $C'(v) < l\_boundQ$ **return T;**
5:      **for each** $u \in N'(v)$
6:          Compute $Upper(u,v)$;
7:          $UP(v) \leftarrow UP(v) + min\{1, Upper(u,v)\}$
8:      **if** $UP(v) < l\_bound$ **continue;**
9:      **for each** $u \in N'(v)$
10:         Compute $F'(u,v)$;
11:         $D'(v) \leftarrow D'(v) + F'(u,v)$;
12:     **if** $D'(v) > l\_bound$ insert v into T
13:     **if** $|T| > K$
14:         remove the last node in T;
15:         $l\_bound \leftarrow$ smallest diversity in T;
16: **return** T;

Algorithm 1. Top-K Diversity Ranking

Since $0 \le S(u,v), S'(v,x) \le 1$ for any nodes u and v, we have for any node x,

$$S(v,x) - S'(v,x) + S(u,x) - S'(u,x)$$
$$\ge (S(v,x) - S'(v,x)) * S(u,x) + (S(u,x) - S'(u,x)) * S'(v,x)$$
$$= S(v,x) * S(u,x) - S'(u,x) * S'(v,x)$$

If we sum up by x for the above inequality, since $S(v,x) = 0$ for $x \notin N(v)$ (resp. for $S(u,x)$), and $S(v,x) * S(u,x) = 0$ for $x \notin (N(v) \bigcap N(u))$, we have

$$C(v) - C'(v) + C(u) - C'(u) \ge S - S' + \sum_{x \in A} S(u,x) * S(v,x) - \sum_{x \in B} S'(u,x) * S'(v,x)$$

where $A = N(u) \cap N(v) - N_{-v}(u) \cap N_{-u}(v)$. $B = N'(u) \cap N'(v) - N'_{-v}(u) \cap N'_{-u}(v)$. As $B \subseteq A$, $S(u,x) \ge S'(u,x)$, $\sum_{x \in A} S(u,x) * S(v,x) - \sum_{x \in B} S'(u,x) * S'(v,x) \ge 0$. Therefore,

$$C(v) - C'(v) + C(u) - C'(u) \ge S - S'$$

So

$$F'(u,v) = 1 - \alpha * \frac{S'}{\sum_{x \in N_{-v}(u)} S'(u,x)}$$
$$\le 1 - \alpha * \frac{(S - (C(u) - C'(u) + C(v) - C'(v)))}{\sum_{x \in N_{-v}(u)} S'(u,x)}$$
$$\le 1 - \alpha * \frac{(S - (C(u) - C'(u) + C(v) - C'(v)))}{C'(u)}$$
$$= Upper(u,v)$$

We thus derived another upper-bound $Upper(u,v)$ for $F'(u,v)$. Thus $F'(u,v) \le min\{1, Upper(u,v)\}$.

To use this upper-bound, we compute S for each pair (u, v) which are each other's r-neighbors in the original network and store these values in the pre-computation stage. Likewise, we also compute and store $C(v)$. When the user inputs a query, we just need to compute $C'(u)$ and $C'(v)$ for the sub-network, which is simply a local neighbor checking, to get $Upper(u, v)$.

The top-k diversity ranking algorithm is as shown in Algorithm 1.

5   Experimental Results

In this section, we did extensive experiments on both synthetic and real data and generated some interesting results. The most diverse nodes on different types of networks are highlighted to illustrate an intuition of diversity. We compare the results of diversity with two classical centrality measures – degree and betweenness centrality and show both the difference and the correlation between them. At last, we implemented our top-k ranking algorithm on dynamic network and demonstrate its efficiency.

5.1   Results on Synthetic Network

We first applied the algorithm to a synthetic network consisting of 92 nodes and 526 edges shown in Fig.2. The network was generated as following: first, we generated three clusters of nodes; in each cluster the nodes only connect with the nodes in the same cluster randomly; then we generated other 10 nodes connecting to any node arbitrarily.

Fig.2 shows the top 20 nodes ranked by degree, betweenness centrality and diversity respectively. The top 10 nodes are highlighted with red color and the sizes of nodes are linear with the ranking (The higher the rank, the larger the size). The second top 10 nodes are highlighted with blue color [1].

This figure demonstrates that the nodes which connect more nodes from different clusters tend to be more diverse. When r increases from 1 to 2, the diverse nodes will further move to the connection points of clusters. It seems that diversity is highly correlated with betweenness centrality on this network. Their correlation coefficients are

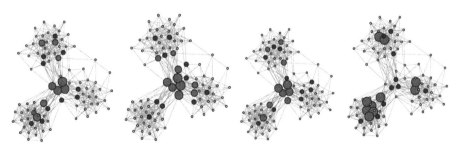

(a) Diversity when r = 1 (b) Diversity when r = 2(c) Betweenness Centrality (d) Degree Centrality

Fig. 2. Synthetic network results

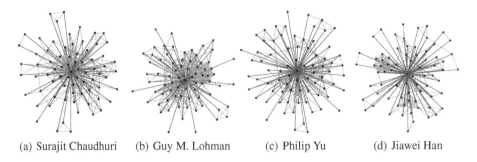

(a) Surajit Chaudhuri   (b) Guy M. Lohman   (c) Philip Yu   (d) Jiawei Han

Fig. 3. Neighborhood of four authors

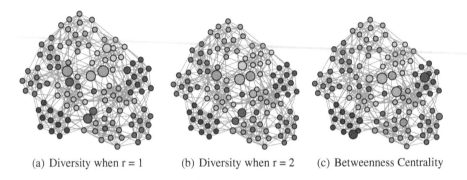

(a) Diversity when r = 1   (b) Diversity when r = 2   (c) Betweenness Centrality

Fig. 4. Network of American football games

shown in Table 5[2]. This large correlation is caused by the characteristic of this network structure. As the network consists of three clusters and some other nodes connecting the clusters, the nodes with high betweenness centrality values also tend to locate on the connection points of clusters. However, diversity is different from betweenness centrality as we analyzed above. And we will show that they are lowly correlated on some networks with different structures.

5.2   Results on DBLP Network

We extracted the network of co-authorship on conference SIGMOD, VLDB and ICDE from DBLP data[3], which means that if two authors cooperated a paper published on these conferences, an edge was generated to link them. Table 3 compares the top 20 author ranked by diversity and betweenness centrality. We set $\alpha = 0.8$, $\delta = 0.5$. As it is proved that on an undirected network degree is consistent to authority (eigenvector centrality) obtained by PageRank [4], we can also treat degree as an authority value and compare it with diversity. Thus Table 3 demonstrates that diversity ranking is different from betweenness centrality ranking as well as authority (degree).

---

[2] SN denotes synthetic network for short.

[3] This network is called as "DB" for short in the remainder of the paper.

Table 3. Author Ranking Results on DB

Diversity when r = 1			Diversity when r = 2		Betweenness Centrality	
Author	DC	Value	Author	Value	Author	Value
Rakesh Agrawal	98	50.94	Rakesh Agrawal	450.84	Rakesh Agrawal	971048.8
David J. DeWitt	118	50.60	David J. DeWitt	434.77	Michael J. Carey	785089.9
Hector Garcia-Molina	98	48.20	Surajit Chaudhuri	402.93	Christos Faloutsos	747502.4
Divesh Srivastava	89	46.75	Michael J. Carey	386.85	David J. DeWitt	746523.0
Surajit Chaudhuri	73	45.53	Divesh Srivastava	373.34	Umeshwar Dayal	737304.2
Raghu Ramakrishnan	90	44.95	Jennifer Widom	367.29	Michael Stonebraker	705067.8
H. V. Jagadish	82	41.53	Hector Garcia-Molina	364.51	Hector Garcia-Molina	685955.0
Hamid Pirahesh	83	41.45	Raghu Ramakrishnan	360.98	Surajit Chaudhuri	631760.8
Michael J. Carey	115	41.05	Michael J. Franklin	360.09	Philip A. Bernstein	628037.5
Michael Stonebraker	113	40.93	Jeffrey F. Naughton	349.62	H. V. Jagadish	604977.7
Jennifer Widom	84	40.29	Hamid Pirahesh	343.99	Divesh Srivastava	562573.6
Christos Faloutsos	94	39.21	H. V. Jagadish	339.80	Raghu Ramakrishnan	555216.0
Jeffrey F. Naughton	95	38.86	Gerhard Weikum	333.76	Gerhard Weikum	540029.5
Guy M. Lohman	73	37.98	Umeshwar Dayal	330.88	Elisa Bertino	533129.3
Michael J. Franklin	76	37.42	Philip A. Bernstein	327.75	Dennis Shasha	526097.3
Nick Koudas	69	37.32	Michael Stonebraker	326.91	Jiawei Han	520527.3
C. Mohan	66	36.19	Abraham Silberschatz	326.70	Michael J. Franklin	518074.6
Gerhard Weikum	80	34.11	C. Mohan	322.23	Gio Wiederhold	517573.1
Philip A. Bernstein	61	33.45	Guy M. Lohman	320.67	Kian-Lee Tan	513349.0
Rajeev Rastogi	75	33.36	Bruce G. Lindsay	312.36	C. Mohan	509267.1

Table 3 demonstrates some interesting results. For example, although the difference between the degrees of R. Agrawal and D. DeWitt is as large as 20, their diversities are nearly the same. The reason should be that R. Agrawal is from industry area and has worked in many companies, e.g., Microsoft, IBM Almaden Research Center, Bell Laboratories, etc. Therefore, Agrawal's cooperators are very diverse. We also compare the diversity of two authors, Surajit Chaudhuri and Guy M. Lohman, who have the same degree. Their neighborhoods as shown in Fig.3(a) and Fig.3(b) demonstrate that Lohman's cooperators connect with each other more closely than Chaudhuri's. Therefore the diversity of Chaudhuri is larger than Lohman as obtained in Table 3.

We can also get similar results on the co-author network of conference KDD and ICDM from DBLP data[4] as shown in Table 4. For example, although Philip S. Yu and Jiawei Han's degrees are roughly the same, their diversities differ a lot, which can also be demonstrated from their neighborhoods as shown in Fig.3(c) and Fig.3(d). The reason should be that Philip S. Yu had worked in industry area and has cooperated with many different persons who have no close relationship. Thus his diversity value is much larger than Jiawei Han.

5.3   Results on Network of American Football Games

We obtained another social network of American football games between Division IA colleges during regular season Fall 2000 [6]. In this data, nodes represent teams and

---

[4] The network is called as "DM" for short in the remainder of the paper.

Table 4. Author Ranking Results on DM

Diversity when r = 1			Diversity when r = 2		Betweenness Centrality	
Author	DC	Value	Author	Value	Author	Value
Philip S. Yu	76	39.72	Philip S. Yu	160.82	Philip S. Yu	544203.3
Jiawei Han	73	26.25	Haixun Wang	107.15	Christos Faloutsos	335598.8
Christos Faloutsos	60	24.77	Jiawei Han	96.85	Heikki Mannila	179383.3
Jian Pei	51	20.37	Christos Faloutsos	93.26	Mohammed Javeed Zaki	158551.1
Haixun Wang	32	19.21	Ke Wang	92.37	Jiawei Han	132043.5
Ke Wang	36	17.30	Jian Pei	91.13	Eamonn J. Keogh	123389.1
Heikki Mannila	39	16.54	Ada Wai-Chee Fu	82.14	Padhraic Smyth	116926.1
Bing Liu	32	15.15	Jianyong Wang	75.56	Jian Pei	112538.7
Mohammed Javeed Zaki	30	14.50	Charu C. Aggarwal	74.11	Charu C. Aggarwal	107042.4
Eamonn J. Keogh	37	14.32	Wei Fan	73.63	Bing Liu	103081.9
Wei Fan	29	14.26	Wei Wang	71.52	Gregory Piatetsky-Shapiro	101267.2
Padhraic Smyth	32	13.89	Bing Liu	70.26	Srinivasan Parthasarathy	95692.4
Wei-Ying Ma	34	13.73	Spiros Papadimitriou	69.17	Ada Wai-Chee Fu	91889.1
Ada Wai-Chee Fu	25	13.70	Hong Cheng	69.14	Ke Wang	90909.1
Qiang Yang	41	13.68	Eamonn J. Keogh	67.69	Haixun Wang	88484.7
Vipin Kumar	29	13.21	Alexander Tuzhilin	64.71	Vipin Kumar	82333.2
Wei Wang	39	13.13	Jiong Yang	63.58	Rakesh Agrawal	80409.2
Hui Xiong	27	13.02	Hongjun Lu	62.50	Huan Liu	79472.5
Huan Liu	28	12.92	David W. Cheung	60.45	Spiros Papadimitriou	78784.6
Alexander Tuzhilin	17	12.16	Michail Vlachos	60.28	Prabhakar Raghavan	77359.7

edges denote that two teams had a game. Fig.4 shows the top 10 nodes with largest diversity and betweenness centrality, which are highlighted by the larger sizes of nodes. The degrees of all the nodes are roughly the same, with the range from 8 to 12. Thus we do not show the degree ranking results. The data also contain the node labels which indicate the conference that each team belongs to. We use different colors to distinguish the labels in the figure. Therefore the results illustrate that the diversity calculated based on network topology is consistent to the diversity based on node labels, which means that the nodes whose neighbors are from more clusters tend to be more diverse. Table 5[5] demonstrates that on this network the diversity is lowly correlated with degree and betweenness centrality.

5.4 Performance Comparison

Fig.5(a) compares the running time of Top-K algorithm with the time of ranking all the nodes on DB and DM networks. It demonstrates that Top-K algorithm is much more efficient and can meet online query needs. We also implemented an efficient betweenness algorithm [3] and compared it with diversity. Fig.5(b) demonstrates that diversity calculation is much faster than betweenness calculation. The reason is that to some extent betweenness centrality is a global measure based on the shortest path calculation between all the pair-nodes which is very time consuming while the diversity measure only needs to count the local neighborhood.

---

[5] FN denotes the social network of American football games for short.

Table 5. Correlation Coefficients of Metrics

Network	#node	#edge	DC vs. BC	DC vs. Diversity r = 1	DC vs. Diversity r = 2	BC vs. Diversity r = 1	BC vs. Diversity r = 2
SN	92	526	0.470	0.874	0.399	0.709	0.828
FN	115	616	0.151	0.345	0.224	0.413	0.463
DB	7640	22309	0.810	0.881	0.819	0.829	0.716
DM	3405	6496	0.665	0.908	0.683	0.701	0.576

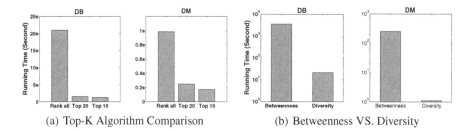

(a) Top-K Algorithm Comparison   (b) Betweenness VS. Diversity

Fig. 5. Performance comparison

6   Discussion

As diversity is a highly subjective concept, we do not think there exists one optimal definition which is applicable for all scenarios. Rather than narrowing ourselves down to one specific definition, we are fully aware of other possible definitions that may be better geared for other applications. For example, a highly intuitive definition can be based on clustering, where nodes are first assigned labels by certain clustering algorithm and then diversity is computed by calculating the information entropy of the cluster distribution of neighbors. This kind of definition needs to at least solve the following issues: (i) The choice of the clustering algorithm dictates the resulting clusters, which in turn determines the diversity computation. The decision on clustering parameters becomes critical and difficult. (ii) The internal cohesion of clusters, which reflects the topology of network, is also an important component for diversity. The diversity of a node connected with a compact cluster should be different from the diversity of a node connected with a loose cluster. Therefore in general still lots of aspects and factors should be exploited for the clustering-based definition. In this paper, we propose a straightforward diversity definition based on the similarity between neighbors instead of solving these problems of clustering.

7   Conclusion

In this paper, we investigated the problem of mining diversity on networks. We gave two criteria to characterize the semantic meaning of diversity and to provide the basis of proposing a reasonable measure definition. Then we studied diversity measure based on network topology and picked a concrete definition to embed the idea. We developed an efficient algorithm to find top-k diverse nodes on dynamic networks. Extensive experiment studies were conducted on synthetic and real data sets. The results are interesting, where individual nodes identified with high diversities are intuitive.

Acknowledgements

The work was supported in part by the U.S. National Science Foundation grants IIS-08-42769 and IIS-09-05215, and the NASA grant NNX08AC35A, and 973 Program of China grant 2006CB303103, and the State Key Program of National Natural Science of China grant 60933013. Any opinions, findings, and conclusions expressed here are those of the authors and do not necessarily reflect the views of the funding agencies.

References

1. http://graphexploration.cond.org/index.html
2. Barabasi, A.-L., Oltvai, Z.N.: Network biology: Understanding the cell's functional organization. Nat. Rev. Genet. 5(2), 101–113 (2004)
3. Brandes, U.: A faster algorithm for betweenness centrality. Journal of Mathematical Sociology 25, 163–177 (2001)
4. Cover, T.M., Thomas, J.A.: Elements of information theory. John Wiley & Sons Inc., Chichester (2006)
5. Faloutsos, M., Faloutsos, P., Faloutsos, C.: On power-law relationships of the internet topology. In: SIGCOMM, pp. 251–262 (1999)
6. Girvan, M., Newman, M.E.J.: Community structure in social and biological networks. Proceedings of the National Academy of Sciences 99(12) (2002)
7. Hwang, W., Kim, T., Ramanathan, M., Zhang, A.: Bridging centrality: graph mining from element level to group level. In: KDD, pp. 336–344 (2008)
8. Kuramochi, M., Karypis, G.: Frequent subgraph discovery. In: ICDM, pp. 313–320 (2001)
9. Lawrence, P., Sergey, B., Motwani, R., Winograd, T.: The pagerank citation ranking: Bringing order to the web. Technical report, Stanford University (1998)
10. Leskovec, J., Kleinberg, J.M., Faloutsos, C.: Graphs over time: densification laws, shrinking diameters and possible explanations. In: KDD, pp. 177–187 (2005)
11. Putnam, R.D.: Bowling Alone: America's Declining Social Capital. Journal of Democracy 6(1) (1995)
12. Rosen-Zvi, M., Griffiths, T., Steyvers, M., Smyth, P.: The author-topic model for authors and documents. In: Proceedings of the 20th conference on Uncertainty in artificial intelligence, Arlington, VA, USA, pp. 487–494. AUAI Press (2004)
13. Stephenson, K., Zelen, M.: Rethinking centrality: Methods and examples. Social Networks 11(1), 1–37 (1989)
14. Wasserman, S., Faust, K.: Social Network Analysis, Methods and Applications. Cambridge University Press, Cambridge (1994)
15. Watts, D.J., Strogatz, S.H.: Collective dynamics of 'small-world' networks. Nature 393(6684), 440–442 (1998)
16. Yan, X., Han, J.: gSpan: Graph-based substructure pattern mining. In: ICDM, pp. 721–724 (2002)

Mining Regular Patterns in Data Streams

Syed Khairuzzaman Tanbeer, Chowdhury Farhan Ahmed, and Byeong-Soo Jeong

Department of Computer Engineering, Kyung Hee University
1 Sochun-dong, Kihung-eup, Youngin-si, Kyonggi-do, Republic of Korea, 446-701
{tanbeer,farhan,jeong}@khu.ac.kr

Abstract. Discovering interesting patterns from high-speed data streams is a challenging problem in data mining. Recently, the support metric-based frequent pattern mining from data stream has achieved a great attention. However, the occurrence frequency of a pattern may not be an appropriate criterion for discovering meaningful patterns. Temporal regularity in occurrence behavior can be a key criterion for assessing the importance of patterns in several online applications such as market basket analysis, gene data analysis, network monitoring, and stock market. A pattern can be said *regular* if its occurrence behavior satisfies a user-given interval in the data steam. Mining *regular* patterns from static databases has recently been addressed. However, even though mining *regular* patterns from stream data is extremely required in online applications, no such algorithm has been proposed yet. Therefore, in this paper we develop a novel tree structure called Regular Pattern Stream tree (RPS-tree), and an efficient mining technique for discovering *regular* patterns over data stream. Using a sliding window method the RPS-tree captures the stream content, and with an efficient tree updating mechanism it constantly processes exact stream data when the stream flows. Extensive experimental analyses show that our RPS-tree is highly efficient in discovering *regular* patterns from a high-speed data stream.

Keywords: Data mining, data stream, pattern mining, regular pattern, sliding window.

1 Introduction

A data stream is a continuous, unbounded, and timely ordered sequence of data elements generated at a rapid rate. Unlike traditional static databases, stream data, in general, has additional processing requirements; i.e., each data element should be examined at most once and processed as fast as possible with the limitation of available memory. Even though mining user-interest based patterns from data stream has become a challenging issue, interests in online stream mining for discovering such patterns dramatically increased [1], [2], [10], [11], [12].

Mining frequent patterns [3], [6], from transactional databases has been actively and widely studied in stream data mining [2], [10], [11], [12] for over a decade. The rationale behind mining frequent patterns is that only patterns occurring at a high frequency in a database are of interest to users. Therefore, a pattern is called frequent if its occurrence frequency (i.e., support) in the database exceeds the user-given

H. Kitagawa et al. (Eds.): DASFAA 2010, Part I, LNCS 5981, pp. 399–413, 2010.

support threshold. However, the occurrence frequency may not always represent the significance of a pattern. The other important criterion for identifying the interestingness of a pattern might be the shape of occurrence i.e., whether the pattern occurs periodically, irregularly, or mostly in a specific time interval.

The significance of patterns with temporal regularity can be revealed in a wide range of applications where users might be interested on the occurrence behavior (*regularity*) of patterns rather than just the occurring frequency. For example, in a retail chain data, some products may be sold more regularly than other products. Thus, even though both of the products are sold frequently over the entire selling history or for a specific time period (e.g., for a year), the products still need to be managed independently. That is, it is necessary to identify a set of items that are sold together at a regular interval for a specified time period. Also, to improve web site design, a site administrator may be interested in regularly visited web page sequences rather than web pages that are heavily hit only for a specific period. As for genetic data analysis, the set of all genes that co-occur at a fixed interval in DNA sequence may carry more significant information to scientists. Again, in stock market the set of stocks indices that rise at a regular interval might be of special interest to stock brokers and traders. The pattern regularity can also be a useful metric among other applications such as network monitoring, telecommunications or the sensor network.

Traditional frequent pattern mining techniques fail to uncover such *regular* patterns because they focus only on the high frequency patterns. Recently, Tanbeer et al. [4] studied the pattern appearance behaviour in static transactional databases. With the help of a *regularity* measure determined by the maximum interval at which a pattern occurs in a database, the study introduced a tree structure called RP-tree to discover *regular* patterns satisfying a user-given *regularity* threshold. The RP-tree requires two database scans and contains the information for only *regular* items in the database. However, with the recent development of technology several online applications require to handle a bulk amount of data in the form of data stream. For example, retail chains record millions of transactions, telecommunications companies connect thousands of calls, and popular web sites log millions of hits at a regular basis. It is, therefore, obvious that, because of the two database scans and the prior knowledge about the *regularity* threshold requirements, the RP-tree is inefficient in discovering *regular* patterns in the above data stream scenarios. Hence, to find *regular* patterns efficiently from data streams we require efficient algorithm that can capture the stream content with one scan and can competently mine the resultant patterns.

Motivated from the above demand, we address a new problem of mining *regular* patterns in data streams. We propose a novel single-pass tree structure, called the RPS-tree (Regular Pattern Stream tree), to capture the stream contents in a compact manner. Using an efficient pattern growth-based mining technique the RPS-tree can mine set of the *regular* patterns in stream data for a user-given *regularity* threshold.

To efficiently handle (or mine) continuously-generated data streams, sliding windows [10], [11], [12] are commonly used because of its flexibility to monitor the stream data at runtime. As new transactions arrive, the oldest transactions in the sliding window expire. Because of the efficient stream handling mechanism, we will exploit the sliding window in our approach.

Main idea of our RPS-tree is to develop a simple, but yet powerful, tree structure that captures the stream content for the current window in full with a single scan in a canonical item order. Such construction feature enables its easy maintenance without any information loss during the slide of window. To the best of our knowledge, RPS-tree is the first effort to mine *regular* patterns from data streams. The experimental analyses on both real and synthetic data show that mining *regular* patterns from data streams with our RPS-tree is more efficient than that with the RP-tree.

The rest of the paper is organized as follows. Section 2 summarizes the existing algorithms related to our work. The detail discussion on RP-tree is also presented here. Section 3 introduces the problem of *regular* pattern mining in data stream. The structure and mining of our proposed RPS-tree are given in Section 4. We report our experimental results in Section 5. Finally, Section 6 concludes the paper.

2   Related Work

Many algorithms have been proposed for mining frequent patterns [3], [8] from static database, since its introduction by Agrawal et al. [6]. Han et al. [3] proposed the frequent pattern tree (FP-tree) and the FP-growth algorithm to mine frequent patterns with a pattern growth approach using only two database scans. Even though FP-growth algorithm has been highly efficient, it is not suitable for mining stream data because of its two database scans requirement.

A large number of techniques have been developed recently to mine frequent patterns from data stream [2], [10], [11], [12]. Algorithms in [10] and [11] use the sliding window concept to capture stream content with the help of a tree-based data structure. To facilitate the efficient mining and tree updating, the DSTree in [11] and the CPS-tree in [10] are constructed for the full window content. Using the FP-growth [3] algorithm both approaches discover the exact set of recent frequent patterns from the data stream with single scan. However, none of the support metric-based frequent pattern mining models is appropriate for discovering the special occurrence (i.e., *periodic* or *cyclic* or *regular*) characteristics of patterns from data stream.

Mining *periodic* patterns [1], [7], *cyclic* patterns [7], [9] and *regular* patterns [4] in static databases have been well-addressed over the last decade. *Periodic* pattern mining problem in time-series data focuses on the cyclic behavior of patterns either in the whole [7] (*full periodic patterns mining*) or at some point [1] (*partial periodic patterns mining*) of time-series. Such pattern mining has also been studied as a wing of sequential pattern mining [5], [9] in recent years. In [9], the authors extended the basic form of sequential patterns to cyclically repeated patterns. A progressive time list-based verification method to mine *periodic* patterns from a sequence of event sets was proposed in [5]. Ozden et al. [7] proposed a method to discover the association rules [6] occurring cyclically in a transactional database. Although mining *periodic* and *cyclic* patterns are closely related to our work, these algorithms cannot be directly applied for finding *regular* patterns from a data stream because they consider time-series or sequential data where the database is static.

Recently, Tanbeer et al. [4] proposed the Regular Pattern tree (RP-tree in short) to exactly mine the *regular* patterns from static transactional databases. The study defines a new *regularity* measure for a pattern determined by the maximum interval at which the same pattern occurs in a database.

Table 1. A transactional data stream (DS)

Id	Transaction	Id	Transaction	Id	Transaction
1	a, c, e, f	4	c, d, e	7	a, c, d, e
2	b, c, f	5	a, b, c, e	8	c, d, e, f
3	b, c, f	6	c, d, e	9	a, c

Construction of an RP-tree requires two database scans: one is for collecting the *regularity* of all distinct items and the other is for building the tree only for the *regular* items in each transaction. To keep track of the occurrence information, RP-tree explicitly maintains the transaction-*ids* (*tid*) of all transactions in the tree structure. It stores the *tid* of a transaction only at the last node of the transaction. The other nodes do not need to carry any occurrence information or support count (as does in FP-tree).

By applying an FP-growth-based [3] efficient pattern growth mining technique and exploiting the *tid*-information kept in the tree structure, RP-tree generates the complete set of *regular* patterns for the user-given *regularity* threshold. While mining *regular* patterns from an RP-tree, the transaction occurrence information maintained in it is used to calculate the *regularity* of each generated pattern.

However, as mentioned before, even though RP-tree efficiently finds *regular* patterns from static transactional databases, it is not suitable for mining *regular* patterns from data streams because of its *regularity* threshold-based tree structure, and two database scans requirement.

3 Problem Definition

Let $L = \{i_1, i_2, \ldots, i_n\}$ be a set of literals, called items that have been used as a unit information of an application domain. A set $X = \{i_j, \ldots, i_k\} \subseteq L$, where $j \leq k$ and $j, k \in [1, n]$, is called a *pattern* (or an *itemset*). A transaction $t = (tid, Y)$ is a tuple where *tid* represents a transaction-*id* (or time of transaction occurrence) and Y is a pattern. If $X \subseteq Y$, it is said that t contains X or X occurs in t. Let $size(t)$ be the size of t, i.e., the number of items in Y.

A data stream DS can formally be defined as an infinite sequence of transactions, $DS = [t_1, t_2, \ldots, t_m)$, where $t_i, i \in [1, m]$ is the i-th arrived transaction. A window W can be referred to as a set of all transactions between the i-th and j-th (where $j > i$) arrival of transactions and the size of W is $|W| = j - i$, i.e., the number of transactions between the i-th and j-th arrival of transactions. Let each slide of window introduce and expire *slide_size*, $1 \leq slide\_size \geq |W|$, transactions into and from the current window.

If X occurs in $t_j, j \in [1, |W|]$, such transaction-*id* is denoted as t_j^X, $j \in [1, |W|]$. Therefore, $T_W^X = \{t_j^X, \ldots, t_k^X\}$, $j, k \in [1, |W|]$ and $j \leq k$ is the set of all transaction-*ids* where X occurs in the current window W.

Definition 1 (a period of X in W). Let t_{j+1}^X and t_j^X $j \in [1, (|W| - 1)]$, be two consecutive transaction-*ids* in T_W^X. The number of transactions (or the time difference) between t_{j+1}^X and t_j^X is defined as a period of X, say p^X (i.e., $p^X = t_{j+1}^X - t_j^X$, $j \in [1, (|W| - 1)]$). For the simplicity of period computation, a '*null*' transaction with no item is considered at the beginning of *W*, i.e., $t_f = 0$ (*null*), where t_f represents the *tid* of the first transaction to be considered. Similarly, t_l, the *tid* of the last transaction to be considered, is the *tid* of the |*W*|-th transaction in the window, i.e., $t_l = t_{|W|}$. For instance, in the stream data in Table 1, consider the window is composed of eight transactions (i.e., *tid* = 1 to *tid* = 8) make the first window, say W_1). Then the set of transactions in W_1 where pattern {*b,c*} appears is $T_{W_1}^{\{b,c\}} = \{2, 3, 5\}$. Therefore, the periods for {*b,c*} are 2 (= 2 - t_f), 1 (= 3 - 2), 2 (= 5 - 3), and 3 (= t_l - 5), where $t_f = 0$ and $t_l = 8$.

The occurrence periods, defined as above, present the exact information about the appearance behavior of a pattern. A pattern will not be *regular* if, at any stage in *W*, it appears after sufficiently large period. The largest occurrence period of a pattern, therefore, can provide the upper limit of its periodic occurrence characteristic. Hence, the measure of the characteristic of a pattern of being *regular* in a *W* (i.e., the *regularity* of that pattern in *W*) can be defined as follows.

Definition 2 (*regularity* of pattern *X* in *W*). Let for a T_W^X, P_W^X be the set of all periods of *X* i.e., $P^X = \{p_1^X, ..., p_s^X\}$, where *s* is the total number of periods of *X* in *W*. Then, the *regularity* of *X* in *W* can be denoted as $reg_W(X) = Max(p_1^X, ..., p_s^X)$. For example, in the *DS* of Table 1 $reg_{W_1}(b,c) = 3$, since $P_{W_1}^{\{b,c\}} = Max(2, 1, 2, 3) = 3$.

Therefore, a pattern is called a *regular* pattern in *W* if its *regularity* in *W* is no more than a user-given maximum *regularity* threshold called *max_reg* λ, with $1 \leq \lambda \leq |W|$. The *regularity* threshold is given as the percentage of window size.

The *regular* patterns in *W*, therefore, satisfy the downward closure property [6], i.e., if a pattern is found to be *regular*, then all of its non-empty subsets will be *regular*. Thus, if a pattern is not *regular*, then none of its supersets can be *regular*. Given *DS*, |*W*|, and a *max_reg*, finding the complete set of *regular* patterns in *W*, R_w that have *regularity* of no greater than the *max_reg* value is the problem of mining *regular* patterns in data stream.

4   RPS-Tree: Design, Construction, and Mining

In this section, we first introduce our RPS-tree for data stream and describe efficient tree update mechanism for RPS-tree. We also discuss the mining of an RPS-tree here.

4.1   Design of an RPS-Tree

The structure of an RPS-tree consists of one *root* node referred to as the "*null*", a set of item-prefix sub-trees (children of the *root*), and an item header table called *regular pattern stream* table (RPS-table in short). Similar to an FP-tree [3] and an RP-tree [4], each node in an RPS-tree represents an itemset in the path from the *root* up to that node.

The RPS-tree maintains the occurrence information of all transactions (in the current window) in the tree structure. To explicitly track such information, it keeps a list of transaction-*id* information only at the last item-node (say, *tail-item*) for a transaction. Such list is called a *tid-list*. Hence, an RPS-tree maintains two types of nodes; say *ordinary nodes* and *tail-nodes*. The former are types of nodes that do not maintain the *tid-list*, whereas the following definition describes the latter type.

Definition 3 (*tail-node*). Let $t = \{i_1, i_2,..., i_n\}$ be a sorted transaction, where i_n is the *tail-item*. If t is inserted into an RPS-tree in this order, then the node of the tree that represents item i_n is defined as the *tail-node* for t. For example, if the first transaction (i.e., *tid* = 1) in the *DS* of Table 1 is inserted into an RPS-tree in lexicographical order, then the node that represents item '*f*' (i.e., the *tail-item* of the transaction) is the *tail-node* in the tree for that transaction.

Nodes of both types explicitly maintain parent, children, and node traversal pointers. In addition, each *tail-node* maintains a *tid-list* and a *tail-node* pointer. The *tail-node* pointer points to either the next *tail-node* in the tree if any, or '*null*'. Irrespective of the node type, no node in the RPS-tree maintains a support count value as does in an FP-tree [3].

The RPS-table consists of each distinct item in the current window with relative *regularity* and a pointer pointing to the first node in the RPS-tree that carries the item. Specifically, the RPS-table of an RPS-tree consists of three fields in sequence (i, r, p); item name (i), *regularity* of i (r), and a pointer to the RPS-tree for i (p). The item name is just a symbol to identify each item. The *regularity* is calculated by traversing the RPS-tree after the construction, which is explained in the next subsection. The item pointer facilitates the fast traversal to the whole tree in the mining phase. In addition, an RPS-tree maintains a *tail-node* pointer (say, tn_p) to point to the first *tail-node* in the tree. These pointers will facilitate fast tree traversal during the *regularity* calculation and tree update operation.

4.2 Construction of an RPS-Tree

The construction of the RPS-tree is featured in such a way that it takes only one scan over the high-speed data stream to capture the full content of the current window. In the RPS-tree, items are arranged according to any canonical order, which can be determined by the user prior to the tree construction. Once the item order is determined (say, for the initial window), items will follow this order in our RPS-tree for subsequent windows.

We use an example to illustrate the step-by-step construction process of an RPS-tree for the *DS* in Table 1. Let us assume that the RPS-tree is constructed in lexicographic order and each window is composed of eight transactions (i.e., the initial window, W_1 contains *tids* from 1 to 8) as shown in Fig. 1(a).

The construction of an RPS-tree is similar to that of an FP-tree [3]. Initially, the RPS-tree is empty (i.e., starts with a '*null*' root node). To simplify the figures, we do not show the node traversal pointers in the trees, although they are maintained as in an FP-tree.

The first transaction to be inserted is $\{a, c, e, f\}$ (i.e., *tid* = 1). As shown in Fig. 1(b), the transaction is inserted in the lexicographic order. Notice that '*f*' is the *tail-item* of the transaction and the *tail-node* "*f*:1" explicitly maintains the *tid* information in its *tid-list*. Also, tn_p points to the first *tail-node* "*f*:1" in the tree (as shown by dotted arrow in the figure). Fig. 1(c) shows the status of the RPS-tree after inserting

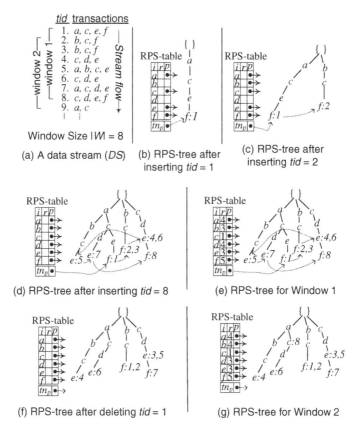

Fig. 1. Construction and update of an RPS-tree for the *DS* in Table 1

the second transaction (i.e., *tid* = 2) in similar fashion. The *tail-node* pointer is updated to point to the next *tail-node* "*f*:2" as shown in the figure. The RPS-tree after capturing all transactions of W_1 is presented in Fig. 1(d). Notice that the RPS-tree in Fig. 1(d) captures the complete information of W_1 in a compact fashion. However, the *regularity* of items in the RPS-table has not been computed yet.

To assist the *regularity* calculation, each item in the RPS-table is assigned a temporary array. Then, starting from the tn_p, and following *tail-node* pointers we visit each *tail-node* and accumulate the *tid*(s) available in its *tid-list* in the respective temporary arrays for every item from that *tail-node* up to the *root*. For example, after visiting the first two *tail-nodes* of "*f*:1" and "*f*:2,3" in the RPS-tree of Fig. 1(d), the contents of the temporary arrays for items '*a*', '*b*', '*c*', '*e*', and '*f*' (i.e., items from *tail-nodes* up to the *root*) are $T^a = \{1\}$, $T^b = \{2, 3\}$, $T^c = \{1, 2, 3\}$, $T^e = \{1\}$, and $T^f = \{1, 2, 3\}$.

Therefore, after finishing the traversal for all *tail-nodes*, we obtain the complete list of *tids* for each item in its respective temporary array. Thus, for instance, the set of transactions for item '*a*' we obtain, $T^a = \{1, 5, 7\}$. Then, it is rather simple calculation to find the P^a from T^a, which gives $reg_{W_1}(a) = 4$. The process of accumulating the *tid*s and calculating the *regularity* of items in the RPS-table is termed as refreshing the RPS-table. Finally, Fig. 1(e) shows the final status of the RPS-tree and the RPS-table with the *regularity* of each item after the RPS-table refreshing operation. The RPS-tree in Fig. 1(e) is ready for mining the set of *regular* patterns from it upon request.

4.3 Updating the RPS-Tree

The simple construction feature of the RPS-tree enables it to delete the oldest and insert new transactions in an efficient manner. Because our RPS-tree keeps the *tail-node* pointers, one can easily locate the transaction(s) to be removed. To illustrate the RPS-tree updating mechanism, we use our running example of RPS-tree construction.

Suppose the window slides transaction-by-transaction (*slide_size* = 1) i.e., each slide of window expires the oldest and inserts one new transactions. Therefore, in this example, *tid* = 1 expires and a new transaction *tid* = 9 appears with the sliding of window.

To reflect the deletion of the oldest transaction we avoid the costly tree traversal operation. Rather following the *tail-node* pointers we visit only the *tail-node*s in the RPS-tree and adjust only the *tid-list*s of each *tail-node* in the tree for deleted transaction(s). We delete the *tid*s in the *tid-list* of each *tail-node* if their values are less than or equal to the *slide_size*; otherwise, we decrement them by *slide_size*. In process, we delete a *tail-node* and its path towards the *root* if its *tid-list* becomes empty. For example, we delete the *tail-node* "*f*:1" and its parent node "*e*", since after adjusting the *tid*s, the *tid-list* of "*f*:1" becomes empty. However, we avoid deleting nodes (toward the *root*) at the parent of "*e*", since it (the parent) has a child other than "*e*". Such operation ensures the deletion of only the expired transactions from the tree. The RPS-tree after deleting the oldest transaction (i.e., *tid* = 1) from the RPS-tree of W_1 and adjusting the *tid-list*s in all *tail-node*s is shown in Fig. 1(f). For the simplicity of figures we avoid showing the *tail-node* pointers in the figures. However, they are maintained as explained above.

Notice that the RPS-tree in Fig. 1(f) is ready to capture the new incoming transaction(s) in the sliding window. New transactions can be easily added to the RPS-tree by using the same technique as illustrated in Figs. 1(b) – (d). Usually, the *regularity* of patterns may change with the sliding of window (i.e., with the deletion and insertion of old and new transactions). For example, with $\lambda = 3$, and $|W| = 8$ for the *DS* in Fig. 1(a) the *regular* patterns {*b*}, and {*b,c*} in W_1 become *irregular* (i.e., a pattern whose *regularity* is greater than *max_reg*) in W_2. Again, the *irregular* patterns {*d*} and {*c,d,e*} in W_1 become *regular* in W_2. Therefore, to reflect the correct *regularity* of each item in the current window, we perform the RPS-table refreshing operation at each window. Fig. 1(g) shows the status of the RPS-tree in W_2 after inserting new transaction and refreshing the RPS-table. Similar to the RPS-tree in Fig. 1(e), the complete set of *regular* patterns for the current window then can be mined from the RPS-tree in Fig. 1(g).

Based on the RPS-tree construction technique discussed above, we have the following property and lemma on the completeness of an RPS-tree. Let for each transaction *t* in a window *W*, *item(t)* be the set of all items in *t* and is called the full item projection of *t*.

Property 1: An RPS-tree contains *item*(*t*) for each transaction in a window only once.

Lemma 1: Given a stream database *DS* and a sliding window *W*, *item*(*t*) of all transactions in *W* can be derived from the RPS-tree for the *W*.

Proof: Based on the RPS-tree construction and updating mechanism and Property 1, *item*(*t*) of each transaction *t* is mapped to only one path in the RPS-tree and any path from the *root* up to a *tail-node* maintains the complete projection for exactly *n* transactions (where *n* is the total number of entries in the *tid-list* of the *tail-node*).   ∎

One may assume that the structure of an RPS-tree may not be memory efficient, since it explicitly maintains *tids* of each transaction in the tree structure. But we argue that the RPS-tree achieves the memory efficiency by keeping such transaction information only at the *tail-nodes* and avoiding the support count field at each node in the tree. Moreover, keeping the *tid* information in tree structure has also been found in literature for efficiently mining frequent patterns [2], [8]. To a certain extent, some of those studies additionally maintain support count and/or the *tid* information [2], [8] in each tree node. Furthermore, with modern technology, main memory space is no longer a big concern. Hence, we made the same realistic assumption as in many studies [11] that we have enough main memory space (in the sense that the trees can fit into the memory).

Since each transaction *t* in *W* contributes at best one path of *size*(*t*) to an RPS-tree, the total size contribution of all transactions in *W* can be at best $\sum_{t \in |W|} | size(t) |$. However, because there are usually many common prefix patterns among the transactions, the size of an RPS-tree is normally much smaller than $\sum_{t \in |W|} | size(t) |$.

It may be assumed that RPS-table refreshing mechanism of RPS-tree may require higher computation cost compared to scanning the stream data twice as in RP-tree. But, we argue that the cost of refreshing the RPS-table by traversing the paths from the *tail-nodes* up to the *root* of the RPS-tree is much less than that by scanning the database a second time, since reading transactions from the memory-resident tree is much faster than scanning them from the database. Also note that, while accumulating the *tids* from a *tail-node* during refreshing the RPS-table, we process as many transactions at a time as the size of its *tid-list*. This multiple transactions processing technique further reduces the RPS-table refreshing cost compared to obtaining the *regularity* of items through a second scan of the stream data. In the next subsection, we discuss the *regular* pattern mining process from the RPS-tree constructed for the current window.

4.4 Mining the RPS-Tree

Similar to the FP-growth [3] mining approach, we recursively mine the RPS-tree of decreasing size to generate *regular* patterns by creating conditional pattern-bases (*PB*) and corresponding conditional trees (*CT*) without additional database scan. Before discussing the mining process we explore the following important property and lemma of an RPS-tree.

Property 2: Each *tail-node* in an RPS-tree maintains the occurrence information of all nodes in the path (from that *tail-node* up to the *root*) in the transactions of its *tid-list*.

Lemma 2: Let $Z = \{a_1, a_2, \ldots, a_n\}$ be a path in an RPS-tree where node a_n, being the *tail-node*, carries the *tid-list* of the path. If the *tid-list* is carried to node a_{n-1}, then node a_{n-1} maintains the occurrence information of path $Z' = \{a_1, a_2, \ldots, a_{n-1}\}$ for the same set of transactions in the *tid-list* without any loss.

Proof: Based on Property 2, the *tid-list* in node a_n explicitly maintains the occurrence information of Z' for the same set of transactions. Therefore, the same *tid-list* at node a_{n-1} exactly maintains the same information for Z' without any loss.   ∎

Using the features revealed by the above property and lemma and based on the downward closure property [6], we proceed to mining the RPS-tree for only *regular* items starting from the bottom up to the top in the RPS-table. If an item i in the RPS-table is an *irregular* item, we ignore mining for it. However, following the node traversal pointers we only visit each node N_i for i in the RPS-tree and carry (i.e, copy) N_i's *tid-list* to its parent N^p. Therefore, the parent node N^p is temporarily converted to a *tail-node* if it was an *ordinary node*; otherwise (i.e., if N^p is a *tail*-node), the *tid-list* is added with its previous *tid-list*. At the same time, from N_i we delete the *tid-list* it borrowed as a parent node from its children (if any). This process of carrying the *tid-list* of a (temporary) *tail-node* to its parent node is termed as *carry-tid* and the set of *tid*(s) carried to the parent is called as *carried-tid*.

We use our running example to illustrate the mining on an RPS-tree. Consider mining the RPS-tree of Fig. 1(e) for $\lambda = 3$. Since 'f', the bottommost item in the RPS-table, is not *regular* (i.e., $reg_{W_i}(f) > 3$), we only perform the *carry-tid* operation for each of its nodes in the RPS-tree. Fig. 2(a) shows the status of the RPS-tree after the *carry-tid* operation for 'f'. The *tid*s shown in dark box in the figure are *carried-tid*s.

Mining for each *regular* item i in the RPS-table, on the other hand, is performed by constructing the conditional pattern-base PB_i for i by projecting only the prefix subpaths of N_i in the RPS-tree with the *tid-list* of N_i. During this projection, we only include *regular* items. Determination of whether an item is *regular* can be easily done by a simple look-up (an $O(1)$ operation) at the RPS-table. There is no worry about possible omission or doubly counting of items. While visiting each N_i, we perform the *carry-tid* operation for the node as well.

To store the *regularity* of items with i, a small RPS-table, say RPS-table$_i$, is maintained for PB_i. While constructing PB_i, to compute the *regularity* of each item j in the RPS-table$_i$, based on Property 2 we map all N_i's *tid-list*s to all items in the respective path explicitly in temporary arrays (one for each item). Once the PB_i is constructed, the contents of the temporary array for j in the RPS-table$_i$ represent the T^{ij} (i.e., set of all *tid*s where items i and j occur together) in PB_i. Therefore, it is a rather simple calculation to compute $reg_W(j)$ from T^{ij} by generating P^{ij}. The conditional tree for i CT_i is, then, constructed from its PB_i by removing all *irregular* items and their respective nodes from the RPS-table$_i$ and PB_i, respectively. If the deleted node is a *tail-node*, based on Lemma 2 its *tid-list* is pushed-up to its parent node.

Let j be the bottommost item in RPS-table$_i$ of CT_i. Then the pattern $\{i,j\}$ is generated as a *regular* pattern with the *regularity* of j in the RPS-table$_i$. The same process of creating a conditional pattern-base and its corresponding conditional tree is repeated for further extensions of pattern $\{i,j\}$.

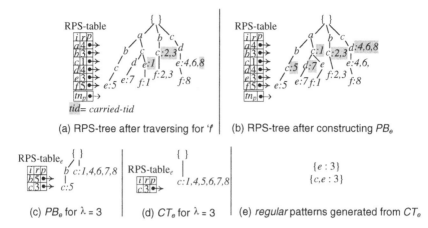

Fig. 2. Mining the RPS-tree of Fig. 1(e) for $\lambda = 3$

The next item in the RPS-table in Fig. 2(a) (i.e., 'e') is a *regular* item (i.e., $reg_{W_i}(e) \leq 3$). Therefore, we construct the PB_e, and then CT_e. We also perform the *carry-tid* operation while constructing the PB_e. The structure of the RPS-tree after the *carry-tid* operation for 'e' is illustrated in Fig. 2(b). Fig. 2(c) shows the structure of the PB_e. The CT_e is constructed by removing all *irregular* items and their respective nodes from the RPS-table$_e$ and PB_e. The CT_e in Fig. 2(d) is, therefore, constructed by deleting all entries for *irregular* item 'b'. The set of all *regular* patterns mined from the CT_e is given in Fig. 2(e). The value after ':' indicates the *regularity* of individual pattern. The whole process is repeated until the top of the RPS-table (i.e., 'a').

Notice that after each successful *carry-tid* operation any node in the RPS-tree retains its original status of either as an *ordinary node* or a *tail-node* (e.g., nodes "e" and "e:4,6" from Fig. 2(a) to Fig. 2(b)). Also, since we start mining from the bottommost item in the RPS-tree, there is no scope of missing any *tid-list* in the whole tree from carrying upward. It can be noticed that, when mining for all items in the RPS-table is completed, the *carry-tid* operations will accumulate a copy of all *tid*s at the *root* node. It is then rather a trivial task to remove them from the *root* to make the tree consistent to be updated for the next window content.

Therefore, from the above mining process we can say that for a given *max_reg* and W the R_W can be generated from an RPS-tree constructed on the window contents. In the next section, we evaluate the performance of our RPS-tree.

5   Experimental Analyses

In this section, we present the experimental results and related analysis on the comparison of proposed RPS-tree with its state-of-the-art counterparts. To the best of our knowledge, the RPS-tree is the first effort to address the problems of *regular* pattern mining in data stream. Therefore, we compare its performance with that of the RP-tree [4], the existing algorithm available for *regular* pattern mining. All programs are

Table 2. Dataset characteristics

Dataset	#Trans.(T)	#Items(I)	MaxTL(MTL)	AvgTL(ATL)
BMS-POS	515,597	1,657	164	6.53
Kosarak	990,002	41,270	2,498	8.10
T10I4D100K	100,000	870	29	10.10

written in Microsoft Visual C++ 6.0 and run with Windows XP on a 2.66 GHz machine with 1GB of main memory. The runtime specifies the total execution time, i.e., CPU and I/Os.

We use several real and synthetic datasets (as in Table 2) which are frequently used in frequent pattern mining experiments, since they maintain the characteristics of transactional data. The first two datasets were obtained from [14]. *BMS-POS* contains several years worth of point-of-sale data from a large electronics retailer. *Kosarak* is a dataset of click-stream data from a Hungarian on-line news portal. *T10I4D100K*, developed by [13], is a synthetic dataset. In all experiments, we consider *slide_size* = 1. In the first experiment, we study the compactness of our RPS-tree in stream data.

5.1 Memory Efficiency

We conducted experiments to verify the memory requirements for our RPS-tree on different datasets by varying the window size. Since RPS-tree is a *regularity* threshold independent tree structure, the *regularity* threshold values do not influence on its memory requirements. Therefore, in this experiment, the reported required memory represents the size of the underlying tree structure after capturing only the complete sliding window content. Because RP-tree is a *regularity* threshold-based tree structure, we do not compare its memory requirement with RPS-tree.

Table 3 reports RPS-tree's memory requirement (on average for all window for a fixed window size) in several datasets with the variation of window size at each case. In *BMS-POS*, for example, when the window size is 100K (i.e., $|W|^1$ = 100K), the required memory is on an average 13.81 MB in each window. Again, in the same dataset RPS-tree consumes on an average 33.51 MB memory when $|W|^4$ = 400K.

Hence, from the data in Table 3 it can be observed that when capturing the stream data of different characteristics, an RPS-tree is memory efficient for the available memory now-a-days. In the next experiment, we compare execution time between our RPS-tree and existing RP-tree.

Table 3. Memory requirement (MB) with window size variation in RPS-tree

Dataset with different window sizes	For window size				
	$|W|^1$	$|W|^2$	$|W|^3$	$|W|^4$	$|W|^5$
BMS-POS ($|W|^1$ = 100K, $|W|^2$ = 200K, $|W|^3$ = 300K, $|W|^4$ = 400K)	13.81	22.26	29.97	33.51	-
Kosarak ($|W|^1$ = 100K, $|W|^2$ = 300K, $|W|^3$ = 500K, $|W|^4$ = 700K, $|W|^5$ = 900K)	55.67	84.92	130.41	159.24	228.97
T10I4D100K ($|W|^1$ = 30K, $|W|^2$ = 50K, $|W|^3$ = 70K, $|W|^4$ = 90K)	3.51	5.09	6.96	8.93	-

5.2 Runtime Efficiency

To study the runtime performance experiments were conducted with a mining request at each window by varying the *max_reg* values for each dataset while the window size |*W*| was kept fixed at reasonably high values. The results of the experiment are shown in Fig. 3. The time shown on the *y*-axes are the total time for scanning the window content, tree construction, tree update and RPS-table refreshing time (only for RPS-tree), and mining. Notice that mining data stream with RP-tree requires scanning each window content twice, since it was originally proposed for static databases.

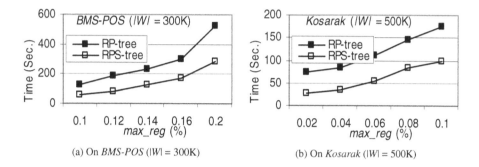

(a) On *BMS-POS* (|*W*| = 300K)   (b) On *Kosarak* (|*W*| = 500K)

Fig. 3. Runtime comparison

As shown in Fig. 3, the higher the *max_reg* values, the longer the overall time re-quired by both trees. The reason is that, the higher the *max_reg* value, the greater the number of *regular* patterns can be generated from the current window. However, the results clearly demonstrate that RPS-tree outperforms RP-tree in terms of overall runtime by multiple orders of magnitude for both high and low *max_reg* values. The key to this performance gain of RPS-tree is its efficient tree updating mechanism that only scans the new incoming transaction(s) once, while RP-tree requires scanning the whole window content twice. The gain of RPS-tree over the RP-tree becomes more prominent when the window size is larger. We also evaluated RPS-tree's performance on the variation of window size, as shown in the next experiment.

5.3 Window Size

Because RPS-tree captures the full window content, its performance may vary de-pending on the window size i.e., |*W*|. Hence, to determine the effect of changes in window size on the runtime of RPS-tree, we analyzed its performance by varying |*W*| over different datasets while keeping the *max_reg* value fixed. The graphs presented in Fig. 4 show the results on *BMS-POS* for *max_reg* = 0.16%, and *Kosarak* for *max_reg* = 0.06%. The *y*-axes in the graphs represent the average total time (including construction time, tree update time for the RPS-tree only, and mining time) required in all active windows.

Larger window sizes resulted in a longer total tree construction time for both trees. However; the overall runtime required by RPS-tree is small enough to handle larger

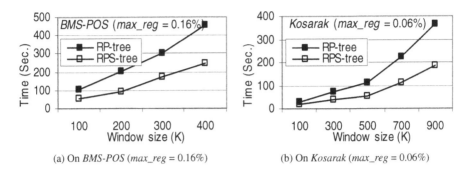

(a) On *BMS-POS* (*max_reg* = 0.16%)   (b) On *Kosarak* (*max_reg* = 0.06%)

Fig. 4. RPS-tree's performance on size of *W*

windows in different datasets. For RP-tree, in contrast, a sharp increase in runtime according to an increase in window size was observed. As a result, the performance gaps between the two tree structures widen for larger windows. For example, in *Kosarak* for *max_reg* = 0.06% when |W| = 100K, RPS-tree's gain is not much prominent (Fig. 4(b)). However, for |W| = 900K RPS-tree achieves a significant improvement in overall runtime. Similar results we obtained in *BMS-POS* as well. Therefore, these results show that RPS-tree is better than RP-tree in handling larger windows and producing the exact set of *regular* patterns within a reasonable amount of time over data streams.

The above experiments demonstrate that RPS-tree outperforms the state-of-the-art algorithms in mining *regular* patterns from data streams of various characteristics (refer to Table 2). The easy and simple maintenance phase of the RPS-tree has been the key to its significant performance gain.

6 Discussions and Conclusions

In this paper, we define the *regularity* of a pattern by its maximum occurrence interval (in a window) calculated from its *tids* (Definition 2) obtained during mining. However, other parameters such as the arithmetic mean or variance of occurrence intervals can also be considered as *regularity* measures for finding interesting patterns from data streams. Since RPS-tree maintains the exact occurrence information for all transactions in the current window, and the mining phase provides the complete *tids* for each pattern, computing such parameters can also be simple similar to computing the maximum occurrence interval for a pattern.

In conclusions, we introduced a new concept of mining interesting patterns (called *regular* patterns) that occur with a temporal *regularity* in high-speed data streams. We proposed a novel tree structure, RPS-tree, to capture the stream content in memory-efficient manner and to enable *regular* pattern mining from it. To obtain the fast and interesting results RPS-tree can be updated efficiently for the current content of the stream. The experimental analysis reveals that RPS-tree is significantly faster than other algorithm that can be used in mining *regular* patterns from a data stream.

References

1. Han, J., Dong, G., Yin, Y.: Efficient Mining of Partial Periodic Patterns in Time Series Database. In: 15th ICDE, pp. 106–115 (1999)
2. Zhi-Jun, X., Hong, C., Li, C.: An Efficient Algorithm for Frequent Itemset Mining on Data Streams. In: ICDM, pp. 474–491 (2006)
3. Han, J., Pei, J., Yin, Y.: Mining Frequent Patterns without Candidate Generation. In: ACM SIGMOD Int. Conf. on Management of Data, pp. 1–12 (2000)
4. Tanbeer, S.K., Ahmed, C.F., Jeong, B.-S., Lee, Y.-K.: Mining Regular Patterns in Transactional Databases. IEICE Trans. on Inf. & Sys. E91-D(11), 2568–2577 (2008)
5. Huang, K.-Y., Chang, C.-H.: Mining Periodic Patterns in Sequence Data. In: Kambayashi, Y., Mohania, M., Wöß, W. (eds.) DaWaK 2004. LNCS, vol. 3181, pp. 401–410. Springer, Heidelberg (2004)
6. Agrawal, R., Srikant, R.: Fast algorithms for Mining Association Rules in Large Databases. In: VLDB, pp. 487–499 (1994)
7. Ozden, B., Ramaswamy, S., Silberschatz, A.: Cyclic Association Rules. In: 14th ICDE, pp. 412–421 (1998)
8. Zaki, M.J., Hsiao, C.-J.: Efficient Algorithms for Mining Closed Itemsets and Their Lattice Structure. IEEE Trans. Knowl. Data Eng. 17(4), 462–478 (2005)
9. Toroslu, I.H., Kantarcioglu, M.: Mining Cyclically Repeated Patterns. In: Kambayashi, Y., Winiwarter, W., Arikawa, M., et al. (eds.) DaWaK 2001. LNCS, vol. 2114, pp. 83–92. Springer, Heidelberg (2001)
10. Tanbeer, S.K., Ahmed, C.F., Jeong, B.-S., Lee, Y.-K.: Sliding Window-based Frequent Pattern Mining over Data Streams. Information Sciences 179, 3843–3865 (2009)
11. Leung, C.K.-S., Khan, Q.I.: DSTree: A Tree Structure for the Mining of Frequent Sets from Data Streams. In: ICDM, pp. 928–932 (2006)
12. Li, H.-F., Lee, S.-Y.: Mining Frequent Itemsets over Data Streams Using Efficient Window Sliding Techniques. Expert Systems with Applications 36, 1466–1477 (2009)
13. IBM, QUEST Data Mining Project, http://www.almaden.ibm.com/cs/quest
14. Frequent Itemset Mining Dataset Repository,
    http://fimi.cs.helsinki.fi/data/

k-ARQ: k-Anonymous Ranking Queries*

Eunjin Jung[1], Sukhyun Ahn[2], and Seung-won Hwang[2]

[1] Dept. of Computer Science
The University of Iowa
ejjung@cs.uiowa.edu
[2] Dept. of Computer Science and Engineering
Pohang University of Science and Technology (POSTECH)
{ashworld,swhwang}@postech.ac.kr

Abstract. With the advent of an unprecedented magnitude of data, top-k queries have gained a lot of attention. However, existing work to date has focused on optimizing efficiency without looking closely at privacy preservation. In this paper, we study how existing approaches have failed to support a combination of accuracy and privacy requirements and we propose a new data publishing framework that supports both areas. We show that satisfying both requirements is an essential problem and propose two comprehensive algorithms. We also validated the correctness and efficiency of our approach using experiments.

1 Introduction

With the advent of data on an unprecedented scale, there has been active research carried out on supporting ranking queries, to effectively narrow down results to a small desired number of matches according to a given ranking criteria [1,2,3,4]. While existing work mostly focused on optimizing the efficiency of computing top-k ranked tuples, there has been less effort made on preserving privacy at the same time.

Privacy preservation research on databases has also been also an active, study area including published work on k-anonymity [5], l-diversity [6], m-invariance [7], and t-closeness [8], but most of these approaches have treated every tuple more or less equally, without taking its ranking into account. As a result, applying these approaches directly on top-k ranked tuples would damage the result quality, by including many non top-k tuples in the results.

To illustrate this problem, consider an example database of job applicants as shown in Fig. 1, where the ranking function is the sum of all three test scores. A recruitment company owns this database of job applicants and, a hiring manager of a company wants to see if this recruitment company has good candidates before paying a service fee. The recruitment company may reveal some information

* This work was supported by Engineering Research Center of Excellence Program of Korea Ministry of Education, Science and Technology (MEST) / Korea Science and Engineering Foundation (KOSEF), grant number R11-2008-007-03003-0.

H. Kitagawa et al. (Eds.): DASFAA 2010, Part I, LNCS 5981, pp. 414–428, 2010.

	Quasi-identifier			Ranking attributes				
Tid	Age	Sex	Zipcode	Course1	Course2	Course3	Score	Rank
1	33	Female	53715	99	99	99	297	1
2	39	Male	53715	96	98	99	293	2
3	23	Female	53715	92	97	95	284	3
4	74	Male	53703	96	96	90	282	4
5	51	Male	53703	98	89	94	281	5
6	37	Female	53703	97	95	88	280	6
7	45	Female	53712	96	95	88	279	7
8	22	Male	53712	98	90	91	279	7
9	33	Female	53712	92	96	90	278	9
10	37	Female	53712	97	89	90	276	10

Fig. 1. An Example Database

about the test scores of their top-k applicants to impress the hiring manager, without compromising the privacy of the applicants. A naive approach would be to publish the table as it is shown in Fig. 1 without any unique identifiers, such as names, SSNs, or passport numbers. However, as shown in [5], revealing quasi-identifiers may seem harmless but when combined with other seemingly harmless public data this may reveal the identities of the applicants. Another approach at the other end of the spectrum of privacy is to publish only the list of total scores, as in [9], but the returned data would only be of limited use for the hiring manager. Our goal is to assist the database owner to publish information on *ranking attributes*, the attributes that most affect tuples' ranking, without compromising too much on privacy.

To consider the privacy, we adopt a widely-known privacy metric called k-anonymity. In a k-anonymous table, the tuples are indistinguishable from the other k-1 tuples. To analyze data quality, we employ a precision metric, *i.e.*, how many returned tuples are top-k ranked, out of the total number of returned tuples. Imagine a scenario where the recruitment company wants to publish the top-6 candidates with 3-anonymity.

Applying the top-k query algorithms to identify the top-6 ranked tuples and then aggregating their values results in Fig. 2(a). This approach includes all of the top-k tuples and, thus achieved *perfect recall*. Each ranking attribute is displayed as a range of values to provide 3-anonymity. Tuples 1, 2, and 3 are not individually identifiable, but are presented as a group of top-3 candidates. The tuple id's in parentheses denote the tuples that are not in the top-6 but were included. Note that the hiring manager does not get to see the Tid column, so he or she can learn about the correlation between the individual course scores and the total score from the range but cannot know which exact tuples are included in the second range. This outcome displays a good result for the top-3 tuples, but the second top-3 tuples suffer exhibit only a 50% accuracy level, *i.e.*, only half of them are in the top-6.

We can also apply a 3-anonymity algorithm to the top-6 tuples. Unfortunately, achieving k-anonymity with a minimal change is NP-hard when $k > 3$ [10], so

Tid	Course1	Course2	Course3	Score	Precision
1,2,3	[92,99]	[97,99]	[95,99]	[284,297]	1
4,5,6,(7,8,10)	[96,98]	[89.96]	[88,94]	[226,282]	0.5

(a) Perfect Recall Example

Tid	Course1	Course2	Course3	Score	Precision
1,5,6,(8,10)	[97,99]	[89,99]	[88,99]	[274,297]	0.6
2,3,4,(9)	[92,96]	[96,98]	[90,99]	[278,293]	0.75

(b) Mondrian Example

Tid	Course1	Course2	Course3	Score	Precision
1,2,3	[92,99]	[97,99]	[95,99]	[284,297]	1
4,6,(7)	[96,97]	[95,96]	[88,90]	[279,282]	0.67

(c) k-ARQ Example

Fig. 2. Results of three different algorithms applied on database in Fig. 1

maximizing data quality while achieving k-anonymity on top-k tuples is not practical. One of the approximation algorithms, Mondrian [11], results in the table shown in Fig. 2(b). This result shows a lower overall precision than Fig. 2(a).

In this paper we propose the use of k-ARQ, a k-Anonymous Ranking Query approach, to ensure both privacy maintenance and precision when ranking attribute publication. We can show that calculating a set of tuples that optimize privacy and precision at the same time is an NP-Complete problem, and introduce two Greedy Approximation Algorithms. Fig. 2(c) shows the result from our Greedy Approximation Algorithms. k-ARQ achieves the same perfect precision as Fig. 2(a) for the top-3 tuples, and achieves better precision for the next top-3. We summarize this comparison below in Table 1.

Table 1. The pros and cons of current state-of-arts

	Privacy	Precision	Ranking Attribute Publication
Privacy top-k [9]	✓	✓	
k-anonymity approximation [11]	✓		✓
k-ARQ	✓	✓	✓

Our key contributions to the research area are as follows:

- To the best of our knowledge, we are the first to study the problem of publishing ranking attributes, while supporting the dual requirements of privacy and accuracy.
- We formalized the problem, and showed that it is an NP-complete problem.
- We develop two Greedy Approximation Algorithms, with optimization heuristics to enhance their efficiency.

– We evaluate our Greedy Algorithms in terms of appropriate use and efficiency, compared to existing approaches. The correctness is defined in this case as satisfying both the privacy maintenance and accuracy requirements, and the efficiency is measured in terms of elapsed time and precision.

The rest of the paper is organized as follows. Section 2 discusses the preliminaries and problem definition, and Section 3 shows that the problem is NP-Complete. Section 4 proposes our Greedy Algorithms. Section 5 presents our evaluation results. Section 6 surveys related work.

2 Preliminaries

As preliminaries, we first formally state our problem (Section 2.1) then discuss why existing solutions (Section 2.2 and 2.3) have failed to address it.

2.1 Problem Definition

The input to our problem is a data table $T = \{t_1, \cdots, t_n\}$, which is a set of m-dimensional n tuples where each tuple t_i is represented by a set of values for m dimensions (attributes), $\{d_1, d_2, \cdots, d_m\}$. Each attribute in T is one of the following: *quasi-identifier*, *ranking attribute*, or *sensitive attribute*, which we will define formally below. Note that our definition is consistent with prior published literatures on this area, as we have indicated with citations.

Definition 1 (Quasi-identifier set QI [12]). *A quasi-identifer set QI is a set of attributes in table T that can be joined with external information to re-identify individual records, such as age, gender, and zip code in Fig. 1.*

Definition 2 (Ranking attribute set RA). *A ranking attribute set RA is a maximal set of attributes that affects the overall ranking (i.e., parameters for $\mathcal{F}$ from the data table and its score), such as course test scores and total score in Fig. 1. Typically ranking attributes have a total order, i.e.between any two values for a ranking attribute, it is defined which one is greater than or equal to the other.*

Definition 3 (Sensitive attribute SA). *A sensitive attribute set SA is a set of attributes associated with the privacy of a user.*

We will now define our proposed problem: the output is a published table T_{pub}, which is derived from the input data table T and includes information on ranking attributes, while satisfying privacy, ranking, and accuracy requirements. For privacy, we use the widely-used k-anonymity method, adopting its definition from [6]. The privacy requirement is specified by k_p for k_p-anonymity.

Definition 4 (K-Anonymity). *A table $T \{t_1, t_2, \cdots, t_n\}$ is a set of tuples where each tuple has dimensions d_j. This table T satisfies k-anonymity if for every tuple t_i there exist $k-1$ other tuples t_1, t_2, t_{k-1} such that $t_i[d_j] = t_1[d_j] = t_2[d_j] = \cdots = t_{k-1}[d_j]$ for any dimension (attribute) d_j.*

The ranking requirement is specified by sending a query for the top k_q tuples based on the user-defined ranking function $\mathcal{F}$. For accuracy, we use the well-known metric of precision and, the ratio of the true positives in the overall results. In our case, this is the ratio of the number of top-k_q tuples in T_{pub} to the number of all tuples in T, and the requirement is specified by the lower bound of this ratio *prec*.

Definition 5 (k-ARQ problem). *For the given data table T and the ranking function $\mathcal{F}$ we compute the score of each tuple and add this as another ranking attribute to T. For the given privacy, ranking, and accuracy requirements (k_p, k_q, prec), calculate a published output table T_{pub} that includes the ranking attribute set RA. (Note that RA includes the $\mathcal{F}$ score now.) T_{pub} is a successful result if and only if*

- *k_p-anonymity: for every tuple t_i in T_{pub} there exist k_p -1 other tuples t_1, t_2, $\cdots$ t_{k_p-1} such that $t_i[d_j] = t_1[d_j] = t_2[d_j] = \cdots = t_{k_p-1}[d_j]$ for any dimension (attribute) d_j in RA.*
- *Ranking and Accuracy $|T_{pub} \cap K_q| \geq prec \times |T_{pub}|$.*

where K_q is a set of top-k_q tuples w.r.t $\mathcal{F}$ scores.

Note that the traditional k-anonymity definition only includes QI. Our definition of k_p-anonymity is only defined over RA, but we can easily achieve k_p anonymity on QI as well. Based on the tuple groupings obtained for our problem, this can be done by simply aggregating the values of QI and SA of each group into ranges as shown in Fig. 3.

Tid	Age	Sex	Zipcode	Course1	Course2	Course3	Score	Precision
1,2,3	[23,39]	human	53715	[92,99]	[97,99]	[95,99]	[284,297]	1
4,6,(7)	[37,74]	human	[53703,53712]	[96,97]	[95,96]	[88,90]	[279,282]	0.67

Fig. 3. Extension to Quasi-identifiers

However, this straightforward adoption is vulnerable to a *homogeneity attack* as discussed in [6], when all SA values in a group happen to be identical. For example, the first group in Fig. 3 shows that all of the top-3 candidates live in the 53715 zipcode area. If a zipcode were a sensitive attribute, then this provides no anonymity for the zipcode even though it's a group of 3 tuples. To avoid this vulnerability, we recommend the application of existing anonymization techniques such as those in [6,7,8] on QI and SA on the groupings obtained for our problem, instead of aggregating values into ranges as discussed above.

2.2   Baseline Approach I: Perfect Recall

We will now discuss a baseline approach of adopting existing solutions for our proposed problem in Section 2.1. We first discuss how to use top-k algorithms

d_1	d_2
B 1.0	C 1.0
A 1.0	A 1.0
E 0.8	L 0.9
K 0.7	D 0.8
F 0.7	G 0.7
D 0.7	H 0.7
J 0.6	F 0.7
I 0.6	E 0.65
H 0.5	I 0.6
C 0.5	B 0.6
L 0.4	J 0.4
G 0.3	K 0.3

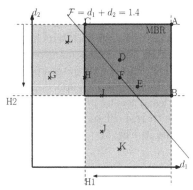

(a) Ordered list of tuples   **(b)** 2-dimensional plot of tuples

Fig. 4. Perfect Recall

such as [1,2,3,4] to distinguish the top-k tuples from the rest and how they, can be leveraged as a baseline solution to our problem.

Fig. 4 presents a geometric illustration of the top-k algorithms. Given that index structure ordering tuples in the order of attribute values d_1 and d_2 as shown in Fig. 4(a) illustrates that these algorithms access objects in a descending order of d_1 and d_2 values, which corresponds to a sweeping hyperplane orthogonal to the d_1 and d_2 axis respectively, downwards towards 0 in Fig. 4(b) as the arrows indicate. These algorithms terminate when the top-k results are guaranteed to exist among the tuples already accessed, *i.e.*, shaded regions in Fig. 4. In other words, when the upper bound score of all tuples in the unshaded region is no higher than k objects in the shaded region with the highest $\mathcal{F}$ scores, these algorithms can safely terminate.

We can use these algorithms to identify a "minimum-bounded rectangle" (which we denote as an MBR) where tightly bounding top k_q tuples are found, and marked as a box as shown in Fig. 4(b) above. When $k_p < k_q$ as in our example, publishing the range of the MBR satisfies both the privacy requirements and ensure all of the actual top_{k_q} are included in T_{pub}, *i.e.*, achieves perfect recall.

While this baseline approach is guaranteed to publish all of the true positives, it does not have any control over how many false positives are included in the results. To illustrates this, when a ranking function is $\mathcal{F} = d_1 + d_2$, a hypersurface $d_1 + d_2 = 1.4$ represents a boundary distinguishing the true positives and negatives. This means that all the object values above this surface are true positives and the rest of object values in the dark-shaded triangular area are true negatives. Depending on data distributions and ranking functions, unbounded number object values may fall into the triangular area, which could lower their precision below the user-specified requirement *prec*.

2.3   Baseline Approach II: Mondrian

An alternative approach is to apply existing anonymization algorithms, that have been typically applied to anonymize both QI and RA. Specifically, we adopted the *mondrian* [11] method, which uses one of the state-of-the-art k-anonymity approximation algorithms, as a baseline approach.

mondrian achieves the user-specified privacy requirement k_p-anonymity by iteratively dividing a data space into equal-sized partitions such that each partition includes at least k_p objects. Mondrian divides the true positive set into half, initially, with respect to d_1 into partitions $P1$ and $P2$, as shown by the vertical division line in Fig. 5.

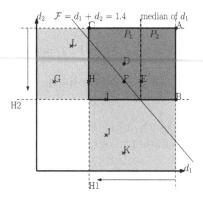

Fig. 5. Mondrian

Fig. 5 illustrates such iterations, based on the same example used in Section 2.2, where $k_q = 6$ and $k_p = 3$, More specifically, we divide the true positive set into half, initially with respect to d_1, as shown by a vertical division line in Fig. 5. This line divides the result set into partitions $P1$ and $P2$ as marked in the figure.

Fig. 6 illustrates each iteration of Mondrian as applied to our current example of $k_p = 3$ and $k_q = 3$ for the database in shown Fig. 1. The *Mondrian* approach

Tid	Course1	Course2	Course3	Score	Tid	Course1	Course2	Course3	Score
1	[97,99]	99	99	297	1	[97,99]	[89,99]	[88,99]	297
5	[97,99]	89	94	281	5	[97,99]	[89,99]	[88,99]	281
6	[97,99]	95	88	280	6	[97,99]	[89,99]	[88,99]	280
2	[92,96]	98	99	293	2	[92,96]	[96,98]	[90,99]	293
3	[92,96]	97	95	284	3	[92,96]	[96,98]	[90,99]	284
4	[92,96]	96	90	282	4	[92,96]	[96,98]	[90,99]	282

(a) Step1 - choosing dimension   (b) Step2 - recoding other dimension

Fig. 6. Step-by-Step Mondrian execution

first divides the values with respect to course 1, by picking the median course 1 score of six true positives, which is 97. Based on this value, we split course 1 into [97, 99] and [92, 96] segments, based on which tuples are divided into the two partitions of tuples 1,5 and 6 and 2,3, and 4, as Fig 6(a) shows. Mondrian then aggregate the values of the course 2 and course 3 scores in each partition into privacy ranges as Fig. 6(b) shows.

Note that this generalization (aggregation) of course 2 and course 3 values results in many false positives, such as for tuple 8,9 and 10 (shown in parentheses to mark false positives) in Fig. 6. Depending on the data distributions and the ranking function, an unbounded number of such false positives can be included in the published table, which makes it difficult to satisfy the user-specified accuracy requirement *prec*. This is due to the nature of the Mondrian method and other k-anonymity approximation algorithms that treat all tuples equally without considering their ranks.

3   Hardness analysis

To show that *k-ARQ* is an NP-Complete problem, first we need to show that the solution to this problem is verifiable in polynomial time. After the verification proof, we will show that a well-known NP-Complete problem, *Subset Sum* problem can be reduced to apply to this problem in polynomial time by showing the polynomial-time reduction and how the solution to one problem is also a solution to the other. For the given parameters $(k_p, k_q, prec)$, we can convert these parameters into (k, p) for a simpler proof. Given the privacy requirement k_p and the ranking requirement k_q, $k = \max(k_p, k_q)$. If $k_p < k_q$, then we can repeat this algorithm multiple times until we get at least k_q tuples. Given k, we can then identify the appropriate p value for the definition below.

Definition 6 (k-ARQ problem). *A table $T \{t_1, t_2, \cdots, t_n\}$ is a set of m-dimensional tuples, and each tuple t_i is labeled as tp if t_i is one of the top k tuples, and as fp otherwise. Each tuple t_i is also associated with count c_i, which indicates how many duplicates of t_i is in T. k-ARQ(T, k, p) returns a split value vector $SV = \langle sv_1, sv_1, \cdots, sv_d \rangle$, which defines a set S as $S = \{t_i : t_i[d_j] \leq sv_j\}$for all $j, 1 \leq j \leq m$, that satisfies the following:*

1. $\sum c_i$ for all $t_i \in S = k$ and
2. $p =$ (the num of tp tuples in S - the num of fp tuples in S)

if such SV exists. If not, k-ARQ(T, k, p) returns a pre-assigned value.

From this definition, the precision of S is $\frac{p+k}{2k}$, thus $p = \frac{prec}{2k} - k$.

Definition 7 (Subset sum problem). *A is a set of integers $\{a_1, a_2, \cdots, a_{pn}\}$ and sum is also an integer. Subsetsum(A, sum) returns $A_s = \{a_{p1}, a_{p2}, \cdots, a_{pt}\}$, where $\sum_{j=1}^{t} a_{pj} = sum$, if such A_s exists. If not, Subsetsum(A,sum) returns a pre-assigned value.*

Polynomial time verification proof: Given a split value vector $SV = \langle sv_1$, $sv_2, \cdots, sv_m \rangle$ for k-ARQ(T, k, p), where T contains n m-dimensional tuples, we can construct S in $O(mn)$ for integer comparisons. By verifying the first condition, $|S| = k$ can be checked in $O(n)$, the worst case being $S = T$, and for the same worst case $O(m)$ for the second condition. Thus, the verification of any solution to k-ARQ(T, k, p) is done in $O(nm)$, polynomial time to its input size.

Polynomial time reduction from a Subset sum problem[1]: Given a subset sum problem *Subsetsum(A, sum)* with an integer set $\{a_1, a_2, \cdots, a_{pn}\}$, for each a_i in A, a pair (t_i, c_i) is created where tuple t_i, $t_i[d_j] = 1$ if $i = j$ and $t_i[d_j] = 0$ otherwise, and the count c_i of t_i is set to be a_i. A pn-dimensional table T is constructed with all t_i. All tuples are labelled in T as tp. Transforming *Subsetsum(A,sum)* into k-ARQ(T, sum, sum) takes polynomial time.

Subsetsum $\Rightarrow k - ARQ$: Imagine that there is a solution A_s for *Subsetsum(A, sum)*, $A_s = \{a_{p1}, a_{p2}, \cdots, a_{pt}\}$. By the definition of the subset sum problem, $\sum_{j=1}^{t} a_{pj} = sum$. Define SV as $sv_i = 1$ if $a_i \in A_s$, 0 otherwise. By definition of S, tuple t_i generated from a_i will be in S. This S satisfies the first condition because $\sum c_i$ for all $t_i \in S = \sum a_j$ for all a_j in $A_s = sum$. This S also satisfies the second condition because the number of tp tuples in S is equal to $\sum_{j=1}^{t} a_{pj} = sum$.

$k - ARQ \Rightarrow$ *Subsetsum*: Given a solution SV for k-ARQ, then compute the set S and construct a subset of A, A_s, such that a_i is in A_s if $sv_i = 1$. $\sum_{a_i \in A_s} a_i = \sum c_i$ for all $t_i \in S = sum$. This satisfies the *Subsetsum(A, sum)* as a solution.

The proof above shows that the k-ARQ problem is an NP-Complete problem. If the desired results is to solve an optimized version of k-ARQ, optimization is possible using p, the precision parameter, or k, the accuracy parameter. Given the same p, as k approaches k_q in the original top-k_q ranking query, the result become more accurate relative to the original result of the top-k_q ranking query. There are only k_q number of candidates for p values, as $1 \leq p \leq k_q$ naturally. Thus, the optimization problem over p is also NP-Complete. Similarly, there are only a limited number of candidates for k value, as $k_p \leq |S| \leq k_p/prec$. As a result, the optimization problem of k-ARQ over k and p is also an NP-Complete Problem.

4   Proposed Solution

This section proposes algorithms to address our proposed problem in Section 2.1. As an exact solution has proven to be NP-complete, we propose the use of two Greedy Approximation Algorithms in Sections 4.1 and 4.2.

4.1   Greedy Algorithm by Deletion

The Greedy by deletion algorithm first initializes the result set as the MBR of true positives (*i.e.*, the output of *PerfectRecall*) and Greedy improves its

---

[1] This reduction technique was inspired by the NP-Completeness proof in [11].

precision toward the optimal value by reducing the boundary on one dimension at a time, until any more deletions would result in a violation of the privacy requirement k_p or reduce the precision. Intuitively, a desirable reduction would eliminate as many false positives as possible while eliminating as few true positives as possible. Our *Greedy Deletion* Algorithm chooses a true positive tuple that has the largest $\Delta_{fp} - \Delta_{tp}$ when removed from the result set, where Δ_{tp} is the number of eliminated true positives and Δ_{fp} is the number of eliminated false positives. To assist in the calculation of $\Delta_{fp} - \Delta_{tp}$, we utilize an ordered list (index) returned by the top-k algorithms.

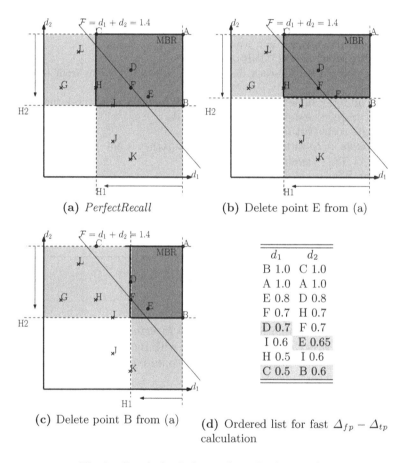

(a) *PerfectRecall*

(b) Delete point E from (a)

(c) Delete point B from (a)

(d) Ordered list for fast $\Delta_{fp} - \Delta_{tp}$ calculation

Fig. 7. Greedy by Deletion from *PerfectRecall*

For example, Fig. 7(a) shows the initial set of results which corresponds to an MBR for true positives. At the first iteration, the *Deletion* algorithm may choose B with the lowest d_2 value or C with the lowest d_1 value. The outcome of each deletion is shown in Fig. 7(b) and (c), respectively. $\Delta_{fp} - \Delta_{tp}$ is calculated by using the sorted lists shown in Fig. 7(d). B is at the bottom of the sorted list for d_2 and the *Deletion* mechanism traverses the list up until the next true positive E. There is one false positive I between B and E, so $\Delta_{fp} - \Delta_{tp} = 1 - 1 = 0$. Similarly, there is H and I between C and D, so $\Delta_{fp} - \Delta_{tp}$ for C is $2 - 1 = 1$. The *Deletion* mechanism chooses C to remove it from the result set and the sorted list is then updated accordingly. The pseudo-code of the *Deletion* algorithm is shown below.

Algorithm 1. Greedy Approach by Deletion

Input: A dataset D divided into tp and fp with m ranking attributes, and user-specified requirements k and p as specified in 6.
Output: The result set RS, MBR of RS
1: Obtain a sorted list for each dimension in descending order from top-k ranking algorithm
2: $RS_t \Leftarrow$ all top-k tuples and false positive tuples in MBR of top-k tuples
3: **while** $|RS_t| \leq k$ and precision of $RS_t \geq p$ **do**
4:     traverse sorted list for each dimension
5:     $candidate \Leftarrow tp$ with the largest $\Delta fp - \Delta tp$ that is not in RS_t
6:     $RS_t \Leftarrow RS_t \setminus \{candidate\}$
7:     $RS \Leftarrow RS_t$
8:     update MBR based on true positive tuples in RS and update RS
9: **end while**
10: **if** MBR of RS satisfy p and k **then**
11:     return RS, MBR
12: **end if**

4.2   Greedy Approach by Insertion

An alternative Greedy algorithm, called the *Insertion Algorithm*, starts with an empty set and inserts the true positive tuple with the largest $\Delta_{tp} - \Delta_{fp}$, breaking ties with ranking, until any more addition would render the precision below the basic requirement.

In our example in Fig. 8(a), after the first iteration, MBR contains only $A = (1.0, 1.0)$. At the second iteration, as all true positives have the same $\Delta_{tp} - \Delta_{fp}$, we pick $B = (1.0, 0.6)$ with the highest ranking. Now the MBR extends to $[1.0, 1.0], [0.6, 1.0]$ and the precision is still 1. At the third iteration, among true positives C,D,E and F, we pick D with the largest $\Delta_{tp} - \Delta_{fp}$. To illustrate, adding C to the result set will result in MBR shown in Fig. 8(b) with $\Delta_{tp} - \Delta_{fp} = |\{C, D, E, F\}| - |\{H, J\}| = 2$, while adding D will result in the MBR shown in Fig. 8(c) with $\Delta_{tp} - \Delta_{fp} = |\{D, E, F\}| - 0 = 3$.

The *Insertion* algorithm is similar to *Deletion* algorithm. The pseudo-code of this algorithm is shown only in [13] due to space limitations.

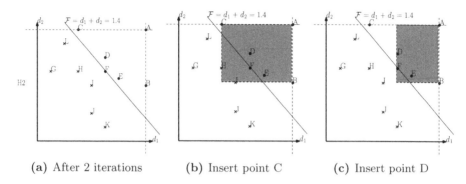

(a) After 2 iterations   (b) Insert point C   (c) Insert point D

Fig. 8. Greedy by Insertion

5   Experimental Results

To validate the effectiveness and efficiency of our algorithms, we generated 100 synthetic datasets, each of which consisted of 10000 tuples with 5 ranking attributes. The ranking function $\mathcal{F}$ is the sum of all of the ranking attribute values. Each dataset was tested 100 times. Our Greedy Algorithms were implemented in C++ using an ODBC connection. All of the performance measurements, our proposed method, Perfect Recall and Mondrian, were performed on a 3.3GHz Intel Dual Core Processor with 2GB RAM running Windows OS. The experiment parameters are summarized below:

Parameter	Result size (k)	precision $(prec)$	dimension (m)	cardinality (n)
Default value	20	0.7	5	10000

To validate the effectiveness of the approach, we can show the success ratio per top-k query as displayed in Fig. 9. We also compared the results from our Greedy Algorithms to optimal solutions when the optimal solutions are known as in Fig. 10. Finally, we can now show the efficiency of our algorithms in terms of time as seen in Fig. 11.

Fig. 9 compares our Greedy Algorithms with *Perfect Recall* and *Mondrian* in terms of the success ratio as the ranking requirement k increases. When a user requests a top-k query, each algorithm may not be able to satisfy both k and p. For each k, the result is an average from 100 executions. Our Greedy Algorithms succeed far more times than *Perfect Recall* and *Mondrian*. This is expected in the case of *Mondrian* as it does not consider the ranking of tuples in partitioning and suffers from the lowest success rate. *PerfectRecall* performs better when k is small, but still only succeeds in about half the amount of times when compared to our Greedy Algorithms. When k becomes as large as 20, *PerfectRecall* always fails to satisfy p.

Fig. 10 shows how our Greedy Algorithms compare to an optimal solution. We generated special datasets where the optimal solution with the highest precision is known for a given k and p, and applied our Greedy Algorithms to them. The

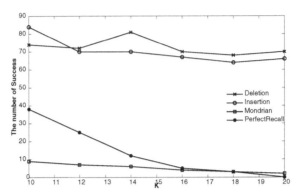

Fig. 9. Success ratio as k increases

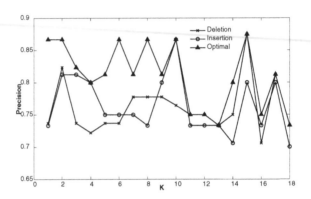

Fig. 10. Precision of results

y-axis shows the precision achieved for each dataset on the X-axis, as specified by experiment ID numbers. Admittedly our solution does not always find a solution, but when it does, the gap between it and the optimal is never lower by more than 0.1, and also in terms of ratio our Greedy Algorithms achieve above 90% relative to the optimal solution.

Fig. 11 compares the efficiency time of our algorithms to others by measuring the execution time as k increases. Since *PerfectRecall* gets its MBR from top-k ranking algorithm, it would not incur any extra cost. For this reason, Fig. 11 does not depict the computational time of *PerfectRecall* but one can imagine this being 0. Each data point is from an average of 100 executions. Since *Mondrian* divides data into equal-sized datasets, the computational time is not affected by k. On the other hand, our Greedy Algorithms need more iterations as k increases and therefore, our execution time increases fast. However, note that this k is limited by the privacy requirement k_p, which usually is much smaller than k_q. For example, if a user requests the top-100 queries with 10-anonymity, our execution time is dominated by 10 and, not by 100. To be precise, it takes less than 1.5 seconds to compute a result set of this type. As future work, we plan to experiment with more heuristics and realistic k_p and k_q values.

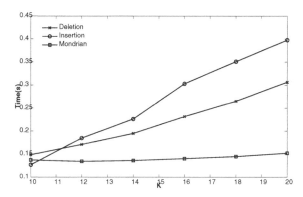

Fig. 11. Computational time

6   Related Work

Ranking queries have been actively studied, as an effective means to narrow down to a relevant data subset in large-scale repositories [1,2,3,4]. Most existing algorithms focus on optimizing the computation of such results and reveals all the information for the top-k results, *i.e.*, identifiers, attribute values, and rank ordering, which may include sensitive data on each individual.

To protect privacy, the notion of k-anonymity [5] has been introduced, to publish an "anonymized" table with the assurance that no individual can be uniquely distinguished from $k - 1$ other tuples. In order to move toward this direction of publishing anonymized tables, the problem is proved to be NP-hard problem for $k > 3$ in [10], followed by efficient approximation partitioning algorithms algorithms [11,12] and clustering algorithms [14,15]. The limitations of k-anonymity were discussed and mitigated in l-diversity [6], m-invariance [7], t-closeness [8], and also in variations of k-anonymity such as (α, k)-anonymity [16] and $(p+, \alpha)$-anonymity [17].

A straightforward solution to our proposed problem with k_p and k_q requirements is to apply these anonymization algorithms over top-k_q tuples returned by top-k query algorithms such as TA-family algorithms [1,2,3,4]. We use one of the algorithms, Mondrian [11], as part of our performance evaluation as a reference point. However, there are limitations in this basic application. Most approximation algorithms assume that there are many more tuples than k_p and use some form of partitioning to ensure k_p-anonymity, so when the input size k_q is already close to k_p, they either cannot partition the input and fail to achieve k_p-anonymity, or have to include too many tuples that are not in the top-k_q results and sacrifice data quality as discussed in Section 2.

To pursue the dual requirements of the privacy and ranking accuracy, the privacy-preserving top-k query algorithm [9] returns the exact top-k results but publishes no other attribute values. For example, if a user wants to find top-100 most popular search keywords is, then the algorithm returns the 100 search keywords but no other information such as how many searches were done for each keyword or in, which region each keyword. Since this algorithm does not publish any attributes, it is very difficult to infer any meaningful statistics from the results, *e.g.*, attributes' sensitivity to rankings or attribute correlations.

Acknowledgement

The authors thank Sriram Pemmaraju for reviewing our NP-completeness proof.

References

1. Fagin, R., Lote, A., Naor, M.: Optimal aggregation algorithms for middleware. In: PODS (2001)
2. Guentzer, U., Balke, W., Kiessling, W.: Optimizing multi-feature queries in image databases. In: VLDB (2000)
3. Bruno, N., Gravano, L., Marian, A.: Evaluating top-k queries over web-accessible databases. In: ICDE (2002)
4. Hwang, S., Chang, K.C.C.: Optimizing access cost for top-k queries over web sources: A unified cost-based approach. In: ICDE (2005)
5. Sweeney, L.: K-anonymity:a model for protecting privacy. International Journal of Uncertainty Fuzziness and Knowledge-based Systems (2002)
6. Machanavajjhala, A., Kifer, D., Gehrke, J., Venkitasubramaniam, M.: L-diversity: Privacy beyond k-anonymity. ACM TKDD 1(1), 3 (2007)
7. Xiao, X., Tao, Y.: M-invariance: towards privacy preserving re-publication of dynamic data sets. In: The 2007 ACM SIGMOD (2007)
8. Li, N., Li, T., Venkatasubramanian, S.: t-closeness: Privacy beyond k-anonymity and l-diversity. In: ICDE (2007)
9. Vaidya, J., Clifton, C.: Privacy-preserving top-k queries. In: ACM SIGMOD (2005)
10. Muthukrishnan, S., Poosala, V., Suel, T.: On rectangular partitionings in two dimensions. In: ICDT (1999)
11. LeFevre, K., DeWitt, D.J., Ramakrishnan, R.: Mondrian multidimensional k-anonymity. In: ICDE (2006)
12. Samarati, P., Sweeney, L.: Protecting privacy when disclosing information: k-anonymity and its enforcement through generalization and suppression. In: IEEE S&P (1998)
13. Ahn, S., Hwang, S.W., Jung, E.: k-anonymous ranking queries. Technical Report TR09-04, Dept. of Computer Science, The University of Iowa (September 2009)
14. Aggarwal, G., Feder, T., Kenthapadi, K., Khuller, S., Panigrahy, R., Thomas, D., Zhu, A.: Achieving anonymity via clustering. In: PODS (2006)
15. Miller, J., Campan, A., Truta, T.M.: Constrained k-anonymity:privacy with generalization boundaries. In: P3DM (2008)
16. Wong, R.C.W., Li, J., Fu, A.W.C., Wang, K.: (alpha, k)-anonymity: an enhanced k-anonymity model for privacy preserving data publishing. In: The 12th ACM SIGKDD (2006)
17. Sun, X., Wang, H., Li, J., Truta, T.M., Li, P.: (p+, α)-sensitive k-anonymity: a new enhanced privacy protection model. In: The 8th IEEE ICCIT (2008)

QSQL: Incorporating Logic-Based Retrieval Conditions into SQL

Sebastian Lehrack and Ingo Schmitt

Brandenburgische Technische Universität Cottbus
Institut für Informatik, Postfach 10 13 44
D-03013 Cottbus, Germany
slehrack@informatik.tu-cottbus.de, schmitt@tu-cottbus.de

Abstract. Evaluating a traditional database query against a data tuple yields `true` on match and `false` on mismatch. Unfortunately, there are many application scenarios where such an evaluation is not possible or does not adequately meet user expectations about vague and uncertain conditions. Thus, there is a need for incorporating impreciseness and proximity into a logic-based query language. The calculus query language CQQL [24] has been developed for such scenarios by exploiting results from quantum logic. In this work we will show how to integrate underlying ideas and concepts of CQQL into SQL.

Keywords: database query language, SQL, information retrieval, DB&IR.

1 Introduction

Evaluating a traditional database query against a data tuple yields `true` on match and `false` on mismatch. Unfortunately, there are many application scenarios where such an evaluation is not possible or does not adequately meet user needs. One problematic application area is information retrieval where finding a complete match in general is hardly possible. Thus, there is a need for incorporating the concepts of impreciseness and proximity into a logic-based query language. To motivate and exemplify the following principles and ideas we introduce a running scenario which is dealing with the sale of TV sets. Amongst other attribute values, the following properties are stored for a single TV set: name, *handling* (ha), *image quality* (iq), existence of an *optical sound port* (osp), *sound quality* (sq) and *status* (st). The three attributes *handling*, *image quality* and *sound quality* contain a rating value for the respective properties decoded as marks from 1 to 6. Whereas, mark 1 stands for an *excellent* test result and mark 6 signals an *inadequate* quality for the tested feature. The domain of the attribute *status* (st) comprehends the three values *available*, *sold* and *ordered*, whereby the underlined abbreviations are used for brevity. Table 1 gives an extracted part of a given data spreadsheet.

The user defines her/his query as: I want to find a device with a handling as easy as possible and the best possible quality of image. If a device is not able to offer a link to a sophisticated sound system via an optical sound port,

H. Kitagawa et al. (Eds.): DASFAA 2010, Part I, LNCS 5981, pp. 429–443, 2010.

the internal sound quality has to be as high as possible. We can formalise the condition as:

$$ha \approx 1 \land iq \approx 1 \land (osp = no \Rightarrow sq \approx 1) \tag{1}$$

The vagueness of the subconditions *handling as easy as possible* ($ha \approx 1$), *best possible quality of image* ($iq \approx 1$) and *sound quality as high as possible* ($sq \approx 1$) cannot be adequately mapped to Boolean truth values. As an inadequate attempt, Table 1 gives the truth values for a Boolean evaluation in round brackets, where the threshold value for an acceptable mark is assumed to be 2. The numeric values in squared brackets and the column scoreval will be discussed later. Obviously, important information are getting lost by the usage of classical Boolean logic. As a consequence, the provided result items are not distinguishable at all. For instance, the TV sets TV1, TV2 and TV3 return all the *same* positive result true for Query 1, in spite of the fact that TV set TV2 has to be acknowledged as the best choice, when the properties handling, image quality and sound quality in conjunction with the considered query are taken into account.

Table 1. Spreadsheet of tested TV sets

TV set						
name	ha	iq	osp	sq	st	scoreval
TV1	**2** (T) [0.8]	**2** (T) [0.8]	**yes** (T) [1.0]	**2** (T) [0.8]	a	0.72
TV2	**2** (T) [0.8]	**1** (T) [1.0]	**yes** (T) [1.0]	**1** (T) [1.0]	s	0.9
TV3	**1** (T) [1.0]	**2** (T) [0.8]	**no** (F) [0.0]	**1** (T) [1.0]	o	0.64
TV4	**3** (F) [0.6]	**1** (T) [1.0]	**no** (F) [0.0]	**4** (F) [0.4]	a	0.32
...	...	...	...	...	...	...

Generally, the user rather wants to see how *near* to her/his vision certain product offers are. Data objects fulfill such queries to a certain degree which can be represented by a value out of the interval $[0,1]$. Based on these *score values* a ranking of all data objects becomes possible which helps to distinguish the result items. Further examples of this kind of uncertain queries can be found in [2].

Our quantitative approach presented in [24] incorporates score values into a logic by exploiting a retrieval model based on quantum logic [3]. An essential advantage of this approach is to preserve the laws of the Boolean logic, given certain syntactical and semantical constraints are respected. Based on quantum logic the calculus query language CQQL, *Commuting Quantum Query Language*, has been developed as an extension of relational domain calculus. The main contribution of this work is the integration of CQQL concepts in combination with a novel weighting approach into SQL.

Next section gives a brief overview about the query language CQQL. Related works and comparable approaches are discussed in Section 3. Section 4 presents the core functionality of QSQL by defining syntax and semantics. In Section 5 we sketch the implementation architecture and present performance experiments with QSQL. Finally, a summary and a outlook is given in Section 6.

2   The Quantum Query Language CQQL

In this section we sketch the basic principles behind evaluating a tuple t against a given CQQL condition c. For readers with special interest in the theoretical foundation we give a more detailed explanation of the retrieval model behind CQQL in [24] and [14]. In comparison to [24] we describe in [14][1] the concepts in a much more intuitive way which does not require knowledge of quantum mechanics.

In general, CQQL enables the logic-based construction of queries starting from traditional Boolean and similarity conditions. The underlying idea is to apply the theory of vector spaces, also known from quantum mechanics and quantum logic, for query processing.

All attribute values of a tuple t are embodied by the direction of a normalised vector. The condition c itself corresponds to a vector subspace also called *condition space*. The evaluation result is then determined by the minimal angle between tuple vector and condition space. The squared cosine of this angle is a value out of the interval $[0, 1]$ and can therefore be interpreted as a similarity measure as well as a score value. A method for a convenient computation of the desired squared cosine of this angle is developed in [24]. It allows to evaluate a tuple t against a *normalised* (see below) condition c constructed by $\wedge, \vee$ and $\neg$ recursively as follows:

$$eval(t, c) = \varphi(t, c) \qquad \text{if } c \text{ is atomic,} \tag{2}$$
$$eval(t, c_1 \wedge c_2) = eval(t, c_1) * eval(t, c_2) \tag{3}$$
$$eval(t, c_1 \vee c_2) = eval(t, c_1) + eval(t, c_2) - \tag{4}$$
$$eval(t, c_1 \wedge c_2)$$
$$eval(t, \neg c) = 1 - eval(t, c) \tag{5}$$

whereby c_1 and c_2 are arbitrary subconditions. The function $\varphi(t, c)$ returns the evaluation of a single similarity predicate c as '$ha \approx 1$'. Its structure depends on the domain of the queried attribute of t. In general, any set of similarity values which can be produced by the scalar product of normalised vectors is supported. That is, the similarity values must form a semi-positive definite correlation matrix.

The defined operations 3 and 4 can only be applied, if the considered condition c is evaluated in a specific *syntactical form*. In this normal form only *mutually exclusive* subconditions or subconditions with *disjoint* sets of restricted attributes are allowed. The algorithm norm [24] transforms an arbitrary condition into the required normal form by using logical transformation rules as idempotence[1], absorption[2] and distributivity[3]. To preserve these logic laws we need following restriction in CQQL: ***In a valid condition any attribute must not be queried by more than one constant in a similarity predicate.*** This restriction will be respected by QSQL. Consequently, the query

---

[1] Idempotence: $A \wedge A \equiv A$ and $A \vee A \equiv A$.
[2] Absorption: $A \vee (A \wedge B) \equiv A$ and $A \wedge (A \vee B) \equiv A$.
[3] Distributivity: $A \wedge (B \vee C) \equiv (A \wedge B) \vee (A \wedge C)$ and $A \vee (B \wedge C) \equiv (A \vee B) \wedge (A \vee C)$.

'$ha \approx 1 \vee ha \approx 3$' is not allowed in CQQL/QSQL. Please notice, that the condition '$(ha \approx 2 \wedge sq \approx 1) \vee (ha \approx 2 \wedge iq \approx 1)$' is valid, because the attribute ha is restricted by the same constant 2 in both occurrences. Furthermore, CQQL/QSQL assume that the evaluations of atomic conditions is based on attributes being independent from each other.

Using more than one similarity conditions on *different* attributes often causes a new problem: which importance on the result should one condition have in comparison to another one? So, in the context of our scenario the user could decide that the impact (weight) of the handling condition should be only a half of the influences of the image condition.

The integration of weights into the CQQL formalism is surprisingly simple, in contrast to the approach from Fagin and Wimmers [10]. At first we assign a weighting variable $\theta_i \in [0, 1]$ to each operand (subcondition) of a conjunction or disjunction, e.g. $c_1 \wedge_{(\theta_1, \theta_2)} c_2$. A weighting variable θ_i controls the influence of the score value produced by evaluating the subcondition c_i. The main idea of our weighting approach is the application of two syntactical substitution rules. They convert a weighted conjunction and a weighted disjunction into unweighted versions of the respective operations. For this purpose, we insert weighting constants as fixed score values into the logical formula q:

$$c_1 \wedge_{(\theta_1, \theta_2)} c_2 \;\rightsquigarrow\; (c_1 \vee \neg\theta_1) \wedge (c_2 \vee \neg\theta_2) \tag{6}$$

$$c_1 \vee_{(\theta_1, \theta_2)} c_2 \;\rightsquigarrow\; (c_1 \wedge \theta_1) \vee (c_2 \wedge \theta_2) \tag{7}$$

To elucidate the mechanism behind the substituted formulas we will examine two extreme cases in more detail. A weighting variable of 0 ($\theta_i = 0$) leads to a behaviour that the corresponding subcondition c_i has no longer any effect on the final evaluation result. On contrary, if both weight variables are equal to 1 ($\theta_1 = \theta_2 = 1$), we achieve the same evaluation result generated by applying the unweighted versions of conjunction and disjunction.

Applying the weighting approach to a weighted version of Query 1 we achieve:

$$(ha \approx 1 \wedge_{0.5,1} iq \approx 1) \wedge (osp = no \Rightarrow sq \approx 1) \equiv$$
$$(ha \approx 1 \wedge_{0.5,1} iq \approx 1) \wedge (\neg(osp = no) \vee sq \approx 1)$$
$$\rightsquigarrow ((ha \approx 1 \vee \neg(0.5)) \wedge (iq \approx 1 \vee \neg(1.0))) \wedge$$
$$(\neg(osp = no) \vee sq \approx 1)$$

The column `scoreval` in Table 1 gives the score values produced by the last query. The single score values for the respective predicates, e.g. '$ha \approx 1$', are given after the queried marks in squared brackets.

3   Related Work

The integration of vague and unprecise conditions into a logic-based query language is a so far not satisfactorily solved research problem [7,15,21,25]. Proposed approaches tackling this problem can be classified into *qualitative* (without score values) and *quantitative* (with score values) methods.

Qualitative approaches: A famous example of a qualitative technique is the *skyline operator* [5] which can be considered as a special case of the winnow operator [6]. It filters out interesting tuples from a potentially large result set. The domination condition of a skyline operator relies on Boolean logic. Therefore, a homogeneous result set is produced once more which cannot sufficiently express different degrees of query matching.

Quantitative approaches: On the contrary, quantitative methods work with score values. For instance, object-relational database systems already use score values for evaluating multimedia conditions. However, score values are outside the logic. Thus, it is required to reduce score values into Boolean truth values[4] before a logic-based query evaluation can take place.

Other quantitative approaches often apply fuzzy logic [4,23,26]. The main principle of fuzzy set theory is to generalise the concept of set membership [27]. In classical set theory a characteristic function $1_A : \Omega \rightarrow \{0,1\}$ defines the memberships of objects $\omega \in \Omega$ to a set $A \subset \Omega$, whereby $1_A(\omega) = 1$, if $\omega \in A$ and $1_A(\omega) = 0$ otherwise. In fuzzy set theory the characteristic function is replaced by a membership function $\mu_M : \Omega \rightarrow [0,1]$, that assigns numbers to objects $\omega \in \Omega$ according to their membership degree to a fuzzy set M. Membership degrees can be used to represent different kinds of imperfect knowledge, including *similarity*, *preference*, and *uncertainty*.

Conjunctions and disjunctions of fuzzy membership degrees are evaluated by special classes of functions called t-norms and t-conorms, respectively. For input values from $\{0,1\}$, all t-norms and t-conorms behave like the Boolean conjunction and disjunction. For the values in between, however, different behaviours are possible. [27] suggests the usage of max for $\vee$, min for $\wedge$ and $(1 - \mu_M(x))$ for $\neg$. Thus, the example condition '$(st = a \vee st = o) \wedge ha \approx 1$' would be evaluated by $min(max(\mu_{[st=a]}(t), \mu_{[st=o]}(t)), \mu_{[ha\approx 1]}(t))$, whereby the fuzzy set $\mu_{[ha\approx 1]}(t)$, for instance, represents the fulfilling of condition '$ha \approx 1$' by tuple t.

The functions min/max are the standard t-norm/t-conorm because it is the only idempotent[2] and first proposed set of functions [27]. Nevertheless, [13] shows that the application of min/max differs from the intuitional understanding of a combination of values, because the binary min/max functions return only one value. This leads to a value dominance of one of the two input values while the other one is completely ignored [12,13].

To exemplify this disadvantage of min/max we consider the condition '$c_1 \equiv ha \approx 1 \wedge iq \approx 1$'. Then, the following equations describe the evaluation by fuzzy logic with min/max (Eq. 8) in comparison to the CQQL/QSQL evaluation (Eq. 9 and 10) introduced in the previous section:

$$eval_{F/min}(t, c_1) = min(\mu_{[ha\approx 1]}(t), \mu_{[iq\approx 1]}(t)) \tag{8}$$

$$eval_Q(t, c_1) = eval_Q(t, \texttt{norm}(ha \approx 1 \wedge iq \approx 1)) \tag{9}$$

$$= \varphi(t, ha \approx 1) * \varphi(t, iq \approx 1) \tag{10}$$

Furthermore, we assume two tuples $t_1[ha, iq] = (1,5)$ and $t_2[ha, iq] = (1,1)$. To evaluate c_1 we need score values for each fuzzy set and predicate, e.g. $\mu_{[ha\approx 1]}(t_1)$ or $\varphi(t_1, ha \approx 1)$. We set the score value 1.0, if the queried property is rated by the best possible mark 1. Thus, $\mu_{[ha\approx 1]}(t_1)$ and $\varphi(t_1, ha \approx 1)$ are evaluated to 1.0,

---

[4] For example, by comparison with a threshold.

since t_1 is rated by mark 1 for handling. On the contrary, mark 5 results to the score value 0.2, i.e. $\mu_{[t_1, iq \approx 1]}(t_1) = \varphi(t_1, iq \approx 1) = 0.2$. If we now apply Equation 8 to t_1 and t_2, i.e. $eval_{F/m}(t_1, c_1) = eval_{F/m}(t_2, c_1) = 1.0$, we can see that t_1 and t_2 cannot be distinguished by the min/max evaluation (in spite of tuple t_2 has an excellent mark in $both$ criteria). Thus, min/max cannot express influences or grades of importance of $both$ values on a result. In contrast, CQQL/QSQL involves both values and therefore avoids the dominance problem of min/max: $eval_Q(t_1, c_1) = 0.2$ and $eval_Q(t_2, c_1) = 1.0$.

The algebraic product $a \cdot b$ for $\wedge$ and the algebraic sum $a + b - a \cdot b$ for $\vee$, which overcomes the dominance problem of min/max, has been also proposed in fuzzy logic [19]. However, a large number of logical laws and semantically equivalences is known from Boolean logic. A user who is intuitively familiar with this equivalences would expect that the same rules are still valid in fuzzy logic and CQQL/QSQL. For instance, the logical combination of a condition with itself, e.g. '$c_2 \equiv iq \approx 1 \wedge iq \approx 1$', should produce the same result as given by evaluating this condition alone[2].

Unfortunately, in fuzzy logic the algebraic product is not idempotent[2] and thus no distributivity[4] holds. This can be easily shown:

$$eval_{F/prod}(t, c_2) = \mu_{[iq \approx 1]}(t) * \mu_{[iq \approx 1]}(t) = (\mu_{[iq \approx 1]}(t))^2 \qquad (11)$$

$$eval_Q(t, c_2) = eval_Q(t, \text{norm}(iq \approx 1 \wedge iq \approx 1)) \qquad (12)$$

$$= \varphi(t, iq \approx 1) \qquad (13)$$

Referring to the user expectation we achieve an incorrect result $eval_{F/prod}(t_1, c_2)$ $= 0.04 \neq 0.2 = eval_{F/prod}(t_1, iq \approx 1)$ in fuzzy logic, if the score values for t_1 are assumed as above. Contrarily, CQQL/QSQL computes the correct result $eval_Q(t_1, c_2) = 0.2$, because of its normalisation algorithm (Eq. 12) recognises that the underlying condition space (Sec. 2, [14]) for '$iq \approx 1$' is intersected by itself. Therefore, the operation '$iq \approx 1 \wedge iq \approx 1$' can be simplified to '$iq \approx 1$' $before$ any evaluation rule is applied (Eq. 13).

In general, CQQL/QSQL is able to differentiate semantical cases by applying Boolean transformation rules on vector spaces during the normalisation. This is impossible in fuzzy logic because required semantics are hidden behind the membership values of the given fuzzy sets [22]. Table 2 summarises the discussed properties.

Table 2. Properties of different evaluation models

Model	Scores	Idempotence	Non-dominating
Boolean logic	no	yes	-
Fuzzy logic (min/max)	yes	yes	no
Fuzzy logic (product)	yes	no	yes
CQQL/QSQL	yes	yes	yes

Probabilistic approaches proposed for example in [20] pan the responsibility for the definition of the correct semantics for conjunction and disjunction to the user. [11] presents several top-k processing techniques using score values computed by score functions. The applied *arithmetic* score functions, however, do not support logic-based conditions.

SQL: An `order` by-clause in SQL also performs a ranking of resulting tuples [9]. However, the order condition is simple based on data values. Arbitrary logical formulas cannot be used in SQL to specify an order condition.

4 Integration of CQQL Concepts into SQL

The Structured Query Language (SQL) is the de facto standard for accessing database systems. Since its introduction in the 70s and its first standardization in 1986 the practical significance of SQL has grown enormously. In this section we will show how we extend SQL's capabilities by CQQL concepts. For this purpose, we establish a new SQL dialect called *Quantum SQL* (QSQL). For defining syntax and semantics we will refer to the core functionality of SQL-92 [9], which covers the well-known relational algebra operations: selection, projection, union, intersection, difference, join and grouping. Object-relational concepts introduced in SQL-99 [1] are not supported by QSQL yet.

We have customised a SQL-92 grammar by inserting QSQL keywords. Obviously, we have to omit a presentation of the whole grammar consisting of 82 rules. Nevertheless, we will state the syntactical characteristics for each basic operation in the following subsections.

For evaluating QSQL queries we develop a mapping between QSQL and SQL-99 [1]. In consequence, every QSQL query can be evaluated against a relational database system, which supports SQL-99, after a normalisation and a further syntactical transformation. In fact, these transformation rules define the semantics of QSQL.

In the remainder, our SQL-99 example statements are ran against the Oracle 11g database system [18]. However, the mapping is designed in a way that a simple adaption to other database systems is possible. We will mention possible marginal differences between the SQL-99 standard syntax and the respective Oracle variant.

4.1 Selection

An important semantical distinction between selecting tuples in QSQL and in SQL is the representation of the result. In Boolean logic a subset of tuples constitutes the outcome of a given query. As already stated, there is a different situation for queries including similarity conditions. In this case each tuple produces a *score value* expressing the degree of fulfilling. Therefore, the query result is achieved by a *list* of *all* tuples ordered by their score values. In our approach, the score value is encoded in an additional attribute `scoreval`, which is automatically generated during the evaluation process.

Syntax in QSQL: A tuple selection in QSQL can be formulated in the same way as in SQL. The logical selection-condition is placed in the `where`-clause. In addition to SQL, there exists the option to apply similarity predicates indicated by the *similarity operator*: $\sim$. As an example we will examine following query: *Determine all* available *TV sets with a handling as easy as possible.* Listing 1.1 gives the QSQL statement for the example.

```
select name
from tv_set
where ( st='a' and ha ~ 1 ) and scoreval > 0
order by scoreval desc
```

Listing 1.1. Selection with similarity condition in QSQL

Mapping to SQL-99: The transformed SQL-99 statement usually includes a huge number of auxiliary functions implemented in the database programing language PL/SQL [16]. Basically, there are two types of supporting PL/SQL-functions: (1) *score functions* computing score values for single predicates and (2) *logical operator functions* connecting subconditions.

As an example for a score function, consider the similarity predicate '$ha \approx 1$' of our current example. Then, the PL/SQL-function HA_TO_SCORE(ha, 1) calculate the score value as normalised distance between the attribute value of ha and the best possible mark 1. The used calculation formula is directly inferred from the CQQL evaluation function $\varphi(t, c)$ for single predicates (Sec. 2). All score functions which are used during the evaluation processing have to be known to the database system in advance. The deployment of them can be automatically accomplished via the construction of the underlying tables.

Besides score functions, the resulting score value also depends on the logical combination of subconditions realised by logical operators. In SQL-99 we apply the PL/SQL-functions LAND, LOR and LNOT to implement the operator semantics given in Section 2 (Eq. (3),(4) and (5)). So, we map the structure of the logical select-conditions to corresponding nested PL/SQL-function calls. Since the computed score value is assigned to the new attribute scoreval we have to transfer the evaluation of logical select-conditions from the where-clause to the select-clause. Using the example in Listing 1.1, Listing 1.2 gives the outcome of mapping QSQL to SQL-99.

```
select name, LAND( EQUAL( st, 'a' ), HA_TO_SCORE( ha, 1 ) ) as scoreval
from tv_set
where scoreval > 0
order by scoreval desc
```

Listing 1.2. Transformed select-statement in SQL-99

The usage of the function EQUAL(st,'a') is necessary because Oracle cannot cast the result of a comparison expression, e.g. 'st='a'', to a numeric value within the select-clause. Please notice, the given statement in Listing 1.2 is simplified in a sense that a repeated calculation of the score value in the where-clause is abbreviated by the attribute name scoreval. Actually, the Oracle

SQL parser expects here the evaluation expression of `scoreval` again. This simplification is used in the remainder of this work.

Next we integrate our weighting approach (Sec. 2). Please regard the query: *Determine all TV sets with a handling as easy as possible and an image quality as good as possible. The rating of the handling property should be twice as important as the image quality benchmark.* According to the linguistic weighting terms we set a weight of 0.5 on the image quality subcondition and a weight of 1.0 on the handling subcondition.

Syntax in QSQL: In QSQL a user can assign a weighting constant from the interval $[0, 1]$ to an arbitrary subcondition. The indicating keyword `weighted by` is written behind the concerned subcondition. If a weighting constant for a subcondition is missing, QSQL implicitly assumes the constant value 1.0 for this subcondition. Listing 1.3 gives the last example expressed as QSQL statement.

Mapping to SQL-99: Before we can map logical select-conditions to PL/SQL-functions we have to dissolve the weighted versions of the logical operators. By applying the substitution rules (Eq. (6) and (7)) we can establish a logical formula on unweighted operators: $(ha \approx 1 \wedge_{(0.5, 1.0)} iq \approx 1) \rightsquigarrow (ha \approx 1 \vee \neg 0.5) \wedge (iq \approx 1 \vee \neg 1.0)$. The SQL-99 version of the last QSQL statement is given in Listing 1.4.

```
select name
from tv_set
where ( ha ~ 1 ) WEIGHTED BY 0.5 and iq ~ 1
```

Listing 1.3. Weighted selection in QSQL

4.2 Projection

In SQL a projection is used to define the column structure of the resulting table by means of a given attribute list.

In contrary to relational algebra, which is *set*-oriented, SQL allows duplicated tuples in tables. If you intend to eliminate these duplicates, you have to use the keyword `distinct` in the `select`-clause. We investigate the *distinct* and the *non-distinct* version of the projection by studying the query: *Determine all producers of TV sets with a handling as easy as possible.*

```
select name, LAND(
  LOR( HA_TO_SCORE( ha, 1 ), LNOT( 0.5 ) ),
  LOR( IQ_TO_SCORE( iq, 1 ), LNOT( 1.0 ) ) ) as scoreval
from tv_set
where scoreval > 0
order by scoreval desc
```

Listing 1.4. Transformed weighted selection in SQL-99

Syntax in QSQL: The syntax for projecting attributes in QSQL does not differ from the usual SQL syntax. The `select`-clause and the keyword `distinct` are used in QSQL in the same way.

Mapping to SQL-99: For the non-distinct version we can state that the projection has no effect on score values. Consequently, we simple copy the projection attribute list from the `select`-clauses of QSQL to SQL-99.

Though, for the `distinct`-case a simple SQL-99 distinct is not sufficient, because duplicated tuples can possess different score values. The question is now: What is the appropriated score value for the condensed tuple in the result table? To answer this question we refer to the strong relationship between the $\exists$-quantifier of the relational domain calculus and the projection operation of the relational algebra. In [17] a mapping between an arbitrary (safe) domain calculus formula and an algebra expression is defined. Thereby, a $\exists$-quantifier is mapped to a projection operation and vice versa, e.g. $\{(Y) \mid \exists X : R(X,Y) \wedge X < 10\} \equiv \pi_Y(\sigma_{X<10}(R))$, whereby X, Y are numeric variables/attributes and R denotes a relation over a relation schema (X, Y).

As already emphasised, the CQQL formalism extends the relational domain calculus. In this sense, we preserve the equivalence between the $\exists$-quantifier and the projection. We choose the *maximum* function, which evaluates the $\exists$-quantifier in CQQL [24], to determine the score value of the resulting tuple. The implementation in SQL-99 utilises a grouping operation over all projected attributes together with the aggregation function `max` (Listing 1.5).

```
select producer, max( HA_TO_SCORE( ha, 1 ) ) as scoreval
from tv_set
group by producer
having scoreval > 0
order by scoreval desc
```

Listing 1.5. Transformed projection in SQL-99

4.3   Union, Intersection and Difference

The union of two tuple sets $E = E_1 \cup E_2$ is one of the classical set operations, besides intersection and difference, provided by QSQL.

In the context of score values we have to revise the traditional semantics of the union operation. First of all, the union of two sets is obviously related to the disjunction of two subconditions in the domain calculus [17]. For instance, following expressions are equivalent: $\{(X) \mid R(X) \wedge (X < 10 \vee X > 15)\} \equiv \sigma_{X<10}(R) \cup \sigma_{X>15}(R)$, when X is a numeric variable/attribute and R denotes a relation over a relational schema (X). Next, we denote the score value of a tuple t contained in a tuple set E_j by the term $sv_{E_j}(t) \in [0, 1]$. The resulting score value $sv_E(t)$ of $E = E_1 \cup E_2$ has to be achieved by combining the two score values $sv_{E_1}(t)$ and $sv_{E_2}(t)$ regarding the same tuple t.

Considering the membership of t to the input sets E_1 and/or E_2 there are three computation cases to deal with:

$$sv_E(t) = \begin{cases} sv_{E_1}(t) + sv_{E_2}(t) - sv_{E_1}(t) * sv_{E_2}(t) & \text{if } t \in E_1 \wedge t \in E_2 \\ sv_{E_1}(t) & \text{if } t \in E_1 \wedge t \notin E_2 \\ sv_{E_2}(t) & \text{if } t \notin E_1 \wedge t \in E_2 \end{cases}$$

For determining $sv_E(t)$ we apparently take advantage of the evaluation rule for the disjunction in CQQL (Eq. (4)).

To exemplify the mapping of a union operation following example query is chosen: *Determine all producers which are capable to offer a TV set or a DVD player. Both devices should be characterised by a handling being as easy as possible.* To process this query we introduce the relation *dvd_player* with the relation schema (name, ha, producer, supplier, delivery_date).

Syntax in QSQL: QSQL uses the keyword union to indicate the union of two tuple sets. We give the example as QSQL statement in Listing 1.6.

```
select producer
from (
    select producer from tv_set where ha ~ 1
  union
    select producer from dvd_player where ha ~ 1 )
```

Listing 1.6. Union of two tables in QSQL

Mapping to SQL-99: For calculating $sv_E(t)$ we split the resulting tuple set E into three disjoint sets whereby each of them corresponds to a single computation case of $sv_E(t)$. The respective SQL statement is shown in Listing 1.7.

```
select producer, max( scoreval )
  as scoreval
from (
    select E1.producer, LOR( E1.scoreval, E2.scoreval ) as scoreval
    from
      ( select producer, HA_TO_SCORE( ha, 1 ) as scoreval
        from tv_set ) E1,
      ( select producer, HA_TO_SCORE( ha, 1 ) as scoreval
        from dvd_player ) E2
    where E1.producer = E2.producer
  union all
    select producer, HA_TO_SCORE( ha, 1 ) as scoreval
    from tv_set
    where producer not in ( select producer from dvd_player )
  union all
    select producer, HA_TO_SCORE( ha, 1 ) as scoreval
    from dvd_player
    where producer not in ( select producer from tv_set ) )
group by producer
having scoreval > 0
order by scoreval desc
```

Listing 1.7. Transformed union of two tables

The extended *similarity* semantics of the *intersection* $(E = E_1 \cap E_2)$ and the *difference* operation $(E = E_1 \backslash E_2)$ are analogously specified. Thus, we establish $sv_E(t)$ for the intersection operation as $sv_{E_1}(t) * sv_{E_2}(t)$. Obviously, the computation of the intersection exploits the evaluation rule for the conjunction in CQQL (Eq. (3)). As for the union operation the score value for the difference is based on more than one computation case:

$$sv_E(t) = \begin{cases} sv_{E_1}(t) * (1 - sv_{E_2}(t)) & \text{if } t \in E_1 \land t \in E_2 \\ sv_{E_1}(t) & \text{if } t \in E_1 \land t \notin E_2 \end{cases}$$

Concerning the first case the second score value $sv_{E_2}(t)$ has a negative effect on the resulting score value $sv_E(t)$. We can affirm these semantics by considering a simple Boolean-based evaluation. In this particular context the input score values $sv_{E_1}(t)$ and $sv_{E_2}(t)$ would be 1.0 (true). So, we obtain a resulting score value of 0.0 (false) which is equivalent to the fact that the considered tuple t must be subtracted from E_1 when t is also given in E_2.

Because of the conceptual correspondences between *union, intersection* and *difference* we will omit special example queries for the intersection and difference operation.

4.4 Join and Grouping

From relational algebra we know that each type of a join operation (inner, outer, natural) can be implemented by the already introduced operations (projection, selection, set operations) and a cross product over several input relations. Therefore, we refer to the discussion about these operations, whereas the score value for a combined tuple in a cross product is the product of the two input tuple score values.

The group operation is used to group a set of tuples with related values. It is very often applied together with a SQL aggregate function, e.g. max, min or sum. Merging tuples from a group of tuples is related to the projection operation which we have already discussed above. Due to this correlation we make use of the *maximum* function again to compute the score value of the resulting tuple. Besides the score value produced by the implicit projection, we must also take the optional having-clause into account. Precisely, we must combine both score values conjunctively to get the final score value. For brevity, we omit a special example query for the grouping.

5 Implementation and Experiments

In this chapter we briefly describe the architecture of QSQL (Fig. 1). To avoid a direct manipulation of Oracle's internal processes we have developed a special QSQL library QSQLforOracle extending the usual Oracle JDBC driver [8] for Oracle 11g. It manages the connection and the communication between a Java application and an Oracle database. Mainly, it takes QSQL queries from a Java application, normalises and transforms them into SQL-99 statements which will be finally sent to the Oracle database server.

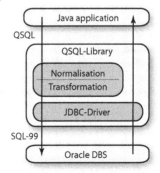

Fig. 1. Implementation architecture of QSQL

By using the QSQL library as a separate normalisation/transformation layer existing databases can also benefit from QSQL. For this purpose, the database administrator must only deploy the score functions (Sec. 4), what can be automatically realised by the adapted QSQL commands `create table` and `alter table`.

In Section 3 we gave a comparison between the *usability* and *expressiveness* of different qualitative and quantitative approaches. In Table 3 we present a *performance* test exploiting Query 1 (Sec. 1) and a set of $100,000$ TV sets. The adapted queries of all approaches are formulated in SQL-99 and are performed on an Oracle 11g database system[5].

Table 3. Performance test in seconds

Model	Time
SQL	0.04
QSQL/unsorted	0.14
Skyline	1.01
Fuzzy/sorted	19.01
QSQL/sorted	19.25

Not surprisingly, the processing times of the Boolean-based approaches SQL with a threshold mark 2 and skyline with a nested select-clause implementation as described in [5] are clearly better than the observed time values for the quantitative approaches Fuzzy logic with min/max as t-norm/t-conorm and QSQL. As already mentioned, in quantitative approaches we sort resulting tuples by their score values. Comparing the unsorting QSQL query (0.14 seconds) with the sorting one (19.25 seconds) reveals that the main contribution to the processing time is caused by the final sorting step. An optimised top-k operator could avoid a sorting of all resulting tuples. Ongoing research activities are dealing with this issue.

6 Summary and Outlook

In this paper we proposed a new SQL dialect called QSQL which integrates the evaluation concepts of the quantum logic-based query language CQQL into SQL. For this purpose, the basic idea of the theoretical model behind CQQL was given. Later, we defined syntax and semantics of QSQL by means of mapping QSQL queries to SQL-99 statements. Various example queries from a running scenario illustrated the introduced principles. Finally, we sketched the implementation architecture of the QSQL and gave an impression about the performance issues of QSQL.

The next step is to extend the functionality of QSQL by further operations, e.g. a top-k operator, and more efficient mapping techniques.

References

1. ANSI/ISO/IEC 9075-1:1999. Information technology - Database languages - SQL - Part 1: Framework (SQL/Framework). ISO (1999)
2. Agrawal, S., Chaudhuri, S., Das, G., Gionis, A.: Automated ranking of database query results. In: CIDR (2003)

---

[5] Hardware: Intel Core 2 Duo 2.5 GHz, 4 GB RAM and Mac OS X 10.5.5.

3. Birkhoff, G., von Neumann, J.: The Logic of Quantum Mechanics. Annals of Mathematics 37, 823–843 (1936)
4. Bosc, P., Pivert, O.: SQLf: A Relational Database Language for Fuzzy Querying. IEEE Transactions on Fuzzy Systems 3(1), 1–17 (1995)
5. Börzsönyi, S., Kossmann, D., Stocker, K.: The Skyline Operator. In: Proceedings of the 17th International Conference on Data Engineering, April 2001, pp. 421–432 (2001)
6. Chomicki, J.: Preference formulas in relational queries. ACM Trans. Database Syst. 28(4), 427–466 (2003)
7. Claremont Workshop. The Claremont Database Research Self Assessment. Technical report (2008)
8. Das, T., Iyer, V., Perry, E.H., Wright, B., Pfaeffle, T.: Oracle Database JDBC Developer's Guide and Reference, 11g Release 1 (11.1). Oracle Corp., Publishers (2007)
9. Date, C.J.: A Guide to the SQL Standard., 4th edn. Addison Wesley, Reading (1997)
10. Fagin, R., Wimmers, E.L.: A Formula for Incorporating Weights into Scoring Rules. Theoretical Computer Science 239(2), 309–338 (2003)
11. Ilyas, I.F., Beskales, G., Soliman, M.A.: A survey of top-k query processing techniques in relational database systems. ACM Computing Surveys 40(4) (2008)
12. Klose, A., Nürnberger, A.: On the properties of prototype-based fuzzy classifiers. IEEE Transactions on Systems, Man, and Cybernetics Part B 37(4), 817–835 (2007)
13. Lee, J.H.: Properties of Extended Boolean Models in Information Retrieval. In: SIGIR (ed.) SIGIR 1994: Proceedings of the 17th annual international ACM SIGIR conference on Research and development in information retrieval, pp. 182–190. Springer, New York (1994)
14. Lehrack, S., Schmitt, I.: The theoretical model behind CQQL. Technical report, BTU Cottbus (2009)
15. Lew, M.S., Sebe, N., Djeraba, C., Jain, R.: Content-based multimedia information retrieval: State of the art and challenges. ACM Trans. Multimedia Comput. Commun. Appl. 2(1), 1–19 (2006)
16. Lorentz, D.: Oracle Database SQL Language Reference, 11g Release 1 (11.1). Oracle Corp., Publishers (2008)
17. Maier, D.: The Theory of Relational Databases. Computer Science Press, Rockville (1983)
18. Morales, T.: Oracle Database Reference, 11g Release 1 (11.1). Oracle Corp., Publishers (2008)
19. Gebhardt, J., Kruse, R., Klawonn, F.: Foundations of Fuzzy Systems. Wiley, Chichester (1994)
20. Rölleke, T., Wu, H., Wang, J., Azzam, H.: Modelling retrieval models in a probabilistic relational algebra with a new operator: the relational bayes. VLDB J. 17(1), 5–37 (2008)
21. Rowe, L.A., Jain, R.: Acm sigmm retreat report on future directions in multimedia research. ACM Trans. Multimedia Comput. Commun. Appl. 1(1), 3–13 (2005)
22. Schmitt, I., Nuernberger, A., Lehrack, S.: On the Relation between Fuzzy and Quantum Logic. In: Views on Fuzzy Sets and Systems from Different Perspectives. Philosophy and Logic, Criticisms and Applications, pp. 421–432 (2009)
23. Schmitt, I., Schulz, N.: Similarity Relational Calculus and its Reduction to a Similarity Algebra. In: Seipel, D., Turull-Torres, J.M.a. (eds.) FoIKS 2004. LNCS, vol. 2942, pp. 252–272. Springer, Heidelberg (2004)

24. Schmitt, I.: QQL: A DB&IR Query Language. The VLDB Journal 17(1), 39–56 (2008)
25. Weikum, G.: DB&IR: both sides now. In: SIGMOD (ed.) SIGMOD 2007: Proceedings of the 2007 ACM SIGMOD international conference on Management of data, pp. 25–30. ACM, New York (2007)
26. Zadeh, L.A.: Fuzzy Logic. IEEE Computer 21(4), 83–93 (1988)
27. Zadeh, L.A.: Fuzzy Sets. Information and Control (8), 338–353 (1965)

k-Selection Query over Uncertain Data

Xingjie Liu[1], Mao Ye[1], Jianliang Xu[2], Yuan Tian[1], and Wang-Chien Lee[1]

[1] Department of Computer Science and Engineering, The Pennsylvania State University,
University Park, PA 16801, USA
{xzl106,mxy177,yxt144,wlee}@cse.psu.edu
[2] Department of Computer Science, Hong Kong Baptist University, Kowloon Tong, Hong Kong
xujl@comp.hkbu.edu.hk

Abstract. This paper studies a new query on uncertain data, called *k-selection query*. Given an uncertain dataset of N objects, where each object is associated with a preference score and a presence probability, a k-selection query returns k objects such that the expected score of the "best available" objects is maximized. This query is useful in many application domains such as entity web search and decision making. In evaluating k-selection queries, we need to compute the expected best score (EBS) for candidate k-selection sets and search for the optimal selection set with the highest EBS. Those operations are costly due to the extremely large search space. In this paper, we identify several important properties of k-selection queries, including *EBS decomposition*, *query recursion*, and *EBS bounding*. Based upon these properties, we first present a dynamic programming (DP) algorithm that answers the query in $O(k \cdot N)$ time. Further, we propose a *Bounding-and-Pruning* (BP) algorithm, that exploits effective search space pruning strategies to find the optimal selection without accessing all objects. We evaluate the DP and BP algorithms using both synthetic and real data. The results show that the proposed algorithms outperform the baseline approach by several orders of magnitude.

1 Introduction

Data uncertainty is pervasive in our world. A web search engine returns a set of pages to a user, but cannot guarantee all the pages are still available. An on-line advertisement site lists many products with nice discounts, but some of them are already sold out in store. A GPS navigator may display nearest restaurants, but some of them may be already full. In presence of data uncertainty, effective queries that facilitate the retrieval of desired data items are urgently needed. In the past few years, several Top-k query semantics [14,10,11,7] for uncertain data have been proposed, trying to capture the possibly "good" items. However, these proposals do not address a very common problem, i.e., a user is typically only interested in the "best" item that is "available". For example, a used car shopper buys the car of his preference that is still for sale. Based on this observation, in this study, we present a new and novel query operator over uncertain data, namely, *k-selection query*. Given that the uncertain availability of data items is captured as a probability, the k-selection query returns a set of k candidate items, such that the expected score of the best available item in the candidate set is optimized.

H. Kitagawa et al. (Eds.): DASFAA 2010, Part I, LNCS 5981, pp. 444–459, 2010.

To illustrate the k-selection query semantics, let us consider a scenario where John plans to purchase a used car. Given an on-line used car database $D = \{d_1, d_2, \cdots, d_N\}$ where each d_i $(1 \leq i \leq N)$ represents a car and d_i ranks higher than d_{i+1}.[1] A top-k query may help by returning k cars based on John's preference. However, contacting all these sellers to find the best available car is time consuming since popular cars with a good deal may be sold quickly. Assuming that the available probabilities of vehicles can be obtained (e.g., a freshly posted used car has a higher available probability than a car posted weeks ago), a k-selection query on D takes into account the scores and availability of the cars to return an *ordered* list of k candidates that maximizes the expected preference score of the *best available* car. Thus, John can be more efficient in finding the best available car by following the list to contact sellers.

Data Object	Scores	Probability
d_1	4	0.3
d_2	3	0.4
d_3	2	0.8
d_4	1	0.6

(a) Dataset Example

Strategies	Results	Expected Best Score
Top-2 Scores	$\{d_1, d_2\}$	2.04
Weighted Score	$\{d_2, d_3\}$	2.16
2-Selection	$\{d_1, d_3\}$	2.32

(b) Query Result Example

Fig. 1. A 2-Selection Query Example

Finding the optimal k-selection to maximize the expected preference score of the *best available* object is not trivial. Consider a toy example in Fig. 1(a), where a set of 4 cars $D = \{d_1, d_2, d_3, d_4\}$ along with their available probabilities are shown. Suppose a user is interested in obtaining an ordered set of two candidates, $\{d_i, d_j\}$, where d_i is ranked higher than d_j. Since d_i has a higher preference score than d_j, the best object of choice is d_i as long as it is available. Only if d_i is unavailable while d_j is available, d_j will become the best choice. Based on the above reasoning, we use *expected best* score (which stands for expected score of the best candidate) to measure the goodness of the returned candidates. First, let us consider $\{d_1, d_2\}$, the candidate set obtained based on the highest scores. Its expected best score is $4 \cdot 0.3 + 3 \cdot (1 - 0.3) \cdot 0.4 = 2.04$. Next, consider $\{d_2, d_3\}$ which is obtained based on the highest weighted score. Its expected best score is $3 \cdot 0.4 + 2 \cdot (1 - 0.4) \cdot 0.8 = 2.16$. The above two strategies look reasonable but they do not yield the best selection because the first strategy does not consider the availability while the second strategy does not consider the ranking order in their selecting processes. As shown, the $\{d_1, d_3\}$, returned by the proposed 2-selection query, yields the highest expected best score $= 4 \cdot 0.3 + 2 \cdot (1 - 0.3) \cdot 0.8 = 2.32$.

Accordingly, the *expected best score (EBS)* for a selection $S \subseteq D$ can be expressed as in Eq. (1).

$$\text{EBS}(S) = \sum_{d_i \in S} f(d_i) \cdot P(d_i \text{ is the best available object in } S) \tag{1}$$

where $f(d_i)$ and $P(d_i)$ denote the preference score and available probability of object d_i, respectively.

---

[1] For simplicity, we assume that the data items have been sorted in the order of John's preference.

Therefore, a k-selection query $Q(k, D)$ aims at returning an ordered subset $S^* \subseteq D$, $|S^*| = k$ such that the EBS of S^* is maximized. S^* can be expressed as shown in Eq. (2):

$$S^* = \arg \max_{S \subseteq D, |S|=k} \text{EBS}(S) \tag{2}$$

The k-selection query is a new type of rank queries on uncertain dara that, to our best knowledge, has not been reported in the literature. Evaluating the k-selection query is very challenging because the candidate objects can not be selected individually to form the optimal solution. As a result, the search space for optimal k-selection is as large as $\binom{N}{k}$, which is significant as N increases. Efficient algorithms for the k-selection query are needed to tackle the challenge.

The contributions made in this paper are summarized as follows:

– We present a new and novel rank query, called k-selection, for uncertain databases.
– Based on the possible world model for uncertain data, we formalize the presentation of *expected best score (EBS)* and propose decomposing techniques to simplify the calculation of EBS.
– We develop a dynamic programming algorithm that solves the k-selection query over sorted data in $O(k \cdot N)$ time (where N is the dataset size).
– A bounding-and-pruning (BP) algorithm is developed based on the EBS bounds and the relationship in their preference scores. Its computational cost is even lower than the DP algorithm for large datasets.
– We conduct a comprehensive performance evaluation on both the synthetic data and real data. The result demonstrates that the proposed DB and BP algorithms outperform the baseline approach by several orders of magnitudes.

The rest of the paper is organized as follows. In Section 2, we review the existing work and formally formulate the problem. Section 3 addresses the problem by decomposing the EBS calculation and identifying the query recursion. Section 4 introduces the dynamic programming algorithm and the bounding-and-pruning algorithm that efficiently processes the k-selection query. Section 5 reports the result obtained from an extensive set of experiments. Section 6 concludes this paper.

2   Preliminaries

In this section, we first briefly review the previous works related to our study. Then, we present the formal definition of the k-selection problem.

2.1   Related Work

The related work involves two major research areas: 1) uncertain data modeling and 2) top-k query processing on uncertain data.

Uncertain data modeling. When inaccurate and incomplete data are considered in a database system, the first issue is how to model the uncertainty. A popular and general model to describe uncertain data is the *possible world semantic* [1,13,9]. Our study in this paper adopts this model. The possible world semantic models the uncertain data as a set of possible instances $D = \{d_1, d_2, \cdots, d_N\}$, and the presence of each instance is associated with a probability $P(d_i)$ $(1 \leq i \leq N)$. The set of possible worlds $\mathcal{PW} = \{W_1, W_2, \cdots, W_M\}$ enumerates all possible combinations of the data instances in D, that may appear at the same time (i.e., in a same possible world). Each possible world W_i has an *appearance probability* which reflects W_i's probability of existence. In this paper, we assume that the appearance of an object is independent from any other objects. Thus, the appearance probability of a possible world W_i can be derived from the membership probabilities of the uncertain objects:

$$P(W_i) = \prod_{d \in W_i} P(d) \cdot \prod_{d \notin W_i} \overline{P(d)} \qquad (3)$$

where the first term represents the probability that all the objects belong to W_i exist, and the second term represents the probability that all objects not in W_i do not exist, with $\overline{P(d)} = 1 - P(d)$.

In [9,8], query processing over independent data presence probabilities is studied, with SQL-like queries on probabilistic databases supported in [8]. Furthermore, since the presence of one object may depend on that of another, this presence dependency is modeled as *generation rules*. The rules define whether an object can present when some others exist and thus model the correlations between data objects. Query processing over correlated objects has been discussed in [17,16].

Top-k query processing over uncertain data. Following the possible world semantics, lots of queries defined in certain databases are revisited in uncertain scenario, such as Top-k queries [14,10,11,7], nearest neighbor queries [6,12,2,5,4,3] and skyline queries [15,18]. Among different queries over uncertain data, top-k queries have received considerable attention [14,10,11,7]. Like the k-selection query, all top-k queries assume a scoring function that assigns a preference score for each object. Because of the data uncertainty, various top-k query semantics and definitions have been explored, including *U-Topk* and *U-kRanks* [14], *PT-Topk* [10], *PK-Topk* [11], and *Expected-kRanks* [7]. However, their goals are essentially different from k-selection query.

The U-Top$k$ introduced in [14] catches the k-object set having the highest accumulated probability of being ranked as top-k objects. Taking the same example in Fig. 1(a), the U-top2 result is $\{d_2, d_3\}$ with top2 probability 0.224. For U-kRanks, the query tries to find the object with the highest probability to be ranked exactly at position $1, 2, \cdots, k$, respectively. Therefore, the object d_3 has the highest probability of being ranked first as well as the second. Thus, a U-2Rank returns the result $\{d_3, d_3\}$. The PT-Top$k$ [10] and PK-Top$k$ [11] define the top-k probabilities for individual objects. Specifically, the top-k probability measures the chance that an object ranks *within* the first k objects. The PT-Top$k$ returns the objects having top-k probability no less than a threshold, and PK-Top$k$ returns the k objects having the largest top-k probabilities. Based on the top-2 probabilities calculated for each object, we can find that, given a threshold of 0.3, PT-Top2 will return objects $\{d_3, d_2, d_1\}$; and PK-Top2 will return $\{d_3, d_2\}$. Most recently, [7] proposed the expected ranking semantics, where each individual object is sorted by its expected rank over all the possible worlds. Based on this definition, the top-2 objects are $\{d_3, d_2\}$. In summary, although existing top-k queries catch the top scored objects with various semantics, because their optimization problems are essentially different with k-selection, none of them can be used to answer the k-selection query.

2.2 Problem Formulation

Given the possible world model, we consider a k-selection query with some possible world W_i. Assume the selection set is S ($S \subseteq D$), since a user will only pick objects from S, the best available object is from the intersection of S and W_i. The *expected best score (EBS)* of S is therefore defined as follows:

Definition 1. *Expected Best Score: The EBS of a candidate answer set S, EBS(S), is defined as the expected best score of S over each possible world $W_i \in \mathcal{PW}$:*

$$EBS(S) = \sum_{W_i \cap S \neq \emptyset} \max_{d \in W_i \cap S} f(d) \cdot P(W_i) \qquad (4)$$

Definition 2. *k-Selection Query: The k-selection query over uncertain dataset D, $Q(k, D)$, is defined as finding the optimal answer set S^* consisting of k objects from D such that the EBS of the k selected objects is maximized (see Eq.(2))*

3 Analysis for k-Selection

To process a k-selection query, one straightforward way is to enumerate all possible selection sets, and for each selection set, enumerate all possible worlds to calculate its EBS. Finally, a set with maximum EBS is returned. This solution is clearly inefficient because it involves a lot of unqualified selection sets, and the EBS calculation accesses a large number of possible worlds one by one. To facilitate the EBS calculation and develop efficient query processing algorithm, in this section, we identify a set of useful properties of EBS and the query.

3.1 Expected Best Score (EBS)

One key step to find the optimal k-selection is to reduce the computation of EBS. From Eqn. (4), we can group the possible worlds based on their best available object. In other words, instead of enumerating all the possible worlds to find EBS of a selection S, we enumerate all the data object $d_i^S \in S$ and then accordingly identify those possible worlds that have the best available object as d_i^S. Therefore, continuing from Eqn. (4), we have:

$$\begin{aligned}
EBS(S) &= \sum_{i=1}^{k} f(d_i^S) \cdot \sum_{W_i \in \mathcal{PW}} P(W_i \mid f(d_i^S) \text{ is the largest in } W_i \cap S) \\
&= \sum_{i=1}^{k} f(d_i^S) \cdot P(d_i^S) \prod_{j=1}^{i-1} \overline{P(d_j^S)}
\end{aligned} \qquad (5)$$

The first step of the above equation eliminates the maximum operator of Eqn. (4) by considering each selected objects one by one. Therefore, for each object d_i^S, the probability that d_i^S is the best available object is the sum of possible world probabilities that d_i^S happens to be the available object with the largest score. In the second step, we further simplify the best available probability as subject to two conditions: 1) d_i^S is available; 2) all the objects within S that have a higher score than d_i^S are unavailable. (Note that objects within S are also numbered in score decreasing order.)

3.2 Query Recursion

For a dataset with N objects, there are a total of $\binom{N}{k}$ possible subsets. Thus, to exhaustively enumerate all possible selections is prohibitively expensive for a large dataset. To develop an efficient algorithm to find S^*, we explore some nice properties of the k-selection query under the data independence assumption.

Theorem 1. EBS Decomposition: *Consider a candidate selection S with objects $\{d_1^S, d_2^S, \cdots, d_k^S\}$ such that d_1^S is the top scored object. We define a partition of selection S as $S = S_{i-} \cup S_{(i+1)+}$, with $S_{i-} = \{d_1^S, d_2^S, \cdots, d_i^S\}$ and $S_{(i+1)+} = \{d_{i+1}^S, d_{i+2}^S, \cdots, d_k^S\}$. The expected best score of S is decomposable as follows:*

$$EBS(S) = EBS(S_{i-}) + \prod_{j=1}^{i} \overline{P(d_j^S)} \cdot EBS(S_{(i+1)+}). \tag{6}$$

Proof. From Eq. (5), we have:

$$EBS(S) = \sum_{j=1}^{i} \prod_{l=1}^{j-1} \overline{P(d_l^S)} P(d_j^S) \cdot f(d_j^S) + \prod_{l=1}^{i} \overline{P(d_l^S)} \cdot \sum_{j=i+1}^{k} \prod_{l=i+1}^{j-1} \overline{P(d_l^S)} P(d_j^S) \cdot f(d_j^S)$$

$$= EBS(S_{i-}) + \prod_{l=1}^{i} \overline{P(d_l^S)} \cdot EBS(S_{(i+1)+}).$$

From Theorem 1, we find that any partition of S can split its EBS into a linear relation as $b_0 + b_1 x$, with b_0 and b_1 only depending on the head partial selection of S_{i-}. And the tail selection $S_{(i+1)+}$ would affect the x value.

Theorem 2. Query Recursion: *For any dataset sorted in descending order of the preference score, $D_{i+} = \{d_i, d_{i+1}, \cdots, d_N\}$, the optimal k-selection set has the maximum EBS as $Opt(k, D_{i+})$. Then, the optimal EBS can be derived recursively as follows:*

$$Opt(k, D_{i+}) = \max \begin{cases} P(d_i)f(d_i) + \overline{P(d_i)} \cdot Opt(k-1, D_{(i+1)+}), \\ Opt(k, D_{(i+1)+}). \end{cases} \tag{7}$$

Proof. Consider the optimal answer set S^* for k-selection query over D_{i+} and $\forall d_i$, it is either included in S^* or not. If $d_i \in S^*$, then because the slope of Eq. (6) is non-negative, S_{i+1+} must also be maximized. By Theorem 1, the corresponding EBS for S^* is $P(d_i)f(d_i) + \overline{P(d_i)} \cdot Opt(k-1, D_{(i+1)+})$. Similarly, if $d_i \notin S^*$, then S^* must also be the optimal set of k-selection query for $D_{(i+1)+}$, with EBS $Opt(k, D_{(i+1)+})$. Thus, the EBS of S^* takes the maximum of the above two cases, as in Eq. (7).

Theorem 2 unleashes the recursion of the k-selection query. Armed with this recursion, we can reduce any k-selection query on D_{i+} to queries with equal to or smaller than k over a smaller data set $D_{(i+1)+}$.

3.3 Bounding Property

The above recursion property helps us to relate the k-selection query with smaller scale queries. It also indicates the dependency of the query to smaller scale queries. In other words, to find the optimal selection for query $Q(k, D_{i+})$, all the queries with smaller k and fewer objects than D_{i+} have to be solved. Since there are still many of these queries, we further explore the bounding property of the k-selection query.

Theorem 3. *EBS Bounding of Optimal Selection: For any dataset D_{i+} with d_i as the top object, the optimal EBS of a k-selection query, $\mathrm{Opt}(k, D_{i+}), k > 0$, is bounded by $[P(d_i)f(d_i), f(d_i)]$.*

Proof. Because all the object scores are positive, any $\mathrm{Opt}(k - 1, D_{(i+1)+})$ is non-negative. Therefore, from the first case in recursion Eq. (7), we must have $\mathrm{Opt}(k, D_{i+}) \geq P(d_i)f(d_i)$. Furthermore, because $f(d_i)$ is the maximum score in D_{i+}, it is also the bound of the maximum score in any set $S \subseteq D_{i+}$. Thus, based on Eq. (4), $\mathrm{Opt}(k, D_{i+}) \leq \max_{d \in S} f(d) \cdot \sum_{W_i \cap S \neq \emptyset} P(W_i) \leq \max_{d \in S} f(d) = f(d_i)$.

4 Query Processing Algorithms

In the previous section, we analyze a set of properties for k-selection queries. Now we present two efficient k-selection processing algorithms based on these properties.

4.1 Dynamic Programming (DP) Algorithm

According to Theorem 2, the k-selection query can be decomposed into queries with smaller dataset size and query size. Specifically, a query $Q(k, D_{i+})$ can be answered in constant time if the sub-queries $Q(k-1, D_{(i+1)+})$ and $Q(k, D_{(i+1)+})$ are solved. Thus, if we link these sub-queries as $Q(k, D_{i+})$'s children, an acyclic recursion graph can be constructed. A sample recursion graph for a 2-selection query over 4 data objects is shown in Fig. 2, with the root node representing $Q(k, D)$. In this graph, there are two scenarios in which the query can be trivially solved: 1) the query size $k = 0$, thus

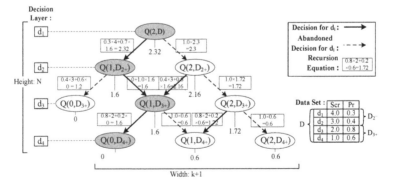

Fig. 2. Example of Dynamic Programming Algorithm (k=2, N=4)

the optimal EBS is also 0 because no object can be selected; 2) the remaining dataset size is 1 (i.e., D_{N+}), which means that the only object must be chosen to maximize the EBS for query size $k > 0$. We call these queries the *base-case queries*. Consider these base-case queries as leaves, the acyclic graph is similar to a tree with height N and width $k+1$ (Fig. 2). Furthermore, the recursion graph can be structured into layers, with the root query on layer 1 and $Q(t, D_{i+})$ ($0 \le t \le k$) queries on layer i. Based on Theorem 2, all the sub-queries in layer i need to decide whether to accept or reject d_i.

Since the evaluation of each sub-query relies on the results of its descendants in the recursion graph, a dynamic programming algorithm is developed to process the queries in a bottom-up fashion in Algorithm 1. Firstly, two types of base-case queries are initialized in line 1. Then, the algorithm recursively determines the optimal selection for each subproblem from bottom up, with queries for smaller datasets evaluated first (lines 2 through 8). For each sub-query $Q(t, D_{i+})$, the variable `accept` gets the optimal EBS assuming that d_i is included; otherwise, `reject` stores the EBS when d_i is excluded. By comparing `accept` and `reject`, the choice of whether to include d_i is stored in the variable `trace`(t, D_{i+}). Finally, the algorithm traces back the `trace` array to find out all accepted objects to obtain the optimal selection (lines 9 through 11). It is not difficult to see that the running time of the dynamic programming algorithm is $O(k \cdot N)$ to traverse the entire recursion graph.

Algorithm 1. Dynamic Programming (DP) Algorithm

Input: Dataset D, query size k
Output: Optimal subset S^* for k-selection over D
1 Initialize $\text{Opt}(0, D_{(0:N)+}) \leftarrow 0$, and $\text{Opt}(1:k, D_{N+}) \leftarrow P(d_N)f(d_N) + 0$;
2 **for** *layer* $i \leftarrow N-1, N-2 \cdots 1$ **do**
3   **for** *query size* $t \leftarrow 1, 2 \cdots k$ **do**
4          $\text{accept} \leftarrow P(d_i)f(d_i) + \overline{P(d_i)} \cdot \text{Opt}(t-1, D_{(i+1)+})$;
5          $\text{reject} \leftarrow \text{Opt}(t, D_{(i+1)+})$;
6          $\text{Opt}(t, D_{i+}) \leftarrow \max(\text{accept}, \text{reject})$;
7          $\text{trace}(t, D_{i+}) \leftarrow \text{accept} > \text{reject}$;

8 Initialize optimal subset $S^* \leftarrow \emptyset$, query size $t \leftarrow k$;
9 **for** *layer* $i \leftarrow 1, 2 \cdots N$ **do**
10   **if** $\text{trace}(t, D_{i+}) = true$ **then**
11         $S^* \leftarrow S^* \cup \{d_i\}, t \leftarrow t - 1$;

12 **return** $S^*, \text{Opt}(k, D)$

Take Fig. 2 as an example. At the beginning, the EBS values for the layer 4 nodes are initialized to $0, 0.6, 0.6$, respectively. Then, the recursion procedure starts from layer 3. Consider the sub-query $Q(1, D_{3+})$ for instance, the left edge from the node represents the case of accepting d_3, with EBS as $0.8 \times 2 + 0.2 \times \text{Opt}(0, D_{4+}) = 1.6$; the right edge represents the choice of rejecting d_3, and the corresponding EBS is $1.0 \times \text{Opt}(1, D_{4+}) = 0.6$. Since accepting d_3 leads to a higher EBS, d_3 will be included in the optimal set for the sub-query $Q(1, D_{3+})$ with solid edge. After the recursion procedure completes, all nodes get their optimal EBS values and decision edge identified. (d_i is selected if the solid edge goes left; otherwise to right). The optimal selection for the root query $Q(2, D)$ can be found by tracing back the decision result of each relevant node. For our running example, the decision path is

$Q(2, D) \rightarrow Q(1, D_{2+}) \rightarrow Q(1, D_{3+}) \rightarrow Q(0, D_{4+})$, and the corresponding decisions along the path are $\langle \texttt{accept}, \texttt{reject}, \texttt{accept} \rangle$ indicating $S^* = \{d_1, d_3\}$.

4.2 Bounding and Pruning Heuristics

The dynamic programming algorithm proposed in the last subsection needs to access all the data objects to find the optimal k-selection. However, the optimal k-selection, after all, tends to include those objects with higher scores. This intuition leads us to consider an algorithm that can stop without solving all the sub-queries.

However, for any sub-query $Q(t, D_{i+})$ $(0 \le t \le k)$, finding pruning rules to stop solving it is not trivial. This is because any of its ancestors above layer i (including the root query) counts on the exact value of $\texttt{Opt}(t, D_{i+})$ to make selection decisions. Thus, to develop efficient pruning heuristics, we start by investigating the relation between the subproblem $Q(t, D_{i+})$ and the root k-selection query $Q(k, D)$. Considering a dataset partition for D as $D_{i-} = \{d_1, d_2, \cdots, d_i\}$ and $D_{(i+1)+} = \{d_{i+1}, d_{i+2}, \cdots, d_N\}$, then a k-selection query over D is also partitioned by selecting t objects from D_{i-} and $k-t$ objects from $D_{(i+1)+}$. We define conditional k-selection queries as follows:

Definition 3. *Conditional k-Selection*: $Q(k, D \mid t, D_{i-})$ *is defined as a conditional k-selection query over* D*, by choosing* t *objects from* D_{i-} *and the other* $k-t$ *objects from* $D_{(i+1)+}$*. The EBS of the entire selection is maximized as* $\texttt{Opt}(k, D \mid t, D_{i-})$*.*

Clearly, the conditional k-selection query is sub-optimal to $Q(k, D)$ because it is restricted to the condition that exactly t objects are selected from D_{i-}. But, the global optimal k-selection can be found by solving a group of conditional k-selection queries.

To find the optimal conditional selection query $\texttt{Opt}(k, D \mid t, D_{i-})$, t objects S_{t-} are chosen from D_{i-} (hereafter S_{t-} is called *head selection*); the remaining $k-t$ objects $S_{(t+1)+}$ will be from $D_{(i+1)+}$ (hereafter $S_{(t+1)+}$ is called *tail selection*). Recall from Theorem 1, the EBS of $S = S_{t-} \cup S_{(t+1)+}$ is a linear function as $b_0 + b_1 x$, where the intercept and slope depend on head selection ($b_0 = \texttt{EBS}(S_{t-})$, $b_1 = \prod_{d \in S_{t-}} \overline{P(d)}$); and x is the EBS of tail selection ($x = \texttt{EBS}(S_{(t+1)+})$). Thus, to find an optimal conditional k-selection, x must be maximized because the slope b_1 is always nonnegative. Furthermore, after x is known, proper head selection shall also be chosen to maximize overall EBS of S.

Since we cannot know the optimal x value without solving $Q(k-t, D_{(i+1)+})$, but from the bounding property, x's bounding can be found without much effort. According to Theorem 3, the optimal $x = \texttt{Opt}(k-t, D_{(i+1)+})$ must fall within $[P(d_{i+1})f(d_{i+1})$, $f(d_{i+1})]$. Combining this value range with all the possible head selection linear functions, we developed two pruning heuristics.

Theorem 4. *Intra-Selection Pruning: For a head selection* S_{t-} *with EBS function represented by* $L(S_{t-}, x) = b_0 + b_1 x$*. For any value* x *within the bounding range* $[P(d_{i+1})f(d_{i+1}), f(d_{i+1})]$*, if there always exists another head selection* S_{t-}^a *having* $L(S_{t-}, x) < L(S_{t-}^a, x)$*,* S_{t-} *will not result in the optimal conditional selection.*

Proof. Suppose the optimal tail selection is $S_{(t+1)+}^*$. By combining the alternative head selection S_{t-}^a, we have the overall EBS as $\texttt{EBS}(S_{t-}^a \cup S_{(t+1)+}^*) = L(S_{t-}^a, x) >$

$L(S_{t-}, x) = \text{EBS}(S_{t-} \cup S^*_{(t+1)+})$. Thus, S_{t-} cannot result in the optimal selection, and is subject to pruning.

Theorem 5. *Inter-Selection Pruning: For a head selection S_{t-} and any value x within the bounding range of $[P(d_{t+1})f(d_{t+1}), f(d_{t+1})]$, if there exists another known k-selection S^a such that $L(S_{t-}, x) < \text{EBS}(S^a)$. Then, the head selection S_{t-} will not result in the optimal selection.*

Theorem 5 is correct because even with the optimal tail selection $S^*_{(t+1)+}$, the linear relation of $L(S_{t-}, x)$ will end up with a lower EBS than $\text{EBS}(S^a)$. This is called inter-selection pruning because the pruning selection S^a may come from any other k-selection without following the restriction of choosing t objects from D_{i-}.

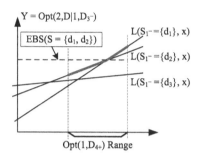

Fig. 3. Example Showing Three Combinations for Partial Dataset D_{4-}

An example illustrating the intra- and inter-selection pruning is shown in Fig. 3, where we consider a conditional query of $Q(2, D \mid 1, D_{3-})$ with $\binom{3}{1} = 3$ different possible head selections: $\{d_1\}, \{d_2\}$, and $\{d_3\}$. Their EBS linear functions as well as the bounding range of $\text{Opt}(1, D_{4+})$ are shown in Fig. 3. Here, we can see that $L(\{d_3\}, x)$ is always lower than $L(\{d_1\}, x)$ and $L(\{d_2\}, x)$ within the value range of $[P(d_4)f(d_4), f(d_4)]$. Thus, according to intra-selection pruning, the head selection $\{d_3\}$ can be safely discarded. In addition, suppose the EBS of $S^a = \{d_1, d_2\}$ is already computed, as in Fig. 3. We can observe that the EBS function of the head selection $\{d_2\}$ is either lower than $\text{EBS}(\{d_1, d_2\})$ or head selection $\{d_1\}$. Therefore, following the inter-selection pruning, $\{d_2\}$ can be safely discarded. After pruning these redundant head selections, the remaining head selections are referred to as effective head selections.

Definition 4. *Effective Head Selections: Considering a sub-query $Q(k, D \mid t, D_{i-})$, the effective head selections $\mathbb{ES}(t, i)$ is head selections remained after the intra- and inter-selection pruning.*

Theorem 6. *Effective Head Selection Recursion: The set of effective head selections $\mathbb{ES}(t, i)$, is a subset of $\mathbb{ES}(t, i - 1) \cup (\mathbb{ES}(t - 1, i - 1) + \{d_i\})$.[2]*

---

[2] The "+" means that d_i is added to each head selection in $\mathbb{ES}(t-1, i-1)$.

Proof. Consider any effective head selection $S_{t-} \in \mathbb{ES}(t, i)$, it either contains d_i or not. If $d_i \in S_{t-}$, then $S_{t-} - \{d_i\}$ must be an effective head selection for $D_{(i-1)-}$ because otherwise S_{t-} will not catch the optimal selection either. Similarly, if $d_i \notin S_{t-}$, S_{t-} must be an effective head selection for $D_{(i-1)-}$.

Theorem 6 points out that we do not need to run a pruning algorithm for all possible head selections from D_{i-} all the times. Instead, it is enough to consider the effective head selections from its predecessors and merging them and apply pruning rules. This recursion significantly reduces the computation of head selection pruning, and motivates a top-down bounding and pruning algorithm.

4.3   Bounding and Pruning (BP) Algorithm

Having introduced the pruning heuristics, now we present the top-down bounding and pruning algorithm to solve the k-selection query (see Algorithm 2).

The main frame of this algorithm is similar to the dynamic programming algorithm 1, except that the sub-queries here are accessed in a top-down fashion. For each sub-query, the algorithm first determines whether the sub-query $Q(k-t, D_{i+})$ is a base-case query (line 7). If so, the optimal EBS for the tail selection $\text{Opt}(k-t, D_{i+})$ is trivially solved (line 9,10). This result can be combined with every effective head selection to find the exact optimal conditional selection EBS. If this conditional selection is better, then the best found k-selection is updated in line 11. For those sub-queries that cannot be trivially solved, we only use constant time to obtain its EBS value bound of $\text{Opt}(k - t, D_{i+})$ in line 13, then all the successors' *effective* head selections are collected and filtered using inter-/intra-selection pruning rules (Theorem 4, 5, 6). Consequently, if no effective head selection is left, the scenario is captured in line 4 and the entire algorithm terminates, asserting the found best k-selection as the global optimal.

Algorithm 2. Bounding and Pruning (BP) Algorithm

Input: Dataset D, query size k
Output: Optimal Subset S^* for k-selection over D
1 Initialize effective head selection $\mathbb{ES}(0 : k, 1 : N) = \emptyset$, $\mathbb{ES}(0, 1) = \{\emptyset\}$;
2 Initialize found best $k$selection as $bestEMS \leftarrow 0, S^* \leftarrow \emptyset$;
3 **for** *layer* $i \leftarrow 1, 2 \cdots N$ **do**
4   **if** $\forall\, \mathbb{ES}(0 : k, i) = \emptyset$ **then**   /* No effective head selections. */
5   break;

6   **for** *query size* $t \leftarrow k, k-1 \cdots 0$ and $\mathbb{ES}(t, i) \neq \emptyset$ **do**
7   **if** $t = k$ **or** $i = N$ **then**   /* Tail selection basecases. */
8   **for** $\forall S_{t-} \in \mathbb{ES}(t, i)$ **do**
9   **if** $t = k$ **then** $\text{EBS}(S) \leftarrow \text{EBS}(S_{t-})$;
10   **if** $i = N$ **then** $\text{EBS}(S) \leftarrow \text{EBS}(S_{t-}) + \prod_{d \in S_{t-}} \overline{P(d)} \cdot P(d_i) f(d_i)$;
11   **if** $EBS(S) > bestEMS$ **then** $bestEMS \leftarrow \text{EBS}(S)$ and $S^* \leftarrow S$

12   **else**   /* Inter-and intra-selection pruning */
13             $\text{TailBound} \leftarrow [p(d_i) f(d_i), f(d_i)]$;
14             $\mathbb{ES}(t, i) \leftarrow \text{MERGEPRUNE}\left(\mathbb{ES}(t, i-1), \mathbb{ES}(t-1, i-1)\right)$;

15 **return** $S^*, bestEMS$

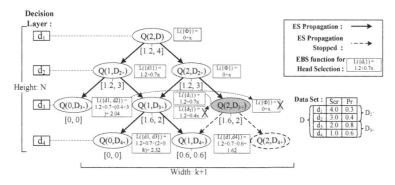

Fig. 4. Example for Bounding and Pruning Algorithm (k=2, N=4)

A running example for the bounding and pruning algorithm is shown in Fig. 4. The EBS bounding range for each tail selection is shown under each node in the figure (e.g. $[1.2, 3]$ for sub-query $Q(2, D_{2+})$ because $f(d_2) * P(d_2) = 1.2$ and $f(d_2) = 3$). For the head selection S_{t-}, the root query $Q(2, D)$ initializes its effective head selection as $\mathbb{ES}(0, 1) = \{\emptyset\}$, and then propagates it to its child nodes $Q(1, D_{2+})$ and $Q(2, D_{2+})$ with $\emptyset \cup \{d_1\}$ and $\emptyset$, respectively. The pruning operation becomes effective on layer 3. For the sub-query $Q(1, D_{3+})$, the before-pruning head selections are $\{d_1\}$ and $\{d_2\}$. Since these two head selections have the EBS functions as $1.2 + 0.7x$ and $1.5 + 0.5x$, respectively, and given tail selection EBS range as $[1.6, 2]$, the head selection $\{d_1\}$ always has a higher EBS (Theorem 4). Thus, $\{d_2\}$ is pruned according to intra-selection pruning. The inter-pruning happens at the node $Q(2, D_{3+})$. Since the only head selection for this node is $\emptyset$ with EBS function $0 + 1x$ and tail selection EBS range $[1.6, 2]$. But at the time, node $Q(0, D_{3+})$ already found a conditional 2-selection $S = \{d_1, d_2\}$ with bestEMS $= 2.04$. This 2-selection is always superior to $Q(2, D \mid 2, D_{3+})$ because the maximum value of $0 + 1x$ over $[1.6, 2]$ is only 2. Thus, after this only head selection $\emptyset$ is pruned, the node has no remaining head selection propagated. Therefore, its child node $Q(2, D_{4+})$ is ignored during the next layer's processing for empty effective front selection. The stopping node is highlighted in orange in Figure 4, and the optimal selection is actually found at node $Q(0, D_{4+})$ with its head selection $\{d_1, d_3\}$.

The bound and pruning algorithm is guaranteed to find the optimal k-selection because all the head selections are kept until it is guaranteed not to catch the optimal selection. And solving the tail selection has always been postponed by using the bounding properties until it is trivial base case.

5   Performance Evaluation

To test k-selection queries, we use both synthetic data and real-world dataset. Real datasets include three data sets named FUEL, NBA and HOU, respectively. FUEL is a 24k 6-dimensional dataset, in which each point stands for the performance of a vehicle. NBA contains around 17k 13-dimensional data points corresponding to the statistics of NBA players' performance in 13 aspects. And HOU consists of 127k 6-dimensional data points, each representing the percentage of an American familys annual expense on 6 types of expenditures.[3] In our experiments, we arbitrarily choose two dimensions from each data set for testing, and assign uniform distributed probability between $[0, 1]$ for each data point. These real-world datasets are used to validate the k-selection performance effectiveness for practical applications. On the other hand, we use the synthetic data to learn insights of k-selection queries and proposed algorithms. For synthetic data, the membership probability of an object is modeled by $P(d_i) = \mu + \delta u(i)$, where $\mu > 0, \delta > 0$, and $u(i)$ uniformly distributed between $[-0.5, 0.5]$. Thus, with default value $\mu = 0.5$ and $\delta = 1$, $P(d_i)$ is a uniform distribution between $[0, 1]$. To generate the ranking score for each object, we set $f(d_i) = 0.5 + \beta u(i) + (1 - |\beta|)u'(i)$, with $u(i)$ the same random variable as in $P(d_i)$ but $u'(i)$ another identical independent random variable. Therefore, β models the *covariance* between $f(d_i)$ and $P(d_i)$ as $\eta = \text{Cov}(f(d_i), P(d_i)) = \delta\beta$. In addition to data modeling, we model the data access I/Os by concerning the object size. We set the page size at 4 KB. Therefore assuming one object occupies θ bytes, it will need one I/O operation every $4000/\theta$ object access. The experiment parameters are summarized in Figure 5. All the experiments are implemented and conducted on a Window Server with Intel Xeon 3.2 GHz CPU and 4 GB RAM. The results presented in this section are averaged over 100 independent runs.

Parameter	Setting	Default
Dataset Size (N)	$100 \sim 100,000$	$10,000$
Query Selection Size (k)	$1 \sim 1,000$	100
Probability Mean (μ)	$0.2 \sim 0.8$	0.5
Probability Range (δ)	$0.2 \sim 1$	1
Probability Score Covariance (η)	$-0.8 \sim 0.8$	0
Object Size (θ)	$10 \sim 1000$	100

Fig. 5. Experiment Parameters

We evaluate the performance of different k-selection algorithms, including dynamic programming algorithm (DP), the bounding and pruning algorithm (BP), and a naive algorithm that examines all $\binom{N}{k}$ candidate selections. Since the naive algorithm cannot scale up to a large dataset, we first compare these algorithms under small datasets. As shown in Fig. 6(a), with 100 objects and 5-selection, the execution time of the naive algorithm is already raised to 100 seconds, which is 10^6 times to the DP algorithm, and 10^5 times to the BP algorithm. For these small datasets, we also find that the BP has a worse performance than the DP algorithm. This is because building a small k by N recursion graph for DP is fast, but the pruning overhead for BP is relatively costly while not much objects can be pruned for a small dataset.

We now proceed to compare the performance of DP and BP under larger datasets. Fig. 6(b) and (c) shows the results using real data. In Fig. 6(b), we find DP is better than BP for a small dataset, which is consistent with what we observed in Fig. 6(a). However, for all other large dataset settings, BP performs much better than DP because of its effective pruning: most of the sub-queries are pruned and left out of computation.

---

[3] Those datasets are collected from www.nba.com, www.ipums.org and www.fueleconomy.gov

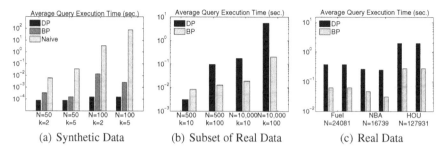

(a) Synthetic Data　　　　　(b) Subset of Real Data　　　　　(c) Real Data

Fig. 6. Algorithm Performance

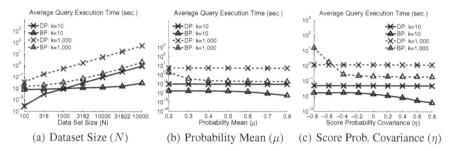

(a) Dataset Size (N)　　(b) Probability Mean (μ)　　(c) Score Prob. Covariance (η)

Fig. 7. Query Response Time under Synthetic Data

To gain more insight into these two algorithms, we further evaluate them using synthetic data under different workload settings. Fig. 7(a) plots the performance results for two series of tests ($k = 10$ and $k = 1,000$) as the dataset size varies from 100 to 10,000. For both of the algorithms, it is found that the query execution time increases with increasing the dataset size or the selection size. This is because the larger is N or k, more sub-queries need to be solved before finding the optimal selections (see Fig. 8(a)). Comparing these two algorithms, again, only when the selection size is small and the dataset size is small, DP outperforms BP. On the other hand, BP outperforms DP by more than an order of magnitude for most of the larger-scale cases. Next, we examine the performance of the two algorithms under different data distribution settings. As for the setting of the membership probabilities, Fig. 7(b) shows that the DP algorithm, again, exhibits a similar performance with various settings. However, the BP algorithm shows a significant decrease in cost when the probability mean is high. The reason is that with a higher probability mean, the optimal k-selection favors to include high-score objects such that the inter-selection pruning could terminate the BP algorithm earlier (as observed in Fig. 8(b)). In Fig. 7(c), a similar trend is found when the score and probability covariance are increased. This is because when the score and probability are correlated, many high-score objects will be selected, thereby making the BP algorithm to explore less objects before termination.

Finally, the impact of the object size is shown in Fig. 9. Although both algorithms need to access more data when a larger dataset is concerned, the BP algorithm performs similarly when the object size is varied. This is because BP only accesses a small portion

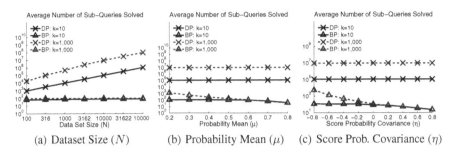

(a) Dataset Size (N)   (b) Probability Mean (μ)   (c) Score Prob. Covariance (η)

Fig. 8. Number of Sub-Queries Solved under Synthetic Data

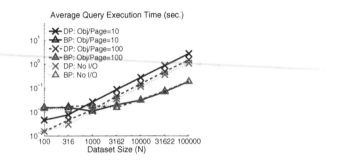

Fig. 9. Impact of Object Size under Various Dataset Sizes

of objects with few I/Os and, hence, even if the object size is large, BP does not suffer from it. On the other hand, however, it is shown that the query execution time of DP increases significantly when the object size increases. This, from another angle, implies that more objects are accessed in DP than BP.

6 Conclusion

In this paper, we introduce a new k-selection query operation over uncertain data, which finds a subset of objects, that yields the highest expected best score (EBS). While the query is very useful for various applications, it may incur very high processing cost due to the extremely large search space. To address this problem, we first analyze the characteristics of the k-selection query and identify a number of properties. Then, we propose two efficient k-selection query processing algorithms. The Dynamic Programming (DP) algorithm, which employs the EBS decomposition and query recursion properties, evaluates sub-queries in a bottom-up fashion recursively. Bound-and-Pruning (BP) algorithm, however, utilize a linear relation of EBS decomposition and bounding, to efficiently reduce the problem search space. Through a set of comprehensive evaluations, we demonstrate that our proposed algorithms are superior to the naive brute-force approach, and are efficient for on-line applications.

Acknowledgement

This research was supported in part by the National Science Foundation under Grant No. CNS-0626709. Jianliang Xu, was supported in part by HK RGC grants HKBU211307 and HKBU210808.

References

1. Abiteboul, S., Kanellakis, P., Grahne, G.: On the Representation and Querying of Sets of Possible Worlds. In: Proceedings of SIGMOD 1987 (1987)
2. Beskales, G., Soliman, M.A., Ilyas, I.F.: Efficient search for the top-k probable nearest neighbors in uncertain databases. In: Proceedings of VLDB 2008 (2008)
3. Cheema, M.A., Lin, X., Wang, W., Zhang, W., Pei, J.: Probabilistic Reverse Nearest Neighbor Queries on Uncertain Data. TKDE 99(1)
4. Cheng, R., Chen 0002, L., Chen, J., Xie, X.: Evaluating probability threshold k-nearest-neighbor queries over uncertain data. In: Proceedings of EDBT 2009 (2009)
5. Cheng, R., Chen, J., Mokbel, M.F., Chow, C.-Y.: Probabilistic Verifiers: Evaluating Constrained Nearest-Neighbor Queries over Uncertain Data. In: Proceedings of ICDE 2008 (2008)
6. Cheng, R., Kalashnikov, D.V., Prabhakar, S.: Querying Imprecise Data in Moving Object Environments. TKDE 16(9)
7. Cormode, G., Li, F., Yi, K.: Semantics of ranking queries for probabilistic data and expected ranks. In: Proceedings of ICDE 2009 (2009)
8. Dalvi, N., Suciu, D.: Efficient Query Evaluation on Probabilistic Databases. In: Proceedings of VLDB 2004 (2004)
9. Fuhr, N., Rölleke, T.: A Probabilistic Relational Algebra for the Integration of Information Retrieval and Database Systems. ACM Transaction on Information System 15(1)
10. Hua, M., Pei, J., Zhang, W., Lin, X.: Ranking queries on uncertain data: a probabilistic threshold approach. In: Proceedings of SIGMOD 2008 (2008)
11. Jin, C., Yi, K., Chen, L., Yu, J.X., Lin, X.: Sliding-window top-k queries on uncertain streams. Proceedings of the VLDB Endowment 1(1)
12. Kriegel, H.-P., Kunath, P., Renz, M.: Probabilistic Nearest-Neighbor Query on Uncertain Objects. In: Proceedings of DSFAA 2007 (2007)
13. Lakshmanan, L.V.S., Leone, N., Ross, R., Subrahmanian, V.S.: ProbView: a Flexible Probabilistic Database System. ACM Transaction on Database System 22(3)
14. Mohamed, I.F.I., Soliman, A., Chang, K.C.-C.: Top-k Query Processing in Uncertain Databases. In: Proceedings of ICDE 2007 (2007)
15. Pei, J., Jiang, B., Lin, X., Yuan, Y.: Probabilistic skylines on uncertain data. In: Proceedings of VLDB 2007 (2007)
16. Prithviraj, S., Deshpande, A.: Representing and Querying Correlated Tuples in Probabilistic Databases. In: Proceedings of ICDE 2007 (2007)
17. Sarma, A.D., Benjelloun, O., Halevy, A., Widom, J.: Working Models for Uncertain Data. In: Proceedings of ICDE 2006 (2006)
18. Zhang, W., Lin, X., Zhang, Y., Wang, W., Yu, J.X.: Probabilistic Skyline Operator over Sliding Windows. In: Proceedings of ICDE 2009 (2009)

Analysis of Implicit Relations on Wikipedia: Measuring Strength through Mining Elucidatory Objects

Xinpeng Zhang, Yasuhito Asano, and Masatoshi Yoshikawa

Kyoto University, Kyoto, Japan 606-8501
{xinpeng.zhang@db.soc.,asano@,yoshikawa@}i.kyoto-u.ac.jp

Abstract. We focus on measuring relations between pairs of objects in Wikipedia whose pages can be regarded as individual objects. Two kinds of relations between two objects exist: in Wikipedia, an explicit relation is represented by a single link between the two pages for the objects, and an implicit relation is represented by a link structure containing the two pages. Previously proposed methods are inadequate for measuring implicit relations because they use only one or two of the following three important factors: distance, connectivity, and co-citation. We propose a new method reflecting all the three factors by using a generalized maximum flow. We confirm that our method can measure the strength of a relation more appropriately than these previously proposed methods do. Another remarkable aspect of our method is mining elucidatory objects, that is, objects constituting a relation. We explain that mining elucidatory objects opens a novel way to deeply understand a relation.

Keywords: link analysis, generalized flow, Wikipedia mining, relation.

1 Introduction

Searching Web pages containing a keyword has grown in this decade, while knowledge search has recently been researched to obtain knowledge of a single object and relations between multiple objects, such as humans, places or events. Searching knowledge of objects using Wikipedia is one of the hottest topics in the field of knowledge search. In Wikipedia, the knowledge of an object is gathered in a single page updated constantly by a number of volunteers. Wikipedia also covers objects in a number of categories, such as people, science, geography, politic, and history. Therefore, searching Wikipedia is usually a better choice for a user to obtain knowledge of a single object than typical search engines.

A user also might desire to discover a relation between two objects. For example, a user might desire to know which countries are strongly related to petroleum, or to know why one country has a stronger relation to petroleum than another country. Typical keyword search engines can neither measure nor explain the strength of a relation. The main issue for measuring relations arises from the fact that two kinds of relations exist: "explicit relations" and "implicit relations." In Wikipedia, an explicit relation is represented by a link. For example, an explicit relation between petroleum and Iraq might be

H. Kitagawa et al. (Eds.): DASFAA 2010, Part I, LNCS 5981, pp. 460–475, 2010.

represented by a link from page "Iraq" to page "Petroleum." An implicit relation is represented by multiple links and pages. For example, as depicted in Figure 8, an implicit relation between petroleum and the USA might be represented by two links: one between "Petroleum" and "Gulf of Mexico" and the other one between "Gulf of Mexico" and the "USA." For an implicit relation between two objects, the objects, except the two objects, constituting the relation is named *elucidatory objects* because such objects enable us to explain the relation. For the example described above, "Gulf of Mexico" is the elucidatory object. The user can understand an explicit relation between two objects easily by reading the pages for the two objects in Wikipedia. By contrast, it is difficult for the user to discover an implicit relation and elucidatory objects without investigating a number of pages and links. Therefore, it is an interesting problem to measure and explain the strength of an implicit relation between two objects in Wikipedia.

Several methods have been proposed for measuring the strength of a relation between two objects on an *information network* (V, E), a directed graph where V is a set of objects; an edge $(u, v) \in E$ exists if and only if object $u \in V$ has an explicit relation to $v \in V$. We can define a *Wikipedia information network* whose vertices are pages of Wikipedia and whose edges are links between pages. Previously proposed methods then can be applied to Wikipedia by using a Wikipedia information network. Most of these methods use only one or two of the three representative concepts for measuring a relation: distance, connectivity, and co-citation, although all the concepts are important factors for implicit relations. Using all the three concepts together would be appropriate for measuring an implicit relation and mining elucidatory objects. Another concept "cohesion," exists for measuring the strength of an implicit relation. CFEC proposed by Koren et al. [1] and PFIBF proposed by Nakayama et al. [2][3] are based on cohesion. We do not adopt cohesion because it has a property unsuitable for measuring "3-hop implicit relations," as we will explain in Section 2.2. In an information network, an implicit relation between two objects s and t is represented by a subgraph containing s and t. We say that the implicit relation is a *k-hop implicit relation* if the subgraph contains a path from s to t whose length is at least $k > 1$. Figure 8 depicts an example of a 3-hop implicit relation between "Petroleum" and the "USA." We observe that a number of 3-hop implicit relations play important roles in Wikipedia.

We propose a new method for measuring a relation on Wikipedia by reflecting all the three concepts: distance, connectivity, and co-citation. Our method uses a "generalized maximum flow" [4][5] on an information network to compute the strength of a relation from object s to object t using the value of the flow whose source is s and destination is t. It introduces a *gain* for every edge on the network. The value of a flow sent along an edge is multiplied by the gain of the edge. Assignment of the gain to each edge is important for measuring a relation using a generalized maximum flow. We propose a heuristic gain function utilizing the category structure in Wikipedia. We confirm through experiments that the gain function is sufficient to measure relations appropriately.

We evaluate our method using computational experiments on Wikipedia. We first select several pages from Wikipedia as our source objects; and for each source object, we select several pages as the destination objects. We then compute the strength of the relation between a source object and each of its destination objects, and rank the destination objects by the strength. By comparing the rankings obtained by our method with those obtained by the "Google Similarity Distance" (GSD) proposed by Cilibrasi and Vitányi [6], PFIBF and CFEC, we ascertain that the rankings obtained by our method are the closest to the rankings obtained by human subjects. Especially, we ascertain that only our method can appropriately measure the strength of 3-hop implicit relations.

Our method can mine elucidatory objects constituting a relation by outputting paths contributing to the generalized maximum flow, that is, paths along which a large amount of flow is sent. We will explain in Section 4.4 that mining elucidatory objects opens a novel way to deeply understand a relation.

Several semantic search engines [7] seem to be used for searching relations between two objects, using a semantic knowledge base [8] extracted from Web or Wikipedia. However, the semantics in these knowledge bases, such as "isCalled," "type" and "sub-ClassOf," are mainly used to construct an ontology for objects. Such semantic knowledge bases are still far from covering relations existing in Wikipedia, such as "Gulf of Mexico" is a major "petroleum" producer. We do not assuming semantics in this paper.

The main contributions of this paper are as follows. (1) A detailed and methodical survey of related work for measuring relations (Section 2). (2) A new method using generalized maximum flow for measuring the strength of a relation between two objects on Wikipedia, reflects the three concepts: distance, connectivity and co-citation (Section 3). (3) Experiments on Wikipedia show that our method is the most appropriate one (Section 4.2). (4) Case studies of mining elucidatory objects for deeply understanding a relation (Section 4.4).

2   Related Work

2.1   Distance, Connectivity, Co-citation

The Erdös number [9] used by mathematicians is based on distance and co-authorships. The legendary mathematician Paul Erdös has a number 0, and the people who co-wrote a paper with Erdös have a number 1; the people who co-wrote a paper with a person with a number 1 have a number 2, and so on. The Erdös number is the distance, or the length of the shortest path, from a person to Erdös on an information network whose edge represents co-authorship; a shorter path represents a stronger relation. However, the Erdös number is inadequate to represent the implicit relation between a person and Erdös because the number does not estimate the connectivity between them.

The connectivity, more precisely the vertex connectivity, from vertex s to vertex t on a network is the minimum number of vertices such that no path exists from s to t if the vertices are removed. The connectivity is also used to measure the fault-tolerant robustness of a network [4]. s has a strong relation to t if the connectivity from s to t is large. The connectivity from s to t is equal to the value of a maximum flow from s to t, where every edge and vertex has capacity 1. The value is equal to the number of vertex-disjoint paths from s to t. However, the distance cannot be estimated by the maximum flow because the amount of a flow along a path is independent of the path length. Lu et al. [10] proposed a method for computing the strength of a relation using a maximum flow. They tried to estimate the distance between two objects using a maximum flow by setting edge capacities. However, the value of a maximum flow does not necessarily decrease by setting only capacities even if the distance becomes larger. Therefore, their method cannot estimate the distance successfully by the value of the maximum flow. Instead of setting capacities, we use a generalized maximum flow by setting every gain to a value less than one. Therefore, the value of a maximum flow in our method decreases if the distance becomes larger.

Co-citation based methods assume that two objects have a strong relation if the number of objects linked by both the two objects is large [11]. On the other hand, co-occurrence is a concept by which the strength is represented by the number of objects linking to both objects. The "Google Similarity Distance" (GSD) proposed by Cilibrasi and Vitányi [6] can be regarded as a co-occurrence based method; it measures the strength of a relation between two words by counting of Web pages containing both words. That is, it implicitly regards the Web pages as the objects linking to the two objects representing the two words. In an information network, an object linked by both objects becomes an object linking to the both if the direction of every edge is reversed. Therefore, co-occurrence can be regarded as the reverse of the co-citation. We then include co-occurrence based methods among co-citation based methods in this paper. Milne and Witten [12] also proposed methods measuring relations between objects in Wikipedia using Wikipedia links based on co-citation. Co-citation based methods cannot deal with a typical implicit relation, such as "person w is regarded as a friend by person v who is regarded as a friend by person u." This relation is represented by the path formed by two edges (u, v) and (v, w). In contrast, co-citation based methods can deal with two edges going into the same vertex, such as edges (u, v) and (w, v). Therefore, co-citation based methods are inadequate for measuring an implicit relation. Furthermore, co-citation based methods cannot deal with 3-hop implicit relations defined in Section 1 because these methods estimate only two edges between the two objects, as explained above. SimRank, proposed by Jeh and Widom [13], is an extension of co-citation based methods. SimRank employs recursive computation of co-cited objects, therefore it can deal with a path whose length is greater than two, although it cannot deal with a typical implicit relation "a friend of a friend" similarly to co-citation based methods.

2.2 Cohesion

In the field of social network analysis, cohesion based methods are known to measure the strength of a relation by counting all paths between two objects. The original cohesion was proposed by Hubbell and Katz [14][15][16]. It has a property that its value greatly increases if a *popular object*, an object linked from or to many objects, exists. As pointed out in other researches [17][1][2], this property is a defect for measuring the strength of a relation. Several cohesion based methods, such as PFIBF and CFEC explained below, were proposed to dissolve this property.

Nakayama et al. [3][2] proposed a cohesion based method named PFIBF. Instead of enumerating all paths, PFIBF approximately counts paths whose length is at most $k > 0$ using the k-th power of the adjacency matrix of an information network. However, in the k-th power of the matrix, a path containing a cycle whose length is at most $k - 1$ would appear. PFIBF cannot distinguish a path containing a cycle from a path containing no cycle. For example, if $k \geq 3$ and two edges (u, v) and (v, u) exist, then PFIBF counts

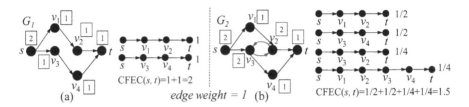

Fig. 1. CFEC on two networks

both path (u, v) and path (u, v, u, v) containing a cycle (u, v, u). Consequently, PFIBF has a property that it estimates a single path, e.g., (u, v) in the above example, for multiple times. The length of a cycle is at least two. No path containing a cycle appears if $k \leq 2$. In fact, PFIBF usually sets $k = 2$. Therefore, PFIBF is inappropriate for measuring a 3-hop implicit relation. The "Effective Conductance" (EC) proposed by Doyle et al. [18] is a cohesion based method also. EC has the same deficit as PFIBF: it counts a path containing a cycle redundantly. Koren et al. [1] proposed cycle-free effective conductance (CFEC) based on EC by solving this deficit. For a positive integer k, CFEC enumerates only the k-shortest paths between s and t, instead of computing all paths. CFEC does not use a path containing a cycle, although it cannot count all paths.

In contrast to the original cohesion, PFIBF and CFEC underestimate a popular object. CFEC defines the weight of path $p = (s = v_1, v_2, ..., v_\ell = t)$ from s to t as $w_{sum}(v_1) \cdot \prod_{i=1}^{\ell-1} \frac{w(v_i, v_{i+1})}{w_{sum}(v_i)}$, where $w(u, v)$ is the weight of edge (u, v) and $w_{sum}(v)$ is the sum of the weights of the edges going from vertex v. Therefore, the weight of a path becomes extremely small if a popular object exists in the path. The strength $C(s, t)$ of the relation between s and t is the sum of the weights of all paths from s to t. Figure 1 depicts two networks and all the paths between s and t. For simplicity, let the weight of every edge be one. The w_{sum} of each vertex is written in the rectangle near the vertex. The weight of each path is presented at the right side of the path. For the network G_1 depicted in Figure 1(a), the w_{sum} of s is 2, and the weight of path (s, v_1, v_2, t) is 1. $C(s, t)$ for G_1 is 2, which is equal to the connectivity between s and t. If we add two edges (v_2, v_3) and (v_3, v_2) to G_1, then we obtain network G_2 in Figure 1(b). Two vertices v_2 and v_3 become more popular in G_2 than they are in G_1, and $C(s, t)$ decreases from 2 in G_1 to 1.5 in G_2. Consequently, CFEC has the property that it could estimate the strength of a relation smaller if popular objects exist. Similarly, PFIBF has the same property. This property seems strange because the connectivity between two vertices on a network never decreases when an edge is added to the network. We ascertain that the property is unsuitable for measuring a relation by comparing our method with PFIBF and CFEC in Section 4.2.

2.3   Explanation of a Relation

Faloutsos et al. [17], Tong and Faloutsos [19] proposed methods for visualizing a sub-network explaining a relation by utilizing the ideas of EC; Koren et al. [1] also proposed such a method based on CFEC. The subnetwork is constructed from a set of paths having the highest weights. These methods always underestimate a popular object. Therefore, a popular object is hardly displayed in a subnetwork explaining a relation in these

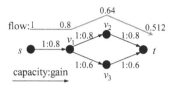

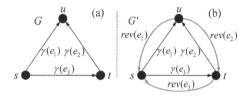

Fig. 2. Generalized maximum flow   **Fig. 3.** A doubled network

methods. However, a popular object constitutes a relation between two objects in some cases. For example, "George W. Bush" is a popular object in the dataset of Wikipedia, and it constitutes a relation between "Junichiro Koizumi" and "Condoleezza Rice." Zhu et al. [20] extract explicit relations between pairs of people from the Web. They then visualize a graph whose vertices are people, edges are explicit relations between two people. They do not explain an implicit relation. Therefore, these methods are inadequate for explaining a relation. We ascertain in Section 4.4 that a more comprehensive alternative is obtainable using our method.

3   Our Method Using Generalized Flow

3.1   Generalized Maximum Flow

The generalized maximum flow problem is identical to the classical maximum flow problem except that every edge e has a gain $\gamma(e) > 0$; the value of a flow sent along edge e is multiplied by $\gamma(e)$. Let $f(e) \geq 0$ be the flow f on edge e, and $\mu(e) \geq 0$ be the capacity of edge e. The capacity constraint $f(e) \leq \mu(e)$ must hold for every edge e. The goal of the problem is to send a flow emanating from the source vertex s into the destination vertex t to the greatest extent possible, subject to the capacity constraints. Let *generalized network* $G = (V, E, s, t, \mu, \gamma)$ be information network (V, E) with the source $s \in V$, the destination $t \in V$, the capacity μ, and the gain γ. Figure 2 depicts an example of a generalized maximum flow on a generalized network. One unit of flow is sent from the source s to v_1, i.e., $f(s, v_1) = 1$, the amount of the flow is multiplied by $\gamma(s, v_1)$ when the flow arrives at v_1. Consequently, only 0.8 units arrive at v_1. In this way, only 0.512 units arrive at the destination t. The capacity constraint for edge $e = (u, v)$ must hold before the gain is multiplied. $f(s, v_1) = 1 \leq \mu(s, v_1)$ must hold, for example.

We propose a new method for measuring the strength of a relation using the generalized maximum flow. The value of flow f is defined as the total amount of f arriving at destination t. To measure the strength of a relation from object s to object t, we use the value of a generalized maximum flow emanating from s as the source into t as the destination; a larger value signifies a stronger relation. We regard the vertices in the paths composing the generalized maximum flow as the objects constituting the relation. We qualitatively ascertain the claim that our method can reflect the three representative concepts explained in Section 2: distance, connectivity, and co-citation.

We first discuss the distance. In the methods based on distance, a shorter path represents a stronger relation. For our method, we set $\gamma(e) < 1$ for every edge e; then a flow considerably decreases along a long path. A short path usually contributes to the generalized maximum flow by a greater amount than a long path does. Therefore, a shorter path means a stronger relation in our method also.

We then discuss the connectivity. In methods based on connectivity, a strong relation is represented by many vertex disjoint paths from the source to the destination. The number of vertex disjoint paths can be computed by solving a classical maximum flow problem. The generalized maximum flow problem is a natural extension of the classical maximum flow problem. Therefore, it also can be used to estimate the connectivity.

We discuss the co-citation at last. A flow emanates from the source into the destination, and therefore the flow seldom uses an edge whose direction is opposite that from the source to the destination. On the other hand, we require use of both directions to estimate the co-citation of two objects. We consider the relation between two objects s and t in the network presented in Figure 3(a). Object u is co-cited by s and t. This co-citation is represented by two edges (s, u) and (t, u). However, we were unable to send a flow from s to t along the two edges, unless we reverse the direction of the edge (t, u) to (u, t). Therefore, we construct a doubled network by adding to every original edge in G a reversed edge whose direction is opposite to the original one. For example, Figure 3(b) depicts the doubled network for the network presented in Figure 3(a). We present the definition of a doubled network.

Definition 1. *Let* $G = (V, E, s, t, \mu, \gamma)$ *be a generalized network, and rev* $: E \rightarrow (0, 1]$ *be a reversed edge gain function for G. The doubled network* $G_{rev} = (V, E', s, t, \mu', \gamma')$ *of G for rev is defined as follows. E' consists of two types of edges: (1) every edge* $e(u, v) \in E$ *with* $\mu'(e(u, v)) = \mu(e(u, v))$ *and* $\gamma'(e(u, v)) = \gamma(e(u, v))$; *and (2) one reversed edge* $e_{rev}(v, u)$ *for every edge* $e(u, v) \in E$ *with* $\mu'(e_{rev}(v, u)) = \mu(e(u, v))$ *and* $\gamma'(e_{rev}(v, u)) = rev(e(u, v))$.

A flow on the original network satisfies the capacity constraint, that is, the flow is send along each (u, v) by at most $\mu(e(u, v))$. The constraint is satisfied on the doubled network if we introduce a new constraint $f(e(u, v))f(e_{rev}(v, u)) = 0$ for flow f. Fortunately, we proved that the value of the generalized maximum flow on a doubled network is unchanged even if the new constraint is introduced. Therefore, we can estimate co-citation using a generalized maximum flow on the doubled network. The proof is omitted because of space limitations.

3.2   Gain Function for Wikipedia

It is desired to assign a larger gain to an important edge because a path composed of edges with large gains can contribute to the value of a flow. The gain of an edge should be determined depending on what kinds of objects the source and the destinations are. For example, if we measure a relation between a Japanese politician and an American politician, then we should assign a larger gain to the primarily important edges connecting Japanese and American politicians, than probably unimportant edges connecting Japanese politicians to baseball players. Edges connecting Japanese politicians would be secondarily important, in the example. To realize such a gain assignment, we need to construct groups of objects in Wikipedia, such as "Japanese politicians" and "baseball

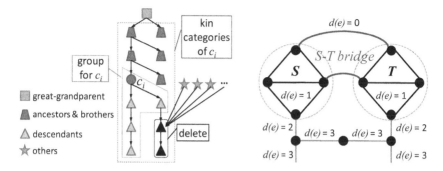

Fig. 4. Grouping for Category c_i **Fig. 5.** Gain function

players". In Wikipedia, the page corresponding to an object belongs to at least one category. For example, the Japanese politician "Junichiro Koizumi" belongs to the category "Members of the Diet of Japan." However, categories cannot be used as groups directly because the category structure of Wikipedia is too fractionalized.

A category representing a concept might have descendant categories each representing a sub concept. The group for category c_i should contain its descendant categories too, as depicted in Figure 4. However, a part of descendant categories represent no sub concepts. For example, "The Pacific War" category is a descendant category of the "Thailand" category. We observed that such irrelevant descendant categories, depicted as black triangles in Figure 4, are usually linked from more than three categories other than kin categories of c_i. We remove them from the group. We omit the detail of constructing a group because of space limitations.

We now propose the gain function for Wikipedia. Let S be the set of objects belonging to a category in group for a category of the source. Similarly, let T be the set of objects for the destination. Then, the primarily important edges explained above are defined as the set of edges (u, v) such that $u \in S \wedge v \in T$ or $u \in T \wedge v \in S$, named an $S - T$ $bridge$. The secondarily important edges are inside S or T, and the unimportant edges would be outside S and T, as illustrated in Figure 5. We assign the gain for an edge $e = (u, v)$ depending on a distance function $d(e)$ between e and $S - T$ $bridge$, defined as follows: if $e \in S - T$ $bridge$, then $d(e) = 0$; if $u \in S \wedge v \in S$ or $u \in T \wedge v \in T$, then $d(e) = 1$; otherwise, $d(e)$ is set to 1 plus the number of edges, including e itself, in the shortest path from e to arbitrary vertex in S or T, computed by ignoring the directions of edges. Figure 5 depicts the definition of $d(e)$. We express the gain function for edge e depending on $d(e)$ with two parameters α and β as

$$\gamma(e) = \alpha * \beta^{d(e)}, 0 < \alpha < 1, 0 < \beta \leq 1,$$

and the reverse gain function is represented with parameter λ as

$$rev(e) = \lambda \times \gamma(e), 0 \leq \lambda \leq 1.$$

If the value of α is fixed, a smaller β produces larger differences between the gains for edges in $S - T$ $bridge$ and those for other edges. λ is used to adjust the importance of a reversed edge. We conduct experiments to determine α, β and λ in Section 4.3.

3.3   Summary of Our Method

We summarize our method for measuring a relation from s to t as follows. (1) Construct a generalized network $G = (V, E, s, t, \mu, \gamma)$ containing s and t from Wikipedia, by determining the parameters α and β explained in Section 3.2. We set the capacity of every edge to one. (2) Determine the parameter λ explained in Section 3.2 for reversed edge gain rev for G, and construct the doubled network G_{rev} of G for rev. (3) Compute a generalized maximum flow g in G_{rev}. (4) Output the value of the flow divided by the square root of $deg(t)$, the number of objects linked from or to t in the dataset, as the strength of the relation. (5) As those constituting the relation, output several paths contributing to the flow.

Computation on a large network is practically impossible. As discussions in [1][13], only a part of the network is significant for measuring a relation. For Wikipedia, we construct G at (1) using pages and links within at most k hop links from s or t in Wikipedia. Careful observation of pages in Wikipedia revealed that several paths composed of three links are interesting for understanding a relation, although we were able to find few interesting paths composed of four links. Furthermore, in preliminary experiments, we constructed G using three and four hop links, separately, and obtained the ranking according to the strength of relations computed by our method. However, the ranking obtained using four hop links is almost identical to that obtained using three hop links. Therefore, we usually set $k = 3$ at (1).

Our method can be applied to both directed network and undirected network. For an undirected network, we set $\lambda = 1$ to use both directions of an edge equally.

We use only a subset of links. The generalized network becomes large if $deg(t)$ is large, and vice versa. The value of the generalized maximum flow becomes large if the generalized network is large. Consequently, the value becomes large for any source if $deg(t)$ is large. On the other hand, the relation between s and t is expected to be independent of $deg(t)$. We decide to divide the value of the flow by $\sqrt{deg(t)}$ at (4). We tried several functions other than $\sqrt{deg(t)}$, such as $deg(t)$ itself or $log(deg(t))$, although $\sqrt{deg(t)}$ is the best among them. Similarly, we can estimate $deg(s)$. However, our main purpose is construction of a ranking according to the strength of relations from a fixed source s to several destinations. Estimating $deg(s)$ does not affect the ranking. Therefore, we do not estimate that.

4   Experiments and Evaluation

4.1   Dataset and Environment

We perform experiments on a Japanese Wikipedia dataset (20090513 snapshot). We first extract 27,380,916 links that appeared in all pages. We then remove pages that are not corresponding to objects, such as each day, month, category, person list, and portal. We also remove the links to such pages, and obtain 11,504,720 remaining links.

We implemented our program in Java and performed experiments on a PC with four 3.0 GHz CPUs (Xeon), 64 GB of RAM, and a 64-bit operating system (Windows Vista).

4.2   Evaluation of Rankings

A good evaluation of methods for measuring the strength of a relation always requires human subjects, as performed in [2][21][22]. There are several benchmark datasets for similarity of words, such as 'WordSimilarity-353" test collection [23]. However, to the best of our knowledge, there is no benchmark dataset for the strength of a relation between objects in Wikipedia, such as people or countries as we used in the experiments. Therefore, we compare the rankings according to the strength of a relation obtained by our method, GSD, PFIBF and CFEC, with those obtained by human subjects. For our method, we set the gain function with $\alpha = 0.8$, $\beta = 0.8$, and $\lambda = 0.8$. The parameters are determined by the estimation of gain function described in the next subsection.

Relations between People: For the source and the destination objects, we select famous person known by the students creating the rankings by their subjects. We first select 10 famous Japanese and American politicians as source objects from Japanese Wikipedia, in order to enables the students to investigate relations among the persons on Wikipedia and create appropriate rankings. As the destination objects for each source, we then select four famous persons related to the source. For each of the 40 obtained pairs of a source and a destination, we compute the strength of the relation from the source to the destination using our method, GSD, PFIBF and CFEC, on the same data set explained in Section 4.1. We then obtain rankings according to the strength. We search Web pages in the domain of Japanese Wikipedia using keywords of the full names of these persons to compute GSD. For PFIBF, edge weight is assigned using FB weighting method of PFIBF [3]. For CFEC, we set the weight of every edge to one. We compare the rankings with those obtained by human subjects. For examining each of the 40 relations, 10 students read the Wikipedia pages corresponding to the source and the destination, and pages related to the source and the destination. Each student gives an integer score of 0–10, independently to the others, as the strength of the relation between each source and destination; a larger score represents a stronger relation. We then obtain rankings according to the average of the scores given by the 10 students.

Table 1 presents the rankings for only five sources because of space limitations. Similar results are obtained for the remaining five sources. For each source, the ranking and the average score obtained by human subjects are written in the column "Human;" an integer 1-4 is assigned as the ranking of the destination; a real number in parentheses is the score. Similarly, the ranking and the strength obtained by our method, GSD, PFIBF and CFEC, are written in the column "Ours 3 hop," "GSD," "PFIBF 2 hop" and "CFEC 3 hop," respectively. "k hop" written behind the name of a method indicates that the method measures a relation between source s and destination t on the network constructed using at most k hop links from s and t. A real number in parentheses is the obtained strength. Note that, GSD uses a smaller real number in parentheses to represent a stronger relation. The shadowed cells for each method emphasize the difference between the ranking obtained by human subjects and that obtained by the method.

The rankings obtained by PFIBF (3 hop) are much worse than those obtained by PFIBF (2 hop). Therefore, we describe the rankings of PFIBF (2 hop) only. Regarding CFEC (3 hop), we use 1000 shortest paths. The rankings obtained by our method are the closest to those obtained by human subjects. However, some rankings by other methods

Table 1. Rankings of persons

Source	Destinations	Human	Ours 3 hop	GSD	PFIBF 2 hop	CFEC 3 hop
Donald Henry Rumsfeld	Dick Cheney	1 (7.7)	1 (2.05)	1 (0.17)	2 (3.38)	2 (1.08)
	Condoleezza Rice	2 (6.9)	2 (1.47)	2 (0.22)	3 (2.58)	4 (0.02)
	Ronald Reagan	3 (5.5)	3 (1.07)	3 (0.35)	1 (3.47)	1 (1.20)
	Junichiro Koizumi	4 (3.8)	4 (0.46)	4 (0.53)	4 (1.63)	3 (0.06)
Nobuta Machimura	Yasuo Fukuda	1 (8.4)	1 (1.67)	1 (0.19)	1 (9.39)	1 (1.38)
	Condoleezza Rice	2 (5.3)	2 (0.82)	2 (0.41)	3 (0.75)	3 (0.01)
	George W. Bush	3 (4.1)	3 (0.64)	4 (0.56)	2 (1.14)	2 (0.02)
	Hillary Clinton	4 (2.6)	4 (0.61)	3 (0.48)	4 (0.27)	4 (0.00)
Kiichi Miyazawa	Noboru Takeshita	1 (8.4)	1 (3.71)	1 (0.09)	1 (12.1)	1 (1.49)
	George H. W. Bush	2 (4.9)	2 (1.07)	4 (0.58)	3 (0.86)	3 (1.04)
	Robert Rubin	3 (4.0)	4 (0.71)	2 (0.49)	4 (0.46)	4 (0.01)
	Bill Clinton	4 (3.9)	3 (1.05)	2 (0.49)	2 (1.74)	2 (1.07)
Junichiro Koizumi	Shinzo Abe	1 (9.1)	1 (5.30)	1 (0.18)	1 (29.6)	1 (1.97)
	Donald Rumsfeld	2 (5.3)	2 (1.99)	2 (0.53)	2 (2.32)	3 (0.12)
	Wen Jiabao	3 (4.5)	4 (1.66)	2 (0.53)	4 (2.00)	2 (1.03)
	Condoleezza Rice	4 (4.1)	3 (1.83)	4 (0.55)	3 (2.17)	4 (0.06)
Yasuo Fukuda	Takeo Fukuda	1 (9.7)	1 (4.04)	1 (0.16)	1 (11.7)	1 (2.12)
	Tony Blair	2 (4.7)	3 (1.43)	4 (0.52)	3 (1.30)	3 (0.06)
	Nicolas Sarkozy	3 (4.6)	2 (1.75)	2 (0.50)	2 (2.07)	2 (1.03)
	Mamoru Mohri	4 (2.8)	4 (0.73)	2 (0.50)	4 (0.47)	4 (0.01)

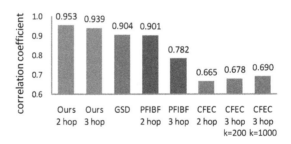

Fig. 6. Average correlation coefficients for all methods

are far from those by human subjects. For example, for "Donald Henry Rumsfeld," PFIBF and CFEC rank "Ronald Reagan" as the first destination, although the students rank him as the third. Similarly, for "Kiichi Miyazawa," GSD ranks "George H. W. Bush" as the fourth, although the students rank him as the second.

We also compute the Pearson product-moment correlation coefficient between the obtained strength and the score given by the students. For each method, Figure 6 depicts the average correlation coefficient for the 10 sources. Note that, the bar "GSD" indicates the absolute value of the coefficient for GSD; the original coefficient for GSD

Table 2. Rankings of states for Petroleum

Ranking	statistics-based	Ours 3 hop	GSD	PFIBF 2 hop	CFEC 3 hop
1	USA	Japan	Iraq	Iran	Saudi Arabia
2	Russia	USA	Iran	Saudi Arabia	Kuwait
3	China	Russia	Saudi Arabia	Iraq	Iraq
4	Saudi Arabia	Saudi Arabia	Kuwait	Japan	Iran
5	Iran	China	Indonesia	Brazil	Egypt
6	Canada	Libya	Libya	Indonesia	Brazil
7	Mexico	Kuwait	UAE	Egypt	Libya
8	Japan	UK	Pakistan	Turkey	UAE
9	Brazil	Iran	Afghanistan	Libya	Indonesia
10	India	Bahrain	Singapore	UAE	Norway

is negative because GSD gives smaller value to represent a stronger relation. For CFEC (3 hop), we present both the coefficients of using $k = 200$ and $k = 1000$ shortest paths.

Our method (2 hop) and our method (3 hop) have the best two correlation coefficients: 0.953 and 0.939. The respective coefficients of GSD and PFIBF 2 hop are fairly good: 0.904 and 0.901. However, GSD cannot use three hop links by nature as explained in Section 2. The coefficient of PFIBF (3 hop) is fairly worse than that of PFIBF (2 hop). Therefore, GSD and PFIBF are unsuitable for measuring the strength of 3-hop implicit relations. Moreover, GSD and PFIBF were unable to mine elucidatory objects constituting an implicit relation, although our method can do so. The coefficients of CFEC are much worse than those of other methods; if the number of shortest paths becomes smaller, then its coefficient becomes smaller.

It took 102s to compute the generalized maximum flow using three hop links for the 40 relations described above. The time for computing PFIBF (3 hop) is 400s, which is about four times longer than our method. For computing CFEC (3 hop), using 200 shortest paths and 1000 shortest paths took 91s and 5631s, respectively.

Relations between Petorleum and Countries: As another experiment, we obtain the rankings according to the strength of the relations from "Petroleum" to each of the 192 states using each method. We also create a *statistics-based ranking* of the 192 states according to the scores computed by the following equation using the statistics about the oil production and consumption of the states [24].

$$score = \frac{oil\ production\ of\ a\ state}{oil\ production\ of\ the\ world} + \frac{oil\ consumption\ of\ a\ state}{oil\ consumption\ of\ the\ world}$$

Although the relation between petroleum to a state is not only dependent on its production and consumption of petroleum, the statistics-based ranking offers an objective way for evaluating the rankings obtained by several methods. The top 10 states in the ranking obtained by each method are presented in Table 2. CFEC is computed using 1000 shortest paths here. Referring to the statistics-based ranking, we can see that our method yielded the most reasonable ranking. Especially, except our method, the two largest consumer "USA" and "China" are not ranked in the top 10 by other methods.

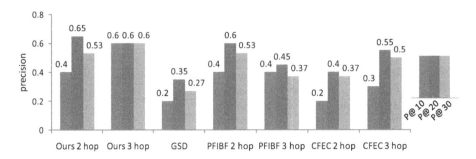

Fig. 7. P@10, P@20 and P@30 of Rankings of states for Petroleum

We then evalute the precision at the top n states of a ranking, abbreviated to P@n, computed by $\frac{|S_n|}{n}$, where S_n is the set of states appeared in both the ranking and the statistics-based ranking. Figure 7 depicts P@10, P@20 and P@30 of all rankings. Similarly to the results of the first experiment depicted in Figure 6, our method (3 hop) and our method (2 hop) generate the highest precisions. The precisions of PFIBF (2 hop) are secondly highest, although those of PFIBF (3 hop) are fairly worse. CFEC (3 hop) performed better than CFEC (2 hop), similar to the first experiment. However, the precision of GSD is the worst in this experiment.

The results of the two experiments imply that our method is the most appropriate one for measuring the strength of a relation. Particularly, our method is the only choice for measuring 3-hop implicit relations.

We also conduct an experiment setting the weight of edge e for CFEC to $\gamma(e)$, our edge gain. However, the obtained rankings are much worse than those obtained by setting edge weight to one. We do not apply the gain function to PFIBF, because PFIBF has its own method for weighting edges.

4.3 Estimation of Gain Function

In this subsection, we evaluate the parameters α, β and λ for our gain function explained in Section 3.2. Let $\rho(\alpha, \beta, \lambda)$ be the correlation coefficient, averaged for the 40 relations among politicians described in Section 4.2, depending on the values of parameters. We set the values of the parameters as $\alpha \in \{0.1, 0.2, ..., 0.9\}$, $\beta \in \{0.1, 0.2, ..., 1.0\}$ and $\lambda \in \{0, 0.1, ..., 1.0\}$. We compute $\rho(\alpha, \beta, \lambda)$ for all the possible $9 \times 10 \times 11 = 990$ combinations of values. Let $\bar{\rho}(\alpha = \chi)$ be the average of $\rho(\alpha, \beta, \lambda)$ obtained by the combinations of fixing $\alpha = \chi$ and varying β and λ. $\bar{\rho}(\beta = \chi)$ and $\bar{\rho}(\lambda = \chi)$ are similarly defined. Table 3 presents the averages $\bar{\rho}(\alpha = \chi)$, $\bar{\rho}(\beta = \chi)$ and $\bar{\rho}(\lambda = \chi)$.

Table 3. Average of correlation coefficients with a fixed parameter

average ╲ χ	0	0.1	0.2	0.3	0.4	0.5	0.6	0.7	0.8	0.9	1
$\bar{\rho}(\alpha = \chi)$	-	0.705	0.811	0.855	0.878	0.891	0.901	0.908	0.914	0.920	-
$\bar{\rho}(\beta = \chi)$	-	0.778	0.805	0.829	0.850	0.870	0.889	0.905	0.913	0.910	0.899
$\bar{\rho}(\lambda = \chi)$	0.810	0.826	0.842	0.855	0.866	0.874	0.880	0.885	0.888	0.891	0.893

The differences between the averages are relatively small when χ is large. Therefore, our method is fairly robust against varying values of the parameters. The highest average for a fixed α is $\bar{\rho}(\alpha = 0.9) = 0.920$, that for β is $\bar{\rho}(\beta = 0.8) = 0.913$, and that for λ is $\bar{\rho}(\lambda = 1.0) = 0.893$. The shadowed cells in the row "$\bar{\rho}(\alpha = \chi)$" indicate that we could find no statistical significance among the distributions of $\rho(\alpha, \beta, \lambda)$ obtained by the combinations of fixing $\alpha = 0.7, 0.8$ or 0.9, by setting the significance level to 0.05. The shadowed cells in the two bottom rows have similar indication. Therefore, candidate combinations producing good results are $\alpha \in \{0.7, 0.8, 0.9\}$, $\beta \in \{0.7, 0.8, 0.9\}$ and $\lambda \in \{0.6, 0.7, ..., 1.0\}$. Similar candidate combinations are obtained by evaluating the P@n of the ranking of states for the 990 combinations of the parameters. We finally choose the combination $\alpha = 0.8$, $\beta = 0.8$ and $\lambda = 0.8$ which produces a medium result among the candidates.

In addition, we obtain the following observations. (1) If $\beta = 1$, then the gain function is insensitive to groups, constructed from the category structure of Wikipedia as explained in Section 3.2. $\bar{\rho}(\beta = 1) = 0.899$ is worse than the best average. Therefore, the category structure is essential to our gain function. (2) If $\lambda = 0$, then no reversed edges are used for measuring a relation. $\bar{\rho}(\lambda = 0) = 0.810$ is the worst value in the bottom row. Therefore, reversed edges used for reflecting co-citation are effective in measuring a relation. Conversely, using no reversed edges would be a deficit of CFEC.

4.4 Case Studies of Elucidatory Objects

For each relation, our method outputs the top-k paths, say top-30 paths, primarily contributing to the generalized maximum flow, that is, paths along which a large amount of the flow is sent. We call objects in such paths *elucidatory objects* affecting the relation. In this subsection, we conduct case studies to demonstrate that elucidatory objects are useful for explaining an implicit relation.

Figure 8 portrays five paths (A)–(E) contributing to the flow emanating from "Petroleum" into the "USA." Each vertex represents a page in Wikipedia, and each edge represents a link from a page to another page. The vertices except "Petroleum" and "USA" are elucidatory objects. We analyze these paths based on the contents of pages in the paths. Path (A) corresponds to the fact that the USA has exploited petroleum in the Gulf of Mexico. John F. Kennedy in path (B), the thirty-fifth President of the USA, considered reducing or abolishing the oil depletion allowance. Path (C) would correspond to the fact that the USA attacked Iraq, which has many oil fields. Alabama, in path (D) of the USA, produces large quantities of plastic from petroleum. USS Gridley in path(E) is a U.S. Navy ship that struck against Iranian oil platforms. The ship also escorted Kuwaiti oil tankers through the Strait of Hormuz. Although paths (B) and (E) would not clearly represent the implicit relation between petroleum and the USA, the other paths are interesting for elucidating the relation. We were unable to find the relation represented by these paths from only two pages "Petroleum" and the "USA." Therefore, our method might help a user to understand a relation. We also investigated the elucidatory objects in the top-30 paths for each of many relations evaluated in the experiments described in Section 4.2, and found that over half of them are meaningful for explaining the relations. One interesting subject of future work is to find a method for filtering out unclear paths. Additionally, several visualization techniques would

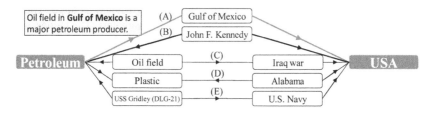

Fig. 8. Explaning an implicit relation

effective for explaining an implicit relation. For example, as depicted in Figure 8, we could represent a path contributing more to a flow by a thicker line to show the importance of a path; displaying a snippet on each edge would also help users to understand the meaning of each path without reading Wikipedia pages.

The methods proposed by Faloutsos et al. [17], Tong and Faloutsos [19] and Koren et al. [1] visualize a subgraph for explaining a relation. However, a user still must investigate important paths in the subgraph to understand the relation. It is easier in usual for a user to understand a relation explained by simple paths rather than a complicated subgraph. Therefore, our method would be better for understanding a relation deeply.

5   Conclusion

We have proposed a new method of measuring the strength of a relation between two objects on Wikipedia. By using a generalized maximum flow, the three representative concepts, distance, connectivity, and co-citation, can be reflected in our method.

We have ascertained that we can obtain a fairly reasonable ranking according to the strength of relations by our method compared with those by GSD [6], PFIBF [3][2] and CFEC [1]. Particularly, our method is the only choice for measuring 3-hop implicit relations. We have also confirmed that elucidatory objects are helpful to deeply understand a relation.

Some future challenges remain. We plan to apply our generalized flow based method to social networks by determining another gain function. Unlike in Wikipedia, objects in most social networks represent people only. A gain function for a social network would be simpler than our function for Wikipedia. We are also interested in seeking possibilities of the elucidatory objects constituting a relation mined by our method. We plan to quantitatively evaluate the elucidatory objects. We are developing a tool for deeply understanding relations by utilizing elucidatory objects.

Acknowledgment

This work was supported in part by the National Institute of Information and Communications Technology.

References

1. Koren, Y., North, S.C., Volinsky, C.: Measuring and extracting proximity in networks. In: Proc. of 12th ACM SIGKDD Conference, pp. 245–255 (2006)
2. Ito, M., Nakayama, K., Hara, T., Nishio, S.: Association thesaurus construction methods based on link co-occurrence analysis for wikipedia. In: CIKM, pp. 817–826 (2008)
3. Nakayama, K., Hara, T., Nishio, S.: Wikipedia mining for an association web thesaurus construction. In: Benatallah, B., Casati, F., Georgakopoulos, D., Bartolini, C., Sadiq, W., Godart, C. (eds.) WISE 2007. LNCS, vol. 4831, pp. 322–334. Springer, Heidelberg (2007)
4. Ahuja, R.K., Magnanti, T.L., Orlin, J.B.: Network Flows: Theory, Algorithms, and Applications. Prentice Hall, New Jersey (1993)
5. Wayne, K.D.: Generalized Maximum Flow Algorithm. PhD thesis, Cornell University, New York, U.S. (January 1999)
6. Cilibrasi, R.L., Vitányi, P.M.B.: The Google similarity distance. IEEE Transactions on Knowledge and Data Engineering 19(3), 370–383 (2007)
7. Kasneci, G., Suchanek, F.M., Ifrim, G., Ramanath, M., Weikum, G.: Naga: Searching and ranking knowledge. In: Proc. of 24th ICDE, pp. 953–962 (2008)
8. Suchanek, F.M., Kasneci, G., Weikum, G.: Yago: a core of semantic knowledge. In: Proc. of 16th WWW, pp. 697–706 (2007)
9. Erdös Number: The Erdös number project, http://www.oakland.edu/enp/
10. Lu, W., Janssen, J., Milios, E., Japkowicz, N., Zhang, Y.: Node similarity in the citation graph. Knowledge and Information Systems 11(1), 105–129 (2006)
11. White, H.D., Griffith, B.C.: Author cocitation: A literature measure of intellectual structure. JASIST 32(3), 163–171 (1981)
12. Milne, D., Witten, I.H.: An effective, low-cost measure of semantic relatedness obtained from wikipedia links (2008)
13. Jeh, G., Widom, J.: Simrank: a measure of structural-context similarity. In: Proc. of 8th ACM SIGKDD Conference, pp. 538–543 (2002)
14. Hubbell, C.H.: An input-output approach to clique identification. Sociolmetry 28, 277–299 (1965)
15. Katz, L.: A new status index derived from sociometric analysis. Psychometrika 18(1), 39–43 (1953)
16. Wasserman, S., Faust, K.: Social Network Analysis: Methods and Application (Structural Analysis in the Social Sciences). Cambridge University Press, New York (1994)
17. Faloutsos, C., Mccurley, K.S., Tomkins, A.: Fast discovery of connection subgraphs. In: Proc. of 10th ACM SIGKDD Conference, pp. 118–127 (2004)
18. Doyle, P.G., Snell, J.L.: Random Walks and Electric Networks, vol. 22. Mathematical Association America, New York (1984)
19. Tong, H., Faloutsos, C.: Center-piece subgraphs: Problem definition and fast solutions. In: Proc. of 12th ACM SIGKDD Conference, pp. 404–413 (2006)
20. Zhu, J., Nie, Z., Liu, X., Zhang, B., Wen, J.R.: Statsnowball: a statistical approach to extracting entity relationships. In: WWW, pp. 101–110 (2009)
21. Xi, W., Fox, E.A., Fan, W., Zhang, B., Chen, Z., Yan, J., Zhuang, D.: Simfusion: measuring similarity using unified relationship matrix. In: Proc. of 28th SIGIR, pp. 130–137 (2005)
22. Gracia, J., Mena, E.: Web-based measure of semantic relatedness. In: Bailey, J., Maier, D., Schewe, K.-D., Thalheim, B., Wang, X.S. (eds.) WISE 2008. LNCS, vol. 5175, pp. 136–150. Springer, Heidelberg (2008)
23. Finkelstein, L., Gabrilovich, E., Matias, Y., Rivlin, E., Solan, Z., Wolfman, G., Ruppin, E.: The WordSimilarity-353 Test Collection (2002)
24. Coutsoukis, P.: Country ranks (2009), http://www.photius.com/rankings/index.html

Summarizing and Extracting Online Public Opinion
from Blog Search Results

Shi Feng[1,2], Daling Wang[1], Ge Yu[1], Binyang Li[2], and Kam-Fai Wong[2]

[1] Northeastern University, Shenyang, China
{fengshi,wangdaling,yuge}@ise.neu.edu.cn
[2] The Chinese University of Hong Kong, Shatin, N.T., Hong Kong, China
{sfeng,byli,kfwong}@se.cuhk.edu.hk

Abstract. As more and more people are willing to publish their attitudes and feelings in blogs, how to provide an efficient way to summarize and extract public opinion in blogosphere has become a major concern for both compute science researchers and sociologist. Different from existing literatures on opinion retrieval and summarization, the major issue of online public opinion monitoring is to find out people's typical opinions and their corresponding distributions on the Web. We observe that blog search results could provide a very useful source for topic-coherent and authoritative opinions of the given query word. In this paper, a lexicon based method is proposed to enrich the representation of blog search results and a spectral clustering algorithm is introduced to partition blog search results into opinion groups, which help us to find out opinion distributions on the Web. A mutual reinforcement random walk model is proposed to rank result items and extract key sentiment words simultaneously, which facilitates user to quickly get the typical opinions of a given topic. Extensive experiments with different query words were conducted based on a real world blog search engine and the experiments results verify the efficiency and effectiveness of our proposed model and methods.

1 Introduction

Online public opinion can be defined as the collection of opinions of many different people on the Web and the sum of all their views [14]. Governments have increasingly found public opinion to be useful tools for guiding their public information and propaganda programs and occasionally for helping in the formulation of other kinds of policies. For individual users, public opinion can help them when making decisions.

Nowadays, people are willing to write about their lives and thoughts in blogs, which are often online diaries published and maintained by bloggers, reporting on their daily activities and feelings. The contents of the blogs include commentaries or discussions on a particular subject, ranging from mainstream topics (e.g., food, music, products, politics, etc.), to highly personal interests [5]. According to statistics, there are more than 100 million blogs on the Internet, which has provided us a rich source for extracting public opinion online.

Fig.1 shows the public opinion extraction results in blogosphere for Liu Xiang's withdrawal from Olympic Games [15]. Different from traditional opinion mining

H. Kitagawa et al. (Eds.): DASFAA 2010, Part I, LNCS 5981, pp. 476–490, 2010.

task, the major issue of online public opinion monitoring is to find out the typical opinions and their corresponding distributions on the Web. In Fig.1 there are nine kinds of opinions, and each one reflects a typical point of view toward Liu's withdrawal. For example, about 22% bloggers support Liu Xiang's decision and about 16% bloggers feel disappointed. However, it's tough work for analysts to get this report, because most of the data collecting and typical opinion summarizing tasks can only be done manually.

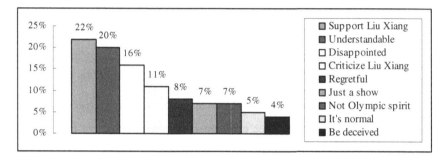

Fig. 1. Online public opinion for Liu Xiang's withdrawing from Beijing Olympic Games

From the discussion above we know that online public opinion extraction is not just a sentiment classification task or the same as opinion summarization task. The three major challenges include:

How to sample the blogosphere. Since there are huge amount of blogs on the Web, given a topic, it is unrealistic to analyze all the topic relevant blogs in the blogosphere. A sampling strategy need to be designed so that we can use a small dataset which could represent as many bloggers' opinions as possible.

How to find the typical opinions. A typical opinion is a point of view held by many people. Public opinion monitoring in blogosphere should aggregate individual attitudes or beliefs and extract typical opinions in the sample dataset.

How to quantitatively measure the distribution of typical opinions. As the example in Fig.1, we should know how many people "*support Liu Xiang*", and how many people "*feel disappointed*" in the dataset, so that we can get a macro view of people's attitudes toward the given topic on the Web.

Most recently, opinion mining techniques have been used to find people's attitudes in blogosphere. Previous studies on opinion retrieval in blogs usually focus on finding the topic relevant opinionated blog entries [22][23], but not the opinion relatedness between the retrieval results. The existing studies on opinion summarization can generate a short abstract of the major opinions in a close blog dataset on a given topic [6]. However, for public opinion monitoring task, the extracted results should not only contain the summary of opinions, but also should include the distribution of each typical opinions. Moreover, previous studies on opinion retrieval and summarization equally treat each blog document, but in real world, blogs from opinion leaders, who have a greater influence on the Web, should be given priority during the summarization. Therefore, there are still some defects in existing methods, which could not totally meet the need for public opinion monitoring task in blogosphere.

In this paper, we propose a new method to summarize and extract public opinion from blog search results. We use the titles and snippets of blog search results (BSRs for short) to summarize public opinion based on the following considerations:

(1) In many cases, bloggers do not confine themselves to one topic in a blog article. But sophisticated Web search techniques could guarantee that BSRs are highly topic relevant to the query word;

(2) Usually, these titles and snippets in BSRs contain bloggers' opinion about the given query word;

(3) Due to the algorithms of blog search engines, the top ranked BSRs are from popular or opinion leader's blog sites. So we can get the public opinion of the whole blogosphere using a relative small BSRs dataset.

(4) We need not to crawl and index the huge amount of blog entries on the Web.

In order to tackle the above three challenges, several hundreds of the top ranked BSRs are used to sample the blogosphere about the given topic. Then a lexicon based method is proposed to measure the underlying opinion relatedness between BSR items and a spectral clustering method is employed to aggregate the opinions into groups, which reflect the opinion distributions on the Web. Finally, a mutual reinforcement random walk model is proposed to rank BSRs and extract key sentiment words in each opinion cluster, which facilitates user to quickly get the typical opinions of the given topic in blogosphere.

To our best knowledge, this is the first paper trying to summarize and extract online public opinion from blog search results. The rest of the paper is organized as follows. Section 2 analyzes the sentiment characteristics of BSR items and discusses the new sentiment representation for BSRs. Section 3 describes opinion clustering algorithm. In Section 4, we will propose a random walk model to rank BSR items and extract key sentiment words simultaneously. Section 5 provides experimental results on real world blog search engine. Section 6 introduces the related work. Finally we present concluding remarks and future work in Section 7.

2   Blog Search Results Sentiment Representation

In this section, we attempt to give BSRs a new representation in order to measure the opinion relatedness between BSR items.

2.1   Characteristics of Blog Search Results

Some commercial blog search engines have been published on the Web [4] [16]. These services usually use Web search techniques and rank the results by their topic relevance and the popularity of the blog entry. To demonstrate the sentiment characteristics of BSRs, we issue the movie name *"Hancock"* in Google Blog Search, then several results are collected and shown in Table 1.

Table 1. The titles and snippets for query word *"Hancock"*

Title	Snippet
Hancock	Go see Hancock. Much respect to Will Smith and the directors behind the film, truly inspirational.
It's a comedy, It's a fantasy, Yes, It's 'Hancock'	Hancock was enjoyable, but no without it problems thanks to many unanswered questions.
Hancock 2008 DVDRip direct links	Hancock 2008 Language: English Runtime: 92 min Country: USA Release Date: 2 July 2008.

We can see that the BSR items in Table 1 have the following characteristics:

(1) The title and snippet of each item are very short, may be just one or two sentences, and sometimes may be just a word.

(2) The titles and snippets are highly relevant to the given query word. That's because the blog search engine employs sophisticated and mature Web search techniques to get the most topic relevant articles and snippets;

(3) Some of the titles and snippets contain the bloggers' sentiments and opinions. As the search results are highly topic-coherent, the sentiment words in titles and snippets mainly reflect the bloggers' own opinions about the given query key word;

(4) Not all the blogs contain authors' emotions, there are some informative results mixed up with affective ones. For example, the last result item in Table 1 tells us the download information of Hancock movie DVDRip;

According to [12], there are two kinds of blog articles in the blogosphere, namely informative blogs and affective blogs, and Table 1 also confirms this point of view. Here we give our definition of Affective BSR and Non-affective BSR.

Affective BSR. An Affective BSR is the BSR item that contains bloggers' sentiments and opinions.

Non-affective BSR. The contents of this genre of BSR include (1) the informative BSR that providing or conveying information and (2) short snippet that do not contain any personal feelings and emotions.

An opinion usually includes opinion holder, opinion target and sentiments. According to the properties of blogs, the opinion holder of a BSR item is the blogger himself/herself. The opinion target is usually the query word or the subtopic related to the query word. From this observation, we submit topic words to a blog search engine, collect the BSR items and aggregate the sentiments in BSRs, so as to summarize and extract the public opinion of the bloggers about the given topic.

2.2 BSRs Sentiment Representation

BSRs are usually very short and opinion words usually do not converge like topic words. So directly applying traditional similarity measure based on term matching to BSRs often produces inadequate results. In this section, we propose a new sentiment representation for BSRs based on WordNet gloss.

WordNet is a large lexical database of English. Nouns, verbs, adjectives and adverbs are grouped into sets of cognitive synonyms (synsets), each expressing a distinct concept [19]. Each synset in WordNet has a gloss that defines the concept that it represents. The intuition of this paper is that the terms with similar sentiments have similar glosses. For example, the synset **A** contains the words *"amusing"*, *"amusive"*, *"diverting"* and *"fun"* and it has the gloss *"providing enjoyment; pleasantly entertaining"*; Synset **B** contains the words *"amused"*, *"diverted"*, *"entertained"* and it has the gloss *"pleasantly occupied"*. The words in synset **A** and **B** are quite different. However, their glosses share the same word *"pleasantly"*. Therefore, synset **A** and **B** have similar sentiment meanings, i.e. when people use the words in **A** and **B**, they tend to express similar state of emotions. In this paper, we attempt to remove non-sentiment words and add the glosses of each emotion-bearing word into BSRs to give them a new sentiment representation. The details of each step are discussed as follows:

Step 1. Lemmatization. We convert the words into their basic lemma form. We do not conduct stemming algorithm to the words because we must keep their original sentiment meanings.

Step 2. Negation Processing. Each word in the negation sentences is replaced by its antonym in WordNet and the words that do not have antonyms in WordNet are given a new prefix *"not-"*.

Step 3. Sentiment Words Tagging. In this paper, we use SentiWordNet as the sentiment lexicon. Extensive experiments show that SentiWordNet is a very effective lexicon tool for finding emotion-bearing words [1][7]. Words with positive or negative strength above a threshold in SentiWordNet are picked out and corresponding words in BSRs are tagged as sentiment words.

According to definition in Section 2.1, suppose R_a represents the set of Affective BSR and R_{na} represents the set of Non-affective BSR. So we get $BSRs = R_a \cup R_{na}$. Let sw denote sentiment word and r, r_i, $r_j \in$ BSRs. If r contains sentiment word, we say $sw \in r$, so we employ the following way to classify BSRs:

$$R_a = \{r_i \mid (\exists sw, sw \in r_i)\}, R_{na} = \{r_j \mid (\forall sw, sw \notin r_j)\} \qquad (1)$$

The above formulas indicate that if r contains at least one sentiment word, we classify it into Affective BSR category; otherwise, we classify it into Non-affective category. It must be emphasized that since our goal is to group BSRs by their opinions, we do not care about the sentiment orientations of each word in BSRs. After this step, we eliminate search result items in R_{na}, and the words in R_a that don't have sentiment tags are also removed. Only sentiment words in R_a are brought to the next processing steps.

Step 4. BSRs Sentiment Representation. After prior processing steps, each remaining BSR item can be represented by a set of sentiment words. Give a BSR item r containing n sentiment words, we have $r = \{sw_1, sw_2, \ldots, sw_n\}$. Synsets and glosses in WordNet are used to expand sentiment words representations, so we have $E(sw) = \{Synset(sw), Gloss(sw)\}$. $Synset(sw)$ denotes all words in the synset of sw; $Gloss(sw)$ denotes the gloss of sw. Therefore, we expand a BSR item r as $Er = \{E(sw_1), E(sw_2), \ldots, E(sw_n)\}$.

We concatenate the expansions of sentiment words in r together, and vector space model is used to represent BSRs. Suppose r_j is a BSR item:

$$\vec{r}_j = (TFIRF(t_1), TFIRF(t_2), ..., TFIRF(t_m)) \tag{2}$$

where t_i (i=1, 2, ..., m) is a term in the new representation, i.e. $t_i \in Er$, TF represents term frequency in the BSRs and IRF denotes the inverse BSR item frequency. We call the new representation of BSRs as Sentiment Vectors (SV), because it can reflect bloggers' original emotions and opinions.

3 Aggregate Opinions in BSRs Based on Spectral Clustering

Blogger's opinions about a certain topic may be opposite or quite different. Our intention is to not only summarize typical opinions, but also find out their corresponding distributions, i.e. how many people hold similar opinions in the blogosphere. So in this section we attempt to aggregate BSRs into opinion clusters.

3.1 Sentiment Similarity Computing

In Section 2.2, we introduce SV to represent bloggers' emotions in BSRs. SV have enriched the representations of the emotion-bearing words in each BSR item, and we employ traditional text similarity measurement algorithm to compute the sentiment similarity between BSR items. Evaluating a variety of similarity measurement algorithms on SV, however, is not the aim of this paper. Rather we simply want to find out whether the new expanded representations of BSRs are effective in reflecting bloggers' opinions. In this paper, we consider BSR items as nodes, and the BSRs collection can be modeled as an affinity graph in which each link denotes the sentiment similarity between BSR items. Formally, given the BSRs set R, let $G=(V,ED)$ be an affinity graph to reflect the relations between items in R. V is the set of vertices and each vertex $v \in V$ is an item in the BSRs set. ED is the set of edges. Each candidate edge e_{ij} in ED is associated with a similarity weight between item v_i and v_j. The similarity weight function $SentiSim$ is defined as:

$$SentiSim(\vec{v}_i, \vec{v}_j) = \frac{\vec{v}_i \cdot \vec{v}_j}{\| \vec{v}_i \| \times | \vec{v}_j |} \tag{3}$$

We use adjacency matrix M to describe the structure of G and the value of M_{ij} represents the weight of an edge in the graph. So we have

$$M_{ij} = \begin{cases} SentiSim(\vec{v}_i, \vec{v}_j), & \text{if } i \neq j \\ 0, & \text{otherwise} \end{cases} \tag{4}$$

Our intention is to aggregate similar opinions in BSRs, namely we have to partition the graph G into several subgraphs and each subgraph should reflect coherent opinions of the bloggers. This is not an easy task because we do not know the structure of the graph G. Moreover, the number of clusters could not be easily predicted in advance. In the next section, we employ a spectral clustering algorithm to partition graph G which does not need to make any assumptions on the form of the clusters and a heuristic method to determine the number of clusters is introduced.

3.2 Spectral Clustering for BSRs

Spectral clustering is an effective algorithm based on graph partitioning. The basic idea of the algorithm is to map the raw data space into eigenspace. In this paper, we choose the MS [8] method with the computation of the Laplacian matrix as follows. Let D be the diagonal degree matrix of M, i.e. $D_{ii} = \sum_j M_{ij}$. The Laplacian matrix L is defined as $L = I - D^{-1}M$. The first k generalized eigenvectors of L is found to compose a new matrix M' and the traditional clustering method such as K-Means can be used on M' to find clusters.

The number of clusters. We could not know the cluster number k in advance. In this paper, we employ a heuristic algorithm to auto determine k by computing eigengap of the matrix L. Matrix perturbation theory indicates that the stability of the eigenvectors of a matrix is determined by the eigengap. However, sometimes the cluster structure of the data is not so obvious, or there may be several big eigengap candidates, i.e. there are several eigenvalues λ_k where $|\lambda_{k+1} - \lambda_k|$ is large. So we use candidate eigengaps to heuristically set the value of k and evaluate the quality of the clustering results to get the best k. Based on the assumption that the best partitioning will have most edges within the subgraphs and little edges between subgraphs, the quality of graph partitioning is defined as [11]:

$$Q(C) = \sum_{i=1}^{k} (e_{ii}/c - (a_i/c)^2) \tag{5}$$

where C is a candidate clustering result, k represents the number of clusters, e_{ii} is the number of edges with both vertices within cluster i, a_i is the number of edges with one or both nodes in cluster i, and c is the total number of edges. The heuristic method to determine k is as follows: (1) Compute the eigenvalues of L; (2) Find the biggest three eigengaps, and set the candidate number of clusters k_1, k_2, k_3; (3) For each candidate k, we employ the MS spectral clustering method [8] to partition G into subgraphs; (4) The $Q(C)$ function in Formula 5 is used to evaluate each candidate clustering results. The best $Q(C)$ is chosen, so the final clustering result and the number of cluster are confirmed.

We call this opinion clustering algorithm as OC algorithm. Using OC algorithm, we can generate opinion coherent clusters of a given BSRs dataset. At the same time, we hope that each BSR item could be ranked by its sentiment coherence to the semantic meanings of the cluster. We will discuss this opinion ranking and keywords extraction method in the next section.

4 Opinion Ranking and Keywords Extraction

When browsing the Web search results, people used to read several top ranked items. Based on this intuition, the proposed algorithm should not only group BSRs into opinion clusters, but also should rank the results in each cluster according to certain metric. Considering the intention of people exploring the blogosphere, the words with higher sentiment strength are better indicator for bloggers' emotions and the BSRs contain definite sentiment orientation and strong emotion meanings will attract more attention. Therefore, the proposed algorithm should rank these BSRs in higher position. And we also hope that the key sentiment words are extracted for each cluster, which could facilitate users' quick browsing through the public opinion summarization results.

Inspired by the work of Wan [18], in this paper we propose a mutual reinforcement random walk model to rank BSRs and extract key sentiment words simultaneously. Our basic assumption is that a BSR item is important if it includes important sentiment words and is heavily linked with other important BSR items. And also, a sentiment word is important if it has higher sentiment strength; it appears in many important BSR items and has relation with many other important sentiment words. This mutual reinforcement relationship of BSRs and sentiment words is shown in Fig.2.

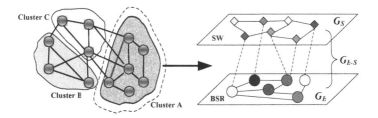

Fig. 2. The mutual reinforcement of BSRs and sentiment words

In Fig.2, given a BSRs dataset, OC algorithm has partitioned the graph into several clusters. In a cluster, BSRs represents the blog search results in a cluster; SW denotes the sentiment word set of the given cluster. We build three graphs G_B, G_S and G_{B-S} to reflect the BSR-BSR, SW-SW, BSR-SW relationship. For bipartite G_{B-S} graph, if a sentiment word sw_j appears in BSR r_i, an edge will be created between r_i and sw_j. Each node in these graphs is associated with a sentiment strength value (shown as different grayscale in the right part of Fig. 2), and based on the random walk on these graphs, this strength is diffused in the three graphs. After several mutual reinforcement iteration steps, the important BSR items could be ranked in higher position and simultaneously we also get the most salient sentiment words in each cluster. The detail of the algorithm is described as follows.

Given a cluster C_o that is the subgraph of G, we have the new adjacency matrix $\{B \mid B_{ij} = M_{ij}$ where $i, j \in C_o\}$ to represent G_B. We use S to denote the adjacency matrix of G_S and the similarity between SW is calculated by cosine similarity of Word-Net expansion representation $E(sw)$. The adjacency matrix of BSR-SW relationship is represented by W, and the weight is computed as:

$$W_{ij} = \frac{TFIRF(sw_j)}{\sum_{sw \in r_i} TFIRF(sw)} \qquad (6)$$

where given a sentiment word sw_j in BSR item r_i. If sw_j appears frequently in r_i and seldom appears in other BSR item, there is a higher weight between sw_j and r_i.

B, S and W is normalized to $\tilde{B}$, $\tilde{S}$ and $\tilde{W}$ respectively and the normalized transpose of W is represented by $\hat{W}$. Let R_{BSR}, R_{SW} denote the ranking scores of BSR and SW. The mutual reinforcement random walk approach can be formulated as follows:

$$\begin{cases} R_{BSR}^{(k+1)} = \alpha\tilde{B}^T R_{BSR}^{(k)} + (1-\alpha)\hat{W}^T R_{SW}^{(k)} \\ R_{SW}^{(k+1)} = \beta\tilde{W}^T R_{BSR}^{(k)} + (1-\beta)\tilde{S}^T R_{SW}^{(k)} \end{cases} \tag{7}$$

Suppose we have:

$$Y = \begin{bmatrix} \alpha\tilde{B}^T & (1-\alpha)\hat{W}^T \\ \beta\tilde{W}^T & (1-\beta)\tilde{S}^T \end{bmatrix}, \quad R = \begin{bmatrix} R_{BSR} \\ R_{SW} \end{bmatrix} \tag{8}$$

In matrix form, we have the equation $YR=\lambda R$. Similar to the idea of PageRank [13], we add links from one node to any other nodes in G_B and G_S graph, so we have:

$$Y = \begin{bmatrix} \alpha((1-d)E/n_1 + d\tilde{B}^T) & (1-\alpha)\hat{W}^T \\ \beta\tilde{W}^T & (1-\beta)((1-d)E/n_2 + d\tilde{S}^T) \end{bmatrix} \tag{9}$$

where E is a square matrix with each element equal 1. We can prove that the transpose of Y is stochastic and irreducible.

Lemma: Y^T is irreducible and when $\alpha + \beta = 1$, it is stochastic.

Proof: There is a link between each node in G_B and G_S, so they are strong connected. Because $G_{B\text{-}S}$ has connected the nodes in G_B and G_S graph, for each pair of nodes u, v in these three graphs, there is a path from u to v. Therefore, the new graph G_{All} composed by G_B, G_S, $G_{B\text{-}S}$ is strong connected. And also there will be more than one path for any pair of nodes in G_{All}, so G_{All} is aperiodic and the matrix Y^T is irreducible. For any column in the left part of Y:

$$\sum_i Y_{ij} = \alpha(\sum_{i=1}^{n_1}\frac{1-d}{n_1} + d\sum_{i=1}^{n_1}\tilde{B}_{ij}) + \beta\sum_{i=1}^{n_3}\tilde{W}_{ij} = \alpha + \beta \tag{10}$$

The same conclusion can be deduced in the right part of Y. So when $\alpha + \beta = 1$, the sum of elements in each column in Y is 1, and the matrix Y^T is stochastic. □

E/n in Formula 9 means that each node in the graph has an equal weight. Recall our assumption that the word with higher sentiment strength is a better indicator for bloggers' emotions, we give each node a different weight during the iteration steps. For graph G_S, the weight of a node is defined by the sentiment strength of the word sw, and we have the weight vector $\{q|q_i = f(sw)\}$, where $f(sw)$ is the sw's sentiment strength in SentiWordNet. For graph G_B, the weight is defined as the average strength of the word in each BSR item, and we have $\{p | p_i = \sum_{j=1}^{N} f(sw_j) / N\}$. p and q is normalized to $\tilde{p}$ and $\tilde{q}$. The matrix Y can be reformulated as follows:

$$Y' = \begin{bmatrix} \alpha((1-d)e\tilde{p}^T + d\tilde{B}^T) & (1-\alpha)\hat{W}^T \\ \beta\tilde{W}^T & (1-\beta)((1-d)e\tilde{q}^T + d\tilde{S}^T) \end{bmatrix} \tag{11}$$

Using Formula 11, we incorporate sentiment strength information into mutual reinforcement random walk model and it can be proved that Y^T is stochastic and irreducible. The power method is used to iteratively find the solution of the equation $YR=R$. It is guaranteed that R will converge to a steady state, which we use as final ranking results of BSR items. Finally, the top 5 ranked sentiment words are extracted as key sentiment words of the cluster. This ranking and summarization method is applied on each cluster.

5 Experiments

5.1 Experiment Setup

Our experiment is conducted on a commodity PC with Windows XP, Core2 Duo CPU and 4GB RAM. Given a query key word, we use Google Blog Search to find the topic relevant blog entries. Titles and snippets are parsed and extracted for further processing steps.

Web search results clustering usually focus on informative, polysemous and poor query words, such as "*java*", "*jaguar*", "*apple*". With different purpose, we pay more attention to the entities' and events' name which can arouse people's interest to publish opinions in blogosphere. The different types of query words used in this paper are shown in Table 2.

Table 2. The query words used in the experiments

ID	Type	Query Words	Data Range
Hancock	Movie	hancock movie	2008.7.1-2008.7.31
Obama	People	president obama	2009.1.1-2009.1.31
Opening	Event	beijing olympic opening	2008.8.8-2008.8.12
IPod	Product	ipod touch	2007.9.1-2007.9.31

Public opinions are highly relevant to the published time. If there is a hot topic emerging on the Internet, people are eager to write down their own opinions about the topic in blogs. However, as the time passing by, the blogs on original topic become fewer and fewer and there maybe a succeeding topic or a new story emerging in the blogosphere. In this paper, we restrict the publishing date of blogs for the query to find the most topic coherent story in the searching space. For example, we restrict the query "*hancock movie*" within one month period since the movie was released.

Usually blog search engines have already ranked results by blogs' authorities, and opinion leaders' blogs and popular blogs can be returned in the higher position. Therefore, we use a relatively small dataset to reflect the major opinions in the whole blogosphere. As a result, less than 1000 BSR items are collected for each query.

Evaluation Measure. There is no ground truth for the clustering results. Since we have model BSRs as graph, the best partitioning results would have most edges with the cluster and little edges between the clusters. So we use Formula 5 in Section 3.2 to evaluate clustering performance.

We use precision (P) at top N results to measure the performance of the ranking algorithm. Since no golden standard is available for these search results, we ask three graduate students major in opinion mining to evaluate the opinion coherence between ranked search results and extracted key sentiment words in each cluster. However, it is very subjective to measure this opinion coherence. The three evaluators are asked to browse through the search results, compare key sentiment words to each result and give a score ranging from 0 to 10 to show how well the extracted key sentiment words match the opinion contained in each search result. If the score is high, it means that the extracted words are good opinion summarization for the given search result item. For each cluster we have:

$$p@N = \frac{\sum_{i=1}^{N} Score_i/10}{N} \tag{12}$$

where N denotes the top N result items in the cluster, and $Score_i$ is the value given by evaluators. Notice that $p@N$ represents the precision of one cluster, and the average precision of all clusters is calculated as $P@N = \left(\sum_k p@N\right)/k$. Finally, we use the average $P@N$ value of the three evaluators to measure the performance of proposed BSRs public opinion summarization and extraction algorithm.

Using Sentiment Vectors and OC algorithm, we hope that most of the Non-affective BSR items are filtered out and the affective ones are ranked in high position. Thus we use Non-affective BSR (I) at top N results to measure this performance:

$$I@N = \frac{I \cap N}{N} \tag{13}$$

where I is the number of Non-affective BSR items in top N results. Three human annotators are asked to label BSR item with A/NA tag, i.e. Affective/Non-affective result. If there is a disagreement between the first two annotators, the third one will decide the final tag of the BSR item.

5.2 Experiment Results

Opinion Clustering Performance. We compare the proposed opinion clustering method with the basic K-Means method. To validate the effectiveness of the cluster number determination algorithm, we manually set k from 2 to 12.

Fig.3 (a) and (b) show the clustering performance and eigenvalues of the query "*hancock movie*". In Fig.3 (a), the Y axis denotes the $Q(C)$ function value. The bigger triangle represents the auto-determined cluster number of OC algorithm. It can be seen from Fig.3 (b) that the "Hancock" dataset has a very obvious eigengap, and the proposed OC algorithm could find the best clustering performance using this eigengap when $k=7$. The "Beijing Olympic opening" and "IPod Touch" dataset also validate the proposed method (due to space limitation, we do not list the figures here). However, in Fig.3 (c) OC algorithm does not find the best partition number (OC predicts $k=10$, but the best Q is achieved when $k=8$). Fig.3 (d) illustrates that the difference between all eigenvalues are approximately the same for the "President Obama" dataset. This indicates that there is no clear cluster structure in "Obama" dataset, and this may because the sentiment words used in this dataset are quite scattered. In this situation, we hope that our ranking algorithm could help us to figure out typical opinions in BSR items.

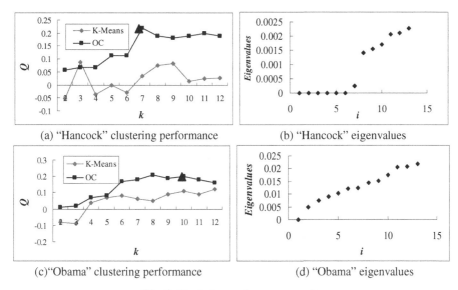

(a) "Hancock" clustering performance   (b) "Hancock" eigenvalues

(c)"Obama" clustering performance   (d) "Obama" eigenvalues

Fig. 3. Clustering performance results

Opinion Ranking Precisions. Here we compare the proposed mutual reinforcement random walk based opinion ranking algorithm (MR algorithm) with two different ranking methods. The first method is directly rank BSR items by average sentiment strength, and we call it SR. The second one is directly employ basic PageRank algorithm on the BSR graph, which does not consider sentiment strength and we call it PR. The key sentiment words of these two methods are extracted by *TFIRF* function, i.e. the words with top *TFIRF* values are extracted as key sentiment words. Here we set $\alpha=\beta=0.5$ of MR algorithm to equally treat the weights of BSRs and sentiment words. The parameter d is set to be 0.85. We use MR-NE represents the mutual reinforcement algorithm without WordNet gloss expansion, i.e. the similarity is computed directly based on sentiment words vectors. The comparison of opinion ranking performance is shown in Fig.4.

We can see from Fig.4 that generally the proposed MR method outperform the other method. Note that the precision is calculated by volunteers' comparing ranked BSRs with extracted key sentiment words. If the key sentiment words could reflect the major opinions expressed in the BSRs, a higher score will be given. Fig.4 validates that the mutual reinforcement method could effectively rank opinions and extract key sentiment words for each cluster and the extracted words could provide a very brief summarization of the major opinions in each cluster. And also we can conclude that the WordNet synset and gloss expansion are effective in finding the underlying opinion relatedness between short BSR texts.

The best performance is achieved using the query *"beijing olympic opening"*. After analyzing the search results, we find that the words that people used to express their opinions on Beijing Olympic open ceremony are really converging. On the other hand, the words reflecting people's opinions on political figures are more complex and scattered. Thus the precision of our algorithm is decreased.

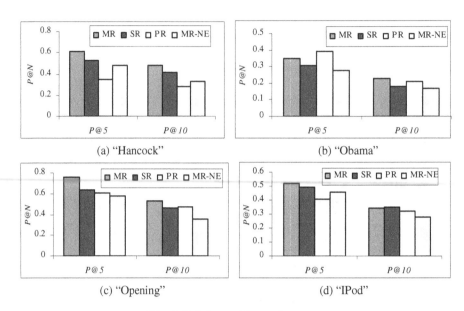

(a) "Hancock"   (b) "Obama"

(c) "Opening"   (d) "IPod"

Fig. 4. Opinion ranking performance

Affective vs Non-affective. The $I@N$ performance using query "*hancock movie*" and "*president obama*" is shown in Fig. 5. It can be seen from Fig. 5 that $I@5$ is relatively small. Using opinion ranking algorithm, there is average 0.6 items in the top 5 BSRs are non-affective for query "*hancock movie*", compared with average 4.6 non-affective items in top 15 BSRs. Therefore, it can be concluded that the proposed algorithm can effectively rank the affective BSR in high position.

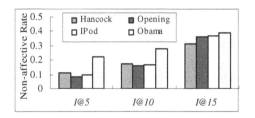
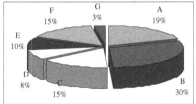

Fig. 5. Performance of $I@N$ **Fig. 6.** Opinion distribution for "*hancock movie*"

Fig. 6 illustrates the clustering results for the query "*hancock movie*". 924 BSR items are parsed for the further clustering steps. Key sentiment words are extracted for each cluster. Take cluster A for example, the extracted words include "*good*" and "*pretty*" and the top rank BSR items contain the sentences such as "*I think overall is good*" and "*saw Hancock yesterday and that was pretty good*". Generally, the proposed method can extract sentiment words which reflect the major opinions expressed in the top ranked results. However, human evaluation for these results is very subjective. The new evaluation method without human involvement will be presented in the future work.

6 Related Work

6.1 Search Results Clustering

Determining the similarity of short text snippets works poorly with traditional document similarity measures. Some lexicon-based and language modeling-based have been proposed to solve the text snippets similarity measurement and clustering problem [9] [20]. But there are still some obstacles to measure the sentiment similarity between the short texts.

Rich literatures have been published on search results clustering. Zeng et al. [21] reformalize the search result clustering problem as a supervised salient phrase ranking problem. Ferragina et al. [3] develop an open-source system which offers both hierarchical clustering and folder labeling with variable-length sentences. There are already some industrial Web search results clustering systems on the Internet [17] [10], which are especially useful for informative, polysemous and poor queries.

We have proposed a sentiment clustering method for blog search results [2]. However, in [2] the sentiment similarity between BSRs is only considered at word level and key words are extracted only by sentiment strength.

6.2 Opinion Mining

The task of opinion retrieval is to find relevant and opinionate documents according to a user's query [22] [23]. TREC started a special track on blog data in 2006 with a main task of retrieving personal opinions towards various topics, and it has been the track that has the most participants in 2007 [7]. However, people could have various opinions on the same topic, and opinion retrieval can not provide users with overall summarization of opinions expressed in blog articles.

TAC 2008 has launched a task on opinion question answering and summarization. Given a list of questions, an exact string answer or several sentences containing the answer should be returned [6]. Different from that task, our intention is to find bloggers' typical opinions and their corresponding distribution in blogosphere.

7 Conclusion and Future Work

Blog has provided a good platform for people to express their opinions and attitudes. In this paper, we propose a method to summarize and extract public opinion based on blog search results. Opinions are aggregated into clusters. A mutual reinforcement random walk model is proposed to rank blog search result items and extract key sentiment words. Experimental results demonstrate that the proposed method can effectively extract public opinion in the blogosphere.

In this study, only sentiment words are used to represent bloggers' opinions in BSRs. In future work, more linguistic information may be considered in the new representation of sentiment vectors.

Acknowledgments. This work is partially supported by National Natural Science Foundation of China (No.60973019, 60973021), CUHK Direct Grant Scheme (No. 2050443, 2050417) and HKSAR's ITF (No. ITS/182/08).

References

1. Esuli, A., Sebastiani, F.: SentiWordNet: A Publicly Available Lexical Resource for Opinion Mining. In: Proceedings of LREC, pp. 417–422 (2006)
2. Feng, S., Wang, D., Yu, G., Yang, C., Yang, N.: Sentiment Clustering: A Novel Method to Explore in the Blogosphere. In: Li, Q., Feng, L., Pei, J., Wang, S.X., Zhou, X., Zhu, Q.-M. (eds.) APWeb/WAIM 2009. LNCS, vol. 5446, pp. 332–344. Springer, Heidelberg (2009)
3. Ferragina, P., Gulli, A.: A Personalized Search Engine based on Web-snippet Hierarchical Clustering. In: Proceedings of WWW, pp. 801–810 (2005)
4. Google Blog Search, http://blogsearch.google.com
5. Kumar, R., Novak, J., Raghavan, P., Tomkins, A.: Structure and Evolution of Blogspace. Commun. ACM 47(12), 35–39 (2004)
6. Li, F., Tang, Y., Huang, M., Zhu, X.: Answering Opinion Questions with Random Walks on Graphs. In: Proceedings of ACL, pp. 737–745 (2009)
7. Macdonald, C., Ounis, I., Soboro, I.: Overview of the TREC-2007 blog track. In: Proceedings of TREC 2007 (2007)
8. Meila, M., Shi, J.: Learning Segmentation by Random Walks. In: NIPS, pp. 873–879 (2000)
9. Metzler, D., Dumais, S., Meek, C.: Similarity Measures for Short Segments of Text. In: Proceedings of ECIR, pp. 16–27 (2007)
10. Mooter, http://www.mooter.com
11. Newman, M., Girvan, M.: Finding and Evaluating Community Structure in Networks. Phys. Rev. E 69(6), 026113 (2004)
12. Ni, X., Xue, G., Ling, X., Yu, Y., Yang, Q.: Exploring in the Weblog Space by Detecting Informative and Affective Articles. In: Proceedings of WWW, pp. 281–290 (2007)
13. Page, L., Brin, S., Motwani, R., Winograd, T.: The PageRank Citation Ranking: Bringing Order to the Web. Technical report, Stanford University (1998)
14. Public Opinion, http://en.wikipedia.org/wiki/Public_opinion
15. Public Opinion Channel, http://yq.people.com.cn/CaseLib.htm
16. Technorati, http://technorati.com
17. Vivisimo, http://vivisimo.com
18. Wan, X., Yang, J., Xiao, J.: Towards an Iterative Reinforcement Approach for Simultaneous Document Summarization and Keyword Extraction. In: Proceedings of ACL, pp. 552–559 (2007)
19. WordNet, http://wordnet.princeton.edu
20. Yih, W., Meek, C.: Improving Similarity Measures for Short Segments of Text. In: Proceedings of AAAI, pp. 1489–1494 (2007)
21. Zeng, H., He, Q., Chen, Z., Ma, W., Ma, J.: Learning to Cluster Web Search Results. In: Proceedings of SIGIR, pp. 210–217 (2004)
22. Zhang, M., Ye, X.: A Generation Model to Unify Topic Relevance and Lexicon-based Sentiment for Opinion Retrieval. In: Proceedings of SIGIR, pp. 411–418 (2008)
23. Zhang, W., Yu, C., Meng, W.: Opinion Retrieval from Blogs. In: Proceedings of CIKM, pp. 831–840 (2007)

Cloud as Virtual Databases:
Bridging Private Databases and Web Services

Hiroaki Ohshima[1], Satoshi Oyama[2], and Katsumi Tanaka[1]

[1] Graduate School of Informatics, Kyoto University, Japan
{ohshima,tanaka}@dl.kuis.kyoto-u.ac.jp
[2] Graduate School of Information Science and Technology,
Hokkaido University, Japan
oyama@ist.hokudai.ac.jp

Abstract. We propose an extensible platform for bridging private databases and Web services. Our main idea is to make a Web service and its results be a set of virtual tables in a relational database (RDB) environment. As private data that cannot be disclosed is stored in private RDBs, these virtual tables realize a bridge between private RDBs and Web services.

1 Introduction

Many Web services are currently offered, and the number of services continues to increase. The *cloud* is so powerful that people will think new services must be made on the cloud. It is, however, very reluctant for people and organization to upload their important private data to the cloud because it is the worst if the private data is leaked out by any chance. Even though the cloud provides private secure space such as Amazon VPC, people still keep to feel uneasy.

People and organization often store their private data in their own relational databases (RDBs). It is necessary to bridge the local databases and Web services when they want to make applications that need both private data and data from Web services. Our approach is to make a Web service and its results be a set of virtual tables in local RDB environments. As private data that cannot be disclosed is stored in private RDBs, these virtual tables realize the bridge to Web services on the cloud. Virtual tables are realized by using table-valued functions. As the platform does not extend the syntax of SQL, it can be extended by users.

2 Related Work

This section describes several related works and compares our work to them. WSQ/DSQ [4] is a relational database environment where Web search engines can be treated as virtual tables. A user can obtain a list of search items with a search ranking, URL, and date from a virtual table. WebQL [1] is a software for Web mining, data extraction, and data integration. It uses an SQL-like language to access different data sources and to extract data from them. They treat several

H. Kitagawa et al. (Eds.): DASFAA 2010, Part I, LNCS 5981, pp. 491–497, 2010.

kinds of data sources and one of them is Web pages. Cafarella et al. [3] created a search system that allows users to use a syntactic pattern as a query. KnowItAll [2] is a Web service using their system. The system can be used to make a natural language application, such as discovering hypernyms/hyponyms of a given word within a huge number of Web pages.

Here, we compare WSQ/DSQ, WebQL, KnowItAll, and our platform. First, except for KnowItAll, none need to store any data initially. They can collect text data both from Web services and from local databases. Second, WSQ/DSQ and WebQL use new languages; users need to pay cost to learn them. KnowItAll does not have a language; it accepts queries that contain a syntactic pattern. Our platform does not have any language extension from existing SQL. Third, WSQ/DSQ does not have natural language processing (NLP) functions such as term extraction using regular expressions. Some processing for text data is usually necessary to join text from the Web and text from local databases. Fourth, our system can store procedures to obtain knowledge from text data. They are stored by making a user-defined function by using the `CREATE FUNCTION` syntax. Last, our system has much more extensibility than others. We implemented some functions for Web services and processing functions such as NLP functions. If users require other functions, they can add their own user-defined functions. It would be difficult to extend the other platforms.

3   Web Services and Functions as Virtual Tables

A large amount of useful information are provided by Web services on the *cloud*. If organizations and people stores their private data on the cloud, it is possible to create many private applications useful for them. It is, however, not acceptable in many cases because the private data must not be stolen by others in all cases. Another way to make such applications is that a private database has capability to connect to Web services. Out approach takes this way, and it is safer for privacy. The concept of the approach is to regard the cloud as virtual databases.

Figure 1 shows the concept of the platform. It bridge local databases and many Web services by regarding the cloud as virtual databases. APIs provided by Web services send data to the platform. A wrapper is made for each of them, and it is treated as a table. These tables are not physical but rather *virtual* ones whose values change according to given arguments. Several functions are prepared, and the outputs of them are also regarded as virtual tables. Processing text data from both local databases and Web services is useful to join data from both of them. Users of the platform can operate all of these by SQL.

Because they all are in an RDB environment, local databases can easily be connected to Web services. Suppose that a private database has a table about researchers, and it contains some data not to be disclosed. A Web service that has publication data for researchers, such as CiteSeer or DBLP, can be joined with the table on the local database, and aggregation of the data will provide the number of research papers for each of the researchers.

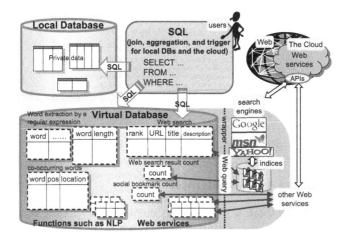

Fig. 1. Concept of the proposed platform

3.1   Virtual Tables Using Table-Valued Functions

Virtual tables are tables that do not store any data but will return table data when they are executed. A **table-valued function** is a function that returns a table. Virtual tables are realized by using table-valued functions. We implemented the platform on Microsoft SQL Server 2005, and SQL CLR on it allows us to make functions by using programming languages supported by the .Net Framework, such as C# and Visual Basic. We do not extend the Transact-SQL.

For example, we created a virtual table that returns Web search results. The table does not *store* any Web search result. When the virtual table is used in SQL statements, it accesses a Web search engine and generates a table data from the returned search results at that time. Many other Web services are also implemented as virtual tables in the same manner. Table-valued functions are also used to implement functions, such as morphological analysis, word extraction, and finding co-occurring words. Conventional RDBMSs have some string functions, but these are too low level to extract meaningful information from text data. The implemented functions return rich results as table data.

As virtual tables has substantial data when they are executed, joining virtual tables and normal tables is easily done. However, because a virtual table changes the result by a given argument value, normal fashion to join virtual tables cannot be used in some cases. APPLY is a proprietary operator in SQL Server 2005. An example usage of the APPLY operator is as follows.

```
SELECT * FROM tbl CROSS APPLY tvf(tbl.col);
```

tbl is a normal table name, and tfv is a table-valued function, and table tbl's column col is given as an argument of tfv. The results columns of this SELECT statement are all tbl's and tfv's. The APPLY operator is used as CROSS APPLY or OUTER APPLY, and they works like INNER JOIN or LEFT JOIN respectively.

Fig. 2. Example of executing `WebSearch`

3.2   Web Services as Virtual Tables

We describe how to treat a Web search engine as a virtual table as an example. A result from a conventional Web search engine normally contains information about searched pages. The information on each page consists of the title, URL, and snippet. The table data of the virtual table is made of them. `WebSearch(query, num)` is the name of the virtual table that returns a set of Web search result items, where the argument `query` is a query and `num` is the maximum number of search results. Each row in the returned table contains a search ranking, URL, title, and snippet. Figure 2 shows an example of executing `WebSearch` with the query "DASFAA2010", where the top 50 search results are retuned as a table. The platform supports many kinds of Web search engines. Search results are obtained through APIs provided by the services. For example, the virtual table `WebSearch_Yahoo` uses Yahoo! Web Search.

Users can also treat many other Web services as virtual tables in the same manner as conventional Web search engines. A Web image search service such as Yahoo! Image Search returns surround text of searched images for a given query word, and a bibliography web service such as CiteSeer and DBLP returns publication information for a given author or document name.

When a Web service provides just a scalar value, the result does not have to be a virtual table. Such Web services are implemented as scalar-valued functions. For example, `WebCount(query)` gives the hit count from a conventional Web search engine for a given query. `SocialBookmarkCount(url)` is another scalar-valued functions that returns the number of social bookmarks in a social bookmark service, such as delicious.com.

3.3   Functions as Virtual Tables

Many Web services return text data. To join local databases and Web services, it is often necessary to extract term-level data from text data. We implemented many functions for information extraction, which are mainly NLP functions. Some of them are briefly introduced as follows.

`ExtractPattern(text, extractPattern)` is a virtual table that extracts words from a given text by using regular expressions. There are two arguments that `text` is a piece of text and `extractPattern` is a regular expression. The function is a virtual table, so the result is as table that contains all of the matched parts in the given regular expression. The `word` column in the returned table has a matched word. There are some other columns for users to specify the matched parts in the regular expression. A simpler function to extract term is `ExtractWord(text, extractPattern)`, where `extractPattern` is a syntactic pattern with a tag `<word>`. `CooccurringWords(text, targetWord)` is a virtual table that returns words co-occurrent with a given target word.

4   Applications

4.1   Joining Local Databases and Web Services

Private data that must not be shared by others is mainly personal information such as salary, birthday, address, and the number of children. On the other hand, Web services provide public information. It sometimes contributes to evaluate the private data. An example to join private data and data in a certain Web service is shown as follows.

```
SELECT Researcher.Fullname, COUNT(*) AS n
FROM Researcher
  CROSS APPLY Publication_DBLP(Researcher.Fullname) AS dblp
GROUP BY Researcher.Fullname
ORDER BY n DESC;
```

There is private information about researchers in a local table `Researcher`. The column `Fullname` stores a list of full names of researchers. `Publication_DBLP` is publication data for researchers. The above SQL lists researcher's name sorted by the number of publications.

The former example joins data from Web services directly to local databases. Another usage is to extract knowledge from data from the Web and to use the knowledge on private data. Here, we introduce Web aggregation, which is a way to acquire knowledge from the Web. It is done by obtaining text resource from Web services, extracting information from the obtained text, and aggregate the extracted information.

We devised a method to obtain coordinate terms of a given word. This method uses syntactic patterns "x or y" and "y or x" where x and y are coordinate terms. The detail of the method is described in [5], our platform has capability to realize the method very easily. Figure 3 shows the definition of the user-defined function to obtain coordinate terms. This can be used as follows.

```
SELECT * FROM CoordinateTerms('Porsche');
```

4.2   Web Trigger

The platform allows setting triggers on virtual tables by providing some stored procedures. To make applications that set triggers on virtual tables, we first need

```
 1: CREATE FUNCTION CoordinateTerms (@q varchar(max))
 2: RETURNS @results TABLE (word NVARCHAR(max), v FLOAT)
 3: AS
 4: BEGIN
 5: DECLARE @wsa TABLE (word NVARCHAR(max), ca int);
 6: DECLARE @wsb TABLE (word NVARCHAR(max), cb int);
 7: INSERT INTO @wsa (word, ca)
 8: SELECT word, count(word) FROM WebSearch('"' + @q + ' or"', 100)
 9:   CROSS APPLY ExtractWord(description, @q + ' or <word>')
10: GROUP BY word;
11: INSERT INTO @wsb (word, cb)
12: SELECT word, count(word) FROM WebSearch('"or ' + @q + '"', 100)
13:   CROSS APPLY ExtractWord(description, '<word> or ' + @q)
14: GROUP BY word;
15: INSERT @results
16: SELECT ta.word, SQRT(ta.ca * tb.cb) v FROM @wsa AS ta, @wsb AS tb
17: WHERE ta.word = tb.word
18: AND ta.word NOT LIKE '%' + @q + '%'
19: ORDER BY v DESC
20: RETURN
21: END
```

Fig. 3. User-defined function to find coordinate terms

```
 1: CREATE FUNCTION CheapHotelInTsukuba()
 2: RETURNS @results TABLE (rest INT, average INT)
 3: AS
 4: BEGIN
 5: INSERT INTO @results (rest, average)
 6: SELECT COUNT(*), AVG(sumCharge)
 7: FROM VacantHotelSearch_Rakuten ('Tsukuba', 'Ibaraki', '2010/4/1', '2010/4/4', 1)
 8: WHERE sumCharge < 30000;
 9: RETURN
10: END
```

Fig. 4. Functions to obtain the number of available hotel rooms and the average charge

to keep Web search service results as a snapshot table. And then the platform checks information in the Web at regular time intervals. When the difference between the previous snapshot and the current information is determined, triggers set on the table are fired.

An example of applications using triggers on virtual tables is to check for available hotel rooms. The virtual table **VacantHotelSearch** provides the results of the Web service. Suppose he has to go to Tsukuba in Ibaraki prefecture in Japan from April 1st to 4th, 2010 and he reserves a hotel later. He wishes to be notified if the available options change. First, he makes a user-defined table-valued function **CheapHotelInTsukuba** that returns the number of available rooms whose charges are less than 30,000 JPY and their average charge. Its definition is shown in Figure 4.

The stored procedure **CreateSnapshotTable** is prepared on the platform, and it is used as follows, where the snapshot table **snap_t_hotel** is updated once an hour (every 60 minutes).

```
CreateSnapshotTable 'CheapHotelInTsukuba()', 'snap_t_hotel', 60
```

```
 1: CREATE TRIGGER tri_tsukuba_hotel
 2: ON snap_t_hotel AFTER UPDATE
 3: AS
 4: BEGIN
 5: DECLARE @rest INT;
 6: SELECT @rest = (SELECT rest FROM snap_tsukuba_hotel);
 7: DECLARE @avg INT;
 8: SELECT @avg = (SELECT average FROM snap_tsukuba_hotel);
 9: DECLARE @message NVARCHAR(max);
10: SET @message = 'The number of hotel options is ' + CAST(@rest AS NVARCHAR(5)) + '.'
11:    + CHAR(13) + CHAR(10) + 'The average charge is ' + CAST(@avg AS NVARCHAR(7)) + '.';
12: EXEC MailSend 'ohshima@dl.kuis.kyoto-u.ac.jp', @message;
13: END
```

Fig. 5. Trigger for checking available hotel rooms

The trigger is written in Figure 5. An e-mail is sent every time the result of CheapHotelInTsukuba changes. The message sent by the system contains the number of hotel options and their average charge.

5   Conclusions

We proposed a platform for bridging local databases and the cloud. It is developed in an existing RDB environment, Microsoft SQL Server 2005. Each Web service is treated as a virtual table. Although it is not easy to brought private data out of local databases, the platform realize to create applications using private data and Web data.

Acknowledgments. This work was supported in part by the following projects and institutions: Grants-in-Aid for Scientific Research (Nos. 18049041 and 21700105) from MEXT of Japan, a Kyoto University GCOE Program entitled "Informatics Education and Research for Knowledge- Circulating Society," and the National Institute of Information and Communications Technology, Japan.

References

1. WebQL on QL2 Software, http://www.ql2.com/
2. KnowItAll, http://www.cs.washington.edu/research/knowitall/
3. Cafarella, M.J., Etzioni, O.: A search engine for natural language applications. In: Proc. of the 14th International Conference on World Wide Web (WWW 2005), May 2005, pp. 442–452 (2005)
4. Goldman, R., Widom, J.: WSQ/DSQ: a practical approach for combined querying of databases and the Web. In: Proc. of the 2000 ACM SIGMOD International Conference on Management of Data (SIGMOD 2000), May 2000, pp. 285–296 (2000)
5. Ohshima, H., Oyama, S., Tanaka, K.: Searching coordinate terms with their context from the Web. In: Aberer, K., Peng, Z., Rundensteiner, E.A., Zhang, Y., Li, X. (eds.) WISE 2006. LNCS, vol. 4255, pp. 40–47. Springer, Heidelberg (2006)

Temporal Top-k Search in Social Tagging Sites Using Multiple Social Networks

Wenyu Huo and Vassilis J. Tsotras

Department of Computer Science and Engineering,
University of California, Riverside, CA, 92521, USA
{whuo,tsotras}@cs.ucr.edu

Abstract. In social tagging sites, users are provided easy ways to create social networks, to post and share items like bookmarks, videos, photos and articles, along with comments and tags. In this paper, we present a study of top-k search in social tagging sites by utilizing multiple social networks and temporal information. In particular, besides the global connection, we consider two main social networks, namely the friendship and the common interest networks in our scoring functions. Based on the degree of participation in various networks, users can be categorized into specific classes that differ in their weights on each scoring component. Temporal information, usually ignored by previous works, can enhance the popularity and freshness of the ranking results. Experiments and evaluations on real social tagging datasets show that our framework works well in practice and give useful and intuitive results.

1 Introduction

The advent of Web 2.0 has facilitated the growth of online communities and applications such as blogs, wikis, online social networks and social tagging sites. In social tagging sites, such as del.icio.us, Flickr, and CiteULike, once a user is logged in, he can easily edit his own personal profile, build social networks with friends, and contribute content by posting bookmarks, videos, photos, or articles. He can also annotate those items with arbitrary tags.

Social tagging sites are free, fun, and functional, attracting more and more people to register as users. Moreover, social tagging sites have formed and stored plenty of valuable information like user-generated items, user social networks, and user tags. How to make good use of this information to improve services such as hot-lists, recommendations and web search is an open and attractive challenge for both academia and industry.

In this paper, we focus on temporal ranking and personal search in social tagging sites. When compared to other related work such as [1,6], our contributions are: First, we apply multiple components to score an item with respect to a particular user's different social networks and assign weights based on the classification of that user's participation in those networks. Then, we take temporal information into account, to enhance popularity and freshness of the top-k

H. Kitagawa et al. (Eds.): DASFAA 2010, Part I, LNCS 5981, pp. 498–504, 2010.

results. We provide a variation of the classic top-k algorithm which works efficiently for our user-dependent temporal scoring functions. Last, experimental evaluations on real social tagging datasets show that our framework works well in practice.

2   Data Model

Previous work in social tagging mostly ignores temporal information, only considering three factors: *User*, *Item*, and *Tags*. We extend the tagging behavior by adding timestamps: <*User, Item, Tags, Timestamp*>, which indicates that a user annotated one item with arbitrary tags at some time. In the following, we first demonstrate the model of social networks and static scoring functions without timestamps, and then explore a method to incorporate temporal information into ranking.

In social tagging sites, users are generally participating in multiple social networks. Aside from the global connection (*Global*), meaning everyone is connecting with anyone else on the whole web, we consider two other main kinds of social networks, namely, friendship (*Friends*) and common interest networks (*Links*).

Friendship is a kind of *explicit* social network. One user can choose to add any other users as friends. Most of them could be acquaintances in real life—friends, schoolmates, business contacts, etc; some may be known through the internet. We use *Friends*(u) to represent all users in a friendship with user u. Social tagging sites enable users to create and join special groups. This is also an explicit social network, since group members have direct connections with each other. We categorize groups into *Friends* as well.

We also consider another kind of social network called common interest network [1]. Different from the traditional explicit social networks built up by adding friends or joining groups, the common interest network is *implicit* in nature, formed based on similar tagging behaviors. The items posted and the tags used by a user can be considered indicators of that person's interests. Linking people together whose tagging behaviors overlap significantly can implicitly form common interest networks.

For example, Let *Items*(u) be the set of items tagged by the user u with any tag. Using *Links*(u) to represent the common interest network for the user u, we could define that another user v is in *Links*(u) iff a large fraction of the items tagged by u are also tagged by v, as $|Items(u) \cap Items(v)| > \theta$, where θ is a given threshold.

Given a query $Q = t_1, \ldots, t_n$ with n tags, issued by user u, and a number k, we want to efficiently return the top-k items with the highest overall scores. Our search strategy is user-focused, giving different results to different users. Our scoring functions consider the user's multiple social networks. Moreover, the top-k results returned take into account tagging behaviors' temporal information.

3   Scoring Functions

The static scoring functions for each social network component and overall combined scores are initially described. A method of weight assignments based on user classification is then discussed. Finally, temporal information of tagging behaviors is added and temporal scoring functions are examined.

3.1   Multiple Social Network Components

The overall static scoring function needs to aggregate three social network components: friendship, common interest network, and global connection.

Given a user u, the friendship component score of an item i for a tag t is defined as the number of users u's friends who tagged i with tag t:

$$Sc_F(i, u, t) = |Friends(u) \cap \{v|Tagging(v, i, t)\}| \qquad (1)$$

Similarly, the score from common interest network is defined as the number of users in u's *Links* who tagged i with tag t:

$$Sc_L(i, u, t) = |Links(u) \cap \{v|Tagging(v, i, t)\}| \qquad (2)$$

Besides the above two scoring component from a user's social networks, we also consider the global effect on scoring. Not everyone is an active participant; if we only use the local social network scoring, the search effectiveness may decrease. The *Global* score, which is user-independant, is defined as the total number of users in the whole website tagged item i with tag t:

$$Sc_G(i, t) = |\{v|Tagging(v, i, t)\}| \qquad (3)$$

As a result, the static overall score of item i for user u with one tag t is an aggregate function of the weighted scores from the three components:

$$Sc_O(i, u, t) = w_1 * Sc_G(i, t) + w_2 * Sc_F(i, u, t) + w_3 * Sc_L(i, u, t) \qquad (4)$$

where w_i is the weight of each component and $\sum_{i=1}^{3} w_i = 1$

Since a query contains multiple tags, we also define the static overall $SCORE$ of item i for user u with the whole query $Q = t_1, \ldots, t_n$ as the sum of the scores from individual tags, which is a monotone aggregation function:

$$SCORE(i, u) = \sum_{j=1}^{n} Sc_O(i, u, t_j) \qquad (5)$$

3.2   User Classification

Different weight assignments of components can generate different overall scores. There are several ways to assign component weights using machine learning or statistics methods. However, those need a large amount of data such as user feedbacks and log records, which are not easy to access. For simplicity, we use a user classification method based on the social networks size and recommend weight assignments for each class.

Users in social tagging sites have different usage patterns and degrees of participation in their social networks. Some users have many friends, while some may only have few. Also, for tagging, some users do frequent tagging and thus have a lot of tagged items; while others may not tag as much.

In our general framework, we use three categories for *Friends* and *Links* social network component, described as: *many, some* and *few*; so there are nine classes totally. Within this classification, we assume that users in the same class have similar degree of trust on each social network scoring component. Then we can give a recommendation of weight assignments for users in each class.

3.3   Temporal Scoring Functions

We believe that ranking results will be more attractive to users not only based on their relevance, but also on popularity and freshness. For example, one item may be more interesting if it is recently added. In this case, a simple interpretation of freshness is the first date the item was posted. However, a more subtle way may consider how many recent tagging behaviors have targeted an item.

Our basic approach is to divide the tagging behaviors into multiple time slices, based on their time stamps. We use m to denote the number of time slices and adjust the weights of different time slices based on their recency. A decay factor a $(0 < a < 1)$ is used to penalize the count score from old time slices. Thus, the temporal score of Global component of item i with tag t can be defined as:

$$TSc_G(i,t) = \sum_{s=1}^{m} Sc_G(i,t,s) * a^{m-s} \tag{6}$$

where $Sc_G(i,t,s)$ is the global score of item i with tag t at time slice s, with $s = m$ being the current time slice.

The temporal scoring functions for *Friends* and *Links* components are defined similarly with the same temporal factors in *Global*:

$$TSc_F(i,u,t) = \sum_{s=1}^{m} Sc_F(i,u,t,s) * a^{m-s} \tag{7}$$

$$TSc_L(i,t) = \sum_{s=1}^{m} Sc_L(i,u,t,s) * a^{m-s} \tag{8}$$

The temporal overall scoring function of item i for user u with tag t is:

$$TSc_O(i,u,t) = w_1 * TSc_G(i,t) + w_2 * TSc_F(i,u,t) + w_3 * TSc_L(i,u,t) \tag{9}$$

Therefore, the temporal scoring for whole query is:

$$TSCORE(i,u) = \sum_{j=1}^{n} TSc_O(i,u,t_j) \tag{10}$$

4   Temporal Ranking Algorithm

Typically, one inverted list is created for each keyword and each entry contains the identifier of a document along with its score for that keyword [2]. For our framework, when the query is composed of multiple tags, we need to access multiple lists and apply the top-k processing algorithms.

One straightforward method is to have one inverted list for each (tag, user) pair and sort items in each list according to the temporal overall score (TSc_O) for the tag t and user u. However, there are too many users registered (del.icio.us has over 5 million users). If we create inverted lists per keyword for each user, there will be a huge amount of inverted lists and thus large space is required.

Another solution is to factor out the user from each inverted list by using upper-bound scores [1]. Since we use the number of users as the static score without normalization and set all three social network component with the same temporal factors for a query, for the same item i with the same tag t, no matter which user, we have $TSc_F <= TSc_G$ and $TSc_L <= TSc_G$. As a result, temporal global score is an upper-bound of temporal overall score for all users. Since the global component scoring is user-independent, we can create only one list for each keyword along with the temporal global scores (TSc_G) as an upper-bound of the user-based temporal overall scores (TSc_O).

The temporal factor can be designed as adjustable for users, so the temporal factors may also need to be factored out from the inverted lists. The static global scores (Sc_G) is an upper-bound for the temporal global scores (TSc_G), since the static scores correspond to the temporal ones with $a = 1$. Therefore, the final upper-bound scores used in the inverted lists are the static global scores (Sc_G).

We can thus extend Fagin's classic top-k TA algorithm [3] to rank the items listed in the order of static global scores (Sc_G) as the upper bound. When a new item is seen for the first time, we compute its exact temporal overall score (TSc_G) with a "local" aggregation function of three component temporal scores. The Algorithm stops whenever the score of the kth item in the heap is no less than the sum of bottom bounds of all lists. More details are covered in [4].

5   Experimental Evaluation

To evaluate the effectiveness of our scoring functions and algorithms, we collected real datasets from CiteULike (http://www.citeulike.org), an academic article social tagging site. In CiteULike, articles are stored with their metadata, abstracts, and links, and users can add tags and personal comments. CiteULike provides some datasets from their core database. However, to get more recent data, we further crawled datasets before 2009.7.1. An extended collection of our experimental evaluations appears in [4].

Here we use the NDCG (normalized discounted cumulated gain) measurement [5] to evaluate the performance of our experiments. Every item in top-k lists is given a corresponding human judgment scoring from 0 to 3 (0=Bad, 1=Fair, 2=Good, 3=Excellent) based on relevance and attractiveness (popularity and freshness) for particular query tags.

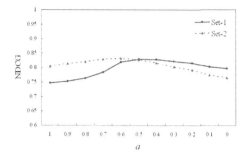

Fig. 1. Average NDCG results for different decay factor a

User		Friends		
		many	*some*	*few*
Links	*many*	Class 1	Class 2	Class 3
	some	Class 4	Class 5	Class 6
	few	Class 7	Class 8	Class 9

Class #	Recommendation
Class 1	r1: $w_1 = 0.1$, $w_2 = 0.45$, $w_3 = 0.45$;
Class 2	r2: $w_1 = 0.1$, $w_2 = 0.3$, $w_3 = 0.6$;
Class 3	r3: $w_1 = 0.1$, $w_2 = 0.1$, $w_3 = 0.8$;
Class 5	r5: $w_1 = 0.2$, $w_2 = 0.4$, $w_3 = 0.4$;
Class 6	r6: $w_1 = 0.2$, $w_2 = 0.3$, $w_3 = 0.5$;
Class 9	r9: $w_1 = 0.4$, $w_2 = 0.3$, $w_3 = 0.3$;

Fig. 2. User classification and weights recommendation for representative classes

Different queries may prefer different temporal factor settings, thus we use two different sets of popular query tags. For set-1, the queries are "social-network" and "tagging". These are popular and very hot recently. For set-2, we use "algorithm" and "database" separately as popular and classic queries.

We divide the time range of our datasets into six-month periods, starting from the most recent 2009.1.1 - 2009.6.30 to earlier time slices, which will remain the same throughout this paper. We change the decay factor a from 1 to 0, which means setting recency priority from low to high. and only evaluate the global temporal scoring (TSc_G) to factor out user diversity.

From the results in Fig. 1, we observe that different kinds of queries have different preferences. Hot queries may prefer recent tagging behaviors much more than classic queries. But for both sets, the average NDCG peaks when a is set 0.5 or 0.6, neither too high to miss the temporal information nor too low to lose classic items.

Then we evaluate the NDCG of different user classes with different weight assignments for each social network and we set *few* as 0-5, *some* as 6-15, and *many* as 15+ for both Friends and Links.

Based on the user classification, we provide an example recommendation of weight assignments for six representative classes in Fig. 2. The decay factor is set as $a = 0.5$ and the time slices are six-months. We tested two queries— "tagging" and "algorithm", picked up two users from our dataset for each class, and extracted the average NDCG. As shown in Fig. 3, in all six representative classes, our multiple-component method produced better NDCG than any other methods considering only one type of social network.

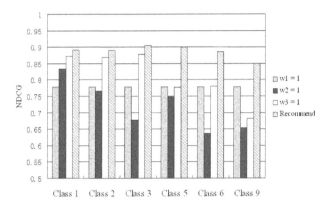

Fig. 3. Average NDCG for weight assignments across six representative classes

6   Conclusions

In this paper, we presented a study of temporal top-k search in social tagging sites using three main types of social networks, friendship, common interest networks, and global connections. To set the weights of each scoring component for different users, a classification method is proposed based on the size of users' social networks. To improve the popularity and freshness of ranking results, the timestamps of tagging behaviors are recorded and divided into multiple time slices and temporal scoring functions are formed by giving higher weights to more recent time slices. In addition, an efficient temporal top-k algorithm for ranking is proposed with upper-bound scores. Experimental evaluation on real datasets shows that our framework and methodology work well in practice.

Acknowledgments. This work is partly supported by NSF grants IIS-0705916 and IIS-0910859.

References

1. Amer-Yahia, S., Benedikt, M., Lakshmanan, L., Stoyanovich, J.: Efficient Network-Aware search in Collaborative Tagging Sites. In: VLDB (2008)
2. Baeza-Yates, R., Ribeiro-Neto, B.: Modern Information Retrieval. Addison-Wesley, Reading (1999)
3. Fagin, R., Lotem, A., Naor, M.: Optimal aggregation algorithms for middleware. In: PODS (2001)
4. Huo, W.: Temporal Top-k Search in Social Tagging Sites Using Multiple Social Networks. Master's thesis, University of California-Riverside (2009)
5. Jarvelin, K., Kekalainen, J.: IR evaluation methods for retrieving highly relevant documents. In: SIGIR (2000)
6. Schenkel, R., Crecelius, T., Kacimi, M., Michel, Neumann, T., Parreira, J. X., Weikum, G.: Efficient top-k querying over social-tagging networks. In: SIGIR (2008)

Air-Indexing on Error Prone Communication Channels

Emmanuel Müller, Philipp Kranen, Michael Nett, Felix Reidl, and Thomas Seidl

Data Management and Data Exploration Group, RWTH Aachen University, Germany
{mueller,kranen,nett,reidl,seidl}@cs.rwth-aachen.de

Abstract. Air-Indexing aims at efficient data dissemination via a wireless broadcast channel to a multitude of mobile clients. As mobile devices have very limited resources, energy efficiency is crucial in such scenarios. Moreover, one has to cope with high error rates on wireless transmissions resulting in packet losses which lead to high energy consumption and long waiting times until a query result is available. We propose a novel cross-layer scheduling with adaptive error correction which enables flexible query optimization on mobile clients. RepAir ensures efficiency by adapting its query processing according to the individual error rate of each client. Thorough experiments show that RepAir yields substantial efficiency improvements in terms of access latency and tuning time compared to competing Air-Indexing approaches.

1 Introduction

Mobile devices are ubiquitous. Novel applications emerge from many different areas such as location based services, mobile information systems, entertainment and multimedia application. A huge variety of data broadcast scenarios have been proposed including traffic information systems, broadcasting of weather or news clips, audio or video guides for museums etc. and broadcasting of small video clips for major events such as the Olympic games. In general, all of these scenarios address a large group of clients using standard mobile devices requiring efficient broadcast techniques. Recent air indexing techniques have been proposed to perform an efficient query processing in such broadcasting scenarios.

In general, air indexing aims at a trade-off between energy efficiency and access efficiency, where the former is measured as the *tuning time* and the latter as *access latency*. The tuning time is the amount of time that the client has to actively receive data, i.e. it is in the *active mode*. For the rest of the time the client switches to the less energy consuming *doze mode*, i.e. it does not download data from the broadcast. The access latency constitutes the total time needed to answer a query, i.e. from the moment the client first tunes into the broadcast until all requested data has been successfully downloaded. To enable selective tuning on mobile clients, several air indexing approaches have been proposed throughout the last years. The general idea is to send additional meta information about the data items and their arrival time within the broadcast such that the clients can switch to doze mode until the desired data arrives. Traditional indexing schemes from the database domain cannot be used because of the missing random access in broadcast scenarios. The presence of communication errors poses an additional obstacle.

H. Kitagawa et al. (Eds.): DASFAA 2010, Part I, LNCS 5981, pp. 505–519, 2010.

Wireless transmission in such mobile scenarios has to cope with error prone communication channels. In todays mobile communication like 3G [7] error correction is always required as communication has to cope with high error rates. For speech transmission these high error rates are typically compensated by transmitting redundant information by special codings. The remaining problem is that still errors occur, resulting in lower speech quality (interrupts, high noise rates, etc.). In data transmission, we rely on receiving data without any errors. Error correction could guarantee this only with tremendous redundancy in transmissions leading to low transmission rates which are not accepted by users due to impractically large transmission times.

Summarizing these facts, we have to cope with several challenges in broadcasting data to mobile devices namely energy efficiency, access efficiency and high error rates. Traditional air indexing approaches have concentrated on the application layer. They have to perform costly reaccess of the broadcast stream in case of an error occurrence. On the other side, the technical layer is not aware of the content it is supposed to transmit and thus error correction techniques are not able to adapt to air indexing requirements.

In this paper we propose a novel air indexing techniques that we call RepAir. We develop an adaptive index scheduling on the technical layer and a flexible error correction on the application layer. As a cross-layer approach, RepAir is aware of the semantic of each packet and can incorporate error correction for improving both tuning time and access latency. For our RepAir scheduling, we employ error correction separately on the index and on the data items to achieve a reliable air indexing and to overcome high penalties on the access latency. Our flexible query optimization can adapt to the individual error rates of each mobile client and yields faster access and lower energy consumption. We show improved access latency and tuning time greatly independent of the error probability.

2   Related Work

Air Indexing techniques that have been proposed throughout the last years can be divided into two major groups: distributed indexing techniques and $(1, m)$ indexing techniques. Both are based on the same underlying assumption that the packet size is fixed during transmission. For air indexing packets are comparable to pages in traditional data bases. Packets are used to transmit both index and data as *index packets* and *data packets*. The differences between distributed and $(1, m)$ indexing lies first in the type of information stored in the index packets and second in the scheduling, i.e. how index and data packets are interleaved to form the broadcast. However, all air indexing techniques focus only on specialized index information and index transmission. Non of these techniques considers forward error correction neither for index nor for data packets. Especially, they all work on the application layer ignoring possible improvements through cross-layer aspects induced by the semantics of index and data packets.

The basic idea underlying (1, m) indexing techniques [5,6,10,2] is to take a traditional index, usually a hierarchical index from the R-Tree family [3,1], and to serialize it for dissemination. The broadcast is then scheduled by sending the entire index and a fraction of $1/m$ of the data in turns yielding m replications of the index per broadcast, hence $(1, m)$. To reduce the total length of a broadcast cycle, other approaches do not send the entire index m times. They split the index into one replicated and several non-replicated parts. Query processing in $(1, m)$ first downloads interesting and available index information (dependent on the approach), then downloads candidate items and finally refines the result set until the exact result is present.

For distributed indexing the data is divided into *frames*, where each frame contains several data items or rather their corresponding data packets. In the broadcast an index packet is sent before each frame. To facilitate navigation within the broadcast during query processing, the index packet contains pointers to other frames and optionally additional information about the local or global data distribution (depending on the actual approach). By using local pointers that index an exponentially increasing object range, distributed indexing schemes allow multiple different search paths whereas $(1, m)$ only allows multiple entry points (repetitions) for the same search path (index). Several approaches have been published using Hilbert Curves [9,4], Voronoi diagrams [17] and D-Trees [15,14]. We will go into more detail describing DSI since we compare against their approaches in our experiments. DSI divides the feature space into a grid such that each grid cell contains at most one data item [9]. After that, the Hilbert Curve (HC) value for each data item is calculated w.r.t. the grid and the items are ordered according to their HC value. The index packets contain a table of pairs (HC_i, p_i), where the $p_0, \ldots, p_t$ are pointers indexing an exponentially increasing number of data items. Two approaches are presented for query processing, the aggressive approach is supposed to optimize the tuning time while the conservative approach tries to optimize the access latency. However, we will show in our experiments in Section 4 that both fail in case of high error rates due to the reloading of lost packets.

Summing up the features of existing approaches, we observe several drawbacks. As none of these techniques includes error correction in the index scheduling, they are forced to costly reaccess the broadcast stream to retrieve lost packets. Furthermore, their query processing is not able to adapt to the varying error rates in each individual mobile client. Overall, this leads to an increase of both access latency and tuning time for these approaches.

3   RepAir Approach

We show that neither recent data scheduling approaches nor error correction techniques on technical layers alone can achieve both an efficient and fault tolerant data dissemination at the same time. Our RepAir scheduling as a cross-layer approach achieves such a solution. The general idea is to trigger different coding for index and data packets at the transmitter side. Each mobile client can use this error correction codes for repairing lost packets. By using different amount of error correction in query processing, a mobile client can adapt to its individual error rate.

3.1   Scheduling in RepAir

As mobile communication uses a packet based transmission, all information is given in packets and the scheduler has only to decide the packet ordering. A packet p is the atomic unit in data transmission. In addition to the carried data or index information it contains header information such as the packet identification number and a pointer to the next index segment. The schedule, which constitutes one broadcast cycle, is first constructed by the transmitter and then repeatedly broadcasted to all receivers. It contains packets to encapsulate data items and index segments. A common assumption for such a broadcast is that index and data remain unchanged. While we keep this assumption for the following sections, we discuss data and index updates in Section 5.

Definition 1. *Air-Indexing Schedule: An Air-Indexing Schedule AIS is a sequence $p_1 \ldots p_N$ of packets containing all data packets $DP = \bigcup_{item \in DB} DP_{item}$ given by the items in the application data base DB and all index packets $IP = \bigcup_{i=1}^m IP_{Index_i}$ in an interleaved fashion.*

As main characteristic of redundant transmission we observe that different index mappings are proposed in recent approaches. The index mapping specified by $f : IP_I \rightarrow DP_I$ gives us the subset of the data base indexed by a set of index packets. The main property of the two variants complete indexing (as in basic $(1, m)$) and distributed indexing is the cardinality of data base items that is indexed by index packets: $|f(\text{complete index})| = |DB|$, $|f(\text{distributed index})| \ll |DB|$. These two variants perform a trade-off between two aspects: A bigger broadcast size (for complete index) and more broadcast cycle passes (for distributed index). However, for error prone communication both variants require multiple broadcast cycle access.

In contrast to both variants of broadcast scheduling (complete and distributed), we reduce the number of required broadcast cycles during query processing through a different form of information redundancy. By our novel RepAir approach we will show that broadcasts, enriched by error correcting codes, yield better access latencies. Especially for error prone communication it is important to ensure a correct index transmission. In contrast to a data segment (interesting only for a few clients) the index is essential for an energy efficient query processing in all mobile clients. Different handling for both types of packets is thus essential for energy efficient query processing.

In a packet based transmission we consider a packet as the atomic unit of communication. Furthermore we assume that due to checksum techniques we can detect an error during transmission. We thus use the common binary erasure model for modeling the errors during communication.

Definition 2. *Binary Erasure Model: A communication layer behaving according to a binary erasure model with error parameter ξ transmits a packet correctly from the transmitter to a receiver with a probability of $(1 - \xi)$.*

As we assume transmission from one transmitter to many receivers, the error probability varies for each receiver. This separate modeling for each mobile client is meaningful as errors are not a global phenomenon affecting all mobile clients in the same manner. There might be various types of errors during the transmission like buildings reducing radio signal quality because of reflections and absorption, interfering signals from multiple transmitters in the same region, low battery and thus low signal strength on the mobile device or even atmospheric noise which disturbs the radio signal. Please note that typical error rates for mobile communication range from 10% up to 70% packet losses even after forward error correction in 3G [7].

Assuming this underlying communication model we have now to consider techniques that handle errors to achieve a correct transmission. On technical layers this problem has already been addressed by error correction techniques: For a fault tolerant transmission one uses coding techniques which cannot only detect errors but also reconstruct the original signal even if some error occurred during the transmission. In general this means that one has to blow-up the transmitted information by redundant data such that errors can be automatically corrected.

Since there is no fixed error rate, RepAir uses a flexible technique which supports a multitude of mobile clients. Rateless "Fountain Codes" [12] support such a scenario for one-to-many communication (broadcast/multicast). They have been designed for fault tolerant transmission in broadcast scenarios like file transmission in satellite television broadcasting. With fountain codes producing an endless coding, we can use the coding packets in the next broadcast for reconstruction as well. Intuitively, a perfect error correcting coding scheme has to code a source packet such that it can not only reconstruct this packet but also help to reconstruct errors in other packets. Belief propagation aims at such a coding scheme. The solition distribution used in our approach ensures that each received packet can be decoded. Furthermore, each packet has high probability to participate in reconstruction of another packet.

Technically, we use a given set SP of source packets and build a coding packet cp by combining a number of source packets which are chosen equally at random, using a simple XOR operation $\oplus$. We refer to the number of source packets included in cp as its degree deg_{cp}. To restore a single source packet sp that is included in cp we need to know all other source packets included in cp. If this is the case, we call cp a resolvable code packet. Once we have a code packet cp with $deg_{cp} = 1$, i.e. $cp = sp$ for a certain source packet sp, we can restore another source packet sp' from cp' if $cp' = sp \oplus sp'$. A similar scenario applies for two known source packets and a corresponding coding packet of degree three etc.

Two major questions arise: First, during scheduling, how do we determine the degree for a coding packet such that we will be able to restore all source packets from a reasonable number of coding packets with high probability? And second, during query processing, how do we determine whether the next coding packet is useful for us, i.e. whether we should spend tuning time to download it? As for the first question there are many approaches in the literature. We adapt the robust soliton distribution from [11] for the degree deg of the coding packets.

Definition 3. *The Robust Soliton Distribution gives the probability $\mu(deg)$ for a coding packet to be of degree deg and is defined by $c > 0$ and δ as follows. Let SP be a set of source packets to be coded and $|SP|$ the size of SP. Then*

$$\mu(i) = (\rho(i) + \tau(i)) / \beta \quad \forall i = 1, \ldots, |SP|$$

where

$$\rho(i) = \begin{cases} 1/|SP| & , \text{if } i = 1 \\ 1/(i \cdot (i-1)) , & \forall i = 2, \ldots, |SP| \end{cases}$$

$$\tau(i) = \begin{cases} R/i \cdot |SP| & , \forall i = 1, \ldots, |SP|/R - 1 \\ R \cdot \ln(R/\delta)/|SP| , & \text{if } i = |SP|/R \\ 0 & , \text{otherwise} \end{cases}$$

$$R = c \cdot \ln(|SP|/\delta)\sqrt{|SP|}$$

and

$$\beta = \sum_{i=1}^{|SP|} \rho(i) + \tau(i)$$

For more details on the robust soliton distribution and its properties please refer to [11]. In the following we will describe how we adapt robust soliton for our RepAir scheduling.

RepAir belongs to the group of $(1, m)$ air indexing techniques. In this work we use a standard R-Tree [3] to index the data items. However, RepAir can be used with any indexing structure as most $(1, m)$ techniques. We do not employ replicated and non-replicated parts, i.e. we index all data items in each of the m index segments. We order the data items according to the leafs of the R-Tree (from left to right) and split them into m groups such that the groups form segments of equal length (as equal as possible; length in terms of byte).

To generate coding packets for the RepAir schedule we employ a pseudo-random number generator $\zeta(seed)$ that produces equally distributed random numbers $rand \in [0, 1)$, which are pseudo-random and recomputable when the $seed$ is known. RepAir computes coding packets for each index segment and each data item separately. We will describe the procedure for an abstract set SP^0 of source packets which can either be the set of index packets constituting an index segment or the set of data packets constituting one data item. First we pick a random seed $seep_{SP^0}$ and compute a set CP^0 of coding packets with $|CP^0| = |SP^0|$. The computation is done in three steps:

1. Determine the degree deg_{cp} for each $cp \in CP^0$ using the robust soliton distribution and $\zeta(seed_{SP^0})$.
2. Compute cp_i by choosing $deg_{cp_i} - 1$ source packets $sp \in SP^0$ equally at random using $\zeta(seed_{SP^0})$ and combining them with sp_i through $\oplus$.
3. Check whether SP^0 is completely resolvable from CP^0. If not, go back to 2.

In the first step we use $\zeta(seed_{SP^0})$ to determine a random value $rand_i$ for each coding packet cp_i. We calculate the degree for cp_i by $deg_{cp_i} = \max\{deg|g(deg) < rand_i\}$ with $g(deg) = \sum_{j=1}^{deg-1} \mu(deg)$. Since $\mu(deg)$ according to Definition 3 is a probability distribution, $g(deg) \in [0, 1)$ holds for all $deg \in \mathbb{N}$.

Having CP^0 we compute additionally a set CP^+ of *blowup packets*, where $|CP^+| = |CP^0| \cdot BU$. BU is a blowup factor that can be set individually for index and data (c.f. Section 4). To provide sufficient yet resolvable information in the blowup packets, we aim at a degree of 8 for all $cp^+ \in CP^+$. However, we restrict their degree to be at most $\lceil |SP^0|/2 \rceil$ to facilitate resolving of coding packets for small sets of source packets. Hence, we set $deg(cp^+) = \min\{\lceil |SP^0|/2 \rceil, 8\}$ for all $cp^+ \in CP^+$. For construction we once again combine $deg(cp^+)$ many source packets $sp \in SP^0$ chosen equally at random using $\zeta(seed_{SP^0})$.

Finally, a RepAir broadcast is built by constructing a fixed number of air indexing schedules, e.g. four schedules AIS_0 to AIS_3 and sending them repeatedly in turns. (We discuss data and index updates in Section 5.) For each AIS a new $seed$ is chosen for each index segment and data item. The advantage of four different AIS lies in

Fig. 1. Abstract query model in mobile data dissemination

the fact that the content of the coding packets differs randomly and thus supports fast resolution with high probability. The header of each packet additionally contains the corresponding seeds $seed_v$ and $seed_{(v+1)\bmod 4}$ for the current and the next AIS. The seed values are crucial for query processing on the mobile client.

3.2   Query Processing in RepAir

Given the scheduling of data and index packets which are sent by the transmitter in an endless broadcast we have to consider query processing for the receiving mobile client. In Figure 1 we give an abstraction from recent query processing schemes. The aim is to describe the abstract steps of query processing together with the associated cost in terms of access latency and tuning time to identify potential for improvement.

1. Initialize Query State: find the next index packet using the pointer contained in the header informnation.
2. Index Found State: receive some index information to compute at least the first candidate for the query result.
3. Necessary Index Loaded State: compute the time slots of candidate data objects.
4. Next Data Item Loaded State: determine whether the query is fully answered. If there are still candidates or unpruned parts of the index left, go back to state 2 and find the next required index information. Otherwise the query is finished.
5. All Data Items Loaded State: the query is finished.

We focus on the marked (*) properties in Figure 1, as these are essential for an efficient query processing on error prone communication channels. Access latency and tuning time between state 3 and state 4 are highly depending on the error probability: For each data item that could not be received without errors the client has to wait for the next broadcast cycle and then tunes in again. The access latency while receiving the index information between state 2 and state 3 is also highly dependent on the error probability as each of the index packets has to be received without errors for a correct query processing. Our RepAir approach enhances both steps (3-4 and 2-3) by the novel cross-layer scheduling. As RepAir is able to reconstruct lost packets instead of waiting for the next broadcast cycle we largely reduce access latency. Tuning time is reduced by avoiding the download of false candidates during query processing (states 3-4).

To answer a query on a mobile device given a RepAir broadcast, we first tune into the broadcast to determine the arrival of the next index segment as described above. Next, we download coding packets from the broadcast that contain index information until we can reconstruct the entire index. After using the index to determine the correct set of data items answering our query, we download for each item corresponding coding

```
01 boolean packetIsUseful( list<Integer> encodedPackets ){
02    removeKnownPackets( encodedPackets );
03    if ( encodedPackets.size() == 1 ) return true;
04    if ( encodedPackets.size() == 0 ) return false;
05    // for all downloaded and not yet resolved coding packets ...
06    for ( CodePacket cpp : packetPool) {
07      if ( encodedPackets.size() < cpp.degree() ) {
08        if ( isSubset( encodedPackets, cpp ) ) return true;
09      } else if ( encodedPackets.size() == cpp.degree() ) {
10        if ( isEqual( encodedPackets, cpp ) ) return false;
11      } else {
12        // a superset is useful if the remainder is useful
13        if ( isSuperset( encodedPackets, cpp) )
14          return packetIsUseful( setDiff( encodedPackets, cpp );
15      }
16    } return true;
17 }
```

Fig. 2. Pseudo code of the method to decide whether a coding packet is useful

packets from the broadcast until we can fully reconstruct the item. This is done in parallel, i.e. we continue downloading *useful* coding packets for other data items even if we did not yet fully restore the previous ones due to too many packet losses.

Hence, we have to determine whether an arriving coding packet is *useful* for us. We describe the process again using a general set of source packets SP^0 constituting either an index segment or a data item of interest. Let $CP = CP^0 \cup CP^+$ be the next set of coding packets corresponding to SP^0. We know the packet id of the first coding packet in CP. After a packet loss, we can determine whether we are still reading a packet belonging to CP using the packet id and the number of corresponding coding packets which is contained in its header.

From the first coding packet that we successfully download we derive the seed $seed_{SP^0}$ for SP^0 in the current AIS as well as the seed for the next AIS. For any $cp \in CP$ we can then calculate the set $SP_{cp} \subset SP^0$ of source packets that are encoded in cp. Figure 2 shows our pseudo code to determine whether cp is interesting based on SP_{cp} (encodedPackets).

We first remove all known source packets from SP_{cp} to receive SP^*_{cp} in (line 2). If there is only one source packet left in SP^*_{cp}, cp is definitely interesting, because we can compute a previously unknown source packet right away (line 3). If there is no new information, we discard cp (line 4). From line 6 to 15 we look at all coding packets cpp corresponding to SP^0 that we have downloaded already but not yet fully resolved (packet pool). Therein we perform three simple checks for each of these coding packets cpp: If SP^*_{cp} is a subset of SP_{cpp}, cp is useful because we can decrease the degree of cpp by adding cp using $\oplus$ (line 08). If SP^*_{cp} and SP_{cpp} contain the same source packets, cp is redundant and therefore not useful (line 10). If SP^*_{cp} is a superset of SP_{cpp}, we can decrease the degree of cp by adding cpp as above. cp is then considered useful, if the remainder $SP_{cp} \setminus SP_{cpp}$ is useful (line 14).

By the three simple checks we just described, we can avoid downloading useless coding packets in many cases. However, there are still cases where a coding packet can be resolved via a linear combination of several known coding packets. To test all possible combinations would be too costly whereas the operations above can be done in linear time. Since we cannot preclude usefulness, we return `true` in case of doubts (line 16). When a coding packet cp is considered useful and downloaded successfully we decrease its degree as well as the degree of all coding packets in the packet pool as much as possible. If thereafter cp is not fully resolved we add it to the packet pool for further computation. Query answering stops when all requested data items are fully restored on the mobile client.

3.3 Discussion

In most of the scenarios introduced in Section 1 the size of one data item is usually larger than 10 kByte, e.g. audio or video clips or enhanced descriptive content for restaurants or touristic places. Moreover, a reasonably short access time is crucial for user acceptance. As an example, consider a very fast wireless connection of 300Kb/sec [7], a user who is willing to wait between 30 seconds and five minutes and an error prone channel that causes the access latency to be in the range of three broadcast cycles. This would restrict the size of a single broadcast cycle to range from 3Mb to 15Mb, e.g. 300 to 1500 data items of 10K each. This is a reasonable number of items for many scenarios such as information on restaurants or touristic places. With a possibly faster wireless connections in the future, the content and size of the single data items will grow as well due to multimedia enhancement.

A look at the size of the resulting index structure reveals a great opportunity to decrease both access latency and tuning time. Assuming 1000 objects of 15 Kb each, indexed in two dimensions using 4 byte floating point numbers, an R-Tree of fanout 10 and pointers of 4 byte, then the resulting R-Tree is of similar size as one data item, i.e. 15 Kb. Now the solution to reduce the tuning time is as simple as can be. If the complete index is received first, the true data objects can exactly be determined. Hence, receiving the complete index improves energy efficiency in comparison to approaches that have only distributed index information at hand. This concept will be strengthened in future even more when the item size grows due to media enhancement.

We consider the above mentioned assumptions as highly plausible and relevant for real application. However, we will show the performance of our approach for many different parameter settings in the next section to showcase the power of our solution.

4 Experiments

We test our RepAir technique against $(1, m)$ indexing with replicated and non-replicated index parts and against both the conservative and aggressive distributed air indexing approaches from [9]. We demonstrate the benefits of RepAir against the well established $(1, m)$ approaches which are achieved due to our cross layer scheduling incorporating forward error correction. For repeatability we used our evaluation framework [13], which can be downloaded from our website[1].

4.1 Setup

Table 1 summarizes the parameters that we investigate in the experiments and shows their variation as well as their standard value, i.e. if not mentioned explicitly, we use

---

[1] http://dme.rwth-aachen.de/repair

Table 1. Parameter variation and standard values

Parameter	Variation	std value	Parameter	Variation	std value
item size	64B - 1MB	10KB	m (RepAir & $(1,m)$)	4,16,32,64	16
#items	100 - 23000	1000	Index blow-up	0-50%	30%
data distribution	as described above	Greece	Data blow-up	0-50%	20%
item sizes	equal or skewed	equal	k **nearest neighbors**	1-20	5
packet size	64-2048 Byte	512 Byte	packet loss probability	0 - 70%	20%
			(corresp. Bit Error Rate)	$0 - 10^{-3}$	(BER $\approx 5 \cdot 10^{-5}$)

the standard values from Table 1. In each experiment we vary one parameter on the x-axis and report access latency and tuning time (plus sometimes the resulting broadcast size) on the y-axis. We use synthetic binary data of the required byte size to investigate equally sized data items. For skewed item sizes we took 1000 images from a gallery containing roughly 600 categories with 100 images each and resized them such that their average size fitted the required byte size. For indexing we use features from three different data distributions. First, we extracted from each image a 3-dimensional HLS color histogram. To investigate equally distributed data we generated 1000 coordinates that were equally random distributed in 2d space. Finally, we use real data representing 23,000 locations in Greece (if less items are used we take a random sample). As RepAir provides a general approach, higher dimensional features could also be used with an appropriate high-dimensional index structure. We use the R-tree as an exemplary index.

Since the resulting measures access latency (AL) and tuning time (TT) can vary depending on the random error, the tune-in time when the query is started, the query region etc., we conduct 1000 queries per parameter setting. We report the median of those 1000 values and provide boxplots on selected experiments. We report results of k nearest neighbor queries. As RepAir downloads the entire index anyway, the presented results can be extrapolated for range queries, since they just need a different descent in the received index structure.

4.2   Fault Tolerance

The main property of an Air-Indexing approach on error prone communication channel is its fault tolerance depending on the error rate of the underlying communication channel. We vary the packet error rate from 0 to 70 percent, which corresponds to a Bit Error Rate (BER) from 0 to 10^{-3} (realistic BER range from 10^{-6} to 10^{-3} [7]). As depicted in Figure 3(a), both variants of DSI do not perform well with increasing error rates. Especially the conservative approach has to load a lot of packets and thus spends the most energy for the query processing. Both variants of (1,m) perform better as they have a more complete index given and thus are more selective in tuning into the broadcast. With high error rates (1,m) has still to receive both index and data segments multiple times as packets are lost more frequently. Our RepAir approach overcomes this drawback by using the incorporated fountain codes to reconstruct lost packets.

Further differences between the approaches are depicted in Figure 3(b) where we measure access latency for each query. We see the contrary effect for the two DSI variants: The aggressive approach answers the query with reasonable low energy consumption (TT) but far to high waiting times (AL). As described in [9], the conservative

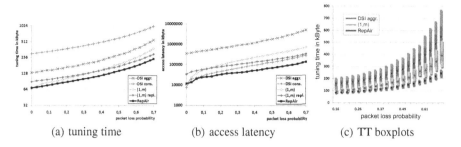

(a) tuning time   (b) access latency   (c) TT boxplots

Fig. 3. a-b: Variation of error probability. RepAir is outperformed only at zero error rate by $(1, m)$ and the conservative DSI due to the blow up overhead. c: Tuning Time boxplots.

approach shows exactly the contrary effect on the two measurements. For (1,m) variants we see similar performance on TT while reduced AL only for the replicated variant. Our RepAir approach shows best performance for both TT and AL with one exception. If no error occurs, the access latency of (1,m) and the conservative DSI is lower than the access latency of the RepAir approach, because RepAir has a larger broadcast size due to the blowup packets. (1,m) replicated and the aggressive DSI both have a higher access latency, because they miss correct candidates during the first broadcast due to their query processing approach.

For the 1000 runs mentioned before, we compare in Figure 3(c) maximum, 95%, median, 5% and minimum values of the tuning time for RepAir, DSI aggressive and (1,m). We can see that not only the median is increasing for higher error rates but also the variance of the results are dramatically increasing. Our RepAir approach shows not only better performance in the median but also the worst case tuning time is far better than for the competing approaches. Due to the cross-layer scheduling, RepAir achieves efficient query processing for both AL and TT and thus outperforms existing approaches.

4.3 Scalability

A general challenge for all index structures is their scalability w.r.t. the broadcast size. For Air-Indexing there are three major aspects that have an effect on the broadcast length: First, the size of each data item. Second, the number of data items, which also effects the size of the index. And third, the amount of index information, e.g. number of pointers in DSI or number of repetitions in $(1, m)$, as the index is transmitted in segments interleaved within the data base. In Figure 4 we see TT, AL and broadcast size (BCS). Only RepAir, (1,m) replicated and DSI aggressive show good scalability for bigger data bases resulting in low TT. DSI even outperforms RepAir for 10000 data items. However, the AL of this technique is unacceptably high as depicted in Figure 4(b) (please notice the logarithmic scale for AL).

Overall RepAir outperforms existing algorithms in terms of scalability w.r.t. the number of data items. The broadcast size of our RepAir approach is slightly bigger than for all other approaches (cf. Fig. 4(c)), as we increase the broadcast size by coding packets. However, as one can see in all other experiments, this overhead results in

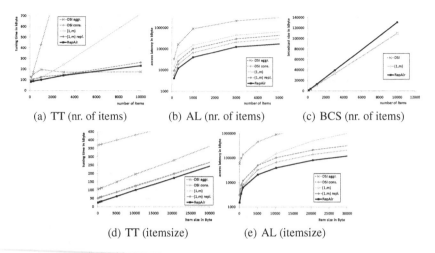

(a) TT (nr. of items)   (b) AL (nr. of items)   (c) BCS (nr. of items)

(d) TT (itemsize)   (e) AL (itemsize)

Fig. 4. Varying the data base size (number of items) (a-c) and the item size (d-e). (cf. Table 1 for remaining parameters)

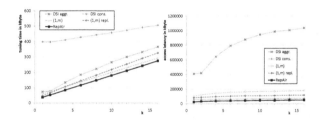

Fig. 5. Tuning time (left) and access latency (right) when varying the query size. (cf. Table 1 for remaining parameters).

a far more efficient query processing as depicted also w.r.t. variable data item size in Figure 4(d-e). Intelligent incorporation of coding packets thus ensures a significant improvement in terms of access latency and tuning time also for larger data bases and data items.

4.4   Variation of Query Size and Scenarios

Next we investigate the effect of the query size on AL and TT for the different approaches by varying k from 1 to 20. As stated above, we leave all other parameters at their standard value given in Table 1. Figure 5 shows the results for tuning time (left) and access latency (right). The tuning time increases nearly linear with each additional data item for all approaches. The increase of (1,m) replicated is a little steeper than (1,m) and RepAir's, since it receives more index information only if needed. The conservative DSI starts with a significantly higher offset at $k = 1$, because it downloads all candidate frames in broadcast order. Since other neighbors might be contained in those

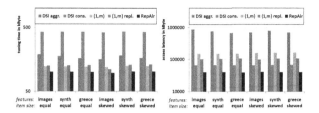

Fig. 6. Tuning time (left) and access latency (right) when varying the scenario. (cf. Table 1 for remaining parameters).

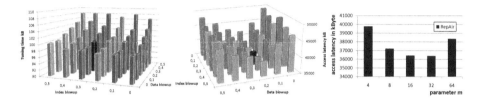

Fig. 7. Effect of blow-up parameters and m (right)

frames, the slope of its graph is significantly lower over k. Inspecting the performance w.r.t. k we show that RepAir outperforms all approaches in AL as well as in TT for any query size.

To test the approaches in different scenarios we tested equally sized items and skewed item sizes as well as three different data distributions in terms of the indexed features: 1) 3d image features (images), 2) equally random distributed 2d features (synth) and 3) 2d locations from greek cities (greece). For the skewed item sizes we used the original binary image data (images resized to 10kB average size) and for the equally sized items we generated 10kB of synthetical binary data (and kept the features as above). Figure 6 shows the resulting TT and AL for all six combinations. As can be seen, RepAir performs better than the competing approaches in all settings.

4.5   Parameter Discussion for RepAir

As we have seen in the previous experiments RepAir outperforms existing approaches due to its cross-layer scheduling. In this section we give a more detailed discussion about this cross-layer scheduling and its parametrization. As described in Section 3.1 our RepAir approach performs fountain coding with individual coding rates for index and data segments. In Figure 7 we show how AL and TT are affected by index and data blow-up. We highlighted the default setting (index blow-up | data blow-up) = $(0.3|0.2)$ which yields best TT and AL for RepAir.

Repair achieves best performance with 20% data blow-up. For lower blow-up rates we achieve a smaller broadcast cycle as we encode less redundant packet information. However, having higher blow-up rates RepAir ensures to receive data items the first time they pass by. Thus the overall access latency is optimal for a medium data blow up with small broadcast cycle but still enough blow-up to reconstruct lost data items.

The index blow-up is more robust as typically the index is relatively small comared to the data base. Thus we can set a higher index blow-up rate of 30% to ensure that the complete index is received without any packet loss.

We observe a similar effect (cf. Fig. 7 right) for the parameter m which denotes how many index segments are interleaved into the data base. For increasing m the broadcast size increases and thus also the AL. For low interleaving rates there are too few index segments and thus the mobile clients have to wait longer time until they can compute the query result. Both effects of setting the blow-ups and m show that one has to ensure small broadcast cycles that contain enough index information for an efficient query processing. Our cross-layer scheduling in the RepAir approach achieves this small but also fault tolerant broadcast by incorporating different blow-up levels for index and data segments.

5   Updates and Further Applications

In general if an update occurs, a new air indexing schedule has to be computed on the server side and broadcasted instead of the old one. However, there are two possible update scenarios with major differences, data updates and index updates. Data updates do only affect the content of the data item, e.g. a stock value or a temperature value. The index is not affected by such changes, i.e. the stock-id or the location-id of the weather measurement stays unchanged. Version bits have been proposed for such scenarios in mobile data dissemination [10] and can be used along with RepAir.

If the version of the index changes during query processing, the query has to be started over again. However, in most broadcasting scenarios (cf. Section 1) the intervals of updates are very long compared to the time needed to answer a user query. If an update occurs once a day or even once per hour but a query is answered within one or two minutes, it is very unlikely that those two actions coincide. In such scenarios one can safely assume static index information. However, for frequently changing index information such assumptions do not hold and further research has to be done in this area. In ongoing work we focus on an update-aware air index structure based on our general RepAir approach.

Air indexing in general is an upcoming research area. Recent publications investigate more specialized tasks such as approximate or continuous kNN queries [16] or valid scope computation for mobile application [8]. Our RepAir approach could be beneficial also in these specialized areas, as RepAir ensures to receive the complete index and thus has information about the overall data distribution.

6   Conclusion

We introduced RepAir, the first cross-layer air indexing approach. Our novel index scheduling includes error correction and adds individual redundancy according to the semantics of the content. With RepAir we propose an enhanced query processing avoiding false candidate downloads. It efficiently determines useful coding packets for reconstruction and is thus flexible for a broad range of error rates. Our simple yet effective approach outperforms competing techniques in both access latency and tuning time.

Thorough experiments show that our novel approach guarantees an energy efficient and fault tolerant data dissemination even for high error rates common in mobile communication. For repeatability and testing of further scenarios, we provide our evaluation software on our website.

Acknowledgments. This research was funded by the cluster of excellence on Ultrahigh speed Mobile Information and Communication (UMIC).

References

1. Beckmann, N., Kriegel, H.P., Schneider, R., Seeger, B.: The R*-tree: an efficient and robust access method for points and rectangles. In: SIGMOD, pp. 322–331 (1990)
2. Gedik, B., Singh, A., Liu, L.: Energy efficient exact KNN search in wireless broadcast environments. In: GIS, pp. 137–146 (2004)
3. Guttman, A.: R-trees: A dynamic index structure for spatial searching. In: SIGMOD, pp. 47–57 (1984)
4. Im, S., Song, M., Hwang, C.S.: An error-resilient cell-based distributed index for location-based wireless broadcast services. In: MobiDE, pp. 59–66 (2006)
5. Imielinski, T., Viswanathan, S., Badrinath, B.R.: Energy efficient indexing on air. In: SIGMOD, pp. 25–36 (1994)
6. Imielinski, T., Viswanathan, S., Badrinath, B.R.: Data on air: Organization and access. TKDE 9(3), 353–372 (1997)
7. Laiho, J., Wacker, A., Novosad, T.: Radio Network Planning and Optimisation for UMTS. Wiley, Chichester (2005)
8. Lee, K.C.K., Schiffman, J., Zheng, B., Lee, W.C.: Valid scope computation for location-dependent spatial query in mobile broadcast environments. In: CIKM (2008)
9. Lee, W.C., Zheng, B.: DSI: A fully distributed spatial index for location-based wireless broadcast services. In: ICDCS, pp. 349–358 (2005)
10. Lo, S.C., Chen, A.L.P.: An adaptive access method for broadcast data under an error-prone mobile environment. TKDE 12(4), 609–620 (2000)
11. Luby, M.: LT codes. In: FOCS, pp. 271–280 (2002)
12. MacKay, D.: Fountain codes. Communications, IEE 152(6), 1062–1068 (2005)
13. Müller, E., Kranen, P., Nett, M., Reidl, F., Seidl, T.: A general framework for data dissemination simulation for real world scenarios. In: MOBICOM (2008)
14. Xu, J., Zheng, B., Lee, W.-C., Lee, D.L.: Energy efficient index for querying location-dependent data in mobile broadcast environments. In: ICDE, pp. 239–250 (2003)
15. Xu, J., Zheng, B., Lee, W.C., Lee, D.L.: The D-Tree: An index structure for planar point queries in location-based wireless services. TKDE 16(12), 1526–1542 (2004)
16. Zhang, X., Lee, W.C., Mitra, P., Zheng, B.: Processing transitive nearest-neighbor queries in multi-channel access environments. In: EDBT, pp. 452–463 (2008)
17. Zheng, B., Xu, J., Lee, W.C., Lee, D.L.: Grid-partition index: a hybrid method for nearest-neighbor queries in wireless location-based services. VLDB J. 15(1), 21–39 (2006)

Content-Based Multipath Routing for Sensor Networks

Inchul Song, Yohan J. Roh, and Myoung Ho Kim

Computer Science, KAIST, Korea
{icsong,yhroh,mhkim}@dbserver.kaist.ac.kr

Abstract. In wireless sensor networks, in-network processing of aggregation queries has been an important technique to reduce energy consumption in wireless communication, which is a main source of energy consumption in sensor devices. In-network processing is typically guided by an *aggregation tree*, where each node forwards partially computed aggregates to its parent. In this paper we consider a routing method for grouped aggregation queries, where sensor readings are divided into disjoint groups according to their values and aggregates are computed for each group. For this type of queries, multipath routing, where each node forwards different aggregates to different nodes, can lead to more efficient in-network processing. However, no multipath routing protocol for efficient in-network processing of grouped aggregation queries has been proposed thus far. In this paper we propose a new routing protocol, called Content-based Multipath Routing (CMR), for efficient in-network processing of grouped aggregation queries. CMR employs multipath routing, and each node forwards partially computed aggregates along different paths based on the *contents* of the aggregates. The experimental results show that CMR outperforms the existing aggregation tree-based routing protocols.

1 Introduction

The advent of wireless sensor networks gives us opportunities unattainable before [1]. They can be deployed in many places such as buildings, manufacturing plants and habitats, and can provide timely and accurate information about environmental conditions. Wireless sensor nodes acquire the status of the environment from different kinds of sensors, and through wireless multihop networking, deliver it to the base station, where users request queries and receive answers.

Energy efficiency is of utmost importance in sensor networks because sensor nodes have limited power. It is known that wireless communication among sensor nodes is a main source of energy consumption [2]. In-network processing is a widely accepted technique to reduce energy consumption in wireless communication in many sensor network applications [3][4]. For aggregation queries such as SUM, AVERAGE, MAX, etc., aggregates are computed *in-network* whenever possible. For example, for a MAX aggregation query, an intermediate node may forward only the maximum value among the sensor readings received from its neighbors, to the next hop node.

H. Kitagawa et al. (Eds.): DASFAA 2010, Part I, LNCS 5981, pp. 520–534, 2010.

This paper is particularly concerned with *grouped aggregation queries*, which divide sensor readings into disjoint groups according to their values and compute aggregates for each group. Consider the following query, "Report the average volume of each room in a building", for monitoring the occupancy of the rooms. This query partitions sensor readings into different groups according to the locations where they are acquired and reports the average volume of each group, i.e., of each room.

For efficient in-network processing of grouped aggregation queries, in-network processing is typically guided by an *aggregation tree*, which is a spanning tree rooted at the base station. In the aggregation tree each node is assigned the distance from the base station, or *level*. In-network processing proceeds *level-by-level* from the farthest nodes toward the nodes near the base station: When the nodes at level l are sending messages, those at level $l - 1$ are receiving. Before sending a message, a node performs in-network processing with its own sensor readings and the partially computed aggregates received from its child nodes and produces new aggregates, one for each group. It then sends these aggregates to its parent.

Although in-network processing based on the aggregation tree has been commonly used because of its simplicity, it may not be the best way to forward aggregates for grouped aggregation queries. For example, consider three different routing approaches for collecting sensor readings that are divided into two groups g_1 and g_2, shown in Figure 1. In the figure, an edge between two nodes indicates that the two can communicate with each other. Black- and grey-colored nodes are the nodes that need to send their sensor readings to the base station, which is connected to node h. The nodes a and d produce sensor readings belonging to group g_1 and the nodes b and e in group g_2. Arrows indicate message transmissions and explosion shapes indicate the places where in-network processing occurs.

In the three routing approaches, the difference lies in the way that node c forwards its aggregates. In the routing approaches of Figure 1a and 1b, which are based on the aggregation tree, node c forwards all its aggregates to its parent, i.e., to node d in Figure 1a and to node e in Figure 1b, respectively. In Figure 1c, however, node c forwards the two aggregates to different nodes: The aggregate in group g_1 to node d and that in group g_2 to node e. Let us, for the sake of brevity, assume that a single message can contain only a single aggregate. Then, the routing approaches in Figure 1a and 1b require 10 message transmissions (count the number of arrows), whereas the routing approach in Figure 1c needs only 8 message transmissions.

As shown in the previous example, the aggregation tree-based routing approach, or single-path routing, where each node forwards all its aggregates to its parent, may not be the best routing strategy for grouped aggregation queries. Nevertheless, as far as we are aware of, no routing protocol that is based on multipath routing has been proposed for efficient in-network processing of grouped aggregation queries.

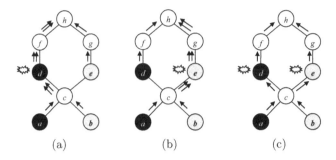

Fig. 1. Three different routing approaches

In this paper we propose a new routing protocol, called Content-based Multi-path Routing (CMR), for efficient in-network processing of grouped aggregation queries. In CMR each node forwards partially computed aggregates along different paths based on the *content* of the aggregates. In other words, it examines the group to which each aggregate belongs and conducts routing in such a way that in-network processing is likely to occur among aggregates in the same group. The experimental study shows that CMR outperforms the existing routing protocols based on the aggregation tree in terms of energy consumption in wireless communication for collecting the results of grouped aggregation queries. Our earlier work [5] was concerned with a routing method for aggregation queries with no grouping. This work extends it for grouped aggregation queries.

The rest of the paper is organized as follows. Section 2 discusses the background of this work and presents related work. Section 3 defines necessary terminology and notations and then describes CMR in detail. In Section 4 we compare the performance of CMR with those of the existing routing protocols that are based on the aggregation tree. Finally, Section 5 concludes the paper.

2   Background

2.1   Query Model

A sensor network can be modelled as a distributed database. Each node generates tuples for the distributed table, named **sensors**, which has one attribute per input of the nodes (e.g., temperature, light, humidity, etc.) A user of a sensor network can query this **sensors** table by using an SQL-like query language.

Consider the following grouped aggregation query that monitors the occupancy of the rooms on the sixth floor in a building, by using microphone sensors attached to sensor nodes and looking for rooms where the average volume is over some threshold [6]:

```
Q1: SELECT AVG(volume), room
    FROM sensors
    WHERE floor = 6
    GROUP BY room
```

```
HAVING AVG(volume) > threshold
EVERY 60s
```

This query first partitions rooms on the sixth floor according to their locations and reports, every 60 seconds, all rooms where the average volume exceeds a specified threshold.

The semantics of grouped aggregation queries is the same as SQL aggregation queries except the EVERY clause (for a detailed description of its semantics, refer to [3]). We consider only standard SQL aggregation functions (AVG, SUM, MIN, MAX, and COUNT) in this paper. In what follows, when we say a 'query', it means a grouped aggregation query, unless otherwise specified.

2.2 In-Network Aggregation Query Processing

Query processing in sensor networks generally consists of two phases. In the *distribution* phase, a query is propagated from the root node, which is connected to the base station, down to all the nodes in the network through some kind of message flooding. In the *collection* phase, sensor readings satisfying the conditions of the query are collected to the base station through some routing protocol. In-network processing is performed whenever possible during the collection phase. The standard SQL aggregation functions can all be computed in-network [7], and how to compute them in-network can be found in [3].

In general, regardless of whether using an aggregation tree or multipath routing, construction of a routing structure that coordinates the operation of in-network processing proceeds similarly as follows:

- *Distance determination*: First, the root node prepares a routing message and records its distance—it is simply zero—on the message and broadcasts it. When a sensor node other than the root node receives a routing message from one of its neighbors and it has not yet decided its distance, it sets its distance to the distance recorded in the routing message plus one and marks the sender as its *candidate parent*. Next, it records its distance on the routing message and broadcasts it. Later, if the node receives a routing message from the node whose distance is one less than its distance, it marks the node as its candidate parent, but this time discards the message. Finally, for other cases, the node just ignores an incoming routing message. In this way, every node in the network discovers its distance from the root node and its candidate parents, which are located one-hop closer to the root node than itself.
- *Structure construction*: In routing protocols based on the aggregation tree, each node selects one of its candidate parents as its parent, to form a routing tree. In multipath routing, this step is omitted.

In-network processing for grouped aggregation queries in the collection phase proceeds as follows. When the nodes at distance d are sending messages, the nodes at distance $d - 1$ are receiving. Before a node sends a message, it combines the aggregates received from its neighbors and its own sensor readings, if available, into new aggregates, producing one aggregate per group. In in-network processing based on the aggregation tree, each node forwards all its newly generated aggregates to its parent. In multipath routing, however, each node decides to which candidate parent to send each aggregate on the fly; thus, different aggregates could be sent to different candidate parents.

2.3   Related Work

It has been shown that the problem of finding the minimum cost routing tree for result collection of grouped aggregation queries is NP-complete [8]. Thus heuristic approaches are generally used for constructing routing trees for efficient in-network processing of grouped aggregation queries in sensor networks.

TAG is one of the most commonly used routing protocols in sensor networks [3][4][7][6]. In TAG, the construction of a routing tree proceeds as described in Section 2.2. The difference is that, in the structure construction step, each node selects the candidate parent from which it received the routing message for the first time as its parent. Note that TAG is a general purpose routing tree that can also be used for grouped aggregation queries.

The Group-aware Network Configuration (GaNC) [9] is specially designed for grouped aggregation queries. When a user poses a query, in the distance determination step, each node additionally records in the message the query and the group to which its sensor readings will likely belongs, based on the specification of the query. In the structure construction step, each node selects as its parent the candidate parent that reported the same group as its group. This may increase the possibility of in-network processing among aggregates belonging to the same group. Unlike GaNC, our proposed method considers the group information of not only its neighbors, but also those residing on the possible paths to the base station.

The Leaves Deletion (LD) algorithm proposed in [8] first constructs a shortest path tree (SPT), where each node connects to the base station in the minimum hops. In fact the SPT constructed in this way is almost the same as the routing tree constructed by TAG. Then, each leaf node changes its parent several times so long as more efficient in-network processing can be achieved.

The routing protocols aforementioned are all tree-based protocols in which each node forwards all its aggregates to its parent. As noted in Section 1, tree-based routing may not be the best routing strategy for grouped aggregation queries. No multipath routing protocol, however, has been proposed for efficient in-network processing of grouped aggregation queries. Although there are some multipath routing protocols, their goal is typically either to increase fault resilience by duplicating aggregates and propagating them along multiple paths [7][10][11], or to support fast recovery by replacing paths in the middle of message propagation in the case of link failures [12].

3   Content-Based Multipath Routing

In this section we present a multipath routing protocol called Content-based Multipath Routing (CMR) that is designed for efficient in-network processing of grouped aggregation queries. CMR decides at each node how to forward aggregates to different candidate parents. The key idea of CMR is that each node forwards aggregates in such a way that in-network processing among aggregates belonging to the same group is performed more frequently and early. Note that frequent in-network processing reduces the data volumes that need to be transferred inside a sensor network. Early in-network processing, where aggregates are soon merged after generation, is preferable than late in-network processing because of its cumulative effect. For example, for a MAX query, suppose that a node needs to forward two values received from its neighbors. Then, it will throw away the smaller one and sends only the larger one to the next hop node. The benefit of not sending the smaller one will be accumulated at every hop up to the base station. The farther the node is located from the base station, the larger the accumulated benefit.

The operation of CMR is based on a distance metric called the *minimum mergeable distance* (MD). Roughly speaking, given a grouped aggregation query, each node computes its MD value per group. The MD value of a node for a group indicates the distance to the closest node that generates sensor readings belonging to the group. The MD values computed are exchanged in a distributed fashion in such a way that each node discovers and stores the MD values of all its candidate parents. In the collection phase, given an aggregate in some group, a node forwards the aggregate to one of the candidate parents that has the smallest MD value for the group. This routing policy will not only increase the occurrences of in-network processing because each node forwards partial aggregates toward some node that generates sensor readings belong to the same group, but also promote early in-network processing because partial aggregates are forwarded toward the closest such node.

In what follows, we first formally define the minimum mergeable distance (MD) and then describe how to compute and exchange MD values in a decentralized manner in a sensor network. Then, we explain how each node forwards partial aggregates to its candidate parents based on their MD values.

3.1   Minimum Mergeable Distance

We model a sensor network as an undirected graph $G = (V, E)$, where V is a set of nodes and E is a set of edges. There is a distinguished node v_0, called the root node, which is an ordinary sensor node or possibly can be the base station. The root node and all other sensor nodes are members of V. An edge (v_i, v_j) is in E if two nodes v_i and v_j can communicate with each other. Figure 2a shows a graph for a sensor network with nine sensor nodes.

The *distance* from v_i to v_j in graph G, denoted by $d_G(v_i, v_j)$, is the length of any shortest path between the two nodes. The distance from the root node v_0 to v_i is simply called the "distance of v_i" and is denoted by $d_G(v_i)$. $d_G(v_0) = 0$

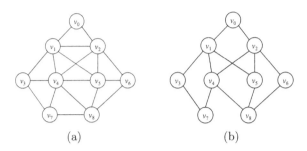

(a)   (b)

Fig. 2. A sensor network graph and its stratified routing graph

by definition. The *stratified routing graph* $S = (V, E')$ of a graph $G = (V, E)$ is a subgraph of G, where an edge $(v_i, v_j) \in E$ is in E' if and only if $|d_G(v_i) - d_G(v_j)| = 1$. The stratified routing graph is also called the *ring topology* in the literature [10][11]. Figure 2a and 2b show a graph and its stratified routing graph, respectively.

Next, we define a support relationship between nodes. Let $S = (V, E)$ be a stratified routing graph. We define a relation $\rightarrow_S$ on V by:

$$\rightarrow_S = \{< v_i, v_j > \mid (v_i, v_j) \in E \text{ and } d_S(v_j) < d_S(v_i)\}.$$

If a pair $< v_i, v_j >$ is in $\rightarrow_S$, we write $v_i \rightarrow_S v_j$. We will omit the subscript S in relation $\rightarrow_S$ when the context is clear. The transitive closure of $\rightarrow$ is denoted by $\rightarrow^+$. We say that v_j is a *support node* of v_i if $v_i \rightarrow^+ v_j$. If $v_i \rightarrow v_j$, we say that v_j is a *direct support node*, or *DS-node*, of v_i. In Figure 2b, node v_0 is the support node of every other node, that is, $v_i \rightarrow^+ v_0$, where $i = 1, \dots, 8$. The support nodes of v_7 are v_0, v_1, v_2, v_3, and v_4, whereas its DS-nodes are v_3 and v_4. There is no support relationship between v_3 and v_8. The support relationship $v_i \rightarrow^+ v_j$ indicates that a path from node v_i to node v_j, through which messages are forwarded toward the root node in the collection phase, can be established.

Given a grouped aggregation query, a node is called a *qualifying node*, or *Q-node*, if it satisfies the conditions in the WHERE clause of the query. For query Q1 in Section 2.1, Q-nodes are the nodes on the sixth floor. Each Q-node is assigned a group ID based on its values of the attributes appeared in the GROUP BY clause of the query. The group ID of a Q-node v is denoted by $group(v)$. For query Q1, each Q-node is assigned the room number where it is placed as its group ID.

We are now ready to define the minimum mergeable distance. Given a grouped aggregation query q and a stratified routing graph $S = (V, E)$, the *minimum mergeable distance* (MD) of a node v_i for a group g, denoted by $\text{MD}_{q,S}(v_i, g)$, is defined as follows:

(1) 0, if node v_i is a Q-node for query q and $group(v_i) = g$,
(2) $\min\{d_G(v_i, v_j) \mid v_i \rightarrow^+ v_j, v_j$ is a Q-node for query q, and $group(v_j) = g\}$, if node v_i is not a Q-node, and there exists one or more such v_j,
(3) $d_G(v_i)$, otherwise.

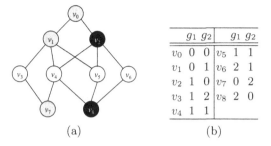

	g_1	g_2		g_1	g_2
v_0	0	0	v_5	1	1
v_1	0	1	v_6	2	1
v_2	1	0	v_7	0	2
v_3	1	2	v_8	2	0
v_4	1	1			

(a)   (b)

Fig. 3. Q-nodes for two groups and minimum mergeable distances

For brevity, we will use $MD(v_i, g)$ instead of $MD_{q,S}(v_i, g)$ if there is no ambiguity. $MD(v_i, g)$ represents the distance from node v_i to the closest Q-node that supports v_i and belongs to group g. Figure 3a shows a stratified routing graph with nine sensor nodes and Q-nodes for some query. In the figure, grey-colored nodes are Q-nodes for group g_1, and black-colored nodes are Q-nodes for group g_2. Figure 3b shows the MD values of each node for groups g_1 and g_2.

Previously, for simplicity of explanation, we defined a Q-node to be a node that satisfies the conditions of a query. However, for a condition such as `temp > 20`, even though a node satisfies this condition at one time, it may not satisfy it at another time. On the other hand, for a condition such as `floor = 2`, if a node satisfies this condition, it will always or likely satisfy this condition during the lifetime of the query. We call a condition of the first type a *dynamic condition*, and that of the second a *static condition*. Because queries may have a dynamic condition in general, we redefine a Q-node to be a node that satisfies all the static conditions of the query. The group of a Q-node is also determined based only on the values of those attributes that may appear in a static condition.

Let v be a node that is not a Q-node for some query. Then, the MD of node v for group g can be calculated as follows:

$$MD(v, g) = \min\{MD(w, g) \,|\, v \to w\} + 1 \tag{1}$$

Equation 1 indicates that the MD of node v for some group can be calculated based solely on the MDs of its DS-nodes for the group. This property enables the implementation of an efficient distributed MD calculation algorithm in a sensor network, which we will describe in the next section.

3.2 Distributed MD Computation

Distributed computation of MD values consists of the following two steps: *construction of stratified routing graph* and *computation of MD*. The construction of stratified routing graph step is the same as the distance determination step described in Section 2.2. Through this step, each node discovers its distance from the root node and its DS-nodes. To reflect the change of network topology,

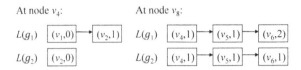

Fig. 4. Group-MD lists at nodes v_4 and v_8

we can execute this step periodically, or whenever the user receives less than a user-specified number of tuples within the specified sample period, as suggested in [4].

The computation of MD step is executed when the user poses a query. In this step, the user query is delivered to every node in the network, the MD of each node is calculated, and each node discovers the MD values of its DS-nodes. This step consists of the following three sub-steps:

– *Collection*. Each node collects the MDs that are broadcast by its DS-nodes and organizes these MDs into a collection of *group-MD lists*, denoted by L. Given a group g, $L(g)$ denotes the list of pairs of a DS-node and its MD for group g, sorted in ascending order of MD. Note that the DS-node that has the smallest MD for group g can be found in the the first entry of the list $L(g)$.
– *Computation*. After receiving the MDs of all its DS-nodes, each node calculates its MDs based on Equation 1. There are two cases to consider. 1) Let us first assume that it is not a Q-node. Then, for each group g in L, its MD for the group is the MD of the DS-node found in the first entry of $L(g)$, plus one. 2) Suppose now that it is a Q-node for some group. Then, its MD for the group is zero, and its MDs for the other groups are calculated as in case 1).
– *Notification*. After calculating its MDs, each node broadcasts them to its neighbors.

The root node initiates the computation of MD step by computing and broadcasting its MD, bypassing the *Collection* sub-step. After that, the execution of this step propagates down to the network. Figure 4 shows how the collections of group-MD lists are populated at nodes v_4 and v_8 after executing this step in Figure 3a.

3.3   The Aggregate Forwarding Algorithm

In this section we describe how each node forwards aggregates to its DS-nodes based on their MD values. The basic idea is that a node forwards each aggregate in some group to the DS-node that has the smallest MD for the group, to promote frequent and early in-network processing.

When forwarding aggregates in this way, a node has to make a good choice about how many number of messages to use. Since a per-message energy cost, regardless of the size of the message, is very high [13], forwarding each aggregate in a separate message may not be the best choice: This may generate too many messages. On the other hand, if a node uses as smaller a number of messages as possible, then some aggregate may end up being contained in a message not targeted to the DS-node that has the smallest MD for the group of the aggregate. Since the optimal number of messages depends on various factors and may not be easily formulated, in this paper we focus only on reducing the energy consumption induced by per-message cost. Thus, in our aggregate forwarding algorithm, a node uses the minimum number of messages. Accordingly, the problem of forwarding aggregates at each node can be formulated as follows.

Suppose that a node has n aggregates to forward, and each message can contain at maximum p aggregates. Then, the node will use $k = \lceil n/p \rceil$ messages to forward its aggregates. A node in our aggregate forwarding algorithm must decide 1) how to distribute n aggregates over k messages and 2) to which DS-node to send each message. Since the number of ways for distributing n aggregates over k messages and deciding the target of each message may be overwhelmingly large, our method tackles this problem heuristically and consists of two steps. In the *cluster generation* step, a node groups aggregates into a number of clusters, each of which contains those aggregates that will go into the same message. In *the cluster assignment* step, the node assigns these clusters to k messages, one by one in some fixed order, and then the node decides to which node to send each message. In what follows, we describe these two steps in more detail.

Cluster generation. Given an aggregate a in some group, the DS-node whose MD for the group is smallest is called the *best DS-node* of the aggregate and is denoted by $best\_DS\_node(a)$. Recall that, given an aggregate a, $best\_DS\_node(a)$ can be found in the first entry of the group-MD list for the group of the aggregate, i.e., $L(group(a))$. $best\_DS\_node(a)$ can be NIL if $L(group(a))$ is empty. In this step a node groups those aggregates with the same best DS-node into a cluster. And the aggregates whose best DS-node is NIL form another cluster. We define the *best DS-node* of a cluster C, denoted by $best\_DS\_node(C)$, to be the best DS-node of any aggregate in the cluster.

Cluster assignment. In this step a node assigns the clusters formed in the previous step to k messages and decides to which DS-node to send each message. Depending on the size of a cluster (the number of aggregates in it), the cluster might need to be assigned to one or more messages. For example, if the size of the cluster is smaller than the size of a message, then we need only one message to accommodate the aggregates in the cluster. On the other hand, if the cluster size exceeds the message size, then we need more than one message for the cluster. Even when the cluster size is smaller than the message size, if the message is already filled with the aggregates from another cluster, then the cluster still might need to be distributed over several messages.

Before we describe how to assign clusters to messages in detail, let us explain how to decide the target of each message first. If a message contains the aggregates from only one cluster, a node simply sends it to the best DS-node of the cluster, because this way the aggregates in the cluster will be sent to their best DS-node. If a message contains the aggregates from more than one cluster, then a node sends it to the best DS-node of the cluster that were first assigned to this message (the order in which clusters are assigned to a message will be explained later). Note that the clusters other than the cluster that were first assigned to the message may not be sent to their best DS-nodes. The target of a message m determined this way is denoted by $target(m)$.

Now we describe how to assign clusters to messages. As noted in the previous paragraph, the order of assigning clusters to messages needs to be carefully determined, because we can end up with some clusters that are not sent to their best DS-nodes. In order to determine the order of assigning clusters to messages, a node examines the benefit of sending each cluster to its best DS-node. If a node sends an aggregate a in a cluster to its best DS-node, the aggregate will arrive at a Q-node that belongs to $group(a)$, after the MD hops of its best DS-node, i.e, after $\text{MD}(best\_DS\_node(a), group(a))$ hops, by the definition of MD. From there on, the aggregate will have a free ride and not have to be transferred due to in-network processing. To capture this notion of free riding, we define the *benefit* of an aggregate a, denoted by $benefit(a)$, to be the number of hops of free riding by sending it to its best DS-node, which is computed as follows:

(1) $benefit(a) = 0$, if $best\_DS\_node(a)$ is NIL,
(2) $benefit(a) = d(v) - \text{MD}(best\_DS\_node(a), group(a))$, otherwise.

In other words, if there exists no best DS-node, then the aggregate will likely arrive at the root node without in-network processing; thus, no free riding and no benefit. On the other hand, if there exists its best DS-node, the aggregate will enjoy free riding, starting after the MD hops up to the root node. Similarly, we define the *benefit* of a cluster C_i, denoted by $benefit(C_i)$, to be the average of the benefits of the aggregates in it:

$$benefit(C_i) = \frac{\sum_{a_i \in C_i} benefit(a_i)}{|C_i|}.$$

Note that the benefit of the cluster whose best DS-node is NIL—the best DS-node of all its aggregates is also NIL—is always the minimum.

In summary, in this step a node processes clusters in decreasing order of their benefits and sets the target of each message to the best DS-node of the cluster that is first assigned to the message. Algorithm 1 describes this step more formally.

In Algorithm 1, a node first sorts clusters in decreasing order of their benefits. It then processes each cluster one by one (line 2), handling the cluster C_{NIL} whose best DS-node is NIL separately at the end. For each cluster, it assigns each aggregate a in the cluster to a message, allocating a new message as needed (line 3). It first attempts to find a message whose target is $best\_DS\_node(a)$, i.e., a message that is already allocated and targeted to $best\_DS\_node(a)$ (line 4). If there is such a message with room (line 5), it places the aggregate in the message. If not, it attempts to allocate a new message (line 8). If this is possible,

---

Algorithm 1. Cluster assignment

Input: C: a set of clusters, C_{NIL}: the cluster whose best DS-node is NIL, k: the number of messages.

Output: M: a set of messages, initially $\emptyset$.

1   Sort C in decreasing order of cluster benefit;

2   **foreach** *cluster* $C_i \in C \setminus C_{\mathrm{NIL}}$ **do**

3   **foreach** *aggregate* $a \in C_i$ **do**

4   Find a message msg from M whose target is *best_DS_node*(a);

5   **if** *Such message with room found* **then**

6 | msg $\leftarrow a$; **return**;

7   **end**

8   **if** $|M| < k$ **then**

9   Allocate a new message msg;

10   *target*(msg) $\leftarrow$ *best_DS_node*(a);

11   msg $\leftarrow a$; $M \leftarrow$ msg; **return**;

12   **end**

13   Find any message msg with room from M; msg $\leftarrow a$;

14   **end**

15 **end**

16 **foreach** *aggregate* $a \in C_{\mathrm{NIL}}$ **do**

17   Find any message msg with room from M; msg $\leftarrow a$;

18 **end**

---

it sets the target of the newly allocated message to *best_DS_node*(a) and puts the aggregate into the message. If it is impossible to allocate a new message, it finds any message with room and places the aggregate there (line 13). Finally, the node puts each aggregate in the cluster C_{NIL} into any message with room (line 16 to 18).

4   Performance Evaluation

We conducted various experiments with our own simulator to compare the performance of our approach with those of the existing routing protocols. The existing routing protocols considered in the evaluation are TAG, GaNC, and LD, described in Section 2.3. We modified LD appropriately to adapt it to our setting. Originally, LD changes the parent of each node several times as long as there is a gain, which can defined differently case by case. In our setting, the parent of each node is changed to only a Q-node and is not changed any more if the changed parent is in the same group as its child node.

As performance metric, we use the amount of energy consumed for wireless communication in collecting the result of a grouped aggregation query in a single sample period. We model per-message energy consumption by the following linear model proposed in [13]:

$$Energy = m \times size + b,$$

Table 1. Default parameters

Parameter	Default value	Parameter	Default value
Network size (m^2)	600×600	Message size	1
Grid size	50×50	Number of groups	8

where b is a fixed component associated with device state changes and channel acquisition overhead, and $m \times size$ is an incremental component which is proportional to the size of the message. We adjusted the coefficients appropriately by linear regression for 802.15.4 radio hardware (such as the CC2420, used in the Telos and micaZ platforms) with 250 kbps data rate. We set m to 11.4 $\mu J/byte$ and b to 487 μJ for sending point-to-point traffic, and 3 $\mu J/byte$ and 414 μJ for receiving point-to-point traffic, respectively. We assume that a message consists of a 11-byte header and 28-byte payload (see TinyOS Enhancement Proposals 111[1]).

In various simulation experiments, sensor nodes are deployed in a rectangular area whose width and height are set to the same length. The default size of the area is $600m \times 600m$. The sensor nodes are arranged into a grid of cells whose default size is 50×50. Each sensor is placed at the center of each grid cell and can communicate with the sensors placed adjacent grid cells: Thus, a sensor node can communicate with at maximum nine sensor nodes. Some nodes at the edges of the grid may have three or five neighbor sensor nodes. The parameter, *message size*, whose default value is one, indicates how many aggregates a single message can contain. A grouped aggregation query divides sensor readings into a number of groups, whose default value is set to 8 groups in the evaluation. Table 1 summarizes the default values of the parameters used in the simulation. We place the base station at the leftmost, uppermost cell of the grid.

In the evaluation, we assume that wireless communication is lossless: That is, a node successfully receives all the messages from other nodes. In addition, we assume that all sensor nodes produce sensor readings that satisfy the conditions of the query. Each sensor reading belongs to a certain group with equal probability of belonging to any group. In all experiments, we have executed each simulation 10 times and computed the average. In all graphs that follow, the y axis indicates the relative cost, i.e., the energy consumption of a routing protocol over that of TAG, which is one of the most commonly used routing protocols. For example, the value 0.8 means the routing protocol consumes only 80% amount of energy compared to TAG, or reduces energy consumption by 20% compared to TAG.

Effect of Number of Groups. In this experiment, we investigate the effect of the number of groups in a query on the performances of the routing protocols. Figure 5a shows the results of the experiment. When the number of groups is one, i.e., all aggregates are in the same group, the three routing protocols show the same performances because in-network processing occurs at every node in

---

[1] http://www.tinyos.net/

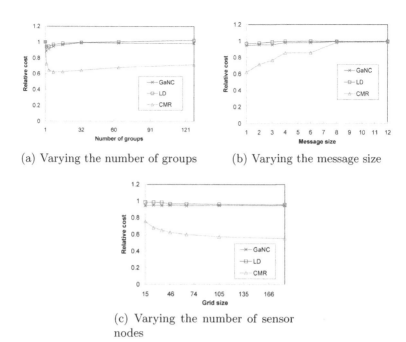

(a) Varying the number of groups   (b) Varying the message size

(c) Varying the number of sensor nodes

Fig. 5. Experimental results

all routing protocols. As the number of groups increases, the proposed method, CMR, outperforms the other routing protocols because of its frequent and early in-network processing effects. As the number of groups becomes larger and larger, its performance slowly degrades but is still better than the other routing protocols. This is because, as the number of groups increases, the number of nearby nodes belonging to the same group decreases, so do the chances of frequent and early in-network processing in CMR.

Effect of Message Size. In this experiment, we evaluate how the message size affects the performances of the routing protocols. The results of this experiment are shown in Figure 5b. When the message size is one, CMR shows the best performance since every aggregate is contained in a separate message, and thus can be sent to its best DS-node. As the message size increases, however, its performance decreases, because some aggregate may end up being contained in a message that is not targeted to its best DS-node.

Effect of Number of Sensor Nodes. This experiment is designed to evaluate the effect of the number of sensor nodes on the performances of the routing protocols by varying the number of sensor nodes from 225 (grid size 15 × 15) to 34225 (grid size 185 × 185). Figure 5c shows the results of the evaluation. As the number of sensor nodes increases, the reduction in energy consumption of the proposed method slowly increases, since the cumulative effect of frequent and early in-network processing in our method also increases.

5   Conclusions

In this paper we have proposed a content-based multipath routing protocol called CMR for efficient in-network processing of grouped aggregation queries. CMR employs multipath routing and forwards aggregates based on the contents of aggregates by means of a distance metric called minimum mergeable distance, to promote frequent and early in-network processing. The experimental evaluation shows that the proposed method outperforms the existing routing protocols. As future work, we plan to perform further experiments to evaluate our proposed method in more realistic sensor network environments.

Acknowledgment

This work was supported by the National Research Foundation of Korea (NRF) grant funded by the Korea government (MEST) (No. 2009-0083055).

References

1. Hill, J.L., Culler, D.E.: Mica: A wireless platform for deeply embedded networks. IEEE Micro 22(6), 12–24 (2002)
2. Zhao, F., Guibas, L.: Wireless Sensor Networks: An Information Processing Approach. The Morgan Kaufmann Series in Networking. Morgan Kaufmann, San Francisco (2004)
3. Madden, S., Franklin, M.J., Hellerstein, J.M., Hong, W.: Tag: a tiny aggregation service for ad-hoc sensor networks. SIGOPS Oper. Syst. Rev. 36(SI), 131–146 (2002)
4. Yao, Y., Gehrke, J.: Query processing in sensor networks. In: Proc. of CIDR (2003)
5. Song, I., Roh, Y., Hyun, D., Kim, M.H.: A query-based routing tree in sensor networks. In: Proc. of Geosensor Networks (2006)
6. Madden, S.R., Franklin, M.J., Hellerstein, J.M., Hong, W.: Tinydb: an acquisitional query processing system for sensor networks. ACM Trans. Database Syst. 30(1), 122–173 (2005)
7. Considine, J., Li, F., Kollios, G., Byers, J.: Approximate aggregation techniques for sensor databases. In: Proc. of ICDE (2004)
8. Cristescu, R., Beferull-Lozano, B., Vetterli, M., Wattenhofer, R.: Network correlated data gathering with explicit communication: Np-completeness and algorithms. IEEE/ACM Transactions on Networking 14(1), 41–54 (2006)
9. Sharaf, A., Beaver, J., Labrinidis, A., Chrysanthis, K.: Balancing energy efficiency and quality of aggregate data in sensor networks. The VLDB Journal 13(4), 384–403 (2004)
10. Manjhi, A., Nath, S., Gibbons, P.B.: Tributaries and deltas: efficient and robust aggregation in sensor network streams. In: Proc. of SIGMOD (2005)
11. Nath, S., Gibbons, P.B., Seshan, S., Anderson, Z.: Synopsis diffusion for robust aggregation in sensor networks. ACM Trans. Sen. Netw. 4(2), 1–40 (2008)
12. Motiwala, M., Elmore, M., Feamster, N., Vempala, S.: Path splicing. SIGCOMM Comput. Commun. Rev. 38(4), 27–38 (2008)
13. Feeney, L., Nilsson, M.: Investigating the energy consumption of a wireless network interface in an ad hoc networking environment. In: Proc. of INFOCOM (2001)

Evaluating Continuous Probabilistic Queries Over Imprecise Sensor Data

Yinuo Zhang[1], Reynold Cheng[1], and Jinchuan Chen[2]

[1] Department of Computer Science, The University of Hong Kong, Pokfulam Road, Hong Kong
{ynzhang,ckcheng}@cs.hku.hk
[2] School of Information, Renmin University of China, Beijing, China
csjcchen@gmail.com

Abstract. Pervasive applications, such as natural habitat monitoring and location-based services, have attracted plenty of research interest. These applications deploy a large number of sensors (e.g. temperature sensors) and positioning devices (e.g. GPS) to collect data from external environments. Very often, these systems have limited network bandwidth and battery resources. The sensors also cannot record accurate values. The uncertainty of these data hence has to been taken into account for query evaluation purposes. In particular, *probabilistic queries*, which consider data impreciseness and provide statistical guarantees in answers, have been recently studied. In this paper, we investigate how to evaluate a long-standing (or *continuous*) probabilistic query. We propose the probabilistic filter protocol, which governs remote sensor devices to decide upon whether values collected should be reported to the query server. This protocol effectively reduces the communication and energy costs of sensor devices. We also introduce the concept of probabilistic tolerance, which allows a query user to relax answer accuracy, in order to further reduce the utilization of resources. Extensive simulations on realistic data show that our method reduces by address more than 99% of savings in communication costs.

1   Introduction

Advances in sensor technologies, mobile positioning and wireless networks have motivated the development of emerging and useful applications [5,10,22,7]. For example, in scientific applications, a vast number of sensors can be deployed in a forest. These values of the sensors are continuously streamed back to the server, which monitors the temperature distribution in the forest for an extensive amount of time. As another example, consider a transportation system, which fetches location information from vehicles' GPS devices periodically. The data collected can be used by mobile commerce and vehicle-traffic pattern analysis applications. For these environments, the notion of the long-standing, or *continuous* queries [21,24,13,15], has been studied. Examples of these queries include: "Report to me the rooms that have their temperature within $[15^o C, 20^o C]$ in the next 24 hours"; "Return the license plate number of vehicles in a designated area within during the next hour". These queries allow users to perform real-time tracking on sensor data, and their query answers are continuously updated to reflect the change in the states of the environments.

H. Kitagawa et al. (Eds.): DASFAA 2010, Part I, LNCS 5981, pp. 535–549, 2010.
© Springer-Verlag Berlin Heidelberg 2010

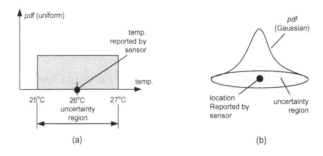

Fig. 1. Uncertainty of (a) temperature and (b) location

An important issue in these applications is *data uncertainty*, which exists due to various factors, such as inaccurate measurements, discrete samplings, and network latency. Services that make use of these data must take uncertainty into consideration, or else the quality and reliability may be affected. To capture data uncertainty, the *attribute uncertainty* model has been proposed in [20,23,3], which assumes that the actual data value is located within a closed region, called the *uncertainty region*. In this region, a non-zero probability density function (pdf) of the value is defined, such that the integration of pdf inside the region is equal to one. Figure 1 (a) illustrates that the uncertain value of a room's temperature in 1D space follows a uniform distribution. In Figure 1 (b) the uncertainty of a mobile object's location in 2D space follows a normalized Gaussian distribution. Notice that the actual temperature or location values may deviate from the ones reported by the sensing devices. Based on attribute uncertainty, the notion of *probabilistic queries* has been recently proposed. These are essentially spatial queries that produce inexact and probabilistic answers [3,4,17,16,14,1].

In this paper, we study the evaluation of the *continuous probabilistic query* (or CPQ in short). These queries produce *probabilistic* guarantees for the answers based on the attribute uncertainty. Moreover, the query answers are constantly updated upon database change. An example of a CPQ can be one that requests the system to report the IDs of rooms whose temperatures are within the range $[26^\circ C, 30^\circ C]$, with a probability higher than 0.7, within the next two hours. Let us suppose there are two rooms: r_1 and r_2, where two sensors are deployed at each room and report their temperature values periodically. At time instant t, their probabilities of being within the specified range are 0.8 and 0.3 respectively. Hence, at time t, $\{r_1\}$ would be the query answer. Now, suppose that the temperature values of the two rooms are reported to the querying server at every t_P time units. Since their temperature values can be changed, the answer to the CPQ can be changed too. For example, at time $t + t_P$, the new probabilities for r_1 and r_2 of satisfying the CPQ are respectively 0.85 and 0.7. Then, the CPQ answer at $t+t_P$ is $\{r_1, r_2\}$. We call the value 0.7, which controls the query answer, the *probability threshold* parameter. This parameter allows a user to specify the level of confidence that he wants to place in the query result.

A simple method for evaluating a CPQ is to allow each sensing device to periodically report their current values, evaluate the probabilities of the new sensor values, and update the query result. This approach is, however, expensive, because a lot of energy and communication resources are drained from the sensing devices. Moreover, when a new value is received, the system has to recompute the CPQ answer. As pointed out in [4], recomputing these probability values require costly numerical integration [4]. It is thus important to control the reporting activities in a careful manner. In this paper, we present a new approach of evaluating a CPQ, which (1) prolongs battery lifetime; (2) saves communication bandwidth; and (3) reduces computation overhead. Specifically, we propose the concept of *probabilistic filter protocol*. A probabilistic filter is essentially a set of conditions deployed to a sensing device, which governs when the device should report its value (e.g., temperature or location) to the system, without violating query correctness requirements [19,5]. Instead of periodically reporting its values, a sensor does so only if this is required by the filter installed on it. In the previous example, the filter would simply be the range $[26^oC, 30^oC]$, and is installed in the sensors in r_1 and r_2. At time $t + t_P$, if the sensor in r_1 checks that its probability for satisfying the CPQ is larger than 0.7, it does not report its updated value. Thus, using the filter protocol, the amount of data sent by the devices, as well the energy spent, can be reduced. Indeed, our experimental results show that the amount of update and energy costs is saved by 99%. Since the server only reacts when it receives data, the computational cost of re-evaluating the CPQ is also smaller.

We also observe that if a user is willing to tolerate some error in her query answer, the performance of the filter protocol can be further improved. In the previous example, suppose that the answer probability of room r_1 has been changed from 0.85 to 0.65. Since the probability threshold is 0.7, r_1's sensor should report its value to the server. However, if the user specifies a "probabilistic tolerance" of 0.1, then, r_1 can choose *not* to report its value to the server. Based on this intuition, we design the *tolerant probabilistic filter*, which exploits the probabilistic tolerance. The new protocol yields more energy and communication cost savings than its non-tolerant counterpart, by around 66%, in our experiments. We will describe the formal definition of probabilistic tolerance, and present the protocol details.

The rest of this paper is organized as follows. Section 2 presents the related work. We describe the problem settings and the query to be studied in Section 3. Then we discuss the probabilistic filter protocol in Section 4. The tolerant probabilistic filter protocol is presented in Section 5. We give our experimental results in Section 6, and conclude the paper in Section 7.

2   Related Work

In the area of continuous query processing, a number of approaches have been proposed to reduce data updates and query computation load. These work include: indexing schemes that can be adapted to handle high update load [21]; incremental algorithms for reducing query re-evaluation costs [24]; the use of adaptive safe regions for reducing update costs [12]; the use of prediction functions for monitoring data streams [13]; and sharing of data changes in multiple-query processing [15].

To reduce system load, researchers have proposed to deploy query processing to remote streaming sources, which are capable of performing some computation. Specifically, the idea of stream filters is studied. Here, each object is installed with some simple conditions, e.g. filter constraints, that are derived from requirements of a continuous query [19,5,22,11,25,8,7]. The remote object sends its data to the server only if this value violates the filter constraints. Since not all values are sent to the server, a substantial amount communication effort can be saved. In this paper, we propose the use of probabilistic filters on data with attribute-uncertainty. To our best knowledge, this has not been addressed before.

There also have been plenty of literature on probabilistic queries. [3] proposed a classification scheme of probabilistic queries based on whether a query returns a numerical value or the identities of the objects that satisfy the query. Many studies focus on reducing the computation time of probabilistic queries since such computing often involves expensive integration operations on the pdfs [4,17].

However, most of the work on probabilistic queries focuses on snapshot queries - queries that only evaluated by the system once. Few studies have addressed the issue of evaluating CPQs. In [1], the problem of updating answers for continuous probabilistic nearest neighbor queries in the server is studied. However, it does not explain how filters can be used to reduce communication and energy costs for this kind of queries. In [9], a tolerance notion for continuous queries has been proposed. However, it does not use the attribute uncertainty model. In [2], we performed a preliminary study of using filters for CPQs. We further improve this method by introducing the probabilistic tolerance, and present an extensive evaluation on our approach.

3   Continuous Probabilistic Queries

In this section, we first explain the details of the system model assumed in this paper (Section 3.1). Then, in Section 3.2, we present the formal definition of CPQ, as well as a simple method of evaluating it.

3.1   System Model

Figure 2 shows the system framework. It consists of a server, where a user can issue her query. The *query manager* evaluates the query based on the data obtained from the *uncertain database* (e.g., [3]), which stores the uncertainty of the data values obtained from external sources. Another important module in the server is the *filter manager*. Its purpose is to instruct a sensor on when to report its updated value, in order to reduce the energy and network bandwidth consumption. In particular, the filter manager derives *filter constraints*, by using the query information and data uncertainty. Then, the filter constraints are sent to the sensors. The server may also request the filter constraints to be removed after the evaluation of a CPQ is completed.

Each sensor is equipped with two components:

- a data collector, which periodically retrieves data values (e.g., temperature or position coordinates) from external environments.
- a set of one or more *filter constraints*, which are boolean expressions for determining whether the value obtained from the data collector is to be sent to the server.

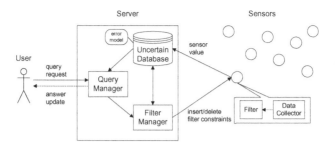

Fig. 2. System Architecture

As discussed in Section 1, we use the *attribute uncertainty* model (i.e., a closed range plus a pdf) [20,23] to represent data impreciseness. The type of uncertainty studied here is the *measurement error* of a sensing device, whose pdf is often in the form of the Gaussian or uniform pdf [20,23,3]. To generate the uncertain data, the uncertain database manager stores two pieces of information: (1) an error model for each type of sensors, for instance, a zero-mean Gaussian pdf with some variance value; and (2) the latest value reported by each sensor. The uncertain data value is then obtained by using the sensor's reported value as the mean, and the uncertainty information (e.g., uncertainty region and the variance) provided by the error model. Figure 1 illustrates the resulting uncertainty model of a sensor's value.

In the sequel, we will assume a one-dimensional data uncertainty model (e.g., Figure 1(a)). However, our method can generally be extended to handle multi-dimensional data. Let us now study how uncertain data is evaluated by a CPQ.

3.2   Evaluation of CPQ

Let $o_1, ..., o_n$ be the IDs of n sensing devices monitored by the system. A CPQ is defined as follows:

Definition 1. *Given a 1D interval R, a time interval $[t_1, t_2]$, a real value $P \in (0, 1]$, a* **continuous probabilistic query** *(or* **CPQ** *in short) returns a set of IDs $\{o_i | p_i(t) \geq P\}$ at every time instant t, where $t \in [t_1, t_2]$, and $p_i(t)$ is the probability that the value of o_i is inside R.*

An example of such a query is: "During the time interval $[1PM, 2PM]$, what are the IDs of sensors, whose probabilities of having temperature values within $R = [10^\circ C, 13^\circ C]$ are more than $P = 0.8$, at each point of time?" Notice that the answer can be changed whenever a new value is reported. For convenience, we call R and P respectively the *query region* and the *probability threshold* of a CPQ. We also name p_i the *qualification probability* of sensor o_i.

At any time t, the qualification probability of a sensor o_i can be computed by performing the following operation:

$$p_i(t) = \int_{u_i(t) \cap R} f_i(x, t) dx \tag{1}$$

In Equation 1, $u_i(t)$ is the uncertainty region of the value of o_i, and $u_i(t) \cap R$ is the overlapping part of $u_i(t)$ and the query region R. Also, x is a vector that denotes a possible value of o_i, and $f_i(x, t)$ is the probability density function (pdf) of x.

Basic CPQ Execution. A simple way of answering a CPQ is to first assume that each sensor's filter has no constraints. When a sensor's value is generated at time t', its new value is immediately sent to the server, and the qualification probabilities of all database sensors are re-evaluated. Then, after all $p_i(t')$ have been computed, the IDs of devices whose qualification probabilities are not smaller than P are returned to the user. The query answer is constantly recomputed during t_1 and t_2.

This approach is expensive, however, because:

1. Every sensor has to report its value to the server periodically, which wastes a lot of energy and network bandwidth;
2. Whenever an update is received, the server has to compute the qualification probability of each sensor in the database, using Equation 1, and this process can be slow.

Let us now the probabilistic filter protocol can tackle these problems.

4   The Probabilistic Filter Protocol

Before presenting the protocol, let us explain the intuition behind its design. Figure 3 shows a range R (the pair of solid-line intervals) and the uncertainty information of two sensors, o_1 and o_2, at current time t_c, represented as gray-colored bars. Let us assume that the probability threshold P is equal to one. Also, the current values extracted from the data collectors of o_1 and o_2 are $v_1(t_c)$ and $v_2(t_c)$ respectively. We can see that o_1's uncertainty region, $u_1(t_c)$, is totally inside R. Also, $v_1(t_c) \in u_1(t_c)$. Hence, o_1 has a qualification probability of one, and o_1 should be included in the current query result. Suppose that the next value of u_1 is still inside R. Then, it is still not necessary for the query result to be updated. More importantly, if o_1 knows about the information of R (the query region of a CPQ), o_1 can check by itself whether it needs to send the update to the server. A "filter constraint" for o_1, when it is inside R, can then be defined as follows:

$$\textbf{if } u_1(t_c) - R \neq \Phi \textbf{ then } \texttt{send } v_1(t_c) \tag{2}$$

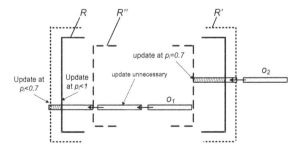

Fig. 3. Illustrating probabilistic filter constraints

which means: "When v_1 has a chance to be outside R, report v_1 to the server". Thus, the server can first compute constraint 2 and send it to o_1. As long as constraint 2 is not satisfied, no update is produced by o_1.

The above technique can be generalized to handle any probability threshold P. Let us consider Figure 3 again, where $P = 0.7$. Suppose that o_1 continues to move towards the left boundary of R, such that a fraction of more than 0.3 of its uncertainty region (shaded) lies outside R. At this point, v_1 must be reported, so that the ID o_1 can be removed from the query result. This is equivalent to using constraint 2, except that R is replaced by R'. Here, R' is derived by using the maximum amount of o_1's uncertainty region allowed on the outside of R, which is equal to 0.3 of o_1's uncertainty region.

Figure 3 also shows that o_2 is currently outside R. For $P = 0.7$, the following constraint can be used:

$$\textbf{if } u_2(t_c) \texttt{ touch } R'' \textbf{ then } \texttt{send } v_2(t_c) \tag{3}$$

When u_2 touches R'', it has a fraction of exactly 0.7 inside R. Upon receiving the update from o_2, the server should insert o_2 to the query result. Notice that while R' is outside R, the region R'' is enclosed by R. In general, for every CPQ with P, two constraints are need, to handle the cases when a value's uncertainty is outside or inside R. An additional advantage of this approach is that a sensor does not need to compute its qualification probability, which can be complicated for a sensor with low computational power.

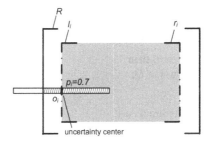

Fig. 4. Checking filter constraints at the sensor

Simple Constraint Verification. In practice, a sensor may not keep the detailed uncertainty information to perform filter constraint checking. Also, since a sensor can have low computational power, it is worthwhile to further simplify the constraint verification process. Observe that the uniform/Gaussian pdf assumed in our uncertainty model has a *symmetric* shape, and is centered around the value sensed from the data collector (c.f. Figure 1). It is then sufficient for the sensor to test the constraints by using *only* its sensed value. Figure 4 illustrates a CPQ with $P = 0.7$. When the sensed value v_i of o_i touches the line l_i, o_i has *exactly* a qualification probability of 0.7. Thus, if v_i is on the left of l_i, its qualification probability must be less than 0.7. Similarly, if v_i is on the right of r_i, its qualification probability is also less than 0.7. Hence, $v_i \in [l_i, r_i]$ if and only if $p_i \geq P$.

The values of l_i and r_i can be obtained by using the pdf information to derive the distance from the boundaries of R. For uniform pdf, the distance can be obtained easily; for Gaussian pdf, the value can be derived by performing table-lookup. This approach is desirable for a sensor with low processing power. Moreover, only one interval ($[l_i, r_i]$) needs to be stored, as opposed to the two intervals presented earlier (e.g., R' and R''). Hence, the precious memory required by a sensor for storing the constraints is also saved.

4.1 Protocol Design

We are now ready to discuss the probabilistic filter protocol. Algorithm 1 below shows the algorithm employed by the server's filter manager.

```
 1  Initialization:
 2  Request data from sensors o₁, . . . , oₘ;
 3  for each sensor oᵢ do
 4  │   UpdateDB(oᵢ);
 5  │   Compute new filter constraint [lᵢ, rᵢ];
 6  └   Send(addFilterConstraint, [lᵢ, rᵢ], oᵢ);

 7  Maintenance:
 8  while t₁ ≤ currentTime ≤ t₂ do
 9  │   Wait for update from oᵢ;
10  │   UpdateDB(oᵢ);
11  │   if update == (oᵢ, delete) then
12  │   │   remove oᵢ from answer of Q;
13  │   if update == (oᵢ, insert) then
14  └   └   insert oᵢ to answer of Q;

15  for each sensor oᵢ do
16  │   Send(deleteFilterConstraint, oᵢ);
```

Algorithm 1. Probabilistic filter protocol (at filter manager)

In this algorithm, after a continuous query Q is registered, the server collects information from all sensors. Based on these values, the server evaluates the filter constraint for each of them. Afterwards, the constraints are installed in the sensors (lines 2-6). These constraints, in the form of $[l_i, r_i]$, are computed by using the method described in the previous section.

When Q is being executed (between times t_1 and t_2), the server continuously listens to updates from all sensors. If it receives an update, it will update the uncertain database (lines 9-10). Then, instead of recomputing the whole query answer of Q, an *incremental update* approach is adopted: the server refreshes the query result according to the update command received (lines 11-14). This is possible, because the update of o_i only affects its own qualification probability, but not other sensors. After the query is completed, the filter constraints for query Q on all sensors are removed (Steps 15-16).

```
 1  currState = FALSE;
 2  while true do
 3      command = receive(server);
 4      switch command do
 5          case addFilterConstraint
 6              Add new filter constraint to o_i;
 7              Stop;
 8          case deleteFilterConstraint
 9              Delete filter constraint from o_i;
10              Stop;
11          result = checkFilterConstraints(v_i,currState);
12          if result == include then
13              currState = TRUE;
14              sendUpdate(o_i,insert);
15          else if result == exclude then
16              currState = FALSE;
17              sendUpdate(o_i,delete);
```

Algorithm 2. Probabilistic filter protocol (at sensor)

Sensor side. Each sensor o_i retrieves data value periodically from the external environment. It also uses a variable called `currState` to store its current state with respect to Q: if o_i is currently included in Q's result, then `currState` has a `true` value, or `false` otherwise. As shown in Algorithm 2, `currState` is initially `FALSE` (line 1). The sensor then continuously listens to the commands from the server (lines 2-3). If the server requests to add or delete filter constraints for a CPQ, it will do so accordingly (lines 4-10). Then, it will check the filter constraints by using its latest sensed value v_i, the `currState` value, and the checking method in the previous section (line 11). If o_i should be included in the query result, o_i changes `currState` to `TRUE`, and notifies the server (lines 12-14). Otherwise, o_i is removed from the query result (lines 15-17).

These algorithms alleviate the problems of the basic protocol discussed in Section 3.2. At the sensor side (Algorithm 2), update is only sent to the server if the filter constraint is violated, not periodically. At the server side (Algorithm 1), since the query answer is updated incrementally, there is no need to compute the qualification probability of each sensor. Moreover, for both the server and sensors, no qualification probabilities are computed. Hence, a significant amount of computational effort at both the server and the sensors is reduced.

5   Tolerant Probabilistic Filters

In this section, we investigate how the performance of the probabilistic filter protocol can be further improved, if the user is willing to sacrifice some degree of accuracy (or equivalently, specify a *tolerance*) in her query answers. We present a definition of tolerance designed for CPQs in Section 5.1. Then, we study how the filter protocol should be modified in order to exploit this tolerance, in Section 5.2.

5.1 Probabilistic Tolerance

The probabilistic tolerance, specified with a real value $\Delta \in [0, 1]$, is defined as follows.

Definition 2. *Given a CPQ Q, and $\Delta \in [0, min(P, 1 - P)]$, a Δ-CPQ returns results $S \cup T$ at every time instant t during the lifetime of Q, where $S = \{o_i | p_i(t) \geq P + \Delta\}$ and $T \subseteq \{o_i | p_i(t) \geq P - \Delta\}$.*

Essentially, the result of Δ-CPQ has the following requirements:

- It contain IDs of all sensors with qualification probabilities not less than $P + \Delta$;
- It does not contain the ID of any sensor whose qualification probability is less than $P - \Delta$;
- It *may* contain a sensor with qualification probability less than $P + \Delta$ but not smaller than $P - \Delta$.

Example. Consider three sensors, o_1, o_2 and o_3, and a CPQ with $P = 0.7$ and $\Delta = 0.1$. Suppose at some time instant, the qualification probabilities p_i's of o_1, o_2, and o_3 are respectively 0.85, 0.55 and 0.71. Since $p_1 \geq 0.7 + 0.1$, o_1 is included in the result of this 0.1-CPQ. On the other hand, $p_2 < 0.7 - 0.1$, and so o_2 is excluded from the query result. For o_3, its probability p_3 is between $[0.6, 0.8]$, and whether o_3 is included in the result does not affect the correctness of the query. Notice that if p_3 was previously greater than 0.8, there is *no need* for o_3 to be removed from the query result, even though its probability is now below 0.7. Hence, o_3 does not have to report its newest value to the server.

5.2 Protocol Design

Given a Δ-CPQ, we first derive two pairs of filter constraints for each sensor. Specifically, we consider the same CPQ, with probability $P + \Delta$, and compute the constraint $[l_i^+, r_i^+]$ for each sensor o_i, using the techniques in Section 4. Recall that if the sensed value v_i is within this range, p_i must be no less than $P + \Delta$. For the same CPQ, we

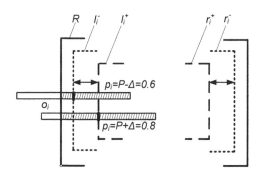

Fig. 5. Filter constraints for enforcing probabilistic tolerance

derive another filter constraint $[l_i^-, u_i^-]$, for probability $P - \Delta$. This means v_i is located in this range, if and only if $p_i \geq P - \Delta$. For example, in Figure 5, $P = 0.7$ and $\Delta = 0.1$. If $v_i \in [l_i^+, r_i^+]$, then p_i must exceed 0.8. On the other hand, if $v_i \notin [l_i^-, r_i^-]$, then $p_i < 0.6$.

The probabilistic tolerance can be enforced by making changes to Algorithms 1 and 2. For the filter manager (Algorithm 1), the maintenance phase (lines 7-16) is the same as before, and so we only display the new initialization phase, as shown in Algorithm 3. The main idea of this algorithm is that for a sensor o_i whose qualification probability p_i is not less than P at initial time t_0, it only needs to report its value v_i if its $p_i < P - \Delta$, or equivalently, $v_i \notin [l_i^-, u_i^-]$. In lines 5-6, all sensors of this type ($R(t_0)$) are assigned the $[l_i^-, u_i^-]$ filters. We say that this filter is *active*, meaning that it is currently employed by the sensor to decide whether to send an update. The other filter, $[l_i^+, u_i^+]$, is not used (or *inactive*) in this moment. However, both filters are sent to the sensor. On the other hand, if $p_i < P$, then o_i has to report v_i when $p_i \geq P + \Delta$, which is equivalent to $v_i \in [l_i^+, u_i^+]$. For this kind of sensors, the roles of the $[l_i^+, u_i^+]$ and $[l_i^-, u_i^-]$ are switched, as shown in lines 8-11.

---

1 **Initialization:**
2 Receive data from all sensors $o_1, \cdots, o_m$;
3 Let $R(t_0)$ be the set $\{o_i | p_i(t_0) \geq P\}$;
4 **for** *each sensor o_i in $R(t_0)$* **do**
5   Compute filter constraints $[l_i^-, u_i^-]$ and $[l_i^+, u_i^+]$;
6   Assign $[l_i^-, u_i^-]$ as *active* filter and $[l_i^+, u_i^+]$ as *inactive* filter;
7   Send the 2 filter constraints to o_i;

8 **for** *each sensor o_i not in $R(t_0)$* **do**
9   Compute filter constraints $[l_i^-, u_i^-]$ and $[l_i^+, u_i^+]$;
10   Assign filter $[l_i^+, u_i^+]$ as *active* filter and $[l_i^-, u_i^-]$ as *inactive* filter;
11   Send the 2 filter constraints to o_i;

Algorithm 3. New initialization phase for filter manager

At the sensor side, Algorithm 2 can generally still be used, except with the following differences. First, two filter constraints are stored. Second, only the *active* filter is used for determining whether to send an update. Third, once an update is sent, the *active* and *inactive* states of the two filters stored in the sensors are swapped.

6   Experimental Evaluation

We now evaluate the performance of our protocols. Section 6.1 presents the experimental setup, and Section 6.2 discusses the results.

6.1   Experimental Setup

We use the temperature sensor readings captured by 54 sensors, deployed in the Intel Berkeley Research lab. The temperature values are collected every 30 seconds. The lowest and the highest temperature values are $+13^{\circ}C$ and $+35^{\circ}C$ respectively. So, we set the domain space as $+10^{\circ}C$ to $+40^{\circ}C$. The uncertainty region of a sensor value is in the range of $\pm 1^{\circ}C$ [18], and we assume that uncertainty pdf is uniform. (We also experiment with Gaussian pdf). We use 54 sensors to generate 155520 records over one day, with a sampling time interval of 30s. For energy consumption, the energy for sending an uplink message is 77.4mJ and the energy for receiving a downlink message is 25.2mJ [6].

Each data point is obtained by averaging over the results of 100 random queries. The size of each query is $5^{\circ}C$. The centers of the queries are randomly selected within $[12.5^{\circ}C, 37.5^{\circ}C]$. All the queries has same duration as simulation period 1 day. By default, $P = 0.6$. Since P cannot be 0 as stated in the definition, so in our experiment we use $P + \epsilon$ where $\epsilon = 10^{-4}$ to substitute $P = 0$ case.

6.2 Experimental Result

Probabilistic Filters. We first evaluate the effectiveness of introducing probabilistic filters. We focus on both communication cost and computation costs. From Figures 6(a) and (b), the use of probabilistic filters reduce the update frequency and energy consumption rate by more than 99%. In detail, the average update frequency for probabilistic filters is around 0.074 per sampling interval, which is much less than when filters are not used. The average energy consumption for the probabilistic filters is 7.6 mJ per sampling interval. Moreover, using our protocol, the server does not need to do any probability computation. Hence, the computational time for handling an update is also significantly reduced (Figure 6(c)).

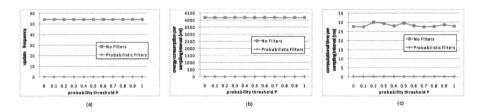

Fig. 6. Probabilistic Filters

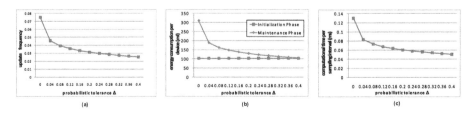

Fig. 7. Probabilistic tolerance ($P = 0.6$)

Probabilistic Tolerance. Next, we evaluate the performance of our tolerant protocol, under different values of Δ. From Figures 7(a) and (b) we can see that the improvement is about a 66% reduction over update frequency and energy consumption (in maintainance phase), when $\Delta = 0.4$. The reason for this improvement is that the increase on the probabilistic tolerance gives more chances for sensors to avoid violating the constraints as well as sending updates. Figure 7(c) also shows that the computational time on the server side is reduced by around 60% at $\Delta = 0.4$. This is the consequence of fewer updates received at the server.

Gaussian Distribution. We also evaluate our protocol for uncertainty pdfs that follow Gaussian distribution. Figure 8 shows that given the same tolerance value, more updates are saved when the Gaussian pdf has a larger variance. For example, if $\Delta = 0.4$, the reduction using variance of 10 units over that of 0.2 units is around 20%. This reflects that the filter constraints (e.g., l_i^+) tend to be further away from the current sensed value under a larger variance. Hence, our protocol works better for Gaussian pdf with a larger variance.

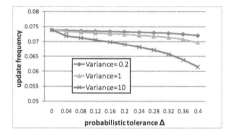 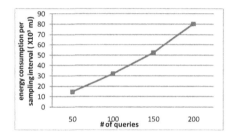

Fig. 8. Probabilistic tolerance using Gaussian distribution (P=0.6)

Fig. 9. Multiple Queries

Multiple Queries. Finally, we evaluate the performance of running multiple queries in the system. We use a number of queries with random sizes and starting times. The lifetime of each CPQ follows a uniform distribution of $[2, 2880]$ sampling intervals. The probability threshold and probabilistic tolerance are also randomly selected. In Figure 9, we can see that the energy consumption rate scales linearly with the number of queries. When we increase the number of queries, the increment on the energy per sampling interval is around 438mJ.

7   Conclusions

Uncertainty management is an important and emerging topic in sensor-monitoring applications. In order to reduce update and energy consumption, we study a protocol for processing continuous probabilistic queries over imprecise sensor data. We further present the concept of probabilistic tolerance, and a protocol which enforces this tolerance, to yield more savings. In the future, we will study how other CPQs (e.g., nearest-neighbor queries) can be supported.

Acknowledgments

Reynold Cheng was supported by the Research Grants Council of Hong Kong (Projects HKU 513307, HKU 513508, HKU 711309E), and the Seed Funding Programme of the University of Hong Kong (grant no. 200808159002). We would like to thank Prof. Kurt Rothermel and Mr. Tobias Farrell (University of Stuttgart) for providing support on the data simulator. We also thank the reviewers for their insightful comments.

References

1. Chen, J., Cheng, R., Mokbel, M., Chow, C.: Scalable processing of snapshot and continuous nearest-neighbor queries over one-dimensional uncertain data. In: VLDBJ (2009)
2. Chen, J., Cheng, R., Zhang, Y., Jin, J.: A probabilistic filter protocol for continuous queries. In: Rothermel, K., Fritsch, D., Blochinger, W., Dürr, F. (eds.) QuaCon 2009. LNCS, vol. 5786, pp. 88–97. Springer, Heidelberg (2009)
3. Cheng, R., Kalashnikov, D.V., Prabhakar, S.: Evaluating probabilistic queries over imprecise data. In: SIGMOD (2003)
4. Cheng, R., Kalashnikov, D.V., Prabhakar, S.: Querying imprecise data in moving object environments. IEEE Trans. on Knowl. and Data Eng. 16(9) (2004)
5. Cheng, R., Kao, B., Prabhakar, S., Kwan, A., Tu, Y.-C.: Adaptive stream filters for entity-based queries with non-value tolerance. In: VLDB (2005)
6. Crossbow Inc. MPR-Mote Processor Radio Board User's Manual
7. Deshpande, A., Khuller, S., Malekian, A., Toossi, M.: Energy efficient monitoring in sensor networks. In: Laber, E.S., Bornstein, C., Nogueira, L.T., Faria, L. (eds.) LATIN 2008. LNCS, vol. 4957, pp. 436–448. Springer, Heidelberg (2008)
8. Elmeleegy, H., Elmagarmid, A.K., Cecchet, E., Arefs, W.G., Zwaenepoel, W.: Online piecewise linear approximation of numerical streams with precision guarantees. In: VLDB (2009)
9. Farrell, T., Cheng, R., Rothermel, K.: Energy-efficient monitoring of mobile objects with uncertainty-aware tolerances. In: IDEAS (2007)
10. Gedik, B., Liu, L.: Mobieyes: Distributed processing of continuously moving queries on moving objects in a mobile system. In: EDBT (2004)
11. Gedik, B., Wu, K.-L., Yu, P.S.: Efficient construction of compact shedding filters for data stream processing. In: ICDE (2008)
12. Hsueh, Y.-L., Zimmermann, R., Ku, W.-S.: Adaptive safe regions for continuous spatial queries over moving objects. In: Zhou, X., et al. (eds.) DASFAA 2009. LNCS, vol. 5463, pp. 71–76. Springer, Heidelberg (2009)
13. Ilarri, S., Wolfson, O., Mena, E.: A query processor for prediction-based monitoring of data streams. In: EDBT (2009)
14. Ishikawa, Y., Iijima, Y., Yu, J.X.: Spatial range querying for gaussian-based imprecise query objects. In: ICDE (2009)
15. Li, J., Deshpande, A., Khuller, S.: Minimizing communication cost in distributed multi-query processing. In: ICDE (2009)
16. Lian, X., Chen, L.: Monochromatic and bichromatic reverse skyline search over uncertain databases. In: SIGMOD (2008)
17. Ljosa, V., Singh, A.K.: Apla: Indexing arbitrary probability distributions. In: ICDE (2007)
18. Microchip Technology Inc. MCP9800/1/2/3 Data Sheet
19. Olston, C., Jiang, J., Widom, J.: Adaptive filters for continuous queries over distributed data streams. In: SIGMOD (2003)

20. Pfoser, D., Jensen, C.S.: Capturing the uncertainty of moving-object representations. In: Güting, R.H., Papadias, D., Lochovsky, F.H. (eds.) SSD 1999. LNCS, vol. 1651, p. 111. Springer, Heidelberg (1999)
21. Prabhakar, S., Xia, Y., Kalashnikov, D.V., Aref, W.G., Hambrusch, S.E.: Query indexing and velocity constrained indexing: Scalable techniques for continuous queries on moving objects. IEEE Trans. Comput. 51(10) (2002)
22. Silberstein, A., Munagala, K., Yang, J.: Energy-efficient monitoring of extreme values in sensor networks. In: SIGMOD (2006)
23. Sistla, P.A., Wolfson, O., Chamberlain, S., Dao, S.: Querying the uncertain position of moving objects. In: Etzion, O., Jajodia, S., Sripada, S. (eds.) Dagstuhl Seminar 1997. LNCS, vol. 1399, p. 310. Springer, Heidelberg (1998)
24. Xiong, X., Mokbel, M.F., Aref, W.G.: Sea-cnn: Scalable processing of continuous k-nearest neighbor queries in spatio-temporal databases. In: ICDE (2005)
25. Zhang, Z., Cheng, R., Papadias, D., Tung, A.K.: Minimizing the communication cost for continuous skyline maintenance. In: SIGMOD (2009)

BPMN Process Views Construction

Sira Yongchareon[1], Chengfei Liu[1], Xiaohui Zhao[1],
and Marek Kowalkiewicz[2]

[1] Centre for Complex Software Systems and Services
Swinburne University of Technology
Melbourne, Victoria, Australia
{syongchareon,cliu,xzhao}@swin.edu.au
[2] SAP Research
Brisbane, Australia
marek.kowalkiewicz@sap.com

Abstract. Process view technology is catching more attentions in modern business process management, as it enables the customisation of business process representation. This capability helps improve the privacy protection, authority control, flexible display, etc., in business process modelling. One of approaches to generate process views is to allow users to construct an aggregate on their underlying processes. However, most aggregation approaches stick to a strong assumption that business processes are always well-structured, which is over strict to BPMN. Aiming to build process views for non-well-structured BPMN processes, this paper investigates the characteristics of BPMN structures, tasks, events, gateways, etc., and proposes a formal process view aggregation approach to facilitate BPMN process view creation. A set of consistency rules and construction rules are defined to regulate the aggregation and guarantee the order preservation, structural and behaviour correctness and a novel aggregation technique, called *EP-Fragment*, is developed to tackle non-well-structured BPMN processes.

1 Introduction

Workflow/process view technologies have been recognised as an important capability for better granularity control of process representation [5, 8-12]. A process view represents a partial view of an actual business process, and therefore separates the process representation from the executable processes. This feature highlights the benefits of process views in the areas of authority control, process visualisation, collaborative business process modelling etc.

Reluctantly, most current research on workflow/process views assumes that business processes are well structured, yet this assumption confronts a lot of conflicts when Business Process Modelling Notations (BPMN) [1] is getting popular. As a graphical modelling tool, BPMN allows users to design business processes arbitrarily, and therefore many practical BPMN processes are not strictly well structured [13]. For example, a BPMN process may have unpaired Fork and Merge or Join gateways. To apply process view technology to BPMN processes, the non-well-structured characteristics of BPMN processes have to be taken into account.

H. Kitagawa et al. (Eds.): DASFAA 2010, Part I, LNCS 5981, pp. 550–564, 2010.

Some research efforts have been put to formalise the construction of process views, mostly by aggregating activities in the corresponding base process [3, 6]; however, their works only focus on construction from basic or compound activities without concerning related events and exceptions, which are common elements in BPMN process designing. These elements help BPMN capture more details of business processes, and they should be considered as well when creating process views for BPMN processes. Furthermore, current approaches do not provide the selective aggregation of branches in Split and Join gateways.

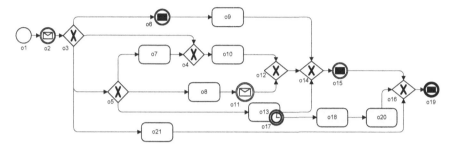

Fig. 1. Motivating example

Figure 1 shows our motivating example of BPMN process. As we can see that some parts of the structure are non-well-structured. For example, the split branches from the Fork gateway o_3 flow to different Join gateways o_4, o_{12}, and o_{14}. The timer-event o_{17} attached to the task o_{13} indicates that the subsequent execution will bypass the Join gateway o_{14} and flow through o_{18}, o_{20}, to o_{16} if the event occurs. For the given process, users may specify the requirement for aggregating tasks o_7, o_8, o_{10}, o_{13}, and event o_{11} in a process view. Two main questions are required to be answered: (1) Is the specified set of objects able to be aggregated? (2) If it is not, then what is the minimal set of objects, including the pre-specified set, for an aggregation?

Aiming at supporting process views generation for BPMN processes, we propose a BPMN process view construction approach that covers the main BPMN elements and characteristics. A set of rules is defined to regulate the view generation in compliance with structural and behavioural consistencies and correctness. Related algorithms are also developed for view checking and construction. Particularly, our approach makes the following contributions to process view research:

- Present an aggregate construction technique, called *EP-Fragment*, to tackle non-well-structured processes and selective aggregation of branches.
- Propose an algorithm for finding minimal aggregate from a set of user-specified tasks. This algorithm helps the automatic aggregation for process views.
- Consider BPMN elements, such as events, exception paths, etc., in our model.

The remainder of this paper is organised as follows. Section 2 provides a formal model of BPMN processes, syntaxes, and components for process view. Section 3 provides a process view construction methodology based on construction rules and consistency constraints; the prototype is also implemented for the proof of our approach. Section 4 reviews the related works. Finally, the concluding remarks are given in Section 5 with an indication on future work.

2 Formal Model of BPMN Processes

In this section, syntaxes, components, and structure of BPMN processes and process views are introduced and defined. A process view constructed on its underlying BPMN process is itself represented by the BPMN diagram. While a full range of BPMN elements are developed and proposed in BPMN 1.2 specification [1] to capture more detailed behaviour of business process, it is adequate to select only a core subset of them for the discussion on BPMN process views. This includes Tasks, Events, Gateways, Control flows, Message flows, Exception flows, and Pools.

Definition 1 (Private process or Process). A private BPMN process bp contains a set of tasks, events, and gateways connected together to represent the execution behaviour of the whole process. We model it as an extended directed-graph which is represented as a tuple (O, T, T^E, G, E, F), where,

- O is a finite set of BPMN element objects divided into disjoint sets of T, G, and E
- T is a finite set of tasks in bp
- G is a finite set of gateways in bp
- E is a finite set of events in bp; $event\_type$: $E \rightarrow \{Start, Catching\text{-}Intermediate, Throwing\text{-}Intermediate, End\}$ is a function used to specify the type of event.
- $F \subseteq O \times O$ is a finite set of control flow relations represented by a directed edge in bp. A control flow $f = (o_i, o_j) \in F$ corresponds to the unique control flow relation between o_i and o_j, where $o_i, o_j \in O$
- $T^E \subseteq E \times T$ is non-injective and non-surjective defining a finite set of attachment relations of intermediate events on tasks, called *Event-attached task relation*. An attachment relation of event e on task t, $t^e = (e, t) \in T^E$ corresponds to the intermediate trigger condition of event e for task t, where $t \in T$, $e \in E$ and $event\_type(e) = Catching\text{-}Intermediate$.
- F^* is reflexive transitive closure of F, written $o_i F^* o_j$, if there exists a path from o_i to o_j. In addition, we can write $o_i (F \cup T^E)^* o_j$ if there exists a path from object o_i to o_j via control flow relations F and event-attached task relations T^E.

Note that the exception flow of the task is a flow leading from an event e in T^E, and there can be one or more events attached to the task defining multiple exception flows. We also define necessary functions that will be used in the paper.

- $in(x) = |\{y \in O \mid \exists y, (y, x) \in F\}|$ returns the in-degree of node x, and $out(x) = |\{y \in O \mid \exists y, (x, y) \in F\}|$ returns the out-degree of node x
- $path\ (o_i \xrightarrow{f_i, f_j} o_j)$ returns a set of all objects in all possible paths leading from o_i via a control flow f_i to o_j via a control flow f_j, such that $\exists o_i, o_j \in O$, $\exists f_i, f_j \in F$, $o_i (f_i F^* f_j) o_j$. A set of objects in *normal path*, denoted as O_{NP}, defines a set of all objects in all possible paths from start event to the end event of a process, i.e., $O_{NP} = path(e_s \xrightarrow{F^*} e_e)$, such that $e_s \in \{E \mid event\_type(E) = Start\}$ and $e_e \in \{E \mid event\_type(E) = End\}$.

It is also conceived that a *well-structured* (opposite to *non-well-structured*) process must contain structures of correct pairs of Fork and Merge or Join gateways, and there must be no branch going out or coming in between the structure [15]. The *non-well-structured* process can be detected by using graph reduction [19] or SESE decomposition technique [20].

Definition 2 (Least Common Predecessor and Least Common Successors). Given a set of objects $N \subseteq O$ in a process, we define a set of least common predecessors and successors of N, denoted as $lcp(N)$ and $lcs(N)$, respectively.

$$lcp(N) = \{ o^p \in O \backslash N \mid \forall o \in N \, (o^p F^* o \wedge (\neg \exists o^q \in O \backslash N \, (o^q F^* o \wedge o^q F^* o^p)))\}$$
$$lcs(N) = \{ o^s \in O \backslash N \mid \forall o \in N \, (o F^* o^s \wedge (\neg \exists o^q \in O \backslash N \, (o F^* o^q \wedge o^s F^* o^q)))\}$$

For the purpose of identifying which flow going out of the least common predecessors and which flow coming into the least common successors, we define two functions $lcpF(N)$ and $lcsF(N)$ as the subset of outgoing flows of $lcp(N)$ and incoming flows of $lcs(N)$, respectively. These subsets only contain the flows in F that flow into or out from the set N.

$$lcpF(N) = \{f_p \in F \mid \forall o^p \in lcp(N), \forall o^s \in lcs(N), \exists o \in N, (o^p, o) \in F \wedge \mid path(o^p \xrightarrow{f_p,\ F^*} o^s) \mid > 0\}$$
$$lcsF(N) = \{f_s \in F \mid \forall o^s \in lcs(N), \forall o^p \in lcp(N), \exists o \in N, (o, o^s) \in F \wedge \mid path(o^p \xrightarrow{F^*,\ f_s} o^s) \mid > 0\}$$

From the lcs and lcp defined above, we can see that if any object does not exists in the *normal path* of the process, but other objects do, then lcs and lcp will not be found. For example, we can determine that lcp and lcs of a set of objects $\{o_7, o_9\}$ in Figure 1 are o_3 and o_{14}, respectively. Correspondingly, the set of flows according to lcp and lcs, i.e., $lcpF$ and $lcsF$, are $\{(o_3, o_6), (o_3, o_5)\}$ and $\{(o_9, o_{14}), (o_{12}, o_{14})\}$, respectively. However, if we include o_{18} into the set, the functions lcp, lcs, $lcpF$, and $lcsF$ will return an empty set as o_{18} does not exist in the *normal path* as same as the others.

Figure 2 illustrates an example of complex scenario showing multiple flows of multiple least common predecessors and successors in a process. Assume that $N=\{t_2, t_3\}$, we find $lcp(N) = \{g_1\}$ and $lcs(N) = \{g_4, g_5\}$; correspondingly, $lcpF(N) = \{(g_1, t_2), (g_1, t_3)\}$ and $lcsF(N) = \{(g_2, g_4), (g_2, g_5), (g_3, g_4), (g_3, g_5)\}$. Similarly, if we assume $N=\{t_4, t_5\}$, then $lcp(N) = \{g_2, g_3\}$ and $lcs(N) = \{g_6\}$. Therefore, we can find that $lcpF(N) = \{(g_2, g_4), (g_2, g_5), (g_3, g_4), (g_3, g_5)\}$ and $lcsF(N) = \{(t_4, g_6), (t_5, g_6)\}$.

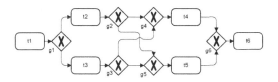

Fig. 2. An example of *lcpF* and *lcsF*

Definition 3 (Exception path). Given a process bp (O, T, T^E, G, E, F), an *exception path* is a set of the paths leading from a catching-intermediate event e in *Event-attached task relation* $(e, t) \in T^E$ to any object in the *normal path* or the end event of the process.

Let *teObject*(e, t) denote the set of objects lying on the *exception path* of $(e, t) \in T^E$.

$$teObject(e, t) = \begin{cases} o \mid \exists o_n \in O_{NP}, eF^* o \wedge oF^* o_n, & \text{if} \mid path(e \xrightarrow{F^*} o_n) \mid > 0 \\ o \mid \exists e_e \in \{E \mid event\_type(E)=End\}, eF^* o \wedge oF^* e_e\}, & \text{otherwise} \end{cases}$$

As shown in Figure 1, we want to find the objects on the exception path of timer event o_{17} which attached to the task o_{13}. As we can find that $lcs(\{o_{17}, o_1\}) = \{o_{16}\}$ in which it exists in both *normal path* from start event o_1 and *exception path* of o_{17}, so the set of objects in exception path $teObject(o_{17}, o_{13}) = \{o_{18}, o_{20}\}$.

Definition 4 (Collaboration Process). A collaboration process is a set of private processes that interacts each other by interchanging messages. Let *cbp* denote a BPMN collaboration process and it is a tuple (BP, M, δ), where

- $BP = \{bp_1, bp_2, \ldots, bp_n\}$, $bp_i \in BP(1 \leq i \leq n)$ is a process existing in *cbp*
- δ: $BP.O \rightarrow P$ is a bijective function describing the object-pool relations between objects in private processes and pools $P = \{p_1, p_2, \ldots, p_k\}$, where pool $p_i \in P(1 \leq i \leq k)$ is used as a container of private process. Correspondingly δ^1: $P \rightarrow BP.O$ is an inverse function
- $M \subseteq \bigcup_{bp \in BP}(bp.T \cup bp.E) \times \bigcup_{bp \in BP}(bp.T \cup bp.E)$, $m = \{(o_i, o_j) \in M \mid \delta(o_i) \neq \delta(o_j)\}$ is a message of the interaction between source o_i and target o_j of tasks or events such that the source and the target must be on different private processes or pools

We define *process view* as an abstract representation of its base collaboration process. The detailed construction process of a view will be introduced in the next section.

3   Process View Construction

Process views are constructed by a set of process view operations in which recent works on process views have summarised two primary operations: Aggregation and Hiding [2, 3]. Aggregation operation provides users to define a set of objects in the base process that has to be aggregated and replace such objects with the aggregate object, while hiding operation will simply hide the specified objects. In this paper we do not consider the hiding operation. The aggregation operation can be iterated in order to achieve the preferred process view. As such, this section will firstly define a set of consistency rules that the constructed process view and its underlying process must comply to maintain the structural and behaviour correctness between them.

3.1   Preliminaries

In this section, we define some necessary terms, definitions and functions that will be used in the process view construction.

Definition 5 (Process fragment or P-fragment). Process fragment represents a partial structure of a private process. Let P-fragment *Pf* denote a nonempty connected sub-graph of a process $bp \in cbp.BP$ and it is a tuple $(O', T', T^{E'}, G', E', F', F_{in}, F_{out})$ where $O' \subseteq O, T' \subseteq T, T^{E'} \subseteq T^E, G' \subseteq G, E' \subseteq E, F' \subseteq O' \times O' \subseteq F$, such that,

- $\forall e_s \in \{E \mid event\_type(E)=Start\}, \forall e_e \in \{E \mid event\_type(E)=End\}, e_s \notin E' \wedge e_e \notin E'$, i.e., *Pf* cannot contain any start or end event of *bp*
- $\exists F_{in}, F_{out} \subseteq F, F \cap ((O \backslash O') \times O') = F_{in} \wedge F \cap (O' \times (O \backslash O')) = F_{out}$; F_{in} and F_{out} are the set of entry flows and exit flows of *Pf*, respectively

- $\forall o_i \in O'$, $\exists o_m$, $o_n \in O'$, $\exists o_x \in O\backslash O'$, $\exists o_y \in O\backslash O'$, $\exists (o_x, o_m) \in F_{in}$, $\exists (o_n, o_y) \in F_{out}$, $o_x F'^* o_i$ $\wedge\, o_i F'^* o_y$, i.e., for every object o_i in $Pf.O'$ there exists a path from entry flow to o_i and from o_i to exit flow
- for every object $o \in O'$ there exists a path $p=(e_s, ..., f_i, ..., o, ..., f_o, ..., e_e)$ starting from e_s to e_e via $f_i \in F_{in}$, o, and $f_o \in F_{out}$

Let *boundary objects* of *Pf* be a set of entry and exit objects of *Pf* which all objects O' in *Pf* are bounded by boundary objects, such that,

- $\exists o_x \in O\backslash O'$, $\exists o_y \in O'$, $(o_x, o_y) \in F_{in}$; o_x is the *entry object* of *Pf*
- $\exists o_y \in O\backslash O'$, $\exists o_x \in O'$, $(o_x, o_y) \in F_{out}$; o_y is the *exit object* of *Pf*

Figure 3 depicts an example of various P-fragments of the process in the motivating example shown in Figure 1. The biggest P-fragment Pf_4 has only one entry object o_3 and one exit object o_{16}. P-fragment Pf_3 has two entry objects o_3 and o_9, and one exit object o_{16}. Similarly, Pf_2 has two entry objects o_3 and o_5, and one exit object o_{14}. Pf_1 has one entry object o_5 but it has two exit objects o_{12} and o_{14}. From the Definition 5, o_{18} and o_{20} are not accounted for exit objects of any P-fragment because they are not in the *normal path*.

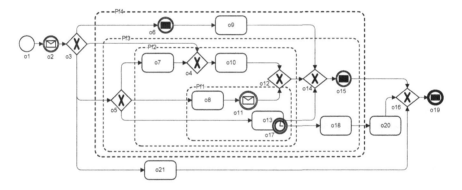

Fig. 3. P-fragments of the motivating example

3.2 Process View Consistency Rules

As stated before, every generated process view must preserve the structural and behaviour correctness when deriving its underlying process which can be the base business process or even inherited process views. In order to preserve such properties, a comprehensive set of *Process view consistency rules* for BPMN processes are defined. Since our previous work [4, 2] defines a set of consistency rules based on BPEL processes, we adapt and extend it as to comply with BPMN.

Assume that v_1 is a process view based on underlying collaboration process *cbp*, and v_2 is a process view constructed by applying an aggregation operation on process view v_1, then v_1 and v_2 must satisfy all consistency rules defined below.

Rule 1 (Order preservation). For any two objects belonging to process views v_1 and v_2, their execution order must be consistent if such objects exists in v_1 and v_2, i.e.,

If o_1, $o_2 \in \bigcup_{bp \in v_1.BP} bp.O \cap \bigcup_{bp \in v_2.BP} bp.O$, such that $o_1 Fo_2$ in v_1, then $o_1 Fo_2$ in v_2

Rule 2 (Branch preservation). For any two objects belonging to process views v_1 and v_2, the branch subjection relationship of them must be consistent, i.e.,

If o_1, $o_2 \in \bigcup_{bp \in v_1.BP} bp.O \cap \bigcup_{bp \in v_2.BP} bp.O$, such that $\neg(o_1 F^* o_2 \vee o_2 F^* o_1)$ then $lcp(\{o_1, o_2\})$ in $v_1 = lcp(\{o_1, o_2\})$ in v_2 and $lcs(\{o_1, o_2\})$ in $v_1 = lcs(\{o_1, o_2\})$ in v_2

Rule 3 (Event-attached task preservation). For any event-attached task relation belonging to v_1 and v_2, an existence of all coherence objects on the exception path led from such attached event must be consistent, i.e.,

If $(e, t) \in \bigcup_{bp \in v_1.BP} bp.T^E \cap \bigcup_{bp \in v_2.BP} bp.T^E$, such that $teObject(e, t)$ exists in v_1, then $teObject(e, t)$ exists in v_2.

Rule 4 (Message flow preservation). For any message flow exists in v_1 and v_2, the message flow relation of its source and target objects must be consistent, i.e.,

If o_1, $o_2 \in \bigcup_{bp \in v_1.BP}(bp.T \cup bp.E) \cap \bigcup_{bp \in v_2.BP}(bp.T \cup bp.E)$, such that $(o_1, o_2) \in v_1.M$ then $(o_1, o_2) \in v_2.M$.

3.3 Constructing an Aggregate

In this section, we define a set of aggregation rules and introduce a formal approach by extending the concept of *P-fragment* to validate the specified set of objects in the process whether it is able to be aggregated. If it is valid, then the result of aggregation is constructed and represented by single atomic task.

3.3.1 Aggregation Rules

Aggregation rules specify the requirements when constructing an aggregate. Let $O^A \subseteq O$ denote a set of objects in process view v_1 that have to be aggregated and let $agg(O^A)$ return an aggregate task in process view v_2 constructed from O^A such that every object in O^A exists in the *normal path* in v_1 and the aggregate satisfies every aggregation rule. We also demonstrate that this proposed set of aggregation rules conforms to *Process view consistency rules*, thus the aggregation operation maintain structural and behaviour correctness between v_1 and v_2.

Aggregation Rule 1 (Atomicity of aggregate). An aggregate behaves as an atomic unit of processing (task); therefore, it must preserve the execution order for every task and event within it, as well as between itself and the process.

It is conceived that the structure and behaviour of every object to be aggregated the aggregate remain internally unchanged. However, the relation and behaviour among those objects in O^A and the other objects $O \backslash O^A$ that are not in the aggregate need to be considered such that there must exist only one in-degree and out-degree of the aggregate which are the least common predecessor of O^A and the least common successor of O^A, respectively, i.e.,

$\forall o \in O^A$ in v_1, $lcp(agg(O^A)) = lcp(O^A) \wedge lcs(agg(O^A)) = lcs(O^A) \wedge | in(agg(O^A)) | = | out(agg(O^A)) | = 1$

This rule demonstrates the conformance to *Process view consistency rules*: (1) Order preservation and (2) Branch preservation.

Aggregation Rule 2 (Objects in exception path). If the task in event-attached task relation is in the aggregate then every object in its exception path must be hidden in the process view; thus, it is not considered to be in the aggregate, i.e.,

If there exists task $t \in O^A \cap T$ and event $e \in E$ such that $(e, t) \in T^E$, then every object $o \in teObject(e, t)$ must be hidden.

The concept behind this rule is that every object in the exception path is treated as an internal behaviour of a task having an event attached to, if the task is to be aggregated then, consequently, such event is to be hidden. Figure 4 shows an example of an application of this rule. If a set of objects $\{t_1, t_2, t_5\}$ is to be aggregated, then the set $\{e_1, e_2, e_3, t_3, t_4, t_6\}$ resulted from $teObject(e_1, t_1) \cup teObject(e_2, t_1)$ must be hidden.

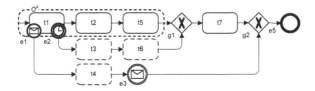

Fig. 4. Aggregating tasks with event-attached task

This rule demonstrates the conformance to *Process view consistency rules*: (3) Event-attached task preservation.

3.3.2 Structure Validation

In this section we propose an approach for structure validation of a given set of objects to be aggregated, called *Enclosed P-fragment*. This approach mainly checks the atomicity of the structure according to Aggregation rule 1. If it is valid, then the aggregate is able to be constructed. However, if it is not valid, we also propose the technique to find the minimum set of objects based on a given set in the next section.

Definition 6 (Enclosed P-fragment or EP-Fragment). Let $Pf(O', T', T^{E'}, G', E', F', F_{in}, F_{out})$ define a P-fragment of a process by the Definition 5. If Pf has only one entry object and one exit object as its boundary, then it is *enclosed*, called *Enclosed P-fragment* or *EP-Fragment*. We can claim that any *EP-Fragment* itself guarantees the atomicity of its whole structure.

Revisiting our motivating example in Figure 3, we can see that Pf_1 is not enclosed since there are two exit objects o_{12} and o_{14}; while o_{18} is not accounted for exit object as it is on the exception path from event o_{17} attached to task o_{13}. Similarly, Pf_2 and Pf_3 are unenclosed. The former has two entry objects o_3 and o_5, and one exit object o_{16}; likewise, the latter has two entry objects o_3 and o_9.

From the Definition 6, multiple entries and multiple exits are allowed for defining *EP-Fragment*. This also enables the selective aggregation of branches feature as illustrated by Pf_4 in Figure 3. The fragment Pf_4 is enclosed because it has only one entry object o_3 and one exit object o_{16}, although there are multiple branches coming in and going out from its fragment.

In order to validate the structure of the given set of objects to be aggregated, we have to find whether the given set of objects is able to form an *EP-Fragment*. To do so, two auxiliary functions are required: *forward walk* and *backward walk*.

Given any two flows in a process: $f_s = (o_x, o_s) \in F$ as an entry flow and $f_e = (o_e, o_y) \in F$ as an exit flow, we want to find two sets of objects, denoted as $\rho Fwd(f_s, o_y)$ and $\rho Bwd(f_e, o_x)$, by walking forward along all possible paths starting from f_s to o_y and by walking backward along all possible paths from f_e to o_x, respectively.

- A *forward walk* function $\rho Fwd(f_s, o_y)$ returns a set of objects by walking forward from f_s to o_y as well as from f_s to the end event of the process
- A *backward walk* function $\rho Bwd(f_e, o_x)$ returns a set of objects by walking backward from f_e to o_x as well as from f_e to the start event of the process.

These two functions can be implemented by extending the depth-first search algorithm so we do not detail them in this paper. Apart from them, we also require two functions to identify a set of objects that does not exist in forward walk but it is found in backward walk, and vice versa. Such functions will help us to validate the *EP-Fragment* as the technique will be described later.

Let function $objOutBwd(f_e, o_x)$ return a set of objects $O^{OB} \subseteq O$ such that it does not exist in forward walk but exists in backward walk and each of such object's flow directly links to the object which exists in both forward and backward walks, i.e., $\forall o_b \in O^{OB}, \exists o \in \rho Fwd(f_s, o_y) \cap \rho Bwd(f_e, o_x), (o_b, o) \in F$. Inversely, function $objOutFwd(f_s, o_y)$ returns a set of objects $O^{OF} \subseteq O$ such that $\forall o_f \in O^{OF}, \exists o \in \rho Fwd(f_s, o_y) \cap \rho Bwd(f_e, o_x), (o, o_f) \in F$.

We can see that if $\rho Fwd(f_s, o_y) = \rho Bwd(f_e, o_x)$, then $objOutFwd(f_s, o_y)$ and $objOutBwd(f_e, o_x)$ return $\varnothing$. This also implies that there exists only one entry flow to the forward walk from f_s to f_e and only one exit flow from the backward walk f_e to f_s.

From Figure 5, we want to find objects in forward and backward walks between an entry flow $f_s = (t_2, g_2)$ and an exit flow $f_e = (g_5, t_7)$. The result of $\rho Fwd(f_s, t_7)$ is $\{g_2, t_3, t_4, t_5, t_6, g_5, g_4, g_6, t_8\}$ and $\rho Bwd(f_e, t_2)$ is $\{g_2, t_3, t_4, t_5, t_6, g_5, g_4, g_1, t_1\}$. Consequently, $objOutFwd(f_s, t_7)$ returns $\{g_6\}$ since it exists in forward walk but does not exist in backward walk. Correspondingly, $objOutBwd(f_e, t_2)$ returns $\{g_1\}$.

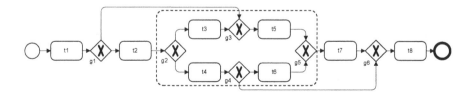

Fig. 5. An example of forward and backward walk in a process

Lemma 1: *Given a set of objects $N \subseteq O$ in a process bp (O, T, T^E, G, E, F), an EP-Fragment Pf $(O', T', T^{E'}, G', E', F', F_{in}, F_{out})$ can be formed by N, if and only if,*

$$- \bigcup_{(f_s, \, o_y) \in lcpF(N) \times lcs(N)} \rho Fwd(f_s, o_y) = \bigcup_{(f_e, \, o_x) \in lcpF(N) \times lcs(N)} \rho Bwd(f_e, o_x)$$

, i.e., the forward walks and backward walks of all combinations of $lcpF$ and $lcsF$
flows return the same result set identical to N in bp (1)

- $\forall f_p \in lcpF(N), \exists o \in N, f_p = (o_x, o)$, i.e., there exists only one entry object o_x (2)
- $\forall f_s \in lcsF(N), \exists o \in N, f_s = (o, o_y)$, i.e., there exists only one exit object o_y (3)

From Figure 3, assuming that $N = \{o_8, o_{11}, o_{13}\}$, then we can find that $lcp(N) = \{o_5\}$
and $lcs(N) = \{o_{14}\}$. Correspondingly, we will find $lcpF(N) = \{(o_5, o_8), (o_5, o_{13})\}$ and
$lcsF(N) = \{(o_{12}, o_{14}), (o_{13}, o_{14})\}$. Because $lcpF$ returns entry flows with only one entry
object o_5 and $lcsF$ returns exit flows with only one exit object o_{14}, therefore N satisfies
condition (2) and (3). After having applied both functions for every combination of
$lcpF(N)$ and $lcsF(N)$, the result sets of ρFwd and ρBwd are $\{o_8, o_{11}, o_{12}, o_{13}\}$ and $\{o_8,$
$o_{11}, o_{12}, o_{13}, o_{10}, o_7, o_4, o_3, o_2, o_1\}$, respectively. As we can see that only the result
from ρFwd is identical to N, but ρBwd is not (condition (1) is not satisfied), thus N
cannot be formed as an *EP-Fragment*.

Figure 6 shows a process with P-fragment Pf_1 in the loop structure. Assume that
$N=\{t_1, t_2, t_3\}$. Since we find that $lcpF(N) = \{(g_1, t_1)\}$ and $lcsF(N) = \{(t_3, g_1)\}$, then
conditions (2) and (3) are satisfied. However, when applying ρFwd and ρBwd functions, $\{t_1, t_2, t_3, g_2, t_4\}$ and $\{t_1, t_2, t_3, g_2\}$ are returned, respectively. The non-identical
results from both functions prove that objects N in Pf_1 can not form an *EP-Fragment*
by not satisfying condition (1). In contrast and clearly, objects in Pf_2 can form an *EP-
Fragment*.

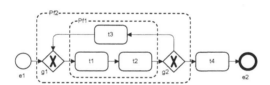

Fig. 6. P-fragments in a loop structure

Theorem 1: *A P-fragment Pf (O', T', $T^{E'}$, G', E', F', F_{in}, F_{out}) in a process bp (O, T,
T^E, G, E, F) can be aggregated if and only if it is enclosed.*

Proof: We prove the claim in two steps: (1) we present that the *EP-Fragment* can be
aggregated and it complies with the Aggregation rules 1 and 2; (2) we show that the
aggregate can form an *EP-Fragment*.

(1) Let Pf (O', T', $T^{E'}$, G', E', F', F_{in}, F_{out}) be an *EP-Fragment* and every object O' in
   Pf is to be aggregated. Aggregation Rule 1 is naturally satisfied by *EP-Fragment*
   (by the Definition 6). Similarly, Aggregation Rule 2 is satisfied by P-fragment
   (by the Definition 5).
(2) Let object $agg(O^A)$ be an aggregate represented as a single atomic task with one
   incoming flow (o_x, $agg(O^A)$) and one outgoing flow ($agg(O^A)$, o_y). Let
   $N=\{agg(O^A)\}$ and then we find $lcp(N) = \{o_x\}$ and $lcs(N) = \{o_y\}$; therefore

condition (2) and (3) of Lemma 1 are satisfied. Apply forward and backward walk functions will return identical result { $agg(O^A)$ }, therefore condition (1) of Lemma 1 is satisfied. So we can conclude that $agg(O^A)$ is an *EP-Fragment*.

3.3.3 Minimal Aggregate

As aforementioned, if a given set of objects cannot be aggregated, i.e., not able to form an *EP-Fragment* by Lemma 1, we facilitate users to be able to do so, by using our proposed minimal aggregate function. For a given set O^A, we can find the minimal aggregate of O^A which satisfies every Aggregation rule. We define $minAgg(O^A)$ as the function that returns a minimal set of objects that can be aggregated, and hides every object on exception paths. The implementation of $minAgg(O^A)$ is illustrated in Algorithm 1.

Algorithm 1. Finding minimal set of objects for the aggregation

---

$minAgg: O^A \to O$
1 **let** $O_{min} = \{\}, O^F = \{\}, O^B = \{\}, O^{OF} = \{\}, O^{OB} = \{\}$
2 **let** $O^{FT} = \{\}, O^{BT} = \{\}$
3 **let** $O_{temp} = O^A$
4 **do**
5   **for each** $(f_s: (o_x, o_s), f_e: (o_e, o_y)) \in lcpF(O_{temp}) \times lcsF(O_{temp})$
6          $O^F = \rho Fwd(f_s, o_y)$
7          $O^B = \rho Bwd (f_e, o_x)$
8          $O^{FT} = O^{FT} \cup (O^F \setminus O^{FT})$
9          $O^{BT} = O^{BT} \cup (O^B \setminus O^{BT})$
10         $O^{OF} = O^{OF} \cup (objOutBwd(f_e, o_x) \setminus O^{OF})$ // find the adjacent exit object
11         $O^{OB} = O^{OB} \cup (objOutFwd(f_s, o_y) \setminus O^{OB})$ // find the adjacent entry object
12   **end for**
13   **if** $(O^{FT} = O^{BT}) \wedge (\forall f_p \in lcpF(O^{FT}), \exists o \in O^{FT}, f_p = (o_x, o) \wedge (\forall f_s \in lcsF(O^{FT}), \exists o \in O^{FT}, f_s = (o, o_y))$ **then break** //break the loop if Agg Rule1 is satisfied
14     $O_{temp} = O_{temp} \cup O^{OF} \cup O^{OB}$
15 **while** $O^{FT} \neq O^{BT}$
16 $O_{min} = O^{FT}$
17 **for each** $t \in O_{min} \cap T$
18   **if** $\exists e \in E, (e, t) \in T^E$ **then**
19   //hide all objects belonging to path of event-attached task(Agg Rule 2)
20   **for each** $o \in teObject(e, t)$
21   **hide object** o and its corresponding flows
22   **end for**
23   **end if**
24 **end for**
25 **return** O_{min}

We explain why $minAgg(O^A)$ returns a minimal set of objects of O^A. Firstly, the given set of objects are validated whether it can form an *EP-Fragment* or not by applying Lemma 1. If it is able to form an *EP-Fragment*, then it is returned which initially satisfies Lemma 1 without extending its boundary. However, if it cannot form an *EP-Fragment*, then a set of adjacent objects resulted from *objOutFwd* and *objOutBwd* functions are added to the object set for each loop (lines 10-11). Since these two functions return only a set of direct adjacent objects which is necessary required intuitively; thus the additional set is minimal then we conclude that this algorithm guarantees the minimum expansion of the object set to form an *EP-Fragment*.

Theorem 2: *A set of objects* $O^A \subseteq O$ *in a process bp* (O, T, T^E, G, E, F) *satisfies all aggregation rules if and only if* $O^A = minAgg(O^A)$.

Proof: To prove this theorem, we need to construct an aggregate by $minAgg(O^A)$ that satisfies both Aggregation Rule 1 and Rule 2.

Aggregation Rule 1: From the Algorithm 1 for *minAgg*, initially we find the *lcpF* and *lcsF* of O^A. Then, all objects within the paths between *lcpF* and *lcsF* are found by ρFwd and ρBwd (lines 5-7). The while loop check if ρFwd does not return the result as identical to the result of ρBwd (line 4), then *Pf* is not enclosed (by Lemma 1, condition 1), then the adjacent objects resulted from *objOutFwd* and *objOutBwd* functions are added to O^A (line 14). Then O^A is repetitively computed for finding the *lcpF* and *lcsF* again finding the entry and exit flows that will be the boundary of the enlarged O^A. ρFwd and ρBwd are used to compare the result and then O^A is validated by *lcpF* and *lcsF* again to check whether it can form an *EP-Fragment* (lines 4-15). If it concludes that such new result set of both ρFwd and ρBwd are identical, and *lcpF* returns flows with one entry object and *lcsF* returns flows with one exit object (line 13), then O^A can form an *EP-Fragment* (by Lemma 1). This concludes that the result aggregate satisfies Aggregation Rule 1.

Aggregation Rule 2: For each object in the result aggregate set that satisfies Aggregation Rule 1, if there exists event e attached to task t in O^A, then such event and every object $o \in teObject(e, t)$ is not included into the aggregate and it is also hidden (lines 17-22). Thus, it satisfies Aggregation Rule 2.

If a given set O^A initially satisfies every condition in Lemma 1 (line 13) and every object in the exception path of every event that attached to task in O^A, then the result set will return the same as an original given set. Therefore, this concludes that $O^A = minAgg(O^A)$. In contrast, if the O^A is not able to satisfy Lemma 1, then the result set will be expanded; hence $|minAgg(O^A)| > |O^A|$. Thus, this concludes that O^A is not able to be aggregated.   □

3.3.4 Effect to Message Flows of an Aggregate

The aggregation has to preserve the consistency of message flow interactions among the process and its participants in the collaboration process. For every incoming and outgoing message flow of the object that have to be aggregated, it also remains for the aggregate, such that,

- $\forall m_x \in v_1.M, \exists o_j \in O^A, \exists o_k \in \bigcup_{bp \in v_1.BP} bp. O \backslash O^A, m_x = (o_j, o_k) \rightarrow (t_{agg}, o_k) \in v_2.M$
- $\forall m_y \in v_1.M, \exists o_m \in \bigcup_{bp \in v_1.BP} bp. O \backslash O^A, \exists o_n \in O^A, m_y = (o_m, o_n) \rightarrow (o_m, t_{agg}) \in v_2.M$

Figure 7 illustrates an example of object aggregation with message flows. All incoming and outgoing messages (a) are rearranged to the aggregate task (b).

This conditional effect of the aggregation satisfies *Process view consistency rules*: (4) Message flow preservation.

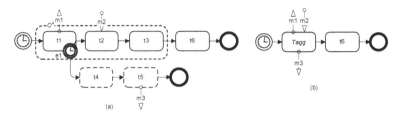

Fig. 7. Aggregation with messages

By considering the Aggregation Rules (1 and 2) and the conditional effect of message flows, we therefore can see that the aggregate task resulted from our aggregation approach satisfies to all Process view consistency rules.

3.4 Prototype

The prototype implementation, named *FlexView*, is currently being developed to support process view construction for BPMN process based on the approach proposed in this paper. The system initially loads the base BPMN file, and then allows users to specify which elements in the process will be aggregated. When the operation is completed, the process view is generated as an output of the system. Figure 8 shows the main screen of the system and an example of process view constructed from the base BPMN process displayed on the BizAgi Process Modeller [17]. Due to limited space, we do not show much detail in the prototype here.

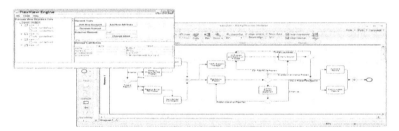

Fig. 8. *FlexView* engine (left) and result process view

4   Related Work and Discussion

Zhao, Liu, Sadiq, and Kowalkiewicz [4] proposed the process view approach based on the perspective of role-based perception control. A set of rules on consistency and validity between the constructed process views and their underlying process view is defined. Compared with our work, they neither provide how each process view is constructed nor consider non-well-structured processes.

Liu and Shen [6] presented an algorithm to construct a process view with an ordering-preserved approach from a given set of conceptual-level activities in a base process. In their approach, the aggregate activities called virtual activities requires to conform membership rule, atomicity rule, and order preservation rule. Compare with our work, they only focus on basic activity aggregation while they do not consider non-well-structured processes and the relation setting of activity in a collaborative process such as messages and event attachments (exception). We extend their work to allow such relations.

Van der Aalst, Dumas et al [16] proposed the framework for staged correctness-preserving configuring reference process models regarding the correctness of syntax and behavioural semantic captured by propositional logic formula. The proposed framework is based on *WF*-net and a set of transition variants used for the configuration: allowed, hidden, and blocked. Compare with our work, they do not provide an aggregation approach to construct the abstracted process model.

Bobrik, Reichert et al [18] presented a visualization approach to tackle inflexibility of building and visualizing personalized views of managed processes. They introduced two basic view operations: reducing and aggregating, and properties of process views. Graph reduction and graph aggregation techniques (by defining SESE region) are used for such operations. This work has some similarities compared with our P-fragment; however, the *EP-Fragment* allows multiple entries and exits to be applicable for selective aggregation of branches. In addition, their work focuses on process visualizing thus relaxing the preservation of structural and behaviour consistencies between base process and its resulted view, while our work is based on the comprehensive set of consistency rules. Their work also does not consider other aspects of BPMN properties, such as exception, but ours does.

Grefen and Eshuis [3] proposed a formal approach to construct a customized process view on a business process. The approach consists of two main phases: a process provider constructs a process view that hides private internal details of the underlying business process, and second phase let a consumer constructs a customized process view tailored to its needs to filter out unwanted process information. However, their approach focuses on block-structured process model represented by hierarchy tree model only and it does not take a graph structure into account. While it is too restrictive and unlikely to see those well-structured process in BPMN process, the approach presented in this paper adapted and extended from their work and our previous work [2] by considering non-well-structured process and event attachments features of BPMN.

Vanhatalo, Volzer, and Koehler [14] proposed a technique for decomposing workflow graphs into a modular and fine fragment by finding Canonical Fragments, and generate the Refine Process Structure Tree. In short, we aim at proposing an aggregate approach that satisfies aggregation rules specifically for BPMN process, while they only focus on finding the finest fragment of graphs.

5   Conclusion and Future Work

This paper presented a novel approach for constructing an aggregate for BPMN process views. The main contribution of this approach is that the core subset of current BPMN standard is taken into account in order to define a comprehensive set of construction rules and consistency rules. Since BPMN is likely to allow processes to be non-well-structured unlike some other standards such as BPEL which are strictly well-structured (block-structure), it is necessary to validate its structure using the *EP-Fragment* validation technique proposed in this paper. Our future work is to support process views for choreography processes in the BPMN 2.0.

Acknowledgement. The research work reported in this paper is supported by Australian Research Council and SAP Research under Linkage Grant LP0669660.

References

1. Object Management Group: Business Process Modeling Notation, V1.2 (January 2009), `http://www.omg.org/spec/BPMN/1.2`
2. Zhao, X., Liu, C., Sadiq, W., Kowalkiewicz, M., Yongchareon, S.: On Supporting Abstraction and Concretisation for WS-BPEL Business Processes. In: Zhou, X., et al. (eds.) DASFAA 2009. LNCS, vol. 5463, pp. 405–420. Springer, Heidelberg (2009)
3. Eshuis, R., Grefen, P.: Constructing customized process views. Data & Knowledge Engineering 64, 419–438 (2008)
4. Zhao, X., Liu, C., Sadiq, W., Kowalkiewicz, M.: Process View Derivation and Composition in a Dynamic Collaboration Environment. In: CoopIS 2008, pp. 82–99 (2008)
5. Liu, C., Li, Q., Zhao, X.: Challenges and opportunities in collaborative business process management. Information System Frontiers (May 21, 2008)
6. Liu, D., Shen, M.: Workflow modeling for virtual processes: an order-preserving process-view approach. Information Systems (28), 505–532 (2003)
7. Liu, D., Shen, M.: Business-to-Business workflow interoperation based on process-views. Decision Support Systems (38), 399–419 (2004)
8. Chiu, D.K.W., Cheung, S.C., Till, S., Karlapalem, K., Li, Q., Kafeza, E.: Workflow View Driven Cross-Organizational Interoperability in a Web Service Environment. Information Technology and Management 5, 221–250 (2004)
9. Chebbi, I., Dustdar, S., Tata, S.: The view-based approach to dynamic inter-organizational workflow cooperation. Data & Knowledge Engineering 56, 139–173 (2006)
10. Zhao, X., Liu, C., Yang, Y.: An Organisational Perspective on Collaborative Business Processes. In: van der Aalst, W.M.P., Benatallah, B., Casati, F., Curbera, F. (eds.) BPM 2005. LNCS, vol. 3649, pp. 17–31. Springer, Heidelberg (2005)
11. Schulz, L.A., Orlowska, M.E.: Facilitating cross-organisational workflows with a workflow view approach. Data & Knowledge Engineering 51, 109–147 (2004)
12. Van der Aalst, W.M.P., Weske, M.: The P2p Approach to Interorganizational Workflows. In: Dittrich, K.R., Geppert, A., Norrie, M.C. (eds.) CAiSE 2001. LNCS, vol. 2068, pp. 140–156. Springer, Heidelberg (2001)
13. Van der Aalst, W.M.P., Ouyang, C., Dumas, M., ter Hofstede, A.H.M.: Pattern-Based Translation of BPMN Process Models to BPEL Web Services. International Journal of Web Services Research (2007)
14. Vanhatalo, J., Volzer, H., Koehler, J.: The Refined Process Structure Tree. In: Dumas, M., Reichert, M., Shan, M.-C. (eds.) BPM 2008. LNCS, vol. 5240, pp. 100–115. Springer, Heidelberg (2008)
15. Liu, R., Kumar, A.: An Analysis and Taxonomy of Unstructured Workflows. In: van der Aalst, W.M.P., Benatallah, B., Casati, F., Curbera, F. (eds.) BPM 2005. LNCS, vol. 3649, pp. 268–284. Springer, Heidelberg (2005)
16. Van der Aalst, W.M.P., Dumas, M., Gottschalk, F.H.M., ter Hofstede, A., La Rosa, M., Mendling, J.: Correctness-Preserving Configuration of Business Process Models. In: FASE 2009, pp. 46–61 (2009)
17. BizAgi Process Modeller, BizAgi, `http://www.bizagi.com`
18. Bobrik, R., Reichert, M., Bauer, T.: View-Based Process Visualization. In: Alonso, G., Dadam, P., Rosemann, M. (eds.) BPM 2007. LNCS, vol. 4714, pp. 88–95. Springer, Heidelberg (2007)
19. Sadiq, W., Orlowska, M.E.: Analyzing process models using graph reduction techniques. In: CAiSE, pp. 117–134 (2000)
20. Vanhatalo, J., Volzer, H., Leymann, F.: Faster and More Focused Control-Flow Analysis for Business Process Models Through SESE Decomposition. In: Krämer, B.J., Lin, K.-J., Narasimhan, P. (eds.) ICSOC 2007. LNCS, vol. 4749, pp. 43–55. Springer, Heidelberg (2007)

Active Duplicate Detection

Ke Deng[1], Liwei Wang[2], Xiaofang Zhou[1],
Shazia Sadiq[1], and Gabriel Pui Cheong Fung[1]

[1] The University of Queensland, Australia
{dengke,zxf,shazia,pcfung}@itee.uq.edu.au
[2] Wuhan University, China
liwei.wang@whu.edu.cn

Abstract. The aim of duplicate detection is to group records in a relation which refer to the same entity in the real world such as a person or business. Most existing works require user specified parameters such as similarity threshold in order to conduct duplicate detection. These methods are called user-first in this paper. However, in many scenarios, pre-specification from the user is very hard and often unreliable, thus limiting applicability of user-first methods. In this paper, we propose a user-last method, called Active Duplicate Detection (ADD), where an initial solution is returned without forcing user to specify such parameters and then user is involved to refine the initial solution. Different from user-first methods where user makes decision before any processing, ADD allows user to make decision based on an initial solution. The identified initial solution in ADD enjoys comparatively high quality and is easy to be refined in a systematic way (at almost zero cost).

1 Introduction

The problem of data quality deservedly attracts significant attention from industry and research communities. An obvious reason is that the value of information fundamentally relies on the quality of the data from which it is derived. One widely studied data quality problem is duplicate detection which identifies the records not identical in representation but referring to the same real world entity. Duplicate data extensively exist in various information systems and impose evident impact in our daily life. In the database community, this problem has been studied for decades and described as merge-purge [1,2], data deduplication [3], instance identification [4] and entity resolution [5]; the same task has also been known as record linkage or record matching [6,7,8,9] in statistics community.

Given a set of records where their pairwise similarities are known, duplicate detection aims to identify duplicates and groups them together. We say a group is correct if it contains only and all records referring to the same entity. Previous works partition records into groups where the highly similar records are in the same group. Such approaches require a *global similarity threshold* to decide whether records should be grouped together. The records that are grouped together are regarded as duplicated records [2,10,11,12]. However, these approaches usually lead to results with poor *recall* and *precision* [13]. [13] indicates that the local structural properties are very important consideration in duplicate detection. [13] proposes two criteria to capture the

H. Kitagawa et al. (Eds.): DASFAA 2010, Part I, LNCS 5981, pp. 565–579, 2010.

local structural properties. The idea is that duplicate records should be close to each other and each duplicate record should have sparse neighbors, i.e. limited neighbors in proximity. The neighbor limit in proximity can be viewed as a *relative global threshold* because it is applied to entire dataset but the size of the proximity area is respective to each record (i.e. decided by the distance from the record to its nearest neighbor). The method proposed in [13] is called DE (i.e. duplicate elimination) in the rest of this paper.

Regardless of which of the existing methods we use, a common limitation is that users need to estimate the *relative/absolute global threshold* in advance. Unfortunately, without understanding the distribution of the data clearly, it is very unlikely to achieve a good estimation. We call these methods *user-first*.

Unlike the existing work, this work does not require user to specify such thresholds for duplicate detection. Specifically, we propose a solution called Active Duplicate Detection (ADD). ADD is based on a new concept of *duplicate principle*. With the duplicate principle, the aggregation characteristics of the duplicate records can be quantitatively evaluated without setting parameters. It follows the observation that the duplicate records tend to be closer to each other and far from others in proximity. For example, if a record is close to a group, we have less confidence on the correctness of this group no matter how far away other records are because this record may belong to this group but was missed. In DE, the similar principle is implied by the two criterion as discussed above. However, DE is different from our method. The criterion are qualitative because it cannot distinguish which one is more likely to be correct if two groups both satisfy the criterion. In contrast, our method is based on quantitative measures of duplicate principle over groups. For a group, it equals to the rate by comparing the group diameter (i.e. the maximum similarity between records in this group) and the distances from this group to other records. With support of the quantitative measures, ADD first finds an initial solution without forcing user to specify parameters, and then user is involved to refine the initial solution. We say that ADD is *user-last*.

In the user-first methods, the final solution is shaped by the user specified parameters which are assumed to be globally applied. But this assumption is not true in many scenarios. The flexible nature of this problem indicates that correct groups usually have very different parameter settings. While it is hard for user-first methods to handle this situation, the robustness can be provided by user-last ADD. The initial solution of ADD is only based on the quantitative measures over the duplicate principle and no parameter is required. This prevents applying improper parameter settings and falsely pruning some correct groups. We argue that it is favorable for user to judge the correctness of groups in the initial solution returned by ADD compared to setting parameters as in the user-first methods.

ADD follows the observation that the duplicate records tend to be closer to each other and far from others in proximity. ADD is different from the typical clustering problem. The goal of clustering is to separate a dataset into discrete subsets by considering the internal homogeneity and the external separation. Once a similarity measure is chosen, the construction of a clustering criterion function usually makes the partition of clusters a global optimization problem. In our problem, we don't have global optimization objective, and we are interested to find the individual optimal groups, i.e. the most isolated. Note that ADD is orthogonal with the similarity measure between records[1]. That is, any distance function can be chosen, such as edit distance or cosine similarity.

In [13], in addition to the *relative global threshold*, there are two more optional parameters, i.e. a solution can be provided without setting of these parameters. One is the maximum number k of records in a correct group and the other is the maximum similarity distance θ between records in a correct group. If user has knowledge on k and θ, the solution can be improved noticeably. In ADD, the settings of k and θ are optional as well. If user has the ability to specify k and θ, ADD generally provides the initial solution with better quality; otherwise, the initial solution needs more refinement. Fortunately, ADD has the mechanism to easily perform refinement. Without the loss of generality, we suppose the k and θ can be specified by user in this work since the effect of no settings of k and θ can be achieved by simply setting k and θ large enough values. ADD is denoted as $ADD(k, \theta)$ when k and θ are specified.

The remainder of this paper is organized as follows. The work related to this problem is reviewed in section 2 and the duplicate principle is discussed in section 3. After that, we introduce the active duplicate detection in section 4 where the duplicate detection tree and the algorithm for search initial solution are introduced. The experimental results are reported in section 5 and the paper is concluded in section 6.

2   Related Work

Duplicate detection has practical significance in data management in particular in large scale information systems. This problem has received a lot of attention from research communities, see [14] for a comprehensive survey. Previous duplicate detection techniques can be classified into supervised and unsupervised approaches [15]. Supervised approaches learn rules from training data which contains known duplicates with various errors observed in practice [12,11]. However, these approaches depend on the training data which is not always available or comprehensive enough even with some active learning approaches [3,16].

The unsupervised methods detect duplicates by applying distance functions to measure the similarity between records. Various similarity functions have been proposed in literature such as edit distance and cosine similarity. In addition, some efforts have been put on learning similarity functions from training datasets [12,11]. The adaptive string similarity metrics is proposed by [11] for different database and application domains in order to improve the accuracy of the similarity measure. [12] develops a general framework of similarity matching by learning from the training dataset consisting of errors and respective solutions. Based on the pairwise similarities, previous duplicate detection techniques apply clustering algorithms to partition records into groups of duplicates based on the pairwise similarity. The highly similar records are in the same group. The single linkage clustering [17] is a widely accepted choice to partition records into groups. Using this method, the dissimilarity between two groups is the minimum dissimilarity between members of the two groups. It works well when the groups of duplicates are very small (of size 2 or 3) [13,18]. If more records belong

---

[1] If two records are similar, they have high similarity, and the distance (or similarity distance) between them is small.

to the same entity, the groups using single linkage may produce a long chain. That is, two records in the same group may be very dissimilar compared to some records in two different groups. Using clustering techniques, duplicate detection requires an *absolute global threshold* to decide whether two records are duplicates [2,10]. However, the global distance threshold leads to a solution with poor recall and precision [13].

To overcome this drawback, [13] points out that local structural properties are very important in duplicate detection. The insight is that if several records are duplicates they should be similar to each other and each duplicate has limited number of neighbors around. This observation is captured by two criteria, i.e. *compact set* and *sparse neighborhood*. The *compact set* means that any records in the group are closer to each other than to any other record, and the sparse neighborhood means that the neighbors of each record in the group are less than a specified number in proximity. These criteria are qualitative since two groups satisfying them are no different. If a group has more than two subgroups and they all meet the requirement of the criteria, it is hard to distinguish which one is more likely to be correct. Thus, the subgroups are always returned in the final solution in [13]. While the compact set criterion is straightforward, the number of neighbors allowed in proximity in sparse neighborhood criterion is a user specified *relative global threshold* since the number of neighbor applies to all groups but the definition of proximity is determined by each record and its nearest neighbor. User needs to estimate the percentage of duplicates and understand the distribution of the dataset.

3   Duplicate Principle

Given a set of records R where no two records in R are identical, duplication detection is to identify all of the records in R that refer to the same entity. In other words, we are trying to group the records in R such that each group maps to one and only one real world entity and no two groups refer to the same entity. Based on the properties of each group, we can define two concepts, namely, *correctness* and *incorrectness*:

Definition 1: Correctness and Incorrectness. Given a group g, we say that its grouping is correct if it contains all records referring to the same entity; otherwise, if it contains some records from other entities or does not contain all records from the same entity, we say that its grouping is incorrect.   □

According to Definition 1, we have the following definition called Confidence on Grouping:

Definition 2: Confidence on Grouping. Given a group g, if g is more likely to be correct, we say that we have more confidence on its grouping.   □

It is intuitive that the duplicated records usually share more common information and thus are more similar to each other. Let us take an example to illustrate the idea of confidence on the grouping. Suppose there is a group g which is formulated by grouping two records: $r_i \in R$ and $r_j \in R$. If there are many records around g such that the distance from any point to r_i is even shorter than the distance between r_i and r_j, then it is obviously not intuitive to group r_i and r_j together. Hence, the confidence of grouping is very low in this situation. As such, the confidence on grouping is related to the distance

Matrices of pairwise similarities

	a	b	c	d	e	f	i	h
a	0	1	8	6.5	9	10	10	9.5
b	-	0	8	7	8	8.5	9	8.5
c	-	-	0	4.5	6.5	7	6.8	6
d	-	-	-	0	10	10.5	12.5	10
e	-	-	-	-	0	0.5	2.1	1
f	-	-	-	-	-	0	1	1
i	-	-	-	-	-	-	0	0.8
h	-	-	-	-	-	-	-	0

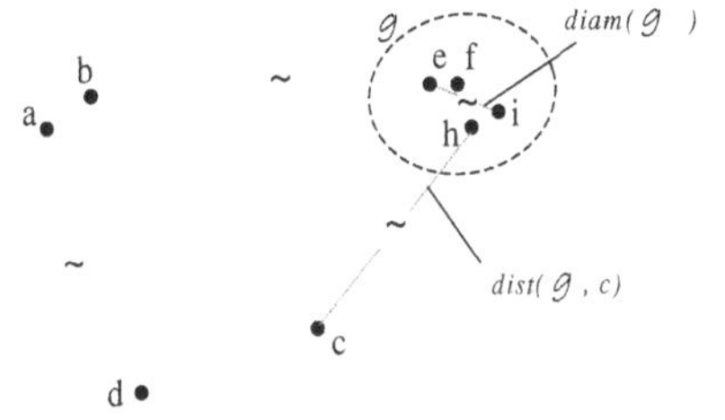

Fig. 1. A schematic example of duplicate principle measure

of the points around a group. This observation is similar to the criteria of compact set and sparse neighborhood in [13]. But we further indicate that the closer point to a group will have a higher impact on our confidence than the farther one to the same group. This point enables us to quantitatively measure the confidence of grouping by considering an increasingly wider proximity so as to avoid the parameter settings as in [13]. We formally define a concept called Duplicate Principle:

Definition 3: Duplicate Principle (DP). Given a set of records R and a group $g \subset R$, our confidence that the records in g are more likely to refer to the same entity is decided by three factors: 1) the records in g tend to be closer to each other, and 2) the other records tend to be farther to g, and 3) the record (not in g) closer to g will have a higher impact to our confidence than the record (not in g) farther to g. □

For two record pairs (r_k, r_f) and (r_i, r_j), suppose they have the same similarity distance. (r_i, r_j) is surrounded closely by many other records while (r_k, r_f) is isolated. Intuitively, (r_k, r_j) is more likely to form a correct group than (r_i, r_j). In other words, (r_k, r_f) has some common textual characteristics which are rarer in other records comparing to (r_i, r_j).

The duplicate principle of a group can be quantitatively appraised to demonstrate how likely this group is correct. For a group $g\{r_1, .., r_n\} \subset R$, let $\max_{\forall r_i, r_j \in g}$ $(dist(r_i, r_j))$ be the *diameter* of g, denoted as *diam(g)*. The diameter is the maximum similarity distance among records in g which describes the *tightness* in g. For any record $r \in R - g$, the similarity distance to g is defined as $min_{\forall r_i \in g} dist(r_i, r), r \in R - g$, the distance to the most similar records in g, denoted as $dist(g, r)$. The greater the $dist(g, r)$ is, the greater the reverse effect of r will be on confidence of g's correctness. According to the duplicate principle, records in g are more likely to be duplicates if $diam(g)$ is smaller and $dist(g, r)$ is greater. The ratio between $diam(g)$ and $dist(g, r)$ is used to measure the duplicate principle of g, denoted as $g.DP$, when a single record $r \in R - g$ is considered.

$$g.DP = \frac{diam(g)}{dist(g,r)}. \tag{1}$$

The smaller DP means that we have more confidence on the correctness of g and the greater DP means that g is more likely to be incorrect. Two situations may cause g to be incorrect. First, g may contain * records referring to some others entities, and the second situation is that some other records should be included in g but are not. In any of these two situations, g is not a correct group. Figure 1 shows a schematic example where h, i, e, f form a group g. The diameter of g is the similarity distance between e and i. For the record c, its distance to g is $dist(g,c)$. We want to point out that the proposed method does not use any spatial techniques and thus it is applicable to various similarity distances functions including edit distance and cosine similarity.

In formula 1, only the reverse effect of a single record $r \in R - g$ is considered in the DP measure of g. If there are more records in $R-g$, a simple method is to aggregate the reverse effects of all records in $R-g$ together, i.e. $g.DP = diam(g) \sum_{\forall r \in R-g} \frac{1}{dist(g,r)}$. To some extent, this simple aggregation reveals the overall reverse effect of all records in $R - g$ to our confidence to the correctness of the group g. But the reverse effect is not gradient descent along with the distance away from g as pointed out in the definition of duplicate principle. Consider $r \in R - g$ is very close to g, it is a strong implication that g is less likely to be correct no matter whether or not all other records are far away from g. That is, r in closer region of g has dominant reverse effect to our confidence of g's correctness.

Thus, the reverse effects of records in $R-g$ should be considered at different distance ranges. A distance range is represented as $[lowbound, upperbound]$. To g, objects with similarity distance falling in $[lowbound, upperbound]$ are in the same distance range. In different distance ranges, the objects are processed separately. An example of distance range is shown in figure 2. In this work, distance ranges are consecutive and equal in length. For each group, the unit length of the distance range is different and the group diameter is a natural choice since the reverse effects are relative to the tightness of g. For example, for a group g with a small diameter, a record $R - g$ has strong reverse effect if its distance to g less than the diameter of g. In this situation, g is very likely to be incorrect. Given a group g, the first distance range is $[0, diam(g))$ (no less than 0 and less than $diam(g)$) and the second distance range is $[diam(g), 2 * diam(g))$, and so on. If a record $r \in R - g$ has $dist(g,r)$ in between $[i * diam(g), (i+1) * diam(g)), i \geq 0$, we say r is in the ith distance range.

For the records in the same range, their reverse effects are similar and simply aggregated as the reverse effect of this distance range, denoted as $g.dp_i$. The following is the formula for calculating $g.dp_i$.

$$g.dp_i = \frac{diam(g)}{dist(g,r)}, \quad i = 0, 1, 2.., \quad for \; \forall r \in R - g \; and$$

$$i * diam(g) \leq dist(g,r) < (i+1) * diam(g) \tag{2}$$

Usually, $g.dp_0$ is computed first since it has the most dominant effect to our confidence to g's correctness; if necessary, the dp in next range is computed. When a set of dp, $(g.dp_0, g.dp_1, ..)$ are computed, we say the DP measure of g, $g.DP$, is

$(g.dp_0, g.dp_1, ..)$. For each given $(g.dp_0, g.dp_1, ..)$, it has practical meaning. For example, $(g.dp_0 = 0, g.dp_1 = 0, g.dp_2 \neq 0..)$ indicates that no record has distance to g less than $2 * diam(g)$.

Given two groups g, g', their DP measures can be compared to evaluate which one is more likely to be correct. We first discuss the situation that $dp_0 = 0$. So, the comparison of DP measures starts from dp_1. If $g.dp_1 < g'.dp_1$, g is more likely to be correct than g' no matter the dp in next range, and we say g dominates g'. If $g.dp_1 = g'.dp_1$, $g.dp_2$ and $g'.dp_2$ needs to be computed and compared. This operation is repeated until the first range where their dps are different is identified. In practise, the number of ranges are usually very small to distinguish two groups (1-2 ranges). The relationship between two groups based on the DP at different ranges can be defined as follows.

Definition 4: Domination. One group g dominates another group g' if $g.dp_i < g'.dp_i$ and $g.dp_j = g'.dp_j, (i = 0, 1, 2.., j = 0, 1, 2, .., 1 \leq j < i)$. This relation is denoted as $g \succ g'$ or $g.DP \succ g'.DP$.   □

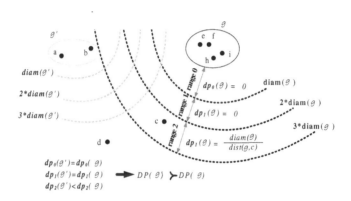

Fig. 2. A schematic example of domination relationship between groups

We just discussed the situation that $g.dp_0 = 0$. In the situation that $g.dp_0 \neq 0$, it means some record(s) in g are closer to some other record(s) not in g than to some record(s) in g. This is a strong indication that g is less likely to be correct. Thus, a group likely to be correct must have $dp_0 = 0$. This requirement is same as the concept of compact set [13] introduced in section 2. Figure 2 shows the domination relationship between two groups g, g' based on $g.DP, g'.DP$. Since $g.dp_0 = g'.dp_0 = 0$, we compare $g.dp_1$ and $d.dp_1$. Since $g.dp_1 = g.dp_1 = 0$, $g.dp_2 = 0$ and $g'.dp_2 = 0.35$ are compared. Since $g.dp_2 < g'.dp_2$, $g \succ g'$, i.e. we have more confidence on g's correctness than g'. Based on the domination relationship between records, the active duplicate detection method is developed.

4   Active Duplicate Detection

Given a set R of records, the active duplicate detection identifies an initial solution of record groupings without any user specified parameter settings. The objective is to return disjoint groups in the ascending order of DP measure. The task can be fulfilled in concept as follows. The group g with the most dominant DP measure is found first. If g also satisfies the additional conditions, i.e. $g.dp_1 = 0$, $|g| < k$ and $diam(g) < \theta$, g is inserted into an initial solution and the included records are removed from R; otherwise, this group is not dropped and the included records remain in R. As aforementioned, k, θ are two global constraints. k is the maximum number of records allowed in a correct group and θ is the maximum similarity distance allowed for records in a correct group. Both k, θ are optional settings provided by user. Then in remaining records in R, the group g' with the most dominant DP measure is found. If the additional conditions are satisfied, g' is insert into the initial solution; otherwise it is dropped. A similar operation is repeated until all qualified groups have been inserted in the initial solution. The cost to do that is prohibitively high due to the combinatorial nature of this problem (NP problem). To reduce the search space, we introduce the duplicate principle tree.

4.1   Duplicate Principle Tree

The duplicate principle tree (DP-tree) for R is a bottom-up process. Initially, each record is in a single group. Each time we find the pair which can form a new group with the most dominant DP measure. That is, we have the strongest confidence that the newly formed group is correct compared to the group formed by merging any other pairs. Let (r_i, r_j) be such a pair. The new group formed is $\{r_i, r_j\}$. The following construction of DP-tree is based on the observation that r_i, r_j should be in the same group. In other words, they cannot be split into two different groups. To prove that, we assume that r_i, r_j are split into two groups $r_i \in g_i$, $r_j \in g_j$ in some stage of the DP-tree construction. The group $\{r_i, r\}$ ($r \in g_i$) must be dominated by $\{r_i, r_j\}$ since $\{r_i, r_j\}$ has the most dominant DP. This is same for r_j and other records in g_j. This means that r_i, r_j are more likely to be duplicate than distinct. It is not reasonable to split r_i, r_j into two groups. This observation can be recursively extended to each group that consists of more than two records. In this work, we call the situation that r_i and r_j will stay in the same group once they are merged *unbreakable group rule*.

Unbreakable Group Rule. Among current groups $G = \{g_1, .., g_n\}$, if a group formed by pair of groups $g_i, g_j \in G$ dominates any group formed by any other pair of groups in G, g_i, g_j are merged to create a new group $\{g_i, g_j\} = g_i \cup g_j$. In the following construction of DP-tree, this new group may merge with other group to form a super group. $\{g_i, g_j\}$ is termed unbreakable, that is, g_i, g_j are always in the same group from now on. □

Following the unbreakable group rule, the groups are merged repeatedly until all records in R are in a single group. The duplicate principle tree is constructed. An example is shown in Figure 4.

In DP-tree, a non-leaf node contains a group formed by merging its two child nodes, and the DP measure is attached. When two groups are merged, the group diameter tends to increase. Note that the records around the resultant group can be close or far away. Thus, the DP measure of the resultant group may increase or decrease. That is, DP does not change monotonically along the path from leaf nodes up to the root in the tree. For this reason, all groups need to be generated in order to find the most dominant groups, i.e. satisfying the duplicate principle the most. The groups at leaf nodes are individual records. The diameter of each such group is 0 and thus $dp = 0$ at all ranges. The group at root contains all records, $dp = 0$ at all ranges as well. Since the root and leaf nodes should be excluded from the solution, their dp_0s are set to be ∞.

Figure 3 is the pseudo-code to construct the DP-tree. When a new group is generated, some operations are performed (line 11-23). From line 11-17, DP of the new group is computed. The DP is compared with the DPs of the groups in its child nodes. The most dominant DP, denoted as $minDP$, is tagged with the new group. It is also marked whether this new group itself has the most dominant DP using a sign $flag$. This is necessary since DP does not change monotonically along the path from leaf node to the root in tree. When browsing the tree, if a node with $flag = 1$, it means that DP of this node dominates that of its child nodes. By doing that, unnecessary tree access can be avoided. In line 18-21, the parent-child relationship is set up by assigning pointers to relevant nodes.

Figure 4 is the duplicate tree of the same example in figure 1. The records which are close to each other and isolated from other records in proximity have dominant DP

Algorithm DP-tree(R)
Input: R a set of records $\{r_1, .., r_{|R|}\}$
Output: DP-tree of R
1. $Output = \emptyset, Tree = \emptyset$; //initialize group sets
2. for each $r_i \in R$
3. $g_i = \{t_i\}$; $g_i.flag = 1$;
4. $g_i.diamter = 0$; $g_i.DP = \infty$;
5. $g_1, , g_{|R|}$ are leaf nodes of $Tree$;
6. $c = 1$;
7. while $|R| > 0$
8.   compute DP for all pairs in G;
9.   if $\{g_i, g_j\}.DP$ dominates DP of all other pairs
10.   create a new group $g_{|T|+c} = \{g_i, g_j\}$;
11. $g_{|T|+c}.DP = \{g_i, g_j\}.DP$;
12. $g_{|T|+c}.flag = 0$;
13.   remove g_i, g_j from G, record $g_{|T|+c}$ in $Tree$;
14. $g_{|T|+c}.minDP = min(g_i.minDP, g_j.minDP)$;
15.   if $g_{|T|+c}.DP \succ min(g_i.minDP, g_j.minDP)$
16. $g_{|T|+c}.flag = 1$;
17. $g_{|T|+c}.minDP = g_{|T|+c}.DP$;
18. $g_i.parent = g_j.parent = g_{|T|+c}$;
19. $g_{|T|+c}.leftChild = g_i$;
20. $g_{|T|+c}.righChild = g_j$;
21.   c=c+1;
22. end while
23. return $Tree$;
END DP-tree

Fig. 3. Pseudo-code for DP-tree construction

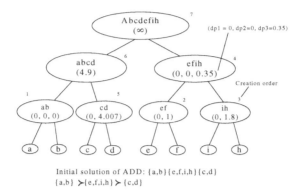

Initial solution of ADD: {a,b}{e,f,i,h}{c,d}
{a,b} ≻{e,f,i,h} ≻ {c,d}

Fig. 4. An example of duplicate tree

measure such as a, b. So, a, b are selected and merged to form a new group $\{a, b\}$. According to the Unbreakable Group Rule, they are not separated in subsequent grouping operations. Compared to group $\{a, b\}$, the distance of c, d is greater and many other records are close. So, the DP of $\{c, d\}$ is clearly dominated by a, b. For the group $\{e, f\}$ and $\{i, h\}$, their DPs are $(0, 1)$ and $(0, 1.8)$. When $\{e, f\}, \{i, h\}$ are merged to be a single group $\{e, f, i, h\}$, the DP decreases even though the diameter increases compared to $\{e, f\}, \{i, h\}$. The reason is that $\{e, f, i, h\}$ is farther away from other records. The DP-tree will then be searched to find the initial solution of ADD.

4.2 Initial Solution and Refinement

Active duplicate detection $ADD(k, \theta)$ first aims to identify an initial solution which can be formally defined as follows:

Initial Solution: given a set R of records, k and θ, $ADD(k, \theta)$ retrieves a set G of disjoint groups from the DP-tree of R. Each group $g \in G$ meets three conditions: 1) $g.DP$ dominates the DP measure of any other groups in DP-tree but not in G, and 2) $g.dp_0 = 0$, and 3) $|g| < k, diam(g) < \theta$.   □

The initial solution is obtained by browsing DP-tree. The search starts from the root. All child nodes are read in a heap H. The node in the heap with the most dominant DP measure is visited and replaced in H by its child nodes. Each time when a node n is visited, we examine whether its DP measure is less than that of all its child nodes in the subtree under n (not only direct child nodes), n is picked up and checked for condition 2 and 3 which are straightforward. Once all conditions are met, this node is inserted into G and its subtree will be pruned. If the conditions are not met, the child nodes are replaced in H. When a leaf node is visited, it is dropped. The search terminates until all nodes in the tree are processed or pruned.

The pseudo code for searching initial solution of $ADD(k, \theta)$ is presented in Figure 5. $ADD(k, \theta)$ invokes function *SEARCH* to recursively visit the nodes in the DP-tree which is constructed by using algorithm *DP-TREE*. In the tree, a node is marked $g.flag = 1$ if its DP dominates its child nodes's DPs. $ADD(k, \theta)$ returns a set of disjoint groups satisfying the conditions of initial solution.

Algorithm $ADD(R, k, \theta)$
Input: R a set of records $\{r_1, .., r_{|R|}\}$
Output: the initial solution
1. $Tree{=}DP\text{-tree}(R)$;
2. $cnode{=}Tree.root$;
3. $output = \emptyset$;
4. $SEARCH(cnode, output, k, \theta)$;
5.   return $output$;
END ADD

Algorithm $SEARCH(node, S, k, \theta)$
Input: θ-diameter threshold, k-group size threshold
Output: S, initial solution
6. $g = node.g$; //group in this node
7.   if $node.flag = 1$ and $diam(g) < \theta$ and $|g| < k$
8.   insert g into output;
9.   return;
10. $SEARCH(node.leftChild, InitialS, k, \theta)$;
11. $SEARCH(node.rightChild, InitialS, k, \theta)$;
END SEARCH

Fig. 5. Pseudo-code for finding the initial solution

Refinement: once the initial solution is returned, the groups in G can be sorted on the dominance relationship of DP measures. The groups with the dominant DP measures are more likely to be correct compared to others. A threshold τ of DP measure can be specified to divide G into two subsets G_1, G_2. If $g \in G_1$, $g.DP \prec \tau$ and $g.DP \succ \tau$ if $g \in G_2$. The groups in G_1 have the most dominant DP measure and thus we have strong confidence of their correctness. So, the groups in G can be returned to the final solution directly. The high precision of G_1 is expected (precision measures the percentage of correct groups in the returned solution). The groups in G_2 need to be examined and refined by user. For a group $g \in G_2$, if user suspects the correctness of g, an exploratory solution can be provided according to which of the two situations (discussed in section 3) are potentially causing the incorrectness of g. If user suspects that g contains records referring to different entities, the subtree under g can be searched to find subgroups in the way as finding initial solution from the tree. If user suspects that g should contain more records, its parent group will be provided. In this case, k and θ are ignored as these global settings can be a reason behind the potentially incorrect group.

5   Experiments

The experimental study covers two aspects of the problem, quality (i.e. precision and recall) and efficiency. In the quality aspect, we compare the solutions using our approach and that using DE developed in [13]. Even though active duplicate detection (ADD) allows user to explore potentially better solution, we still use the initial solution in the comparison for fairness. The impact of reference set to the quality of initial solution is also studied. In the efficiency aspect, several large datasets are tested using ADD in terms of scalability.

5.1   Experiment Setup and Evaluation Metrics

The experiments are conducted using a PC (Intel Core2 6600 CPU 2.4GHz, 2GB memory). We consider the real datasets publicly available from the Riddle reposi-tory: $Restaurants, BirdScott, Parks, Census$. These four datasets are also used in the experiment in [13]. $Restaurant$ is a database containing 864 restaurant records that contain 112 duplicates (i.e. 112 entities have duplicates). Each record is com-posed of four fields: $Name, Address, City$ and $Cuisine$. Based on $Restaurant$, we create two datasets $Short\ Restaurants$ ($Restaurants[Name]$) and $Long\ Restau-rants(Restaurants[Name, Address, City, Cuisine])$ in order to test the effect of record length to the solution. $BirdScott$ is a collection of 719 bird records that contain 155 duplicates. $Parks$ contains 254 duplicates in 654 records. $Census$ is a collection of 833 records that contain 333 duplicates. In each dataset, the similarity between two records are measured using the normalized edit distance (divided by the length of the longer record).

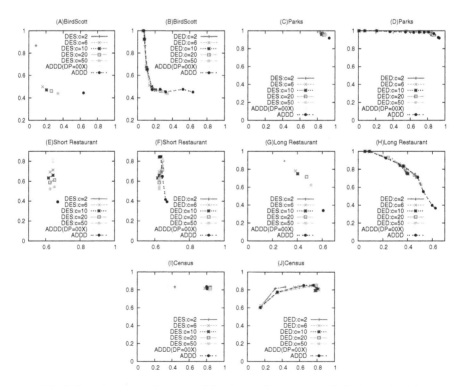

Fig. 6. Experiments results on *precision (vertical axis) vs. recall (horizontal axis)*

The *precision* and *recall* metrics are used to evaluate the quality. In our experiments, precision measures the percentage of correct groups in the returned solution and recall is the percentage of correct groups returned from all correct groups. Higher recall and precision values indicate better quality. We use *precision* vs. *recall* graph to show the experiment results where the $(recall, precision)$ values for various parameter settings are demonstrated.

For DE, the minimum compact set is always selected since it is required in theory by [13]. That is, if one group satisfying the criteria (i.e. compact set and sparse neighborhood) has more than two disjoint subgroups satisfying the criteria as well, the subgroups are selected in the final solution. However, when conducting experiments in [13], the minimum compact set is not strictly required in the final solution. The reason as explained by [13] is that the non minimum compact set is rare and so it is not necessary to check it. In our experiment, we strictly require that the groups in the final solution of DE are minimum compact sets to make the experiment fair. For the parameter c of sparse neighborhood criterion, we use the same settings as that used in [13] ($c = 6$) for the same datasets $Restaurants$, $BirdScott$, $Parks$ and $Census$. We also test the situation when c is set at $[2, 6, 10, 20, 50]$ in order to examine the impact of c to the precision and recall of the final solution using DE. For the proximity associated with c, we use the same definition as that in [13], i.e. the double of the distance from each record to its nearest neighbor.

For ADD, the initial solution is returned first. A fraction of initial solution with the best duplicate principle measure can be reported directly due to high confidence. We tested the accuracy and recall of this fraction by setting the DP measure as $(dp_0 = 0, dp_1 = 0, dp_2 \neq 0)$ in the experiments, denoted by $DP = 00x$. Any group with the DP dominating $DP = 00x$ is in the first fraction of the initial solution. The meaning of $(dp_0 = 0, dp_1 = 0, dp_2 \neq 0)$ is that no record has distance to the group less than the double of the group diameter.

Two optional parameters tested are k, θ where k is the size of a correct group (the maximum number of records) and θ is the diameter of a correct group (the maximum similarity distance of records). As discussed in section 1, they are optional parameters specified by user to indicate the condition of a correct group. To demonstrate the impact of k, θ, they are tested separately. When the setting of k changes from 2 to 5 with step length 1, θ is not specified. Similarly, when the setting of θ changes from 0 to 1 with step length 0.05, k is not specified. In all figures, $DES, ADDS$ denote algorithm DE, ADD running at different setting of k, while $DED, ADDD$ donate algorithm DE, ADD running at different settings of θ. $ADDS(DP = 00x)$ denotes the algorithm ADD by only selecting the first fraction of the initial solution.

5.2   Precision vs. Recall

In figure 6, the experiment results for five different datasets are demonstrated. For each dataset, the impact of k, θ are presented in two separate figures. This depicts the situation that the given threshold $(dp_0 = 0, dp_1 = 0, dp_2 \neq 0)$ is usually reliable for the first fraction of initial solution with a high precision compared to the final solution of DES. Thus, the groups in the first fraction of the initial solution can be directly returned to the final solution. In the same figures, it is also clear the corresponding recall of $ADD(DP = 00x)$ is relatively low. But, this is not a problem for ADD since the first fraction is only a part of the final solution and the other part is found by refining the rest of the initial solution. Thus, the initial solution should contain most of the correct groups, i.e. high recall. In the experiments, the entire initial solution returned by $ADDS$ demonstrates a clearly better recall than that of DES in all datasets at all settings of k.

In figure 6(b)(d)(f)(h)(j), the algorithms are tested for various θ. Compared to k, the impact of θ is significant. The reason is that most datasets have a large percentage of groups that are of size 2. Therefore, once $k > 2$, the effect of k is hardly noticed. In contrast, θ is more robust. Even though two tuples have the best DP measure, if their similarity distance is greater than θ, they can not be in the same group. Thus, both $ADDD$ and DED are sensitive to the settings of θ. In all settings of θ, the $ADDD(DP = 00x)$ always has the reliable precision comparing to DED while the $ADDD$ always has the higher recall for all datasets.

Effect of c. A good estimation of parameter c requires user to understand the dataset before any processing and have a proper knowledge of statistics. In the experiments, the impact of various estimations of c to the precision and recall is tested. The results are presented in figure 6.

We first look at the situation that k is given 6(a)(c)(e)(g)(i). When the setting of c changes from 2 to 20, we observe that the precision decreases and the recall increases. Since c is upper bound of the sparse neighborhood criterion, the greater c means it can be satisfied by more groups and, as a consequence, the recall of the final solution increases as illustrated. The increase of recall means that many groups that should be in the final solution are pruned if the setting of c is small. At the same time, the precision decreases. If precision decreases sharply and recall increases slowly like in 6(e), it means the small setting of c is proper for this dataset. If precision decreases slowly and recall increase sharply like in 6(a)(i), it means the greater setting of c is proper. If precision decreases slowly and recall increase slowly like in 6(c), it means the setting of c is irrelevant.

In the situation that θ is given 6(b)(d)(f)(h)(j). The setting of c also changes from 2 to 50, we observe the same situation that the precision decreases and the recall increases. Similarly, the greater c allows more groups to qualify to be in the final solution and several patterns are observed. The small c is proper if precision decreases sharply and recall increases slowly like in 6(f); the greater c is proper if precision decreases slowly and recall increase sharply like in 6(b)(j); c is irrelevant if precision decreases slowly and recall increase slowly like in 6(c).

We notice that most duplicates in the datasets studied here is of size 2-3. Thus, the significant impact of changing c to the final solution suggests that the distribution in proximity of different records can be very different. This situation makes it hard for user to estimate a proper c. The expertise and knowledge in statistics are critical.

6   Conclusion

In this paper, we have presented a new method for duplicate detection, namely 'Active Duplicate Detection' or ADD. ADD provides a means of creating an initial solution (or grouping of duplicate records) without requiring any user input. This is in contrast with most existing approaches to duplicate detection which require user to provide certain parameters up front. Assuming that user has knowledge of the underlying datasets so as to provide correct parameter settings is not a sound assumption. As such, ADD provides a much more globally applicable solution. Further, we have demonstrated that ADD performance both in quality of results (precision vs. recall) as well as efficiency is highly competitive and in some case superior to existing solutions.

References

1. Hernandez, M., Stolfo, S.: The merge/purge problem for large databases. In: SIGMOD (1995)
2. Hernandez, M., Stolfo, S.: Real-world data is dirty: data cleansing and the merge/purge problem for large databases. Data mining and knowledge discovery 2(1), 9–37 (1998)
3. Sarawagi, S., Bhamidipaty, A.: Interactive deduplication using axtive learning. In: SIGKDD (2002)
4. Wang, Y., Madnick, S.: The inter-database instance identification problem in integrating autonomous systems. In: ICDE (1989)
5. Benjelloun, O., Garcia-Molina, H., Menestrina, D., Su, Q., Whang, S.E., Widom, J.: Swoosh: A generic approach to entity resolution. The VLDB Journal (2008)
6. Newcombe, H.: Record linking: The design of efficient systems for linking records into individual and family histories. Am. J. Human Genetics 19(3), 335–359 (1967)
7. Tepping, B.: A model for optimum linkage of records. J. Am. Statistical Assoc. 63(324), 1321–1332 (1968)
8. Felligi, I., Sunter, A.: A theory for record linkage. Journal of the Amercian Statistical Society 64, 1183–1210 (1969)
9. Newcombe, H.: Handbook of Record Linkage. Oxford Univ. Press, Oxford (1988)
10. Monge, A., Elkan, C.: An efficient domain independent algorithm for detecting approacimatly duplicate database records. In: SIGKDD (1997)
11. Bilenko, M., Mooney, R.: Adaptive duplicate detection using learnable string similarity measures. In: SIGKDD (2003)
12. Cohen, W., Richman, J.: Learing to match and cluster large hihg-dimensional data sets for data integration. In: SIGKDD (2002)
13. Chaudhuri, S., Ganti, V., Motwani, R.: Robust identification of fuzzy duplicates. In: ICDE (2005)
14. Elmagarmid, A.K., Ipeirotis, P.G., Verykios, V.S.: Duplicate record detection: A survey. TKDE 19(1), 1–16 (2007)
15. Winkler, W.: Data cleaning methods. In: SIGMOD workshop on data cleaning, record linkage, and object identification (2003)
16. Tejada, S., Knoblosk, C., Minton, S.: Learing domain-independent string tranformation weights for high accuracy object identification. In: SIGKDD (2002)
17. Jain, A., Dubes, R.: Algorithms for clustering data. Prentice Hall, Englewood Cliffs (1988)
18. Chaudhuri, S., Sarma, A.D., Ganti, V., Kaushik, R.: Leveraging aggregate constraints for deduplication. In: SIGMOD (2007)

FlexTable: Using a Dynamic Relation Model to Store RDF Data

Yan Wang, Xiaoyong Du, Jiaheng Lu, and Xiaofang Wang

Key Labs of Data Engineering and Knowledge Engineering, Ministry of Education, China
Information School, Renmin University of China, China
{wangyan15,duyong,jiahenglu,wangxiaofang}@ruc.edu.cn

Abstract. Efficient management of RDF data is an important factor in realizing the Semantic Web vision. The existing approaches store RDF data based on triples instead of a relation model. In this paper, we propose a system called FlexTable, where all triples of an instance are coalesced into one tuple and all tuples are stored in relation schemas. The main technical challenge is how to partition all the triples into several tables, i.e. it is needed to design an effective and dynamic schema structure to store RDF triples. To deal with this challenge, we firstly propose a schema evolution method called LBA, which is based on a lattice structure to automatically evolve schemas while new triples are inserted. Secondly, we propose a novel page layout with an interpreted storage format to reduce the physical adjustment cost during schema evolution. Finally we perform comprehensive experiments on two practical RDF data sets to demonstrate that FlexTable is superior to the state-of-the-art approaches.

Keywords: FlexTable, RDF data, dynamic relational model, Lattice.

1 Introduction

The Resource Description Framework(RDF) is a flexible model for representing information about resources in the World Wide Web (WWW). When an increasing amount of RDF data is becoming available, RDF model appears to have a great momentum on the Web. The RDF model has also attracted attentions in the database community. Recently, many database researchers have proposed some solutions to store and query RDF data efficiently.

The popular solutions are called TripleStore[4] and VerPart[2], which could be seen in Fig.1. The former one uses one table to store all the RDF triples. This table only has three attributes which separately corresponds to subject, predicate and value of triples. The weak point of this solution lies in that it stores predicates as values in the table, not attributes, such as in traditional relation database. Then statistics of attributes is useless for RDF queries. In RDF queries, a key operation is to find instance with a given predicate, however, the statistics of attribute 'predicate' in TripleStore contain summary of all the predicates. So statistics of a predicate could not be obtained from TripleStore directly, which will deteriorate performance of RDF queries.

H. Kitagawa et al. (Eds.): DASFAA 2010, Part I, LNCS 5981, pp. 580–594, 2010.
© Springer-Verlag Berlin Heidelberg 2010

For example, there is a query to find names of graduates from Zurich University who won Nobel Prize in Physics. To answer this query, users need to scan the only one table in TripleStore twice and join the results of these scans. When the scans are more than two, the join order will be very important for query performance. However, in mainstream databases, the statistics for each predicate could not be gained directly, that will be harmful to join performance.

The latter one uses N two-column tables to store triples, where each table corresponds to one predicate and N is the number of all the predicates in an RDF data set. In such table, the first column stores subjects of triples and the second column contains values of those subjects in corresponding predicate. This solution is better than TripleStore in statistics collecting and query performance, where statistics of each predicate could be extracted with traditional methods in DBMS. But, in VertPart, values of each instance are stored separately, where statistics of predicates correlation are lost. Then DBMS may recommend a low-efficient query path for an RDF query.

For instance, there is another RDF data set extracted from WWW, where each triple represents that a book is tagged by a user with a tag as a predicate. A triple <"Harry Potter and the Philosopher's Stone", "young adult", "David"> means that user "David" uses "young adult" to tag "Harry Potter and the Philosopher's Stone". Assuming that there is a query issued for retrieving those books that having three tags simultaneously, such as "fantasy", "fiction", and "children". VertPart's performance advantage, comparing with TripleStore, lies in that the scan of these predicates is more efficient. However, there still leaves a problem for joining results of these scans in a high speed. Unfortunately, this kind of queries in RDF data applications is popular, such as finding groups or building a taxonomic structure among instances. If a book tagged with "fantasy", "fiction" is inserted into RDF data set, we could classify it to a class of "fantastic fiction". When another book is inserted with more tags, including "fantasy" and "fiction", we can conclude that it belongs to a subclass of "fantastic fiction" and could build relations between these classes. In this kind of applications, while each instance is inserted, system must search the class in which its tags are contained and classify it to that class. Because the join condition for these intermediate results requires final results to have same subjects. If all the triples describing one instance are organized together in one page, the join cost will be reduced dramatically while the number of join predicates is very big.

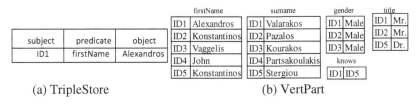

Fig. 1. An Example of TripleStore and VertPart

These requirements urge that triples should be organized as triple groups, and a query could be restricted in each group to reduce scan cost. Secondly, all triples sharing same subjects should be stored in one page to reduce join cost. However, there exist some problems to do that. First question is how to group the triples to reduce query cost. And the second one is how to support this process dynamically. As we know that in real applications, data sets are growing over time. If triple grouping is redone while each new triple is inserted, the cost will be prohibitively high. So using a dynamic model will be very important for triple grouping. The last one is when supporting dynamic triple grouping, new inserted triple will induce merging or adjusting of the original groups. And the changes of original groups will cause changes of triple positions, which is a costly operation. To resolve these problems, we propose to use a dynamic relation model called FlexTable to support dynamical triple grouping.

Providing a relation model for RDF data storage is necessary and significant. As written in [1], EF Codd has pointed out that a relation model of data is superior in several respects to the graph or network model. The reason why current solutions do not use relation model to store RDF data lies in that RDF's logical model is a graph model. Naturally, researchers use a physical graph model to store RDF data, such as TripleStore and VertPart. However, for the applications pointed above, we could find that it is time to come back to relation model to store RDF data. Our idea is something like that in [12], called clustered property table, both of these solutions prefer to manage RDF data in relation model. And the difference of our methods lies in that they focus on a static circumstance, and our focus is on a dynamic circumstance.

From the following experiments, a conclusion could be drawn that clustered property table will be better than VertPart. In [2], although Abadi had shown that, in column-store database, the performance of VertPart was superior to clustered property table, he also pointed out that this phenomenon was observed in column-store database and the performance of VertPart and clustered property table differed litter in row-store database. In [8], Sidirourgos showed that even in column-store database, the performance of VertPart was not always better than clustered property table, which depends on different data set. So, we continue to extend the method of clustered property table to a dynamic circumstance. Now our implementation is realized on row-store database, in the future we will implement them on column-store database.

In this paper, we propose a system called FlexTable to store RDF data with a dynamic relation model. To the best of our knowledge, there is no work today storing RDF data with a dynamic relation model.

There are some contributions of FlexTable. Firstly, a method based on lattice-structure is proposed to design evolving triple groups. In this method, each triple group is stored in a table and predicates in a triple group correspond to attributes in that table. In the following sections, the attribute set of a table is called a schema. Whether or not to adjust a schema depends on the similarity between the schema and its most similar schema. If similarity is high enough, a new schema is produced to contain tuples and attributes of the original two. In the following sections, the process of dynamical schema adjustment is called schema evolution.

Secondly, a new data page layout is designed for reducing cost of schema evolution. As we know, in mainstream database, reorganizing a tuple requires correctly extracting values according to the order of attributes in schema. When a new schema is produced to merge the original two, the attribute order in the new schema is different with that of the original two. So it is necessary to reorganize values from old schemas to insert in the new one, which causes large cost when this process happens frequently. Unfortunately, in many data applications, they are such cases, such as tagging system, Yago[1]. In this paper, a new type of data page is introduced to contain

---

[1] Yago, http://www.mpi-inf.mpg.de/yago-naga/yago/

interpreted information of attributes, which will reduce the cost of rewritten data pages.

The rest of this paper is organized as follows. In section 2, we introduce preliminary of FlexTable. Section 3 we present an algorithm for designing a dynamic relation model. In section 4, we propose a new page layout to reduce adjustment cost of tables. Section 5 gives some detail experiments to show the improvement of FlexTable. In section 6, we discuss the state of the art of storing RDF data in RDBMS. Finally, we conclude in section 7.

2 Preliminary

An RDF term is an URI or a literal or a blank node. A triple $(s,p,v) \in (U \cup B) \times U \times (U \cup B \cup L)$ is called an RDF triple, where U is a set of URIs, B is a set of blank node and L is a set of literals. In a triple, s is called subject, p is predicate and v is value.

Definition 1 -- RDF tuple. An RDF tuple is a tuple coalesced with a set of triples having a same subject.

An RDF tuple is structured as a conventional tuple in DBMS. The subject of those triples is stored in an attribute named by "URI". Values of each predicate are stored in an attribute named by the predicate. In an RDF tuple, the number of values in a predicate is always more than one, so we use an array to store those values. In Table1, the first tuple is an example of an RDF tuple describing "Albert_Einstein", edges in Fig.2 correspond to triples composing this RDF tuple.

Definition 2 -- RDF schema. An RDF schema is a set of RDF tuples.

An RDF schema is stored as a table in FlexTable. Each row corresponds to an RDF tuple and each column corresponds to an attributes of RDF tuples. Table1 is an example of an RDF schema.

Table 1. An Example of RDF schema

URI	GraduatedFrom	bornIn	diedIn
Albert_Einstein	University_of_Zurich	Ulm	Princeton,_New_Jersey
Leonhard_Euler	University_of_Basel	Basel	Saint_Petersburg

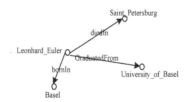

Fig. 2. An example of triples about "Leonhard_Euler"

3   Schema Evolution

As we know, it is efficient to manage chaotic things with classifying them into several groups. This principle is also useful in managing RDF data. However, all of current solutions for storing RDF data, including TripleStore and VertPart, logically regard all the triples as a whole, which disobeys the above principle. Although VertPart could physically store triples in several tables, it also logically treats all the triples as one schema. When triples are considered as a whole, the correlations of all predicates are difficult to compute. As we have pointed out in the first section, many applications need to issue queries with joins. In this case, join order and predicate correlation statistics would have a great effect on query performance. However, there is no such functionality in current solutions. So, classification of triples should be added into RDF data management to reduce query cost. By using classification, predicates could be clustered into several classes which could help to improve query performance with adjustment of join order.

To resolve these problems listed above, a new approach is introduced to organize RDF data in schemas as done in RDBMS. A naïve method is to extract an RDF schema from an RDF tuple. However, with this approach, there will be a drawback that few RDF tuples will share a same RDF schema. In this paper, the rigid constraint is relaxed, if two schemas are most similar, these two could be merged into one. In other words, similar schemas are merged automatically according to their similarity, not manually.

In the following subsections, firstly a formula to measure similarity of two schemas is introduced. Secondly a lattice-based algorithm (LBA) is presented to make certain which schema pair to merge. This method is better than the brute force approach, which computes all the pairs of existing schemas. Thirdly, a control parameter is proposed to judge when to stop evolve.

3.1   Similarity Measurement

As we have pointed out, two schemas with maximum similarity value will be merged while a new RDF tuple is inserted. To compute the similarity of these two schemas, a cosine-distance measure is introduced. Here each schema is represented as an attribute set, e.g. { URI, 'cityInState', 'geopoliticalSubdivision', 'majorCityInState', 'conceptuallyRelated'} is a schema with six attributes as listed above. To compute cosine-distance of two schemas, only the latter four attributes are needed. Formula 1 is given to compute importance of an attribute in one schema.

$$\mu(s_i, a_j) = \sup(s_i, a_j) \cdot \log \frac{|\{s\}|}{|\{s \mid a_j \in s\}|} \tag{1}$$

In Formula 1, "s_i" represents a schema and "a_j" represents an attribute. This formula references classical tf/idf formula used in Information Retrieval. The former part $\sup(s_i, a_j)$ is to compute the importance of attribute "a_j" in schema "s_i", and the latter part is to compute the importance of attribute "a_j" in all the schemas. In Formula 1, $\sup(s_i, a_j)$ is computed as ratio of RDF tuples which have values in attribute "a_j" to all

RDF tuples contained in "s_i". While $sup(s_i, a_j) = 1$, all the RDF tuples in "s_i" have values in attribute "a_j", it means that the importance of "a_j" to "s_i" is much higher than that of "a_i" to other schemas.

The latter part is to compute the logarithm of the ratio of all schemas to those schemas which have attribute "a_j". This part measures the distribution of attribute "a_j" in all schemas. While more schemas containing attribute "a_j", the importance of "a_j" to "s_i" would be less. For example, if attribute "inUniversity" existes in less schemas than attribute "name", two schemas sharing the attribute "inUniversity" are more similar than those only sharing attribute "name". Because the former two is likely to be students or teachers in a university and the latter two could be anything else.

With Formula 1, similarity of two schemas could be computed as a cosine-distance formula written as follows:

$$sim(s_i, s_j) = \frac{\sum_k \mu(s_i, a_k) \bullet \mu(s_j, a_k)}{\sqrt{\sum_k \mu(s_i, a_k)^2} \sqrt{\sum_k \mu(s_j, a_k)^2}} \tag{2}$$

3.2 Lattice-Based Algorithm(LBA)

A straightforward method used in schema evolution, named as brute force approach, is to compute all the similarity of schema pairs with formula 2 and pick up the most similar pair to merge. With brute force approach, each time schema evolution is invoked, the similarity between new schema and existing schemas should be computed and similarity of all the schema pairs should be saved. It could be inferred that time complexity of this approach is $O(n)$ and space complexity is $O(n^2)$, the cost of which is prohibitively expensive.

To reduce the cost of schema evolution, inspired with [10], a lattice-based algorithm (LBA) is introduced to prune the unnecessary computation. Here a lattice is built as follows. Firstly, each RDF schema is corresponded to a node in the lattice. Secondly, while all the attributes of schema A is contained in attribute set of schema B, then A is an ancestor of B and B correspondingly is a descendant of A. When there is no other schema that is both an ancestor of B and a descendant of A, A is called a parent of B and B is a child of A. The relationship between A and B corresponds to a line from A to B. When a top node, which has no attribute, and a bottom node, which has all the attributes as its attribute, is added, a lattice could be drawn. An example is shown in Fig.3, where there are four nodes, such as {P2,P3}, {P1,P5}, {P2,P3,P4}, {P1,P2,P3,P5}. Solid lines in Fig.3 represent that the upper node is a parent of the lower node. Dashed lines in Fig.3 represent that these nodes are brother nodes. Here, only the nodes that have same parents are concerned as brother nodes. In our algorithms, only the similarities between parent-child schema pair or brother schema pair are computed. Inferred from Fig.3, similarity between {P1,P5} and {P2,P3,P4} are unnecessary to compute, compared with brute force approach. And in a sparse data set, while the bigger the number of schema is, the more unnecessary computation cost is pruned.

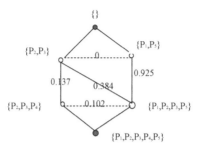

Fig. 3. An Example of RDF Schema Lattice

In Algorithm EvolutionLattice, when an RDF tuple is inserted, a schema S_{new} is extracted from it. After adding S_{new} into lattice, similarity values δ_i between S_{new} and its parents, brothers S_i are computed. Assuming that the biggest similarity is δ', say between schema S_m and S_n. While exists $\delta_i > \delta'$, then merge the schema S_i with S_{new}. Otherwise a new schema is created for S_{new} and S_m is merged with S_n.

```
Algorithm EvolutionLattice(tuple, lattice)
Input: tuple - An RDF tuple
       lattice - An RDF schema lattice
Output: lattice
1:   schema ← ExtractSchema(tuple);
2:   AddSchema(schema, lattice);
3:   schemaPair ← GetMaxSimPair(lattice);
4:   if(NeedMerge(schemaPair))
5:       newSchema=MergeSchema(schemaPair);
6:       AddSchema(newSchema,lattice)
7:   InsertTuple(tuple);
8:   return lattice;
```

The input parameters of *EvolutionLattice* are two, *tuple* corresponds to a new inserted RDF tuple, and *lattice* corresponds to the schema lattice structure described above. The aim of *ExtractSchema* in line 1 is to extract an RDF schema from the inserted RDF tuple. Line 2 invokes *AddSchema* to add the extracted schema into *lattice* and compute similarity between this schema and its related schemas connected with lines in *lattice*. In Line 3, *GetMaxSimPair* is used to select a schema pair with max similarity from *lattice*. When *NeedMerge* returns true, *MergeSchema* is invoked to merge these schemas. In line 5, the merged schema *newSchema* is inserted into the *lattice*. At last, *InserTuple* is invoked to insert the inputting RDF tuple into FlexTable. The keys of this algorithm are *AddSchema* and *NeedMerge*, the detailed descriptions of the former one is shown as follows and the latter is described in 3.3.

```
Algorithm AddSchema(schema, lattice)
Input: schema -- A new schema
       lattice - An RDF schema lattice
Output: lattice
1: bottom ← getBottomNode(lattice);
2: stack ← new Stack(bottom);
```

```
3:   while(!isEmpty(stack))
4:       temp ← pop(stack);
5:       if (schema is ancestor of temp)
6:           push all parents of temp into stack;
7:       else
8:           AddChildren(temp's children, schema);
9:            compute similarity between temp's children and
schema;
10: top ← getTopNode(lattice);
11: push top in stack;
12: while(!isEmpty(stack))
13:      temp ← pop(stack);
14:      if (temp is ancestor of schema)
15:          push all children of temp into stack;
16:      else
17:          AddParents(temp's parents, schema);
18:           compute similarity between temp's parents and
schema;
19:          compute similarity between temp and schema;
20:           compute similarity between temp's brothers and
schema;
21: return lattice;
```

Algorithm *AddSchema* adds lines between new schema and existing schemas in *lattice*. It is divided into two parts, first part corresponds to line 1 to line 9, which finds all the children of a new schema, add relations between the schema and its children, and compute similarity between these pairs. The second part corresponds to line 10 to line 20, which finds all the parents of the new schema, and add relations, compute similarities between them. In addition, the brother relations are ascertained only when parents of schema are found, then similarities between *schema* and its brothers are computed after *AddParents* is invoked. Note that, in line 20 where *temp* is also a brother of schema, the similarity between these two schemas should be computed here.

3.3 Control Parameter

On implementation of Schema Evolution, there is another problem, which is when to stop merge. The strategy used here is to compute the storage gain of schema evolution. Firstly, storage cost of two schemas and a new schema is computed. If storage cost of a new schema is smaller than that of existing two schemas, then merge these two schemas into the new one. Otherwise, there is no need for actions.

The storage cost of a schema includes two parts, one is to store schema meta-information, such as the name of schema and names of attributes in schema, and the other is to store tuples in schema such as null value bitmap of a tuple and actual values. In RDBMS, the storage cost of a schema could be expressed as formula 3.

$$C = \alpha + \beta \bullet |A| + \gamma \bullet |A| \bullet |N| + C_{val} \tag{3}$$

Here, α is storage cost of schema name and other information that only stored one time for a schema. And β is storage cost of each attribute in one schema, $|A|$ is the number of attributes in corresponding schema, $\beta\bullet|A|$ is total storage cost of all attributes of one schema. The third part is storage cost of null value bitmap for all tuples which is used to reduce storage cost for null value, where $|N|$ is number of RDF tuples in the schema, γ is storage cost of each bitmap. The fourth part is storage cost of actual values.

With Formula 3, storage gain for schema merging could be computed as Formula 4:

$$
\begin{aligned}
C_{new} &= \gamma\bullet|A_{s1\cup s2}|\bullet(|N_{s1}|+|N_{s2}|)+\beta\bullet|A_{s1\cup s2}|+\alpha+C_{val1}+C_{val2} \\
C_{gain} &= C_{new}-C_{s1}-C_{s2}=\gamma\bullet((|A_{s1}|-|A_{s1\cap s2}|)\bullet|N_{s2}|+(|A_{s2}|-|A_{s1\cap s2}|)\bullet|N_{s1}|) \qquad (4)\\
&\quad -\alpha-\beta\bullet|A_{s1\cap s2}|
\end{aligned}
$$

While C_{gain} is bigger than zero, the function NeedMerge in EvolutionLattice returns "true", otherwise returns "false". If the result is "false", merging is not happened.

In summary, we propose a formula to compute similarity between two schemas firstly. Secondly, a lattice-based algorithm (LBA) to design dynamic relational schemas for RDF data is introduced. At last, a formula is used to determine when to merge two schemas.

4   Modification of Physical Storage

In traditional databases, such as PostgreSQL, a tuple's values are stored in the same order as order of attributes in schema. This method has a benefit to reduce storage space because system can use a bitmap to indicate whether the value of an attribute in one tuple is null or not. So with this approach, storage of null values is avoided. For example, a tuple with a not-null value bitmap "101" means that the value of the second attribute is null.

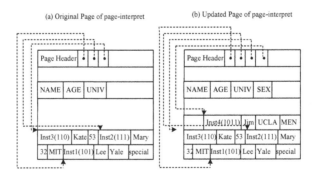

Fig. 4. An Example of FlexTable data page layout

However the above method is inefficient when schema evolution happens frequently. When the system merges two existing schemas, at least data pages of one schema are required to be rewritten correspondingly. For example, on assumption that attribute set of SchemaA is {name, age, univ} and attribute set of SchemaB is {name, sex, univ}. There is an RDF tuple tupA in SchemaA having a value {'Kate', 53} with a not-null bitmap '110'. And another RDF tuple tupB in SchemaB has a value {'Jim', 'MEN', 'UCLA'} with a not-null bitmap '111'. After SchemaA merges with SchemaB, the resulting attribute set of new schema changes to {name, age, univ, sex}, there exists a problem that not-null bitmap of tupB should be explicitly rewritten to '1011'. Therefore, the cost of schema merging is prohibitively high in the traditional page layout. We tackle this problem with introducing a novel page layout which uses an interpreted storage format. Specifically, with interpreted storage, the system must "interpret" the attribute names and their values for each tuple at query access time, whereas with positional storage like bitmap in PostgreSQL, the position information of the attributes is pre-compiled. In the scenario that schema changes frequently, the benefits of not storing null values become less, while the performance gain in avoiding the data pages rewritten becomes more desirable.

Inspired by [11], we introduce a new approach called page-interpret to divide data page into three regions: the first is page header area, the second is attribute interpreted area, and the third is data value area. As shown in Fig.4(a), behind "Page Header" and pointers of tuples in current page, we record attribute information such as attributes' ids, which area is called attribute interpreted area. Next we give an example to show how to process schema merging between schemaA (name, age, univ) and schemaB (name, sex, univ). When a new tuple (Jim, 'MALE', 'UCLA') is inserted, it is important to note that only the last non-full page needs to be rewritten (shown in Fig.4(b)), while previous full pages in original schemaA and schemaB could remain unchanged. Note that our approach does not affect the query performance. With page layout of FlexTable, the only additional cost is to interpret a tuple organized according to page-level's attribute set into a tuple organized according to table-level's attribute set, which is trivial comparing with I/O cost.

5   Experiment and Analysis

Setting. The experiments are carried out on a PC running Windows XP, with Intel Pentium-(R) Dual CPU T2390@1.86GHz and 1GB of RAM. We use a 160GB SATA hard drive. The data set for experiments are two. One is FreeToGovCyc[1], called govcyc in the following sub-sections, a practical and sparse Semantic Web data set, with 45823 triples, 10905 instances and 441 properties. The other one is triples randomly extracted from Yago, called yago in the following sub-sections, which occupies near 1/100 of Yago dataset, with 1,000,000 triples, 152,362 instances and 87 properties. The modifications of physical storage are implemented on PostgreSQL 8.1.3. The algorithm of SchemaEvolution is programmed with JAVA.

5.1   Analysis of Triples Import

Time cost of triples import on FlexTable is composed of two parts. One is the time for execution of LBA, which corresponds to the time for logical adjustment of RDF tuples. The other one is the time for inserting triple set into its corresponding table, which corresponds to the time for physical adjustment of RDF tuples' positions.

---

[1] FreeToGovCyc,
http://semweb.mcdonaldbradley.com/OWL/Cyc/FreeToGov/60704/FreeTo GovCyc.owl

Experiments are executed to compare FlexTable with VertPart and TripleStore. Here each table in FlexTable has an index on URI to accelerate queries executed in the next sub-section. To make experiments equality, each table of VertPart also has an index on instance. Our TripleStore simulates Hexastore[6], where if an instance has some values on one predicate, these values are organized as an array to store in database, so does the implementation of VertPart. Because our queries are nothing to do with the column 'value', only two indexes are built on TripleStore. They are (instance, predicate) and (predicate, instance).

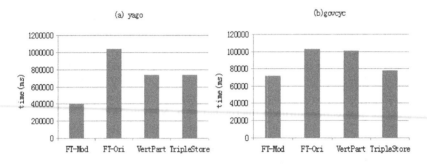

Fig. 5. Comparisons of Time Cost for Triples Import

Fig.5 shows two experiments implemented on yago and govcyc. FT-mod corresponds to FlexTable implemented on data page layout introduced in section4. And FT-Ori corresponds to FlexTable implemented on data page layout of PostgreSQL. From Fig.5, a conclusion could be drawn that no matter in what data set, FT-Mod is the solution with least time cost. And FT-Ori is the most expensive solution for triple import. In yago, the difference of cost between FT-Mod and other solutions is more evident than that in govcyc. It results from the cause that data in govcyc is sparser than that in yago. So the time of LBA executed on govcyc is bigger than that on yago. Because cost of schema evolution occupies most part of whole cost in FT-Mod, the advantages of FT-Mod is clearer in yago.

5.2 Analysis of Storage Cost

Table2 shows comparisons of storage cost in FlexTable with other solutions. Here FlexTable means FT-Mod in sub-section 5.1. From Table2, we could find that storage cost of FlexTable is the smallest. There are two reasons. The first one is that the role of control parameter described in sub-section 3.2 makes storage cost smaller. The second one is that in FlexTable values of one instance are organized as one tuple which corresponds to several tuples in other solutions, so it could avoid storing redundant tuple headers.

In Table2, a phenomenon could be found that, in govcyc, storage cost of VertPart is much bigger than that of TripleStore. However, in yago, it is on the contrary. This phenomenon results from the fact that data of govcyc is sparser than that of yago. For the reason that VertPart creates one table for each predicate, so there are many free space in pages of VertPart.

Table 2. Comparisons of Storage Cost

DataSet	FlexTable(KB)	VertPart(KB)	TripleStore(KB)
govcyc	4,592	15,112	9,576
yago	47,400	56,168	76,808

5.3 Analysis of Query Performance

Our test queries are represented in SPARQL syntax. The pattern of test query is in the form "select ?x where {?x pred1 ?val1. {?x pred2 ?val2}.... {?x predN ?valN}}". This kind of query is to search all the instances simultaneously having predicates in the query. It is an important kind of query in Semantic Web construction. Assuming that pred1, pred2, ... , predN are predicates describing instance A. With this query, users could find which instances belong to subclass of instance A's class. Then they could build a taxonomy structure between instances. These predicates are shown in Table3. Fig.6 shows our experimental results by comparing it with VertPart and TripleStore. We add predicates to the query pattern one by one. Firstly, we use the first two predicates in Table3. Secondly, the third predicate is added to join with results of last query. Then the third, the fourth, the fifth, ... , the Nth predicate are added to the query pattern. With increase of N, we find that our approach is superior to the existing solutions in terms of query performance, especially in yago. For example, when the attribute number to join in yago is 8, VertPart takes 407ms and TripleStore takes 312ms, but FlexTable takes only 172ms. In Fig.6, changes of FlexTable's performance could be explained that the number of tables need to be scanned in FlexTable significantly decreases with increase of N. And at the same time, the number of joins in data dictionary to find the correct table to scan will becomes bigger, which deteriorates test query's performance. So, in Fig.6(a), query cost of N=1 is biggest, then in most cases it decreases with increase of N. When N is 7, the query performance is the best. And when N is eight, the cost becomes bigger than the case that N=7. VertPart suffers from scanning and joining more tables, so its cost is always bigger tan FlexTable. TripleStore scans data from one index, the next join could use buffer of last join. So, in yago, performance of the query executed in TripleStore does not change much. In govcyc, because the data set is too small, the tendency of query performance are not so evident. However, we could still find that performance of FlexTable is better than other two in many cases.

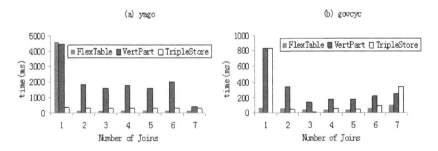

Fig. 6. Comparisons of query performance while increasing number of joins in a query

Table 3. Filtered Predicates

Sequence	govcyc	yago
0	comment	type
1	definingmt	Iscalled
2	conceptuallyrelated	bornondate
3	positivevestedinterest	livesin
4	allies	diedondate
5	possesses	influences
6	economicinterestin	iscitizenof
7	capitalcity	diedin

6 State of the Art

In this section, we discuss the state of the art of storing RDF data in RDBMS, with an extended look at comparing them with FlexTable.

6.1 TripleStore

Although there have been some non-relational RDBMS approaches for storing RDF data, such as [3]. The majority of RDF data storage solutions uses RDBMSs, such as Jena[4], Sesame[5], Oracle[14]. While storing RDF data in a relational database, there is a problem to translate triples into tuples. The straightforward way is to translate a triple into a tuple, i.e. a table with three attributes is created for storing triples, whose attributes separately correspond to subject, predicate and object of triples. This approach is called TripleStore, which is shown in Fig.1(a).

In TripleStore, the cost of table scanning and joining is much expensive. So researchers need to add many indexes to improve query performance of Triple Store. For example, in Hexastore[6], there are at least 6 indexes needed to add, which are all possible permutations of attributes subject(s), predicate(p) and value(v), i.e. {s, p, v}, {s, v, p}, {p, s, v}, {p, v, s}, {v, s, p} and {v, p, s}. To reduce storage cost of so many indexes, in rdf-3x[7], a compressed method is presented. From their experiments, using the compressed method, storage cost of 6 indexes which is six times of original cost could be reduced to double of original cost. In [13], Neumann proposes some optimization skills for rdf-3x optimizer.

These optimizations for TripleStore are orthogonal to ours. The focus of FlexTable is on how to partition triples into several correlated groups to accelerate query performance. So these optimizations also could be added to FlexTable to improve query performance and reduce storage cost.

6.2 VertPart

VertPart is a variant of TripleStore. Because in TripleStore, storing all triples in one table is a bottleneck for query performance, researchers consider some methods to partition triples into different tables. In an RDF data set, there is an assumption that the number of predicates is far less than that of subjects and objects. According to this assumption, partitioning triples based on predicates is a feasible way to improve performance of TripleStore. In [2], triples with same predicates are partitioned into a table named by the predicate, whose attributes separately correspond to subject and object of triples. This approach is called VertPart, shown in Figure1(b). With the method of VertPart, these tables with two attributes could be stored in column-store database. In [2], a detail experiment is executed to show that storing RDF data in column-store database is better than that of row-store database. In [8], experiments are shown that gain of performance in column-store database depends on the number of predicates in a data set. There exist potential scalability problems for VertPart when the number of predicates in an RDF data set is high. In [9], an experimental comparison of RDF data management approaches is done to show that none of current approaches can compete with a purely relational model.

As proved in some papers, VertPart is effective in some applications, such as traditional library systems, where predicates are predefined by administrators or content managers. However it is infeasible for WWW applications, where predicates are defined freely by users. There are several reasons. Firstly, in these applications the number of predicates is much bigger than that of subjects. So VertPart will produce many tables with few triples, which increases storage cost. Secondly, in practice each subject is described with many predicates. If users need to search all predicates of a given subject, system needs to scan all predicate tables, which is poor in query performance. So it could be concluded that VertPart is not suitable for storing triples produced in WWW applications.

From above analyses, we could find that the best approach for storing RDF date is to store them in a relation model. And our work is to manage RDF data with a dynamic relation model, which could reduce the cost of schema adjustment during incremental production of RDF triples.

7 Conclusion

In this paper, we present an RDF storage system, called FlexTable, which designs a dynamic relation model to support efficient storage and query for RDF data. Firstly, we introduce a mechanism to support dynamic schema evolution. We also propose a novel page layout to avoid physical data rewritten during schema evolution. Finally, comprehensive experiments are performed to demonstrate the advantages of Flex-Table over existing methods in terms of triple import, storage and query performance. In the future, we will extend FlexTable to column-store database.

Acknowledgments. Yan Wang, Xiaoyong Du and Xiaofang Wang are partially supported by National Natural Science Foundation of China under Grant 60873017. Jiaheng Lu is partially supported by 863 National High-Tech Research Plan of China (No: 2009AA01Z133, 2009AA01Z149), National Natural Science Foundation of China (NSFC) (No.60903056), Key Project in Ministry of Education (No: 109004) and SRFDP Fund for the Doctoral Program(No.20090004120002).

References

1. Code, E.F.: A relational model of data for large shared data banks. Communications of the ACM 26, 64–69 (1983)
2. Abadi, D.J., Marcus, A., Madden, S.R., Hollenbach, K.: Scalable Semantic Web Data Management Using Vertical Partitioning. In: VLDB 2007, pp. 411–422 (2007)
3. Bonstrom, V., Hinze, A., Schweppe, H.: Storing RDF as a graph. In: LA-WEB 2003, pp. 27–36 (2003)
4. Carroll, J.J., Dickinson, I., Dollin, C.: Jena: implementing the semantic web recommendations. Technical Report, HP Labs (2004)
5. Broekstra, J., Kampman, A., Harmelen, F.: Sesame: A Generic Architecture for Storing and Querying RDF and RDF Schema. In: Horrocks, I., Hendler, J. (eds.) ISWC 2002. LNCS, vol. 2342, pp. 54–68. Springer, Heidelberg (2002)
6. Weiss, C., Karras, P., Bernstein, A.: Hexastore: Sextuple Indexing for Semantic Web Data Management. In: VLDB 2008, pp. 1008–1019 (2008)
7. Neumann, T., Weikum, G.: RDF-3X: a RISC-style Engine for RDF. In: VLDB 2008, pp. 647–659 (2008)
8. Sidirourgos, L., Goncalves, R., Kersten, M., Nes, N., Manegold, S.: Column-Store Support for RDF Data Management: not all swans are white. In: VLDB 2008, pp. 1553–1563 (2008)
9. Schmidt, M., Hornung, T., et al.: An Experimental Comparison of RDF Data Management Approaches in a SPARQL Benchmark Scenario. In: Sheth, A.P., Staab, S., Dean, M., Paolucci, M., Maynard, D., Finin, T., Thirunarayan, K. (eds.) ISWC 2008. LNCS, vol. 5318, pp. 82–97. Springer, Heidelberg (2008)
10. Quan, T.T., Hui, S.C., Cao, T.H.: A Fuzzy FCA-based Approach to Conceptual Clustering for Automatic Generation of Concept Hierarchy on Uncertainty Data. In: CLA 2004, pp. 1–12 (2004)
11. Chu, E., Beckmann, J., Naughton, J.: The Case for a Wide-Table Approach to Manage Sparse Relational Data Sets. In: SIGMOD 2007, pp. 821–832 (2007)
12. Wilkinson, K.: Jena Property Table Implementation. Technical Report, HP Labs (2006)
13. Neumann, T., Weikum, G.: Scalable Join Processing on Very Large RDF Graphs. In: SIGMOD 2009, pp. 627–640 (2009)
14. Chong, E.I., Das, S., Eadon, G., Srinivasan, J.: An Efficient SQL-based RDF Querying Scheme. In: VLDB 2005, pp. 1216–1227 (2005)

An MDL Approach to Efficiently Discover Communities in Bipartite Network*

Kaikuo Xu, Changjie Tang, Chuan Li, Yexi Jiang, and Rong Tang

School of Computer Science, Sichuan University, China
kaikuoxu@gmail.com, cjtang@scu.edu.cn

Abstract. Bipartite network is a branch of complex network. It is widely used in many applications such as social network analysis, collaborative filtering and information retrieval. Partitioning a bipartite network into smaller modules helps to get insight of the structure of the bipartite network. The main contributions of this paper include: (1) proposing an MDL 21 criterion for identifying a good partition of a bipartite network. (2) presenting a greedy algorithm based on combination theory, named as MDL-greedy, to approach the optimal partition of a bipartite network. The greedy algorithm automatically searches for the number of partitions, and requires no user intervention. (3) conducting experiments on synthetic datasets and the southern women dataset. The results show that our method generates higher quality results than the state-of-art methods Cross-Association and Information-theoretic co-clustering. Experiment results also show the good scalability of the proposed algorithm. The highest improvement could be up to about 14% for the precision, 40% for the ratio and 70% for the running time.

Keywords: Community Detection, Bipartite Network, Minimum Description Length, Information Theory.

1 Introduction

To understand the structure of a complex system, a common approach is to map the interconnected objects in the complex system to a complex network and study the structure of the complex network. During the mapping, the interacted objects are mapped into highly connected modules that are only weakly connected to one other. The modules are considered to embody the basic functions of the complex network. Thus people can identify the modules or communities of which the complex network is composed in order to comprehend the structure of the complex system [1-5]. One classical application is the recommendation system for E-Commence like Amazon. Let's look at an example.

Example 1. Fig 1 shows an example about books and customers. In (a), there are eleven customers and four books. To recommend books to a given customer 'C', a

---

* This work was supported by NSFC Grant Number: 60773169 and 11-th Five Years Key Programs for Sci. &Tech. Development of China under grant No. 2006BAI05A01.

H. Kitagawa et al. (Eds.): DASFAA 2010, Part I, LNCS 5981, pp. 595–611, 2010.

recommendation system needs to search customers with similar tastes of 'C' (or books having the same topic with the books bought by 'C') from the sample data. To do so, the data is transformed into a two-mode network first, as shown in (b). The recommendation system could detect communities in the two-mode network directly. The result for community detection on (b) is shown in (c). Another choice is to project the two-mode network into one mode network and detect communities in the one-mode network. In Fig 1, (b) is projected to (d) and the result for community detection on (d) is shown in (e). Customers in the same group are considered to share the similar tastes and books in the same group are considered to share the same topic. In Fig 1, 'A','B','D', and 'E' are considered to share the similar tastes with 'C' and '2' is considered to have the same topic with '1'. At last, the results are selectively applied to recommendation according to the accuracy of different methods.

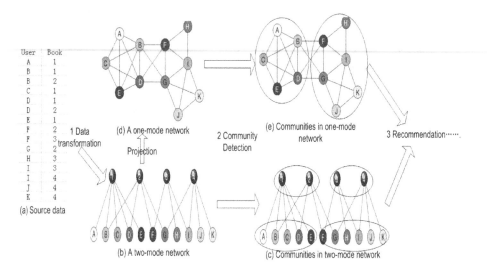

Fig. 1. The flowchart of recommendation for some sample data

A complex network is called one-mode network if it is mapped from a complex system composed of one type of objects. There are extended works on detecting communities in one-mode network. Among these works, a popular approach is to search for partitions that maximize the modularity [6, 7]. However, as a criterion of community quality, modularity is considered to have two disadvantages [8]: a resolution limit and a bias to equal-sized communities. Compression-based method that searches for partitions minimizing the description length is a recent alternation [9, 10]. Compression-based method usually makes use of the MDL principle in the field of information theory [11].

A complex network is called two-mode network or bipartite network if it is mapped from a complex system composed of two types of objects. In this paper, bipartite network refers to two-mode network. Fig. 1 (d) and (b) show examples of both one-mode network and two-mode network [12]. According to Fig.1, one-mode network has only one node set while two-mode network has two disjoint node sets. For one-mode network, node connects to each other in the same set. We only need to analyze the connections to detect the communities in one-mode network. For two-mode network, nodes in the same set are not connected. To detect the communities of one node set, we need study its connections to the other node set and the community structure of the other node set simultaneously. Thus communities of two node sets can be detected simultaneously.

As discussed in the section of related works, the existing methods for bipartite network is still far from sound comparing with the methods for one-mode network. This paper proposes a compression-based method to resolve communities in bipartite network as: (a) defines an MDL criterion which is extended from [9]; and (b) proposes a greedy algorithm which is adapted from [13]. The criterion is based on binomial coefficient and the proposed greedy algorithm takes advantage of the properties of binomial coefficient. The proposed algorithm automatically discovers the communities of two node sets simultaneously. Experiment results show that the newly proposed method successfully finds communities that CA (cross-association) method [13] fails to find and it is more accurate than the well known ITCC (Information-theoretic co-clustering) method [14] while detecting communities.

The rest of the paper is organized as follows. Section 2 gives some related work. Section 3 introduces the basic idea of our method and the MDL criterion. Section 4 proposes a greedy algorithm that finds communities according to the proposed MDL criterion. Section 5 evaluates the proposed algorithms on four datasets. Section 6 concludes the paper.

2 Related Works

The earliest significant work on this topic is the biclustering/co-clustering [15] which simultaneously searches row/column communities. A row community and a column community together with the connections between them form a submatrix (bicluster). Cheng [16] proposed a biclustering algorithm for gene expression data analysis, using a greedy algorithm that identifies one bicluster at a time by minimizing the sum squared residue. Dhillon [14] proposed an information-theoretic co-clustering, which searches communities by minimizing the KL-divergence between the biclusters and the original matrix. According to the experiments, the results generated by biclustering/co-clustering are highly accurate. However, the row/column community number needs to be specified before running the algorithm.

Recently, Guimera [17] proposed a projection based method. It transforms the bipartite network to one-mode network and uses method for one-mode network to discover communities. However, the projection process is usually considered to cause information loss, which will leads to bad result. Barber [18] extended the modularity [6] to bipartite network and proposed a method searching communities by minimizing the modularity. Barber's method requires a serious constraint: the numbers of row community and column community need to be equal. Lehmann [19] proposed a method detecting biclique communities. However, its result highly depends on the user's specification on the relaxation of biclique community. All methods mentioned above require manual specification on some parameters.

At the same time, some parameter-free methods were proposed. Chakrabarti [13] proposed a CA(cross-association) method. It automatically discovers row/column communities simultaneously. Sun [20] extended CA to deal with time-evolving bipartite network and its ability to discover communities for static bipartite network is no better than CA. Papadimitriou[21] extended CA to search not only global communities but also local communities and its ability to discover global communities for static bipartite network is also no better than CA. Nevertheless, the result generated by CA is still unsatisfactory as shown in section 5.

3 An MDL Criterion

The MDL principle has been successfully used in many applications such as universal coding, linear regression and density estimation [11]. The basic idea of MDL principle is to use the regularity of given data to compress the data. The regularity that compresses the data the most is considered to describe the given data the best. In our method, the community structure is treated as the regularity of a bipartite network. Thus according to the MDL principle, to search a good community structure is actually to search the community structure that compresses the bipartite network the most. The sketch of the proposed method for bipartite network community detection is as follows.

1) Define a formula to compute the bits for expressing a bipartite network directly. Let l be the bit length to directly express the bipartite network;
2) Divide the process to express the bipartite network into two parts: one part to express its community structure and the other part to express the extra information describing the network given the community structure. Let l_c be the bit length. Note that if all vertices are put into one community, l_c is equal to l;
3) Search for the optimum partition that minimizes l_c. This method is called a compression-based method because l_c is expected to be less than l. l_c is called the MDL criterion since it is the optimizing target.

Table 1. Table of main symbols

Symbol	Definition
A	Binary data matrix
n, m	Dimensions of A (rows, columns)
k, e	Number of row and column communities
k^*, e^*	Optimal number of communities
(Q_r, Q_c)	A partition
$A_{i,j}$	Submatrix of A composed of rows in community i and columns in community j
a_i, b_j	Dimensions of $A_{i,j}$
$n(A_{i,j})$	Number of elements in $A_{i,j}$ $n(A_{i,j}) = a_i b_j$
$n_1(A_{i,j})$	Number of 1 in $A_{i,j}$
$C(A_{i,j})$	Code cost for $A_{i,j}$
$T(A; k, e, Q_r, Q_c)$	Total cost for A
$Ar_{i,j}$	Sub-matrix of A composed of row i and columns in community j
$n(Ar_{i,j})$	Number of elements in $Ar_{i,j}$ $n(Ar_{i,j}) = b_j$
$n_1(Ar_{i,j})$	Number of 1 in $Ar_{i,j}$

3.1 Terminologies

Let X be an unweighted and undirected bipartite network. Suppose X is composed of two disjoint node sets S_r and S_c, the size of S_r is n and the size of S_c is m, then X can be described as a $n \times m$ (n, $m \geqslant 1$) adjacency matrix A. Let us index the rows as 1, 2,..., n and columns as 1, 2,..., m. Let k denote the number of row communities and let e denote the number of column communities. Let us index the row communities by 1, 2, ... , k and the column communities by 1, 2, ... , e. Let

$$Q_r: \{1, 2, \ldots , n\} \rightarrow \{1, 2,\ldots, k\}$$

$$Q_c: \{1, 2, \ldots , m\} \rightarrow \{1, 2,\ldots, e\}$$

denotes the assignments of rows to row communities and columns to column communities, respectively. The pair (Q_r, Q_c) is referred as a partition. We denote submatrix of A composed of rows in community i and columns in community j as $A_{i,j}$, $i = 1$, 2, ..., k and $j = 1, 2, ..., e$. Let the dimensions of $A_{i,j}$ be (a_i, b_j).

Example 2. Fig 1 (c) shows a partition example for Fig1 (b). In this example, $S_r = \{1,2,3,4\}$, $S_c = \{A,B,C,D,E,F,G,H,I,J,K\}$; $n = 4$, $m = 11$; $k = 2$, $e = 2$; $Q_r = \{1,2,3,4\} \rightarrow \{1,1,2,2\}$, $Q_c = \{A,B,C,D,E,F,G,H,I,J,K\} \rightarrow \{1,1,1,1,1,2,2,2,2,2,2\}$.

3.2 A Two-Part Coding

We now describe a two-part coding for matrix A. The first part is the *partition complexity* that describes the partition (Q_r, Q_c). The second part is the *conditional complexity* that describes the matrix given the partition.

3.2.1 Partition Complexity
The partition complexity consists of the following terms:

1) Bits needed to describe the number of rows, columns and number of 1: n, m, and $n_1(A)$. Since this term does not vary with different partitions, it makes no sense to the result so that it can be ignored.
2) Bits needed to describe the number of communities: $\log(n)$ for k and $\log(m)$ for e. Since these two terms do not vary with different partitions, it is ignored as well.
3) Bits needed to describe the community to which the rows and columns belong: $n\log(k)$ and $m\log(e)$.
4) Bits needed to describe all the ke submatrix $A_{i,j}$, i.e. the number of 1 between pairs of row community and column community: $ke\log(n_1(A))$

3.2.2 Conditional Complexity
Example 3. Assume there is a matrix A with only 1 row community and 1 column community. Given $n(A)$ and $n_1(A)$, the left information of A is expressed as follows: (1) Each combination of the $n_1(A)$ 1's is mapped to an integer index . Thus it requires

$\binom{n(A)}{n_1(A)}$ integers to map all the combinations. (2) All the integers are encoded into bits.
Consider the following matrix

$$A = \begin{bmatrix} 1 & 0 & 0 & 0 \\ 0 & 0 & 1 & 0 \\ 0 & 1 & 0 & 0 \\ 0 & 0 & 0 & 1 \end{bmatrix}$$

It requires $\binom{n(A)}{n_1(A)} = \binom{16}{4} = 1820$ integers. The matrix could be encoded using
$\log(1820)$ bits.

Given the partition complexity, the information to fully express A can be expressed
using $\sum_{1 \le i \le k} \sum_{1 \le j \le e} C(A_{i,j})$ bits where

$$\sum_{1 \le i \le k} \sum_{1 \le j \le e} C(A_{i,j}) = \log\left[\prod_{i=1}^{k} \prod_{j=1}^{e} \binom{n(A_{i,j})}{n_1(A_{i,j})} \right] \quad (1)$$

Then the total bits to describe A with respect to a given partition is

$$T(A; k, e, Q_r, Q_c) = n\log(k) + m\log(e) + ke\log(n_1(A))$$
$$+ \log\left[\prod_{i=1}^{k} \prod_{j=1}^{e} \binom{n(A_{i,j})}{n_1(A_{i,j})} \right] \quad (2)$$

3.3 Problem Formulation

By the MDL principle, smaller $T(A; k, e, Q_r, Q_c)$ leads to higher quality partition.
Thus in order to resolve the communities of A, we need to find the optimal partition
that minimizes $T(A; k, e, Q_r, Q_c)$. The optimal partition correspond to k^*, e^*, Q_r^*, Q_c^*
and $T(A; k^*, e^*, Q_r^*, Q_c^*)$. Typically, this problem is a combination optimization
problem, thus is a NP-hard problem [22]. A common approach to conquer such problem is the evolutionary algorithm [23, 24]. In order to obtain a deterministic result, we
use an approximate algorithm instead.

4   A Split-Refine Greedy Algorithm

4.1   Sketch of the Algorithm

The proposed algorithm is a greedy algorithm, as shown in Algorithm 1. It is the algorithm for the case when n is bigger than or equal to m. And it has a counterpart for the
case when n is smaller than m. It starts from $k = 1$ and $e = 1$. For each iteration, it
performs a row and column splitting (line 6), a column splitting only (line 7), and a
row splitting only (line 8). Each step above guarantees that the total cost is non-
increasing, so it is considered to be a greedy algorithm. At last, it adjusts the search

step size for the three steps above. The procedure terminates if the failure continues $3*\sigma$ times, where σ is a threshold. Here σ is set to be 3.

Algorithm 1. MDL-greedy

Input: A
Output: k^*, e^*, Q_r^*, Q_c^*
1. $k = 1; e = 1; Q_r = \{1,1,\ldots,1\}; Q_c = \{1,1,\ldots,1\}; T = 0; F=0, R.rowstep = 1, R.columnstep = 1$
                //start as one community; F is the flag for no improvement;
                R is a counter for repeated splitting, i.e. the search step size
2.   currentcost = CostComputation (k, e, Q_r, Q_c)
3. $F = 0;$
4.   if $n >= m$
5.   do
6.   **Row-column-greedy**$(k, e, Q_r, Q_c, F, R);$
7.   **Column-greedy**$(k, e, Q_r, Q_c, F, R);$
8.   **Row-greedy**$(k, e, Q_r, Q_c, F, R);$
9.   **Adjust-searchstep**$(F, R);$
10.   while $(F < 3\sigma)$
11.   end
12. $k^* = k, e^* = e, Q_r^* = Q_r, Q_c^* = Q_c$
13. **return**
Procedure **CostComputation** (k, e, Q_r, Q_c)
    //compute the cost of a partition according to formula 2

Algorithm 2 shows the row and column searching. It splits the row and column first. During the splitting, a new row community and a new column splitting are generated. After the splitting, it reassigns each row/column to a new community that has the least cost. At last, the algorithm checks whether the splitting leads to a better compression. If the compression has been improved, the new partition is assigned to the current partition. The check guarantees the non-increasing of the total cost. Algorithm 2 has its counterparts for **Column-greedy** and **Row-greedy.** There are two sub-procedures SplitRow() and ReassginRowCommunity(), which are described in Algorithm 4 and Algortihm 5, respectively. SplitColumn() and ReassginColumnCommunity() are fundamentally the same as the former two except that they are for column.

Algorithm 2. Row-column-greedy

Input: k, e, Q_r, Q_c, F, R
Output: k, e, Q_r, Q_c, F, R
1. $[tmpk, tmpQ_r]$ = SplitRow(k, Q_r, R); $[tmpe, tmpQ_c]$ = SplitColumn$(tmpe, tmpQ_c, R)$;
2. $[tmpk, tmpe, tmpQ_r, tmpQ_c]$ = ReassginRowCommunity$(tmpk, e, tmpQ_r, Q_c)$;
3. $[tmpk, tmpe, tmpQ_r, tmpQ_c]$ = ReassginColumnCommunity$(tmpk, tmpe, tmpQ_r, tmpQ_c))$;
4.   newcost = CostComputation$(tmpk, tmpe, tmpQ_r, tmpQ_c)$;
5.   if newcost >= currentcost
6. $F = F + 1;$
7.   else
8.   currentcost = newcost; $k = tmpk$; $Q_r = tmpQ_r$ $e = tmpe$; $Q_c = tmpQ_c$; $F = 0$;
9.   end

Algorithm 1 adjusts the search size R according to the failure during the search. Algorithm 3 shows the detail. For each consecutive three times failure, the search step R will be increased. If there is no such failure, the search step size is kept as 1.

Algorithm 3. Adjust-searchstep

Input: F, R
Output: F, R
1.   if $F\%3 == 0$
2.   if $k > e$ $R.rowstep++$;
3.   else if $e > k$
4. $Rcolumnstep++$;
5.   else
6. $R.rowstep++$;
7. $Rcolumnstep++$;
8.   end
9.   end
10.   else
11. $R.rowstep=1$;
12. $Rcolumnstep=1$;
13.   end

4.2 Split the Submatrix

The algorithm splits the submatrix by splitting the row/column respectively. The detail of function SplitRow() is as follows: line 3 picks up the submatrix whose cost is the highest; line 11 judges whether the reduced cost after removing a row and regarding it as a community is higher than or equal to a threshold. In our work, the threshold is set as the maximum reduced cost. The splitting step searches for the right value for k and e.

Lemma 4.1. if $A = \begin{bmatrix} A_1 \\ A_2 \end{bmatrix}$, then $C(A_1) + C(A_2) \leqslant C(A)$.

Proof

$$C(A) = \log[\binom{n(A)}{n_1(A)}] = \log[\binom{n(A_1)+n(A_2)}{n_1(A_1)+n_1(A_2)}]$$

$$= \log[\sum_{x=0}^{n_1(A)} \binom{n(A_1)}{n_1(A)-x}\binom{n(A_2)}{x}]$$

$$\geq \log[\binom{n(A_1)}{n_1(A_1)}\binom{n(A_2)}{n_1(A_2)}]$$

$$= \log[\binom{n(A_1)}{n_1(A_1)}]+\log[\binom{n(A_2)}{n_1(A_2)}]$$

$$= C(A_1)+C(A_2)$$

where the second equality follows the Vandermondes's Identity [25] and the inequality follows the monotonic of the log function.

Corollary 4.1. For any $k_2 \geqslant k_1$ and $e_2 \geqslant e_1$, there exists a partition such that

$$\sum_{1\leq i\leq k_2}\sum_{1\leq j\leq e_2} C(A_{i,j}) \leq \sum_{1\leq i\leq k_1}\sum_{1\leq j\leq e_1} C(A_{i,j})$$

Proof. This simply follows Lemma 4.1.
According to Corollary 4.1, the function SplitRow() will never increase the conditional complexity.

Algorithm 4. SplitRow

Input: k, Q_r, R
Output: k, Q_r
1.   Repeat line 2~line14 $R.rowsize$ times
2. $k = k + 1$;

3. $r = \arg\max\limits_{1 \leq i \leq k} \sum\limits_{1 \leq j \leq e} \binom{n(A_{i,j})}{n_1(A_{i,j})}$

4.   for each row i $(1 \leq i \leq n)$
5.   if $Q_r(i) == r$

6. $\mathrm{costr}(i) = \sum\limits_{1 \leq j \leq e} \left(\binom{n(A_{r,j})}{n_1(A_{r,j})} - \binom{n(A_{r,j}) - n(Ar_{i,j})}{n_1(A_{r,j}) - n_1(Ar_{i,j})} \right) - \binom{n(Ar_{i,j})}{n_1(Ar_{i,j})} \right)$

      //record the reduced cost when removing row i from community r and regard it
        as a community
7.   end
8.   end
9. $threshold$ = max(costr);
10.   for each row i $(1 \leq i \leq n)$
11.   if costr(i) >= threshold
12. $Q_r(i) == k$;
13.   end
14.   end
15.   **return**

4.3 Reassign the Rows and Columns to Community

The reassign step is a refinement of the result from the split step. It not only refines the splitted row communities, but also refines the column communities so as to fit the change of row communities. Line 5 picks up the community to which assigning a row will add the least cost to the total. Line 7 judges whether to reassign a row will reduce the total cost. The symbols before reassignment are denoted as k^0 and $C(A_{i,j}^0)$, and the symbols after the reassignment are denoted as k^1 and $C(A_{i,j}^1)$.

Theorem 4.1. $\sum\limits_{1 \leq i \leq k^0} \sum\limits_{1 \leq j \leq e} C(A_{i,j}^0) \geq \sum\limits_{1 \leq i \leq k^1} \sum\limits_{1 \leq j \leq e} C(A_{i,j}^1)$

Proof. This simply follows the reassignment judgment.

By Theorem 4.1, the function ReassginRowCommunity() will never increase the conditional complexity.

4.4 Computational Complexity

Line 5 of Algorithm 5 computes ke binomial coefficient. The same computation is also required for the column reassignment. Thus the computational complexity of Algorithm 5 is $O(I(n+m)ke)$, where I is the count of the outer loop.

    Line 2 of Algorithm 4 computes ke binomial coefficient, and the loop on line 3 computes ne binomial coefficient. Thus the computational complexity of Algorithm 4 is $O((k+n)e)$. Here the loop time R is ignored since its maximum value is 3.

Considering the computation on columns, the complexity of each iteration in Algorithm 1 is $O(2I(n+m)ke + 2ke+ne+mk) = O(I(n+m)ke)$. Thus the complexity of Algorithm 1 is $O((k^*+e^*) I(n + m)k^*e^*)$. Since k^*, e^*, and I are small according to our experiments, Algorithm 1 is linear.

Algorithm 5. ReassignRowCommunity

Input: k, e, Q_r, Q_c
Output: k, e, Q_r, Q_c
1.   do
2.   currentcost = CostComputation (k, e, Q_r, Q_c);
3. $tmpQ_r = Q_r$;
4.   for each row r $(1 \leq r \leq n)$

5. $\quad i = \arg\min_{1 \leq i \leq k} \sum_{1 \leq j \leq e} ((^{n(A_{i,j})+n(Ar_{r,j})}_{n_1(A_{i,j})+n_1(Ar_{r,j})}) - (^{n(A_{i,j})}_{n_1(A_{i,j})}))$;

6. $\quad ipos = Q_r(r)$;
       //the current community row r belong to

7. $\quad$ if $\quad \sum_{1 \leq j \leq e} ((^{n(A_{ipos,j})}_{n_1(A_{ipos,j})}) - (^{n(A_{ipos,j})-n(Ar_{r,j})}_{n_1(A_{ipos,j})-n_1(Ar_{r,j})}))$

$\geq \sum_{1 \leq j \leq e} ((^{n(A_{ipos,j})+n(Ar_{r,j})}_{n_1(A_{ipos,j})+n_1(Ar_{r,j})}) - (^{n(A_{ipos,j})}_{n_1(A_{ipos,j})}))$

8. $\quad\quad tmpQ_r(r) = i$
9. $\quad\quad Q_r = tmpQ_r$;
10. $\quad$ end
11. $\quad$ end
12. $\quad$ for each column c $(1 \leq c \leq m)$
13. $\quad\quad$ do the same as done from each row
14. $\quad$ end
15. $\quad$ newcost = CostComputation (k, e, Q_r, Q_c)
16. while newcost < currentcost;
17. **return**

---

5 Performance Study

All experiments are conducted on an INTEL core 2DuoProcessorE2160 with 2G memory, running Windows XP. All algorithms are implemented in Matlab R2007b. The program is run on J2SE 5.0.

5.1 The Synthetic Datasets

To check the performance of the proposed algorithm, the algorithm is examined over four synthetic datasets, which parallel the datasets for one-mode network in [26].

Dataset1: The sizes of S_r and S_c for each graph are set as 192 and 192, respectively. S_r and S_c are divided into three communities of 64 vertices, denoted as $\{S_{r1}, S_{r2}, S_{r3}\}$ and $\{S_{c1}, S_{c2}, S_{c3}\}$ respectively. Edges are placed between vertex pairs of different types independently and randomly. The distribution of the edges for a graph is shown in Table 2. Here Z_{ij} is the average degree of the vertices in S_{ri}. Several constraints are added on the generated graphs: For a given i, $\sum_{1 \le j \le 3} Z_{ij} = 24$; $Z_{ii} > \sum_{j!=i} Z_{ij}$. All $Z_{ij}^{i!=j}$ are the same.

Table 2. Edge distribution for dataset 1

Graph	S_{c1}	S_{c2}	S_{c3}
S_{r1}	Z_{11}	Z_{12}	Z_{13}
S_{r2}	Z_{21}	Z_{22}	Z_{23}
S_{r3}	Z_{31}	Z_{32}	Z_{33}

Dataset 2: the number of graphs is the same as that of datasets 1. Each graph in dataset 1 has a corresponding subgraph in dataset 2, as shown in Table 3. All graphs in dataset 2 are generated by removing S_{r3} and its related edges from graphs in dataset 1.

Dataset 3: Let y be an integer and $1 \le y \le 21$. The sizes of S_r and S_c for each graph are set as 192y and192y, respectively. S_r and S_c are divided into three communities of 64y vertices, denoted as $\{S_{r1}, S_{r2}, S_{r3}\}$ and $\{S_{c1}, S_{c2}, S_{c3}\}$ respectively. Edges are placed between vertex pairs of different types independently and randomly. The edge distribution for a graph is shown in Table 2. Here Z_{ij} is the average degree of the vertices in S_{ri}. Two constraints are added on the generated graphs: For a given i,

$$\sum_{1 \le j \le 3} Z_{ij} = 24y, \quad Z_{ij}^{i!=j} = 2y.$$

Table 3. Edge distribution for dataset 2

Graph	S_{c1}	S_{c2}	S_{c3}
S_{r1}	Z_{11}	Z_{12}	Z_{13}
S_{r2}	Z_{21}	Z_{22}	Z_{23}

Dataset 4: the number of graphs is the same as that of datasets 3. Each graph in dataset 3 has a corresponding subgraph in dataset 2, as shown in Table 3. All the graphs in dataset 4 are generated by removing S_{r3} and its related edges from graphs in dataset 3.

5.2 Evaluation and Results

There is no 'standard' criterion to measure the quality of the results. Therefore, the precision, i.e. the fraction of vertices classified correctly, is computed as in [26], and it is adapted to bipartite network. The vertices in S_r that has been correctly classified is denoted as $P(S_r)$, the vertices in S_c that has been correctly classified is denoted as $P(S_c)$. Then precision can be computed as follows:

$$P(V) = \frac{|S_r|}{|S_r| + |S_c|} * \frac{|P(S_r)|}{|S_r|} + \frac{|S_c|}{|S_r| + |S_c|} * \frac{|P(S_c)|}{|S_c|}$$

Another criterion is the ratio between the number of discovered communities and the correct community number. The number of correct row communities is denoted as N_r and the number of correct column communities is denoted as N_c. Then ratio can be computed as follows:

$$Ratio = \frac{N_r}{N_r + N_c} * \frac{k}{N_r} + \frac{N_c}{N_r + N_c} * \frac{l}{N_c}$$

When *Ratio* is equal to 1, the method discovers the correct community number. When *Ratio* is less than 1, the method discovers less than the correct community number. And when *Ratio* is bigger than 1, the method discovers more than the correct community number.

Table 4. Precision&Ratio for dataset 1

	$i \stackrel{!}{=} j$ Z_{ij}	MDL greedy	CA	ITCC
$P(V)$	1	0.9765±0.0525	0.9100±0.0665	0.9744±0.0529
	2	0.9944±0.0277	0.9394±0.0569	0.9746±0.0566
	3	0.9897±0.0373	0.9414±0.0583	0.9776±0.0595
	4	0.9787±0.0507	0.9262±0.0642	0.9719±0.0663
	5	0.9470±0.0577	0.8794±0.0797	0.8553±0.1233
Ratio	1	1.0567±0.1258	1.1900±0.1276	-
	2	1.0133±0.0656	1.1600±0.1296	-
	3	1.0233±0.0855	1.1450±0.1415	-
	4	1.0433±0.1127	1.1650±0.1411	-
	5	1.0733±0.1577	1.2667±0.2247	-

Table 5. Precision&Ratio for dataset 2

	$i \stackrel{!}{=} j$ Z_{ij}	MDL greedy	CA	ITCC
$P(V)$	1	0.9958±0.0181	0.8745±0.0585	0.9329±0.0166
	2	0.9978±0.0130	0.8804±0.0631	0.9350±0.0523
	3	0.9939±0.0188	0.8896±0.0693	0.9388±0.0395
	4	0.9853±0.0207	0.8922±0.0671	0.9048±0.0789
	5	0.9430±0.0431	0.8339±0.0753	0.8097±0.1009
Ratio	1	1.0100±0.0438	1.3420±0.1372	-
	2	1.0040±0.0281	1.3300±0.1738	-
	3	1.0080±0.0394	1.2780±0.1703	-
	4	1.0100±0.0438	1.3060±0.1874	-
	5	1.0540±0.1749	1.4580±0.2417	-

The algorithms MDL-greedy, CA and ITCC are tested on dataset 1 and dataset 2. Although the 'farthest' initialization is reported to be the best in [14], 'random' initialization gains better result than 'farthest' initialization for ITCC in this experiment.

Therefore, 'random' initialization is adopted for ITCC. Furthermore, the community number is specified manually as the correct number since ITCC cannot search it automatically. Therefore, there is no *Ratio* for ITCC. The results are the average precision/ratio and their standard deviation over 100 random network generations, which are shown in Table 4 and Table 5. On both datasets MDL-greedy outperforms CA and ITCC. MDL-greedy gains a higher precision and is more stable than the other two methods. MDL-greedy scarcely discovers wrong community number while CA tends to generate more communities. When $Z_{ij}^{i!=j}$ increase to 5, the performance of all methods drops. The precision for ITCC drops the most, which is even worse than CA. The precision drop of MDL-greedy is acceptable because the variation is small. Comparing with CA, the *Ratio* for MDL-greedy is hardly affected. Obviously, CA gets trapped in the noise produced by the big $Z_{ij}^{i!=j}$ when searching for the number of communities. A sample result on a specified network is shown in Fig 2 to explain this phenomenon.

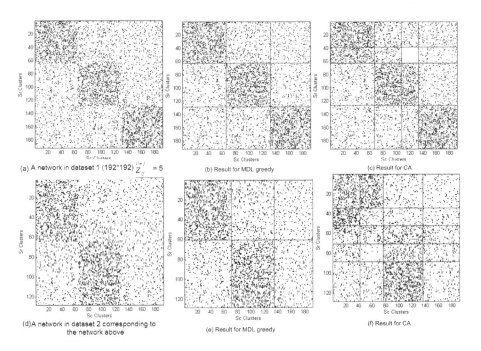

(a) A network in dataset 1 (192*192) $Z_{ij}^{i!=j}$ = 5

(b) Result for MDL greedy

(c) Result for CA

(d)A network in dataset 2 corresponding to the network above

(e) Result for MDL greedy

(f) Result for CA

Fig. 2. A sample result for MDL-greedy & CA on dataset 1&2

5.3 Scalability

The algorithms MDL-greedy and CA are tested on dataset 3 and dataset 4. ITCC is not tested here since it requires manual specification on community number. The time cost for finding the optimum partition is recorded in our testing. The results are shown in Fig 3. The results show that the cost of MDL-greedy increases linearly along the increasing of y and is very stable. MDL-greedy outperforms CA on both datasets.

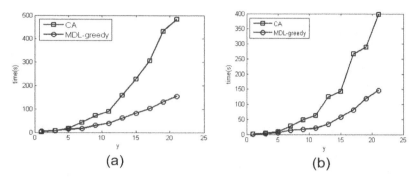

Fig. 3. Time cost for dataset 3(a)&dataset 4(b)

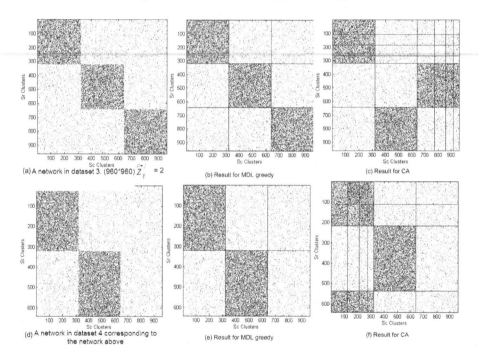

Fig. 4. A sample result for MDL-greedy & CA on dataset 3&4

When y is small, the performances of MDL-greedy and CA are close. But along the increasing of *y*, the performance difference between MDL-greedy and CA becomes larger and larger. According to the *Ratio* value in Table 4 and Table 5, the difference is attributed to the extra burden from CA's tendency to discover the community number more than the correct one. This phenomenon is very obvious in the result for dataset 4. Since the networks in dataset 4 are asymmetric, CA sometimes consumes much more time than expected. There are two jumps in Fig 3(b): from *y*=15 to *y*= 17 and from *y*=19 to *y*=21. A sample result for both methods on a specified network is shown in Fig 4. The number of communities that CA generates is much larger than the actual number: (k_{CA}*=6, l_{CA}*=6) vs (k*=3, l*=3) in Fig 4(c) and (k_{CA}*=4, l_{CA}*=6) vs (k*=2, l*=3) in Fig 4(f). Generating more communities will decrease both the precision and scalability of CA. This may make MDL-greedy more practical than CA, since practical problems often provide large datasets.

5.4 The Southern Women Data Set

The southern women dataset is commonly used to evaluate the performance of bipartite network community detection methods [27]. It records eighteen women's attendance on fourteen events. Freeman has described it as "...a touchstone for comparing analytic methods in social network analysis[28]". MDL-greedy, CA and ITCC are run on this dataset. Again 'random' initialization is adopted for ITCC. ITCC is run twice and the community number are set as ($k = 2, l = 3$) and ($k = 2, l = 4$) respectively. The results are shown in Fig 5, where the women are labeled as 'W*' and the events are labeled as 'E*'. According to Freeman[28], the 'perfect' community for the southern women are 'W1-W9' and 'W10-W18'. There is no focus on the events. It is shown from Fig 5 that both MDL-greedy and CA detect the correct community number for the southern women. The result of MDL-greedy is exactly the 'perfect' community while the result of CA is quite different from the 'perfect' one. And no method in [28] generates the same result as CA. No method in [28] generates the same result as ITCC ($k = 2, l = 3$) either, which is also far from the 'perfect' one. However, the result of ITCC ($k = 2, l = 4$) is very close to the 'perfect' one in which only 'w8' is misplaced.

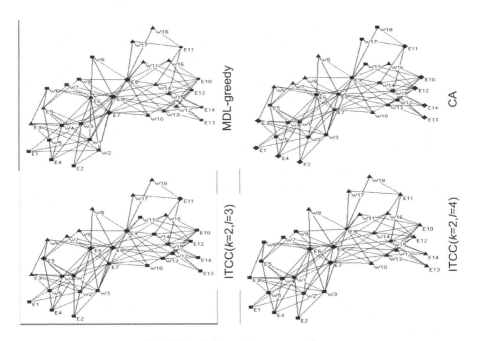

Fig. 5. Result for southern women dataset

6 Conclusions

To resolve communities in bipartite networks, the compression-based method is proposed for bipartite networks. A binomial coefficient based formula is applied as the description length of a partition. A greedy algorithm to minimizing the description length is also proposed under a split-refine framework. It successfully searches communities in automatic style for different types of nodes simultaneously. The experiment results show the high accuracy of the proposed method in the perspectives of both manually defined measures: precision and ratio. The experiment results also show its almost linear scale with the node size of the bipartite network, which fits the computational complexity analysis well.

References

[1] Newman, M.E.J.: The Structure and Function of Complex Networks. SIAM Review 45, 167–256 (2003)
[2] Guimerà, R., Amaral, L.: Functional cartography of complex metabolic networks. Nature 433, 895–900 (2005)
[3] Danon, L., Duch, J., Diaz-Guilera, A., Arenas, A.: Comparing community structure identification. J. Stat. Mech. P09008 (2005)
[4] Newman, M.E.J., Girvan, M.: Finding and Evaluating Community Structure in Networks. Physical Review E 69, 026113 (2004)
[5] Linden, G., Smith, B., York, J.: Amazon.com Recommendations: Item-to-Item Collaborative Filtering. IEEE Internet Computing 7, 76–80 (2003)
[6] Newman, M.E.J.: Modularity and community structure in networks. Proceedings of the National Academy of Sciences 103, 8577–8582 (2006)
[7] Pujol, J., Béjar, J., Delgado, J.: Clustering algorithm for determining community structure in large networks. Physical Review E 74, 9 (2006)
[8] Fortunato, S., Barthelemy, M.: Resolution Limit in Community Detection. Proceedings of the National Academy of Sciences 104, 36–41 (2007)
[9] Rosvall, M., Bergstrom, C.T.: An information-theoretic framework for resolving community structure in complex networks. Proc. Natl. Acad. Sci. USA 104, 7327–7331 (2007)
[10] Rosvall, M., Bergstrom, C.T.: Maps of random walks on complex networks reveal community structure. Proc. Natl. Acad. Sci. USA 105, 1118–1123 (2008)
[11] Barron, A., Rissanen, J., Yu, B.: The minimum description principle in coding and modeling. IEEE Transactions on Information Theory 44, 2743–2760 (1998)
[12] Strogatz, S.H.: Exploring complex networks. Nature 410, 268–276 (2001)
[13] Chakrabarti, D., Papadimitriou, S., Modha, D.S., Faloutsos, C.: Fully automatic cross-associations. In: Proc. Tenth ACM SIGKDD Int. Conf. Knowl. Discov. Data Min., pp. 79–88 (2004)
[14] Dhillon, I.S., Mallela, S., Modha, D.S.: Information-theoretic co-clustering. In: KDD (2003)
[15] Madeira, S.C., Oliveira, A.L.: Biclustering algorithms for biological data analysis: a survey. IEEE/ACM TCBB 1, 24–45 (2004)
[16] Cheng, Y., Church, G.M.: Biclustering of expression data. In: ISMB (2000)
[17] Guimerà, R., Sales-Pardo, M., Lan, A.: Module identification in bipartite and directed networks. Physical Review E 76 (2007)

[18] Barber, M.J.: Modularity and community detection in bipartite network. Physical Review E 76 (2007)

[19] Lehmann, S., Schwartz, M., Hansen, L.K.: Biclique communities. Phys. Rev. E 78, 016108 (2008)

[20] Sun, J., Faloutsos, C., Papadimitriou, S., Yu, P.: GraphScope: Parameter-free mining of large time-evolving graphs. In: KDD (2007)

[21] Papadimitriou, S., Sun, J., Faloutsos, C., Yu, P.: Hierarchical, parameter-free community discovery. In: Daelemans, W., Goethals, B., Morik, K. (eds.) ECML PKDD 2008, Part II. LNCS (LNAI), vol. 5212, pp. 170–187. Springer, Heidelberg (2008)

[22] Sipser, M.: Introduction to the Theory of Computation. PWS Publishing Company (1997)

[23] Ashlock, D.: Evolutionary computation for modeling and optimization. Springer, New York (2005)

[24] Xu, K.K., Liu, Y.T., Tang, R.e.a.: A novel method for real parameter optimization based on Gene Expression Programming. Applies Soft Computing 9, 725–737 (2009)

[25] Rosen, K.H.: Discrete mathematics and its applications, 4th edn. WCB/McGraw-Hill, Boston (1999)

[26] Girvan, M., Neuman, M.E.J.: Community structure in social and biological networks. Proc. Natl. Acad. Sci. USA 99, 7821–7826 (2002)

[27] Davis, A., Gardner, B.B., Gardner, M.R.: Deep South. University of Chicago Press, Chicago (1941)

[28] Freeman, L.: Dynamic Social Network Modeling and Analysis. The National Academies Press, Washington (2003)

Fires on the Web: Towards Efficient Exploring Historical Web Graphs

Zhenglu Yang[1], Jeffrey Xu Yu[2], Zheng Liu[2], and Masaru Kitsuregawa[1]

[1] Institute of Industrial Science, The University of Tokyo, Japan
{yangzl,kitsure}@tkl.iis.u-tokyo.ac.jp
[2] Chinese University of Hongkong, China
{yu,zliu}@se.cuhk.edu.hk

Abstract. Discovery of evolving regions in large graphs is an important issue because it is the basis of many applications such as spam websites detection in the Web, community lifecycle exploration in social networks, and so forth. In this paper, we aim to study a new problem, which explores the evolution process between two historic snapshots of an evolving graph. A formal definition of this problem is presented. The evolution process is simulated as a fire propagation scenario based on the Forest Fire Model (FFM) [17]. We propose two efficient solutions to tackle the issue which are grounded on the probabilistic guarantee. The experimental results show that our solutions are efficient with regard to the performance and effective on the well fitness of the major characteristics of evolving graphs.

1 Introduction

Graphs represent the complex structural relationships among objects in various domains in the real world. While these structural relationships are not static, graphs evolve as time goes by. Evolving graphs are usually in the form of a set of graphs at discontinuous time stamps, where the period between two adjacent time stamps may be quite long. Take the Web archive for example. Due to its large size, the Web or a part of it is periodically archived by months or even by years. Mining evolving graphs is important in many applications including the detection of spam websites on the Internet [7], exploration of community lifecycle in social networks [20], and identification of co-evolution relationships between structure and function in bio-informatics [18].

While many of the existing studies have paid attentions to finding stable or changing regions in evolving graphs [1,19,10], only a few of them are about how graphs evolve. The researchers have proposed various generative models to capture the statistical properties of the graph evolution such as Power Law distribution [5], effective diameter [21], and so forth. In this paper, however, we study a new problem, which is to model the evolving process between two historical snapshots of an evolving graph. Fig. 1 briefly shows our idea. G is an evolving graph which evolves from time t to t'. Suppose we have the graph snapshots at time t and t', and the real evolution details between these two

H. Kitagawa et al. (Eds.): DASFAA 2010, Part I, LNCS 5981, pp. 612–626, 2010.

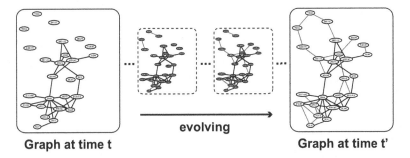

Graph at time t   evolving   **Graph at time t'**

Fig. 1. Example Graphs (All the figures in this paper are made colorful for clarity)

snapshots are unknown. We would like to generate a series of virtual graph snapshots, as shown in the dotted parts in Fig. 1. What is important is that the statistical properties of the evolution must be maintained in these virtual graph snapshots. By doing this, we decompose the macro graph change of the two real snapshots into several micro changes. The benefits are twofold.

Firstly, by parsing the graphs into historical snapshots, we can learn the evolution of the different parts, and thus the future trends of these regions can be predicted. Fig. 2 shows a concrete example, which is conducted on the DBLP dataset[1]. The extracted virtual historical snapshot steps can help us understand the evolution of co-authorship relations. It seems that Web community evolution detection [2] can do the same thing, but the work in this paper is different from that. We aim to detect the changes throughout the whole graph and do not constrain the work on the boundary subgraph (community) that may be defined by the users (i.e., with keywords).

Secondly, successfully tackling the issue proposed in this paper can address the critical issue of the lack of intermediate order-based Web data between two historical Web graphs. Many existing studies on static/dynamic graph mining can profit from restoring these historical graphs such as frequent subgraph discovery [9], temporal graph cluster detection [1], micro view on social networks [13], and so forth. As such, this research work is orthogonal to the existing issues on graph mining in a complementary manner.

For ease of exposition and without loss of generality, we only consider the node/edge insertion scenario in this paper. It should be noted that our approaches can handle the scenario where both the insertion and the deletion of nodes/edges occur. The difficulty in generating these virtual graph snapshots is that there are numerous number of possibilities.

Our approach adopts the Forest Fire Model (FFM) [17], which has been demonstrated as successfully explaining the mechanism of dynamic systems [8]. The new edge linkage action can be thought of as fire propagation, and the nodes on the new edges are the origins of the fire. The virtual historical graph is then to be thought of as the snapshot on tracking how the fire propagates on the whole graph. We will give the formal definition of the problem shortly.

---

[1] www.informatik.uni-trier.de/~ley/db

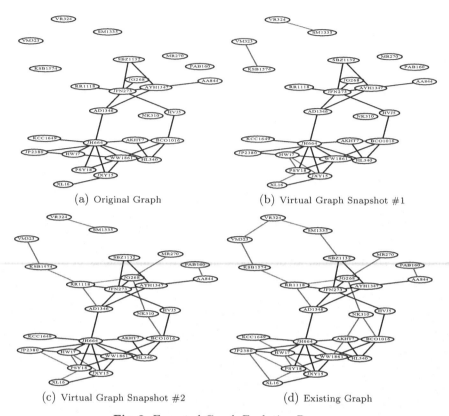

(a) Original Graph

(b) Virtual Graph Snapshot #1

(c) Virtual Graph Snapshot #2

(d) Existing Graph

Fig. 2. Expected Graph Evolution Process

The contributions of our paper are as follows:

- We propose a new problem of how to trace back the virtual snapshots of evolving graphs. The process can be simulated based on the FFM. The historical graph is deemed as a snapshot of tracking the fire propagation situation on the graph.
- We propose two approaches, *bottom-up* and *leap-search*. The *bottom-up* strategy examines the candidates from scratch in a global view, while the *lead-search* method applies a density-oriented candidate selection mechanism. We also explore the heuristics based on the properties of evolving graphs to improve the efficiency of the two approaches.
- We conduct comprehensive experiments on real large evolving graphs. The evaluation results demonstrate the effectiveness and efficiency of the proposed solutions.

The remainder of this paper is organized as follows. We introduce the preliminaries in Section 2. The bottom-up and leap-search solutions are presented in Sections 3 and 4, respectively. Section 5 reports the experimental results and Section 6 concludes the paper.

Table 1. The summary of notation

Notation	Definition
G	An evolving graph or a graph adjacent matrix
G_t	An evolving graph at time t
V_t	The set of vertices in the graph G_t
E_t	The set of edges in the graph G_t
$|V_t|$	Number of vertices in the graph G_t
$|E_t|$	Number of edges in the graph G_t
i, j, k	Vertices in a graph
$e= (i, j)$	Edges in a graph
$deg(u)$	Degree of vertex u
$v_{i,j}$	Fire propagation velocity between vertices i and j
$t_{i,j}$	Fire propagation time between vertices i and j
$d_{i,j}$	Distance or length between vertices i and j
c	Backward burning probability

2 Preliminary

In this paper, we deal with undirected evolving graphs. Let G denote an evolving graph. A snapshot of the graph G at time t is represented as $G_t = (V_t, E_t)$, where V_t is the set of vertices at time t; and $E_t \subseteq V_t \times V_t$ is the set of edges at time t. The notations used in this paper is summarized in Table 1. To trace back the historical snapshots of an evolving graph, we will introduce novel strategies based on the FFM [17].

2.1 Forest Fire Model

The FFM was first proposed in ecology [17], where the scholars were interested in how to control and predict wildfire in the real world environment. Henley [8] first introduced the FFM in studying the characteristics of self-organized systems. From then on, the FFM has succeeded in explaining many real dynamical systems such as Geographic Information Systems (GIS) [6], Affiliation Network [12], arXiv citation [12], and so forth. Most especially, the FFM fits in many properties of real Web data we would like to explore in this paper: (1) the rich get richer, which is called the attachment process (heavy-tailed degrees) [3]; (2) leads to community structure [11]; (3) densification [12]; and (4) shrinking effective diameter.

Note that the FFM is related to random walk and electric currents [4] with regard to the issue of evolving graphs. The difference is that the latter two have not taken all the aforementioned four properties of real Web graph into account. As far as we know, there is only one work [12] on studying evolving Web graph based on the FFM. There are three main differences between this work and [12]: (1) [12] aims to generate synthetic evolving graphs from scratch, while in this work we trace back the historical snapshots between two real graphs; (2) [12] randomly selects the initial fired nodes, while in this work we deliberately choose the initial fired nodes with probabilistic guarantee; and (3) we introduce the fire propagation velocity into our framework, while [12] does not consider this property. These distinct issues are due to the different purposes of the two works; [12] intended to generate synthetic evolving graph from scratch on the fly, while we aim to study the whole history of how an old graph evolves to a

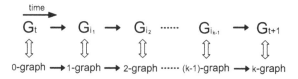

Fig. 3. Graph evolving process

newer graph. By parsing the evolving process into virtual historical snapshots, we can learn the micro evolution of the different regions in a large graph to make a navigational prediction on the evolving trend of the graph.

2.2 Discovery of the Historical Snapshot Graph Problem

Given two graphs, G_t and G_{t+1}, at time t and time $t+1$, the problem of discovering the historical graph snapshots involves tracing back the virtual graphs after inserting n new edges[2], where $1 \leq n \leq (|E(G_{t+1})| - |E(G_t)|) = k$, into the old graph G_t. Let n-*graph* denote the graph snapshot which has n new edges. Fig. 3 illustrates the evolving process of the graph. G_t and G_{t+1} can be mapped at 0-*graph* and k-*graph*, respectively. **The issue of discovering the historical graph snapshots is then equivalent to finding the n-graphs**, where $0 < n < k$ and $(|E(G_{t+1})| - |E(G_t)|) = k$.

For example, Fig. 2 (b)-(c) are the graph snapshots of Fig. 2 (a) by inserting 6 and 14 new edges, denoted as 6-*graph* and 14-*graph*, respectively. The new edge linkage action can be thought of as fire propagation, and the nodes on the new edges are the origins of the fire. The virtual historical graph is then thought of as the snapshot on tracking how the fire propagates on the whole graph. In this paper, we introduce how to set the initial fire energy and how fast fire propagates on the graph. As far as we know, this work is the first one to study these issues. **The problem of finding the n-graphs is then equivalent to discovering the burning out n-graphs**, which is formally defined in Definition 1.

Definition 1. Burning Out n-Graph (BOG) Problem
Given two undirected graphs $G_t = (V_t, E_t)$ and $G_{t+1} = (V_{t+1}, E_{t+1})$; a cost function $CF(i,j)$ on edge (i,j) where $i \in V_t \cup V_{t+1}$ and $j \in V_t \cup V_{t+1}$; a user preferred number n of new edges, find the subgraph H of $G_t \cup G_{t+1}$ such that

- *The number of new edges on H is n, and*
- $\sum_{(i,j) \in E(H)} CF(i,j)$ *is minimized.*

$E(H)$ is the edge set of graph H. The cost function CF can be considered as the time necessary to construct the subgraph H (or burning out it), as will be well defined shortly with the help of the FFM [17].

---

[2] New vertices are accompanied with new edges.

2.3   Discovery of the Burning Out n-Graph (BOG) Problem

We aim to address the issue of finding the first burning out n-graphs, which is equivalent to extracting the historical graph snapshots. In this section, we introduce the basic configuration of the fire model such as the initial fire energy, the velocity and time for fire propagation, and the update of the fire energy.

Initial fire energy: Consider an FFM, where the fires are caused by those changing edges (with vertex insertion), and each changing edge introduces one unit fire energy. The initial fire energy of a vertex is the accumulated energy of all its adjacent new edges.

Fire propagation: The cost function $CF(i,j)$ introduced in Section 2.2 is defined as the time $t_{i,j}$ of the fire propagation consumed on the edge between the two adjacent vertices i and j. We have

$$CF(i,j) = t_{i,j} = \frac{d_{i,j}}{\bar{v}_{i,j} + \bar{v}_{j,i}}, \tag{1}$$

where $d_{i,j}$ denotes the distance between the two adjacent nodes (i and j), and $\bar{v}_{i,j}$ and $\bar{v}_{j,i}$ denote the average fire propagation velocity from vertice i to j, and vice versa, respectively. It is interesting to note that the fire propagation has direction, which conforms to the common intuition that the effects of changing edges/vertices spread from the origins to the distant edges/vertices[3]. The average velocity $\bar{v}_{i,j}$ is mainly dependent on the initial burning energy[4] ℓ_i. Hence, simplify the model without loss of generality, we have $\bar{v}_{i,j} = \gamma_i * \ell_i$, where γ_i is a constant. In this paper, we assume $\gamma_i = \gamma_j = \ldots = 1$, and Eq. (1) can be deduced as

$$CF(i,j) = t_{i,j} = \frac{d_{i,j}}{\ell_i + \ell_j}. \tag{2}$$

Fire energy update: The fire energy ℓ should be updated after each successful propagation by using the following equation.

$$\ell_{j_{t'}} = \ell_{j_t} + \sum_{i \in Neigh_{suc}(j)} (1 - c)\, \ell_{i \to j}, \tag{3}$$

where ℓ_{j_t} and $\ell_{j_{t'}}$ are the fire energy of j at time t and time t' respectively, c is a *backward burning probability* [12], and $\ell_{i \to j}$ is the fire energy transferred from i to j. Thus we have $\ell_{i \to j} = \ell_i/deg(i)$. $Neigh_{suc}(j)$ denotes the neighbour nodes of j that successfully transfer their fire energy to j between time t and time t'. Note that once a node successfully propagates fires to its neighbours, its own fire energy is reset to zero at the time.

In this paper, we propose two approaches to discover the fastest burning out n-graphs. The *bottom-up* approach examines the candidates from scratch

---

[3] This mechanism will help to address the issue on directed graphs, which however, is out scope of this paper.

[4] In a real ecological environment, other effects such as wind, topography slope, and so forth, should also be taken into account.

Algorithm 1. Bottom-up algorithm

Input: The graph G_{i-1} and G_i at time t_{i-1} and t_i, a user preferred number n
Output: The fastest burning out n-graph(s)

1   $H = G_{i-1} \cup G_i$;
2   **for each** vertex $v_i \in H$ **do** //initial fire energy
3       $fe[v_i]$ =num_of_adjacent_changing_edges;
4   can_graph_list=store v_i as a graph;
5   num_of_new_edge=0;
6   **while** num_of_new_edge $< n$ **do**
7   **for each** graph $g \in$ can_graph_list **do**
8   g'= appending g with new edge e of minimal spreading time;
9   t_{local}=fire propagation time on e;
10   **if** $t_{local} < t_{min_{local}}$ **do**
11              $t_{min_{local}}$=t_{local};
12   update g' in can_graph_list;
13   update $fe[v_i]$ if new edge introduces a new vertex v_i;
14   num_of_new_edge++;
15 output the n-graphs with minimal fire propagation time;

---

in a global view with the dynamic threshold guaranteed, while the *leap-search* approach proposes a density-oriented candidate selection strategy.

3   The Bottom-Up Approach

We develop a bottom-up greedy algorithm to extract the burning out n-graphs. The algorithm follows the candidate generation and verification iteration. To accelerate the process, the threshold-based technique is proposed to prune the candidates early.

Candidate generation: In each iteration, a k-graph g is grown up to a candidate $(k+1)$-graph g' by introducing a new edge e_{new}, where the following conditions hold: (1) e_{new} is connected to some vertices in g; and (2) the fire spreading time t from g to v_{new} is minimized[5].

Verification: If the burning out time of the candidate $(k+1)$-graph g' is greater by far than the best one, we turn to the next candidate subgraph. Otherwise, we update the fire energy of the new vertex, which is introduced by the new edge based on Eq. 3, and continually grow g' to a larger candidate graph with another candiate-generation-and-test iteration.

The basic bottom-up greedy algorithm is shown in Algorithm 1. We will introduce the pruning techniques in the next section. In the initial phase (line 1-5), the two graphs are joined. The changing edges of each vertex are counted while joining, stored in an array fe. The vertex is put into the candidate list as a graph. The candidate generation and test iteration is from line 6 to line 14. Given a k-graph g, we generate its candidate (k+1)-graph, by finding a nearest edge to g, which may come from its nearest neighbor vertex or internal unlinked

---

[5] Specifically, t is computed based on Eq. 2. By default, $d_{i,j}$ is set to 1 for the unweighted graph in this paper. For the weighted graph, $d_{i,j}$ can be set according to the weight of the edge.

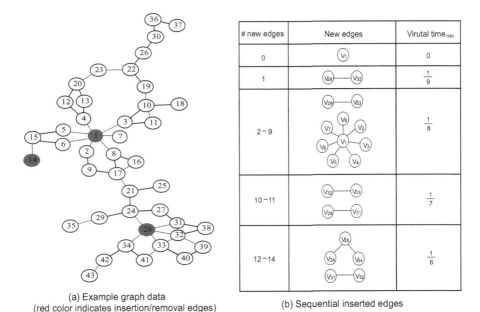

(a) Example graph data
(red color indicates insertion/removal edges)

(b) Sequential inserted edges

Fig. 4. Example graph

edge. The fire propagation time is computed based on Eq. 3. If the time is smaller than the least time of burning out graph time so far, we update the new graph information (line 12) and also update the fire energy of a new vertex if it exists (line 13). The iteration will be completed when the burning out n-graph is found.

3.1   Implementation Details

Pruning Techniques. We can prune many candidate graphs early based on the threshold-based technique. The threshold time t_{th} is dynamically updated based on the most optimal graph by far. Before a candidate graph g grows up to a larger candidate one, if we find the construction (burning out) time t_g of g is already greater than t_{th}, this round can be terminated and continued to the next round. The reason why we jump to the next round instead of the remaining candidates is that we can rank all the candidate graphs based on their burning out time in ascending order (by using minimal heap). If t_g is greater than t_{th}, then the time of all the remaining candidates (in the heap) should be greater than t_{th}; hence, this round can be safely terminated. The early pruning rule is justified based on the following lemma.

Lemma 1 (Anti-monotone). *Let g be a graph with k edges and g' be a connected supergraph of g with $(k+1)$ edges. The burning out times of g and g' are t and t', respectively. Let t_{th} be a threshold time. If $t > t_{th}$, then $t' > t_{th}$.*

Algorithm 2. Pruning strategy for the bottom-up algorithm

```
1   sort fe in descending order;
2   min_heap ← store vⱼ ∈ fe with maximal value as a subgraph;
3   while min_heap is not empty do
4       g=subgraph with minimal value tₘ in min_heap;
5       if #new edge(g) ≥ n do
6           break;
7       g′ = appending g with edge of minimal spreading time;
8       t_local=fires propagation time of the new edge;
9       min_heap ← store g′ with t_local_max;
10      update fe[vᵢ] if new edge introduces a new vertex vᵢ;
11      vⱼ=unvisited node in fe whose energy is the largest;
12      for each neighbor vₖ of vⱼ do
13          if t(vₖ, vⱼ) < t_local do
14              min_heap ← store vⱼ as a subgraph;
15              break;
```

Proof Sketch. (By contradiction) Suppose we have $t' \leq t_{th}$. We reduce g' to g'' by removing the edge not existing in g. Let t'' be the burning out time of g''. We have $t'' \leq t_{th}$. As t'' is equal to t, it results in $t \leq t_{th}$, which is contradictory to the assumption; thus the lemma holds.   □

Lemma 1 can efficiently prune many candidate subgraphs, as will be demonstrated in the experimental results. The reason for this is that due to the Densification Power Law [12] property, the threshold of the most optimal subgraph (a "rich" one) by far will have high probability greater than most of the other subgraphs ("poor" ones); thus, the latter ones can be pruned earlier without testing. The optimized algorithm is shown in Algorithm 2, which replaces line 5-14 in Algorithm 1. The algorithm is self-explanatory, and we provide a concrete example to illustrate the process.

Example 1. Suppose we want to determine the snapshot with an insertion of 14 new edges. The graph in Fig. 4 (a) is our example graph, where the red lines indicate the changing edges. We first scan the graph to accumulate the number of the changing adjacent edges of each vertex and sort them. Therefore, we obtain a list $V_1 : 7, V_{28} : 5, V_{32} : 4, V_{33} : 3, \ldots$. Note that we only record the vertex which has at least one changing adjacent edge. Next, we push V_1 into the heap (because it has the largest initial fire energy and may propagate the fire faster). We traverse the adjacent vertices of V_1 and compute the time that the fire can be spread to the nearest neighbor, resulting in $1/8$ unit time (w.r.t Eq. 2, where $\ell_{V_1} = 7$ and $\ell_{V_2} = 1$). Note that there are multiple nearest neighbors, e.g., V_2, V_3, V_4, etc. The new subgraph is pushed into the heap, and the fire energy of each vertex involved is updated (w.r.t Eq. 3). We also push the node V_{28} into the heap because its initial energy is the largest among the unvisited nodes. Recursively, we execute the process until we find that the total number of new edges is greater than or equal to the threshold (i.e., 14). The results are listed in Fig. 4 (b). Note that the historical snapshot should be the union of graph G' of the new edges g_n with the old graph G_t, where G'=$G_t \cup g_n$.

Algorithm 3. Leap-search algorithm

Input: The graph G_{i-1} and G_i at time t_{i-1} and t_i, a user preferred number n
Output: The fastest burning out n-subgraph(s)

1   $H = G_{i-1} \cup G_i$;
2   **for each** vertex $v_i \in H$ **do** //*initial fire energy*
3       $fe[v_i]$ =num_of_adjacent_changing_edges;
4   sort fe in descending order;
5   **for each** $v_j \in fe$ **do** //*generate candidate core subgraphs*
6   g=store v_j with energy value as a subgraph;
7   **while** #new edge$(g) < n$ **do**
8   g'= appending g with edge of minimal spreading time;
9   t_{local}=fires propagation time of the new edge;
10   t_{local_max}=maximal value of t_{local};
11          $cand\_graph\_list$=store g' with t_{local_max};
12 **for each** subgraph $g \in cand\_graph\_list$ **do** //*test the candidate graph*
        //*update g if possible*
13   **for each** edge $e(i, j) \in g$ **do**
14   update $fe[i]$ and $fe[j]$ by checking neighbors of i and j;
15   update fire propagation time of $e(i, j)$;
16   update least time t_{least} necessary to fire out g;
17 Output the n-graphs(s) with minimal fire propagation time;

---

4   The Leap Search Approach

In this section, we present a leap-search based method for the extraction of burning out n-graphs. The method is efficient in processing graphs with a large n, where growing up the candidate subgraphs from scratch by using bottom-up growth can be time consuming. The approach is based on the density-oriented candidate selection strategy. The intuitive idea is that fire transfers fast in those regions where the energy density is high. The extraction process is also composed of candidate generation and verification iteration.

Candidate generation: Starting from the nodes with the most fire energy, we grow them by selecting their nearest neighbors (w.r.t. the fire propagation time) until the number of the new edges in the subgraph graph is equal to n. In other words, we do not wait for other possible candidates to grow up. During the growing process, we record the least time necessary to transfer the fire.

Verification: We check the bottleneck nodes of the fire propagation in these subgraphs (as indicated by the least time), greedily find the neighbors which can remedy the weak edges on spreading the fire, and then update the least time value. Through a recursive process, we finally determine the first burning out regions with the least n new edges.

The leap-search algorithm is shown in Algorithm 3. The initial phase (line 1-4) is similar to that of Algorithm 1. The candidate core subgraphs are generated first (line 5-11). Starting from the nodes with most fire energy, we grow them by linking to their nearest neighbors (w.r.t. the fire propagation time) or their internal unlinked edges (line 8) until the number of the new edges in the subgraph graph is equal to or greater than n. Different from the bottom-up algorithm, we do not wait for other possible candidates to grow up. During the growing process, we record the least time necessary to transfer the fire among them (line 9). In

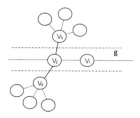

Fig. 5. Update mechanism for weak links (red line indicates the edge has changed and this figure is best viewed in color)

Algorithm 4. Pruning strategy for the leap-search algorithm

---

1   t_{th} =by far min time to burn out a n-candidate graph;
2   max_heap=candidate graphs with fire propagation time t;
3   **while** max_heap is not empty **do**
4   g=subgraph with maximal value in max_heap;
5   **for** all edges of g
6           $e(i,j)$=g's unvisited slowest edge on propagating fire;
7   update $fe[i]$ and $fe[j]$ by checking neighbors of i and j;
8   update fire propagation time $t_{e_{i,j}}$ of $e(i,j)$;
9   if($t_{e_{i,j}} > t_{th}$)
10   break;

---

the candidate test phase (line 12-16), we check the bottleneck nodes of the fire propagation in these subgraphs (as indicated by the least time), greedily find the neighbors which can remedy the weak edges on spreading the fire (line 14-15), and then update the least time value[6] (line 16). The detail of the updating mechanism will be described shortly. Finally, we determine the first burning out regions with the least n new edges.

4.1   Implementation Details

Weak Link Updating Mechanism. We introduce how to update the fire propagation time of the weak links. Suppose we have a subgraph as shown in Fig. 5. Nodes V_1 and V_2 have been included in the candidate graph g, but nodes V_3 and V_4 are outside of g. If we know that edge $e(V_1, V_2)$ is a weak edge of g (i.e., the fire propagation time is slow), then we start from nodes V_1 and V_2, and check whether their neighbors can transfer fire energy to them. For this example, node V_2 has two neighbors, V_3 and V_4, with a large fire energy and can propagate fire to V_2. Therefore, we update $fe[V_2]$ (line 14 in Algorithm 3) by using Eq. 3. The fire spreading time of $e(V_1, V_2)$ is also updated (line 15).

Pruning Strategy. When testing the candidate graphs (line 12-16 in Algorithm 3), we can prune many candidates early by using a time threshold, t_{th}, which is by far the fastest time to burn out an n-graph. Given a candidate graph, if after we update it by using the mechanism introduced in the last section the burning time is still smaller than t_{th}, then we can safely prune this candidate graph. To

---

[6] The replacement should guarantee the number of new edge will not decrease.

Data set name	DBLP		Enron		Web	
Classifed type	old	new	old	new	old	new
Recorded time	2001	2007	2001-10-1	2001-12-31	2004-5	2005-7
Nodes	1849	5289	6310	10008	2446029	3078826
Edges	4732	16667	30637	67777	57312778	71400464

Fig. 6. Statistics of the three data sets

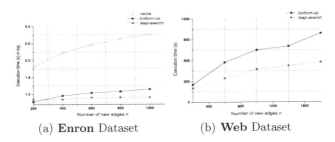

(a) **Enron** Dataset   (b) **Web** Dataset

Fig. 7. Efficiency of the proposed solutions

further improve the efficiency, the slowest edges on propagating fires in candidate graphs are tested first. The optimized algorithm is shown in Algorithm 4.

5   Performance Evaluation

To evaluate our strategies, we conducted extensive experiments. We performed the experiments using a Itanium2 CPU (1.5GHz) server with a 128G memory[7], running Redhat linux. All the algorithms were written in C++. We conducted experiments on three real life datasets, *DBLP*, *Enron*, *Web*.

The first dataset, *DBLP*, is extracted from DBLP website[8] and focuses on the bibliography information from database community. It contains the co-authorship information of major database conferences from 2001 to 2007. The second dataset, *Enron*, records the email communication information of each day from 2001-10-01 to 2001-12-31. For detail of these two datasets refer [15]. The third dataset, *Web*, records two snapshots of Japanese web pages (in jp domain) in May 2004 and July 2005, respectively. Part of this dataset is reported in [22]. The basic statistics of the datasets are shown in Fig. 6. Refer [23] for more experimental results.

5.1   Efficiency of Our Solutions

We compare our two algorithms, *bottom-up* and *leap-search*, with a naive method [23]. The result is shown in Fig. 7. Note that the execution time on the *Enron* dataset is in logarithm format. We can see that our solutions are much faster than the naive one, about one to two orders of magnitude (as shown in Fig. 7 (a)). For the huge Web dataset, the naive algorithm can not finish in

---

[7] The actual used memory is smaller than 8G.
[8] www.informatik.uni-trier.de/~ley/db

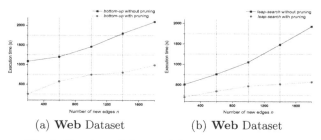

(a) **Web** Dataset   (b) **Web** Dataset

Fig. 8. Pruning efficiency of the proposed strategies

reasonable time. Between our two approaches, the *leap-search* performs better than the *bottom-up* when the number of new edges n becomes larger. The reason is that for large value of n, growing graphs from scratch by evaluating all the neighbors is time consuming. However, for a small value of n, the cost for growing graphs becomes smaller and the cost for updating weak links becomes larger; thus, the *bottom-up* algorithm performs better.

5.2 Pruning Efficiency of Our Solutions

In this section, we evaluate the efficiency of the proposed pruning techniques. The result is illustrated in Fig. 8. We can see that with pruning techniques, the *bottom-up* algorithm can perform much better, as shown in Fig. 8 (a). The reason is that many of the candidate graphs can be pruned sharply with the anti-monotone rule and the refinement strategy. For the *leap-search* approach, the early pruning technique can remove many candidate graphs from the heap. Thus, the overall performance is improved. In summary, with the pruning techniques, both algorithms only need to test and update a small number of candidate graphs, which lead to good scalability with respect to the cardinality of the datasets and the value of n.

5.3 Effectiveness of Our Solutions

We examine whether the discovered virtual historical snapshots restore the real data with high precision and follow the important properties of evolving graphs.

- **Precision.** We compare three algorithms, *random*, *bottom-up* and *leap-search* on the precision metric[9]. The *random* method is implemented by randomly selecting n new edges combined with the old graph as the virtual snapshot. The quantitative metric is defined as $precision = \frac{\Delta E_R \cap \Delta E_V}{\Delta E_R}$, where ΔE_R denotes the set of actual new edges and ΔE_V represents the set of virtual generated new edges. Our methods can get rather high precision as illustrated in the figure. The reason why the precision decreases on restoring the older snapshots (i.e., 10/15), is due to the small number of the new edges, which leads to difficulty in locating the new edges. For our two algorithms, there is a trade-off between efficiency and effectiveness.

---

[9] Due to its prohibitive cost on execution compared with others, the *naive* algorithm has not been considered here.

Table 2. Precision evaluation

(a) Precision on **Enron** dataset

Date=	10/15	11/1	11/15	12/1	12/15
random	0.180	0.545	0.699	0.767	0.816
bottom-up	0.620	0.743	0.858	0.911	0.935
leap-search	0.517	0.654	0.779	0.863	0.907

(b) Precision on **DBLP** dataset

Year=	2002	2003	2004	2005	2006
random	0.129	0.271	0.415	0.622	0.731
bottom-up	0.546	0.644	0.700	0.773	0.902
leap-search	0.414	0.526	0.611	0.725	0.846

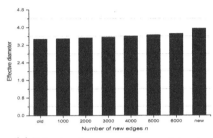
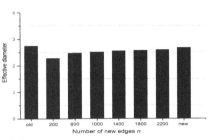

(a) Effective diameter of **Enron** dataset   (b) Effective diameter of **DBLP** dataset

Fig. 9. Effective diameter

- **Effective Diameter.** Fig. 9 shows the values of effective diameters [21] for the historical snapshots. We can see that the virtual snapshots mainly express the transition characteristic between the two real graphs. In Fig. 9 (b), the effective diameter drops after inserting a few edges. We argue that the reason is due to the relative small community of the DB scholars. The first few new edges may link to many others because these insertion edges may be caused by those influential people (with more initial fire energy).
- **Degree Distribution.** We also evaluate the degree distributions of the virtual historical snapshots. The temporal degree distribution follows the power law distribution and the changing edge degree distribution obeys the densification power law distribution. Refer [23] for detail.

6 Conclusion

In this paper we have studied a new problem of tracing back the virtual historical snapshots. Two solutions have been proposed, the *bottom-up* and the *leap-search*. We have conducted extensive experiments and the results show that our approaches can restore the historical graph snapshots efficiently while maintaining the evolution properties. In the future, we will evaluate some other predictors such as those proposed in [14,16].

Acknowledgements

The work was done when the first author visited the Chinese University of Hongkong. The work was supported by grants of the Research Grants Council of the Hong Kong SAR, China No. 419008 and 419109.

References

1. Bansal, N., Chiang, F., Koudas, N., Tompa, F.W.: Seeking stable clusters in the blogosphere. In: VLDB (2007)
2. Breiger, R., Carley, K., Pattison, P.: Dynamic social network modeling and analysis: Workshop summary and papers (2003)
3. Broder, A., Kumar, R., Maghoul, F., Raghavan, P., Rajagopalan, S., Stata, R., Tomkins, A., Wiener, J.: Graph structure in the web. Comput. Netw. 33(1-6) (2000)
4. Doyle, P.G., Snell, J.L.: Random walks and electric networks. Carus Mathematical Monographs, Mathematical Association of America 22 (1984)
5. Faloutsos, M., Faloutsos, P., Faloutsos, C.: On power-law relationships of the internet topology. In: SIGCOMM (1999)
6. Goncalves, P., Diogo, P.: Geographic information systems and cellular automata: A new approach to forest fire simulation. In: EGIS (1994)
7. Gyöngyi, Z., Garcia-Molina, H., Pedersen, J.: Combating web spam with trustrank. In: VLDB (2004)
8. Henley, C.: Self-organized percolation: A simpler model. Bull. Am. Phys. Soc. 34 (1989)
9. Inokuchi, A., Washio, T., Motoda, H.: An apriori-based algorithm for mining frequent substructures from graph data. In: Zighed, D.A., Komorowski, J., Żytkow, J.M. (eds.) PKDD 2000. LNCS (LNAI), vol. 1910, pp. 13–23. Springer, Heidelberg (2000)
10. Inokuchi, A., Washio, T.: A fast method to mine frequent subsequences from graph sequence data. In: ICDM (2008)
11. Kumar, R., Raghavan, P., Rajagopalan, S., Sivakumar, D., Tomkins, A., Upfal, E.: Stochastic models for the web graph. In: FOCS (2000)
12. Leskovec, J., Kleinberg, J., Faloutsos, C.: Graphs over time: densification laws, shrinking diameters and possible explanations. In: KDD (2005)
13. Leskovec, J., Backstrom, L., Kumar, R., Tomkins, A.: Microscopic evolution of social networks. In: KDD (2008)
14. Liben-Nowell, D., Kleinberg, J.: The link-prediction problem for social networks. JASIST 58(7) (2007)
15. Liu, Z., Yu, J.X., Ke, Y., Lin, X., Chen, L.: Spotting significant changing subgraphs in evolving graphs. In: ICDM (2008)
16. Newman, M.E.J.: The structure and function of complex networks. SIAM Review 45(2) (2003)
17. Rothermel, R.C.: A mathematical model for predicting fire spread in wildland fuels. USDA Forest Service, Ogden, UT, Tech. Rep. (1972)
18. Shakhnovich, B.E., Harvey, J.M.: Quantifying structure-function uncertainty: a graph theoretical exploration into the origins and limitations of protein annotation. Jounal of Molecular Biology 4(337) (2004)
19. Sun, J., Faloutsos, C., Papadimitriou, S., Yu, P.S.: Graphscope: parameter-free mining of large time-evolving graphs. In: KDD (2007)
20. Tantipathananandh, C., Berger-Wolf, T., Kempe, D.: A framework for community identification in dynamic social networks. In: KDD (2007)
21. Tauro, S., Palmer, C., Siganos, G., Faloutsos, M.: A simple conceptual model for the Internet topology. In: GLOBECOM (2001)
22. Toyoda, M., Kitsuregawa, M.: What's really new on the web? identifying new pages from a series of unstable web snapshots. In: WWW (2006)
23. Yang, Z., Yu, J.X., Liu, Z., Kitsuregawa, M.: Fires on the Web: Towards Efficient Exploring Historical Web Graphs. University of Tokyo, Tech. Rep. (2009)

Identifying Community Structures in Networks with Seed Expansion

Fang Wei[1,2], Weining Qian[3], Zhongchao Fei[1,2], and Aoying Zhou[3]

[1] Portfolio Strategy&Technology Leadership CTO Group, CPG,
Alcatel-Lucent Shanghai Bell
[2] School of Computer Science, Fudan University,
Shanghai, China
{Fang.Wei,Zhongchao.Fei}@alcatel-sbell.com.cn
[3] Software Engineering Institute, East China Normal University, Shanghai, China
{wnqian,ayzhou}@sei.ecnu.edu.cn

Abstract. Real-world networks naturally contain a lot of communities. Identifying the community structures is a crucial endeavor to analyze the networks. Here, we propose a novel algorithm which finds the community structures from seed expansion. Its expansion process bases on the transmissive probabilities coming from seed vertices and the modularity Q function which is firstly defined by Newman et al.. The experimental evaluation is conducted on real-world networks. The evaluation shows that our algorithm has good results in quality.

1 Introduction

Real-world networks naturally contain a lot of communities. Generally, the community is some tightly-linked entities. It often represents a set of common interests or functional interactions members. Hence, how to identify the community structures from complex networks is an interesting problem. Recently, it has attracted wide attention in many domains and disciplines.

In this paper, we propose a novel algorithm which identifies the community structures from seed expansion. The algorithm is based on transmissive probabilities and the modularity Q function which is a new metric to measure the community structures and firstly defined in [1].

The algorithm begins with a set of seed vertices. From the initial probabilities of seed information, we compute the transmissive probabilities of newly- expanded vertices. The probabilities reflect the possibility that seed vertices extend to new vertices. Sorting the probabilities at each time step, we get a descending order of newly-expanded vertices. According to the order, we compute the change value on modularity Q for current community candidate. If the change value is larger than zero, it means that adding the vertex brings good community structure to the seed set and it has a chance of further expansion.Otherwise, the vertex is deleted.

H. Kitagawa et al. (Eds.): DASFAA 2010, Part I, LNCS 5981, pp. 627–634, 2010.

The contributions of this paper are summarized as follows:

1. Vertices' transmissive probabilities are defined and the change values on modularity Q are chosen to decide the expansion. The algorithm doesn't need to predefine the threshold of probability.
2. In five real-world networks, we evaluate our algorithm in two expansion cases which are overlapping and non-overlapping. Both cases show that our method has good performances.

2 Preliminaries

2.1 Network Model and Community

A network can be modeled as a graph. Let G = (V, E) be a graph with n vertices and m edges, V and E are respectively the vertex set and edge set. The networks discussed in this paper are unweighted and undirected graphs.

The community structure in network often refers to a subgraph of tightly-linked vertices. Its connections among community vertices are relatively denser than the ones crossing to the rest of the network.

We introduce one of its definitions, which is described in [2].

Definition 1. *If a network owning m edges has been divided into k communities, its modularity Q function is:*

$$Q = \sum_{c=1}^{k} [\frac{l_c}{m} - (\frac{d_c}{2m})^2]$$

where c is the label of community. l_c is the total number of edges whose ends both lie in the same community c, and d_c is the sum of the degrees of vertices in community c.

When the community structure is updated by adding new vertices, we consider the change of Q value. If adding a vertex to community structure, the links pointed to the community members are called as **inlinks** and the links pointed to the outside vertices which not belong to the community structure are named as **outlinks**. Their total number are respectively denoted by $|IL|$ and $|OL|$.

2.2 Definitions

In our algorithm, the probability for the transition from a vertex to another is also considered. It is defined as:

Definition 2. *The probability for the transmission from vertex j to vertex i is:*

$$P(v_j \rightarrow v_i) = \frac{1}{d(v_j)}$$

where vertex j is vertex i's neighbor and $d(v_j)$ is the degree of vertex j.

The higher vertex's transmissive probability is, the tighter connection between the vertex with its neighbor is.

3   Our Algorithm

Our algorithm is like the aggregative process. The aggregative criterion is the change value of modularity Q.

Its expansion process will be introduced in following subsections.

3.1   The Change Value on Modularity Function Q

From the seed vertices, we begin the expansion process. For every newly-expanded vertex, we will consider whether it brings good community structure to seed group.

When community c absorbs a new vertex w, it gets a new structure and its initial Q becomes Q'. Then, the change value on Q corresponds to:

$$\Delta Q = Q' - Q = \frac{|IL_w|}{m} - \frac{2d_c \times d_w + d_w^2}{4m^2}$$

Here, $|IL_w|$ is the total number of inlinks, d_w is the degree of vertex w.

The value is decided not only by the links of new vertex, but also by the total degrees of current seed set.

3.2   Expansion Step and Transmissive Probability

For the vertex v in seed S, its initial probability at the beginning of expansion can be measured by:

$$P_0(v) = \frac{d(v)}{\sum_{u \in S} d(u)}$$

where v and u are the members of seed S.

After the initial state, the probability of vertex at each expansion step can be computed based on its links information. Summarizing the amount of probabilities transferred from its neighbors and itself, we get the probability of vertex i at time step t:

Definition 3. *The probability of vertex i at time step t is:*

$$P_t(v_i) = P_{t-1}(v_i) + \sum_{v_k \in N(v_i)} P_{t-1}(v_k) \times P(v_k \rightarrow v_i)$$

where $N(v_k)$ are the neighbors of vertex k.

By the above formula, we get vertices' probabilities at each expansion state and sort these values in descending order. Then, every newly-expanded vertex is scanned by this order for the locally-optimal expansion and the Q change value corresponding to each scanning state is measured.

If the change value is larger than zero, it will be added into the seed for further expansion. If ΔQ is smaller than zero, the vertex will be discarded. The next vertex in probability order will repeat the computing process. The computing seeks the locally-optimal value.

After the newly-expanded vertices are scanned, the seed group is updated and the expansion process reaches new time step. The process is repeated until the Q value of seed structure has no better changes.

3.3   The Bound of Expansion Process

In expansion process, the spread of seed information can be viewed as informa-tion's random walk. At each step, the walker focuses on a vertex and moves to its neighborhood vertex at random. The sequence of walkers forms a Markov chain. From given vertex, the random walks tend to scan more often the vertices that are tightly connected with seed vertices.

After a series of expansions, the Markov chain approaches the steady distri-bution which is at the mixing time of random walk. The extent of convergence about walks is measured by the L1-distance [3]. Based on the vertex's probabil-ity defined in our algorithm, the L1-distance can be computed. By the distance, the mixing time of our expansion is estimated.

4   Experimental Evaluation

The experiments are conducted on a single processor of a 3.2 GHz Pentium Xeon, which has 2GB RAM and runs with Window 2K. The experimental datasets come from five real-world networks which are described in Table 1.

Table 1. The features of our experimental datasets

Data Name	Vertices	Edges	Data Source
Zachary's karate club	34	78	http://www-personal.umich.edu/~mejn/netdata/
Football games	115	613	http://www-personal.umich.edu/~mejn/netdata/
NIPS coauthorships	1,063	2,083	http://www.cs.toronto.edu/~roweis/data.html
Protein interactions	1,458	1,948	http://www.nd.edu/ networks/resources.htm
KDD citations	27,400	352,504	http://www.cs.cornell.edu/projects/kddcup/

In expansion process, the newly-expanded vertices face the fact that whether the locally-optimal goal permits them to belong to several communities, which corresponds to overlapping or non-overlapping expansion approach. Both the cases have practical requirements. We will perform our algorithm on two cases and compare their modularity Q with other algorithms.

4.1   Selecting Seeds

Selecting good seeds is significant to expansion process. But the analysis of seeds choosing is not our emphasis, the paper focuses on expansion steps.

Our seeds chosen in experiments come from the seeds described in [4]. The con-crete approach is coarsening the origin graph into a series of smaller graphs. They find the partition clues from coarsening graphs. By the recursive partition, the final results are our seed sets. They are often a set of tightly-connected vertices.

4.2   The Experimental Analysis of Five Datasets

In experiments, the average Q value for all discovering communities is adopted to measure the expansion process. It is denoted as avg_Q.

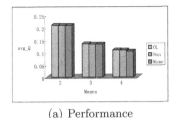

(a) Performance

Fig. 1. The experimental results of Karate Club at various means

The Mome algorithm in [5] is one of the latest methods for the community discovery. We compare our algorithm denoted as PQ(P comes from vertex's probabilities and Q is the modularity function) with it at the same conditions.

The following subsections are the evaluation with same number of means. In all figures, PQ algorithm with overlapping expansion approach is abbreviated as "OL" , the non-overlapping one is "Non" and the Mome algorithms is denoted as "Mome".

Zachary' Karate Club: It is one of the classic datasets in social network analysis. The karate club belonged to an American university in 1970s. Wayne Zachary concluded the social interactions among club members and constructed this network. It has 34 vertices represented the number of club members and 78 edges expressed their social interactions.

For the performance on avg_Q, our algorithm has slight superiority which is shown in Figure 1.

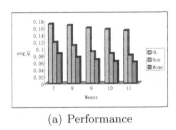

(a) Performance

Fig. 2. The experimental results of Football Games at various means

American college football teams: This network comes from the games played between 115 American college football teams in 2000. Each team is represented by a vertex in the network and the games between two teams are denoted as edges. Each team belongs to a conference, and the games between inter-conference are played more than intra-conference ones.

From Figure 2, we can clearly see that the expansion with overlapping is better than the one with non-overlapping approach, and the non-overlapping one is better than Mome. The overlapping approach has distinct superiority over Mome.

632   F. Wei et al.

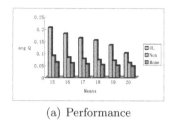

(a) Performance

Fig. 3. The experimental results of NIPS Coauthorship at various means

The Co-authorships of NIPS conference: This data is drawn from the co-authorship network about the NIPS conference papers. The vertices represent the authors and edges are the co-authorships among authors.

In Figure 3, the overlapping expansion is obviously better than non-overlapping approach and Mome, the non-overlapping approach has a little superiority over Mome.

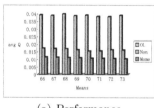

(a) Performance

Fig. 4. The experimental results of Protein interactions at various means

Protein interactions: The dataset is the protein network which indicates the interactions between proteins. Figure 4 is its experimental evaluation. The overlapping approach and non-overlapping one are obviously better than Mome.

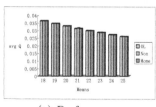

(a) Performance

Fig. 5. The experimental results of KDD Coauthorship at various means

The citationships of KDD papers: The last dataset is about the paper citations of KDD conference. Its links come from the citationships among conference papers.

In Figure 5, three approaches have similar results. Our methods is not worse than Mome.

From the experimental analysis, our algorithm has better results than Mome in both overlapping and non-overlapping expansion cases. We also find that the overlapping expansion makes the single community structure more complete.

5   Related Work

Andersen and Teng in [6] describe a method to find communities from seed set. They use the lazy random walks technique to sweep each vertex. At the sweep cut, the method decides whether there is a jump in the expanded probabilities of vertices. Furthermore, they need to seek the help of some additional operations to improve the local properties of sweep cut.

There is also a algorithm based on seed expansion in [4]. It is designed for the problem of detecting overlapping community structures. Its expansion process is limited by the thresholds of vertices degree-normalized probabilities and overlapping rates. Our PQ algorithm doesn't need to predefine the threshold. The expansion process is guided by the change value of modularity Q.

6   Conclusion

In this paper, we present a new method for identifying the community structures in networks. Our method is based on seed expansion. The change value of modularity function is used to evaluate the contribution of newly-expanded vertex to current seed group. The transmissive probabilities are used to deduce the relationships between neighborhood vertices.

In five real-world datasets, the algorithm is evaluated from overlapping and non-overlapping cases. Both of them show that our method has good quality than others.

Acknowledgments

Thanks to IBM China Research Laboratory for providing the comparing data in experiments.

The project is partially supported by National Basic Research Program of China (973 Program,No.2010CB731402), National Nature Science Foundation of China(No.60833003) and Shanghai International Collaboration Foundation Project (No. 09530708400).

References

1. Newman, M.E.J., Girvan, M.: Finding and evaluating community structure in networks. Physical Review 69, 026113 (2004)
2. Fortunato, S., Castellano, C.: Community Structure in Graphs. Physics and Society (2007)

3. Montenegro, R., Tetali, P.: Mathematical Aspects of Mixing Times in Markov Chains. Foundations and Trends in Theoretical Computer Science, vol. 1(3) (2006)
4. Wei, F., Qian, W., Wang, C., Zhou, A.: Detecting overlapping community structures in networks. World Wide Web Journal 12(2), 235 (2009)
5. Zhu, Z., Wang, C., Ma, L., Pan, Y., Ding, Z.: Scalable Community Discovery of Large Networks. In: The Ninth International Conference on Web-Aged Information Management (2008)
6. Andersen, R., Lang, K.J.: Communities from seed sets. In: Proceedings of the 15th International World Wide Web Conference (2006)

Dynamic Agglomerative-Divisive Clustering of Clickthrough Data for Collaborative Web Search

Kenneth Wai-Ting Leung and Dik Lun Lee

Department of Computer Science and Engineering,
The Hong Kong University of Science and Technology, Hong Kong
{kwtleung,dlee}@cse.ust.hk

Abstract. In this paper, we model clickthroughs as a tripartite graph involving users, queries and concepts embodied in the clicked pages. We develop the Dynamic Agglomerative-Divisive Clustering (*DADC*) algorithm for clustering the tripartite clickthrough graph to identify groups of similar users, queries and concepts to support collaborative web search. Since the clickthrough graph is updated frequently, DADC clusters the graph incrementally, whereas most of the traditional agglomerative methods cluster the whole graph all over again. Moreover, clickthroughs are usually noisy and reflect diverse interests of the users. Thus, traditional agglomerative clustering methods tend to generate large clusters when the clickthrough graph is large. DADC avoids generating large clusters using two interleaving phases: the *agglomerative* and *divisive* phases. The agglomerative phase iteratively merges similar clusters together to avoid generating sparse clusters. On the other hand, the divisive phase iteratively splits large clusters into smaller clusters to maintain the coherence of the clusters and restructures the existing clusters to allow DADC to incrementally update the affected clusters as new clickthrough data arrives.

1 Introduction

The exponential growth of information on the Internet has created great demands on highly effective search engines. Finding relevant information in such a large volume of data to satisfy users' information needs becomes a difficult and challenging task. To improve retrieval effectiveness, personalized search engines create user profiles recording the users' preferences, which are used to adjust the search results to suit the users' preferences. Clickthrough data, which contains a user's queries, the retrieved results and the results that the user has clicked on, is an important implicit relevance feedback available on a search engine. A user clicks on documents mostly because the user believes they satisfy his/her information needs. Thus, most personalized systems [4], [6] rely on analyzing the clickthrough data to extract the users' preferences.

In order to fully utilize the clickthrough data to improve retrieval effectiveness, we propose a Community Clickthrough Model (CCM) which incorporates multiple-type objects, namely, users, queries, and concepts embodied in documents returned by the search engine and those clicked by the users, in a tripartite graph model. Based on CCM, we develop the Dynamic Agglomerative-Divisive Clustering (*DADC*) algorithm to generate clusters of similar users, similar queries and similar concepts. The user

H. Kitagawa et al. (Eds.): DASFAA 2010, Part I, LNCS 5981, pp. 635–642, 2010.

clusters form user communities that are useful in collaborative filtering to predict the interests of a user. The query clusters are useful in providing query suggestions for users to formula more effective queries, while the concept clusters are useful in categorizing result pages.

The main contributions of this paper are:

- We develop a tripartite Community Clickthrough Model (CCM) which alleviate the click sparsity problem by considering in CCM concepts embodied in the documents instead of the the documents themselves.
- We developed the Dynamic Agglomerative-Divisive Clustering (DADC) algorithm to produce clusters of similar users, similar queries and similar concepts based on CCM. DADC is able to resolve semantic ambiguities and hence produces better precision and recall compared to the existing state-of-the-art clustering methods.
- Partitional clustering methods, such as K-Means, are fast but lack accuracy. Moreover, in most cases, the number of clusters K has to be determined ahead of time as an input to the clustering algorithm. On the other hand, hierarchical clustering methods (divisive or agglomerative) are more accurate, but they are slow especially when the data set is large. DADC retains the accuracy of hierarchical clustering methods and allows new incoming data to be clustered online dynamically.

The rest of the paper is organized as follows. Section 2 discusses the related work. In Section 3, we present our community clickthrough model and the method to extract concepts embodied in documents for constructing the model. In Section 4, we describe our DADC method. Experimental results evaluating the performance of DADC against three state-of-the-art methods (BB, CubeSVD, M-LSA) and a baseline method (K-Means) are presented in Section 5. Section 6 concludes the paper.

2   Related Work

In this section, we review a few state-of-the-art techniques for query clustering.

2.1   BB's Graph-Based Clustering

In Beeferman and Berger's agglomerative clustering algorithm [1] (or simply called BB's algorithm in this paper), a query-document bipartite graph is firstly constructed with with one set of nodes corresponds to the set of the submitted queries, while the other set of nodes corresponds to the set of clicked documents. When a use submits a query and clicks on a document, the corresponding query and the clicked document are linked together with an edge on the bipartite graph. During the clustering process, the algorithm iteratively merges the two most similar query into one query node, then the two most similar documents into one document node, and the process of alternative merging is repeated until the termination condition is satisfied.

2.2   CubeSVD

In order to model the relationships between users, queries, and documents, CubeSVD [9] models the clickthrough data as a 3-order tensor $\mathcal{A}$. After the tensor $\mathcal{A}$ is constructed, a Higher-Order Singular Value Decomposition (*HOSVD*) technique is employed to simplify $\mathcal{A}$ into lower-order matrix, in order to apply latent relationships analysis to discover latent relationships among users, queries, and documents in the tensor $\mathcal{A}$.

2.3 M-LSA

In the web domain, M-LSA [10] represents the relationships between users, queries, and documents with three co-occurrence matrices ($M_{u \times q}$, $M_{u \times d}$, and $M_{q \times d}$), where u, q, and d are the users, queries and documents respectively. A unified co-occurrence matrix R is constructed using the co-occurrence matrices. Similar to LSA, M-LSA also employs Eigen Value Decomposition (*EVD*) to discover important objects from the object collections from R.

2.4 Divisive-Agglomerative Clustering

The Divisive-Agglomerative clustering [7] is a top-down, incremental algorithm for clustering stream data into a tree-shape structure [8]. The root node of the tree contains the complete set of data. The algorithm splits the root node into two smaller clusters according to a heuristic condition on the diameters of the clusters. The two new clusters are linked to the root node as it's child nodes. The algorithm iterates the splitting process on the leave nodes until no new cluster (new child node) can be formed. Apart from splitting, the algorithm also has an agglomerative phase to re-aggregate the leave nodes to adapt the tree structure as the data are updated.

3 Community Clickthrough Model

The three clustering methods (BB, CubeSVD, and M-LSA) discussed in Section 2 are content-ignorance in that two queries are related if they induce clicks on the same document. They completely ignore the content of the documents. One major problem with content-ignorance model is that the number of common clicks on documents induced by different queries is very small. Beeferman and Berger [1] reported that the chance for two random queries to have a common click is merely 6.38×10^{-5} in a large clickthrough data set from a commercial search engine. Thus, the bipartite graph or the co-occurrence matrix would be too sparse for obtaining useful clustering results.

To alleviate the click sparsity problem, we introduce a content-aware clickthrough model, called Community Clickthrough Model (CCM), which replaces *clicked documents* with *concepts embodied in the clicked documents*. When a user u_i submits a query q_j, an edge is created between u_i and q_j representing the relationship between u_i and q_j. Similarly, if a query q_i retrieves a document that embodies concept c_j, an edge is created between q_i and c_j. When a user u_i clicks on a document that embodies concept c_k, an edge is created between u_i and c_k. For clarity, when a user u clicks on a document that embodies a concept c, we simply say u *clicks* on c; if q retrieves a document that embodies concept c, we say u *retrieves* c. Thus, CCM is a tripartite graph relating *users*, their submitted *queries*, the *retrieved concepts* and the *clicked concepts*, which are a subset of the retrieved concepts.

To identify concepts embodied in a document, we define a concept as a sequence of one or more words that occur frequently in the web-snippets[1] of a particular query. These word sequences represent important concepts related to the query because they co-exist in close proximity with the query in the top documents. The following support formula, which is inspired by the well-known problem of finding frequent item sets in data mining [2], is employed to measure the *interestingness* of a particular keyword/phrase c_i extracted from the web-snippets arising from q:

$$support(c_i) = \frac{sf(c_i)}{n} \cdot |c_i| \tag{1}$$

where $sf(c_i)$ is the snippet frequency of the keyword/phrase c_i (i.e. the number of web-snippets containing c_i), n is the number of web-snippets returned and $|c_i|$ is the number of terms in the keyword/phrase c_i. If the support of a keyword/phrase c_i is greater than the threshold s ($s = 0.07$ in our experiments), we treat c_i as a concept for the query q.

4 Dynamic Agglomerative-Divisive Clustering

Our Dynamic Agglomerative-Divisive Clustering (*DADC*) algorithm performs updates efficiently by incrementally updating the tripartite graph as new data arrives. It consists of two phases, namely, the **agglomerative phase** and **divisive phase**. The agglomerative phase is based on Beeferman and Berger's agglomerative clustering algorithm [1], which iteratively merges similar clusters. The divisive phase splits large clusters into small ones using the Hoeffding bound [3] as a criterion. It prevents clusters from growing without bound when new data arrives. The clickthrough data is first converted into a tripartite graph as described in Section 3, and DADC would iteratively merge and split nodes in the tripartite graph until the termination condition is reached.

4.1 Agglomerative Phase

The agglomerative phase is based on the tripartite graph described in Section 3 with the following assumptions: 1) Two users are similar if they submit similar queries and click on similar concepts, 2) Two queries are similar if they are submitted by similar users and retrieve similar concepts, and 3) Two concepts are similar if they are clicked by similar users and are retrieved by similar queries.

Based on the above assumptions, we propose the following similarity functions to compute the similarity between pair of users, pair of queries, and pair of concepts.

$$sim(u_i, u_j) = \alpha_1 \cdot \frac{Q_{u_i} \cdot Q_{u_j}}{\| Q_{u_i} \| \| Q_{u_j} \|} + \beta_1 \cdot \frac{C_{u_i} \cdot C_{u_j}}{\| C_{u_i} \| \| C_{u_j} \|} \tag{2}$$

$$sim(q_i, q_j) = \alpha_2 \cdot \frac{U_{q_i} \cdot U_{q_j}}{\| U_{q_i} \| \| U_{q_j} \|} + \beta_2 \cdot \frac{C_{q_i} \cdot C_{q_j}}{\| C_{q_i} \| \| C_{q_j} \|} \tag{3}$$

$$sim(c_i, c_j) = \alpha_3 \cdot \frac{U_{c_i} \cdot U_{c_j}}{\| U_{c_i} \| \| U_{c_j} \|} + \beta_3 \cdot \frac{Q_{c_i} \cdot Q_{c_j}}{\| Q_{c_i} \| \| Q_{c_j} \|} \tag{4}$$

---

[1] "Web-snippet" denotes the title, summary and URL of a Web page returned by search engines.

where Q_{u_i} is a weight vector for the set of neighbor query nodes of the user node u_i in the tripartite graph G_3, the weight of a query neighbor node $q_{(k,u_i)}$ in the weight vector Q_{u_i} is the weight of the link connecting u_i and $q_{(k,q_i)}$ in G_3. C_{u_i} is a weight vector for the set of neighbor concept nodes of the user node u_i in G_3, and the weight of a query neighbor node $c_{(k,u_i)}$ in C_{u_i} is the weight of the link connecting u_i and $c_{(k,u_i)}$ in G_3. Similarly, U_{q_i} is a weight vector for the set of neighbor user nodes of the query node q_i, C_{q_i} is a weight vector for the set of neighbor concept nodes of the query node q_i, U_{c_i} is a weight vector for the set of neighbor user nodes of the concept node c_i, and Q_{c_j} is a weight vector for the set of neighbor query nodes of the concept node c_i.

The condition $\alpha + \beta = 1$ is imposed on Equations (2), (3) and (4) to make the similarities lie between $[0, 1]$. The similarity of two nodes is 0 if they do not share any common node in G_3, while the similarity of two nodes is 1 if they share exactly the same set of neighboring nodes.

In the agglomerative phase, the algorithm merges the two most similar users based on Equations (2), then the two most similar queries are merged based on Equations (3), and finally the two most similar concepts are merged based on Equations (4), and so on. The procedure repeats until no new cluster (user, query or document cluster) can be formed by merging.

4.2 Divisive Phase

The divisive phase employs a hierarchical clustering technique, which is an inverse of the agglomerative phase (splitting instead of merging). It iteratively splits large clusters into two smaller clusters until no new clusters can be formed by splitting. One major problem in the divisive phase is to determine the minimum number of observations necessary for the phase to converge. To resolve the problem, the Hoeffding bound [3] is employed to ensure that after n independent observations of a real-valued random variable r with range R, and with confidence $1 - \delta$ (where δ is the split threshold), the true mean of r is at least $\bar{r} - \epsilon$, where $\bar{r}$ is the observed mean of the samples and

$$\epsilon = \sqrt{\frac{R^2 \ln(1/\delta)}{2n}} \tag{5}$$

In the divisive phase, each cluster is assigned with a different ϵ, namely, ϵ_k. Assume that the distances between pair of users, pair of queries, and pair of concepts are defined as follows.

$$d(u_i, u_j) = \sqrt{\sum_{k=1}^{n}(q_{(k,u_i)} - q_{(k,u_j)})^2 + \sum_{k=1}^{m}(c_{(k,u_i)} - c_{(k,u_j)})^2} \tag{6}$$

$$d(q_i, q_j) = \sqrt{\sum_{k=1}^{n}(u_{(k,q_i)} - u_{(k,q_j)})^2 + \sum_{k=1}^{m}(c_{(k,q_i)} - c_{(k,q_j)})^2} \tag{7}$$

$$d(c_i, c_j) = \sqrt{\sum_{k=1}^{n}(u_{(k,c_i)} - u_{(k,c_j)})^2 + \sum_{k=1}^{m}(q_{(k,c_i)} - q_{(k,c_j)})^2} \tag{8}$$

$q_{(k,u_i)} \in Q_{u_i}$ is the weight of the link connecting u_i and $q_{(k,u_i)}$, and $c_{(k,u_i)} \in C_{u_i}$ is the weight of the link connecting u_i and $c_{(k,u_i)}$. Similarly, $u_{(k,q_i)} \in U_{q_i}$, $c_{(k,q_i)} \in C_{q_i}$, $u_{(k,c_i)} \in U_{c_i}$, and $q_{(k,c_i)} \in Q_{c_i}$.

Assume that two pairs of nodes $(d1_n = d(n_i, n_j)$ and $d2_n = d(n_k, n_l))$ are the top-most and second top-most dissimilar nodes in a cluster (based on the distance Equation 6), (7), or (8). Assume that $\triangle d = d(n_i, n_j) - d(n_k, n_l)$, if $\triangle d > \epsilon_k$, with probability $1 - \delta$, the differences between $d(n_i, n_j)$ and $d(n_k, n_l)$ is large than zero, and pick (n_i, n_j) as the boundary of the cluster when applying Hoeffding bound with $\triangle d$. In the divisive phase, n_i and n_j are selected as the pivots for the splitting, and the clusters are split according to the statistical confidence given by Hoeffding bound.

5 Experimental Results

In this Section, we compare and analyze the clustering results from the five clustering algorithms (K-Means, CubeSVD, M-LSA, BB, and DADC). In Section 5.1, we describe the setup for collecting the clickthrough data. In Section 5.2, The five clustering algorithms are compared and evaluated in Section 5.2.

5.1 Experimental Setup

To evaluate the performance of the five clustering algorithms, we developed a middle-ware to serve as Google's frontend for collecting clickthrough data. 64 users are invited to use our middleware to search for the answers of 239 test queries. We ask human judges to determine a standard cluster for each of the users, queries, and concepts. The clusters obtained form the five clustering algorithms are compared against the standard cluster for their correctness.

5.2 Performance Comparison

We have already shown in [5] that clustering algorithms that employ concepts achieve better precisions comparing to content-ignorance methods that consider document only. Thus, all of the five methods compared in this section are based on the *CCM* model as described in Section 3, which is based on the relationships between users, queries, and concepts. A tripartite graph as described in Section 3 is used for BB and DADC methods. CubeSVD is based a 3-order tensor (with user, query and document dimensions) to model the relationships between users, queries, and concepts. For M-LSA, it is based on the unified matrix $R_{u,q,c}$.

Table 1 shows the average precision, recall, and F-measure values of K-Means, CubeSVD, M-LSA, BB, and DADC methods. K-Means is served as the baseline in the comparison. As discussed in Section 1, hierarchical clustering methods are slower, but more accurate comparing to partitional clustering methods. As expected, we observe that DADC yields the best average precision, recall, and F-measure values (0.9622, 0.7700, and 0.8479), while BB yields the second best average precision, recall, and F-measure values (0.9141, 0.6732, and 0.7531). DADC gains 32.3% average precision, 144% average recall, and 127% average F-measure, while BB gains 25.6% average

Table 1. Precisions of K-Means, CubeSVD, M-LSA, BB, and DADC at Optimal Thresholds

	K-Means	CubeSVD	M-LSA	BB	DADC
Avg. Precision	0.7275	0.7612	0.8642	0.9141	0.9622
Avg. Recall	0.3152	0.2998	0.6064	0.6732	0.7700
Avg. F-Measure	0.3733	0.4189	0.6793	0.7531	0.8479

precision, 113% average recall, and 101% average F-measure comparing to the baseline method. The extra divisive phase in DADC helps to separate dissimilar objects into different clusters, resulting in more accurate clusters comparing to BB.

We observe that the average precision, recall, and F-measure values of CubeSVD are very close to those obtained by the baseline method. As discussed in [9], CubeSVD generates new associations among the users, queries, and concepts in the reconstructed tensor $\hat{A}$ through the CubeSVD analysis. The new associations in $\hat{A}$ brings not only similar objects, but also unrelated objects together. We observe that the predicted associations from the CubeSVD analysis may not be always correct, and the slight overlap of the incorrect associations brings the unrelated objects together, forming large clusters containing both similar and dissimilar objects. CubeSVD's large clusters lead to an average precision (0.7612) that is slightly better than the baseline method (0.7275), but a very bad recall (0.2998) comparing to the baseline (0.3152). We observe that M-LSA, which aims at identifying the most important objects among the co-occurrence data, does not suffer from the incorrect prediction problem of CubeSVD. It can successfully identify the important object among the co-occurrence data, and group similar objects into the same cluster. It gains 18.7% average precision, 92% average recall, and 82% average F-measure comparing to the baseline method.

6   Conclusions

We propose a dynamic Agglomerative-Divisive clustering (*DADC*) algorithm to effectively exploit the relationships among the users, queries, and concepts in clickthrough data. We evaluate DADC against four different state of the art methods (BB, CubeSVD, M-LSA, and K-Means), experimental results confirm DADC can accurately determine the user, query, and concept clusters from the clickthrough data, and it significantly outperforms existing cluster strategies designed for multiple-type data objects.

For future work, we will investigate collaborative filtering methods to predict the interests of a user from the user communities to enhance the accuracy of personalization and study methods to provide personalized query suggestions based on the user communities and query clusters to help users formula more effective queries.

References

1. Beeferman, D., Berger, A.: Agglomerative clustering of a search engine query log. In: Proc. of ACM SIGKDD Conference (2000)
2. Church, K.W., Gale, W., Hanks, P., Hindle, D.: Using statistics in lexical analysis. In: Lexical Acquisition: Exploiting On-Line Resources to Build a Lexicon (1991)

3. Hoeffding, W.: Probability inequalities for sums of bounded random variables. JASA 58(301) (1963)
4. Joachims, T.: Optimizing search engines using clickthrough data. In: Proc. of ACM SIGKDD Conference (2002)
5. Leung, K.W.T., Ng, W., Lee, D.L.: Personalized concept-based clustering of search engine queries. IEEE TKDE 20(11) (2008)
6. Ng, W., Deng, L., Lee, D.L.: Mining user preference using spy voting for search engine personalization. ACM TOIT 7(4) (2007)
7. Rodrigues, P.P., Gama, J.: Semi-fuzzy splitting in online divisive-agglomerative clustering. In: Neves, J., Santos, M.F., Machado, J.M. (eds.) EPIA 2007. LNCS (LNAI), vol. 4874, pp. 133–144. Springer, Heidelberg (2007)
8. Rodrigues, P.P., Gama, J., Pedroso, J.P.: Hierarchical clustering of time-series data streams. IEEE TKDE 20(5) (2008)
9. Sun, J.T., Zeng, H.J., Liu, H., Lu, Y.: Cubesvd: A novel approach to personalized web search. In: Proc. of WWW Conference (2005)
10. Wang, X., Sun, J.T., Chen, Z., Zhai, C.: Latent semantic analysis for multiple-type interrelated data objects. In: Proc. of ACM SIGIR Conference (2006)

Author Index